Astronomy at High Angular Resolution 2011

The Central Kiloparsec in Galactic Nuclei

Journal of Physics: Conference Series Volume 372

Bad Honnef, Germany
28 August - 2 September 2011

ISBN: 978-1-62276-806-6
ISSN: 1742-6588

Printed from e-media with permission by:

Curran Associates, Inc.
57 Morehouse Lane
Red Hook, NY 12571

Some format issues inherent in the e-media version may also appear in this print version.

Printed by Curran Associates, Inc. (2013)

Institute of Physics
Dirac House, Temple Back
Bristol BS1 6BE UK

Phone: 44 1 17 929 7481
Fax: 44 1 17 920 0979

techtracking@iop.org

Additional copies of this publication are available from:

Curran Associates, Inc.
57 Morehouse Lane
Red Hook, NY 12571 USA
Phone: 845-758-0400
Fax: 845-758-2634
Email: curran@proceedings.com
Web: www.proceedings.com

TABLE OF CONTENTS

SESSION A: THE NUCLEI OF ACTIVE GALAXIES

SESSION B: THE GALACTIC CENTER

SESSION C: THE IMMEDIATE ENVIRONMENT OF SMBHS

SESSION D: THE PHYSICS OF NUCLEAR JETS AND THEIR INTERACTION WITH THE ISM

SESSION E: THE CENTRAL 100PC OF THE NUCLEAR ENVIRONMENT

SESSION F: STAR FORMATION IN THAT REGION

POSTER PRESENTATIONS

Preface

We are pleased to present the proceedings from the Astronomy at High Angular Resolution 2011: *The central kiloparsec in galactic nuclei* conference. The conference took place in the Physikzentrum of the Deutsche Physikalische Gesellschaft (DPG), Bad Honnef, Germany, from 28 August to 2 September 2011. It was the second conference of this kind, following the Astronomy at High Angular Resolution conference held in Bad Honnef, three years earlier in 2008.

The main objective of the conference was to frame the discussion of the broad range of physical processes that occur in the central 100pc of galactic nuclei. In most cases, this domain is difficult to probe through observations. This is mainly because of the lack of angular resolution, the brightness of the central engine and possible obscurations through dust and gas, which play together in the central regions of host galaxies of galactic nuclei within a broad range of activity. The presence of large amounts of molecular and atomic (both neutral and ionized) gas, dust and central engines with outflows and jets implies that the conditions for star formation in these regions are very special, and probably different from those in the disks of host galaxies.

Numerous presentations covering a broad range of topics, both theoretical and experimental, those related to research on Active Galactic Nuclei and on a wide range of observed wavelengths were submitted to the Scientific Organizing Committee. Presentations have been grouped into six sessions:

- The nuclei of active galaxies
- The Galactic Center
- The immediate environment of Super Massive Black Holes
- The physics of nuclear jets and the interaction of the interstellar medium
- The central 100pc of the nuclear environment
- Star formation in that region

The editors thank all participants of the AHAR 2011 conference for their enthusiasm and their numerous and vivid contributions to this conference. We would especially like to thank John Hugh Seiradakis from the Aristotle University of Thessaloniki in Greece for giving the dinner talk on the most astounding ancient Antikythera mechanism. We would also like to thank Victor Gomer and the staff of the Physikzentrum of the Deutsche Physikalische Gesellschaft in Bad Honnef where the conference took place. Last but not least we would like to thank all unnamed helpers, without whom the organisation of this conference would not have been possible.

Financial support for this conference was granted by the *Deutsche Forschungsgemeinschaft (DFG)* Sonderforschungsbereich project number SFB 956. We also acknowledge support from the European Community Framework Programme 7, Advanced Radio Astronomy in Europe, grant agreement no. 227290.

Christof Iserlohe, Vladimir Karas, Melanie Krips, Andreas Eckart, Silke Britzen and Sebastian Fischer

The Editors

Journal of Physics: Conference Series **372** (2012) 011001 doi:10.1088/1742-6596/372/1/011001

Conference Photographs

Conference Group Photo, September 1 2011

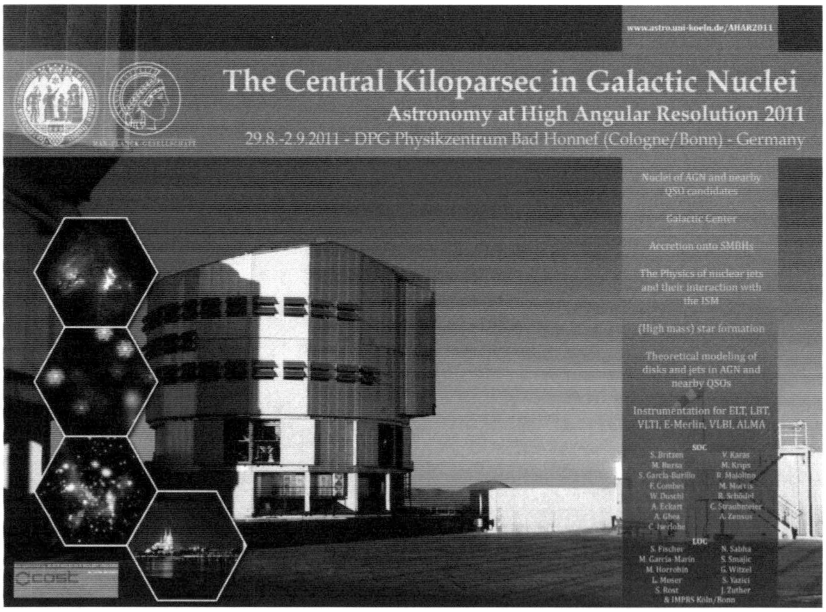

Poster from the announcement of the conference

Journal of Physics: Conference Series **372** (2012) 011001 doi:10.1088/1742-6596/372/1/011001

Conference dinner aboard *Moby Dick*

Contents of the proceedings

This volume is organized in the order of the speakers during the conference and according to the authors' family name for poster presentations. Table 1 gives an overview of the oral presentations and their sessions. Table 2 gives an overview of poster presentations. Not all participants have submitted their contribution in written form. Articles submitted to this volume are highlighted with * behind the name of the author. If an article has been submitted the table lists the title of the article, otherwise the title of the presentation.

Table 1: Sessions and Talks

Continued on next page

– continued from previous page

Ladislav Subr *	Dynamical evolution of the young stellar disc in the Galactic centre
Mark Morris	Gas Dynamics near Sgr A*
Jihane Moultaka *	Ices in the Galactic Center?
Tobias Fritz	The extinction towards the galactic center and new determination of the distance to the Galactic Center
Rainer Schödel (invited) *	SagittariusA* in the Infrared
Rainer Buchholz *	Near-infrared polarization in the central parsec of the galactic center
Andreas Eckart *	Flare emission from Sagittarius A*
Jason Dexter *	Radiative Models of Sagittarius A* and M87 from Relativistic MHD Simulations
Gunther Witzel	Time series of near-infrared flux densities of Sagittarius A*
Michael Kramer	Studying the Galactic Centre using Pulsars: I. Finding pulsars and GC astrophysics
Norbert Wex	Studying the Galactic Centre using Pulsars: II. Probing the spacetime of Sgr A*
Farhad Yusef-Zadeh *	The Origin of Parsec-Scale Gaseous and Stellar Disks in the Galactic Center and AGNs
Donald Figer (invited)	Massive star formation in nuclei of galaxies
Joel Sanchez Bermudez *	NaCo/SAM observations of sources at the Galactic Center
Noriyuki Matsunaga *	The evolution of the Galactic nuclear bulge traced by Cepheid variable stars
Denise Riquelme *	Disk-Halo interaction: The molecular clouds in the Galactic center region

Session C: The immediate environment of SMBHs
Chairs: Avery Broderick and Martin Elvis

Avery Broderick (invited)	Imaging Black Hole Horizons with the Event Horizon Telescope
Vladimir Karas *	Frame-dragging effects on magnetic fields near a rotating black hole
Silke Britzen *	Detecting supermassive binary black holes with VLBI - discovery of a ring-structure in 3C454.3
Helene Flohic	Multiwavelength monitoring of a low-mass AGN: structure of the accretion disk
Markus Boettcher *	Probing the Central Regions of AGN through VHE Gamma-Ray induced Pair cascades
Gabriele Bruni *	The central structure of Broad Absorption Line QSOs: observational characteristics in the cm-mm wavelength domain
Martin Elvis *	Slicing the Torus: Obscuring Structures in Quasars
Makoto Kishimoto *	Probing the innermost dusty structure in AGN with mid-IR and near-IR interferometers
Daniel Asmus *	Nuclear mid-infrared properties of nearby low-luminosity AGN
Konrad Tristram *	The complexity of parsec-scaled dusty tori in AGN
Leonard Burtscher	The MIDI AGN Large Programme: A statistical sample of resolved AGN tori
Rachel Mason *	The Nuclear Infrared Emission of Low-Luminosity AGN
Christian Leipski *	The dust emission of high-redshift quasars

Session D: The physics of nuclear jets and their interaction with the ISM
Chair: Francoise Combes

Melanie Krips (invited) *	Multiple-Line Study of NGC 1068: Hot Molecular Gas Caused by Jet-Gas Interaction in the Central 100pc?
Rebeca Aladro *	Chemical complexity in NGC1068
Mohammad Zamaninasab	The quasar 3C454.3: Results of 1995-2011 VLBI and single dish monitoring

Continued on next page

– continued from previous page

Alexander Tchekhovskoy (invited) *	General Relativistic Modeling of Magnetized Jets from Accreting Black Holes
Tuomas Savolainen	Parsec-scale jet properties and gamma-ray emission in the MOJAVE sample of blazars
Zulema Abraham	The precessing jet in the core of NGC 1275 (3C84)
Jonathan Leon-Tavares	The interplay between the jet, BLR and gamma-ray emission at several parsecs away from the central engine of superluminal AGN.
Oliver Porth	Signatures of synchrotron radiation from the relativistic jet base

Session E: The central 100 pc of the nuclear environment
Chairs: Donald Figer and Melanie Krips

Francoise Combes (invited) *	Dynamical processes in galaxy centers
David Meier *	Resolving Gas Chemistry at Arcsecond Scales in Nearby Spiral Nuclei
Satoki Matsushita *	10 pc Scale Circumnuclear Molecular Gas Imaging of Nearby AGNs
Tessel Van der Laan *	Are circumnuclear (starburst) rings important in the evolution of the central kiloparsec?
Petri Vaisanen *	Central regions of LIRGs: rings, hidden starbursts, Supernovae and star clusters
Richard I. Davies (invited) *	Do stellar winds play a decisive role in feeding AGN?
Julia Scharwaechter	Circum-nuclear molecular hydrogen emission in nearby Seyfert galaxies
Q.Daniel Wang	X-raying the circum-nuclear regions of nearby galaxies at high spatial and spectral resolutions
Anca Constantin *	Linking the Supermassive Black Hole Growth with the Megamaser Emission
Monica Valencia-S. *	Discovery of an Intermediate Mass Black Hole at the center of the starburst/Seyfert composite galaxy IRAS 01072+4954

Session F: Star formation in that region
Chair: Richard I. Davies

Jesus-Martin Pintado (invited)	High mass star formation in nuclei of galaxies and the central regions of the Milky Way
Sungsoo S. Kim *	Simulations of Nuclear Star-Forming Rings: A Case of the Milky Way
Jonathan Exposito	Star Formation and jet-induced coronal lines near the core of NGC 1068
Santiago Garcia-Burillo *	The feeding of activity in galaxies: a molecular line perspective
Rainer Beck *	Magnetic fields and gas flows around circumnuclear starbursts
Stefanie Mühle *	Molecular gas in active environments

Table 2: Poster presentations

Blank, Marvin *	A fresh look at the starburst-AGN connection
Bremer, Markus *	Optical Integral Field Spectroscopy of NGC 5850
Brodatzki, Katharina Anna	Internal gamma-gamma-opacity in the LLAGNs M87 and Cen A
Bursa, Michal	Time-resolved X-ray polarimetry as a tool to probe accretion flow structure
Czerny, Bozena *	QPO in RE J1034+396: model constraints from observed trends
Dovciak, Michal *	Polarization in lamp-post model of black-hole accretion discs
Endo, Wataru	Quasi-Periodic-Oscillations of Sagittarius A* in Short Millimeter-wave
Fischer, Sebastian *	A close look at Seyfert 2 nuclei
Goicoechea, Luis Julian *	Continuum reverberation mapping in a z = 1.41 radio-loud quasar
Goosmann, Rene	The fragmented structure of the X-ray warm absorber in NGC 3783
Grossberger, Christoph	Structural variability of 3C 111 on parsec scales
Haas, Jaroslav *	Dynamical evolution of thin stellar discs
Hryniewicz, Krzysztof *	Probing broad-line region of the weak line quasar SDSS J094533.99+100950.1
Hwang, Chorng-Yuan *	Star Formation in the Central Kiloparsec of Nearby Active Galaxies
Ikhsanov, Nazar *	Magnetically controlled accretion onto a black hole
Kunneriath, Devaky *	The Galactic centre mini-spiral with CARMA
König, Sabine *	CI, [CII] and CO observations towards TNJ 1338?1942: Probing the ISM in a massive proto-cluster galaxy at z=4.11
König, Sabine	Water megamasers in nearby QSOs
Lee, Ansun	Gas Inflow from the Central Few Hundred Parsec to the Few Parsec Regions
Marin, Frederic *	Modeling the optical/UV polarization while flying around the tilted outflows of NGC 1068
Messineo, Maria	Classification of bright evolved stars with 2MASS and GLIMPSE data.
Pan, Hsi-An *	Star Formation Timescale in the Circumnuclear Starburst Ring of Barred Galaxy NGC 7552
Prandoni, Isabella *	Searching for Molecular Gas in Southern Radio Galaxies
Rosenberg, Marissa *	[FeII] as a tracer supernova rate
Ruff, Andrea *	Near infrared hydrogen emission line ratios as diagnostics of the broad emission line region
Sabha, Nadeen	The Galactic Center in Mid-Infrared
Schinzel, Frank *	The long-term evolution of the parsec scale jet of the quasar 3C 345
Schnülle, Kirsten *	Charged Balanced Black Rings in Five Dimensions
Shahzamanian Sichani, Banafsheh *	Spectral Evolution of Field Galaxies
Valencia-S., Monica *	Modeling polarization signatures of NIR radiation from the Sgr A* black hole environs
Vitale, Mariangela *	Classifying Active Galactic Nuclei from the zCOSMOS 20k spectroscopic survey. Looking for a mass trend.
Westhues, Christian	Photometric AGN reverberation mapping

Peer review statement

All papers published in this volume of *Journal of Physics: Conference Series* have been peer reviewed through processes administered by the proceedings Editors. Reviews were conducted by expert referees to the professional and scientific standards expected of a proceedings journal published by IOP Publishing.

The Latest Results from *QUEST*, the *Q*uasar and *U*LIRG *E*volution *St*udy

Sylvain Veilleux[1,2]

[1] Department of Astronomy, University of Maryland, College Park, MD 20742 USA
[2] Astroparticle Physics Laboratory, NASA Goddard Space Flight Center, Code 661, Greenbelt, MD 20771 USA

E-mail: veilleux@astro.umd.edu

Abstract. The latest results from our *Q*uasar-*U*LIRG *E*volution *St*udy (*QUEST*) are described with an emphasis on the recent scientific breakthroughs on the issue of quasar feedback.

1. Introduction

Gas-rich galaxy merging at high redshifts may trigger major starbursts, lead to the formation of elliptical galaxies, and account for the growth of supermassive black holes (BHs; e.g., Sanders et al. 1988; Hopkins et al. 2009). This merger-driven evolutionary scenario starts with a completely obscured ultraluminous infrared galaxy (ULIRG). As the system evolves, the obscuring gas and dust is gradually dispersed, giving rise to dusty QSOs and finally to completely exposed QSOs. Powerful winds, driven by the central quasar or the surrounding starburst, have been invoked to stop the growth of both the BH and spheroidal component and explain the tight BH-spheroid mass relation (e.g., Murray et al. 2005). These winds are purported to quench star formation in the merger remnants ("negative mechanical feedback"), creating a population of red gas-poor ellipticals and explaining the bimodal color distribution observed in galaxy surveys.

Mergers are not the unique explanation for starbursts and quasars at high redshifts and the formation of elliptical galaxies. An increasingly popular alternative is the formation and growth of galaxies via cold gas accretion (e.g., Keres et al. 2005; Dekel & Birnboim 2006, 2008; Genzel et al. 2006, 2008; Dekel et al. 2009; Ceverino et al. 2010). However, these two different scenarios have several characteristics in common: (1) massive gas inflows to the center of the potential well, (2) formation of spheroids via violent relaxation and transformation of the accreted gas into new stars, (3) growth of the central black hole, and (4) the need for galactic winds to remove the left-over gas and produce the gas-poor elliptical galaxies observed in the local universe. Here we briefly review the observational evidence for processes (1)–(4), emphasizing the exciting new results relating to (4). We focus our discussion on the results from our own survey of local systems, *QUEST*, the Quasar-ULIRG Evolution STudy. The ULIRG component of this survey focuses on the 1-Jy sample, a sample of $z < 0.3$ 118 ULIRGs selected at 60 μm from a redshift survey of the *IRAS* faint source catalog (Kim & Sanders 1998). The QSO component focuses on the Palomar-Green (PG) quasars of Schmidt & Green (1983). The *QUEST* sample of $\sim$40 QSOs was carefully selected to match the redshifts of the *QUEST* set of ULIRGs and cover the full range in luminosity, infrared excess [$\log(L_{IR}/L_{BOL})$], and radio loudness of the PG QSOs.

2. Gas Inflows

There is now considerable *indirect* evidence that massive gas inflows have indeed taken place in galaxy mergers. In Rupke et al. (2008), we measured nuclear or near-nuclear oxygen abundances in a sample of 100 star-forming LIRGs and ULIRGs using new, previously published, and archival spectroscopy of strong emission lines (including [O II] $\lambda\lambda$3726, 3729) in galaxies with redshifts $< z > \sim 0.1$. When compared to local emission-line galaxies of similar luminosity and mass (using the near-infrared luminosity-metallicity and mass-metallicity relations), we found that LIRGs and ULIRGs are oxygen underabundant by a factor of 2 on average. As a corollary, LIRGs and ULIRGs were also found to have smaller effective yields. We concluded that the observed underabundance results from the combination of a decrease of abundance with increasing radius in the progenitor galaxies and strong, interaction- or merger-induced gas inflow into the galaxy nucleus. This conclusion demonstrates that local abundance scaling relations are not universal, a fact that must be accounted for when interpreting abundances earlier in the universe's history, when merger-induced star formation may be the dominant mode in massive systems. Kewley et al. (2006) came to a similar conclusion in their study of an optically selected sample of nearby interacting galaxies. Recent spatially-resolved spectroscopic studies of these objects, combined with numerical simulations, have confirmed the importance of gas inflows on the final distribution of metals in merger remnants (e.g., Rupke et al. 2010ab).

3. Formation of Spheroids

Our comprehensive ground- and space-based imaging and spectroscopic program of all ULIRGs and QSOs in the *QUEST* sample has demonstrated that ULIRGs are advanced mergers of gas-rich, disk galaxies sampling the entire Toomre merger sequence beyond the first peri-passage (Veilleux et al. 2002). The NIR light distributions and stellar kinematics in single-nucleus ULIRGs and QSOs are consistent with those of $\sim 0.5-5L^*$ spheroids in formation (e.g., Veilleux et al. 2006, 2009; Dasyra et al. 2006ab, 2007). Numerous other studies have come to similar conclusions (e.g., Rothberg & Joseph 2006; Rothberg & Fischer 2010; Cales et al. 2011 among others).

A brief exercise in numerology provides some perspective on this result. The number density of local ULIRGs is $\sim 2.5 \times 10^{-7}$ Mpc^{-3}. This is only $\sim$(1/7000) $\times$ the number density of SDSS ellipticals. The obvious but long-standing question is whether there are enough "spheroids in formation" at high redshifts to account for all local ellipticals (e.g., Toomre 1977). Bright submm-selected galaxies (SMGs) may fit the bill. They show all of the tell-tale signs of high-z spheroids in formation: obvious mergers, rich in molecular gas, and intense ULIRG-like star formation rates ($>$ 100 M$_\odot$ yr^{-1}; e.g., Greve et al. 2005; Tacconi et al. 2006, 2008). Located at $z \sim 1-4$, they are $\sim$2 orders of magnitude more numerous than local ULIRGs. The expected number of SMG "descendants" is given by n(SMGs) $\times$ τ(SMG era) / τ(SMG lifetime), where n(SMGs) $\sim 3 \times 10^{-5}$ Mpc^{-3} is the number density of SMGs, τ(SMG era) $\sim$ 1500 Myr is a rough estimate of the duration of the SMG era, and τ(SMG lifetime) $\sim$ 300 Myr is the lifetime of individual SMGs. The expected number of descendants is thus similar to the number density of local ellipticals with $L > L^*$. Their measured stellar masses also appear consistent with this idea (e.g., Borys et al. 2005), but they are denser and more compact than local massive ellipticals (e.g., Tacconi et al. 2008). Two possible solutions have been proposed for this discrepancy: (1) the end-result of SMGs are "puffed up" by subsequent dry major mergers and minor mergers. (2) The compactness observed by Tacconi et al. (2008) and others may be misleading: in reality perhaps there is a bright central starburst region dominating the surface brightness distribution, which is surrounded by a much larger, undetected halo of older stars dominating the overall mass distribution. This last possibility should be relatively easy to test with deeper rest-frame NIR observations of these objects.

4. Black Hole Growth

AGN have long been known to be present in ULIRGs (e.g., Sanders et al. 1988), in greater numbers than in LIRGs (e.g., Veilleux et al. 1995, 1999ab). However, the role they play in powering ULIRGs has been much harder to quantify, except in the few cases where a broad-line region is detected in the optical or near-infrared (these objects are invariably dominated by the AGN; e.g., Veilleux et al. 1997, 1999).

Dust obscuration remains a potential problem, so it is important to observe these objects at wavelengths where the AGN diagnostics are less affected by this dust: mid-to-far infrared, radio, and X-rays. Here we focus our discussion on the mid-infrared and X-ray results. This is not to say that the far-infrared is not potentially useful; on the contrary, *Herschel* data have already been used successfully to estimate the AGN contribution in some ULIRGs (e.g., van der werf et al. 2010). However, whether these data can be used for accurate (±15%) estimates remain to be proven at this (admittedly early) stage of investigation. The reason to skip the radio (cm-wave) methods is different: the main problem here is the fact that this energy range is a minor contributor to the overall bolometric luminosity of ULIRGs, so one always needs to make the questionable assumption that results derived from the radio data apply to the mid-far infrared, where the bulk of the power is emitted.

The advent of *Spitzer* has revolutionized our understanding of (U)LIRGs and QSOs at mid-infrared wavelengths. In Schweitzer et al. (2006), we showed that starbursts are responsible for at least ~30%, but likely most, of the FIR luminosity of PG QSOs. We argued in Netzer et al. (2007) that both strong- and weak-FIR emitting sources have the same, or very similar, intrinsic AGN spectral energy distributions (SEDs). In Schweitzer et al. (2008), we found that emission from dust in the innermost part of the narrow-line region is needed in addition to the traditional obscuring torus in order to explain the silicate emission in these QSOs. In Veilleux et al. (2009b), the contribution of nuclear activity to the bolometric luminosity in the *QUEST* ULIRGs and QSOs was quantified using six independent methods that span a range in wavelength and give consistent results within ~ ±10%-15% on average. This agreement suggests that deeply buried active galactic nuclei (AGNs) invisible to the *Spitzer* InfraRed Spectrograph but bright in the far-infrared are not common in this sample. The average derived AGN contribution in ULIRGs was found to be ~35%-40%, ranging from ~15%-35% among "cool" ($f_{25}/f_{60} \leq 0.2$) optically classified HII-like and LINER ULIRGs to ~50 and ~75% among warm Seyfert 2 and Seyfert 1 ULIRGs, respectively. This number exceeds ~80% in PG QSOs. ULIRGs fall in one of three distinct AGN classes: (1) objects with small extinctions and large polycyclic aromatic hydrocarbon (PAH) equivalent widths are highly starburst-dominated; (2) systems with large extinctions and modest PAH equivalent widths have larger AGN contributions, but still tend to be starburst-dominated; and (3) ULIRGs with both small extinctions and small PAH equivalent widths host AGN that are at least as powerful as the starbursts. The AGN contributions in class 2 ULIRGs are more uncertain than in the other objects. A morphological trend is seen along the sequence (1)-(2)-(3), in general agreement with the standard ULIRG-QSO evolution scenario and suggestive of a broad peak in extinction during the intermediate stages of merger evolution. However, the scatter in this sequence, including the presence of a significant number of AGN-dominated systems prior to coalescence and starburst-dominated but fully merged systems, implies that black hole accretion, in addition to depending on the merger phase, also has a strong chaotic/random component, as in local AGNs. See Armus et al. (2007), Farrah et al. (2007), Desai et al. (2007), and Petric et al. (2011) for more *Spitzer* results on (U)LIRGs.

More recently, we carried a similar analysis using new and archival *Chandra* and *XMM-Newton* X-ray data on 40 ULIRGs and 26 QSO from the *QUEST* sample (Teng & Veilleux 2010). A combination of traditional and hardness ratio spectral fitting methods was used to characterize the X-ray properties of these objects. The absorption-corrected 2-10 keV to bolometric luminosity ratios were used as a proxy for the fractional contribution of the AGN to the bolometric luminosity of these systems. The results were in general agreement with those obtained in the mid-infrared with *Spitzer*: the likelihood for dominant nuclear activity increases along the merger sequence from "cool" ULIRGs, "warm" ULIRGs, infrared-bright QSOs, and infrared-faint QSOs. The starburst dominates the total power in ULIRGs prior to the merger, and this is followed by rapid black hole growth during and after coalescence. The main concern with using the X-rays to quantify the AGN contribution is that the 2-10 keV luminosity may not be adequately corrected for absorption. It is therefore crucial to look beyond 10 keV to search for, and quantify, highly absorbed AGN possibly lurking in some of these systems. The discovery of a Compton-thick AGN in the ULIRG/QSO Mrk 231 is a good case in point (Braito et al. 2004). An attempt was made with *Suzaku* to study local ULIRGs beyond 10 keV, but only one object was detected with any significance (Teng et al. 2009). More comprehensive studies will only be possible with the next generation of X-ray facilities (e.g., NuSTAR, Astro-H).

5. Galactic Winds

There is growing observational support for large-scale galactic winds: e.g., most galaxies with star formation rate (SFR) densities in excess of $\sim$0.1 $M_\odot$ yr^{-1} kpc^{-2} show signatures of outflows, both locally and at high redshifts (e.g., Veilleux et al. 2005). The local outflows are often spatially resolved, allowing determination of the mass outflow rates ($\sim$ 0.1-5 x SFR) and kinetic energies ($\sim 10^{56-58}$ ergs in ULIRGs; Rupke et al. 2002, 2005abc; Martin 2005, 2006). These winds are primarily driven by the starburst rather than the AGN, except in late-stage mergers with quasar signatures where velocities $\geq$1000 km s^{-1} (cf. 100-400 km s^{-1} in other systems) are sometimes observed (Rupke et al. 2005c; Krug et al. 2010).

In the past year, there have been three major scientific breakthroughs in this area of research:

- Our *Herschel* PACS GTO survey of ULIRGs (*SHINING*, PI Sturm) has revealed far-infrared (FIR) OH features with P-Cygni profiles indicative of massive *molecular* outflows in most ($\sim$70%) ULIRGs, including the closest quasar known, Mrk 231 (Fischer et al. 2010; Sturm et al. 2011; Fig. 1). In some of these objects the (terminal) outflow velocities exceed 1000 km s^{-1}, and their outflow rates (up to $\sim$1200 $M_\odot$ yr^{-1}) are several times larger than their star formation rates. We compared the outflow signatures in different types of ULIRGs and in starburst galaxies to address the issue of the energy source (AGN or starburst) of these outflows. In Sturm et al. (2011), we reported preliminary evidence that ULIRGs with a higher AGN luminosity (and higher AGN contribution to $L_{\rm IR}$) have higher terminal velocities and shorter gas depletion timescales. Taken at face value, the outflows in the observed ULIRGs are able to expel the cold gas reservoirs from the centers of these objects within $\sim 10^6 - 10^8$ years.
- Feruglio et al. (2010) used the IRAM PdB Interferometer to obtain a very deep map of the CO(1-0) transition in Mrk 231. Thanks to the wide band, broad wings were detected in the profile of the CO line, with velocities of up to 750 km s^{-1} and spatially resolved on kpc scale (Fig. 1). These broad CO wings trace a giant molecular outflow of about $\sim$700 $M_\odot$ yr^{-1}, far larger than the on-going SFR ($\sim$200 $M_\odot$ yr^{-1}) in the host galaxy. Remarkably, this CO outflow coincides spatially with blueshifted optical Na ID 5890, 5896 Å absorption features (Rupke et al. 2005c).
- Our recent Gemini/IFU observations have revealed that the Na ID outflow in Mrk 231 is not only spatially resolved on kpc scale but also wide-angled, thus driven by a QSO wind rather than a jet (Rupke & Veilleux 2011; Fig. 2). In this paper, we showed that the nuclear region hosts an outflow with blueshifted velocities reaching 1100 km s^{-1}, extending 2–3 kpc from the nucleus in all directions in the plane of the sky. A radio jet impacts the outflow north of the nucleus, accelerating it to even higher velocities (up to 1400 km s^{-1}). Finally, 3.5 kpc south of the nucleus, star formation is simultaneously powering an outflow that

reaches more modest velocities of only 570 km s^{-1}. Blueshifted ionized gas is also detected around the nucleus at lower velocities and smaller scales. The mass and energy flux from the outflow are >2.5 times the star formation rate and >0.7% of the active galactic nucleus luminosity. The former suggests strong negative feedback to star formation and the latter is consistent with the coupling efciency required in the AGN feedback model of Hopkins & Elvis (2010). In other words, this powerful outflow may be the long-sought "smoking gun" of quasar mechanical feedback that clears out the neutral + molecular disk formed from dissipative collapse during the merger, resulting in red and dead elliptical galaxies without star formation activity.

Our group is actively following up on these results using *Herschel* (guaranteed time key project *SHINING* and its OT1 and OT2 extensions), *HST*, *Chandra*, and a variety of ground-based facilities including Gemini and IRAM.

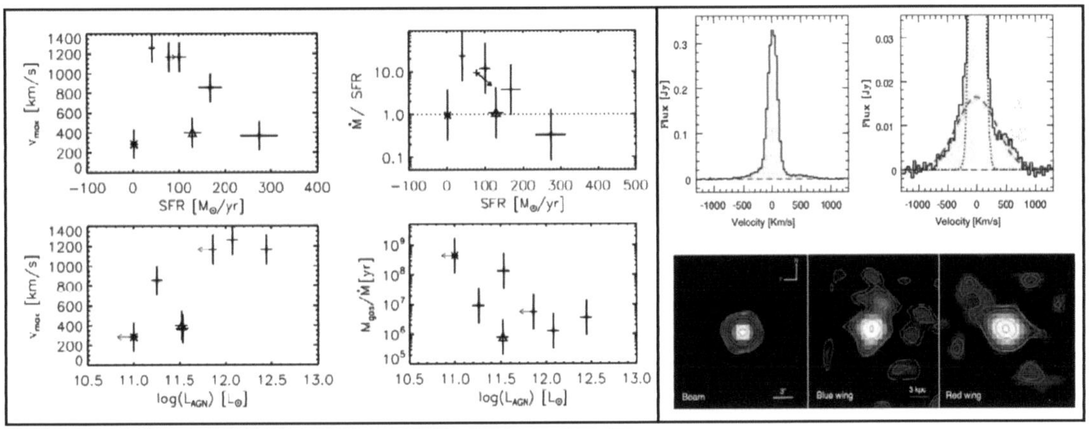

Figure 1. (Left box) Powerful molecular outflows are detected in most (~70%) local (U)LIRGs / FIR-bright QSOs with *Herschel* (from Sturm et al. 2011). The left panels show the maximum (≈ terminal) outflow velocity as a function of SFR and AGN luminosity (asterisk = NGC 253; triangle = Arp 220). The upper right panel shows the ratio of mass outflow rate to SFR as a function of SFR, while the bottom right panel is the depletion time scale as a function of AGN luminosity. Note the possible trends of increasing outflow velocities and decreasing depletion time scales with increasing AGN luminosities, suggestive of increasing influence of QSO feedback. (Right box) Continuum-subtracted spectrum of the CO(1–0) transition in Mrk 231 from Feruglio et al. (2010). Upper left panel shows the full flux scale, while upper right panel is an expanded flux scale to highlight the broad wings. The lower panels show the CO maps of the broad wings: beam profile on the left, the blue wing (from –500 to –700 km s^{-1}) in the middle, and red wing (from 500 to 800 km s^{-1}) on the right. The blue-wing emission is slightly extended. The mass outflow rate is ~700 $M_\odot$ yr^{-1} or ~2.5 × SFR.

6. Summary and Prospects

The latest results from *QUEST* are:

1. Trends seen between merger phase, AGN fractional contribution to the bolometric luminosity and extinction are in general agreement with the standard ULIRG – QSO evolution scenario and suggestive of a broad peak in extinction during the intermediate stages of merger evolution.

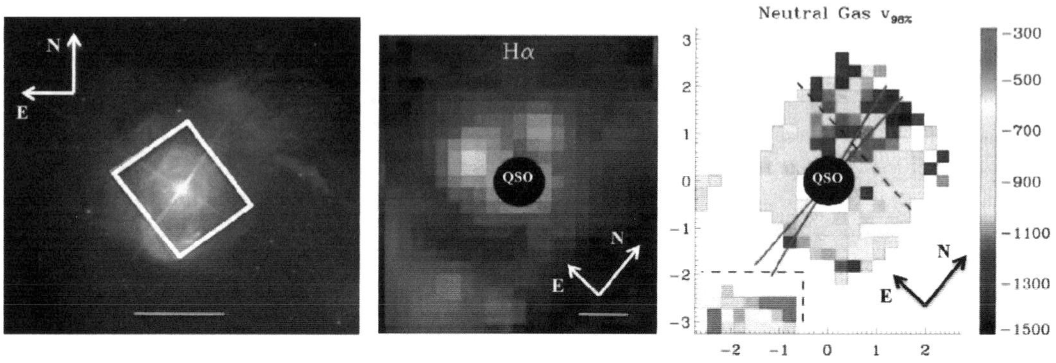

Figure 2.
(*Left*) *HST* / ACS continuum image of Mrk 231 in the 435W and 814W filters. The horizontal bar = 5 kpc (∼6″). The field of view of the Gemini/IFU data (6.3″ × 7.5″) of RV11 is overlaid as a box. Note the tidal arc ∼3.5 kpc (∼4″) south of the nucleus. (*Middle*) Hα emission, in logarithmic flux units. Bright features are unobscured HII regions, while fainter emission is likely shock-excited (elevated [N II]/Hα). The horizontal bar = 1 kpc (∼1.2″). Note the emission coincident with the tidal arc. (*Right*) Velocity field of the Na ID absorbing material in Mrk 231 relative to systemic. The spatial scale of this panel is in kpc and roughly matches that of the middle panel. A nuclear (blueshifted) outflow extends from the nucleus up to 2-3 kpc in all directions (as projected in the plane of the sky). The high velocities (∼1000 km s^{-1}) and derived large mass outflow rate (≳400 $M_\odot$ yr^{-1}) suggest that the AGN powers the nuclear wind. The northern quadrant of the nuclear wind is further accelerated by the narrow radio jet (indicated by the two red lines). A lower-velocity starburst-driven outflow is present in the south, coincident with the tidal arc. (from Rupke & Veilleux 2011)

2. The scatter in this sequence implies that black hole accretion, in addition to depending on merger phase, also has a strong stochastic/random component.
3. Starburst-driven winds help clear dusty cocoons around most ULIRG merger remnants.
4. *Powerful* neutral/molecular quasar-driven winds are seen in a few nearby systems. They are capable of removing all of the gas in these systems in $10^6 - 10^8$ yrs and therefore halting star formation activity on that same timescale.

The prospects are good to independently verify results #1 and #2 using *Herschel* and the soon-to-be-launched next generation of hard X-ray telescopes (e.g., NuSTAR, Astro-H). There is little doubt that *Herschel* and *HST* will provide important new constraints in the near future on #3 and #4. 3D data from 8-meter class telescopes, IRAM, and ALMA will provide the crucial spatial information needed to more accurately derive the important dynamical quantities of the neutral and molecular components of these outflows and evaluate the role they play on the overall evolution of the galaxy hosts. We all look forward to AHAR 2014 to discuss these new results!

Acknowledgments
This paper was written on behalf of the *QUEST* team. The author was supported in part by NASA through *Spitzer* grant 126375, *Chandra* General Observer grant G090125X, *XMM-Newton* General Observer grant NNX09AF05G, the Astrophysics Data Analysis Program (ADP) research grant NNX09AC79G. The author also acknowledges support from a Senior NPP Award

Journal of Physics: Conference Series **372** (2012) 012001 doi:10.1088/1742-6596/372/1/012001

from NASA and thanks the host institution, the Goddard Space Flight Center, where this paper was written.

7. References

Armus L, et al. 2007 *ApJ* **656** 148
Borys C, et al. 2005 *ApJ* **635** 853
Braito V, et al. 2004 *A&A* **420** 79
Cales S L, et al. 2011 *ApJ* **741** 106
Ceverino D, Dekel A and Bournaud F 2010 *MNRAS* **404** 2151
Dasyra K M, et al. 2006a *ApJ* **638** 745
Dasyra K M, et al. 2006b *ApJ* **651** 835
Dasyra K M, et al. 2007 *ApJ* **657** 102
Dekel A and Birnboim Y 2006 *MNRAS* **368** 2
Dekel A and Birnboim Y 2008 *MNRAS* **383** 119
Dekel A, et al. 2009 *Nature* **457** 451
Desai V, et al. 2007 *ApJ* **669** 810
Farrah et al. 2007 *ApJ* **667** 149
Feruglio C, et al. 2010 *A&A* **518** L155
Fischer J, et al. 2010 *A&A* **518** L41
Genzel R, et al. 2006 *Nature* **442** 786
Genzel R, et al. 2008 *ApJ* **687** 59
Greve T R, et al. 2005 *MNRAS* 359 1165
Hopkins P F and Elvis M 2010 *MNRAS* **401** 7
Hopkins P F, et al. 2009 *MNRAS* **398** 303
Keres D, Katz N, Weinberg D H and Davé R 2005 *MNRAS* **363** 2
Kewley L, Geller M J and Barton E J 2006 *AJ* **131** 2004
Kim D C and Sanders D B 1998 *ApJS* **119** 41
Krug H B, Rupke D S N and Veilleux S 2010 *ApJ* **708** 1145
Martin C L 2005 *ApJ* **621** 227
Martin C L 2006 *ApJ* **647** 222
Murray N, Quataert E and Thompson T A 2005 *ApJ* **618** 569
Netzer H, et al. 2007 *ApJ* **666** 806
Petric A O, et al. 2011 *ApJ* **730** 28
Rothberg B and Joseph R D 2006 *AJ* **131** 185
Rothberg B and Fischer J 2010 *ApJ* **712** 318
Rupke D S N, Veilleux S and Sanders D B 2002 *ApJ* **570** 588
Rupke D S N, Kewley L J and Barnes J E 2010a *ApJ* **710** L156
Rupke D S N, Kewley L J and Chien J-H 2010b *ApJ* **723** 1255
Rupke D S N and Veilleux S 2011 *ApJ* **729** L27
Rupke D S N, Veilleux S and Sanders D B 2005a *ApJS* **160** 87
Rupke D S N, Veilleux S and Sanders D B 2005b *ApJS* **160** 115
Rupke D S N, Veilleux S and Sanders D B 2005c *ApJ* **632** 751
Rupke D S N, Veilleux S and Baker A 2008 *ApJ* **674** 172
Sanders D B, et al. 1988 *ApJ* **325** 74
Schmidt M and Green R 1983 *ApJ* **269** 352
Schweitzer M, et al. 2006 *ApJ* **649** 79
Sturm E, et al. 2011 *ApJ* **733** L16
Tacconi L J, et al. 2006 *ApJ* **640** 228
Tacconi L J, et al. 2008 *ApJ* **680** 246
Teng S H, & Veilleux S 2010 *ApJ* **725** 1848
Teng S H, et al. 2009 *ApJ* **710** 2035
Toomre A 1977 *Evolution of Galaxies and Stellar Populations* ed B M Tinsley and R B Larson (New Haven: Yale University Observatory) p 401
van der werf P P, et al. 2010 *A&A* **518** L42
Veilleux S, Cecil G and Bland-Hawthorn J 2005 *ARAA* **43** 769
Veilleux S, Kim D-C and Sanders D B 1997 *ApJ* **477** 631
Veilleux S, Kim D-C and Sanders D B 1999a *ApJ* **522** 113
Veilleux S, Kim D-C and Sanders D B 1999b *ApJ* **522** 139

Veilleux S, Kim D-C and Sanders D B 2002 *ApJS* **143** 315
Veilleux S, et al. 1995 *ApJS* **98** 171
Veilleux S, et al. 2006 *ApJ* **643** 707
Veilleux S, et al. 2009a *ApJS* **182** 628
Veilleux S, et al. 2009b *ApJ* **701** 587

The Cosmogony of Super-Massive Black Holes[1]

Wolfgang J. Duschl[*,**,***], **Peter A. Strittmatter**[**,****]

[*] Astrophysik Kiel, Institut für Theoretische Physik und Astrophysik, Christian-Albrechts-Universität zu Kiel, Leibnizstr. 15, 24118 Kiel, Germany[2]

[**] Steward Observatory, The University of Arizona, 933 N. Cherry Ave., Tucson, AZ 85721, USA

[***] wjd@astrophysik.uni-kiel.de; [****] pstrittmatter@as.arizona.edu

Abstract. We summarize our recent results on the evolution of super-massive black holes in the cores of galaxies. Our models are based on the viscous, Eddington-limited evolution of self-gravitating accretion disks. We find that, within this framework, one can explain the growth time scales and the luminosities of individual objects as well as the variation of these quantities for AGN with different black hole masses.

1. Introduction

There is overwhelming evidence that (almost) all galactic centers contain super-massive black holes (SMBHs) with the vast majority of them in the mass range of $10^{6...10}\mathrm{M}_\odot$ [2,3]. Accretion into these black holes is held responsible for the energy output of (active) galactic nuclei [4,5]. Surveys in the optical and X-ray wavelength regimes find that the lower the BH mass, the later in the evolution of the Universe the co-moving space density peaks [6-13]. This indicates that the evolution of black holes with a smaller (final) mass takes longer than that of more massive ones.

In this contribution, we summarize our results on the evolution of self-gravitating accretion disks in galactic centers and the resulting luminosity and mass evolution of AGN and their SMBHs.

2. The Scenario

The scenario behind our models is that a galaxy-galaxy interaction or merger has driven by tidal forces a considerable amount of gas and dust towards the galactic center. As the tidal forces are not capable of removing all the angular momentum of the gas, a massive disk will form. Viscous forces will then lead to accretion which ultimately will transport the gas into the center of the galaxy and feed a black hole there. Observations, for instance, of ultra-luminous infrared galaxies indicate that the most massive of these disks may have initial masses of $M_{\mathrm{d,i}} = 10^{10}\ \mathrm{M}_\odot$ and, in some cases, even more [14]. Their typical sizes are of order of a few 10^2 pc. These disks are the configurations from where our model calculations start. For numerical reasons a seed black hole with an initial mass $M_{\mathrm{BH,i}} \ll M_{\mathrm{d,i}}$ is added. Our models are (i) rotationally symmetric; (ii) symmetric with respect to their rotational plane; (iii) geometrically thin; and (iv) in vertical hydrostatic equilibrium. Their main differences to standard accretion disk models are that (i) we allow for self-gravity of the disk to be of relevance; (ii)

[1] This contribution is partly based on [1].
[2] Address for correspondence

in order to be able to deal with self-gravitating disks, we use the β-parameterization for the viscosity [15]; (iii) we take the Eddington limit into account [16]; and (iv) we allow the size of the disk to evolve as required by the conservation laws (for further physical and numerical details, see [1]).

3. The Reference Model

We define a reference model with an initial disk mass $M_{\mathrm{d,i}} = 10^{10}\mathrm{M}_{\odot}$, an outer radius of 10^2 pc, an initially constant radial distribution of the surface density Σ, a viscosity parameter $\beta = 10^{-3}$, and an initial central mass $M_{\mathrm{BH,i}} = 10^4\mathrm{M}_{\odot} \ll M_{\mathrm{d,i}}$.

Figure 1 shows the evolution of the mass flow rate $\dot{M}$ and the luminosity L as a function of time, with $\dot{M}_{\mathrm{d}}$ the rate at which the disk delivers mass, $\dot{M}_{\mathrm{Edd}}$ the mass flow rate corresponding to the respective Eddington limit, and $\dot{M}_{BH}$ the growth rate of the black hole, given by

$$\dot{M}_{\mathrm{BH}} = \min(\dot{M}_{\mathrm{d}}, \dot{M}_{\mathrm{Edd}}).$$

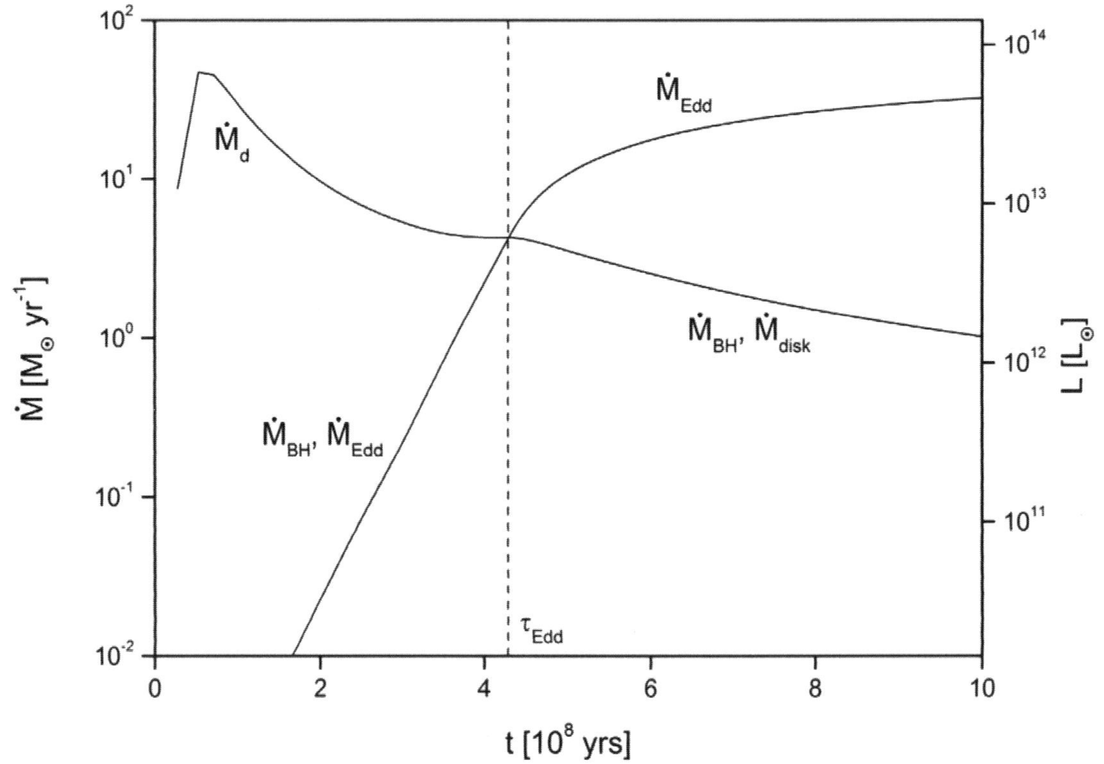

Figure 1: The evolution of the mass flow rates and the luminosity of the Reference Model (adapted from [1]).

We find that initially the evolution is dominated by the Eddington limit. Only a (initially small) fraction of the mass delivered by the disk actually contributes to the growth of the black hole. We assume that whatever falls victim to the Eddington limit is lost from the system, for instance as a jet or outflow. In the present context, however, we do not follow the further fate of this material. After a time τ_{Edd} the mass of the black hole has grown and the mass flow rate from the disk fallen enough for the latter to become smaller than the Eddington limit. From now on the black hole can accrete all the material coming from the disk. During this free accretion phase, the Eddington ratio $\dot{M}_{\mathrm{BH}}/\dot{M}_{\mathrm{Edd}}$ continues to become ever smaller. We expect that at a later time radiation-inefficient accretion will take over and lead to considerably smaller luminosities [17]. In the present context, we have, however, followed the evolution of the disks and black holes for not long enough for that transition to be reached.

4. Variations of the Model Parameters

Starting from the reference model, we have varied its parameters with an emphasis on the initial disk mass $M_{\mathrm{d,i}}$ and the initial disk size. We find that in general, unless the Eddington limit is not reached at all, the evolution of the mass flow rates follows the same pattern: An initially Eddington-limited phase during which the black hole mass grows exponentially is followed be free accretion which still can contribute considerably to the black hole mass, albeit no longer exponentially. Among other results ([1]), we find that the evolution time scale depends strongly on the initial mass and the size of the disk. The smaller and the more massive the initial disk is, the faster it evolves. This translates into the most massive black holes forming the quickest, and is in remarkable agreement with the observational findings of the above-mentioned surveys. Due to the Eddington limit, there is, however, a lower limit to the disk size which is still beneficial to a faster evolution: too small disks are so strongly Eddington-limited that most of their mass is lost rather than accreted into the black hole. Our models also allow for black hole masses $> 10^9 \mathrm{M}_\odot$ in the very young Universe (redshifts $z > 6$), in agreement with the observations of the most distance quasars known to date [18-21].

References

[1] Duschl WJ and Strittmatter PA 2011 *MNRAS* **413** 1495
[2] Kormendy J and Richstone D 1995 *ARA&A* **33** 581
[3] Richstone D et al 1998 *Nat* **395** A14
[4] Salpeter EE 1964 *ApJ* **140** 796
[5] Zeldovich YaB 1964 *SovPhysDokl* **9** 19
[6] Cowie LL et al 2003 *ApJ* **584** L57
[7] Fiore F et al 2003 *A&A* **409** 79
[8] Hasinger G 2003 *AIP ConfS* **666** 227
[9] Ueda Y et al 2003 *ApJ* **598** 886
[10] Wolf C et al 2003 *A&A* **408** 499
[11] Cristiani S et al 2004 *ApJ* **600** L119
[12] Hasinger G et al 2005 *A&A* **441** 417
[13] Fontanot F et al 2007 *A&A* **461** 39
[14] Sanders DB et al 2009 *ASP ConfS* **408** 3
[15] Duschl WJ et al 2000 *A&A* **357** 1123
[16] Heinzeller D and Duschl WJ 2007 *MNRAS* **374** 1176
[17] Beckert T and Duschl WJ 2002 *A&A* **387** 422
[18] Barth AJ et al 2003 *ApJ* **594** L95
[19] Fan X et al 2003 *AJ* **125** 1649
[20] Willott CJ et al 2007 *AJ* **134** 2435
[21] Mortlock DJ et al 2011 *Nat* **474** 616

Growing supermassive black holes: sub-grid modelling and intermediate-scale processes

Alexander Hobbs
Institute for Astronomy, ETH Zurich, Switzerland
E-mail: ahobbs@phys.ethz.ch

Abstract. The sheer range of scales in the Universe makes it impossible to model all at once. It is necessary, therefore, when conducting numerical experiments, that we employ sub-resolution prescriptions that can represent the scales we are unable to model directly. In this article we present a prescription for black hole growth that incorporates a different accretion regime from the standard approach used in the literature, and discuss the results of dedicated simulations of intermediate processes between small-scale accretion flows and large-scale cosmological volumes that can strongly enhance the accretion rate onto the black hole at the centre of a galaxy.

1. Introduction

Modelling galaxy formation requires at least 10 orders of magnitude in scale, as well as a huge variety of physical processes. These range from the physics of relativistic accretion flows around black holes ($\lesssim 10^{-4}$ pc) all the way to gravitational dynamics on the scales of galaxy clusters ($\gtrsim$ Mpc). In-between there is a host of baryonic physics to include - the hydrodynamics of the gas, radiative cooling, star formation, and magnetic fields, as well as any associated feedback processes. Naturally, all of these couple to each other in complex ways, making the task of simulating galaxy formation extremely difficult.

It is, nonetheless, vital that we do so, for numerical simulations are our only practical means of experimenting with the Universe. The basic procedure for these experiments is relatively simple, consisting of 5 distinct steps:

(i) Build models of what we are interested in
(ii) Simulate the time evolution of those models subject to physical laws / processes
(iii) Vary the parameters of the model
(iv) Analyse the effect on the evolution of the simulation
(v) Compare to observations and refine the model

where this process is repeated to achieve the required level of accuracy and as more detailed observations emerge. The largest stumbling block in this sequence is step (ii), as it is impossible to include all of the required physical processes (even the ones that we know about) due to limitations on computational resources and an inability to resolve all of the necessary scales in a simulation. We must therefore employ prescriptions that encapsulate processes occurring below the resolution limit; so-called "sub-grid" models.

One particular process that suffers from these difficulties is the accretion of gas onto a supermassive black hole (SMBH). This is a hugely important aspect of galaxy formation and evolution, as it is now well established that many galaxies in the local Universe harbour SMBHs with masses $10^6 \lesssim M_{\rm bh}/{\rm M}_\odot \lesssim 10^9$ at their centres. Furthermore, SMBHs show a clear relationship with their host galaxy, their masses known to be correlated with observable properties of the galactic bulge. The best established correlations are between the SMBH mass, $M_{\rm bh}$, and the velocity dispersion, σ, i.e., the $M_{\rm bh} - \sigma$ relation [1, 2], and the $M_{\rm bh} - M_{\rm bulge}$ relation [3, 4]. Understanding the formation and growth mechanisms of SMBHs is therefore crucial in determining the evolution of the larger host systems.

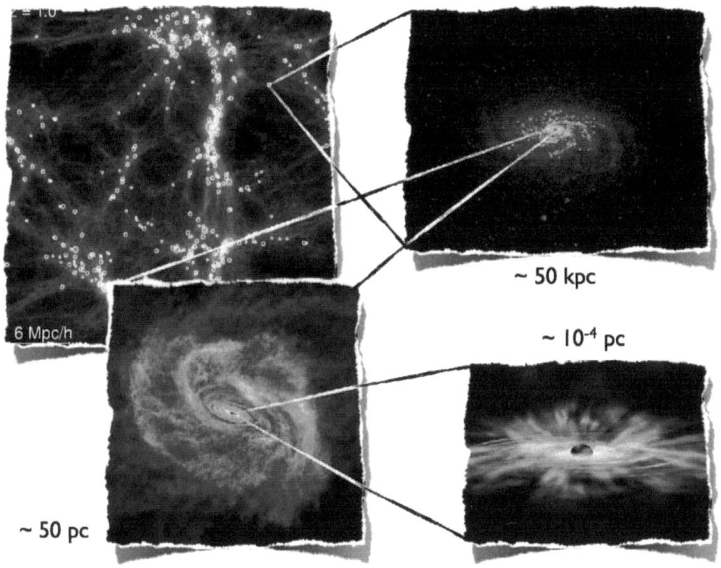

Figure 1. The range of scales required to fully model SMBH growth over cosmic time. Filamentary structures on cosmological scales (top left) funnel gas into forming galaxies (top right), the centres of which harbour SMBHs surrounded by swirling tori of gas (bottom left). As the gas makes its way down to the last stable orbit of the SMBH it forms a dense, hot accretion disc that gives rise to powerful outflows (bottom right). Top left-hand figure from [5].

Figure 1 depicts the range of scales that one must take account of in order to build a complete picture of SMBH growth from the early Universe to the present day. It is clear that this growth is a dynamic process, as surveys of active galactic nuclei (AGN) have uncovered the existence of quasars of mass $M_{\rm bh} \sim 10^9\,{\rm M}_\odot$ at $z \sim 7$, when the Universe was $\lesssim 1/10^{\rm th}$ of its current age; this implies that many SMBHs had already assembled their mass by this time [6].

Of course, our understanding of how SMBHs grow to these masses is incomplete. We know that black holes grow by accreting low angular momentum material from their surroundings, yet the character of the accretion flow onto an SMBH is governed by physical processes as diverse as galaxy mergers [7], turbulence induced by stellar feedback [8] and black hole accretion-driven outflows [9].

The typical masses of "seed" BHs in the early Universe are under debate, but it is still clear that to reach the required $\sim 10^9\,{\rm M}_\odot$ the SMBH must have experienced a high and sustained accretion rate. Even at the higher end of possible seed masses, i.e., $\sim 100\,{\rm M}_\odot$ [10], we would still require Eddington-limited growth in order to reach the observed masses of early Universe quasars [11]. One must therefore conclude that the accretion flow, as well as the evolution of the galaxy as a whole, is likely strongly influenced by AGN feedback that is coupled to the SMBH growth rate [12].

Journal of Physics: Conference Series **372** (2012) 012003 doi:10.1088/1742-6596/372/1/012003

2. Sub-grid modelling of SMBH growth

Black hole growth is now routinely modelled in galaxy formation simulations (for a fiducial work see [13]) and the importance of SMBHs in shaping the properties of galaxies has been demonstrated [14, 15, 5]. Given the difficulty in accounting for the sheer range of scales involved, we are compelled to make use of sub-grid modelling. The majority of galaxy formation simulations published in the literature incorporate what we shall term the "Bondi-Hoyle model" for black hole growth [13, 16, 17, 18, 19, 20], which derives from the work of [21] and [22] - hereafter B&H. This model assumes the simplest possible accretion flow, where the gas is at rest at infinity and accretes steadily onto a black hole, subject only to the (Newtonian) gravity of the latter, which is modelled as a point mass.

In galaxy formation simulations, unfortunately, this idealised picture is far from satisfied, as the gas inflow is complicated considerably by the properties of the flow at larger scales. Owing to the requirement of modelling the inflow and eventual accretion in a sub-grid fashion, we find it is useful to cast the infalling gas into several accretion "regimes", with the relative contribution each determined from the character of the flow at the resolution limit of the simulation. To start with, we have the two extremes:

(i) stationary, where the gas is completely supported and at rest at the resolution limit
(ii) in free-fall, where the gas flow is entirely radial and influenced only by gravity

It is important to recognise that the reality will lie somewhere in between. Specific cases that lie between these extremes are:

- the gas is unsupported at the resolution limit but shocked at smaller radii, after which it may begin to tend to free-fall once again if the cooling time is short
- the gas possesses angular momentum, and is thus supported by rotation but inflows through viscous processes

The mode of Bondi-Hoyle accretion corresponds to option (i), and so employing the Bondi-Hoyle sub-grid prescription in simulations implicity assumes such a regime. The opposite case, that of the gas being in free-fall, is therefore not catered for in this standard model.

2.1. The Bondi-Hoyle accretion regime

The nature of the Bondi-Hoyle regime for accretion is summarised in the first sentence of the [22]'s abstract: "The special accretion problem is investigated in which the motion is steady and spherically symmetrical, the gas being at rest at infinity". Physically, gas can only be at rest if it is not subject to any forces. The only external force acting on the gas in the B&H problem, i.e., the gravitational force, is due to the black hole. Self-gravity of the gas is neglected. The "infinity" in question is a region at a distance large enough from the SMBH that the gravitational force exerted by the latter is negligible when compared to the pressure forces within the gas. This is quantified by defining the Bondi radius,

$$r_B = 2GM_{\mathrm{bh}}/c_\infty^2 \tag{1}$$

where M_{bh} is the mass of the central object and c_∞ is the sound speed of the gas far from the hole.

Applying the Bondi-Hoyle formalism to black hole growth assumes that the accretion rate onto the SMBH is commensurate with the accretion rate through the Bondi radius (i.e., the flow is steady-state) and therefore given by

$$\dot{M}_{\mathrm{bh}} = \pi\lambda(\Gamma) r_B^2 \rho_\infty c_\infty = \frac{4\pi\lambda(\Gamma) G^2 M_{\mathrm{bh}}^2 \rho_\infty}{c_\infty^3} \tag{2}$$

where ρ_∞, c_∞ are the density and sound speed at infinity respectively, and $\lambda(\Gamma)$ contains all the corrections arising due to the finite pressure gradient force in the problem. In sub-grid modelling, however, the relevant gas properties "at infinity" are evaluated at a radius around the SMBH that corresponds to the smallest resolvable scale. Often this is orders of magnitude larger than the Bondi radius. We clearly require the gas to be at rest at this evaluation radius, and for the SMBH to dominate the gravitational potential in order for the Bondi-Hoyle formalism to be applicable, even to spherical flow. Such a situation could be achieved if the gas within the SMBH sphere of influence is at the virial temperature and therefore in hydrostatic balance [23, 24]. This may be the case for low-luminosity SMBHs in giant elliptical galaxies where the gas is rather tenuous and hot due to a long cooling time [25].

2.2. The free-fall accretion regime

However, in cosmological simulations (as in the real Universe) we expect SMBHs to be immersed in stellar bulges and dark matter haloes that are typically $\sim 10^3$ to 10^4 times more massive than the SMBH [4, 26]. Furthermore, in the epoch when SMBHs grow rapidly, their hosts are very gas rich, and the inflow of gas from large scales cannot be easily captured by gradual cooling from a tenuous hot halo [27, 28, 29, 30, 31]. Higher density gas is likely to cool much faster and hence is likely to be much cooler than the virial temperature. We therefore expect an inflow of gas to the centre rather than hydrostatic balance.

Due to the dominant potential energy of the halo, and the fact that the temperature of the gas infalling from large scales cannot be greater than the virial temperature at the halo virial radius, we can see that for spherical flow with a relatively soft equation of state the character of the inflow may reach that of free-fall at the evaluation radius instead of being stationary.

2.3. An interpolation approach

Of course, the two regimes discussed above constitute a simplified picture. In reality gas that has accreted onto a halo from outside the virial radius may shock at smaller radii, heating up to the local $T_{\rm vir}$ at that radius. However, due to the high densities reached for the shocked gas the cooling time is likely to be short [27, 28], and in particular significantly less than the free-fall time [32]. Gas may therefore be stationary temporarily at the shock radius but as it cools and begins to infall it will quickly tend to the free-fall velocity, and certainly by the time a radial inflow reaches the Bondi radius the assumption of being at rest will no longer be satisfied.

The list of regimes above is essentially a range of "supporting mechanisms" for the gas at the evaluation radius, the relative importance of which must be taken account of to infer the appropriate accretion rate. This is very difficult to do in a simulation, and so as a starting point we deal solely with the two extremes - (i) and (ii) as referred to above. We require a formula that interpolates between these two cases, and so we propose, as per [33], the expression

$$\dot{M}_{\rm interp} = \frac{4\pi\lambda(\Gamma)G^2 M_{\rm enc}^2 \rho_\infty}{(c_\infty^2 + \sigma^2)^{3/2}} \tag{3}$$

where σ is the velocity dispersion for the external potential, $\sigma \sim (GM_{\rm enc}(r)/r)^{1/2}$ and $M_{\rm enc}$ is the enclosed mass. This expression tends to the standard Bondi-Hoyle formula (equation 2) in the limit that $c_\infty \gg \sigma$, i.e., the gas internal energy dominates over the potential energy of the halo, and $M_{\rm enc} \rightarrow M_{\rm bh}$, i.e., as we approach the SMBH radius of influence. In the opposite limit, where the halo potential energy dominates, $\sigma \gg c_\infty$, we recover the free-fall estimate, $M_{\rm gas}/t_{\rm ff}$, where $M_{\rm gas}$ is the enclosed gas mass and $t_{\rm ff} = (r^3/2GM(r))^{1/2}$ is the free-fall timescale. This interpolation approach is a minor modification of the full Bondi-Hoyle-Lyttleton

[34, 22] expression for an accretor that is moving relative to the gas, with two changes: (1) the velocity dispersion replaces the relative velocity between gas and accretor, and (2) the total enclosed mass (external potential + gas + SMBH) replaces the black hole mass. Finally, in order to incorporate (in a crude fashion) the effect of AGN feedback on the accretion the rate should be capped at Eddington (as is commonly employed in simulations), namely

$$\dot{M}_{\rm acc} = \texttt{max}(\dot{M}_{\rm interp}, \dot{M}_{\rm Edd}) \tag{4}$$

3. Turbulent feeding on intermediate scales

A major caveat which we must point out with the discussion so far is that the presence of angular momentum in the gas alters things significantly. Indeed, angular momentum can be viewed as a physical "barrier" to accretion, with only the lowest angular momentum material able to flow to smaller scales and accrete onto the SMBH. One should therefore endeavour to take account of angular momentum in a sub-grid model of SMBH growth if at all possible. For this we recommend the "accretion disc particle" method of [35], which tracks and utilises the angular momentum in the gas to infer the rate of accretion and feedback.

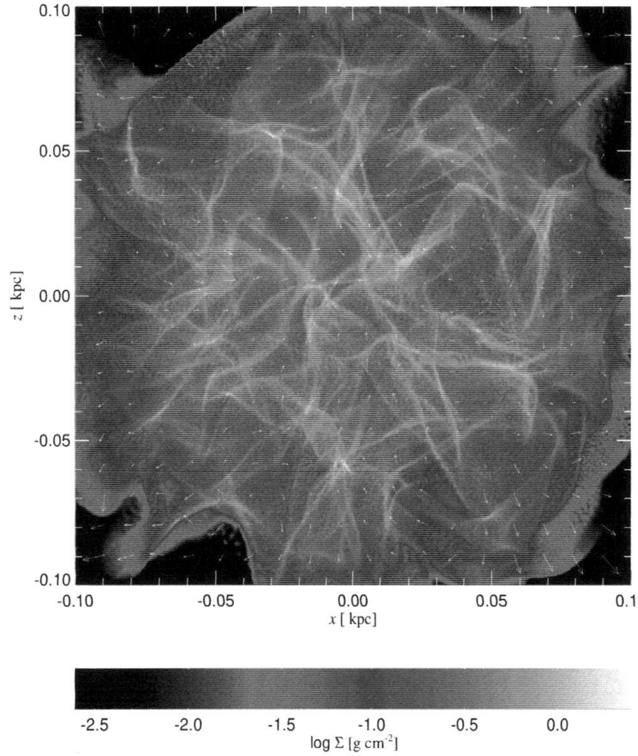

Figure 2. Projected gas column density in the fiducial turbulent feeding simulation, at an early time (10^5 yrs), before gas has had a chance to accrete on the SMBH. Note the formation of multiple thin and dense filaments due to convergent turbulent velocity flows.

Of course, taking account of angular momentum in the sub-grid model will only make it harder for an SMBH to accrete, as viscous timescales, even at parsec scales, are often prohibitively long (with standard assumptions [36], $t_{\rm visc}$ through a $\sim$ 10 pc scale disc is $\sim t_{\rm Hubble}$). We therefore need a process that can allow gas to acquire low angular momentum. A promising candidate is the supernovae-driven turbulent feeding model of [8], which would typically occur at scales intermediate between detailed simulations of black hole accretion discs and large-scale cosmological simulations, and is thus not captured by the majority of galaxy formation models. Indeed, it is important to take account of processes occurring on intermediate scales which may often get missed by this dichotomy of approaches to SMBH growth.

Figure 2, taken from [8], shows a visualisation from a numerical simulation of a mode of turbulent feeding at scales of ~ 100 pc. The turbulent velocity field imposed on the gas (in addition to a net rotation) is tailored to the typical energy and distribution of supernovae explosions in a spherical shell of gas at these scales. The formation of overdense filaments can clearly be seen, due to the action of the supernovae feedback in sweeping up and shocking the gas.

This turbulent feeding model yields some very interesting and potentially useful results in terms of the accretion rate through the central 1 pc (the accretion radius, $r_{\rm acc}$, of the simulations). There is a strong trend of enhanced accretion with levels of increasing turbulence - indeed, with a mean turbulent velocity of ~ 200 km s^{-1}, which is the approximate value of the velocity dispersion at these scales, the accretion rate is $2 - 3$ orders of magnitude higher than in the same simulation with no seeded turbulence. These results can be seen in Figure 3. The explanation for this lies in the angular momentum distribution with and without turbulence. When the velocity field was purely rotational (about a single axis), the angular momentum distribution was highly peaked, corresponding to a thin disc or a ring at a particular radius outside $r_{\rm acc}$. This ring can only accrete into the central parsec by viscous processes, thereby yielding a very low accretion rate. With the introduction of a turbulent velocity field, however, the angular momentum distribution was significantly broadened, putting some of the gas on low enough angular momentum orbits to travel inside the accretion radius *ballistically*, unaffected by hydrodynamical drag due to the high overdensity of the infalling gas clump or filament.

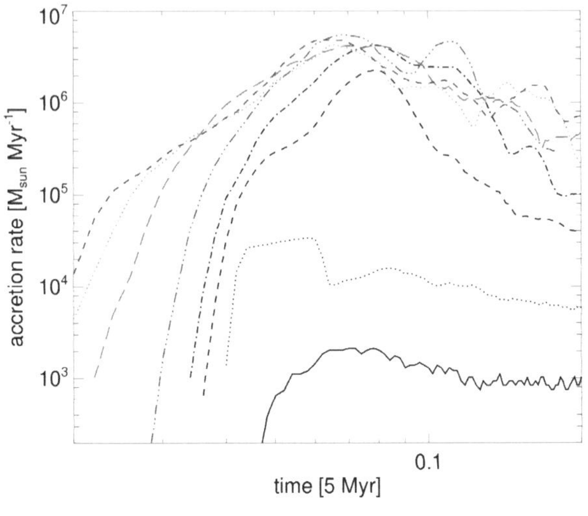

Figure 3. Accretion rate versus time for a fiducial set of simulations, with $v_{\rm rot} = 60$ km s^{-1}. Linestyles correspond to $v_{\rm turb} = 0$ (black solid), 20 km s^{-1} (black dotted), 40 km s^{-1} (black dashed), 60 km s^{-1} (black dot-dash), 100 km s^{-1} (brown dot-dot-dash), 200 km s^{-1} (red long dashed), 300 km s^{-1} (pink dotted), and 400 km s^{-1} (blue dashed) in code units. We can clearly see that the accretion rate on the black hole strongly increases with increasing levels of turbulence when rotation is present.

The dominant mode of feeding is therefore a ballistic one, but to later times an extended (and still quite turbulent) disc forms which must then accrete viscously. Such a structure is however no longer a ring but instead extends over a wider range of radii and accretes more rapidly than the non-turbulent case - this is due to the action of a supersonic turbulent viscosity that is maintained over the lifetime of the infall, as gas from the outer parts of the shell continues to impact upon the disc. Turbulence thus enhances the accretion rate in both modes of feeding. Of course, if the velocity field had no rotational component the opposite effect would be seen, since without turbulent broadening of the distribution all of the gas would simply end up inside the accretion radius. As we have mentioned, however, the gas typically inflows from considerably larger scales and so it is likely that angular momentum will, by the time the gas has reached these intermediate scales, translate into significant rotation.

The supernovae-driven turbulent feeding model can be described analytically by convolving a distribution of turbulent velocities (typically Maxwellian) with the mass of gas in the loss-cone of an isotropically-emitted wind in rotation around the SMBH - see [8] for more details. For a shell of initially constant density, with the outer radius $\approx 3 \times$ the inner radius, this yields an accreted mass fraction of

$$\frac{\Delta M}{M_{\rm shell}} = \frac{3(GM_{\rm bh}r_{\rm acc})^{1/2}}{2\sqrt{2}r_{\rm out}v_{\rm rot}} \cdot \psi + \frac{9(GM_{\rm bh}r_{\rm acc})^{1/2}}{4\sqrt{2}rv_{\rm turb}\pi^{1/2}} \cdot \varphi \tag{5}$$

where $v_{\rm turb}$ is the mean turbulent velocity, and $v_{\rm rot}$ is the rotation velocity. The functions ψ and φ are given by

$$\psi \equiv \texttt{erf}\left(\frac{5v_{\rm rot}}{4v_{\rm turb}}\right) - \texttt{erf}\left(\frac{3v_{\rm rot}}{4v_{\rm turb}}\right)$$

$$\varphi \equiv e^{-9v_{\rm rot}^2/16v_{\rm turb}^2} - e^{-25v_{\rm rot}^2/16v_{\rm turb}^2}$$

which, in the limit that $v_{\rm turb} \gg v_{\rm rot}$, simplifies to

$$\frac{\Delta M}{M_{\rm shell}} \approx \frac{(2GM_{\rm bh}r_{\rm acc})^{1/2}}{r_{\rm out}v_{\rm turb}}\left(1 + \frac{v_{\rm rot}^2}{v_{\rm turb}^2}\right) \tag{6}$$

where we have dropped numerical factors of approximately unity.

This analytical theory reproduces the numerical results from the simulations well [8], and can be used to make a prediction for the surface density of the disc that forms once the main ballistic mode of feeding ends. The prediction of the surface density, $\Sigma(R) \propto R^{-3/2}$, is quite steep. We note that [37] have recently found evidence that for high-luminosity AGN, the surface density distribution of heated dust (a tracer for the AGN disc) inferred from mid-IR interferometry can indeed be quite steep, going as $R^{-1} \lesssim \Sigma(R) \lesssim R^{-3/2}$ (refer to Figure 14 in [37]). The supernovae-driven turbulent feeding model is therefore a promising one both in terms of inferred SMBH accretion rates and observations of AGN tori. Moreover, the implication that the progenitor for the turbulence is feedback from star formation speaks to a connection between the stellar properties of the bulge and the SMBH mass, such as is observed.

4. Conclusions

Sub-grid modelling is a vital aspect of galaxy formation simulations. Processes occurring below the limit of resolution can be encapsulated into sub-grid prescriptions to allow us to explore the formation and evolution of galaxies over cosmic time. It is, however, extremely important that sub-grid models continue to be refined and improved using dedicated simulations as our understanding of the relevant processes improves. In particular, the extension of existing SMBH growth prescriptions to include different accretion regimes will help to shed light on the existence of quasars at high redshift and the observed SMBH-host correlations through the action of feedback. The Bondi-Hoyle/free-fall interpolation model of [33] attempts to account for the opposite extreme from the standard Bondi-Hoyle formalism in models of SMBH growth. Taking an approach to sub-grid modelling that is physically motivated, rather than empirically-motivated, will allow numerical simulators to construct models from first principles.

An important area of sub-grid modelling that is often overlooked is that of "intermediate" scales, which may be too small to be captured at high resolution in simulations of, say, galaxy mergers, but too large to be of interest to simulators studying detailed accretion flows at the last stable orbit around an SMBH. However, processes occurring on these intermediate scales can strongly influence the flow onto the SMBH, and indeed provide a potential solution to the AGN feeding problem. The supernovae-driven turbulent feeding model of [8] may be such a process.

5. References

[1] Gebhardt K, Bender R, Bower G, Dressler A, Faber S M, Filippenko A V, Green R, Grillmair C, Ho L C, Kormendy J, Lauer T R, Magorrian J, Pinkney J, Richstone D and Tremaine S 2000 *ApJL* **539** L13–L16
[2] Ferrarese L and Merritt D 2000 *ApJL* **539** L9–L12
[3] Magorrian J, Tremaine S, Richstone D, Bender R, Bower G, Dressler A, Faber S M, Gebhardt K, Green R, Grillmair C, Kormendy J and Lauer T 1998 *AJ* **115** 2285–2305
[4] Häring N and Rix H W 2004 *ApJL* **604** L89–L92
[5] Di Matteo T, Colberg J, Springel V, Hernquist L and Sijacki D 2008 *ApJ* **676** 33–53 (*Preprint* 0705.2269)
[6] Mortlock D J, Warren S J, Venemans B P, Patel M, Hewett P C, McMahon R G, Simpson C, Theuns T, Gonzáles-Solares E A, Adamson A, Dye S, Hambly N C, Hirst P, Irwin M J, Kuiper E, Lawrence A and Röttgering H J A 2011 *Nature* **474** 616–619 (*Preprint* 1106.6088)
[7] Hopkins P F and Quataert E 2010 *MNRAS* **407** 1529–1564 (*Preprint* 0912.3257)
[8] Hobbs A, Nayakshin S, Power C and King A 2011 *MNRAS* **413** 2633–2650 (*Preprint* 1001.3883)
[9] Nayakshin S and Power C 2010 *MNRAS* **402** 789–802 (*Preprint* 0911.2434)
[10] Madau P and Rees M J 2001 *ApJL* **551** L27–L30 (*Preprint* arXiv:astro-ph/0101223)
[11] King A R and Pringle J E 2007 *MNRAS* **377** L25–L28
[12] Power C, Zubovas K, Nayakshin S and King A R 2011 *MNRAS* **413** L110–L113 (*Preprint* 1103.1702)
[13] Springel V, Di Matteo T and Hernquist L 2005 *MNRAS* **361** 776–794 (*Preprint* astro-ph/0411108)
[14] Croton D J, Springel V, White S D M, De Lucia G, Frenk C S, Gao L, Jenkins A, Kauffmann G, Navarro J F and Yoshida N 2006 *MNRAS* **365** 11–28 (*Preprint* arXiv:astro-ph/0508046)
[15] Bower R G, Benson A J, Malbon R, Helly J C, Frenk C S, Baugh C M, Cole S and Lacey C G 2006 *MNRAS* **370** 645–655 (*Preprint* arXiv:astro-ph/0511338)
[16] Sijacki D, Springel V, Di Matteo T and Hernquist L 2007 *MNRAS* **380** 877–900 (*Preprint* 0705.2238)
[17] Pelupessy F I, Di Matteo T and Ciardi B 2007 *ApJ* **665** 107–119
[18] Di Matteo T, Colberg J, Springel V, Hernquist L and Sijacki D 2008 *ApJ* **676** 33–53 (*Preprint* 0705.2269)
[19] Johansson P H, Naab T and Burkert A 2009 *ApJ* **690** 802–821 (*Preprint* 0802.0210)
[20] Kim J h, Wise J H, Alvarez M A and Abel T 2011 *ArXiv e-prints* (*Preprint* 1106.4007)
[21] Bondi H and Hoyle F 1944 *MNRAS* **104** 273–+
[22] Bondi H 1952 *MNRAS* **112** 195–+
[23] Komatsu E and Seljak U 2001 *MNRAS* **327** 1353–1366 (*Preprint* arXiv:astro-ph/0106151)
[24] Suto Y, Sasaki S and Makino N 1998 *ApJ* **509** 544–550 (*Preprint* arXiv:astro-ph/9807112)
[25] Churazov E, Sazonov S, Sunyaev R, Forman W, Jones C and Böhringer H 2005 *MNRAS* **363** L91–L95 (*Preprint* arXiv:astro-ph/0507073)
[26] Guo Q, White S, Li C and Boylan-Kolchin M 2010 *MNRAS* **404** 1111–1120 (*Preprint* 0909.4305)
[27] Birnboim Y and Dekel A 2003 *MNRAS* **345** 349–364 (*Preprint* arXiv:astro-ph/0302161)
[28] Kereš D, Katz N, Weinberg D H and Davé R 2005 *MNRAS* **363** 2–28 (*Preprint* arXiv:astro-ph/0407095)
[29] Kereš D, Katz N, Fardal M, Davé R and Weinberg D H 2009 *MNRAS* **395** 160–179 (*Preprint* 0809.1430)
[30] Dekel A, Birnboim Y, Engel G, Freundlich J, Goerdt T, Mumcuoglu M, Neistein E, Pichon C, Teyssier R and Zinger E 2009 *Nature* **457** 451–454 (*Preprint* 0808.0553)
[31] Kimm T, Devriendt J, Slyz A, Pichon C, Kassin S A and Dubois Y 2011 *ArXiv e-prints* (*Preprint* 1106.0538)
[32] Nulsen P E J and Fabian A C 2000 *MNRAS* **311** 346–356 (*Preprint* arXiv:astro-ph/9908282)
[33] Hobbs A, Power C, Nayakshin S and King A 2012 *MNRAS*
[34] Hoyle F and Lyttleton R A 1939 The effect of interstellar matter on climatic variation *Proceedings of the Cambridge Philosophical Society* (*Proceedings of the Cambridge Philosophical Society* vol 34) pp 405–+
[35] Power C, Nayakshin S and King A 2010 *MNRAS* 1860–+ (*Preprint* 1003.0605)
[36] Shakura N I and Sunyaev R A 1973 *A&A* **24** 337–355
[37] Kishimoto M, Hönig S F, Antonucci R, Millour F, Tristram K R W and Weigelt G 2011 *A&A* **536** A78 (*Preprint* 1110.4290)

Testing the AGN Unification Model in the Infrared

C. Ramos Almeida[1], N. A. Levenson[2], A. Alonso-Herrero[3,4], A. Asensio Ramos[5,6], J. M. Rodríguez Espinosa[5,6], A. M. Pérez García[5,6], C. Packham[7], R. Mason[8], J. T. Radomski[2], and T. Díaz-Santos[9]

[1] University of Sheffield, Department of Physics & Astronomy, S3 7RH, UK.
[2] Gemini Observatory, Casilla 603, La Serena, Chile.
[3] Instituto de Física de Cantabria, CSIC-UC, Santander, E-39005, Spain.
[4] Augusto González Linares Senior Research Fellow.
[5] Instituto de Astrofísica de Canarias, La Laguna, Tenerife, E-38205, Spain.
[6] Departamento de Astrofísica, Universidad de La Laguna, E-38205, Spain.
[7] Astronomy Department, University of Florida, Gainesville, FL 32611-2055, USA.
[8] Gemini Observatory, Northern Operations Center, Hilo, HI 96720, USA.
[9] Spitzer Science Center, Pasadena, CA 91125, USA.

E-mail: `c.ramos@sheffield.ac.uk`

Abstract. We present near-to-mid-infrared spectral energy distributions (SEDs) for 21 Seyfert galaxies, using subarcsecond resolution imaging data. Our aim is to compare the properties Seyfert 1 (Sy1) and Seyfert 2 (Sy2) tori using clumpy torus models and a Bayesian approach to fit the infrared (IR) nuclear SEDs. These dusty tori have physical sizes smaller than 6 pc radius, as derived from our fits. Active galactic nuclei (AGN) unification schemes account for a variety of observational differences in terms of viewing geometry. However, we find evidence that strong unification may not hold, and that the immediate dusty surroundings of Sy1 and Sy2 nuclei are intrinsically different. The Type 2 tori studied here are broader, have more clumps, and these clumps have lower optical depths than those of Type 1 tori. The larger the covering factor of the torus, the smaller the probability of having direct view of the AGN, and vice-versa. In our sample, Sy2 tori have larger covering factors (C_T=0.95±0.02) and smaller escape probabilities than those of Sy1 (C_T=0.5±0.1). Thus, on the basis of the results presented here, the classification of a Seyfert galaxy may depend more on the intrinsic properties of the torus rather than on its mere inclination, in contradiction with the simplest unification model.

1. Introduction

The unified model for active galaxies [4] is based on the existence of a dusty toroidal structure surrounding the central region of AGN. This geometry of the obscuring material allows the central engines of Type 1 AGN to be seen directly, resulting in typical spectra with both narrow and broad emission lines, whereas in Type 2 AGN the broad-line region (BLR) is obscured.

Pioneering work in modelling the dusty torus [28, 27, 10, 8, 11] assumed a uniform dust density distribution, although from the start, [16] realized that smooth dust distributions cannot survive within the AGN vicinity. In order to solve the discrepancies between the observations and the predictions from smooth torus models, an intensive search for an alternative torus geometry has been carried out in the last decade. The clumpy dusty torus models [22, 23, 13, 34] propose that the dust is distributed in clumps, instead of homogeneously filling the torus volume, and they are making significant progress in accounting for the mid-IR (MIR) emission of AGNs [19, 21, 14, 25, 33, 31, 12, 1].

In this work, we report IR nuclear fluxes for a representative sample of nearby Seyfert galaxies, for which we estimate unresolved nuclear MIR fluxes. We compile near-IR (NIR) nuclear fluxes from the literature of similar resolution to construct nuclear SEDs for all the galaxies, and fit them with clumpy torus models, which we interpolate from the [22, 23] database. Table 1 summarizes key observational properties of the sources in the sample.

Table 1. Basic Galaxy Data.

Galaxy	Seyfert type	z	Distance (Mpc)	Scale (pc arcsec^{-1})	Telescope	Instrument	Filters	Resolution (arcsec)
NGC 1097	1	0.0042	19	92	Gemini-S	T-ReCS	Si-5, Qa	0.41, 0.52
NGC 1566	1	0.0050	20	97	Gemini-S	T-ReCS	Si-2, Qa	0.30, 0.53
NGC 6221	1	0.0050	18	87	Gemini-S	T-ReCS	Si-2, Qa	0.32, 0.55
NGC 7469	1	0.0163	65	315	Gemini-S	T-ReCS	Si-2, Qa	0.31, 0.55
NGC 3227	1.5	0.0039	17	82	Gemini-N	Michelle	N'	0.39
NGC 4151	1.5	0.0033	13	64	Gemini-N	OSCIR	N, IHW18	0.53, 0.58
NGC 6814	1.5	0.0052	21	102	Gemini-S	T-ReCS	Si-2, Qa	0.28, 0.53
NGC 1365	1.8	0.0055	18	107	CTIO 4m	OSCIR	N, IHW18	0.92, 1.03
NGC 2992	1.9	0.0077	31	149	Gemini-N	Michelle	N', Qa	0.32, 0.53
NGC 5506	1.9	0.0062	25	120	Gemini-N	Michelle	N', Qa	0.36, 0.51
Centaurus A	2	0.0018	3.5	17	Gemini-S	T-ReCS	Si-2, Qa	0.30, 0.53
Circinus	2	0.0014	4	20	Gemini-S	T-ReCS	Si-2, Qa	0.33, 0.55
IC 5063	2	0.0113	45	219	Gemini-S	T-ReCS	Si-2, Qa	0.40, 0.62
Mrk 573	2	0.0172	69	334	Gemini-S	T-ReCS	N, Qa	0.36, 0.54
NGC 1386	2	0.0029	11	56	Gemini-S	T-ReCS	N, Qa	0.31, 0.54
NGC 1808	2	0.0033	11	64	CTIO 4m	OSCIR	N, IHW18	0.94, 1.02
NGC 3081	2	0.0080	32	155	Gemini-S	T-ReCS	Si-2, Qa	0.30, 0.56
NGC 3281	2	0.0107	43	208	Gemini-S	T-ReCS	N, Qa	0.34, 0.58
NGC 4388	2	0.0084	34	163	Gemini-N	Michelle	N', Qa	0.34, 0.50
NGC 7172	2	0.0087	35	169	Gemini-S	T-ReCS	N	0.51
NGC 7582	2	0.0053	21	103	CTIO 4m	OSCIR	N, IHW18	0.76, 0.99

2. MIR Observations

Ground-based MIR high-angular resolution observations of 21 nearby active galaxies were carried out over the past years for a variety of science drivers. We make use of this archive of data in this paper. The first set of observations was obtained with the University of Florida MIR camera/spectrometer OSCIR at the 4 m Blanco Telescope at Cerro Tololo Inter-American Observatory (CTIO) and at the Gemini North Telescope. Another set of observations was performed with the MIR camera/spectrograph T-ReCS (Thermal-Region Camera Spectrograph) on the Gemini-South telescope. The last set was obtained with the MIR camera/spectrograph Michelle on the Gemini North Telescope (see Table 1). Observations of PSF and flux standard stars were made for each galaxy through the same filters. The PSF star images were employed to determine the unresolved (i.e., nuclear) component of each galaxy. For further details on the MIR observations of the whole sample we refer the reader to [33, 31].

3. SED Observational Properties

Using the MIR data described in Section 2 and NIR data of similar resolution compiled from the literature, we constructed subarcsecond resolution nuclear SEDs in the wavelength range from ~1 to 18 μm for the galaxies in Table 1. Figures 1 and 2 show the individual SEDs.

Using the highest resolution data only ($\leq$0.55 arcsec) we constructed average Sy1 and Sy2 templates (see Figure 4 in [31]) and measured the 1.265–18.3 μm IR slope ($f_\nu \alpha \nu^{-\alpha_{IR}}$) of the Sy1 template: $\alpha_{IR} = 1.7 \pm 0.3$, as well as the NIR ($\alpha_{NIR} = 1.6 \pm 0.2$) and MIR spectral indexes ($\alpha_{MIR} = 2.0 \pm 0.2$). A flat NIR slope indicates an important contribution of hot dust emission (up to ~1000-1200 K; [6]) that comes from the immediate vicinity of the AGN. The shape of the Sy2 mean SED is very steep ($\alpha_{IR} = 3.1 \pm 0.9$, $\alpha_{NIR} = 3.6 \pm 0.8$, and $\alpha_{MIR} = 2.0 \pm 0.2$) compared with those of the Type 1 Seyferts. In general, Sy2 have steeper 1–10 μm SEDs than Sy1 [9, 3, 2]. On the contrary, the MIR slope is the same for the Sy1 and Sy2 templates ($\alpha_{MIR} = 2.0 \pm 0.2$).

4. SED Modelling

The clumpy dusty torus models of Nenkova et al. (2002) hold that the dust surrounding the central engine of an AGN is distributed in clumps, instead of homogeneously filling the torus volume. The clumpy database now contains 1.2×10^6 models, calculated for a fine grid of model parameters. The inherent degeneracy between these parameters has to be taken into account when fitting the observables. To this end, we developed a Bayesian inference tool (BayesClumpy; [5]).

The priors for the model parameters are assumed to be truncated uniform distributions in the intervals reported in Table 2. Therefore, we give the same weight to all the values in each interval. Apart from the six parameters that characterize the models, there is an additional parameter that accounts for the vertical displacement required to match the fluxes of a chosen model to an observed SED. This vertical shift scales with the AGN bolometric luminosity (see Section 5).

Table 2. Clumpy Model Parameters and Considered Intervals

Parameter	Abbreviation	Interval
Width of the angular distribution of clouds	σ	[15°, 70°]
Radial extent of the torus	Y	[5, 30]
Number of clouds along the radial equatorial direction	N_0	[1, 15]
Power-law index of the radial density profile	q	[0, 3]
Inclination angle of the torus	i	[0°, 90°]
Optical depth per single cloud	τ_V	[5, 150]

The results of the fitting process of the IR SEDs with the interpolated version of the clumpy models of [22, 23] are the posterior distributions for the six free parameters that describe the models and the vertical shift. When the observed data introduce sufficient information into the fit, the resulting posteriors will clearly differ from the input uniform priors, either showing trends or being centered at certain values within the intervals considered.

We fitted the individual IR SEDs with BayesClumpy, modelling the torus emission and the direct AGN contribution (the latter as a broken power law). We also consider the IR extinction curve of [7] to take into account any possible foreground extinction from the host galaxy. Although the solutions to the Bayesian inference problem are the posterior distributions of each parameter, we can translate the results into corresponding spectra (see Figures 1 and 2). The solid lines correspond to the model described by the combination of parameters that maximizes their probability distributions (maximum-a-posteriori; MAP). Dashed lines represent the model computed with the median value of the probability distribution of each parameter. Shaded regions indicate the range of models compatible with the 68% confidence interval for each parameter around the median.

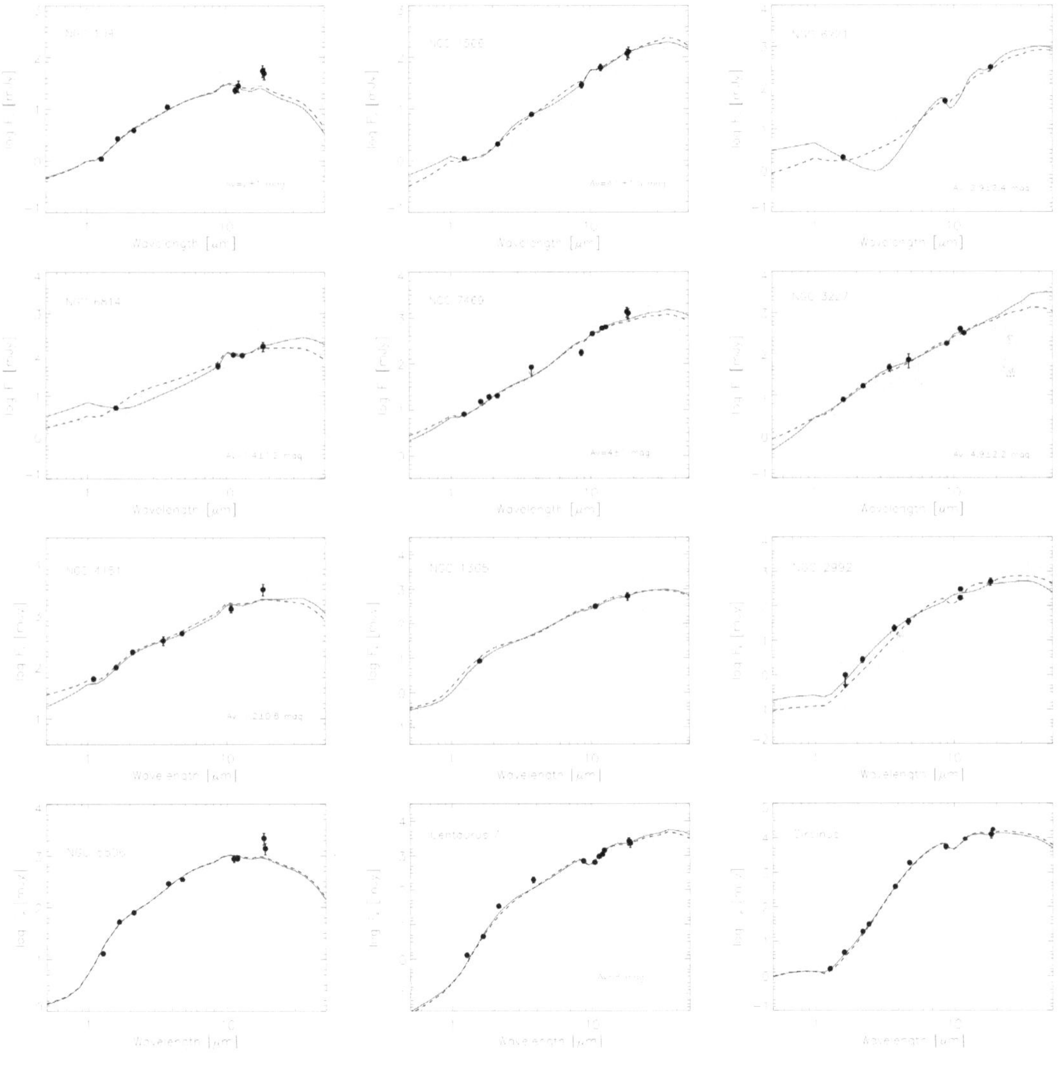

Figure 1. High spatial resolution IR SEDs of the Seyfert galaxies (dots). Solid and dashed lines correspond to the MAP and median models, respectively. Shaded regions indicate the range of models compatible with the 68% confidence interval for each parameter around the median.

The clumpy models successfully reproduce the observed Seyfert SEDs studied here with compatible results among them. This indicates that the NIR and MIR unresolved fluxes employed here are dominated by a combination of reprocessed emission from dust in the torus and direct AGN emission.

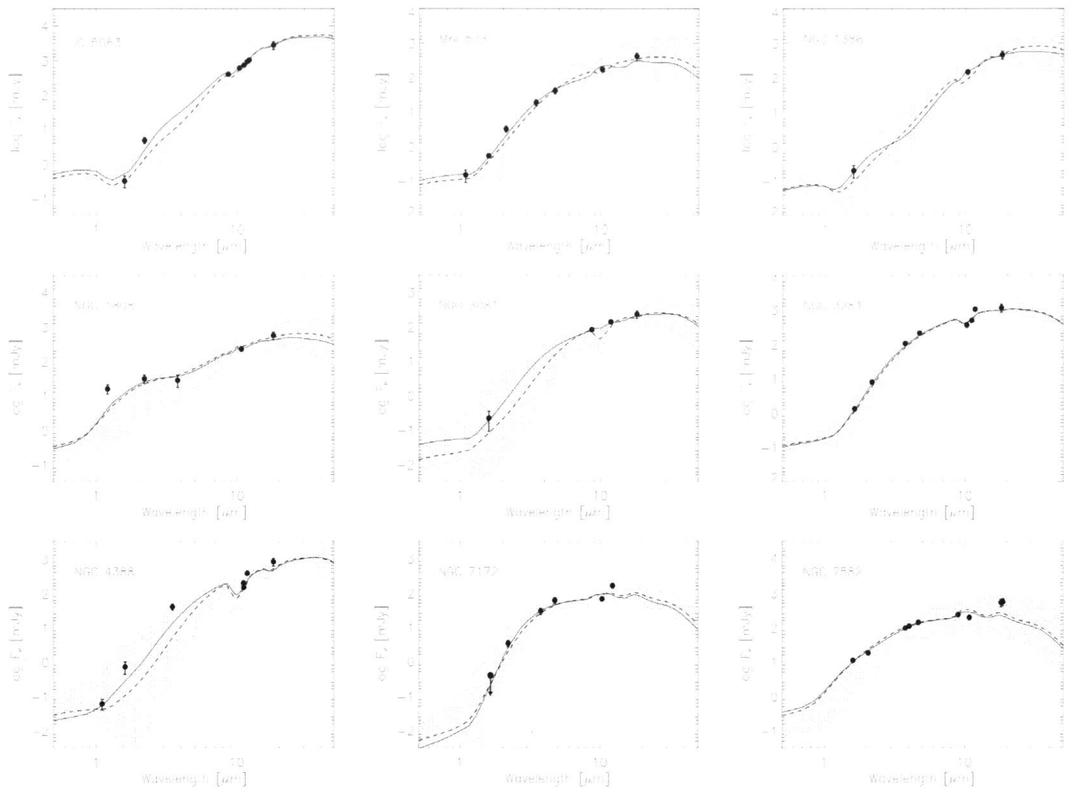

Figure 2. Same as in Figure 1, but for the remaining Sy2 galaxies.

5. Comparison between Type 1 and Type 2 Seyfert tori.

The aim of this work is to compare between the properties of Type 1 and 2 tori under the assumption that the SEDs studied here are torus/AGN dominated. Despite the relatively low number of objects considered (7 Sy1 and 9 Sy2; the unreliable fits of NGC 1808 and NGC 7582 are excluded), we find that some of the parameters are significantly different between Sy1 and Sy2.

To take full advantage of the Bayesian approach, the best way to compare the results for Sy1 and Sy2 galaxies is to derive joint posterior distributions for the full Type 1 and Type 2 datasets respectively. Thus, we normalized all the Sy1 SEDs at 8.74 μm and fitted them together using BayesClumpy, and we did the same for the Sy2. We considered the mean redshift for the Sy1 (z=0.0061$\pm$0.0045) and for the Sy2 (0.0078$\pm$0.0051) in the fits. In Figure 3 we show the Sy1 (left panel) and Sy2 fits (right panel). Note that the MAP and median models predict a flat SED with the silicate feature in weak emission for the Sy1 galaxies, and steeper and with the silicate band in shallow absorption for Sy2.

The comparison between the Sy1 and Sy2 posterior distributions is shown in Figure 4. From a visual inspection it is clear that the joint posteriors of the parameters N_0, q, τ_V, and σ are completely different between Sy1 and Sy2. There is no overlap between the 1-sigma intervals. In Table 3 we report the median and mode values of the histograms in Figure 4.

In order to quantify how different the probability distributions are, we calculated the Kullback-Leibler divergence (KLD; [17]) between the Sy1 and Sy2 posteriors. This divergence takes into account the full shape of the posterior and it is a positive value, and equal to zero when two distributions are identical. Therefore, the larger the value of KLD, the more different

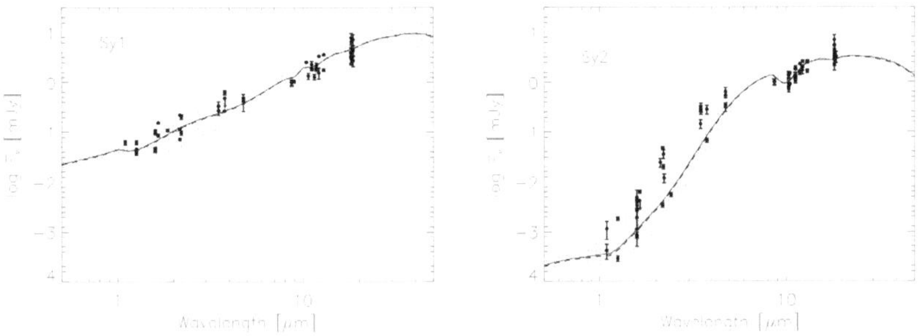

Figure 3. Same as in Figure 1, but for the Sy1 (left) and Sy2 SEDs (right) normalized at 8.74 μm.

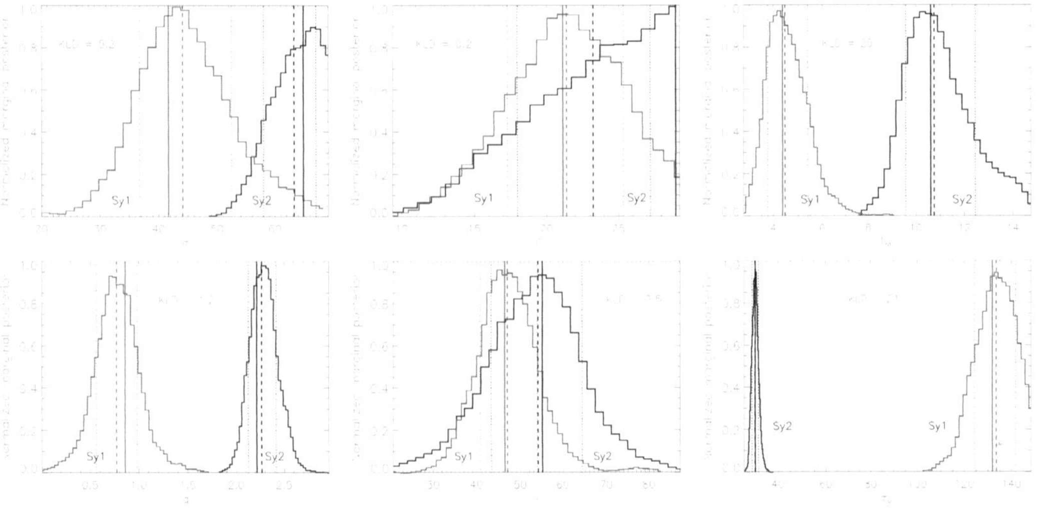

Figure 4. Posterior distributions resulting from the fits of the joint Sy1 and Sy2 SEDs. KLD values derived from the comparison between Sy1 and Sy2 for each parameter are labelled.

Table 3. Statistics of the comparison between Sy1 and Sy2 parameters

Type	σ (deg)		Y		N_0		q		i (deg)		τ_V	
	Sy1	Sy2	Sy1	Sy2	Sy1	Sy2	Sy1	Sy2	Sy1	Sy2	Sy1	Sy2
Medians	$44\pm^{8}_{7}$	$63\pm^{4}_{5}$	21 ± 4	$23\pm^{4}_{5}$	4 ± 1	$11\pm^{2}_{1}$	0.8 ± 0.2	2.3 ± 0.1	$47\pm^{7}_{6}$	$54\pm^{10}_{11}$	$133\pm^{8}_{9}$	30 ± 1
Modes	42	65	21	29	4	11	0.9	2.2	46	55	132	30

the posteriors. We find KLD>1 for σ, N_0, q, and τ_V (see Figure 4). These are indeed the four parameters whose 1-sigma regions do not overlap. Thus, we consider their differences significant between Sy1 and Sy2. For both Y and i we find KLD<1 and similar median values between Sy1 and Sy2.

Sy1 tori are narrower and have fewer clouds ($\sigma = 44^o\pm^{8^o}_{7^o}$; N_0=4±1) than those of Sy2 ($\sigma = 63^o\pm^{4^o}_{5^o}$; $N_0=11\pm^{2}_{1}$). The radial density distribution of the clouds is also different between the two Seyfert types according to this analysis: in Sy2, the majority of the clumps are distributed very close to the nucleus (i.e. steep radial density distribution; q=2.3±0.1) whereas for Sy1 the clouds distribution is flatter (q=0.8±0.2). On the other hand, the optical depth of the clouds in Sy1 tori is larger ($\tau_V=133\pm^{8}_{9}$) than in Sy2 (τ_V=30±1).

Interestingly, we find high as well as low values of the inclination angle of the torus for Sy1 and Sy2 (see Table 3). This variety in the i values translates into the similar median values of the joint Sy1 and Sy2 posteriors ($47^o \pm^{7^o}_{6^o}$ for Sy1 and $54^o \pm^{10^o}_{11^o}$ for Sy2), which are intermediate within the considered prior (i=[0°,90°]). *This is telling us that, in the clumpy torus scenario, the classification of a Seyfert galaxy as a Type 1 or 2 may depend on the intrinsic properties of the torus rather than in mere inclination.*

In Figure 5 we represent the median values of σ and N_0 for the different Seyfert types over the covering factor contours. The covering factor is defined as $C_T = 1 - \int e^{-N_{LOS}(i)} dcos(i)$. Type 1 nuclei tend to be located within lower C_T contours ($C_T \leq 0.6$) than those of Type 2s, for which $C_T \geq 0.5$, with the exceptions of Centaurus A and Mrk 573. We have represented with larger symbols the median values from the joint σ and N_0 posteriors reported in Table 3.

Since the covering factor is a non-linear function of the torus parameters, we took full advantage of our Bayesian approach and generated joint posterior distributions for C_T from those in Figure 4. The median values of the histograms are C_T(Sy2)=0.95±0.02 and C_T(Sy1)=0.5±0.1. The divergence between the Sy1 and Sy2 C_T posteriors is KLD=28, indicating that the difference is significant (1-sigma regions do not overlap). Thus, Sy1 tori in our sample have lower C_Ts than those of Sy2, implying that they are intrinsically different.

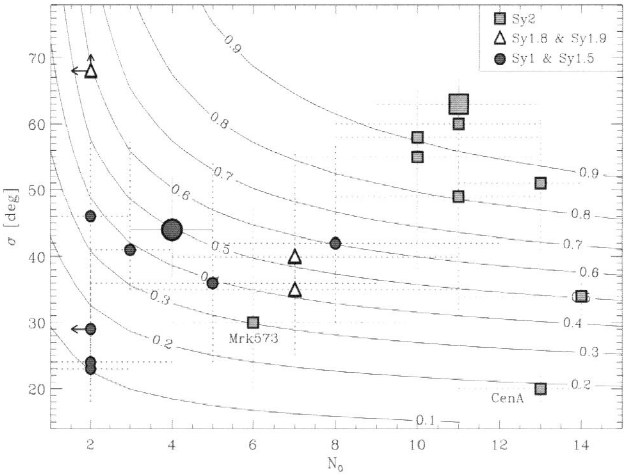

Figure 5. σ versus N_0 for the individual galaxies. Either median values or upper/lower limits are taken from the fits presented here. Error bars indicate 68% confidence level around the median. The big dot and square correspond to the average σ and N_0 values for Sy1 and Sy2 from Table 3.

The clumpy model fits yield the intrinsic bolometric luminosity of AGN (L_{bol}^{AGN}) by means of the vertical shift applied to match the observational data points. Combining this value with the torus luminosity (L_{bol}^{tor}), obtained by integrating the corresponding model torus emission (without the AGN contribution), we derive the reprocessing efficiency (RE) of the torus ($L_{bol}^{tor}/L_{bol}^{AGN}$). The previous values are calculated on the Bayesian framework, by combining the posterior distributions of the model parameters. Sy2 tori in our sample are more efficient reprocessors than Sy1, absorbing and re-emitting the majority of the intrinsic AGN luminosity in the IR: RE(Sy2)=[0.4, 1.0], with a median value of 0.8 and RE(Sy1)=[0.2,0.7], with median of 0.5.

We considered a possible dependency of the RE (or alternatively the covering factor; C_T) on L_{bol}^{AGN}, since the amount of incoming radiation from the AGN could possibly have some influence on the reprocessed energy or even in the torus properties (e.g., receeding torus scenario; [18]). However, we find no relationship between the two quantities in the luminosity range considered.

This means that the reprocessing efficiency depends primarily on the total number of clouds available to absorb the incident radiation, i.e. on the torus covering factor. However, the possible dependence of the torus properties on the AGN luminosity considering a broader luminosity range is further investigated in [1].

In general, the IR SED fitting does not constrain the size of the torus (Y) as well as other model parameters (see Section 5.3 in [33]). The NIR and MIR observations are sensitive to the warm dust (located within ~10 pc of the nucleus), which depends on the combination of model parameters N_0, q, and Y [35]. Far-IR observations are more sensitive to the torus extent independently, as shown in [32]. The outer size of the torus scales with the AGN bolometric luminosity: $R_o = Y R_d$, so assuming a dust sublimation temperature of 1500 K, $R_o = 0.4\ Y\ (L_{bol}^{AGN}/10^{45})^{0.5}$ pc. We derived R_o posterior distributions from those of L_{bol}^{AGN} and Y and find that all tori in our sample have outer radii smaller than 6 pc, in agreement with MIR direct imaging of nearby Seyferts [26, 30] and also interferometric observations [15, 36, 20, 29].

Summarizing, we find tantalizing evidence, albeit for a small sample of Seyfert galaxies and under the clumpy torus hypotesis, that the classification as a Type 1 and Type 2 depends on the intrinsic torus properties, in contradiction with the simplest unification model.

References

[1] Alonso-Herrero, A., et al. 2011, ApJ, 736, 82
[2] Alonso-Herrero, A., Quillen, A. C., Rieke, G. H., Ivanov, V. D., & Efstathiou, A. 2003, AJ, 126, 81
[3] Alonso-Herrero, A., Quillen, A. C., Simpson, C, Efstathiou, A., & Ward, M. J. 2001, AJ, 121, 1369
[4] Antonucci, R. R. J. 1993, ARA&A, 31, 473
[5] Asensio Ramos, A. & Ramos Almeida, C. 2009, ApJ, 696, 2075
[6] Barvainis, R. 1987, ApJ, 320, 537
[7] Chiar, J. E. & Tielens, A. G. G. M. 2006, ApJ, 637, 774
[8] Efstathiou, A., & Rowan-Robinson, M. 1995, MNRAS, 273, 649
[9] Fadda, D., Giuricin, G., Granato, G. L., & Vecchies, D. 1998, ApJ, 496, 117
[10] Granato, G. L., & Danese, L. 1994, MNRAS, 268, 235
[11] Granato, G. L., Danese, L., & Franceschini, A. 1997, ApJ, 486, 147
[12] Hönig, S. F., et al. 2010, A&A, 515, 23
[13] Hönig, S. F., Beckert, T., Ohnaka, K., & Weigelt, G. 2006, A&A, 452, 459
[14] Horst, H., Duschl, W. J., Gandhi, P., & Smette, A. 2009, A&A, 495, 137
[15] Jaffe, W., et al. 2004, Nature, 429, 47
[16] Krolik, J. H. & Begelman, M. C. 1988, ApJ, 329, 702
[17] Kullback, S., & Leibler, A. 1951, Annals Math. Stat., 22, 79
[18] Lawrence, A. 1991, MNRAS, 252, 586
[19] Mason, R. E., et al. 2006, ApJ, 640, 612
[20] Meisenheimer, K., et al. 2007, A&A, 471, 453
[21] Mor, R., Netzer, H., & Elitzur, M. 2009, ApJ, 705, 298
[22] Nenkova, M., Sirocky, M. M., Ivezić, Ž., & Elitzur, M. 2008, Ap
[23] Nenkova, M., Sirocky, M. M., Nikutta, R., Ivezić, Ž., & Elitzur, M. 2008, ApJ, 685, 160
[24] Nenkova, M., Ivezić, Ž., & Elitzur, M., 2002, ApJ, 570, 9
[25] Nikutta, R., Elitzur, M., & Lacy, M. 2009, ApJ, 707, 1550
[26] Packham, C., Radomski, J. T., Roche, P. F., Aitken, D. K., Perlman, E., Alonso-Herrero, A., Colina, L., & Telesco, C. M. 2005, ApJ, 618, L17
[27] Pier, E. A., & Krolik, J. H. 1993, ApJ, 418, 673
[28] Pier, E. A., & Krolik, J. H. 1992, ApJ, 401, 99
[29] Raban, D., Jaffe, W., Röttgering, H., Meisenheimer, K., & Tristram, K. 2009, MNRAS, 394, 1325
[30] Radomski, J. T., et al. 2008, ApJ, 681, 141
[31] Ramos Almeida, C., et al. 2011a, ApJ, 731, 92
[32] Ramos Almeida, C., et al. 2011b, MNRAS, 417, L46
[33] Ramos Almeida, C., et al. 2009, ApJ, 702, 1127
[34] Schartmann, M., et al. 2008, A&A, 482, 67
[35] Thompson, G. D., Levenson, N. A., Uddin, S. A., & Sirocky, M. M. 2009, ApJ, 697, 182
[36] Tristram, K. R. W, et al. 2007, A&A, 474, 837

Astronomy at High Angular Resolution 2011 IOP Publishing
Journal of Physics: Conference Series **372** (2012) 012005 doi:10.1088/1742-6596/372/1/012005

Radio properties of 0.4 < z < 0.8 BAL QSOs

J. Zuther

I. Physikalisches Institut, Universität zu Köln, Zülpicher Straße 77, 50937 Köln, Germany

E-mail: zuther@ph1.uni-koeln.de

S.Komossa

Max-Planck-Institut für Radioastronomie, Auf dem Hügel 69, 53121 Bonn, Germany

E-mail: skomossa@mpifr.de

H. Zhou

Polar Research Institute of China, Jinqiao Rd. 451, Shanghai 200136, China
Department for Astronomy, University of Science and Technology of China, Hefei, Anhui, China

E-mail: zhouhongyan@pric.gov.cn

Abstract. We present MERLIN 18 cm radio continuum observations of three $0.4 < z < 0.8$ broad-absorption-line (BAL) quasi-stellar objects (QSOs). The question, whether BAL QSOs can be interpreted in a pure orientation scheme, or whether evolutionary aspects have to be taken into account, is still a matter of debate. Radio-interferometric observations can provide additional information about both aspects. The presence of jets, their orientation, and beaming can provide clues on the orientation of the nuclear components. On the other hand, the compactness of the nuclear radio emission and its spectral index can provide information about the evolutionary state of the AGN, and of starbursts, if present.

1. Introduction

There appears to be an intimate link between the growth of supermassive black holes (SMBHs) and the galaxy bulges they reside in (e.g., [1, 2] and references therein). This suggests the existence of physical feedback mechanisms to regulate their co-evolution. In the quasi-stellar object (QSO) phase, i.e. during which the bulk of the accretion of matter onto the SMBH occurs, winds/outflows from the innermost region of the quasar can be an important source of feedback [2]. An extreme manifestation of strong outflows is the broad absorption line (BAL) phenomenon in QSOs. BAL QSOs are a rare subclass, comprising about 15% of the QSO population at high redshifts [3]. The outflows in the BAL QSOs are revealed by metal absorption systems with large blueshifted velocities (several thousand km/s). In recent years, two interpretations of this phenomenon emerged:

(1) Unified scheme: Under the assumption of the unified scheme for active galaxies [4], the absorbing clouds have a small covering factor as seen from the QSO nucleus [5]. Thus, the frequency of detection just translates to the rate at which our line of site intercepts the outflow.

(2) Evolution scheme: The evolution model for BAL QSOs represents one aspect of the *merger-driven ultraluminous-infrared galaxy (ULIRG)-to-QSO-evolution* scenario proposed by [6], which describes the process of accumulation of large amounts of molecular gas and dust in the center of the merger remnant, subsequently triggering massive star formation (ULIRG phase) and nuclear activity. In the next phase, which marks the BAL QSOs stage, these young objects are ejecting their gaseous envelopes at very high velocities following the initial turn-on of the active galactic nucleus [7, 8, 9]. This turn-on can also be connected with circumnuclear starbursts which contribute to such outflows [10]. In the final stage the AGN feedback has removed all the gas and dust from the nuclear region and appears as a classical QSO with a red elliptical host [11]. There is quite some controversy about the nature of the BAL phenomenon, since several investigations support one or the other model (cf. discussion in [12]).

Quite recently, studies in the radio domain have been able to shed light on these issues [13, 14, 15, 16]. The radio properties can be used to study orientation and compactness of BAL QSOs. The former revealed that BAL QSOs apparently occur in two varieties, namely in equatorial and polar configurations [17]. The latter, combined with radio-spectral indices, shows that BAL QSOs are similar to young radio sources [15, 16]. It, therefore, appears that the BAL phenomenon is more complex and probably a combination of the two "basic" scenarios.

The origin of the BAL phenomenon is still under debate and maybe linked with the host galaxy in an evolution scenario. BAL QSOs are typically found at high redshifts ($z > 1$) [18] coinciding with the peak of the cosmic QSO space density. Because of the small angular extent of the host galaxy and brightness of the QSO, it is very costly to address this issue. However, when inspecting observations of the handful of low-redshift BAL QSOs, one finds that virtually all of them are associated with galaxy mergers and enhanced star formation [9, 10]. Therefore, it might be instructive to have a closer look at the extended radio emission in BAL QSOs.

2. Observations and results

Here, we focus on the $z < 1$ part of the redshift distribution of BAL QSOs, which will allow us to carry out more detailed structural investigations (Zuther et al. in prep.). In a pilot study, we chose three low-ionization BAL QSOs in the $0.4 < z < 0.8$ rangerange from the sample of SDSS low-ionization BAL QSOs [19].These three objects were detected as unresolved sources with the Very Large Array FIRST survey at a few milli-Jansky (0758+553, 1017+125, 0908+407). Based on the 20 cm and SDSS i-band flux densities, their radio-loudness is of intermediate character, $\log R \equiv \log(S_{20\mathrm{cm}}/S_i) \approx 1$. They have SMBH masses of about $10^{7.5} M_\odot$ and accrete at the Eddington limit (based on their g-band luminosities and a bolometric correction according to [20]). We observed these three BAL QSOs at 18 cm with the Multi-Element-Radio-Linked Interferometer Network (MERLIN). The reconstructed images have an angular resolution of about 300 milli-arcseconds (mas). This corresponds to linear scales of 1.6-2 kpc at the object's redshifts. Maps are presented in Fig. 1.

The objects in the MERLIN maps are unresolved and recover about 60% of the FIRST 20 cm flux density. Most of the flux, therefore, is constrained to the central two kilo-parsecs. These objects show no jet-like large-scale structures, neither in the FIRST images on scales of several 10 kpc, nor our MERLIN images. Taking variability and interstellar scintillation into account, which typically can be of the order of 10-30%, it seems plausible that low-surface-brightness emission due to extended emission by star formation does not play a dominant role - at least at radio wavelengths. We, furthermore, estimated the star-formation rate assuming that all radio emission is due to star formation. This would then predict about $1000 M_\odot \mathrm{yr}^{-1}$. The corresponding far-infrared luminosity would be of the order of $10^{13} L_\odot$, in conflict with the fact that none of the objects has been observed to be a ULIRG. The brightness temperatures of about 10^5 K, estimated from the MERLIN beam size, are on the lower end of brightness temperatures found for typical accretion scenarios ($\sim 10^{5-7}$ K, [21]). Nevertheless, assuming

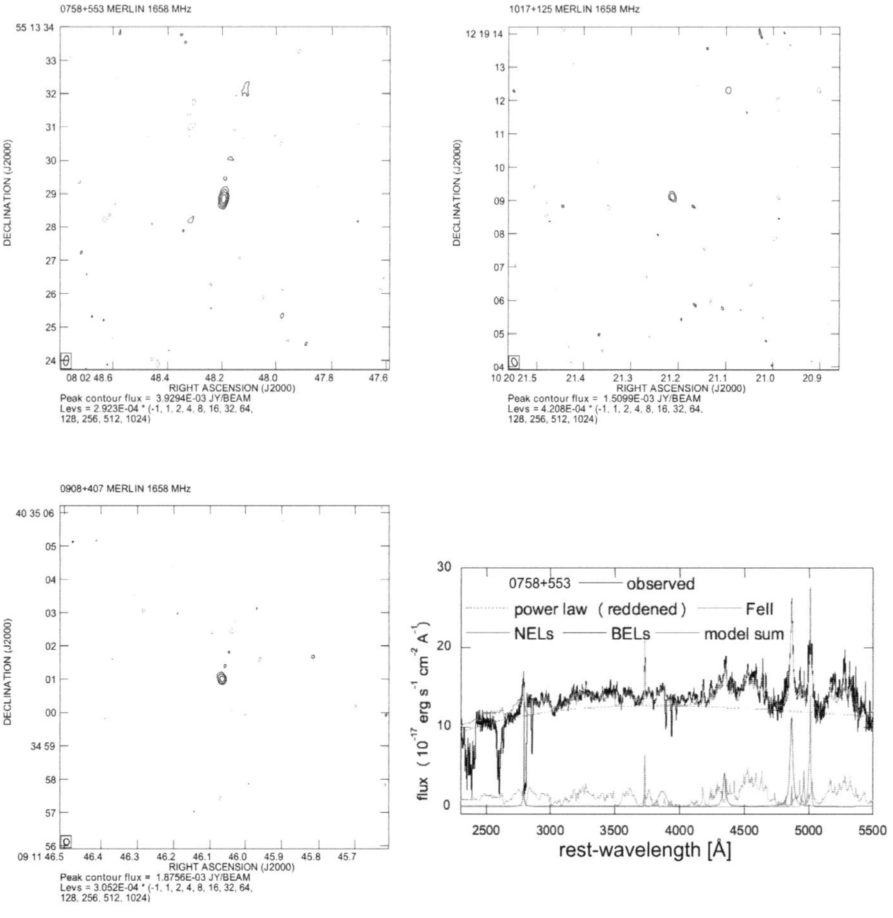

Figure 1. MERLIN 18 cm maps of the three MgII BAL QSOs. Contours are in multiples of three times the off-source noise in the images. Beam sizes are indicated in the lower-left corner of each map. The lower-right panel shows a representative SDSS spectrum. Components of the spectral fit are indicated.

that the emission is originating from an even more compact region (e.g., [13, 22]), one can expect the MERLIN-based brightness temperatures to be a lower limit. We therefore conclude that the radio emission is not dominated by starbursts.

This conclusion is supported by the fact that the SDSS spectra are dominated by the QSO, even though one might expect to recognize the host galaxy in the spectrum because of the strong supression of the AGN emission by the BAL outflow (lower right panel in Fig. 1). However, the [OII] emission is relatively strong and might be related to star formation [23]. Therefore, higher angular resolution observations, preferentially at several wavelengths are needed.

3. Summary

We presented preliminary results from a radio-interferometric study of intermediate-redshift BAL QSOs. Our MERLIN 18 cm data show compact emission on the sub-2 kpc scale. The compactness together with the brightness temperatures and the AGN dominated SDSS spectra indicate that star formation does not play a dominant role in these objects - at least as traced by the radio emission.

Although we did not find jet structures on large scales, we cannot conclude on the orientation properties, because the scales probed with MERLIN at 18 cm are still too large. We showed the feasibility to study intermediate-redshift BAL QSOs at milli-Jy flux-density levels with interferometric techniques. Future follow-up observations at shorter wavelengths, especially with *e*MERLIN, will provide higher angular resolution and additional spectral information, necessary to study the origin of the compact radio emission in these objects.

Acknowledgments

This work has benefited from research funding from the European Community's sixth Framework Programme under RadioNet R113CT 2003 5058187. MERLIN is a National Facility operated by the University of Manchester at Jodrell Bank Observatory on behalf of STFC. Part of this work was supported by the German *Deutsche Forschungsgemeinschaft, DFG* project number SFB 956. JZ acknowledges support from the European project EuroVO DCA under the Research e-Infrastructures area (RI031675), and travel support within the DFG Schwerpunktprogram SPP1177.

References

[1] Alexander, D. M., & Hickox, R. C. 2011, arXiv:1112.1949
[2] Fabian, A. C. 2012, arXiv:1204.4114
[3] Hewett, P. C., & Foltz, C. B. 2003, AJ, 125, 1784
[4] Antonucci, R. 1993, ARA&A, 31, 473
[5] Weymann, R. J., Morris, S. L., Foltz, C. B., & Hewett, P. C. 1991, ApJ, 373, 23
[6] Sanders, D. B., Soifer, B. T., Elias, J. H., et al. 1988, ApJ, 325, 74
[7] Hazard, C., Morton, D. C., Terlevich, R., & McMahon, R. 1984, ApJ, 282, 33
[8] Briggs, F. H., Turnshek, D. A., & Wolfe, A. M. 1984, ApJ, 287, 549
[9] Canalizo, G., & Stockton, A. 2001, ApJ, 555, 719
[10] Lípari, S. L., & Terlevich, R. J. 2006, MNRAS, 368, 1001
[11] Tacconi, L. J., Genzel, R., Lutz, D., et al. 2002, ApJ, 580, 73
[12] Brotherton, M. S. 2007, The Central Engine of Active Galactic Nuclei, 373, 315
[13] Jiang, D. R., & Wang, T. G. 2003, A&A, 397, L13
[14] Gregg, M. D., Becker, R. H., & de Vries, W. 2006, ApJ, 641, 210
[15] Montenegro-Montes, F. M., Mack, K.-H., Vigotti, M., et al. 2008, MNRAS, 388, 1853
[16] DiPompeo, M. A., Brotherton, M. S., De Breuck, C., & Laurent-Muehleisen, S. 2012, arXiv:1201.2916
[17] Zhou, H., Wang, T., Wang, H., et al. 2006, ApJ, 639, 716
[18] Trump, J. R., Hall, P. B., Reichard, T. A., et al. 2006, ApJS, 165, 1
[19] Zhang, S., Wang, T.-G., Wang, H., et al. 2010, ApJ, 714, 367
[20] Vestergaard, M. 2004, ApJ, 601, 676
[21] Narayan, R., Mahadevan, R., Grindlay, J. E., Popham, R. G., & Gammie, C. 1998, ApJ, 492, 554
[22] Zuther, J., Fischer, S., & Eckart, A. 2012, arXiv:1204.3162
[23] Ho, L. C. 2005, ApJ, 629, 680

Journal of Physics: Conference Series **372** (2012) 012006 doi:10.1088/1742-6596/372/1/012006

The SED of Low-Luminosity AGNs at high-spatial resolution

J A Fernández-Ontiveros[1,2,3], M A Prieto[1,2], J A Acosta-Pulido[1,2] and M Montes[1,2]

[1] Max-Planck-Institut für Radioastronomie (MPIfR), Auf dem Hügel 69, Bonn, D–53121, Germany

[2] Instituto de Astrofísica de Canarias (IAC), Vía Láctea s/n, La Laguna, E–38200, Spain

[3] Departamento de Astrofísica, Facultad de Física, Universidad de La Laguna, Astrofísico Fco. Sánchez s/n, La Laguna, E–38207, Spain

E-mail: jafo@mpifr.de

Abstract.
The inner structure of AGNs is expected to change below a certain luminosity limit. The big blue bump, footprint of the accretion disk, is absent for the majority of low-luminosity AGNs (LLAGNs). Moreover, recent simulations suggest that the torus, a keystone in the Unified Model, vanishes for nuclei with $L_{bol} \lesssim 10^{42}\,\mathrm{erg\,s^{-1}}$. However, the study of LLAGN is a complex task due to the contribution of the host galaxy, which light swamps these faint nuclei. This is specially critical in the IR range, at the maximum of the torus emission, due to the contribution of the old stellar population and/or dust in the nuclear region. Adaptive optics imaging in the NIR (VLT/NaCo) together with diffraction limited imaging in the mid-IR (VLT/VISIR) permit us to isolate the nuclear emission for some of the nearest LLAGNs in the Southern Hemisphere. These data were extended to the optical/UV range (*HST*), radio (VLA, VLBI) and X-rays (*Chandra*, *XMM*-Newton, *Integral*), in order to build a genuine spectral energy distribution (SED) for each AGN with a consistent spatial resolution ($< 0\farcs5$) across the whole spectral range. From the individual SEDs, we construct an average SED for LLAGNs sampled in all the wavebands mentioned before. Compared with previous multiwavelength studies of LLAGNs, this work covers the mid-IR and NIR ranges with high-spatial resolution data. The LLAGNs in the sample present a large diversity in terms of SED shapes. Some of them are very well described by a self-absorbed synchrotron (e.g. NGC 1052), while some other present a thermal-like bump at $\sim 1\,\mu$m (NGC 4594). All of them are significantly different when compared with bright Seyferts and quasars, suggesting that the inner structure of AGNs (i.e. the torus and the accretion disk) suffers intrinsic changes at low luminosities.

1. Introduction

The majority of AGNs spend their lives in a low state, characterized by a low accretion rate and a modest luminosity. These are known as low-luminosity AGNs (LLAGNs), the faintest but also the most numerous members of this family, including $\sim 1/3$ of all galaxies in the Local Universe [1]. However, LLAGNs are not just scale-down versions of the bright counterparts, Seyfert galaxies and quasars. Their main characteristics are: *i)* absence of the **big blue bump** [2], footprint of the accretion disk in the Spectral Energy Distribution (SED), *ii)* LLAGNs are **"radio-loud"**, usually showing compact cores and parsec-scale radio jets [3], and *iii)* they are frequently associated with **low ionization** nuclear emission-line regions (LINERs, [4]). The differences between high-luminosity objects and LLAGNs might be associated with the structural changes predicted at low luminosities. Both the broad-line region and the "torus", keystones of the Unified Model [5], are expected to vanish at $L_{bol} \lesssim 10^{42}$ erg/s [6, 7]. Instead of the classical scheme, the nucleus of LLAGNs is described in terms of a **radiatively inefficient** structure in which most of the gravitational energy is not released in the form of electromagnetic radiation [8, 1]. Its main components are: *i)* a radiatively inefficient accretion flow (RIAF), *ii)* a truncated accretion disk, and *iii)* a jet or an outflow. These components contribute to the total energy output and compete at certain wavelength ranges, but the relative importance of each one of them is still not clear (e.g. [9]). Specifically, RIAF-dominated models predict that most of the energy is released at radio and X-ray wavelengths, while high-spatial resolution data suggest an important contribution in the IR (see Fig. 2). On the other hand, some LLAGNs still show broad lines in polarized light [10] together with a high absorption column ($\gtrsim 10^{23}\,\mathrm{cm}^{-2}$, [11]), which claim for the presence of a torus in their nuclei. In brief, LLAGNs permit to explore the boundaries of unified model, the luminosity radiated is not high enough to support the torus and the structure of the accretion disk also seems to be altered at low accretion ratios.

2. Dataset

The dataset on which this work is based consist on a multiwavelength, high-spatial resolution dataset covering the radio, infrared, optical, UV and X-ray ranges. This project is a follow-up of *The central parsecs of the nearest galaxies*[1] [12, 13], a high-spatial resolution study of the brightest and nearest Seyfert galaxies carried out at sub-arcsec scales with the Very Large Telescope (VLT). The objects included in the present work (Table 1, Fig. 1) correspond to AGNs one to two order of magnitude fainter ($L_{bol} \lesssim 10^{42}\,\mathrm{erg\,s^{-1}}$).

The observations include the near (NIR, VLT/NaCo) and mid-infrared (mid-IR, VLT/VISIR) ranges for the inner few kpc of a sample of 6 galaxies (Table 1). These observations were acquired using VLT/NaCo adaptive optics in the NIR and diffraction limited imaging with VLT/VISIR in the mid-IR. The dataset was completed with the highest-spatial resolution data available in the optical/UV range from the *HST* scientific archive. Nuclear fluxes were measured using aperture photometry of the unresolved component in the centre ($\lesssim 0''.1$), subtracting the local background around ($0''.2$–$0''.3$). In the mid-IR, *PSF* photometry was performed (see [12]). An extensive search in the literature has been performed in order to complete the radio and X-ray ranges. Flux measurements in these ranges correspond mainly to Very Large Array (VLA) and Very Long Baseline Interferometry (VLBI) in radio and *Chandra*, *XMM*-Newton and *Integral* at X-rays. The characteristics of this dataset permit us to build, for each LLAGN in the sample, a consistent SED of the same physical region, very well sampled over a wide range in wavelength. This permit us to disentangle the nuclear emission from the host galaxy light, which otherwise dominates the energy output.

3. Multiwavelength SEDs

Multiwavelength SEDs were built at a consistent subarsec resolution based on the aperture photometry data in the IR, optical and UV ranges together with the radio and X-ray measurements from the literature (Section 2). First, we will focus on the cases of NGC 1052 and the Sombrero galaxy (NGC 4594). Both are very well sampled in wavelength. The former is a very good example of a SED described by a power-law in the IR-to-UV range, while the latter present a different behaviour showing a thermal NIR bump at $\sim 1\,\mu$m. The rest of the galaxies in the sample present a SED shape with some degree of similarity to one of these two cases.

[1] `http://www.iac.es/project/parsec`

Table 1. Galaxy sample (objects sorted by increasing distance). FWHM correspond to the size of the most compact object found in the FOV. L_{bol} is estimated from the high-spatial resolution SEDs.

Name	**D** [Mpc]	**Ref.**	**FWHM** [pc]	**1″** [pc]	$\mathbf{L_{bol}}$ [erg s^{-1}]	**Type**	**Class**
NGC 4594	9.08	[14]	3.1	44.0	1.9×10^{41}	SA(s)a	LINER 2
NGC 1097	14.2	[15]	8.9	68.8	1.9×10^{42}	SB(r′l)b	LINER 1
NGC 1386	15.3	[14]	6.7	74.2	6.6×10^{42}	SB(s)a	Sy 2
M87	16.7	[16]	9.2	81.0	1.2×10^{42}	cD	LINER 1
NGC 1052	18.0	[14]	10.5	87.3	9.3×10^{42}	E4	LINER 1.9
NGC 3169	24.7	[17]	10.9	119.7	8.1×10^{41}	SA(s)a	LINER 2

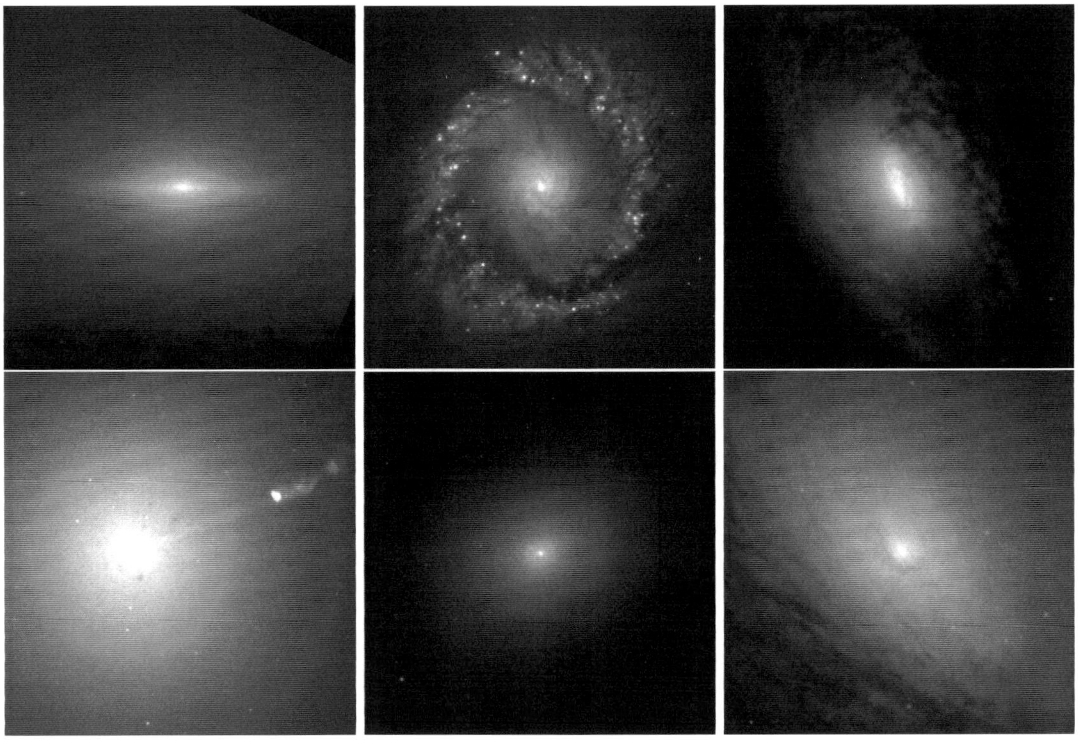

Figure 1. Optical-IR colour composition for the central parsecs of the galaxies in the sample. From left to right, top to bottom, the frames corresponds to NGC 4594 (FOV= 15″ × 15″), NGC 1097 (32″ × 32″), NGC 1386 (26″ × 26″), M87 (29″ × 29″), NGC 1052 (30″ × 30″) and NGC 3169 (10″ × 10″).

3.1. NGC 1052

NGC 1052 is a prototypical example of a LINER type 2 nucleus [4]. Fig. 2 shows the high-spatial resolution SED (black dots), together with large aperture measurements (grey spikes). The nucleus is the brightest source in the radio and mid-IR ranges, and seems to dominate the millimetre and sub-millimetre ranges as well. However, the bulk of the stellar light is nearly two order of magnitude above the nuclear emission for large apertures at optical and IR wavelengths. Only high-spatial resolution measurements permit to isolate the flux of the LLAGN and discern

Journal of Physics: Conference Series **372** (2012) 012006 doi:10.1088/1742-6596/372/1/012006

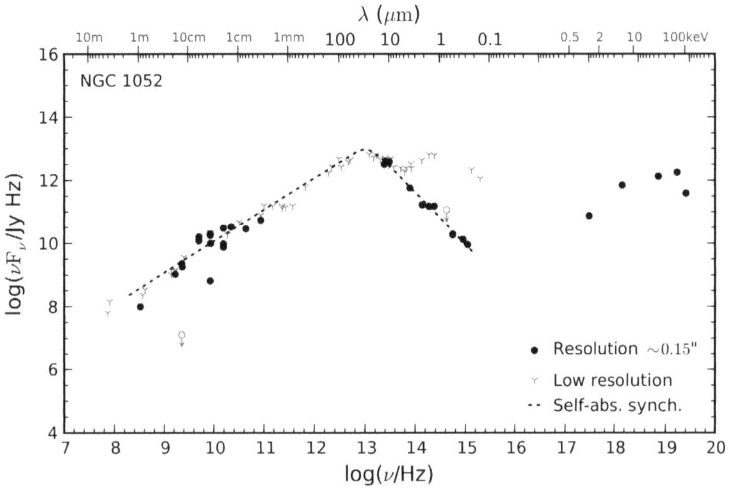

Figure 2. High-spatial resolution SED for NGC 1052 (black dots). Low-spatial resolution data are also shown for comparison (grey spikes). A self-absorbed synchrotron model, representative of a jet emission, has been fitted to the radio, IR, optical and UV range (dashed-line).

the nature of the emission in this range.

Based on the high X-ray absorption column ($N_H \sim 10^{23}\,\mathrm{cm}^{-2}$, [18]) and the presence of a broad Hα line in polarised light [19], one would expect a strongly obscured nucleus for this galaxy in the optical/UV range. However, instead of a continuum affected by a high reddening, the high-spatial resolution SED follows a power law in the mid-IR-to-UV range (Fig. 2). Moreover, a variation in the position of the nucleus with wavelength, common in highly reddened AGNs, is not detected in this case. This power-law behaviour is similar to the case of NGC 4258 [20]. A self-absorbed synchrotron [21], i.e. a very simple model of a jet, describes very well the high-spatial resolution SED of NGC 1052 in the radio-to-UV range (dashed-line in Fig. 2). Radio core fluxes, corresponding to milliarcsec apertures are not included in the fit. The optically-thick part of the self-absorbed synchrotron, from radio to mid-IR, presents a flat slope ($\alpha \approx 0$, $S_\nu \propto \nu^{-\alpha}$), followed by a turnover and the optically-thin emission in the mid-IR-to-UV range ($\alpha \approx 2.7$). Although a detailed modelling is desired, this simple exercise show that the energy output in the nucleus of NGC 1052 is driven the jet. Other LLAGNs also show a similar behaviour (e.g. M81, [22]). In this context, the X-ray flux would be associated to the comptonization of photons emitted by syncrotron radiation.

The IR range is described by the Unified Model [5] in terms of thermal emission from dust in a torus of a few parsecs in size. However, at a spatial resolution of $\sim 10.5\,\mathrm{pc}$, the dominant mechanism seems to be non-thermal. This is supported by the polarisation degree ($\sim$ 3–5%) found in the continuum emission by [23] in the IR and by [24] in the UV range, based on $7''$ and $10''$ diameter apertures, respectively. If a torus is present in the centre of this LLAGN, as suggested by the broad Hα line in the polarised spectrum and the high N_H column density, its contribution to the parsec-scale emission is very low.

3.2. NGC 4594

Also known as the Sombrero galaxy or M104, NGC 4594 hosts the faintest nuclei in the sample ($L_{bol} \sim 1.9 \times 10^{41}\,\mathrm{erg\,s^{-1}}$). Although this LLAGN belongs to the type 2 class, the nucleus is not obscured, as suggested by X-ray ($N_H \sim 10^{21}\,\mathrm{cm}^{-2}$, [25]) and optical observations ($E(B-V) = 0.08$, [26]). Fig. 3 shows the high-spatial resolution SED (black dots), together

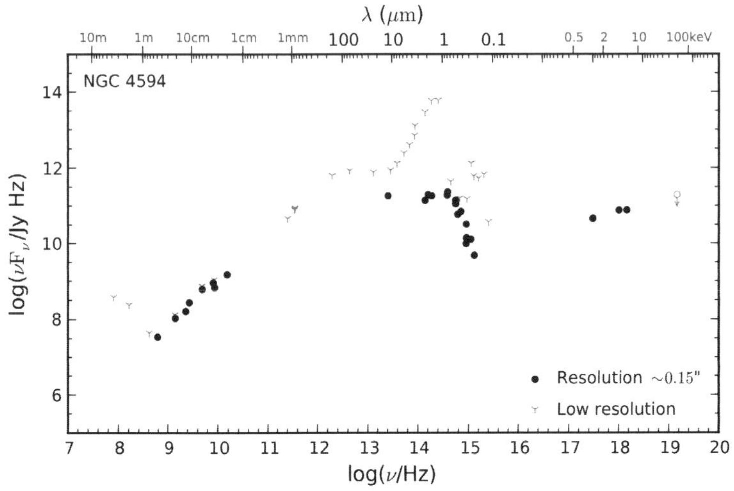

Figure 3. High-spatial resolution SED for NGC 4594 (black dots). Low-spatial resolution data are also shown for comparison (grey spikes).

with the measurements based on large apertures (grey spikes). As in the case of NGC 1052, the optical range is dominated by the stellar emission in low-spatial resolution data. At high-spatial resolution, NGC 4594 shows a thermal-like bump with a maximum at $\sim 1\,\mu$m. This component presents a soft variability in a timescale of a few years, thus cannot have a stellar origin. On the other hand, this thermal-like bump in the IR can be associated to a truncated accretion disk, expected for LLAGNs [8], although the non-detection of the Fe K line emission in X-rays by [27] suggests the absence of an accretion disk in this object. Another possible origin for this emission could be associated with synchrotron emission produced in the base of a jet, as in the case of X-Ray binaries in the low-hard state [28, 22].

4. Average LLAGN at parsec scales

Left panel in Fig. 4 shows the high-spatial resolution SED for all the LLAGNs in the sample. Each SED has been normalized by its mean power $(\overline{\nu F_\nu})$ in order to facilitate their comparison. The dispersion of the SED shapes is large, specially in the mid-IR-to-UV range. However, three main characteristics are common for all the LLAGNs in the sample: *i)* a flatter spectra at radio wavelengths when compared with Seyfert galaxies, associated to the presence of compact cores and jets, *ii)* the bulk of the emission is released in the IR range, and *iii)* the IR luminosity is very similar to the X-ray luminosity. The latter suggests that there might be a relation between both ranges. In Seyferts and quasars, the IR emission is produced by dust in the torus, as a consequence of the absorption and re-emission of the optical and X-ray light. In LLAGNs the presence of the torus component is not so clear. Figs 2 and 3 show that the IR is not apparently dominated by a thermal component, as in the case of Seyfert nuclei. Notwithstanding, the mechanism that drives the energy output in LLAGNs seems to connect both ranges.

In order to compare with previous studies, an average LLAGN template was obtained from the individual SEDs. The method described in [13] was followed to derive the average LLAGN SED (right panel in Fig. 4). The main characteristics mentioned before are reflected in the average template, which shows a flat radio continuum and a maximum in the mid-IR. The diversity of SEDs in the mid-IR-to-UV range translates into a mix, in the average LLAGN, of the main individual features. These are the thermal-like bump around $\sim 1\,\mu$m plus a steep power-law continuum decreasing from the mid-IR to the UV.

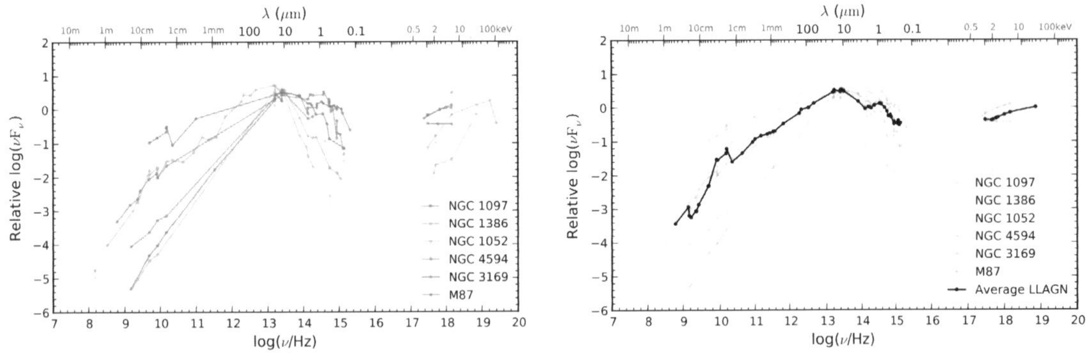

Figure 4. *Left panel:* high-spatial resolution SEDs for the six LLAGNs in the sample.*Right panel:* in black, average SED for LLAGNs obtained from the individual SEDs shown in the left panel.

4.1. Comparison with previous studies

Left panel in Fig. 5 shows the average LLAGN (black dots) together with an average SED distribution obtained from the sample of LINERs in [29] (green dots), using the same method as in our sample. The average LINER SED given by [30] is also shown (grey triangles). Overall, our average SED for LLAGNs is consistent with previous works in the common wavelength ranges covered. LLAGNs are bright at radio wavelengths but do not show the big blue bump, footprint of the accretion disk, in the optica/UV range. In contrast, the spectral slope in this part of the spectrum is very steep, decreasing with wavelength. At X-rays, the SED becomes almost flat for all the templates.

However, subarsec resolution data has been included for the first time in the NIR (VLT/NaCo adptive opctics) and mid-IR (VLT/VISIR diffraction limited data) ranges. This permit us to characterize the region of the spectrum in which these faint nuclei radiate most of their luminosity. As mentioned before, the nature of the emission in this range seems to be associated with non-thermal processes. For comparison, the average SED for "radio-loud" quasars given by [31] is also shown in the left panel of Fig. 5 (orange line). As pointed out by previous authors, the "radio-loud" quasar template presents also a similar shape when compared with LLAGNs in the radio and X-ray ranges.

4.2. Low and high-luminosity AGNs

The average SED for Seyfert 1 (in blue) and Seyfert 2 nuclei (in red) derived from high-spatial resolution data by [13] are shown in the right panel of Fig. 5. Compared with the LLAGN template, Seyferts nuclei exhibit a strong and thermal mid-IR component, associated with the emission from the dust in the torus. At radio wavelengths, both Seyfert SEDs show a steeper decrease, more similar to "radio-quiet" quasars, in contrast with LLAGNs which are "radio loud". In the optical/UV range, the differences between both classes of AGN are larger. Seyfert 1s present a big blue bump, i.e. the emission from the accretion disk, although this is softer when compared with the one in quasars [13]. In contrast, the shape of the Seyfert 2 template is determined by the obscuration in these nuclei. This is in line with the Unified Model, in which the torus hides the central engine for type 2 objects. However, LLAGNs present a significantly different shape when compared with Seyferts. Furthermore, its SED cannot be reproduced by applying a foreground extinction to the Seyfert 1 template (dashed line in left panel of Fig. 5).

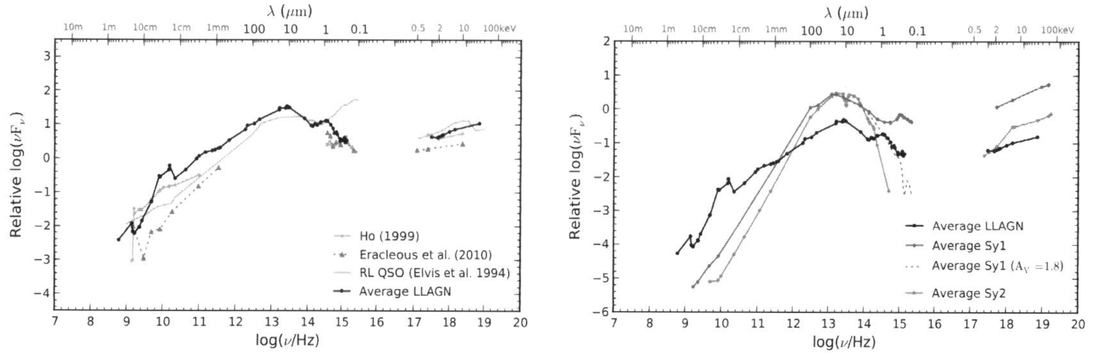

Figure 5. *Left panel:* comparison of the LLAGN average SED obtained in this study (black dots) with previous studies based on high-spatial resolution data, from [29] (green dots) and [30] (grey triangles). In orange, the radio-loud template for QSOs provided by [31]. *Right panel:* high-spatial resolution average SED for LLAGNs (black dots), Seyfert 1s (blue dots, [13]) and Seyfert 2s (red dots, [13]).

5. Summary

The importance of LLAGNs lies in the fact that they permit to explore the limits of the classical picture for active nuclei. At very low luminosities ($L_{bol} \lesssim 10^{42}\,\mathrm{erg\,s^{-1}}$) the broad-line region and the torus are expected to vanish, giving way to a radiatively inefficient structure. As a consequence of the changes in the inner structure, the SED of LLAGNs is expected to present differences with regard to the bright class.

A high-spatial resolution study has been performed for a sample of 6 nearby LLAGNs. This includes sub-arcsec resolution data in the radio, IR, optical, UV and X-ray ranges and allows us to disentangle the nuclear light from that of the host galaxy. For the first time, the mid-IR to NIR range is sampled at high-spatial resolution, a range in which LLAGNs radiate most of their luminosity. The shape of the SED in LLAGNs suggests that non-thermal processes can dominate the emission from radio to UV wavelengths. Self-absorbed synchrotron emission in a jet is a possible mechanism to explain the radio-to-UV continuum [22]. Still, in the case of the Sombrero galaxy this synchrotron continuum is not present in the IR-to-UV range, and the thermal-like emission could be associated to a truncated accretion disk.

Overall, the high-spatial resolution SEDs of the LLAGNs in the sample are intrinsically different when compared with bright AGNs. The big blue bump is absent, even for unobscured nuclei. The strong thermal IR component associated with the torus in Seyfert nuclei is not detected for LLAGNs, in line with numerical simulations that predicts this structure to receed at low-luminosities. If it is still present, the torus does not seem to dominate the emission in faint nuclei. Nevertheless, there are still some similarities with the brighter counterparts. At radio wavelengths, LLAGNs show a similar spectral index as "radio-loud" quasars. Moreover, the behaviour of faint Seyferts (e.g. NGC 1386) is more similar to that of LLAGNs rather than the case of bright Seyferts, suggesting a smooth transition between both classes of AGNs.

References

[1] Ho L C 2008 ARA&A **46** 475–539 (*Preprint* 0803.2268)
[2] Ho L C 1996 PASP **108** 637
[3] Nagar N M, Falcke H and Wilson A S 2005 A&A **435** 521–543 (*Preprint* arXiv:astro-ph/0502551)
[4] Heckman T M 1980 A&A **87** 152–164
[5] Antonucci R 1993 ARA&A **31** 473–521
[6] Laor A 2003 ApJ **590** 86–94 (*Preprint* arXiv:astro-ph/0302541)

[7] Hönig S F and Beckert T 2007 MNRAS **380** 1172–1176
[8] Narayan R 2005 Ap&SS **300** 177–188 (*Preprint* arXiv:astro-ph/0411385)
[9] Yu Z, Yuan F and Ho L C 2011 ApJ **726** 87 (*Preprint* 1011.1962)
[10] Barth A J, Filippenko A V and Moran E C 1999 ApJ **525** 673–684 (*Preprint* arXiv:astro-ph/9905290)
[11] González-Martín O, Masegosa J, Márquez I and Guainazzi M 2009 ApJ **704** 1570–1585 (*Preprint* 0909.4082)
[12] Reunanen J, Prieto M A and Siebenmorgen R 2010 MNRAS **402** 879–894 (*Preprint* 0911.2112)
[13] Prieto M A, Reunanen J, Tristram K R W, Neumayer N, Fernandez-Ontiveros J A, Orienti M and Meisenheimer K 2010 MNRAS **402** 724–744 (*Preprint* 0910.3771)
[14] Jensen J B, Tonry J L, Barris B J, Thompson R I, Liu M C, Rieke M J, Ajhar E A and Blakeslee J P 2003 ApJ **583** 712–726 (*Preprint* arXiv:astro-ph/0210129)
[15] Tully R B, Rizzi L, Shaya E J, Courtois H M, Makarov D I and Jacobs B A 2009 AJ **138** 323–331
[16] Blakeslee J P, Jordán A, Mei S, Côté P, Ferrarese L, Infante L, Peng E W, Tonry J L and West M J 2009 ApJ **694** 556–572 (*Preprint* 0901.1138)
[17] Mandel K S, Wood-Vasey W M, Friedman A S and Kirshner R P 2009 ApJ **704** 629–651 (*Preprint* 0908.0536)
[18] Fukazawa Y, Hiragi K, Mizuno M, Nishino S, Hayashi K, Yamasaki T, Shirai H, Takahashi H and Ohno M 2011 ApJ **727** 19 (*Preprint* 1011.1750)
[19] Barth A J, Filippenko A V and Moran E C 1999 ApJ **515** L61–L64 (*Preprint* arXiv:astro-ph/9902193)
[20] Chary R, Becklin E E, Evans A S, Neugebauer G, Scoville N Z, Matthews K and Ressler M E 2000 ApJ **531** 756–763 (*Preprint* arXiv:astro-ph/9910557)
[21] Türler M, Courvoisier T J L and Paltani S 2000 A&A **361** 850–862 (*Preprint* arXiv:astro-ph/0008480)
[22] Markoff S, Nowak M, Young A, Marshall H L, Canizares C R, Peck A, Krips M, Petitpas G, Schödel R, Bower G C, Chandra P, Ray A, Muno M, Gallagher S, Hornstein S and Cheung C C 2008 ApJ **681** 905–924 (*Preprint* 0804.0344)
[23] Rieke G H, Lebofsky M J and Kemp J C 1982 ApJ **252** L53–L56
[24] Heeschen D S 1973 ApJ **179** L93
[25] González-Martín O, Masegosa J, Márquez I, Guerrero M A and Dultzin-Hacyan D 2006 A&A **460** 45–57 (*Preprint* arXiv:astro-ph/0605629)
[26] Ho L C, Filippenko A V and Sargent W L W 1997 ApJS **112** 315 (*Preprint* arXiv:astro-ph/9704107)
[27] Pellegrini S, Baldi A, Fabbiano G and Kim D W 2003 ApJ **597** 175–185 (*Preprint* arXiv:astro-ph/0307142)
[28] Markoff S, Nowak M A and Wilms J 2005 ApJ **635** 1203–1216 (*Preprint* arXiv:astro-ph/0509028)
[29] Ho L C 1999 ApJ **516** 672–682 (*Preprint* arXiv:astro-ph/9905012)
[30] Eracleous M, Hwang J A and Flohic H M L G 2010 ApJS **187** 135–148 (*Preprint* 1001.2924)
[31] Elvis M, Wilkes B J, McDowell J C, Green R F, Bechtold J, Willner S P, Oey M S, Polomski E and Cutri R 1994 ApJS **95** 1–68

Journal of Physics: Conference Series **372** (2012) 012007 doi:10.1088/1742-6596/372/1/012007

F-GAMMA: On the phenomenological classification of continuum radio spectra variability patterns of *Fermi* blazars

E Angelakis[1], L Fuhrmann[1], I Nestoras[1], C M Fromm[1], M Perucho-Pla[2], R Schmidt[1], J A Zensus[1], N Marchili[1], T P Krichbaum[1], H Ungerechts[3], A Sievers[3], D Riquelme[3] and V Pavlidou[1]

[1] Max-Planck-Institut für Radioastronomie, Auf dem Hügel 69, DE-53121, Bonn, Germany
[2] Department d'Astronomia i Astrofísica, Universitat de València, C/Dr. Moliner 50, 46100 Burjassot, València, Spain
[3] Instituto de Radio Astronoma Milimétrica, Avenida Divina Pastora 7, Local 20, E 18012, Granada, Spain

E-mail: eangelakis@mpifr.de

Abstract. The *F-GAMMA* program is a coordinated effort to investigate the physics of Active Galactic Nuclei (AGNs) via multi-frequency monitoring of *Fermi* blazars. In the current study we show and discuss the evolution of broad-band radio spectra, which are measured at ten frequencies between 2.64 and 142 GHz using the Effelsberg 100-m and the IRAM 30-m telescopes. It is shown that any of the 78 sources studied can be classified in terms of their variability characteristics in merely 5 types of variability. It is argued that these can be attributed to only two classes of variability mechanisms. The first four types are dominated by spectral evolution and can be described by a simple two-component system composed of: (a) a steep quiescent spectral component from a large scale jet and (b) a time evolving flare component following the "Shock-in-Jet" evolutionary path. The fifth type is characterised by an achromatic change of the broad band spectrum, which could be attributed to a different mechanism, likely involving differential Doppler boosting caused by geometrical effects. Here we present the classification, the assumed physical scenario and the results of calculations that have been performed for the spectral evolution of flares.

1. Introduction

Among the most evident characteristics of blazars is their intense variability at all wavelengths. Studies of the variability characteristics, preferably with simultaneous data, shed light on the physics driving the energy production and dissipation in these systems (e.g. [2, 3]). The *F-GAMMA* program [4, 5, 6] explores this possibility by coordinated monthly monitoring of *Fermi* γ-ray monitored blazars. *F-GAMMA* is covering mostly the radio cm to sub-mm bands primarily with the Effelsberg 100-m, the IRAM 30-m and the APEX 12-m telescopes (although optical telescopes are participating as well, Fuhrmann et al. in prep.) for ~ 60 prominent blazars.

The physical cause for the variability itself has been long debated. The "Shock-in-Jet" model suggested by [1], is the most accepted one and attributes the variability to shocks propagating down the jet. The basic assumption is that variations at the jet base (e.g. changes in the injection rate, the magnetic field, bulk Lorentz factor etc.) cause the formation of shocks, which consequently go through first *Compton*, then *synchrotron* and finally *adiabatic* energy losses. This is the main model prediction which is adopted in the following.

An alternative model, the "Internal Shock Model" proposed by [7], suggests that energy is channeled into the jet in an intermittent way. "Plasma shells", may have different bulk Lorentz factors and masses, so that faster shells can catch up with slower ones, collide and relativistic shocks can occur. The shock accelerates electrons to relativistic energies so that they can emit synchrotron and inverse Compton radiation.

Other models explain the variability geometrically. For example, [8] suggest that orbital motion and jet precession, e.g. caused by a binary black hole system, can produce helical jet morphologies which can bent the jet, changing its orientation and shape. This may then cause brightness and spectral variability through changes in the Doppler factor.

2. Observations

The work discussed here is based on observations conducted quasi-simultaneously with the Effelsberg 100-m and the IRAM 30-m telescope within the *F-GAMMA* program which is described in [4, 5, 6]. The 100-m telescope has been observing between 2.64 and 43.05 GHz at 8 frequencies over a baseline of 4.5 years (January 2007 – June 2011). The 30-m telescope has been measuring at 86 and 142 GHz (June 2007 – June 2011). The millimetre observations for *F-GAMMA* are closely coordinated with the more general flux monitoring conducted by IRAM, and data from both programs are included in this paper. The Effelsberg 100-m telescope is equipped with circularly polarised feeds, while the IRAM 30-m with linearly polarised ones. Details are given elsewhere (Fuhrmann et al. in prep., Angelakis et al. in prep., Nestoras et al. in prep.). Measurements at 4.85 GHz, 10.45 GHz, 32.0 GHz, 86.24 GHz and 142 GHz are done differentially either by using multi-feed systems, or, at IRAM 30-m, by wobbler switching.

On average, the time needed for observing an entire spectrum of any given source at Effelsberg alone is of the order of 35 minutes, while at IRAM, roughly 2 minutes. The combined spectra (Effelsberg and IRAM) are observed quasi-simultaneously within approximately one week. That is, neither the single-facility nor the combined spectra are likely to be affected by source variability. In the current study only data collected until June 2011, have been used.

3. Data reduction

The antenna temperature measured for a certain source is subjected to a series of corrections to compute the source flux density outside the terrestrial atmosphere. These operations, are: (a) *Pointing correction*, meant to correct for small pointing residuals. (b) *Elevation dependent gain correction*, correcting for the loss of antenna sensitivity caused by gravitational deformation of the reflector. (c) *Atmospheric opacity correction*, which corrects for the attenuation due to the terrestrial atmospheric absorption. (d) *Absolute calibration (sensitivity correction)*, which does the translation of the previously corrected antenna temperature to SI units. All measurements are subsequently exposed to quality checks. The overall uncertainties are typically of the order of $0.5 - 5\,\%$ for Effelsberg and of the order of $\leq 10\,\%$ for IRAM. More details can be found in [9] as well as in Angelakis et al. (in prep.) and in Nestoras et al. (in prep.).

4. Phenomenological classification of the spectrum variability pattern

The following discussion refers to the phenomenological classification of the variability patterns shown by the broad-band spectra of sources in the *F-GAMMA* sample relative to the observing band-pass. The term "variability pattern" is assigned to the shape traced by the source broad-band spectrum – as that is observed monthly – as a function of time. In total 78 sources have been examined and their spectra span, in total, over 4.5 years of observations.

A visual inspection of the examined sources reveals a wealth of spectral features as well as of variability patterns that different sources exhibit. For instance, the variability amplitude, the frequency of occurrences and the pace at which the variability events evolve, seem to be different for different sources. Interestingly, despite the apparent complexity it appears that the 78 sources that have been studied here can be classified in only five phenomenological classes on the basis of its variability pattern, which are numbered from 1 to 5 (more details will be given by Angelakis et al. in prep.). Four of them show also sub-types which however do not deserve a separate type and are named after the main type followed by the letter "b". The prototype sources are shown in figures 2–10. The main phenomenological characteristics of these types, are:

Type 1 : This variability pattern is clearly dominated by spectral evolution. That is, at an instant of time the spectrum appears convex and its peak $(S_{\mathrm{m}}, \nu_{\mathrm{m}})$ is drifting within the observing band-pass from high towards lower frequencies, covering a significant area in the $S-\nu$ space. Its shape is smoothly changing towards an ultimate flat or mildly steep power law which is then followed by new events. For this type no evidence for the presence of an underlying quiescent steep spectrum is seen. The lowest frequencies in the bandpass are remarkably variable indicating that the activity ceases at frequencies much lower than the lowest in our band-pass. The prototype source is shown in figure. 2.

Type 1b : As a sub-class of the previous one, type 1b shares the same characteristics with type 1 except that the lowest frequency does not show as intense a variability. The activity ceases around this part of the band-pass. A source representative of this type is shown in figure 3.

Type 2 : This type is also dominated by spectral evolution. The basic characteristic of this case is the fact that the flux density at the lowest frequency during the steepest spectrum phase is higher than that during the inverted spectrum phase. Moreover, the maximum flux density reached by the flaring events is significantly above that at the lowest frequency. This implies that the observed steep spectrum is not a quiescent spectrum but rather the relic of an older, yet recent, outburst. The prototype of this type is shown in figure 4.

Type 3 : Type 3, shown in figure 5, is dominated by the spectral evolution as well. The identifying characteristics, are: (a) the fact that the lowest frequency practically does not vary and, (b) the maximum flux density level reached by outbursts is comparable to that at the lowest band-pass frequency. This phenomenology leaves hints that the events cease very close to the lowest frequency of the band-pass and hence a quiescent spectrum is becoming barely evident.

Type 3b : Type 3b, shown in figure 6, is very similar to type 3. Here however the quiescent spectrum is seen clearly at least at the 2 lowest frequencies.

Type 4 : Sources of this type spend most of the time as steep spectrum ones which are sometimes showing an outburst of relatively low power propagating towards low frequencies. A representative case is shown in figure 7.

Type 4b : This type includes persistently steep spectrum cases as it is shown in figure 8.

In all previous classes, type 1 to 4b, the occurring variability is clearly dominated by spectral evolution. However, there exists a class of sources that show a fundamentally different behaviour. The variability happens self-similarly without signs of spectral evolution. Those are grouped in a separate type with two sub-types:

Type 5 : In this case the spectrum is convex and follows an "achromatic" evolution. That is, it shifts its position in the $S-\nu$ space preserving its shape. This is shown clearly in figure 9.

Type 5b : This type shows, in principle, characteristics similar to the previous one but there occurs a mild yet noticeable shift of the peak $(S_{\rm m}, \nu_{\rm m})$ towards lower frequencies as the peak flux density increases. A characteristic case is shown in figure 10.

The distribution of sources over the different variability types are shown in figure 1. The sub-classes with the label "b" are numbered with appending the fraction 0.5 to the numerical tag of the corresponding type (e.g. Type 5b appears as 5.5). As can be seen, the majority of the sources show variability of type 1 – 3b which is the result of the *F-GAMMA* sample selection (i.e. sources variable at all frequencies). Interestingly, a small number of sources (eight) show "achromatic" behaviour calling for a dedicated study as it is discussed later.

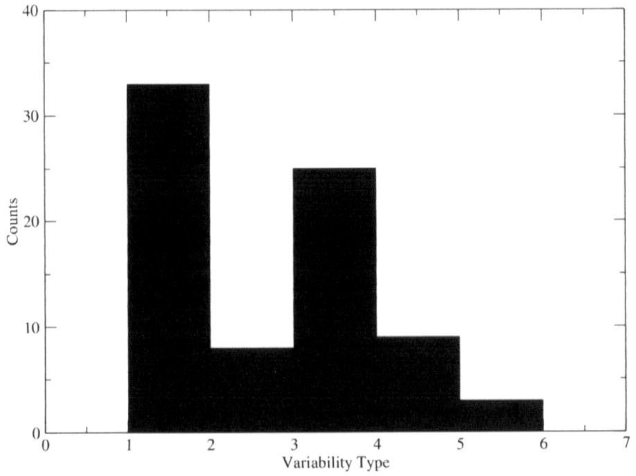

Figure 1. The distribution of sources over spectral variability type. The sub-classes labeled "b" are enumerated with appending .5 to the class number. For example, 5b should appear as 5.5. The number of sources used here are 78.

This classification is done solely on the basis of the phenomenological characteristics of the variability pattern shown by the radio spectra within a given band-pass. As it is discussed in the next section, it appears that all the phenomenology for types 1–4b can be naturally explained by the same physical system observed under different circumstances.

5. A physical interpretation of the variability types

In the following it is suggested that the classification discussed earlier can be reproduced by a simple two-component system observed under different circumstances.

Let us assume a simple two-component system at $z = 0$, consisted of:

(i) A power-law quiescent spectrum with $S \propto \nu^{\alpha}$ (e.g. $\alpha \approx -0.5$). This can be attributed to the optically thin diffuse emission of a large scale jet or even relic recent flaring event or blends of such.

(ii) A convex synchrotron self-absorbed spectrum (hereafter SSA) representative of a recent outburst superimposed on the quiescent part.

The assumed configuration is presented in figure 11 where the shaded areas denote the observing band-pass. The phenomenology shown there captures the system (solid line) at an instant in time. Consequently the spectral shape that would be observed at that instant of time would depend on two parameters of the shaded areas (band-pass) shown there:

(i) The ***position*** of the shaded areas relative to the high and low frequency peak of the assumed system. This parameter denotes the relative position of the centre of our band-pass with respect to the source spectrum.

(ii) The ***width*** of the shaded areas relative to the width of the bridge between the optically thick part of the outburst and the steep part of the quiescent spectrum. This parameter denotes the fraction of the spectrum that the band-pass can sample.

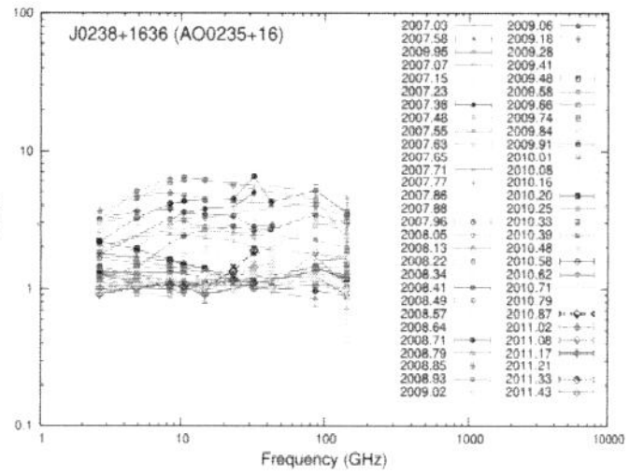

Figure 2. Prototype source for variability type 1.

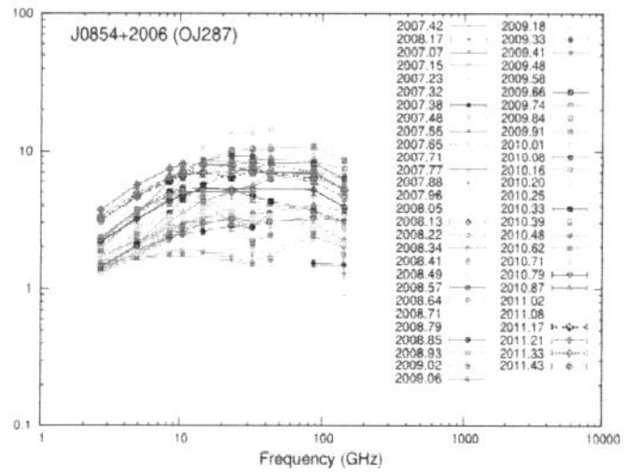

Figure 3. Prototype source for variability type 1b.

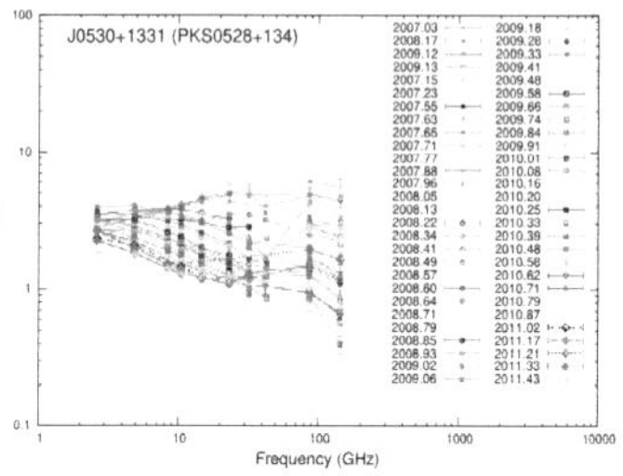

Figure 4. Prototype source for variability type 2.

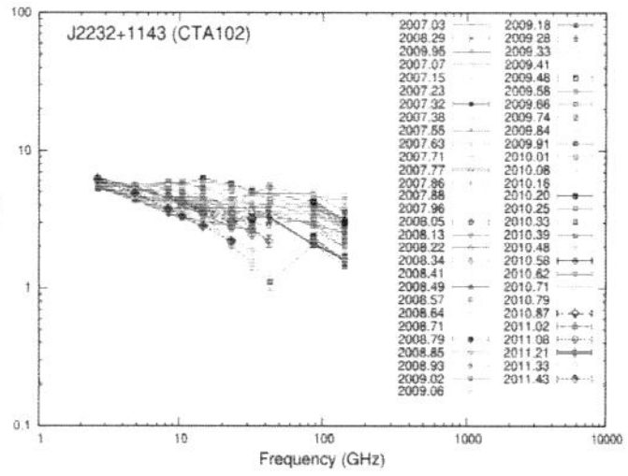

Figure 5. Prototype source for variability type 3.

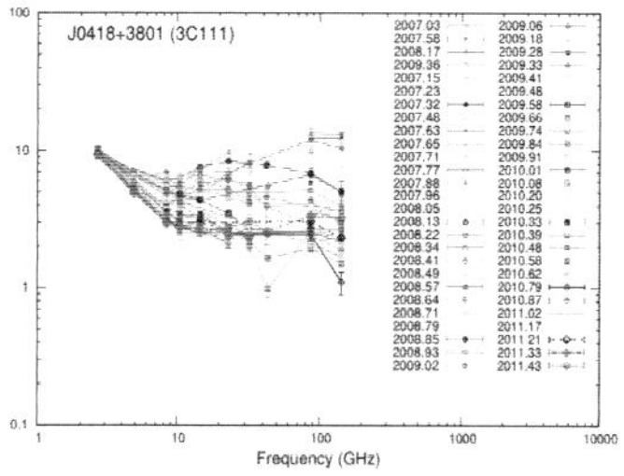

Figure 6. Prototype source for variability type 3b.

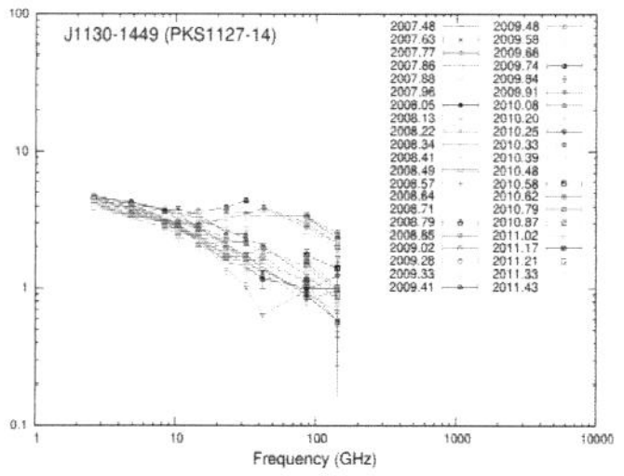

Figure 7. Prototype source for variability type 4.

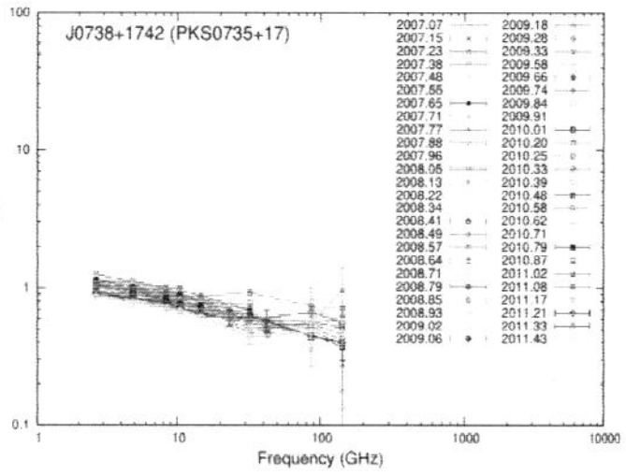

Figure 8. Prototype source for variability type 4b.

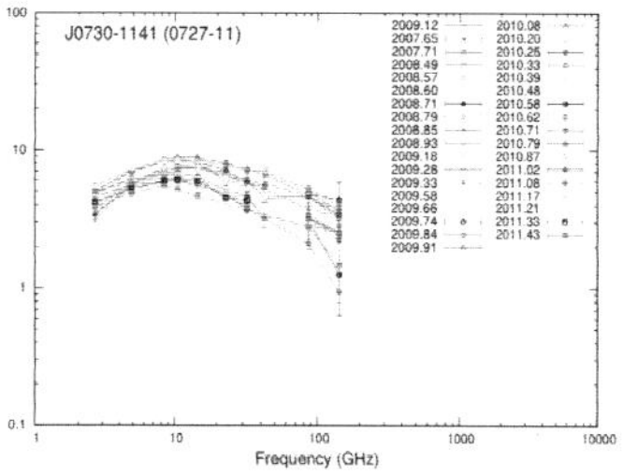

Figure 9. Prototype source for variability type 5.

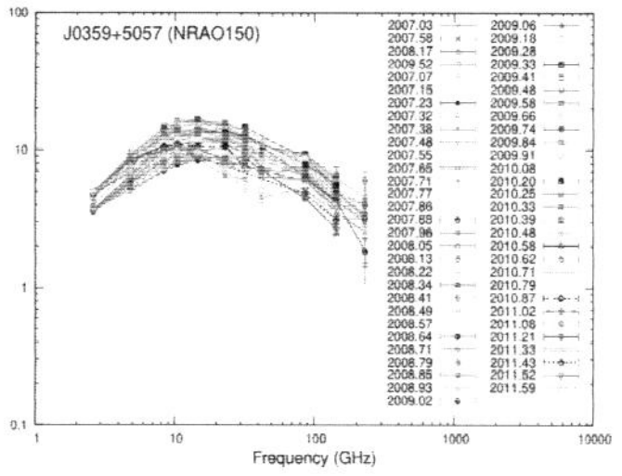

Figure 10. Prototype source for variability type 5b.

Within this scenario it can be said that as the variability type increases the dominance of the steep quiescent spectrum becomes progressively larger and the basic characteristics of the observed variability types 1–4b can be reproduced naturally with the appropriate modulation of these two parameters.

The question that naturally arises then is how can these two quantities be modulated. Qualitatively speaking, this can be formulated in terms of the combination of (a) *redshift* and (b) *source intrinsic properties*. The *redshift* changes the relative *position* of the band-pass allowing a different part of the spectrum to be sampled. The *source intrinsic properties* imply that different sources show different spectral characteristics, such as: peak frequency of the outburst, peak flux density excess of the outburst relative to the quiescent spectrum, different broadness of the valley, different broadness of the SSA spectrum of the outburst etc. Accounting now for the dynamical evolution of a flaring event in the $S_{\rm m} - \nu_{\rm m}$ space, one can introduce a third factor namely (c) the *flare specific properties* which of course are also a function of the *source intrinsic properties* and allow the system to evolve dynamically. While factors (a) and (b) have

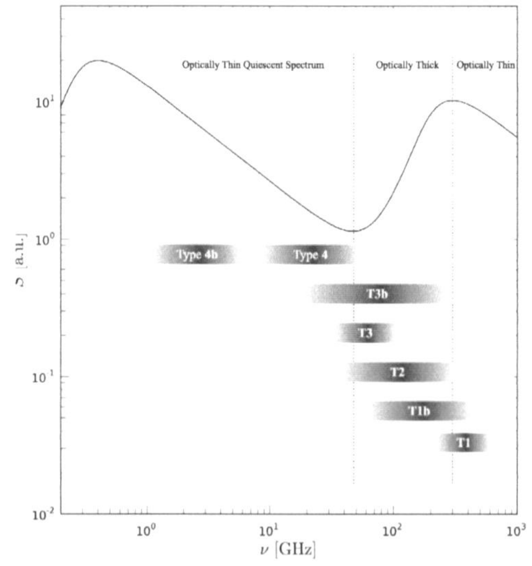

Figure 11. The assumed common underlying two-component system. The different variability types can be reproduced with the appropriate modulation of the relative position and relative broadness of the band-pass denoted by the grey shaded areas.

a static effect and determine the general shape of the observed spectrum, the latter one (c) changes both the relative position and width of the band-pass dynamically shaping the specific characteristics of the variability pattern. In the following we present calculations that have been done to examine whether this scenario can reproduce the observed phenomenologies.

6. Modelling the broad band radio spectra

Following the hypothesis that all the observed events are the reflection of the same process, namely shocks evolving in jets seen with different frequency band-passes at different evolutionary stages, the Shock-in-Jet model [1, 10] has been applied to reproduce their temporal evolution.

Specifically, our approach is presented in [11], and relies on the information extracted from the quiescent spectrum and the flaring event to connect the different radiative evolutionary stages of the shock (*Compton*, *Synchrotron* and *Adiabatic* stage). The distribution of the quiescent spectrum parameters depends on the intrinsic source properties such as the magnetic field in the jet, B, the Doppler factor, D, and the normalisation coefficient of the spectrum, K. The spectral behaviour of the injected component, depends on the evolution of the physical parameters of the shock and is parametrized by the exponents in the relations giving B, K and D as functions of the distance along the jet: $L \propto R^r$, $B \propto L^{-r \cdot b}$, $K \propto L^{-r \cdot k}$, $D \propto L^{-r \cdot d}$, where r is the opening rate of the jet. In the current work it is assumed that the *Synchrotron* stage is very short compared to the other two stages (see [11]).

We have calculated the spectral evolution of flares for three characteristic cases in terms of source *luminosity*: weak, medium and strong and for three *redshifts*: 0, 1.5 and 3. In every case the peak flux of the flaring event at time 0 was set equal to the turnover flux of the quiescent spectrum. Subsequently the flare has been left to evolve according to the evolutionary model discussed previously ([1, 10, 11]). Figures 12–20 show the spectra expected to be observed by a facility with band-pass of 2–140 GHz (*F-GAMMA* program). It is evident from there that the majority of the types discussed earlier are reproduced naturally. Elsewhere (Angelakis et al. in prep.) more cases are simulated increasing the expected variability patterns.

7. Discussion

From the discussion in section 4, it is evident that the variability phenomenologies of the studied blazars can be categorised in two fundamentally different classes. (a) Sources that are dominated by spectral evolution and (b) sources that have a convex spectrum and vary self-similarly with

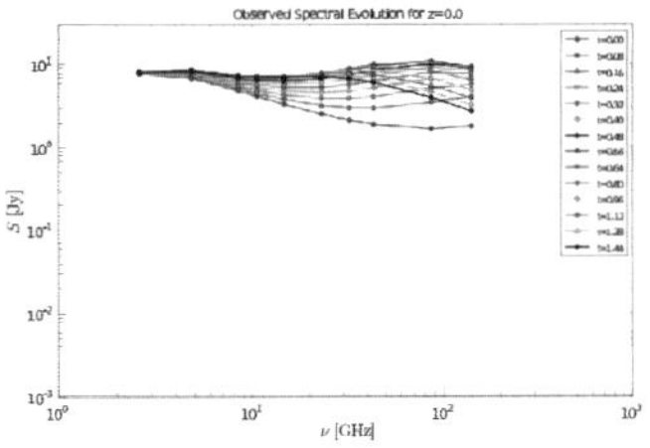

Figure 12. The spectrum expected from a powerful source at $z = 0$.

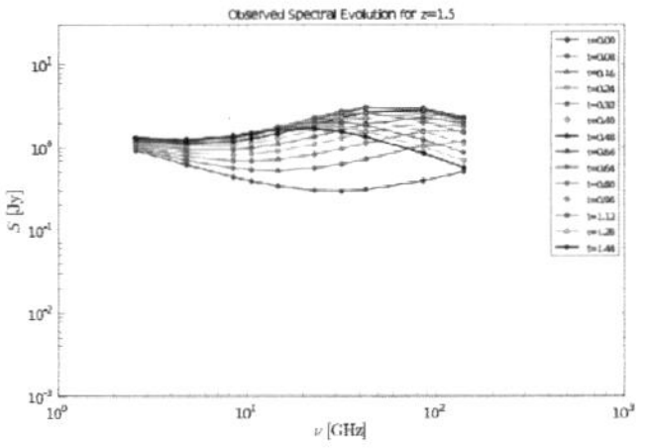

Figure 13. The spectrum expected from a powerful source at $z = 1.5$.

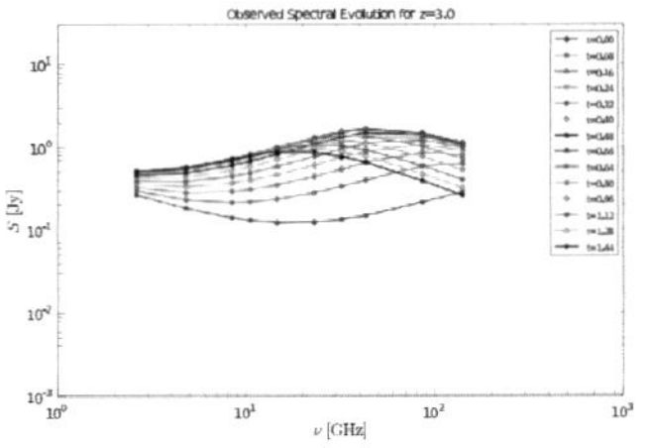

Figure 14. The spectrum expected from a powerful source at $z = 3.0$.

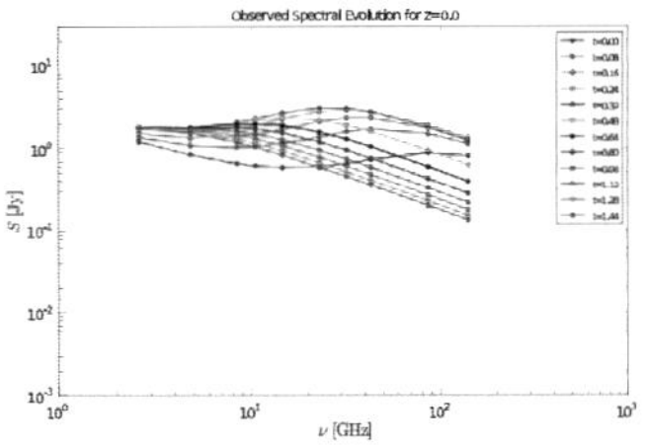

Figure 15. The spectrum expected from a medium source at $z = 0$.

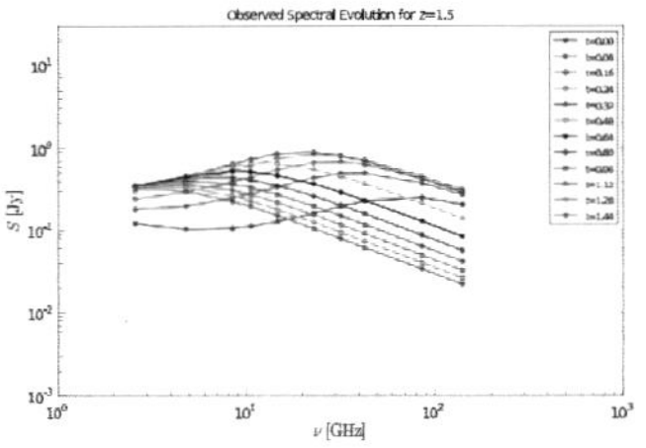

Figure 16. The spectrum expected from a medium source at $z = 1.5$.

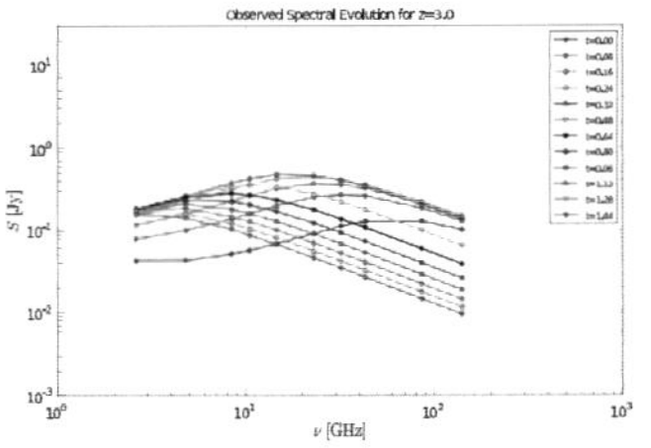

Figure 17. The spectrum expected from a medium source at $z = 3.0$.

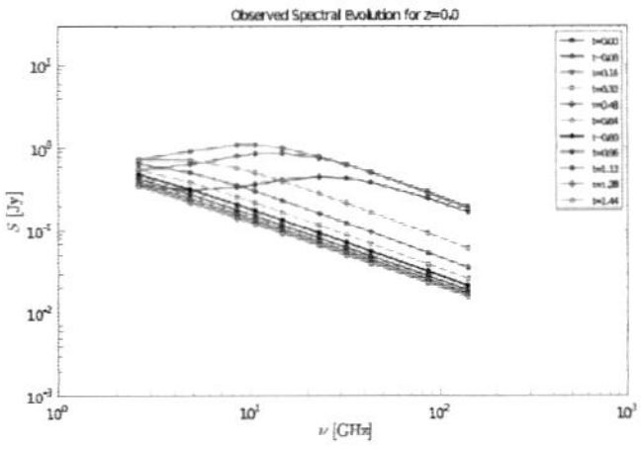

Figure 18. The spectrum expected from a weak source at $z = 0$.

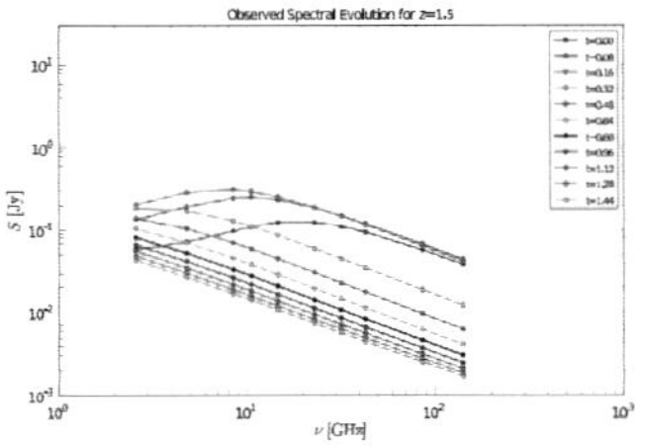

Figure 19. The spectrum expected from a weak source at $z = 1.5$.

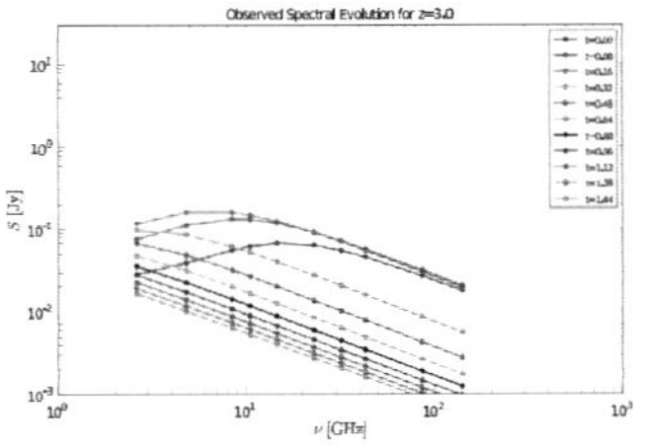

Figure 20. The spectrum expected from a weak source at $z = 3.0$.

only mild if at all spectral evolution. This implies that there must exist two distinct mechanisms causing variability. It must be noted though that this refers to the available baseline (roughly five years) meaning that sources of type 5 and 5b could still show spectral evolution over longer time scales. An additional element that points towards a different variability mechanism is the persistency of the spectral shape, in the case of type 5 and the fashion of change in the case of type 5b which is not seen in cases of clear spectral evolution.

Of the 78 sources that have been examined for the current work, 8 show achromatic variability. The interesting characteristics is that in the cases that show a mild spectral evolution, the turnover flux and frequency S_{m} and ν_{m}, are evolving in an anti-correlated fashion (see e.g. figure 10). Apart from the fact that all of them show the clear presence of a large scale jet even at 2 cm (as it is shown from the MOJAVE images [12]) no other peculiar property has been identified so far. Possible mechanisms that are currently investigated, include: changes in the magnetic field structure, changes in the Doppler factors and geometrical effects.

It is noteworthy that none of the studied sources has shown a switch of type at least over the baseline of the *F-GAMMA* program, neither between types of the same underlying mechanism (i.e. 1–4b) nor between types with different underlying mechanism (i.e. types 1–4b and types 5, 5b). This suggests that the mechanism producing the variability is, either a fingerprint characteristic of the source, or the conditions that determine it change over longer time scales, if at all. Possible switch from achromatic to an evolution dominate behaviour would imply that the suggested dichotomy of the variability mechanisms is not valid. In any case, the persistency of the evolution dominated types implies that the power deposited in each event for a certain source is not varying significantly from one event to the other. Further investigations to explore this statement are underway and will be presented elsewhere.

Concerning the evolution dominated case, it seems that the Marscher & Gear model [1] provides a precise reproduction of the observed phenomenology and most importantly, over a range of intrinsic parameters covered by the *F-GAMMA* sample. Studies to examine whether other variability mechanisms can reproduce the observed phenomenology (shapes, time scales etc.) are needed. In any case, any successful model could be used for extracting the physical parameters from the observed spectra.

Acknowledgments

Based on observations with the 100 m telescope of the MPIfR (Max-Planck-Institut für Radioastronomie). Based on observations carried out with the IRAM 30m Telescope. IRAM is supported by INSU/CNRS (France), MPG (Germany) and IGN (Spain). I. Nestoras and R. Schmidt are members of the International Max Planck Research School (IMPRS) for Astronomy and Astrophysics at the Universities of Bonn and Cologne. The *F-GAMMA* team sincerely thanks the Time Allocation Committee of the 100-m and 30-m telescope for supporting the continuation of the program. E. Angelakis feels obliged to wholeheartedly thank Dr. A. Kraus for the constant support and the very constructive discussions.

References

[1] Marscher A P and Gear W K 1985 *ApJ* **298** 114–127
[2] Boettcher M 2010 *ArXiv e-prints* (*Preprint* 1006.5048)
[3] Boettcher M and Dermer C D 2010 *Bulletin of the American Astronomical Society* (*Bulletin of the American Astronomical Society* vol 42) pp 706–+
[4] Fuhrmann L, Zensus J A, Krichbaum T P, Angelakis E and Readhead A C S 2007 *The First GLAST Symposium* (*American Institute of Physics Conference Series* vol 921) ed Ritz S, Michelson P and Meegan C A pp 249–251
[5] Angelakis E, Fuhrmann L, Marchili N, Krichbaum T P and Zensus J A 2008 *Memorie della Societa Astronomica Italiana* **79** 1042–+ (*Preprint* 0809.3912)
[6] Angelakis E, Fuhrmann L, Nestoras I, Zensus J A, Marchili N, Pavlidou V and Krichbaum T P 2010 *ArXiv e-prints* (*Preprint* 1006.5610)
[7] Spada M, Ghisellini G, Lazzati D and Celotti A 2001 *MNRAS* **325** 1559–1570 (*Preprint* arXiv:astro-ph/0103424)
[8] Villata M and Raiteri C M 1999 *A&A* **347** 30–36
[9] Angelakis E, Kraus A, Readhead A C S, Zensus J A, Bustos R, Krichbaum T P, Witzel A and Pearson T J 2009 *A&A* **501** 801–812 (*Preprint* 0905.2660)
[10] Türler M, Courvoisier T and Paltani S 2000 *A&A* **361** 850–862 (*Preprint* arXiv:astro-ph/0008480)
[11] Fromm C M, Perucho M, Ros E, Savolainen T, Lobanov A P, Zensus J A, Aller M F, Aller H D, Gurwell M A and Lähteenmäki A 2011 *A&A* **531** A95+ (*Preprint* 1105.5024)
[12] Kellermann K I, Lister M L, Homan D C, Vermeulen R C, Cohen M H, Ros E, Kadler M, Zensus J A and Kovalev Y Y 2004 *ApJ* **609** 539–563 (*Preprint* astro-ph/0403320)

Journal of Physics: Conference Series **372** (2012) 012008 doi:10.1088/1742-6596/372/1/012008

Supermassive Black Holes and Their Relationships with Their Host Galaxies

Bradley M. Peterson

Department of Astronomy and Center for Cosmology and AstroParticle Physics, The Ohio State University, 140 West 18th Avenue, Columbus, OH 43210, USA

E-mail: peterson.12@osu.edu

Abstract. We review how the masses of black holes in active galactic nuclei are measured and outline the current limitations and uncertainties. Masses have been measured directly by emission-line reverberation for nearly 50 relatively nearby AGNs, but uncertainties due to the unknown geometry and projection effects limit the accuracy of these masses to ~ 0.3 dex. Reverberation studies show that there is a very tight relationship between the broad-line region radius and the AGN luminosity, with an intrinsic scatter of ~ 0.1 dex, which shows (1) that the largest source of systematic uncertainty in the black hole mass determinations is how the velocity field of the broad-line region is characterized, not the size of the broad-line region, and (2) that the size of the broad-line region can be estimated to fairly high accuracy from the AGN luminosity alone, thus providing a powerful indirect method of estimating black hole masses in even distant AGNs.

1. Introduction

The existence of supermassive black holes at the centers of at least some galaxies has been suspected for nearly 50 years, when it was first suggested that accretion powers quasars. Oddly enough, the first real evidence that supermassive black holes reside in galactic centers was not in quasars, but in quiescent galactic nuclei, where their masses were determined by the kinds of high-angular resolution studies featured at this workshop. High angular resolution is required for most dynamical measurements of the black hole mass $M_{\rm BH}$ since one must resolve (or nearly resolve, depending on who you listen to) the black hole radius of influence,

$$r_* = 2GM_{\rm BH}/\sigma_*^2, \tag{1}$$

where σ_* is the stellar velocity dispersion of the bulge, since it is within this distance that the black hole dominates the dynamics of the stars and gas. In this contribution, however, I'm going to focus instead on a complementary method, reverberation mapping (Blandford & McKee 1982; Peterson 1993), that substitutes time resolution for angular resolution to measure the masses of black holes in active galactic nuclei (AGNs). While application of this method is restricted to Type 1 (broad line) AGNs, which are found in only a small fraction of all luminous galaxies, the importance of reverberation mapping derives from the fact that it is the *only* direct method of measuring the masses of supermassive black holes at large cosmological distances, as illustrated in Figure 1.

It is useful at this point to distinguish between *direct* and *indirect* methods of measuring the masses of black holes in galaxies. *Direct* methods are those which involve measuring the

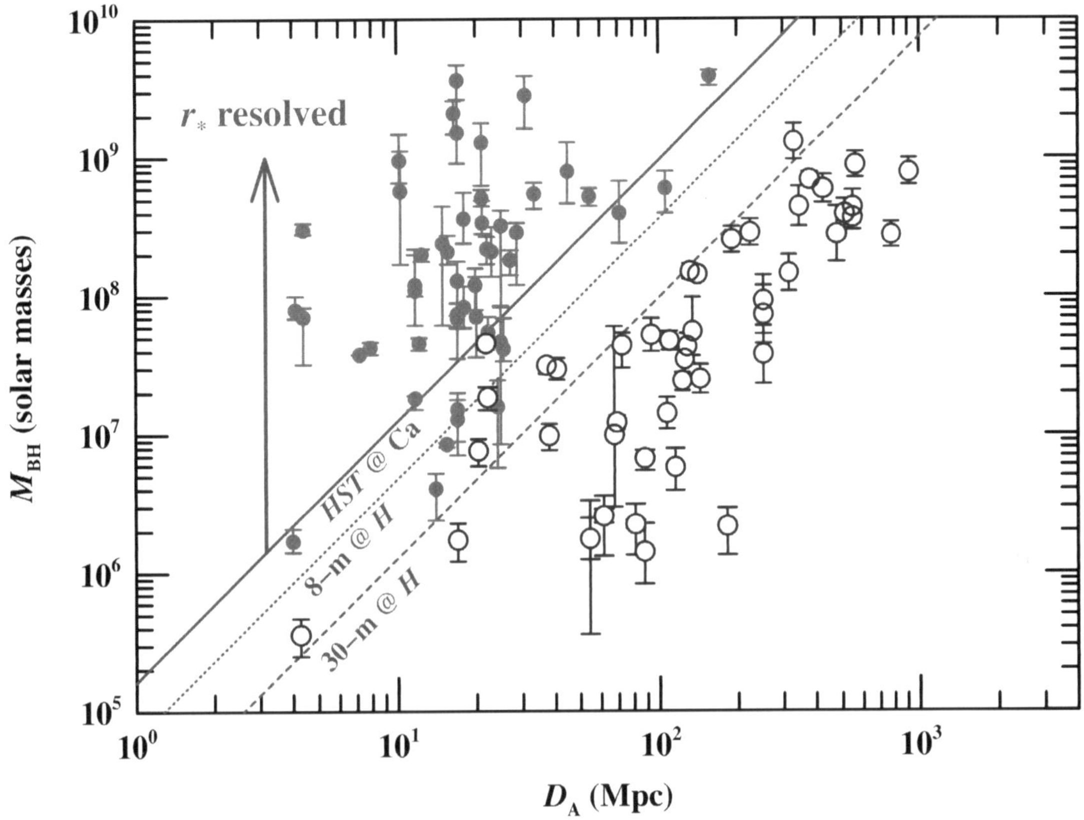

Figure 1. Black hole mass as a function of angular size distance for quiescent galaxies measured by high-angular resolution techniques (red filled circles) and for Type 1 AGNs measured by reverberation mapping (blue open circles). The diagonal lines show the limit at which the black hole radius of influence (equation 1) is resolved for spectral features used to measure stellar velocity dispersion. Even the largest telescopes working at the diffraction limit will not significantly increase the number of AGNs for which it is possible to measure black hole masses using high angular-resolution techniques. At large distances, and even at small distances for low-mass black holes, reverberation mapping is the most promising technique for measuring black hole mass.

motions of gas or stars that are accelerated by the black hole itself; these would include proper motions and radial velocities of stars in the center of the Milky Way and megamasers in Type 2 AGNs, stellar or gas dynamics in nearby galaxies, and AGN reverberation mapping. *Indirect* methods are those in which we measure observable properties that that are *correlated* with the black hole mass. These would include the correlations between (1) black-hole mass and stellar bulge velocity dispersion (the "$M_{\rm BH}$–σ_* relationship") (2) black-hole mass and bulge luminosity (the "$M_{\rm BH}$–$L_{\rm bulge}$ relationship"), the fundamental plane, and, in the case of Type 1 AGNs, reverberation-based scaling relationships (see §3.1).

With the exception of the measurements of masses of the black holes at the center of the Milky Way and the megamaser source NGC 4258, it is probably safe to say that measurements using the direct methods listed above are still probably uncertain by at least a factor of two. The systematic errors associated with the reverberation-based masses are not well-understood because of our lack of understanding of the geometry of the broad-line region (BLR) and how the mass measurements are affected by projection. Reverberation masses are usually computed from

$$M_{\rm BH} = f\left(\frac{\Delta V^2 R}{G}\right) + gL, \tag{2}$$

where ΔV is the emission-line width, R is the size of the BLR as measured for this particular emission line, G is the gravitational constant, L is the AGN luminosity, and f and g are constants. In the first term on the right-hand side, the quantity in parentheses is called the "virial product" which is comprised of the two observables and has units of mass. The systematics associated with geometry and projection are all subsumed into the dimensionless factor f, which is expected to be of order unity, and should be different for each AGN. It is possible to obtain an average value $\langle f \rangle$ by noting that AGNs clearly show the same $M_{\rm BH}$–σ_* relationship as quiescent galaxies, except that the AGN relationship has an undefined zero point, i.e., $\langle f \rangle$ is unknown. If we assume that the zero-points are identical for AGNs and quiescent galaxies, then we find that $\langle f \rangle \approx 5.25$ (Onken et al. 2004; Woo et al. 2010, though see also Graham et al. 2011 and §4); the scatter around the AGN $M_{\rm BH}$–σ_* relationship then suggests that the typical systematic uncertainties are probably no more than ~ 0.3 dex.

The second term on the right-hand side of equation (2) accounts for a possible contribution by radiation pressure (Marconi et al. 2008). There is active debate about whether or not this term is significant. Space limitations preclude further discussion of this interesting possibility, and for the time being I will make the cavalier assumption that $g = 0$.

2. Reverberation Mapping and Black Hole Masses

2.1. Measuring the Size of the Emission-Line Region

The ultraviolet, optical, and near-infrared spectra of Type 1 AGNs are dominated by strong Doppler-broadened emission lines which are driven by the ionizing continuum emission that arises in the accretion disk surrounding the central supermassive black hole. The accretion process has numerous instabilities, so the continuum emission from the central source varies with time in an unpredictable fashion (though it can be statistically well-modeled as a damped random walk; see Kelly et al. 2009). The emission lines respond to these variations in the continuum flux, but with a time delay that is attributable to the light-travel time across the line-emitting region. The goal of reverberation mapping is to measure the time-delayed response of emission lines to the continuum variations and thus determine the structure and kinematics of the BLR (see Peterson 2001 for a tutorial on the method). Without going into detail, with a couple of simple assumptions that are easily justified *post facto* (e.g., that the observable continuum variations track those of the ionizing continuum with a negligible time delay), it is straightforward to measure the mean response time for the line-emitting region by cross-correlating continuum and emission-line light curves; this time delay τ, or lag, gives the typical distance from the accretion disk to where the response of that particular emission line is most pronounced, $R = c\tau$.

Operationally, reverberation mapping requires long-duration, high-precision spectrophotmetric monitoring of AGNs. Despite the difficulty, large allocations of telescope time required, and risk (due primarily to weather and equipment failure), reverberation mapping programs have yielded results for nearly 50 AGNs at the time of this workshop. The major conclusions from these studies are:

(i) Different emission lines respond with different time delays. Lines characteristic of more highly ionized gas (e.g., C IV λ1549, He II λ4686) respond more rapidly than those characteristic of less highly ionized gas (e.g., the hydrogen Balmer lines), providing evidence for ionization stratification of the BLR.

(ii) In every case where time delays for multiple lines are measured, there is a strong anticorrelation between lag and emission-line Doppler width ΔV of the form $\tau \propto \Delta V^{-2}$, justifying the use of equation (2) to obtain the mass of the central source.

(iii) For a given emission line, the lag is longer in more luminous AGNs, consistent with $R \propto L^{1/2}$, as discussed in §3.1. Furthermore, when the response of a particular emission line is measured at different times when the continuum is in different luminosity states, the lag varies as $\tau \propto L_{\rm UV}^{1/2}$, where $L_{\rm UV}$ is the luminosity in the ultraviolet, as close as possible to the ionizing continuum — the optical continuum varies in phase with the UV continuum, to a good approximation, but with a lower amplitude. The emission-line width also changes with luminosity, in a way that is consistent with a constant value of $\Delta V^2\tau$, again justifying use of equation (2).

2.2. Measuring the Velocity Dispersion of the Line-Emitting Region

To calculate the enclosed mass from equation (2), one wants to consider the velocity dispersion of the gas that is actually producing the variability signal. It is common to isolate the variable part of the emission line by combining all of the spectra obtained in the reverberation program into an rms residual spectrum (root mean square flux at each wavelength, with the mean flux subtracted), which removes the non-variable, or slowly varying, components of the spectrum. The emission-line width is usually characterized by either the full-width at half maximum (FWHM) or the line dispersion $\sigma_{\rm line}$ (i.e., the second moment of the line profile). Both have particular advantages and disadvantages: FWHM is easy to measure and is usually not affected by blending with other lines, but it is more sensitive to unremoved narrow-line components, random noise and, in the case of resonance lines, absorption by intervening gas. Line dispersion is well-defined, even in noisy spectra, less sensitive to narrow-line components and self-absorption, and more accurate for low-contrast lines, but is much more sensitive to blending with other features. Choice of line-width measure turns out to be quite important, as described in §3.2.

2.3. Reverberation-Based Masses

Once the lag and line width are measured, equation (2) is used to compute the mass. As described above, the mean scaling factor is currently determined from the $M_{\rm BH}$–σ_* relationship. While the fact that AGNs show an $M_{\rm BH}$–σ_* relationship with a slope that is the same as that for quiescent galaxies is somehow reassuring, we really would like additional assurances that these masses are correct. Fortunately, as can been seen upon careful inspection of Figure 1, there are a bare handful of AGNs that are close enough and massive enough that their black hole masses can be measured by high-angular resolution studies of stellar and gas dynamics. Results for two well-studied AGNs are listed in Table 1. Noting that the uncertainties for the reverberation results do not include any systematic errors (which are likely at the factor of two to three level), the results are all in quite reasonable agreement.

A further consistency check is provided by comparing the $M_{\rm BH}$–$L_{\rm bulge}$ relationships for AGNs and quiescent galaxies. The agreement is found to be very good if the zero point from the $M_{\rm BH}$–σ_* relationship is adopted (Bentz et al. 2009a).

3. Indirect Reverberation-Based Black Hole Masses

Reverberation mapping is costly in terms of telescope time, even for relatively nearby bright AGNs. The problem of measuring masses for a large population of AGNs, including faint high-redshift quasars, becomes daunting. Fortunately, however, their are indirect methods of mass measurement that allow us to estimate masses in AGNs from more easily acquired data. Specifically, in the case of Type 1 AGNs we can use the relationship between BLR region radius and AGN luminosity to side-step the reverberation process: from a single spectrum, we can

Table 1. Black Hole Masses (in units of $10^6\ M_\odot$)

Galaxy	NGC 3227	NGC 4151
Direct methods:		
Stellar dynamics	7–20[a]	< 70[b]
Gas dynamics	20^{+10}_{-4}[c]	$30^{+7.5}_{-22}$[c]
Reverberation	7.63 ± 1.7[d]	46 ± 5[e]
Indirect methods:		
$M_{\rm BH}$–σ_*	25	6.1
R–L scaling	15	65

[a] Davies et al. (2006).
[b] Onken et al. (2007).
[c] Hicks & Malkan (2008).
[d] Denney et al. (2010).
[e] Bentz et al. (2006).

measure both luminosity, from which we infer the BLR radius, and line width and then compute the mass trivially. Of course, there are subtleties, as discussed below.

3.1. *The Radius–Luminosity Relationship*

Long before it was actually confirmed observationally, a relationship between the radius of the BLR and the AGN luminosity was expected on the basis of photoionization physics. Naïve expectations based on the gross similarity of broad-line spectra over several orders of magnitude in luminosity suggested that the relationship should have the form $R \propto L^{1/2}$, which is actually in remarkable agreement with the observations, once care is taken to carefully remove the host-galaxy starlight contamination (Bentz et al. 2009b). The relationship is observationally well-constrained only for Hβ (Figure 2). When only the highest-quality reverberation data are used, the intrinsic scatter in this relationship is very small, approximately 0.1 dex. Thus, measurement of the AGN luminosity from a single spectrum can give an estimate of the size of the Hβ-emitting region that is nearly as good as one can obtain by reverberation!

3.2. *Measuring the Line Width*

When considering the many possible sources of random and systematic error in reverberation-based mass estimates, it is common to focus on uncertainties in the R–L relationship. While there are possible difficulties, this is a misguided concern: proper characterization of the velocity dispersion of the broad-line emitting gas is a far great problem. First, the emission lines are generally broader in the mean spectrum (or a single spectrum) than in the rms spectrum, which effectively requires a larger value of $\langle f \rangle$. Second, the mean spectrum (or a single spectrum) is heavily contaminated by constant or slowly varying components like the narrow-line component, which arises in low-density gas at much larger distances, and broad Fe II emission, which can be particularly troublesome for Hβ in particular. And third, the line shape is a function of the line width, as illustrated in Figure 3, where we use the ratio of FWHM/$\sigma_{\rm line}$ as a rough characterization of the line profile.

This last point is actually quite important: if you consider the masses of the AGNs with the largest (~ 3.5) and the smallest (~ 0.7) values of FWHM/$\sigma_{\rm line}$, their ratio differs by a factor of ~ 25, depending on whether you use FWHM or $\sigma_{\rm line}$ as the line-width measure. This actually

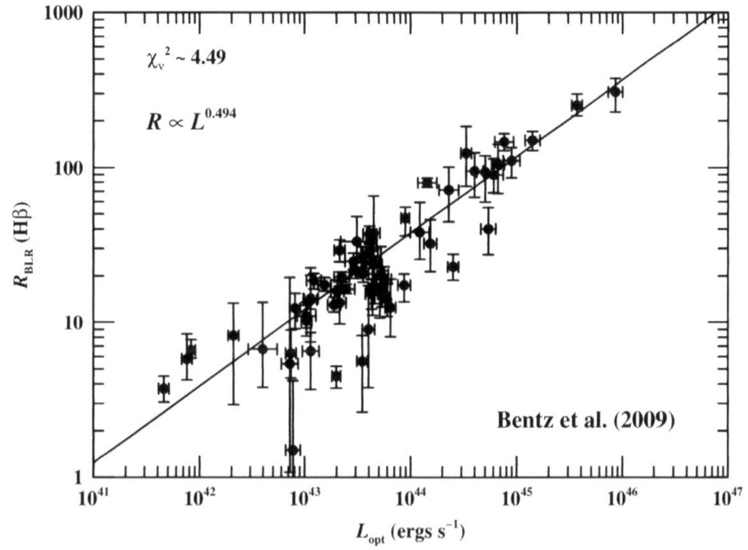

Figure 2. The relationship between the size of the broad-line region as measured for the Hβ emission line as a function of the AGN optical (5100 Å) continuum luminosity. From Bentz et al. (2009b).

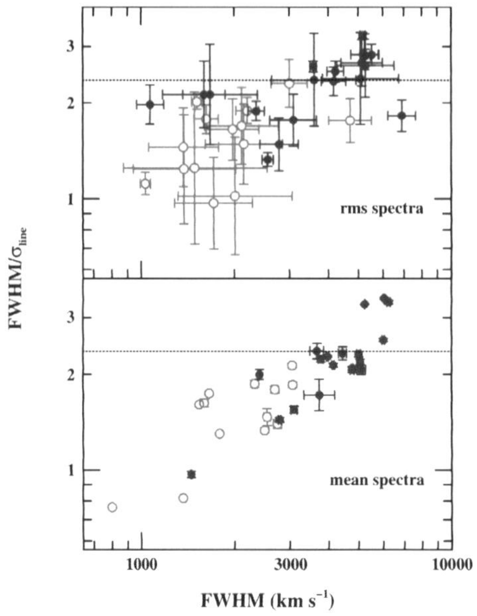

Figure 3. Ratio of FWHM/$\sigma_{\rm line}$ for rms (top) and mean (bottom) as a function of FWHM, demonstrating that line shape and and line width are correlated.

brings out another point: the average value of FWHM/$\sigma_{\rm line}$ is about 2.35, the same as for a Gaussian. So fitting a Gaussian to an emission line, like people often do to better determine the line widths in noisy spectra, works well for an *average* quasar emission line, but it is a terrible approximation for both the widest and narrowest lines. I'll return to this point shortly.

So which of FWHM and $\sigma_{\rm line}$ is the better measure? Which introduces less bias into the mass scale? My first parenthetical comment is to not get your expectations too high: what is remarkable to me is that we can take these two simple parameters, time lag and line width, that in some sense average over what is undoubtedly a very complex system, and compute a mass that's good to a factor of two or three! But, still we need these uncertainies to be random rather than systematic (i.e., not a function of line width or luminosity).

My own best answer to which line-width measure to use is that I have some preference for $\sigma_{\rm line}$ over FWHM. The evidence for favoring $\sigma_{\rm line}$ is threefold, but none of these make the case definitive. First, the virial relationships, i.e., plots of line width versus lag, are rather tighter for $\sigma_{\rm line}$ than for FWHM (Peterson et al. 2004). Second, if we consider the well-studied case of NGC 5548 and plot the reverberation mass measurements as a function of AGN luminosity (Figure 4), we find that the mass is independent of luminosity for $\sigma_{\rm line}$, but shows some (weak) luminosity dependence if FWHM is used: I hasten to add that neither of these statements has a high level of statistical signficance.

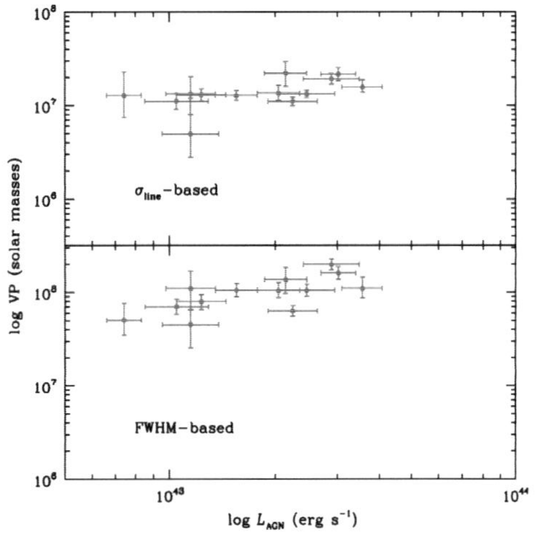

Figure 4. Virial product (i.e., $M_{\rm BH}/\langle f \rangle$) based on $\sigma_{\rm line}$ (top) and FWHM (bottom) as a function of luminosity as NGC 5548 varies in brightness over time. The masses based on $\sigma_{\rm line}$ are independent of luminosity; those based on FWHM show dependence on luminosity, though both of these conclusions are of low statistical significance.

My third argument is more subtle: suppose that $\sigma_{\rm line}$ is the correct, unbiased way to characterize the line width. If one uses FWHM instead, Figure 3 shows that the broadest lines will be overestimated and the narrowest lines will be underestimated. All other things being equal, the largest masses are too high and the lowest masses are too low. This would explain qualitatively the "sub-Eddington boundary" found by Steinhardt & Elvis (2010) using black-hole masses computed from FWHM of Gaussian fits to quasar spectra (Shen et al. 2008). Rafiee & Hall (2011) show that the sub-Eddington limit is significantly weakened when directly measured values of $\sigma_{\rm line}$ are used instead of FWHM of Gaussian fits.

4. Future Considerations

On the whole, the news is pretty good: reverberation mapping can potentially yield masses that are accurate to around a factor of 2–3 or so. Longer, more densely sampled reverberation campaigns are beginning to yield velocity–delay maps (e.g., Bentz et al. 2010; Barth et al. 2011; Denney et al. 2011) and in the cases where dynamical modeling has been done, the black hole masses are consistent with the masses computed from equation (2). It is extremely important to do this for as many AGNs as possible in order to determine $\langle f \rangle$ directly and remove dependence of the reverberation mass scale on relationships between the black hole mass and host galaxy properties. As more and better data are acquired, the simple relationships between black hole mass and host galaxy properties are beginning to look more complex. For example, there is some evidence that narrow-line Seyfert 1 galaxies, which are thought to be high Eddington rate accretors, reside in galaxies with pseudobulges rather than classical bulges (see Orban de Xivry et al. 2011, Mathur et al. 2012, and references therein). The evolutionary paths for black hole growth may well depend on the black hole mass, with higher-mass black holes growing early, largely through mergers, and lower-mass black holes growing later through secular processes (see Kormendy, Bender, & Cornell 2011 and references therein); relationships like the R–L relationship (§3.1) are probably not affected (as once gas gets down to the inner parsec or so, the black hole hardly cares how the gas got there), but the relationships between the black hole mass and the host properties might well be different.

Acknowledgments
I am grateful to the US National Science Foundation for support of this work through grant AST-1008882. I thank the organizers for their kind hospitality and for an enjoyable workshop, and I thank to Smita Mathur for comments on the manuscript.

References

Barth, A., et al. 2011, ApJ, 743: 4
Bentz, M.C., et al. 2006, ApJ, 651, 775
Bentz, M.C., Peterson, B.M., Pogge, R.W., & Vestergaard, M. 2009a, ApJ, 694, L166
Bentz, M.C., Peterson, B.M., Netzer, H., Pogge, R.W., & Vestergaard, M. 2009b, ApJ, 697, 160
Bentz, M.C., et al. 2010, ApJ, 720, L46
Blandford, R.D., & McKee, C.F. 1982, ApJ, 255, 419
Davies, R.I., et al. 2006, ApJ, 646, 754
Denney, K.D., et al. 2010, ApJ, 721, 715
Denney, K.D., et al. 2011, in Proceedings of the Workshop *Narrow-Line Seyfert 1 Galaxies and Their Place in the Universe*, PoS(NLS1) 034
Graham, A.W., Onken, C.A., Athanassoula, E., & Combes, F. 2011, MNRAS, 412, 2211
Hicks, E.K.S., & Malkan, M.A. 2008, ApJS, 174, 31
Kelly, B.C., Bechtold, J., & Siemiginowska, A. 2009, ApJ, 698, 895
Kormendy, J., Bender, R., & Cornell, M.E. 2011, Nature, 469, 374
Marconi, A., et al. 2008, ApJ, 678, 693
Mathur, S., Fields, D., Peterson, B.M., & Drupe, D. 2012, submitted to ApJ [arXiv:1102.0537]
Onken, C.A., Ferrarese, L., Merritt, D., Peterson, B.M., Pogge, R.W., Vestergaard, M., & Wandel, A. 2004, ApJ, 615, 645
Onken, C.A., et al. 2007, ApJ, 670, 105
Orban de Xivry, G., Davies, R., Schartmann, M., Komossa, S., Marconi, A., Hicks, E., Engel, H., & Tacconi, L. 2011, MNRAS, 417, 2721
Peterson, B.M. 1993, PASP, 105, 247
Peterson, B.M. 2001, in Advanced Lectures on the Starburst–AGN Connection, ed. I. Aretxaga, D. Kunth, R. Mújica (Singapore: World Scientific), p. 3
Peterson, B.M., et al. 2004, ApJ, 613, 682
Rafiee, A., & Hall, P.B. 2011, MNRAS, 415, 2932
Shen, Y., Greene, J.E., Strauss, M.A., Richards, G.T., & Schneider, D.P. 2008, ApJ, 680, 169
Steinhardt, C.L., & Elvis, M. 2010, MNRAS, 402, 2637
Woo, J., et al. 2010, ApJ, 716, 269

Astronomy at High Angular Resolution 2011 IOP Publishing
Journal of Physics: Conference Series **372** (2012) 012009 doi:10.1088/1742-6596/372/1/012009

High-resolution infrared observations of active galactic nuclei

Jörg-Uwe Pott
Max-Planck-Institut für Astronomie, Königstuhl 17, D-69117 Heidelberg, Germany
E-mail: jpott@mpia.de

Abstract. Interferometric resolution at IR wavelengths offers for the first time the possibility to zoom into the nuclei of galaxies beyond the circumnuclear stellar structures and spatially resolve gas and dust in the innermost regions (0.05-5pc), dominated by the central black hole. Ultimate goal is to reveal new aspects of AGN feeding, and interaction with its host galaxy. After first successes of resolving AGN with infrared interferometry (VLTI, Keck-IF), the second generation of high-resolution interferometric imagers behind 8m class telescopes is currently being built. I will summarize current aspects and successes of the field, and present our activities to provide extended capabilities for VLTI-Midi and -Matisse, LBT-Linc-Nirvana and Keck-Astra to study a larger sample of AGN in greater detail.

1. Which AGN questions can be (partially) answered by IR observations at interferometric resolution?

Besides nuclear fusion in stars, accretion of matter onto super-massive black holes in active galactic nuclei (AGN) is the second dominating source of radiative energy in the universe. While we already attempt to use nuclear fusion on Earth, the details of the physics in an AGN are much less understood, and a fascinating area of research. High-angular resolution (HAR) infrared (IR) observations help to reveal the involved physical processes in the outer (red) parts of the black hole's accretion disk, as well as its immediate surroundings. The structure and kinematics of the broad and narrow emission line regions (BLR resp. NLR), as well as the distribution and content toroidal obscuration region (TOR aka torus) are being studied. Their properties are connected to feeding the accretion process and how it feeds back to the surroundings. While typical linear torus sizes are well suited for the resolution of current long baseline interferometers, like VLTI and KI, line reverberation mapping observations suggest that resolving the broad line region, being located inside of the dusty torus, requires even higher angular resolution.

Apart from measuring their direct properties, the interaction of the AGN with its host galaxy is an active field of research. The feeding material, finally settling in the accretion disk, might originate in circumnuclear star burst outflows, star formation next to an AGN is often observed. On the other hand, AGN feedback has possibly a significant impact on regulating the star formation in the galaxy. Such host-AGN interaction scenarios require sensitive spectroscopic observations, which typically cannot be provided by interferometric arrays, which, due to their limited collecting area, are optimized to observe the brightest and compact details. However, AO-assisted narrow-band imaging or IFU observations, combined with Fizeau interferometry (Linc-Nirvana), can trace the observations of such AGN-host interaction.

A third area of research, related to HAR-observations of AGN, uses AGN and their properties as tracers for galactic evolution. HAR observations are particularly adequate to study the merger of galaxies in the latest phases. Imaging traces galaxy mergers by resolving distorted shapes of the host. In a few cases, even two AGN are observed, which trace the black hole mass build up via merging.

In this article, I will concentrate mostly on the interferometric observation of AGN physics, using the highest resolution available, provided by Michelson interferometric arrays. The last section will contain a brief outlook to instruments of the next generation interferometers, in the construction of which I am currently involved, and examples of their potential impact on extragalactic HAR observations.

2. Exploring the immediate AGN environment

Given IR broad-band fluxes, and the sizes between 0.1-10 pc, the thermal continuum radiation of dusty clouds of AGN tori can be observed with modern telescope arrays. The last decade has seen quite some groundbreaking work in this field.

First AGN fringes at NIR [Swain et al.(2003), Wittkowski et al.(2004)] and MIR wavelengths [Jaffe et al.(2004)] have been observed, showing the general existence of resolvable dust structures so close to the AGN. The measured sizes are are an order of magnitude smaller than smooth torus models, and can only explained by dusty clumps [Nenkova et al.(2010)]. First interferometric observations of dust heated by an Advection Dominated Accretion Flow (ADAF) in CenA [Meisenheimer et al.(2007)]. While originally several emission mechanism have been proposed to explain the NIR bump in AGN SEDs, interferometric confirmations of the size-luminosity relation [Kishimoto et al.(2009), Kishimoto et al.(2011), Tristram et al.(2009), Tristram & Schartmann(2011)], together with optical-to-NIR reverberation measurements [Suganuma et al.(2006)] strongly support a dusty, thermal origin of the NIR radiation. [Pott et al.(2010a)] pioneered the interferometric version of the dust reverberation measurement, showing suprisingly different results than the spatially unresolved version [Koshida et al.(2009)]. As discussed by the former authors, the interferometrically derived dust size scale is a more robust measurement of the average location of the emitting dust, while, as typically for reverberation measurements, the latter method is sensitive to the innermost dust. Burtscher presented at this conference the first sizeable, systematic, interferometric AGN survey of 14 AGN suitable to probe for differences between the dust morphology in type 1 and type 2 AGN [Burtscher et al.(2012)].

3. What is required to go beyond shear torus size estimates?

Now we search for ways to go into the details of the circumnuclear dust. There is several paths to go forward.

- multi-wavelength studies to identify large-scale structure
- high-precision measurements to identify the small scale structure (clouds)
- spatially resolved multi-λ reverberation measurements to understand the heating throught the entire torus
- spectro-astrometry of the broad-line region.

Currently, it is not clear if the dust structures seen at 2 μm and 10 μm are the same, and physically connected. Multi-wavelength campaigns are rare, and have just started, since typically the interferometric AGN observations have been carried out in different hemispheres: 2 μm with the more sensitive Keck interferometer on Mauna Kea, and the 10 μm observations with the MIDI beamcombiner of the VLTI on Cerro Paranal in Chile. However, recent advances with the VLTI instrumentation will allow for more efficient NIR-AGN observations and facilitate future multi-wavelength studies (ESO press releases ann11031 & ann11021).

Direct detection of small scale structure in the torus (like dusty clouds) will require high-precision measurments. Current MIR visibility measurements suffer from the difficult photometric measurements in the large thermal background. Upcoming NIR instrumentation and fringe tracking possibilities at the VLTI will offer to combine measurements of a few percent precision of the correlated flux with good uv-coverage. Ultimately, the 2nd generation instruments like GRAVITY and MATISSE will provide the tools to distinguish clumpy structure and warps in the dust morphology.

Interferometric NIR imaging of the broadline region would require 0.1 mas resolution, or baselines of order 1 km, at the sensitivity of todays arrays. No currently built or planned array provides these performance numbers. However, spectro-interferometric astrometry has proven on sky to deliver differential astrometric precision up to 3 μas for bright stars [Pott et al.(2010b)]. A few tens of micro-arcsec would be enough to astrometrically resolve the kinematic properties of the BLR with respect to the continuum radiation. We currently work to upgrade the KI-ASTRA spectroscopy [Woillez et al.(2010)] to allow the spectroscopic observation of the brightest AGN.

4. What to expect from the next generation of instruments?

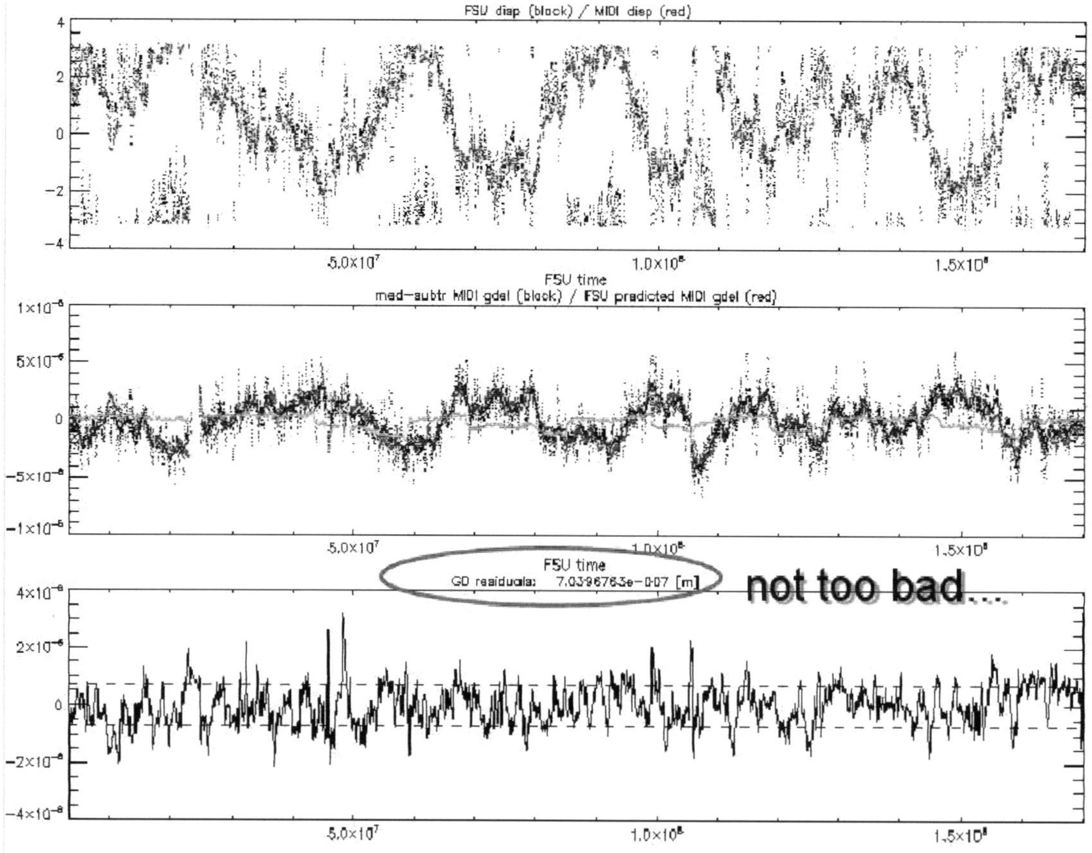

Figure 1. First results from commissioning the PRIMA-FSU as MIDI fringe tracker, showing that the MIDI group delay can be controlled to better than $\lambda/10$.

MPIA contributes to the next generation of infrared high-angular resolution technology. Here, I will briefly report on three of our interferometric instrumentation initiatives for the next years.

As discussed above, the VLTI/MIDI precision is currently limited to typically 5-10%, which is not sufficient to resolve fine-structure details in the dust morphology of AGN surroundings. In collaboration with ESO, we work on commissioning the PRIMA-FSU K-band fringe tracker as external fringe tracker for the MIR measurements. First results are very encouraging (Fig. 1, see also [Müller et al.(2010)]). Key of the fringe sensing unit (FSU) is that low-dispersion fringes are recorded in K-band. Spectrally dispersed FSU fringes allow to distinguish the water vapor seeing from the dry air fringe position, which is a prerequisite to precisely predict the mid-infrared fringe position [Koresko et al.(2006)]. This technology will allow both to improve on the precision *and* the sensitivity of MIDI. While the precision is important for fine structure, the sensitivity improvement is needed to enlarge the number observable targets, and be able to detect significant differences between different type of AGN. This remains difficult, since also within a class the detailed properties of each AGN can be quite different, depending on several physical parameters of the accretion process.

A leap forward regarding the MIR imaging capabilities of the VLTI will be provide by MATISSE, the 2nd generation 4-telescope beam-combiner [Lopez et al.(2008)]. Delivering 6 baselines and 3 closure-phases at a time, MATISSE is dedicated to model-independent interferometric imaging of the dusty environment of AGN. Particularly interesting will be the fact, that MATISSE also offers L/M-band observing capabilities. This will finally close the gap for HAR-AGN observations between 2 μm and 10 μm. This is crucial to get a complete picture of the dust formation in AGN. First light for MATISSE is foreseen for 2016.

A different type of gap will be closed by the Fizeau-interferometric imager at the LBT, LINC-NIRVANA [Herbst et al.(2010)]. This strategic LBT instrument will, from 2014 on, deliver AO-like diffraction limited images of a 23m telescope. While current interferometric arrays already deliver higher resolution, only the complete uv-coverage of such an Fizeau-imager will allow faint images in narrow- and broad-band of the surroundings of AGN at 50-100pc scale. Here, circum-nuclear starbursts, jets, and AGN winds are key targets to investigate the processes of AGN feeding and feedback. LINC-NIRVANA will be the next step in resolution and sensitivity from the current 8m-class A0-imagers towards ELT and JWST science cases.

References

[Swain et al.(2003)] Swain, M., Vasisht, G., Akeson, R., et al. 2003, ApJL, 596, L163
[Wittkowski et al.(2004)] Wittkowski, M., Kervella, P., Arsenault, R., et al. 2004, A&A, 418, L39
[Jaffe et al.(2004)] Jaffe, W., Meisenheimer, K., Röttgering, H. J. A., et al. 2004, Nature, 429, 47
[Meisenheimer et al.(2007)] Meisenheimer, K., Tristram, K. R. W., Jaffe, W., et al. 2007, A&A, 471, 453
[Kishimoto et al.(2009)] Kishimoto, M., Hönig, S. F., Antonucci, R., et al. 2009, A&A, 507, L57
[Kishimoto et al.(2011)] Kishimoto, M., Hönig, S. F., Antonucci, R., et al. 2011, A&A, 527, A121
[Tristram et al.(2009)] Tristram, K. R. W., Raban, D., Meisenheimer, K., et al. 2009, A&A, 502, 67
[Tristram & Schartmann(2011)] Tristram, K. R. W., & Schartmann, M. 2011, A&A, 531, A99
[Pott et al.(2010a)] Pott, J.-U., Malkan, M. A., Elitzur, M., et al. 2010, ApJ, 715, 736
[Burtscher et al.(2012)] Burtscher, L., et al., *MIDI AGN Large Programme: A statistical sample of resolved AGN tori*, in this conference proceedings
[Nenkova et al.(2010)] Nenkova, M., Sirocky, M. M., Nikutta, R., Ivezić, Ž., & Elitzur, M. 2010, ApJ, 723, 1827
[Suganuma et al.(2006)] Suganuma, M., Yoshii, Y., Kobayashi, Y., et al. 2006, ApJ, 639, 46
[Koshida et al.(2009)] Koshida, S., Yoshii, Y., Kobayashi, Y., et al. 2009, ApJL, 700, L109
[Pott et al.(2010b)] Pott, J.-U., Woillez, J., Ragland, S., et al. 2010, ApJ, 721, 802
[Woillez et al.(2010)] Woillez, J., Akeson, R., Colavita, M., et al. 2010, SPIE, 7734,
[Müller et al.(2010)] Müller, A., Pott, J.-U., Morel, S., et al. 2010, SPIE, 7734,
[Koresko et al.(2006)] Koresko, C., Colavita, M. M., Serabyn, E., Booth, A., & Garcia, J. 2006, SPIE, 6268,
[Lopez et al.(2008)] Lopez, B., Antonelli, P., Wolf, S., et al. 2008, SPIE, 7013,
[Herbst et al.(2010)] Herbst, T. M., Ragazzoni, R., Eckart, A., & Weigelt, G. 2010, SPIE, 7734,

Journal of Physics: Conference Series **372** (2012) 012010 doi:10.1088/1742-6596/372/1/012010

Type 1 AGN at low z. I. Emission properties

J Stern and A Laor

Department of Physics, Technion – Israel Institute of Technology, Haifa, Israel

E-mail: stern@physics.technion.ac.il, laor@physics.technion.ac.il

Abstract. We analyze the emission properties of a new sample of 3 596 type 1 AGN, selected from the SDSS DR7 based on the detection of broad Hα emission. The sample extends over a broad Hα luminosity $L_{\rm bH\alpha}$ of $10^{40}-10^{44}$ erg s^{-1} and a broad Hα FWHM of 1 000–25 000 km s^{-1}, which covers the range of black hole mass $10^6 < M_{\rm BH}/{\rm M}_\odot < 10^{9.5}$ and luminosity in Eddington units $10^{-3} < L/L_{\rm Edd} < 1$. We combine ROSAT, GALEX and 2MASS observations to form the SED from 2.2 μm to 2 keV. We find the following: 1. The distribution of the Hα FWHM values is independent of luminosity. 2. The observed mean optical-UV SED is well matched by a fixed shape SED of luminous quasars, which scales linearly with $L_{\rm bH\alpha}$, and a host galaxy contribution. 3. The host galaxy r-band (fibre) luminosity function follows well the luminosity function of inactive non-emission line galaxies (NEG), consistent with a fixed fraction of $\sim 3\%$ of NEG hosting an AGN, regardless of the host luminosity. 4. The optical-UV SED of the more luminous AGN shows a small dispersion, consistent with dust reddening of a blue SED, as expected for thermal thin accretion disc emission. 5. There is a rather tight relation of νL_ν(2 keV) and $L_{\rm bH\alpha}$, which provides a useful probe for unobscured (true) type 2 AGN.

1. Introduction

We examine the emission properties of a new sample of low z broad line AGN, aimed to be an extension of the SDSS quasar catalog (QCV, Schneider et al. 2010) to low luminosities. This contribution summarizes the main results obtained on the distribution of the broad line FWHM (§3), the host galaxies (§4), and the AGN spectral energy distribution (§§5–7). Full details of the methods used, and a discussion of the implication of the results, are given in the main article (Stern & Laor, submitted).

2. Sample Creation

The type 1 (T1) sample is selected from the SDSS 7$^{\rm th}$ data release (DR7, Abazajian et al. 2009), based on the detection of a broad Hα emission line. The parent sample includes 232 837 spectra with S/N > 10 and $0.005 < z < 0.31$. The broad Hα emission is identified by modeling the nearby narrow emission lines, stellar absorption features and featureless continuum. We find broad Hα emission in 3 596 objects (1.5%), which constitute the T1 sample.

The optical spectra in the T1 sample are supplemented by nearly complete near IR photometry from 2MASS (Skrutskie et al. 2006, detection rate 97%) and UV photometry from GALEX (Martin et al. 2005, 93%). We also add X-ray photometry from the ROSAT (Voges et al. 1999) survey, available for 43% of the T1 objects. The measured broad Hα luminosities ($L_{\rm bH\alpha}$) and FWHMs (Δv) and the photometric luminosities are available electronically.

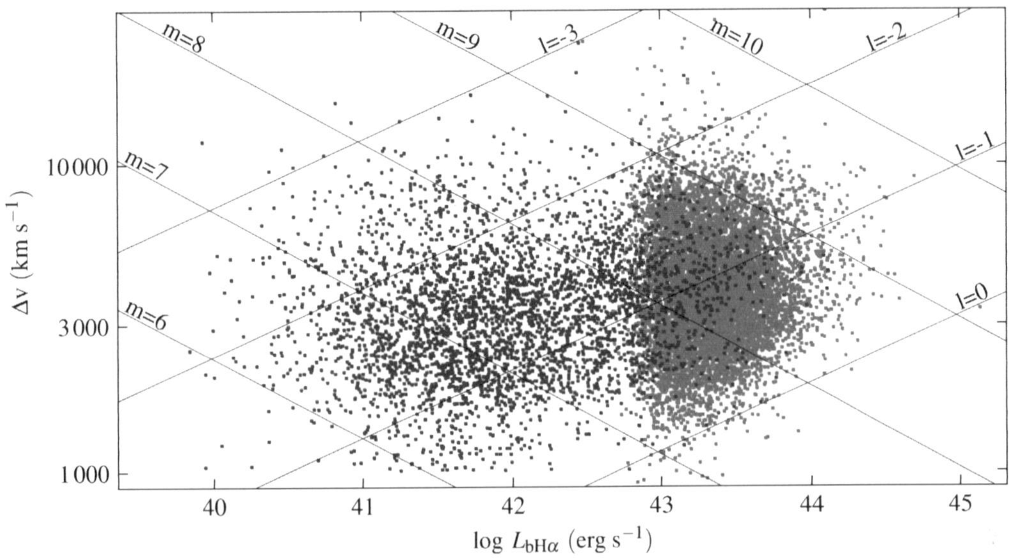

Figure 1. The distributions of the T1 (dark gray) and QCIV (light gray) samples in the broad Hα luminosity vs. FWHM plain. Crossing lines mark decades of the implied $M_{\rm BH}$ and $L/L_{\rm Edd}$.

3. How broad do the broad lines get?

Various theoretical models of the broad line region (BLR) and the AGN ionizing continuum predict an upper limit on Δv, beyond which the BLR does not exist (Nicastro 2000, Elitzur and Shlosman 2006, Laor & Davis 2011). The models differ in the dependence of the maximal Δv on luminosity. In Figure 1, we plot the $L_{\rm bH\alpha}$ vs. Δv distributions of the T1 objects and the 8 185 QCIV objects where the broad Hβ is available (measurements from Shen et al. 2008).

The T1 sample extends down by two orders of magnitude in $L_{\rm bH\alpha}$ compared to QCIV. The Eddington ratio $L/L_{\rm Edd}$ ($\equiv 10^l$) and black hole mass $M_{\rm BH}$ ($\equiv 10^m {\rm M}_\odot$) are derived from $L_{\rm bH\alpha}$ and Δv, in a manner similar to Greene & Ho (2005). It can be seen that the Eddington limit ($l = 0$) sets a minimum Δv with increasing $L_{\rm bH\alpha}$ for $L_{\rm bH\alpha} > 10^{42.5}$ erg s^{-1}. This, together with the rarity of AGN with $m > 9.5$, leads to a decrease in the range of observed Δv values with increasing $L_{\rm bH\alpha}$. A similar convergence of the range of Δv for the Mg II and C IV lines at the highest continuum luminosities was noted by Fine et al. (2008, 2010).

There is a steep decline in the number of objects with Δv > 10 000 km s^{-1}, at all luminosities. This decline is seen clearly in Figure 2, which plots the Δv distribution of the combined T1 + QCIV sample at different $L_{\rm bol}$. The distributions are remarkably similar, showing a roughly linear decline of $\log({\rm d}N/{\rm d}\log\Delta{\rm v})$ vs. Δv, or equivalently ${\rm d}N/{\rm d}\log\Delta{\rm v} \propto e^{-\Delta{\rm v}/\Delta{\rm v}_0}$, with $\Delta{\rm v}_0 \approx 2700$ km s^{-1}. The origin of this similarity is not clear, as the Δv distributions should be set by the distribution of the m and l values. Either these distributions somehow lead to a Δv distribution which is independent of $L_{\rm bH\alpha}$, or it may imply that Δv, rather than m and l, sets the observed Δv distribution through some unknown mechanism. Note the effect of the Eddington limit at Δv < 3 000 km s^{-1}, which increases the minimal Δv and the peak position with increasing $L_{\rm bol}$.

4. How does the AGN fraction change with galaxy luminosity?

Kauffmann et al. (2003), and various follow up studies, found that AGN tend to reside in massive hosts. We therefore examine the T1 fraction as a function of host luminosity.

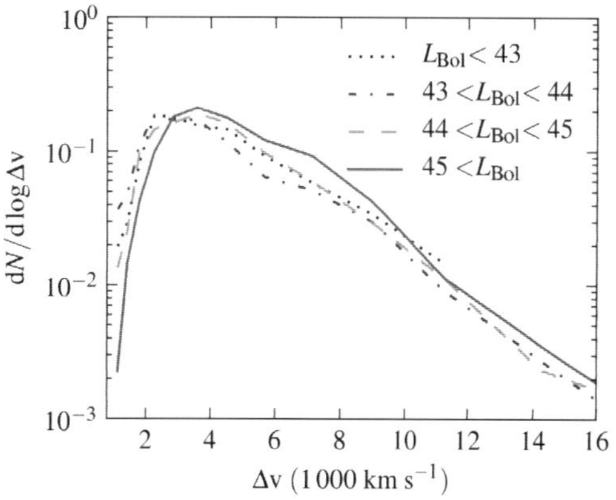

Figure 2. The FWHM distribution of the T1+QCIV sample, at different $L_{\rm bol}$ bins.

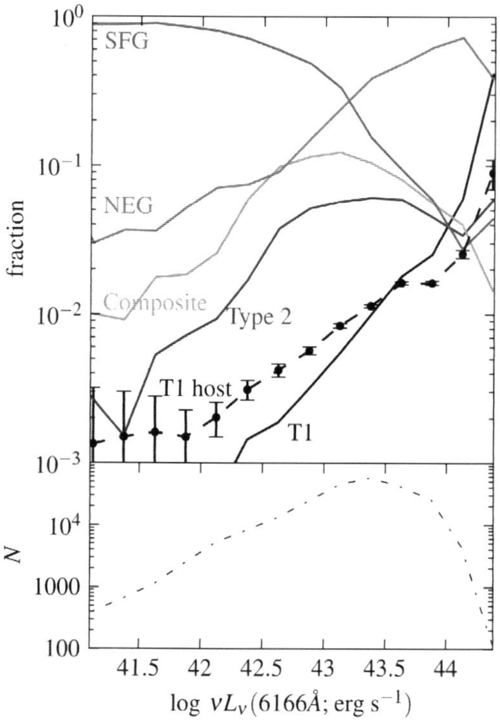

Figure 3. The spectral type distribution of galaxies in the parent sample, as a function of luminosity (r band). Non-T1 classifications are from Brinchmann et al. (2004). The 'T1 host' line is corrected for the AGN contribution. The lower panel shows the number of objects per 0.25 decade in νL_ν.

Figure 3 presents the classification of $z < 0.2$ galaxies in the parent sample (§2) as a function of the spectroscopically measured νL_ν(6166Å). The type 2, Star Forming Galaxy (SFG), Non-Emission line Galaxy (NEG) and composite classifications (based on the equivalent width and BPT ratios of the narrow lines) are taken from Brinchmann et al (2004). At low luminosities >90% of the SDSS galaxies are SFG, while at high luminosities the NEG dominate. The 'T1 host' line is derived from the νL_ν(6166Å) of the T1 objects, corrected for the AGN contribution (see §5 below). Remarkably, the T1 host fraction follows well the NEG fraction, or equivalently, their galaxy luminosity distribution is similar. This may suggest that T1s reside in NEG, as also implied by their similar concentration indices and bulge/total light ratios (Kauffmann et al. 2003). The fraction of NEG hosting broad line AGN at the level detectable in this study is $\sim 3\%$, independent of galaxy luminosity.

5. How does the AGN spectral energy distribution (SED) depend on luminosity?

Figure 4 compares $L_{\rm bH\alpha}$ with the observed 2.2 μm – 2 keV continuum luminosity, in the T1 sample. In the luminous bins in the FUV, NUV and 3940Å panels, the slopes are consistent with a linear relation, in all other bands and luminosities the slopes are flatter. Since the FUV occurs close to the peak position of the AGN SED (e.g. Zheng et al. 1997), the linear relation between νL_ν(FUV) and $L_{\rm bH\alpha}$ suggests that $L_{\rm bH\alpha}$ provides a good estimator of $L_{\rm bol}$. Also, the linear relation suggests that the mean covering factor of the BLR is independent of luminosity.

In all bands excluding X-ray, there is a transition from a steeper slope to a flatter slope with decreasing $L_{\rm bH\alpha}$. The transition $L_{\rm bH\alpha}$ increases with wavelength, up to a maximum of $\simeq 10^{43}$ erg s^{-1} at 1.2μm. The transition occurs due to the host contribution, which peaks at 1.2μm.

The transition between AGN and host dominance of the SED can be seen in Figure 5. The left panel plots the mean SEDs in the $L_{\rm bH\alpha}$ bins, i.e. each line connects the mean values of

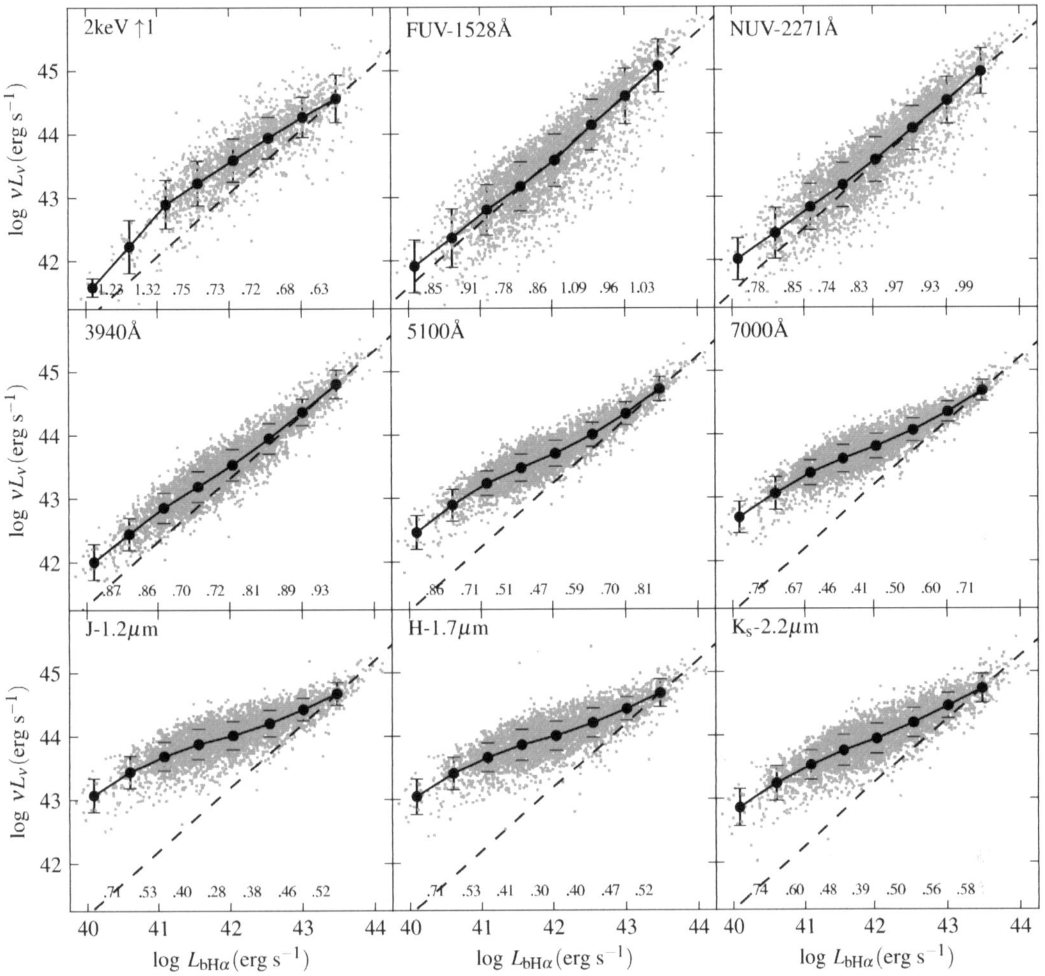

Figure 4. The $L_{\rm bH\alpha}$ vs. observed (AGN+host) νL_ν from the X-ray to the NIR, in the T1 sample. The X-ray νL_ν are shifted upward by 1 decade for presentation purpose. The large black dots and associated error bars mark the mean and dispersion of νL_ν in half-decade wide $L_{\rm bH\alpha}$ bins. The solid lines connect the means of adjacent bins, with the local slopes written underneath. Dashed lines depict a linear relation, normalized to the highest $L_{\rm bH\alpha}$ bin.

different continuum wavelengths for the same $L_{\rm bH\alpha}$ (Fig. 4). Note the transition from an AGN-dominated UV peaked SED to a galaxy-dominated NIR peaked SED with decreasing $L_{\rm bH\alpha}$. As the net AGN SED, we take the Richards et al. (2006, hereafter R06) SED derived from luminous quasars in the SDSS. We scale the R06 SED to the UV in the highest $L_{\rm bH\alpha}$ bin, and rescale by the relative $L_{\rm bH\alpha}$ of the lower seven bins. The scaled R06 SEDs are the putative net AGN SEDs, assuming the mean SED remains fixed with luminosity and scales linearly with $L_{\rm bH\alpha}$. In the middle panel of Fig. 5 we plot the residuals derived by subtracting the AGN from the observed SED, and in the right panel we plot the mean SEDs of inactive SDSS galaxies, with z-distributions matched to the z-distributions of each $L_{\rm bH\alpha}$ bin.

The residuals all appear to have a typical galaxy SED. This justifies (qualitatively) the simple scaling prescription we used for the net AGN SED. Note the significant excess luminosity in the K_s and H bands in the most luminous residuals, and the non-linear νL_ν(X-ray)-$L_{\rm bH\alpha}$ relation

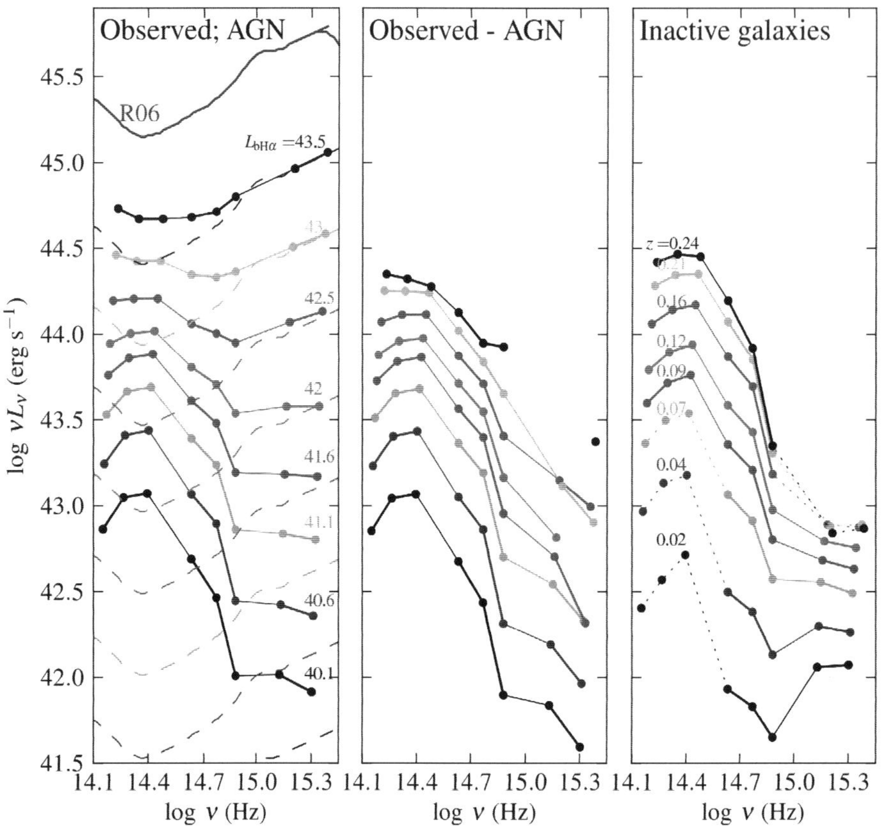

Figure 5. The luminosity dependence of the mean T1 objects SEDs at 2.2μm – 1528Å. **Left** Solid lines mark the observed SEDs for the eight $L_{\rm bH\alpha}$ bins (mean $L_{\rm bH\alpha}$ noted). Dashed lines mark the net AGN SED, derived by scaling the R06 quasar SED by the mean $L_{\rm bH\alpha}$ of the bin with the same color. **Middle** Residuals of the subtraction of the the scaled R06 SEDs from the observed AGN SEDs. **Right** Mean SEDs of z-matched inactive galaxies (mean z noted). Dotted lines mark bands with $< 70\%$ detections.

(Fig. 4), which imply that the linear scaling of the net AGN SED is applicable only to the optical-UV region.

In order to quantitatively evaluate the AGN SED scaling law, we define the scaling index β, where $L_{\rm AGN}(\lambda) \propto L_{\rm bH\alpha}^{\beta}$. A β independent of wavelength λ indicates the mean AGN SED shape remains fixed with luminosity. From the qualitative analysis shown in Fig. 5, we expect $\beta \sim 1$ for λ in the optical or UV. In Figure 6, we plot the implied mean host luminosity $L_{\rm host} \equiv \nu L_\nu - L_{\rm AGN}$ in the SDSS u and z bands, for different $L_{\rm bH\alpha}$ and β. As the expected mean $L_{\rm host}$ of a certain $L_{\rm bH\alpha}$ bin, we use the mean luminosity of type 2 AGN with the same z distribution as the $L_{\rm bH\alpha}$ bin. The type 2s are taken from the sample of Brinchmann et al. (2004).

The top panel of Fig. 6 shows that in the u band, $\beta < 0.9$ is ruled out, as the implied residual is negative for $L_{\rm bH\alpha} > 10^{41}$ erg s^{-1}. It is remarkable that the simplest scaling, $\beta = 1$ in both bands, leads to the closest match of the mean type 1 host to the mean type 2 host. Together with the linear νL_ν(UV)-$L_{\rm bH\alpha}$ relationship at high luminosities (Fig. 4), this implies that the mean optical-UV net AGN SED has a fixed shape, and scales linearly with $L_{\rm bH\alpha}$.

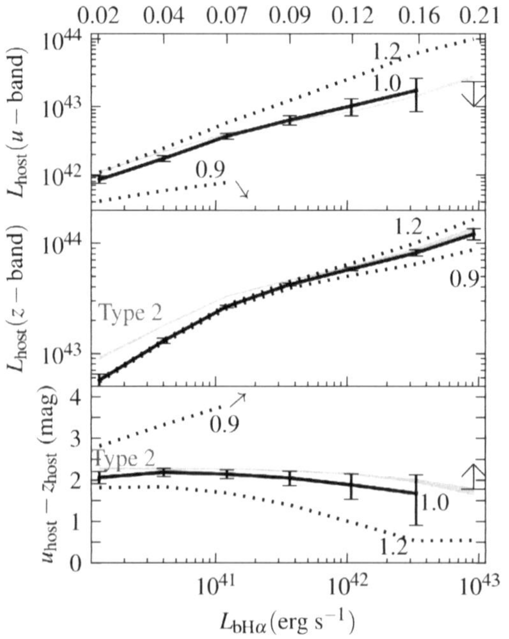

Figure 6. A comparison of the implied mean T1 host luminosity for different $L_{\mathrm{bH}\alpha}$ bins, for different AGN SED scaling laws (black solid/dotted lines, β noted), with z-matched type 2 AGN. The comparison is made in the u band (3551Å), z band (8932Å), and $u-z$ color. The mean z of each $L_{\mathrm{bH}\alpha}$ bin is noted on top. Arrows mark a negative implied host.

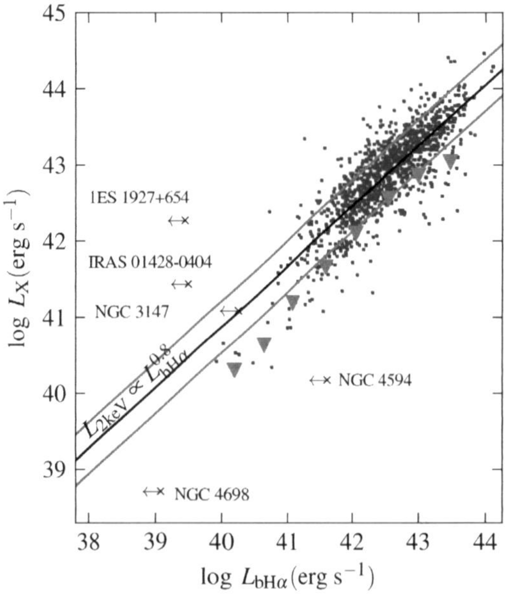

Figure 7. The L_{X} vs. $L_{\mathrm{bH}\alpha}$ relation, as a probe of true type 2 AGN. Dots mark T1 objects with $F_{\mathrm{bH}\alpha} > 10^{-13.5}$ erg s^{-1} cm^{-2}. Red triangles mark mean upper limits in 0.5-dec $L_{\mathrm{bH}\alpha}$ bins. The best power law fit and dispersion are shown as black and gray lines (slope noted). Arrows mark the upper limits on $L_{\mathrm{bH}\alpha}$ of five true type 2 candidates.

6. Is the lack of BLR emission in true type 2 candidates significant?

True type 2 candidates are AGN in which the BLR is not observed, despite apparently unobscured X-ray emission. In Figure 7 we compare the upper limits on $L_{\mathrm{bH}\alpha}$ in five true type 2 candidates from Shi et al. (2010) and Tran et al. (2011), with the T1 sample $L_{\mathrm{X}}(\equiv \nu L_\nu(2\mathrm{keV}))$ vs. $L_{\mathrm{bH}\alpha}$ relation. Upper limits on $L_{\mathrm{bH}\alpha}$ are derived from the expected Δv (based on the published M_{BH} and L/L_{Edd}, §3) and the flux density near Hα. The absence of a broad Hα in NGC 4594, NGC 4698, and NGC 3147 is not significant. In the other two objects the expected $L_{\mathrm{bH}\alpha}$ is well above the upper limits, and these two objects appear to be true type 2 AGN.

7. What drives the dispersion in the optical-UV SED of AGN?

In §5 we have shown that the mean SED of broad line AGN is well reproduced by the sum of the mean SED of luminous quasars, scaled down by $L_{\mathrm{bH}\alpha}$, and a host contribution. Here, we examine the dispersion in the shapes of individual SEDs.

In Figure 8, we show the distributions of $\nu L_\nu(\mathrm{FUV})/L_{\mathrm{bH}\alpha}$ and optical slope α_{opt} as a function of $L_{\mathrm{bH}\alpha}$. The $\nu L_\nu(\mathrm{FUV})/L_{\mathrm{bH}\alpha}$ distributions for $\log L_{\mathrm{bH}\alpha} > 40.5$ have a very similar shape. All distributions have a narrow peak, with an extended tail towards low $\nu L_\nu(\mathrm{FUV})/L_{\mathrm{bH}\alpha}$ values. The narrow peak suggests there is a small dispersion in the covering factor of the BLR. In the $\log L_{\mathrm{bH}\alpha} = 40.5$ bin, host contribution to the UV broadens the distribution and increases the mean value.

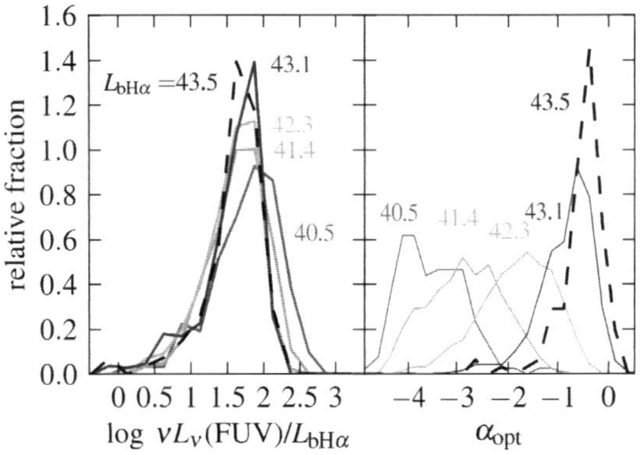

Figure 8. The T1 objects $\nu L_\nu(\mathrm{FUV})/L_{\mathrm{bH}\alpha}$ and α_{opt} distributions as a function of $L_{\mathrm{bH}\alpha}$. Solid (dashed) lines represent bins of width 1 (0.5) dec in $L_{\mathrm{bH}\alpha}$. The mean $L_{\mathrm{bH}\alpha}$ are noted.

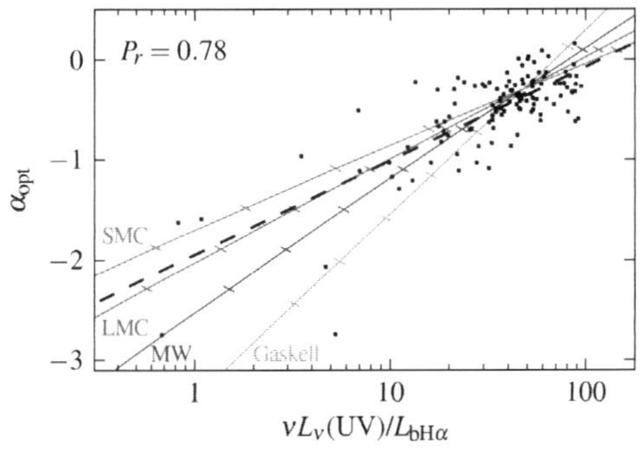

Figure 9. The α_{opt} vs. $\nu L_\nu(\mathrm{UV})/L_{\mathrm{bH}\alpha}$ of the T1 objects in the log $L_{\mathrm{bH}\alpha} = 43.5$ bin. Also shown are the best fit (dashed line) and expected relations for different extinction laws (solid lines, marks every 0.1 in $E_{\mathrm{B-V}}$).

The α_{opt} distributions are redder at lower luminosities, due to the host emission. The log $L_{\mathrm{bH}\alpha} = 43.5$ bin is least affected by the host contribution, and has a narrow peak and extended red tail, similar to the $\nu L_\nu(\mathrm{FUV})/L_{\mathrm{bH}\alpha}$ distribution, and to the continuum slope distribution found by Richards et al. (2003) on luminous quasars. There is therefore a small dispersion in $\nu L_\nu(\mathrm{FUV})/L_{\mathrm{bH}\alpha}$ and α_{opt} in the net AGN emission, based on the objects least affected by the host contribution. The similar shape may indicate a common origin of the dispersion, such as dust reddening.

The dust reddening scenario is explored in Figures 9 and 10. In Fig. 9 we plot α_{opt} vs. $\nu L_\nu(\mathrm{FUV})/L_{\mathrm{bH}\alpha}$ for the $\log L_{\mathrm{bH}\alpha} = 43.5$ bin. The two independent ratios are highly correlated, as expected if dust extinction contributes to the dispersion of the AGN SED. An SMC/LMC extinction law (using the Pei 1992 formulation) is preferred over a Milky Way or Gaskell & Benker (2007) extinction law, as found by Hopkins et al. (2004) on luminous quasars. In Fig. 10, we plot the 2.2 μm – 2 keV mean SED for different α_{opt} bins at the highest $L_{\mathrm{bH}\alpha}$ bins. In the $\log L_{\mathrm{bH}\alpha} = 43.5$ panel, the difference between SEDs increases from the NIR to the UV and disappears in the X-ray, also consistent with dust extinction. A similar behavior is observed in the log $L_{\mathrm{bH}\alpha} = 43$ panel, together with a weak anti-correlation of NIR luminosity and α_{opt}. The latter trend originates from the range of host luminosities, where a relatively luminous host increases $\nu L_\nu(1.2\mu\mathrm{m})$ and decreases α_{opt}.

The reddening origin for the dispersion in the net AGN SED implied by Figs. 9 and 10, already at $\alpha_{\mathrm{opt}} > -1$, indicates that the intrinsic AGN SED dispersion is even smaller than seen in Fig. 8. A rather uniform and blue intrinsic optical-UV SED is expected if AGN are powered by a thin accretion disc, which emits locally close to a blackbody.

8. Conclusions

We present and analyze a new sample (T1) of 3 596 broad Hα selected AGN from the SDSS DR7, with $\log L_{\mathrm{bH}\alpha} = 40 - 44$, which spans $m = 6 - 9$ and $l = -3 - 0$. We add UV (GALEX), IR (2MASS), and X-ray (ROSAT) luminosities to form the mean SED. The main results are:

(i) The Hα FWHM velocity distribution $\mathrm{d}N/\mathrm{d}\log\Delta\mathrm{v}$ is independent of luminosity and falls exponentially with Δv. The origin of this distribution remains to be understood.

(ii) The observed mean 9000Å–1500Å SED, as a function of $L_{\mathrm{bH}\alpha}$, is consistent with a sum of the mean SED of luminous quasars, which scales linearly with $L_{\mathrm{bH}\alpha}$, and a host galaxy

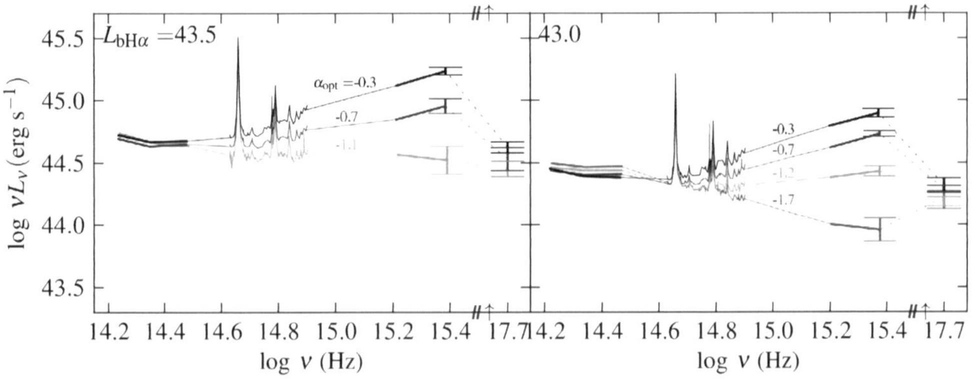

Figure 10. The mean IR to X-ray SEDs and optical spectra as a function of α_{opt}, for T1 objects with $L_{bH\alpha} = 10^{43.5}$ and 10^{43} erg s^{-1}. The value of L_X is shifted upward (+1 dec) for presentation. The error bars denote the errors in the mean L_X and νL_ν(FUV).

contribution.

(iii) The host galaxy r-band luminosity function of T1 objects follows the NEG luminosity function, with a relative normalization of $\sim$ 3%, suggesting that the host of broad line AGN are NEG, and the AGN probability of occurrence is independent of the host mass.

(iv) The dispersion in the optical-UV SED in luminous AGN ($\log L_{bH\alpha} \geq 43$), is consistent with reddening. This indicates the intrinsic SED of AGN is blue, with a small dispersion, as predicted from thermal thin accretion disc models.

(v) The $L_{bH\alpha}$ versus L_X correlation provides a useful probe for unobscured narrow line AGN. It can be used to test if the absence of a broad Hα, in X-ray detected AGN, is significant.

References

Abazajian K N *et al* 2009 *ApJS* **182** 543
Brinchmann J, Charlot S, White S D M, Tremonti C, Kauffmann G, Heckman T and Brinkmann J 2004 *MNRAS* **351** 1151
Elitzur M and Shlosman I 2006 *ApJ* **648** L101
Fine S *et al* 2008 *MNRAS* **390** 1413
Fine S *et al* 2010 *MNRAS* **409** 591
Gaskell C M and Benker A J 2007 *ArXiv e-prints* 0711.1013
Greene J E and Ho L C 2005 *ApJ* **630** 122
Hopkins P F *et al* 2004 *AJ* **128** 1112
Kauffmann G *et al* 2003 *MNRAS* **346** 1055
Laor A and Davis S W 2011 *MNRAS* **417** 681
Martin D C *et al* 2005 *ApJ* **619** L1
Nicastro F 2000 *ApJ* **530** L65
Pei Y C 1992 *ApJ* **395** 130
Richards G T *et al* 2003 *AJ* **126** 1131
Richards G T *et al* 2006 *ApJS* **166** 470 (R06)
Schneider D P *et al* 2010 *AJ* **139** 2360
Shen Y, Greene J E, Strauss M A, Richards G T and Schneider D P 2008 *ApJ* **680** 169
Shi Y, Rieke G H, Smith P, Rigby J, Hines D, Donley J, Schmidt G and Diamond-Stanic A M 2010 *ApJ* **714** 115
Skrutskie M F *et al* 2006 *AJ* **131** 1163
Stern J and Laor A, submitted to *MNRAS*
Tran H D, Lyke J E and Mader J A 2011 *ApJ* **726** L21
Voges W *et al* 1999 *A&A* **349** 389
Zheng W, Kriss G A, Telfer R C, Grimes J P and Davidsen A F 1997 *ApJ* **475** 469

Astronomy at High Angular Resolution 2011 IOP Publishing
Journal of Physics: Conference Series **372** (2012) 012011 doi:10.1088/1742-6596/372/1/012011

Formation and propagation of MHD jets - relativistic jets, radiation pressure, and shock-induced rotation

Christian Fendt, Oliver Porth[1], and Bhargav Vaidya[2]
Max Planck Institute for Astronomy, Königstuhl 17, D-69117 Heidelberg, Germany
E-mail: fendt@mpia.de

Abstract. We present recent results of magnetohydrodynamic (MHD) simulations of jet formation and propagation, discussing a variety of astrophysical setups. In the first approach the role of the disk magnetic flux profile and disk mass flux profile concerning the jet collimation degree is investigated. Our results suggest (and quantify) that in general magnetized outflows launched from a compact region close to the inner disk radius tend to be less collimated. The second setup considers simulation of relativistic MHD jet formation, considering jet launched from the surface of a Keplerian disks, demonstrating - for the first time - the self-collimating ability of relativistic MHD jets. We obtain Lorentz factors up to $\simeq 10$ while acquiring a high degree of collimation of about 1 degree. We then present MHD jet formation simulations taking into account radiation pressure of a central luminous source. We investigate radiative effects on jet collimation and propagation - an environment which is interesting for outflows from massive young stars and active galactic nuclei. Finally, we present a model which explains a possible jet rotation by shock compression of an intrinsic helical magnetic field.

1. Introduction

Astrophysical jets are highly collimated beams of high-velocity material, observed in a variety of astronomical sources - among them young stellar objects (YSO), micro-quasars, or active galactic nuclei (AGN). The current understanding of jet formation is that these outflows are launched by *magnetohydrodynamic* (MHD) processes in the close vicinity of the central object - an accretion disk surrounding a protostar or a compact object [1, 2, 3]. Besides the contribution of the disk wind to the overall jet flow, there will be a further contribution launched by the central object. In case of stellar sources, a stellar wind may contribute additional Poynting flux or pressure [4, 5, 6]. In case of a central black hole, a central Poynting dominated spine jet may exist being driven by the Blandford-Znajek mechanism [7]. The details of all the physical processes involved are, however, not completely understood.

In this proceedings we present results of axisymmetric MHD simulations investigating certain aspects of jet formation and propagation. As the structure and the evolution of the accretion disk as well as the origin and the distribution of the jet-launching magnetic field is not really known, we performed a parameter study exploring how the disk magnetic flux distribution and the disk wind density profile affect the asymptotic collimation degree (Sect. 2).
In Sect. 3 we present simulations of collimating disk winds into relativistic jets applying (special)

[1] Present address: School of Mathematics, University of Leeds, Leeds, LS2 9JT, UK
[2] Present address: School of Physics and Astronomy, University of Leeds, Leeds, LS6 2EN, UK

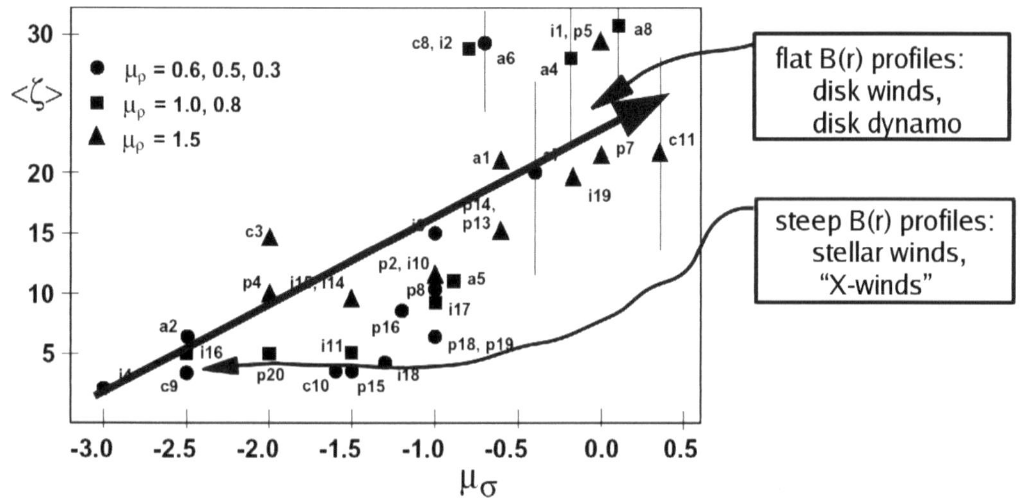

Figure 1. Outflow collimation degree as function of the power law index of the disk magnetic field, resp. magnetization profile. Error bars indicate flows which have not reached a steady state.

relativistic MHD simulations from the surface of a disk which is in Keplerian equilibrium with a central compact object. Essentially, these results prove (for the first time) the self-collimating feasibility of relativistic MHD jets.
Jets may originate in a hot and luminous environment and may be subject to de-collimating radiation pressure. This is the topic of Sect. 4.
Jet rotation is observationally indicated for stellar jets. However, rotation is a natural ingredient for any jet launching. In Sect. 5 we argue that jet rotation can be induced also from jet-internal shocks which compress the helical field and give rise to a toroidal Lorentz force.

2. Jet collimation and the accretion disk magnetic flux profile

Here we show results of axisymmetric jet formation simulations from disks of different magnetic flux profile (see [8]). The physical grid size corresponds to $(r \times z) = (150 \times 300)\, r_{\rm in}$, where $r_{\rm in}$ is the inner disk radius which could be properly scaled to the astrophysical object investigated. Here we have applied the ZEUS-3D code.

We start from an initially force-free magnetic field distribution within a gas in hydrostatic equilibrium. The simulations evolve under the boundary condition of a prescribed mass flux launched from the disk surface into the outflow. We have run models covering a wide range of different profiles for the disk magnetic flux or the disk wind mass flux, both parameterized by power laws, $B_{\rm p,disk}(r) \sim r^{-\mu}$, and $\rho_{\rm wind}(r) \sim r^{-\mu_\rho}$. Both quantities can be combined in the disk wind magnetization parameter (similar to the Michel magnetization parameter [9]), $\sigma_{\rm wind} \sim B_{\rm p,disk}^2 r^4 \Omega^2 / \dot{M}_{\rm wind} \sim r^{\mu_\sigma}$.

We quantify the collimation degree by comparing the axial and lateral mass fluxes (see [8], [10]). Figure 1 shows the collimation degree $\zeta \simeq (\dot{M}_z/\dot{M}_r)$ plotted against the power law exponent of the disk wind magnetization μ_σ. Our main result is that steep magnetization profiles, resp. disk magnetic field profiles, are unlikely to launch highly collimated outflows. However, flat profiles - in general leading to a higher collimation degree - tend to be unstable, i.e. they do not evolve into a steady state.

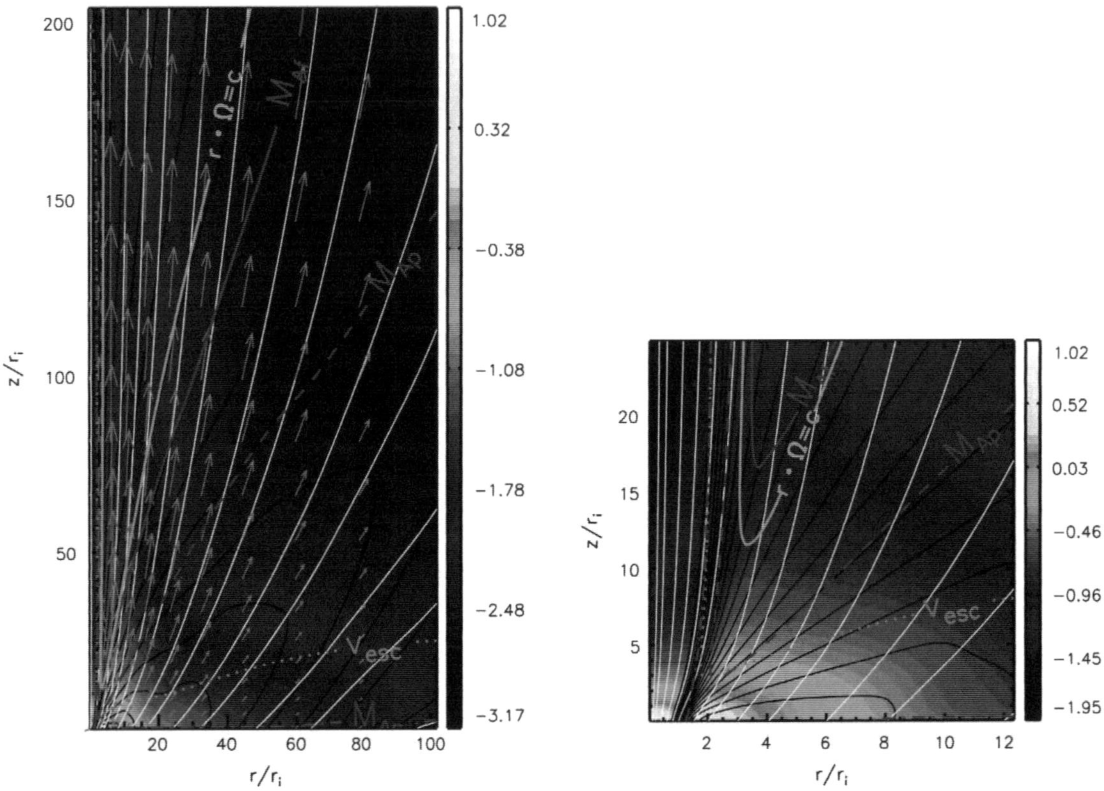

Figure 2. Logarithmic (rest-frame) density of the jet in the steady-state. Shown are poloidal magnetic field lines (solid white), electric current flow lines (solid black), characteristic MHD surfaces (dark grey dotted, dashed, solid), surface of escape velocity (grey dotted), light surface (light grey solid). The arrows in the top plot indicate the velocity field. The right figure is the enlarged central region. Note the location of the light surface $R_{\rm L}(r,z) \equiv c/\Omega$, with the so-called angular velocity of the field lines Ω. (Adapted from [12]).

3. Relativistic MHD jets from accretion disks

We have extended our non-relativistic simulations to the (special) relativistic regime using the PLUTO code [11]. In order to apply an astrophysically motivated Keplerian disk boundary condition for the jet outflow we have added Newtonian gravity to the special relativistic equations (see [12, 13]).

The standard setup involves a poloidal field strength considering a plasma-beta between 0.001 and 1 at the inner disk radius and a poloidal field profile which is fixed in time by the boundary condition. The toroidal magnetic field induced during the flow evolution is floated into the boundary domain. However, in an alternative approach we prescribe a higher Poynting flux by injecting plasma with a higher toroidal field strength (up to a factor 8). Depending on the Poynting flux injected, the resulting Lorentz factors of the jet reach number values between $\Gamma = 1.5 - 10$.

Typical jet opening angles for the relativistic part of the outflow are $1-7$ degrees, however, on the physical scales covered by the simulations of up to 6000^2 Schwarzschild radii the asymptotic jet is not jet fully accelerated or collimated.

Figure 2 shows a typical result of a (mildly) relativistic jet with a Lorentz factor of 1.5. The

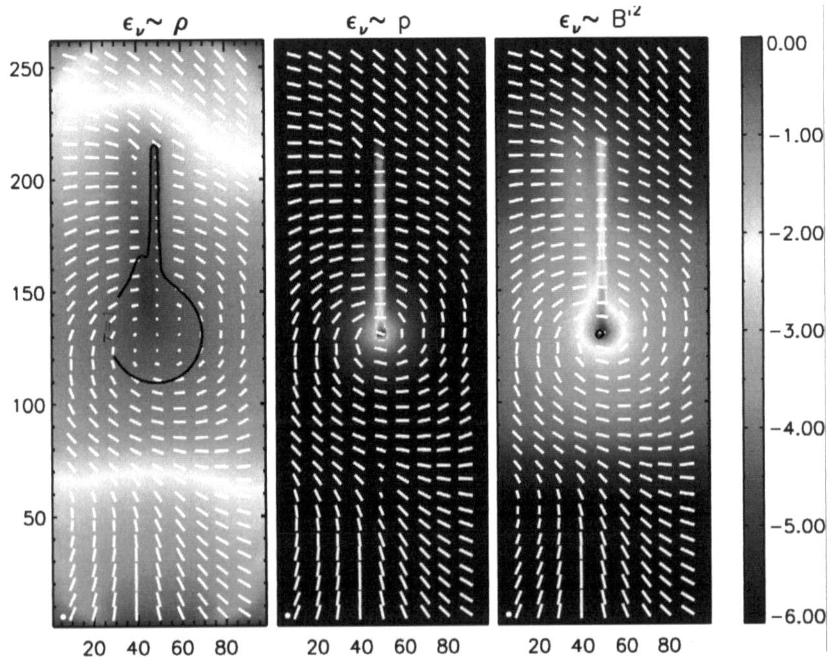

Figure 3. Ideal resolution synchrotron maps considering relativistic MHD simulations of jet formation and polarized radiation transfer. Results for three tracers for the particle acceleration are shown: density (left), thermal pressure (middle), magnetic energy (right). Shown are polarization vectors and the (normalized) 43 GHz intensity distribution for a viewing angle of 30 deg. The radial box size corresponds 400 inner disk radii.

location of the magnetosonic surfaces is shown as also the location of the light surface where the hypothetical rotation of the field lines supercedes the speed of light, $R_{\rm L}(r,z) \equiv c/\Omega$. The asymptotic region of the relativistic jet enclosed by the light surface is the truly relativistic part of the outflow. This part of the outflow originated close to the inner edge of the accretion disk deep in the potential well where the rotation is most rapid.

Having derived a physically consistent distribution of the magnetohydrodynamic variables by our numerical simulations - i.e. the jet magnetic field distribution, the density distribution, and the jet velocity - we may us this information to derive consistent radio synchrotron maps following a relativistic polarized radiation transfer within the jet (Fig. 3). Since particle acceleration is not included in the MHD model, additional assumptions have to be made. We have tested three tracers for the power-law particle acceleration, i.e. density, thermal pressure, or magnetic energy density [12, 13]. All tracers give a similar polarization structure, although the intensity distribution differs. The spiky axial structure results from a high temperature, thin axial outflow from the compact object area and could be considered either as a Blandford-Znajek jet or a coronal wind.

4. De-collimation of MHD jets by radiation pressure

Jets may be launched in a hot and luminous environment - examples are accretion disks around compact objects or massive young stars. In this case the outflow dynamics can be affected by the strong radiation field. In order to quantify this effect numerically, we have run simulations of MHD jet formation using the PLUTO code [11] and where we have implemented the physical effect of radiation pressure. Our model setup so far was applied for a central luminous star

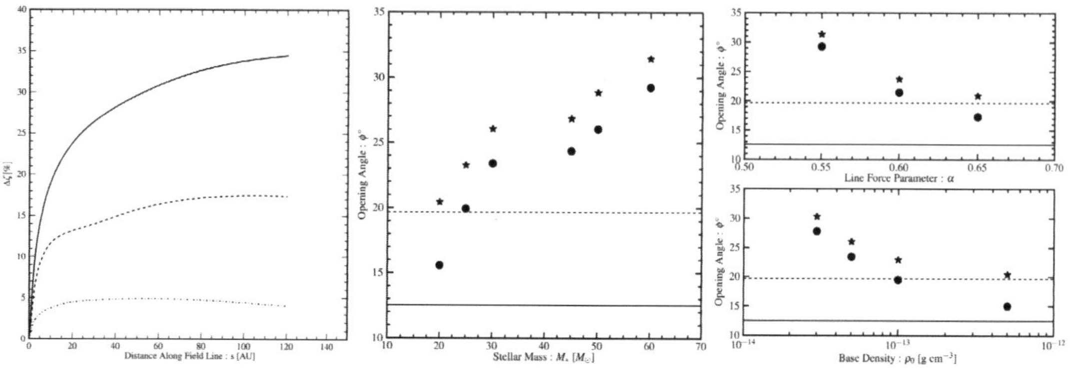

Figure 4. MHD jet formation under radiation pressure. Example of outflows from massive young stars (taken from [14]). Comparison of opening angles for the field line with varying magnetic field strength (left): $B = 5.1\,\mathrm{G}$ (solid), $6.6\,\mathrm{G}$ (dashed) and $11.5\,\mathrm{G}$ (dotted). Jet opening angles (middle) at the Alfvén point (stars) and the fast point (bullets) of a field line rooted at $r = 5\,\mathrm{AU}$ for runs with different stellar mass and thus luminosity. Jet opening angles (right) for runs with different *alpha* (top panel), and different injection density ρ_0.

surrounded by an (also luminous) accretion disk [14]. Thus, both the central object and the surrounding inner hot accretion disk may contribute to the radiative forces. The model could be extrapolated to AGN jets, however, certain constraints on the parameter values must be carefully considered (e.g. the ionization degree, the temperature, the existence of lines).

The radiative force considered in our study

$$\vec{F}^{\rm rad} = \vec{f}_{\rm cont,*} + \vec{f}_{\rm cont,disk} + \vec{f}_{\rm line,*} + \vec{f}_{\rm line,disk} \quad (1)$$

comprises of acceleration due to continuum radiation from the central object, $\vec{f}_{\rm cont,*}$, and the disk, $\vec{f}_{\rm cont,disk}$, and those due to lines forces from the disk, $\vec{f}_{\rm line,disk}$, and the central object $\vec{f}_{\rm line,*}$, For the line forces we apply the well known CAK theory [15], which has been also explored for AGN disk winds by [16]. The line force is expressed as a product of the force due to continuum radiation and a force multiplier $M(t)$. The optical depth parameter t is related to the velocity gradient along the l.o.s., the wind density, and the ion thermal speed $v_{\rm th}$, thus $t = (\rho\sigma_{\rm e} v_{\rm th})/|\hat{\vec{n}} \cdot \nabla(\hat{\vec{n}} \cdot \vec{v})|$. An empirical form for $M(t)$ as a sum over all lines for model atmospheres for massive OB stars has been defined as $M(t) \sim kt^{-\alpha}$, where k and α are line force parameters [17]. Depending on the selection of lines, typical values for k range from 0.4-0.6 and for α between 0.3-0.7. A force multiplier parameterization, independent of arbitrary v_{th}, was introduced by [18].

We start off with a pure MHD simulation similar to [19] or [8]. When a steady outflow is established after about 200 inner orbital periods, we switch-on the radiative forces. The radiative forces disturb the pure MHD jet structure, and a new quasi-steady state is reached after another 200 rotations [14].

Our results are the following (see Fig.4 for a visual summary of these results). The line-driven force from a central object can significantly modify the outflow dynamics in terms of collimation and acceleration. We find that the outflow velocity is increased by a factor of 1.5-2 by radiation forces as compared to the pure MHD flow. Also, the degree of collimation is visibly lowered, e.g., in a 30% change in the magnetic flux profile, thus a wider opening angle of the magnetic field lines. For the example of massive young stars we have investigated different stellar masses thus central luminosity and radiation pressure. We determine the amount of de-collimation by

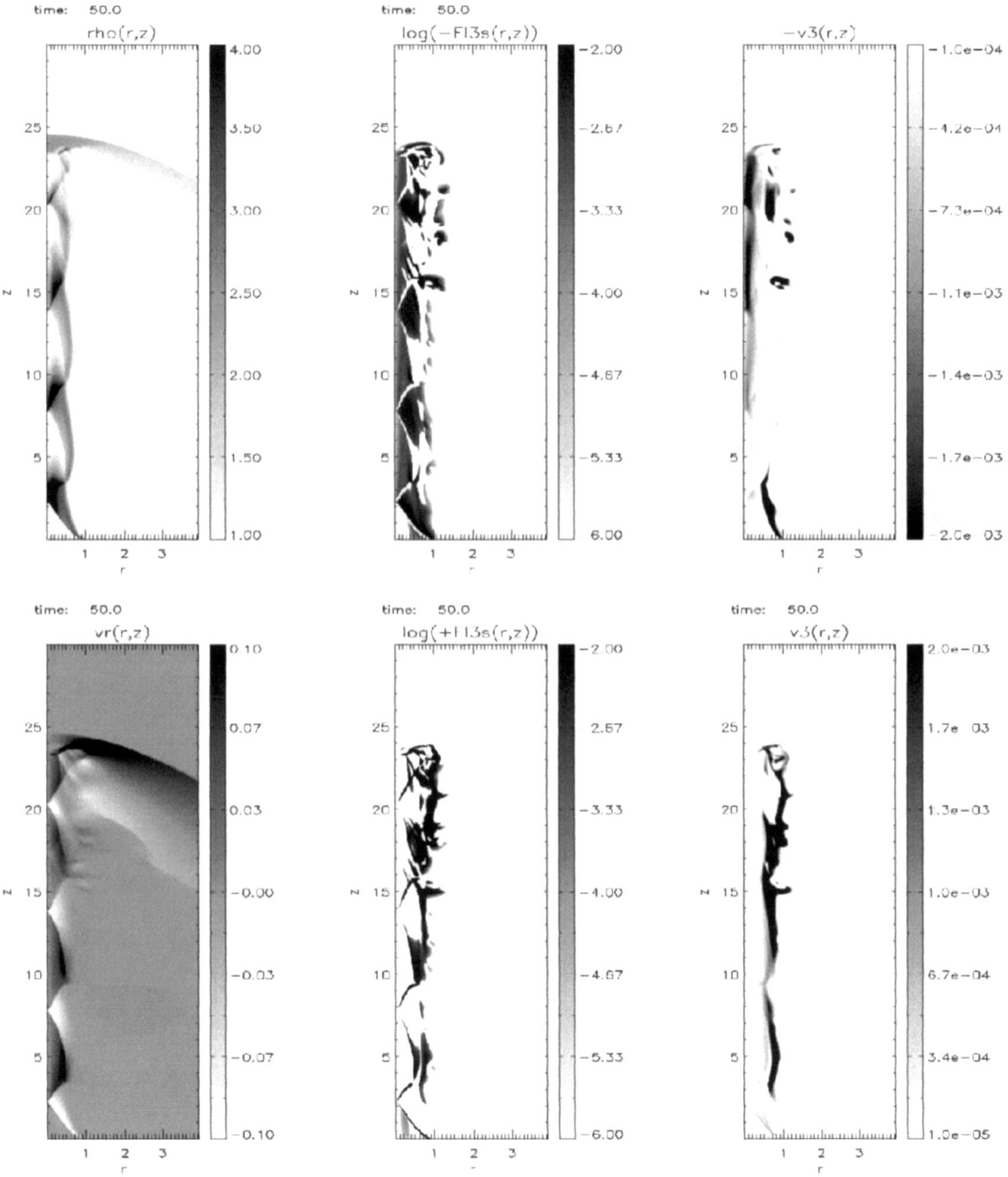

Figure 5. Axisymmetric MHD simulation considering a high (external) Mach number jet propagation with a rather weak magnetic field (taken from [20]. Shown are the density distribution (top left), the radial velocity distribution (lower left), the negative (top) and positive (bottom) toroidal specific Lorentz force components (middle), and negative (top) and positive (bottom) toroidal velocity components (right) at dynamical time $t = 50$ for a sub-grid of $r < 4.0$ of the whole computational domain $(0.0 < r < 10.0, 0.0 < z < 30.0)$, where $r = 1$ is the radius of the jet injected into the ambient gas.

measuring the opening angles of a typical magnetic field line at the magnetosonic points. We find that by increasing the central mass (thus luminosity) from e.g. $20 - 60\,M_\odot$, the outflow opening angle changes from $20 - 32$ degree, indicating a substantial amount of jet de-collimation.

We have carried out simulations with different field strengths considering a plasma-β of $1.0, 3.0, 5.0$. For a flow with high magnetic flux the radiative forces do not considerably affect the collimation degree. For a flow with low magnetic flux, the dynamical effect of radiative forces is greatly increased. We further find that the line-force parameter α is critical in determining the magnitude of the line-driven forces. Lower values of α lead to an efficient radiative force from the central star and thus de-collimate the flow to a larger extent as compared to higher values of α. We find that the radiation forces may also change the mass flow rates in the jet. Even a small change in α may lead to significant changes in mass flux of up to $\simeq 28\%$. In the examples investigated, the line forces due to the hot accretion disk do not play a significant role in controlling the dynamics of the MHD outflow, simply because they are orders of magnitude smaller than all other forces that affect the flow dynamically. This might be different in disk jets from AGN which are less dense and more ionized.

5. Jet rotation driven by MHD shocks in helical magnetic fields

Finally, we propose and numerically investigate the hypothesis that a rotation of astrophysical jets can be caused by magnetohydrodynamic shocks in a helical magnetic field (see [20] for details). Shock compression of the helical magnetic field results in a toroidal Lorentz force component which will accelerate the jet material in toroidal direction. This process transforms magnetic angular momentum (magnetic stress) carried along the jet into kinetic angular momentum (rotation). Hints of such a mechanism were observed in numerical simulations by [21, 22, 23], but were not discussed in great detail, or even applied to observations.

In our simulations the jet is injected into the ambient gas typically with zero kinetic angular momentum (no rotation). Different dynamical parameters for jet propagation are applied such as the jet internal Alfvén Mach number and fast magnetosonic Mach number, the density contrast of jet to ambient medium, or the external sonic Mach number of the jet. Figure 5 show the time evolution of the rotational velocity in a propagating jet with time-dependent mass injection. It is clearly visible the rotational velocity is increased at the (internal and external) shock positions. The enhanced rotational velocity coincides as well with the locations of enhanced toroidal Lorentz force [20]. For the example of protostellar jets, the numerically derived rotation feature looks consistent with the observations, i.e. rotational velocities of 0.1-1% of the jet bulk velocity.

The mechanism we suggest should work for a variety of jet applications, e.g. protostellar or extragalactic jets, and internal jet shocks (jet knots) or external shocks between the jet and ambient gas (entrainment).

Acknowledgments

We acknowledge the LCA team and M. Norman and D. Clarke for the possibility to use the ZEUS code, and Andrea Mignone and the PLUTO team for the possibility to use their code. The projects were partly funded by the *International Max Planck Research School for Astronomy and Cosmic Physics at the University of Heidelberg* (IMPRS-HD), the *Heidelberg Graduate School of Fundamental Physics*, and the *Klaus Tschira Stiftung* (KTS) in Heidelberg.

References

[1] Blandford, R.D. & Payne, D.G. 1982, *MNRAS* 199, 883
[2] Pudritz, R.E., Ouyed, R., Fendt, Ch., Brandenburg, A. 2007, in: B. Reipurth, D. Jewitt, & K. Keil (eds.), *Protostars & Planets V*, University of Arizona Press, Tucson, 2007, p.277
[3] Shang, H., Li, Z.-Y., Hirano, N. 2007, in: B. Reipurth, D. Jewitt, & K. Keil (eds.), *Protostars & Planets V*, University of Arizona Press, Tucson, 2007, p.261
[4] Keppens, R., Goedbloed, J.P. 1999, *A&A* 343, 351

[5] Matt, S., Pudritz, R.E. 2008, ApJ, 678, 1109
[6] Fendt, Ch. 2009, *ApJ* 692, 346
[7] Blandford, R.D. & Znajek, R.L. 1977, *MNRAS* 179, 433
[8] Fendt, Ch. 2006, *ApJ* 651, 272
[9] Michel, F.C. 1969, *ApJ* 158, 727
[10] Fendt, Ch., Čemeljić, M. 2002, *A&A* 395, 1045
[11] Mignone, A., Bodo, G., Massaglia, S., Matsakos, T., Tesileanu, O., Zanni, C., Ferrari, A. 2007 *ApJS* 170, 22
[12] Porth, O., Fendt, C. 2010, *ApJ* 709, 1100
[13] Porth, O., Fendt, C., Meliani, Z., Vaidya, B. 2011 *ApJ* 737, 42
[14] Vaidya, B., Fendt, C., Beuther, H., Porth, O. 2011 *ApJ* 742, 56
[15] Castor, J.I., Abbott, D.C., Klein, R.I. 1975 *ApJ* 195, 157
[16] Proga, D. 2003 *ApJ* 585, 406
[17] Abbott, D.C. 1982 *ApJ* 259, 282
[18] Gayley, K.G. 1995 *ApJ* 454, 410
[19] Ouyed, R., Pudritz, R.E. 1997 *ApJ* 482, 712
[20] Fendt, C. 2011 *ApJ* 737, 43
[21] Kössl, D., Müller, E., Hillebrandt, W. 1990 *A&A* 229, 378
[22] Uchida, Y., Todo, Y., Rosner, R., Shibata, K. 1992 *PASJ* 44, 227
[23] Todo, Y., Uchida, Y., Sato, T., Rosner, R. 1992 *PASJ* 44, 245

Spectral energy distribution template of redshift-zero AGN and the comparison with that of quasars

Almudena Prieto

Instituto de Astrofisica de Canarias, Tenerife, Spain

aprieto at iac dot es

Abstract. Spectral energy distributions (SEDs) of the central few tens of parsecs of some of the nearest active galactic nuclei (AGN) have been used to construct a genuine redshift-zero SED template [1]. This paper presents the comparison between this redshift-zero template and those of the radio loud and radio quiet quasar templates by [2]. Despite the four orders of magnitude difference in power between the redshift-zero AGNs, mostly Seyfert galaxies, and that of quasars, they all show a remarkable similarity. The distinction relays on the importance of the so-called blue bump component - emission from the accretion disk - which is absent in Seyfert type 2 , mildly present in Seyfert type 1, seen in full realm in quasars. Conversely, the characteristic red bump component - dust reprocessed emission in the IR - is present in all cases. The difference between the three AGN classes can be ascribed in terms of central dust obscuration and progressive view angle of the nuclear core.

1. High spatial resolution SEDs: data source for redshift-zero AGNs

A best estimate of the pure AGN light contribution on very nearby targets has been derived from very high spatial resolution data over a wide range of the electromagnetic spectrum. The compiled SEDs represent the central few tens of parsec region of some of the nearest and brightest AGN. The work is motivated by the current possibility to obtain subarcsec resolution data in the near-to-mid-IR - with adaptive optics and interferometry, and thus at comparable resolutions to those obtained in radio interferometry and the optical-to-UV with HST. The near-to mid- IR high resolution data in this work come mostly from the ESO Very Large Telescope (VLT), hence this study relies on Southern targets, all well known objects, mostly Seyfert galaxies and low luminosity AGN -LINER type. Objects studied in this project include Centaurus A, NGC 1068, Circinus, NGC 1097, NGC 5506, NGC 7582, NGC 3783, NGC 1566 and NGC 7469, Mrk 1239, NGC 1365, MCG -6-30-15, NGC 1052, NGC 3169 and M87.

The main data sources are: VLA and ATCA in their extended configurations in radio, VLT diffraction-limited images and interferometry in the mid-infrared (mid-IR), VLT adaptive-optics images in the near-infrared, and HST imaging and spectra in the optical-ultraviolet. Although X-rays and Gamma-rays do not provide such a fine resolution, information when available above 20 keV are also included in the SEDs on the assumption that above that energy we are mostly sampling the AGN core. Most of the data comes from BeppoSAX and INTEGRAL telescopes. When available, Chandra data extracted from the central 1 arcsec resolution beam is also included.

Individual SEDs for each of the targets are presented and discussed in [1,3].

2. The spectral energy distribution "template" of redshift zero AGN

The study sample includes different type of AGN: namely Seyfert type 1 and type 2 with a range in power of 10^{42} erg/s < Lbol ~< 10^{44} erg/s) and Low Luminosity AGN (LLAGN) with Lbol <~ 10^{42} erg/s. Each of theses three classes shows a vey specific SED at these high spatial resolutions. To produce genuine SED templates for each type, the best representative examples of each class were selected [1]. By representative it means that intermediate-type-class objects were excluded.

Figure 1a,b presents the resulting SED templates for the Seyfert type 1- and 2- class, indicating as well the targets used to construct them. That derived for LLAGNs is presented in [3] and will not be discussed here. The method used to construct these templates is described in [1] Briefly, each target SED is normalized to the mean value of its total power, this being derived from integrating its high spatial resolution SED. In doing so, the normalized SEDs result at comparable scale at about the optical-IR region. Then, for each Seyfert type, a median average across the spectrum was produced. The reliability of these templates can be judged by comparing them with the individual SEDs used to construct them, all in each class showing very similar distributions.

Figure 1c shows the Seyfert type 1 and 2 templates plotted on top of each other. There is not further normalization in this plot, note that both averages SEDs show at about the same level in the 10 um region. The two relevant features in both templates is the IR bump and the rising high-energy spectrum.

The IR bump, peaking at about 2 um, can be reconciled with dust emission with an equivalent grey-body temperature of ~300 K [1]. The shape of the bump longward of 2 μm is similar for both Seyfert types; it is shortward of this wavelength where the difference arises: type 2 nuclei are characterized by a sharp decay in power from 2 μm onwards to the optical wavelengths, type 1s present also a decay shortward of 2 μm but this recovers at about 1 μm to give rise to the characteristic blue bump feature seen in Quasars and type 1 sources in general [2]. This dramatic change in the SED suggests a clearer line of sight in type 1s but and obscured one in type 2s. This is in line with the existence of a nuclear parsec-scale torus: the shallower spectrum in type 1s reflects the contribution of much hotter dust from the inner region of the torus which we are able to see directly in this sources; in the type 2s this innermost region remains hidden from our line and we are sampling the enshrouding colder dust only.

The second important feature in the templates is the high energy spectrum. Within the common sampled energy band - 1 to ~100 keV - this region appears rather similar in both AGN types, the general trend being that of a gentle increase in power with increasing frequency.

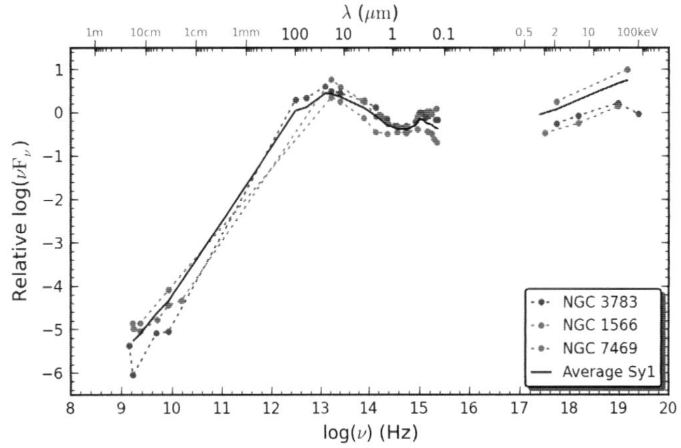

Figure 1a: average SED of Seyfert type 1 objects and individual SEDs –normalised to their total power- used to construct the average. Filled points on each SED are the original data source.

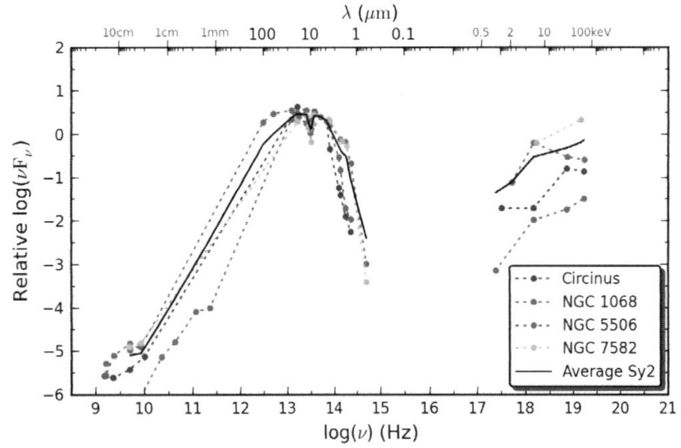

Figure 1b: same as in figure 1a but for Seyfert type 2s.

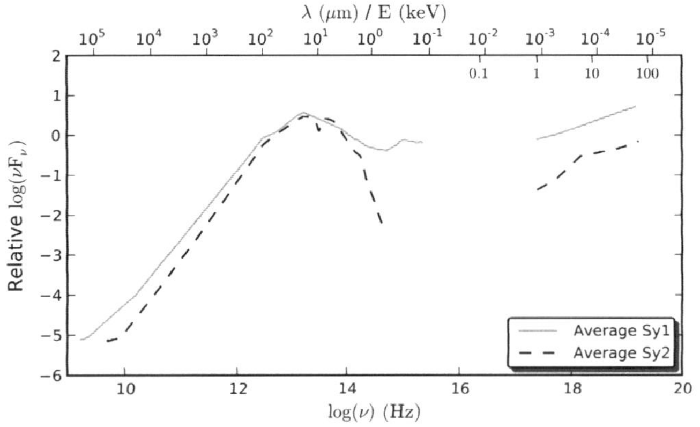

Figure 1c: Comparison of Seyfert type 1 and 2 average SED templates (taken from [1]).

3. Comparison with Quasar's template

Elvis et al [2] constructed the averaged SED template of a sample of radio quiet and of radio loud quasars. Both set of quasars were selected for being strong UV –X-ray emitters. Their average power falls in the 10^{45} erg/s range, i.e. about three to four orders of magnitude above than that of the Seyfert's.

Figure 2 shows the comparison of all the templates: radio loud and radio quiet quasars, Seyfert type 1 and 2. For this comparison, the quasar templates were normalized to their total power integrated over their respective SED in the same manner as done with the Seyfert's. The resulting normalized SEDs are plotted on top of each other in figure 2. It can be seen that all fall at about the same scale at the location of the IR bump.

Despite the difference in power between the Seyfert and the Quasar class, their respective averaged SEDs show a remarkable resemblance. All are characterized by a strong emission bump in the IR, centered at about the same location over the 2 - 10 um region, and with comparable width. As in the Seyfert's case described above, the shape of the bump longward of 2 μm is similar for all types, it being characterized by a smooth fall off towards the radio waves. In the radio loud quasars, the bump starts to depart from the common shape longward of the millimeter region because of their stronger radio emission. Shortward of 2 μm, the difference AGN types start to show their signature:

- The Seyfert type 2 SED is characterized by a sharp decay from 2 μm onwards. Shortward of 1μm, all Seyfert type 2 nuclei are undetected, hence the data gap in their template.
- Seyfert type 1 and quasars, both radio -loud and –quiet, present also a decay shortward of 2 μm but this is progressive: relatively strong in Seyfert type 1, milder in Quasars. The decay ends at a at common depression point at about 1 μm – the characteristic inflexion point associated with dust sublimation temperature- followed by a rising blue bump emission component. The rise up of this component is also progressive, in this case it is mild in Seyfert 1s but very strong in quasars.

This progressive change in the SED shape as one moves from Seyfert type 2s to quasars can be ascribed to the effect produced by a central -- parsec-scale -- dust obscuration effect in the line of sight. Nuclear dust fully obscures our line of sight to Seyfert 2 cores, produces partial obscuration in type 1s and only in quasars dust is off from our line of sight. This is the well known prediction of the AGN unification models, seen for the first time reflected in the spectral energy distributions of Seyfert and quasars alike.

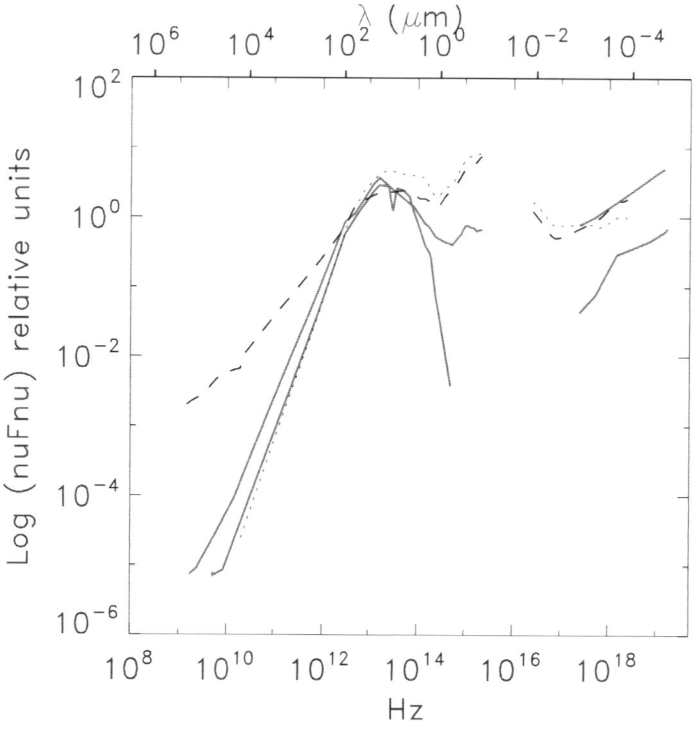

Figure 2: SED templates for four different AGN types, all normalized to the median of their total power: Seyfert type 2 (red), Seyfert type 1 (blue), radio-loud quasar (dash-line), radio quiet quasar (dots). Seyfert SEDs are from [1], that of quasar's from [2].

4. Availability of the Seyfert SED templates

Tabular ascii form of the Seyfert type 1 and 2 SED templates are available at http://www.iac.es/proyecto/parsec/main/seyfert-SED-template.php

References

[1] Prieto et al, 2010, MNRAS
[2] Elvis et al. 1994, ApJ
[3] Fernandez-Ontiveros et al. (this proceeding)

The hypothesis of the dust origin of the Broad Line Region in Active Galactic Nuclei

Bozena Czerny, Krzysztof Hryniewicz
Copernicus Astronomical Center, Bartycka 18, 00-716 Warsaw, Poland
E-mail: bcz@camk.edu.pl

Abstract. Strong broad emission lines are the most important signatures of active galactic nuclei. These lines allowed to discover the cosmological nature of quasars, and at present these lines allow for convenient method of weighting the black holes residing in their nuclei. However, a question remains why such strong lines form there in the first place. Specifically, in the case of Low Ionization Lines, there must be a mechanism which leads to an efficient rise of the material from the surface of the accretion disk surrounding a black hole but at the same time should not give a strong signature of the systematic outflow, as the Balmer lines are not significantly shifted with respect to the Narrow Line Region. We determine the effective temperature of the accretion disk underlying the Hβ line at the basis of the time delay measured from reverberation and the simple Shakura-Sunyaev theory of accretion disks. We obtain that this temperature is universal, and equal 995 ± 74 K, independently from the black hole mass and accretion rate of the source. This result suggests to us that the dust formation in the disk atmosphere is responsible for the strong rise of the material. However, as the material gains height above the disk it becomes irradiated, the dust evaporates, the radiation pressure force suddenly drops and the material fall back again at the disk. Therefore, a failed wind forms. In the simple version of the model the disk irradiation is neglected, but in the present paper we also discuss this irradiation and we use the observed variation of the Broad Line Region in NGC 5548 to constrain the character of this non-local non-stationary phenomenon. The current instruments cannot resolve the Broad Line Region but future instrumentation may allow to test the model directly.

1. Introduction

Broad emission lines in quasars and in other types of active galaxies have lead to the identification of those objects as distant sources with strong non-stellar nuclear emission coming from their nuclei due to accretion onto black holes. Both [1] and [2] in their seminal papers were just studying emission lines. Since that time there was a tremendous progress in our understanding of the nature of active galaxies and in our knowledge of the Broad Line Region (BLR) properties. Emission line measurements are also nowadays a powerful tool to measure the mass of the central black hole [3, 4] and these measurements done for large samples of quasars (e.g. [5]) give the insight into the evolution of galaxies. But one question still remains to be answered: why this region actually forms there.

The properties of the BLR contain a hidden hint. The argument goes through three independent lines of reasoning.

First point: The lines cover a considerable fraction (0.1 - 0.3) of an AGN sky from the point of view of the nucleus, as estimated from the line luminosity (e.g. [6]). At the same time they are almost never along the line of sight to the nucleus although NAL are seen in a significant fraction of objects (BAL are rare; 20% of normal quasars [7]) and both NAL and BAL form at larger distances from the nucleus. There are rare reports of the transient absorption phenomenon seen in X-ray data which are likely to be due to BLR clouds [8]. These observations indicate that the BLR is not spherical but flat and located close to the equatorial plane; in side-viewed objects this region is hidden from us by the dusty-molecular torus and in top-viewed objects the line of sights does not cross BLR. Kinetic arguments also suggest such a flattening (e.g. [9, 10]).

Second point: the line widths are from 1000 to 10 000 km s^{-1}, consistent with the Keplerian disk-like motion. However, lines rarely show two-component profile expected from a flattened disk-like configuration. There are rare objects showing clearly double-peaked lines but these are mostly radio-galaxies [12], and some objects hint for the disk-like profiles in the variable part of their line profiles. The majority of quasars have a single peak. Murray & Chiang have shown that even a disk-like configuration can lead to single profiles but it requires special wind velocity profile [13]. And no systematic shift (outflow velocity signature) with respect to the Narrow Line Region (NLR) is seen in Balmer lines.

Third point: The careful studies of the luminosity and the line ratios, particularly in respects to the Low Ionization Lines like Balmer lines and Mg II [14] suggest that additional collisional heating is needed in addition to pure radiative heating estimated from the shape of the continuum emission from the central region of an active galaxy.

All three problems point towards the BLR as flat configuration roughly in Keplerian motion but with additional turbulence with the turbulence velocities of order of a thousand km s^{-1}. Such velocities, although only a fraction of the Keplerian speed would be 100 times higher than the sound speed, and the development of a such highly supersonic turbulence is not easily explained.

In our paper we offer the plausible mechanism which gives strict quantitative predictions where the LIL part of the BLR would form. Future high resolution observations will possibly resolve the BLR region and offer direct test of the proposed mechanism.

2. Disk temperature underlying the BLR

We determine the effective temperature of the accretion disk atmosphere underlying the BLR using the observational determination of the BLR radius from the reverberation method and combine it with the Shakura-Sunyaev theory of accretion disks [16]. The scaling of the BLR with luminosity discovered by [3] has been studied by several authors (e.g. [4]). Here we use the result of [15], where the starlight contamination was carefully removed. The power-law index in the scaling between the BLR size and the monochromatic luminosity in their analysis is consistent with 0.5 within the errors. We thus fixed this slope at 0.5, and refitted their data again, obtaining the relation

$$\log R_{BLR}[\mathrm{H}\beta] = 1.538 \pm 0.027 + 0.5 \log L_{44,5100}, \tag{1}$$

where $R_{BLR}[\mathrm{H}\beta]$ is in light days and $L_{44,5100}$ is the monochromatic luminosity at 5100 Å measured in units of 10^{44} erg s^{-1}. The value 0.027 is the error in the best-fit vertical normalization; the dispersion around the best fit is larger but still small at 0.21. We assume that the disk is not strongly irradiated, and the outflow from the disk is not too strong, so the simple theory of the alpha-disk [16] applies. In this case, the monochromatic flux at 5100 Å can be calculated from the model, if the black hole mass and the accretion rate are known. Taking the numerical formula of [17] (corrected for a factor-of-two error, Nikolajuk, private communication), we derive

$$\log L_{44,5100} = \frac{2}{3} \log(M\dot{M}) - 43.8820 + \log \cos i, \tag{2}$$

where the black hole mass, M, and the accretion rate, $\dot{M}$, are in g and g s^{-1}, respectively. This formula allows us to calculate the BLR radius from the Eq. 1, if the product of $(M\dot{M})$ is known.

For completeness, we introduce a geometrical factor representing the effect of the inclination angle, i for the measured delay (for details, see [18]). This gives the relation between the actual disk radius, r, and the measured delay of Hβ line:

$$r = \frac{R_{BLR}[\mathrm{H}\beta]}{1 + \sin i} \tag{3}$$

and we fix this factor adopting the average inclination $i = 39.2°$ after [19].

The Shakura-Sunyaev accretion disk theory allows us to calculate the effective temperature of the disk at that radius as a function of the black hole mass and accretion rate

$$T_{eff} = \left(\frac{3GM\dot{M}}{8\pi r^3 \sigma_B}\right)^{0.25}, \tag{4}$$

where σ_B is the Stefan-Boltzmann constant. Since BLR is far from the central black hole, the inner boundary effect can be neglected.

When we combine Eqs. 1, 2, 3, and 4, the dependence on the unknown mass and accretion rate vanishes, and we obtain a single value for all the sources in the sample

$$T_{eff} = 995 \pm 74\mathrm{K}, \tag{5}$$

where the error reflects the error in the constant in Eq. 1. We stress that the resulting value does not depend either on the black hole mass or accretion rate. The value is universal, for all sources.

This value is interesting since it is close to the critical temperature at which the dust can form. The distance is smaller than the distance to the dusty torus $R_{dust} \sim 0.4 L_{45}^{1/2}$ pc [20] where L_{45} stands for the bolometric luminosity in 10^{45} erg s^{-1}) since in our case the temperature results only from the local disk dissipation instead of the irradiation by the central parts of AGN.

3. The mechanism of formation of the BLR

This uniform effective temperature of 1000 K of the disk underlying the onset of the BLR implies that the dust present in the disk atmosphere is responsible for the phenomenon. The radiation pressure acting on dusty atmosphere is very large, as it is known for example for Wolf-Rayet stars and the strong wind develops. However, unlike in stars, the further development of the wind is modified by two factors: gravitational attraction in disk geometry rises with the height and, more importantly, the irradiation by the central parts of the accretion disk (UV) also rises very strongly due to better geometrical exposure. The dust evaporates, the rising material looses the radiation pressure support and falls back onto the disk surface. Such a failed wind can account both for the large additional turbulent velocity without any strong signature of outflow since both outflow and inflow take place with comparable velocities, and for the additional heating, as the outflowing and infalling clumps collide.

This region is separate from the region known as dusty/molecular torus. At large distances the UV irradiation is less important, the dust survives even at large height from the disk and the outflow proceeds forming an appearance of a torus. The schematic representation of the BLR formation mechanism is shown in Fig 1.

4. Discussion

The mechanism of BLR formation proposed in this paper and in [18] is attractive since it is universal for all sources, and the physics of dusty winds is known from stellar evolution.

Journal of Physics: Conference Series **372** (2012) 012013 doi:10.1088/1742-6596/372/1/012013

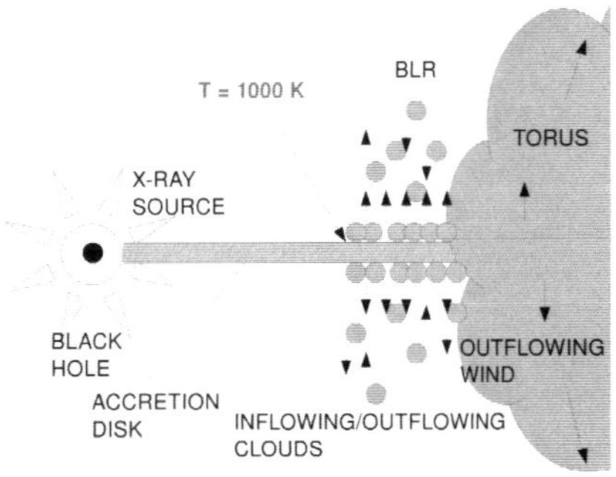

Figure 1. The schematic picture of the formation of BLR. In this region dusty wind rises up, the dust evaporates due to irradiation by the central parts, gravity wins and dustless material falls back onto the disk. Dusty torus is further from the nucleus where irradiation does not destroy the dust.

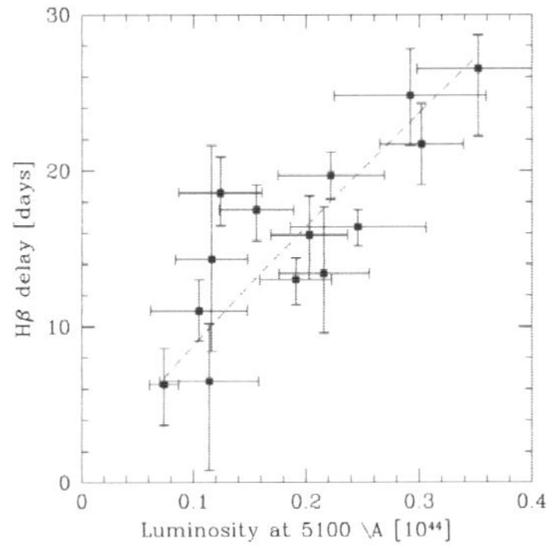

Figure 2. The dependence of the size of the BLR in NGC 5548 on the monochromatic luminosity: data points come from Bentz et al. [15], and the best fit line has the slope 0.90 ± 0.22 in the logarithmic space, instead of 0.5 characteristic for the statistical sample.

Although the possible presence of the dust in the BLR clouds was previously discussed by a number of authors, our idea of the key importance of the dust for BLR is new and should be further tested. The whole idea automatically explains why the size of the BLR region scales with the square root of the monochromatic luminosity, with small dispersion, since no dependence on the second parameter - for example bolometric luminosity is expected. We expect a weak dependence of the inclination angle but this also remains to be tested.

4.1. Determination of the disk temperature in all monitored sources

In the paper [18] we tested additionally the mechanism by determining the disk BLR temperature for a number of sources which were carefully monitored for many years (e.g.. [4]) and their luminosity was corrected for the starlight contribution at the basis of the HST data (Beztz et al. [15]). For all sources we obtain the temperature value close to 1000 K. This is also the case for NGC 5548 monitored for 14 years.

We do not see any correlation between the accretion rate and the disk effective temperature at the onset of BLR in the objects for which the reverberation monitoring was performed. Mechanism simply connecting the BLR distance with irradiation would likely contain the trend with the bolometric luminosity which (in Shakura-Sunyaev model) rises faster with the accretion rate than the monochromatic luminosity.

4.2. The role of the disk irradiation

In our approach to the description of the disk atmosphere underlying the BLR and to the description of the monochromatic flux we used the simple local stationary disk model of Shakura & Sunyaev. However, observations of AGN show that AGN are variable and this variability is a central point of the monitoring approach and the measurement of delays. It means not only that the disk is not stationary, but it implies that non-local phenomena (irradiation of the outer disk parts by the inner disk) are complicated by these time delays.

The measured time delays between the continua measured at UBV bands are typically of order of a day or less in Seyfert galaxies (see e.g. [21]), the major trends in optical luminosity are of order of tens to hundreds of days, and measured time delays of Hβ line are of order of days (see compilation in Bentz et al. [15]). Both an outer and an inner disk are the sources of variability, as argued for example by [22], and the perturbations in the outer disk propagate mechanically inwards through the disk (sound waves, viscous propagation) while perturbations in the inner disk affects the outer disk through irradiation, and the timescale is the light travel time. This complex situation can be roughly represented with two different accretion rates/luminosities: inner accretion rate visible through the monochromatic flux and responsible for the current strength of irradiation and the outer accretion rate (past value, likely closer to the average value) characteristic for the disk structure underlying the BLR.

We can now consider various possibilities depending on the importance of the irradiation on the measured value.

If the whole idea of [18] is incorrect, and the dust formation in the accretion disk atmosphere has nothing to do with the formation of the BLR, then we return to the classical point of view that the onset of BLR requires a specific value of the ionization parameter. In this case the radius of the BLR should scale with the accretion rate at the inner disk as

$$R_{BLR} \propto \dot{M}^{1/2}. \tag{6}$$

If we assume that the monochromatic flux is nevertheless dominated by the local dissipation, i.e. we can use Eq. 2, the final dependence between the monochromatic luminosity and the BLR radius will be

$$R_{LBR} \propto L_{5100}^{0.75} M^{-1/2}, \tag{7}$$

i.e. the slope will be close to the older value and the presence of the additional dependence on the black hole mass would be expected. For a given source, only the dependence on the monochromatic luminosity matters, while in a sample of sources there is a likely bias towards positive correlation between the monochromatic luminosity and the mass, leading to much more shallow dependence.

Therefore, we additionally plotted the results of the monitoring for NGC 5548, taking the data points from Bentz et al. where the starlight was subtracted. The relation is shown in Fig. 2. The slope is indeed steeper, equal to 0.90 ± 0.22. The lower limit is consistent within the error with the expectations based on the ionization parameter, but the value itself is surprisingly high. Therefore, we considered whether the dust based scenario can explain such a steep slope. If we follow the general idea of [18] of dust onset of the BLR but we include the irradiation of the disk surface we also obtain the dependence of the BLR position on the accretion rate at the inner disk. Assuming that the irradiation actually dominates we have

$$F_{disk} = \frac{L_{bol}}{4\pi r^2} \frac{h}{r} \sqrt{\gamma(\gamma - 1)}, \tag{8}$$

where $h(r)$ is the disk thickness, and the radial dependence is in the form of a power law, $h(r) \propto r^{\gamma}$.

[23] considered direct and indirect irradiation of an AGN accretion disk and calculated the disk thickness of the non-irradiated and irradiated disk taking into account the vertical disk structure with convection and realistic opacities, with atomic transitions, molecules and dust. The disk thickness rises slowly with the radius at the inner part of the disk and then flares at larger distances, roughly as predicted by the original Shakura & Sunyaev model. The final rise in Loska et al. [23] is rougly with $h \propto r^{1.25}$ (see their Fig. 3). If we supplement this value of γ into Eq. 8, assume the BLR location at a fixed temperature determined by the dust, and combine it with the expression for the monochromatic flux we obtain the expected scaling

$$R_{BLR} \propto L_{5100}^{0.86} M^{-0.57}, \tag{9}$$

Table 1. Observed size of the BLR in the monitored AGN.

name	redshift	R_{BLR} ly	R_{BLR} μarc sec	name	redshift	R_{BLR} ly	R_{BLR} μarc sec
Mrk335	0.02579	15.7	25.3	PG 0026+129	0.14200	111.0	32.5
PG 0052+251	0.15500	89.8	24.1	Fairall 9	0.04702	17.4	15.4
Mrk 590	0.02639	25.6	40.3	3C 120	0.03301	38.1	48.0
Ark 120	0.03271	39.7	50.5	Mrk 79	0.02219	15.2	28.5
PG 0804+761	0.10000	146.9	61.1	PG 0844+349	0.06400	32.2	20.9
Mrk 110	0.03529	25.5	30.0	PG 0953+414	0.23410	150.1	26.7
NGC 3227	0.00386	7.8	84.0	NGC 3516	0.00884	6.7	31.5
NGC 3783	0.00973	10.2	43.6	NGC 4051	0.00234	5.8	103.1
NGC 4151	0.00332	6.6	82.7	PG 1211+143	0.08090	93.8	48.2
PG 1226+023	0.15834	306.8	80.6	PG 1229+204	0.06301	37.8	24.9
NGC 4593	0.00900	3.7	17.1	PG 1307+085	0.15500	105.6	28.3
IC 4329A	0.01605	1.5	3.9	Mrk 279	0.03045	16.7	22.8
PG 1411+442	0.08960	124.3	57.7	NGC 5548	0.01718	18.0	43.6
PG 1426+015	0.08647	95.0	45.7	Mrk 817	0.03146	21.8	28.8
PG 1613+658	0.12900	40.1	12.9	PG 1617+175	0.11244	1.5	26.4
PG 1700+518	0.29200	251.8	35.9	3C 390.3	0.05610	23.6	17.5
Mrk 509	0.03440	79.6	96.2	PG 2130+099	0.06298	22.9	15.1
NGC 7469	0.01632	4.5	11.5				

i.e. the trend with the black hole mass appears but the trend with the luminosity matches nicely the one seen for NGC 5548.

Instead, if the irradiation dominates at 5100 Å the spectrum has much flatter slope ($F_\nu \propto \nu^{-1}$) than in the Shakura-Sunyaev disk ($F_\nu \propto \nu^{1/3}$), and the monochromatic flux is proportional to the accretion rate, $L_{5100} \propto \dot{M}$. In this case the combination of the irradiation dominating 5100 Å and the assumption of the specific value of the ionization parameter specified BLR gives the required relation for the statistical sample

$$R_{BLR} \propto L_{5100}^{0.5}, \tag{10}$$

but it is inconsistent with the trends observed in NGC 5548.

4.3. Further tests of the model

Better tests of the idea that the dust is the key mechanism behind the formation of the BLR require both more results of the reverberation monitoring and better description of the actual dusty outflow.

Observationally, results from many years for a single source would allow to see whether the steeper relation seen for NGC 5548 is a typical pattern for a *single* source variability. Monitoring of more objects in a still broader range of masses and accretion rates would allow to test the exact form of coupling of the BLR size to the mass and accretion rate.

Direct observational approach to the location of the BLR is not possible at present. The requested spatial resolution is very high: the optimum source is NGC 4051 (100 μ arc sec) due to its low redshift (see Table 1), but in the future even such an extreme resolution will be achieved. Such a possibility was already envisioned by [24].

Theoretical studies are also needed since the actual formation of the dust in the disk atmosphere, dusty outflow, dust/gas coupling, dust evaporation and subsequent inflow are very complex phenomena. There are papers addressing some of the issues (for example, see a recent paper of [25] about the dusty outflow forming the dusty torus), but never all of them. In addition, the dust may acquire charge, and the presence of the large scale magnetic field may also modify the dust particle behaviour, (see e.g. [26] for extreme examples of charges close to a black hole).

Acknowledgments
This work was supported in part by grant NN 203 380136.

References

[1] Seyfert, C. K., 1943, ApJ, 97, 28S
[2] Schmidt, M., 1963, Nature, 197, 1040S
[3] Kaspi, S., et al., 2000, ApJ, 533, 631
[4] Peterson, B.M., et al., 2004, ApJ, 613, 682
[5] Vestergaard, M., Fan, X., Tremonti, C.A., Osmer, P.S., Richards, G.T., 2008, ApJ, 674, L1
[6] Goad, M., Koratkar, A., 1998, ApJ, 495, 718
[7] Knigge, C., Scaringi, S., Goad, M. R., Cottis, C. E., 2008, MNRAS, 386, 1426
[8] Risaliti, G., Nardini, E., Salvati, M., Elvis, M., Fabbiano, G., Maiolino, R., Pietrini, P., Torricelli-Ciamponi, G., 2011, MNRAS, 410, 1027
[9] Nikołajuk, M., Czerny, B., Ziółkowski, J., 2005, AIPC, 801, 220
[10] Collin, S., Kawaguchi, T., Peterson, B. M., Vestergaard, M., 2006, A&A, 456, 75
[11] Collin, S., Hure, J.-M., 2001, A&A, 372, 50
[12] Eracleous, M., Halpern, J. P., 2003, ApJ, 599, 886
[13] Murray, N., Chiang, J., 1997, ApJ, 474, 91
[14] Collin-Souffrin, S., Dyson, J.E., McDowell, J.C., Perry, J.J., 1988, MNRAS, 232, 539
[15] Bentz, M.C., Peterson, B.M., Netzer, H., Pogge, R.W., Vestergaard, M., 2009, ApJ, 705, 199
[16] Shakura, N.I., Sunyeav, R.A., 1973, A&A, 24, 337
[17] Tripp, T.M., Bechold, J., Green, R.F., 1994, ApJ, 433, 533
[18] Czerny, B. & Hryniewicz, K., 2011, A&A, 525, L8
[19] Lawrence, A., Elvis, M., 2010, ApJ, 714, 561
[20] Nenkova, M., Sirocky, M. M., Nikutta, R., Ivezic, Z., Elitzur, M. 2008, ApJ, 685, 160
[21] Sergeev, S. G.; Doroshenko, V. T., Golubinskiy, Yu. V., Merkulova, N. I., Sergeeva, E. A., 2005, ApJ, 622, 129
[22] Gaskell, C.M., 2008, RMxAC, 32, 1
[23] Loska, Z., Czerny, B., Szczerba, R., 2004, MNRAS, 355, 1080
[24] Elvis, M., Karovska, M., 2002, ApJ, 581, L67
[25] Dorodnitsyn, A., Bisnovatyi-Kogan, G. S., Kallman, T., 2011, ApJ, 741, 29
[26] Kovar, J. et al., 2011, Phys.Rev.D., 84, 084002

Astronomy at High Angular Resolution 2011 IOP Publishing
Journal of Physics: Conference Series **372** (2012) 012014 doi:10.1088/1742-6596/372/1/012014

The inner 200 parsecs of NGC4151 observed with OSIRIS

Christof Iserlohe[1], **Alfred Krabbe**[2], **James Larkin**[3]

[1] I. Physikalisches Institut der Universitaet zu Koeln, Zuelpicherstrasse 77, 50937 Koeln, Germany
[2] Deutsches SOFIA Institut (DSI) an der Universitaet Stuttgart, Pfaffenwaldring 31, 70569 Stuttgart, Germany
[3] Physics and Astronomy Building, 430 Portola Plaza, Box 951547, Los Angeles, CA. 90095-1547, USA

E-mail: `ciserlohe@ph1.uni-koeln.de`

Abstract. NGC 4151 is a Seyfert 1.5 galaxy. Due to its proximity of only 13 Mpc it is an ideal testbed and therefore one of the most intensively studied Seyfert galaxies. NGC4151 shows an outflow in the Narrow Line Region (NLR) along the NE and SW direction. This outflow has been modelled by several authors ([9], [10] and [11]) as two hollow bi-cones. The excitation conditions in the NLR have been investigated by [11] and have been attributed to photoionization due to the central engine. [4] also observed a linear radio structure in the inner few parsec with a pronounced S-like curvature which is slightly misaligned with the bi-cones major axes. Here we compare these findings with our results and discuss the role of the radio jet. We also refer the reader to the full paper (Iserlohe et. al., 2012) which will be published soon.

1. Introduction

NGC4151 has been observed with OSIRIS (<u>O</u>H <u>S</u>uppressing <u>I</u>nfra-<u>R</u>ed <u>I</u>maging <u>S</u>pectrograph) at the W. M. Keck Observatory. OSIRIS is an integral field spectrograph [1] for the near-infrared (z, J, H and K band) with a nominal spectral resolution $\lambda/\Delta\lambda = 3700$ that works with the Keck Adaptive Optics System [2]. The design of the instrument is based on concepts developed for the TIGER spectrograph [3] and utilizes an infrared transmissive microlens array that samples a rectangular field-of-view (FoV) of the adaptive optics focal plane. We performed AO assisted spectroscopic H and K band observations of the inner 100 parsec with linear resolution on-site of about 5 parsec. The AO loop was locked on the nucleus.

The K band spectra reveal a wealth of emission lines (see figure 1). Brγ shows a broad and a narrow component on top. Ro-vibrational transitions of molecular hydrogen are observed as well as highly ionized species like [CaVIII] and [SiVI]. HeI appears in emission and absorption. The nuclear spectrum is significantly redder than the spectrum extracted from the whole FoV.

2. Feeding

Radio observations by [4] show in HI an oval on scales of 10 arcminutes. Kinematically this oval has been described as a weak bar with inflow of gas towards the nucleus. The major axis of this oval is aligned roughly from NW to SE. K-band imaging of the inner few arcminutes

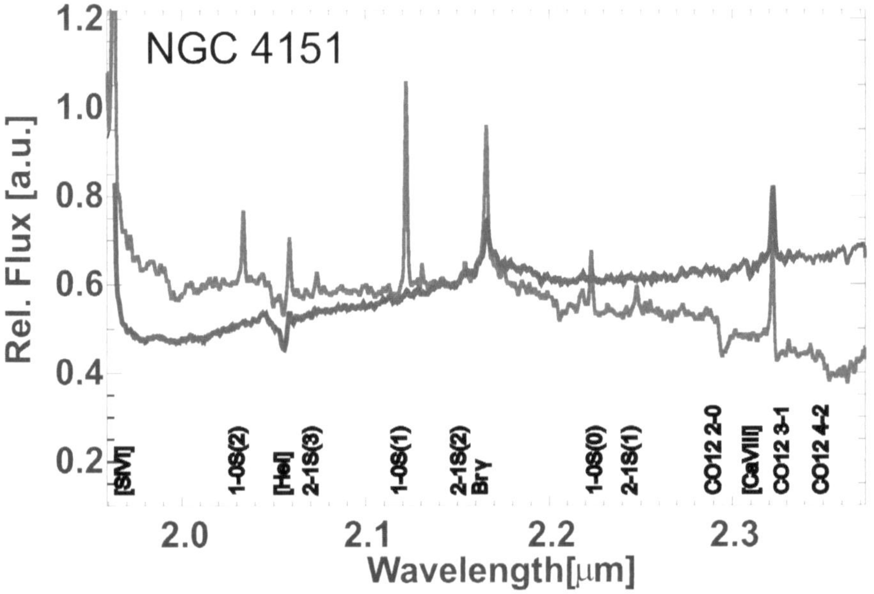

Figure 1. K-band spectra of NGC4151. Nuclear spectrum extracted from a circular aperture with a diameter of 0.3 arcseconds centered on the continuum peak (red) and spectrum extracted from our whole Field-of-View (green).

of NGC4151 show an inner K band continuum bar which is well aligned with the major axis of this oval. Along this bar and on arcsecond scales our spectra are also significantly redder than perpendicular to the bar implying that there is more obscuring material along that bar. Roughly perpendicular to this bar narrow Brγ is observed in the Narrow Line Region (NLR). The dynamics of this gas reveals three components: Rotation in the galactic disc of NGC4151, an outflow into the NLR (roughly from NE to SW) and a velocity signature which resembles a keplerian disc. Latter is centered at the position of the continuum peak and is aligned roughly from NW to SE and may hence not represent an outflow. To investigate this velocity signature we fit a thin, keplerian disc model to it following [6]. Free parameters of the model are the mass of the black hole, the inclination, the scale parameter and the width of the seeing disc. If we assume that the inclination of this disc is the same as the inclination of the galactic disc of NGC4151 (23 degrees) we derive a black hole mass of $5 \pm 2 \times 10^7$ solar masses compatible with measurements by [7] (4.5×10^7 solar masses, reverberation mapping) and [8] (4.5×10^7 solar masses, modelling the outer stellar velocity field). Also the line-of-nodes is aligned with the inner bar with an accuracy of about 5 degrees. We assume that this keplerian disc represents another feeding stage closer to the nucleus.

3. Feedback

The kinematics of the Narrow Line Region (e.g. observed in [OIII]λ500 nm) is compatible with an outflow in two hollow bi-cones ([9], [10] and [12]). The outflow is modelled with constant acceleration (e.g. due to radiation pressure). The corresponding half opening angles are 15 degrees and 33 degrees respectively with a position angle of 60 degrees and an inclination of 45 degrees. A comparison between the velocities of selected clouds measured in [OIII] [5] and [FeII]λ1.664 reveals that the [FeII] kinematics is very similar to that observed in [OIII]. To investigate whether the [FeII]λ1.664 emission we observe is compatible with this model we analyze the

[FeII] emission morphology by overplotting regions from which emission is expected according to the model with channel maps in [FeII]. Most of the emission seen in [FeII] emerges with zero velocity and we assume that this emission emerges from the galactic disc of NGC4151. Velocities and positions of clouds with other than systemic velocities are well compatible with this model.

The S-like shape of the 6 cm radio jet [4] has been puzzling for a long time. It has been speculated that ram pressure due to the rotation of the galactic disc of NGC4151 into the path of the ejected plasma may be the cause. Here we present the proof, that this is indeed the case. Figure 2 shows the 6 cm radio jet overplotted to the coadded emission of ro-vib transitions of molecular hydrogen (1-0S(0) + 1-0S(1) + 1-0S(2)). Most of the molecular emission seen in this figure emerges with systemic velocity and is hence embedded in the disc of NGC4151. The sense of rotation is clockwise. The correlation between the individual radio knots and bright spots in molecular hydrogen is indeed striking. This also shows that the radio jet must have a low inclination and almost runs completely in the disc.

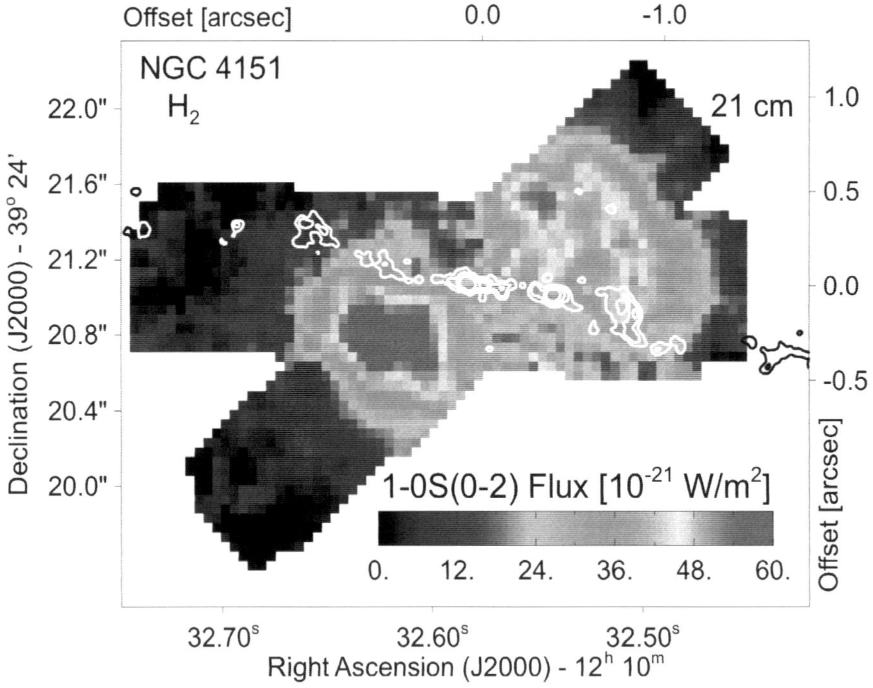

Figure 2. The 6 cm radio jet (contours) with the coadded emission of ro-vib transitions of molecular hydrogen (1-0S(0) + 1-0S(1) + 1-0S(2)).

4. HeI absorption in the nuclear spectrum

Absorption of UV excited species is observed in nuclear spectra of NGC4151 [13]. Here, we observe infrared absorption and emission of HeIλ2.058 where the lower state 2^1S is metastable and has an excitation energy of approximately 21 eV (see the Grothrian diagram in figure 3). The gaussian emission component emerges with systemic velocity while the lorentzian absorption component is redshifted by 280 km/s with a half-width-half-maximum of 400 km/s. The emission component is spatially resolved, especially into the direction of the Narrow Line Region, while the absorption component remains spatially unresolved. Following [14] we derive a column density of 5×10^{12} cm^{-2} for a covering factor of 1. The minimum covering factor derived is 0.1 with a column density of 1×10^{14} cm^{-2}. We may actually see a UV excited gas cloud orbiting closely to the nucleus which is detected for the first time in the near-infrared.

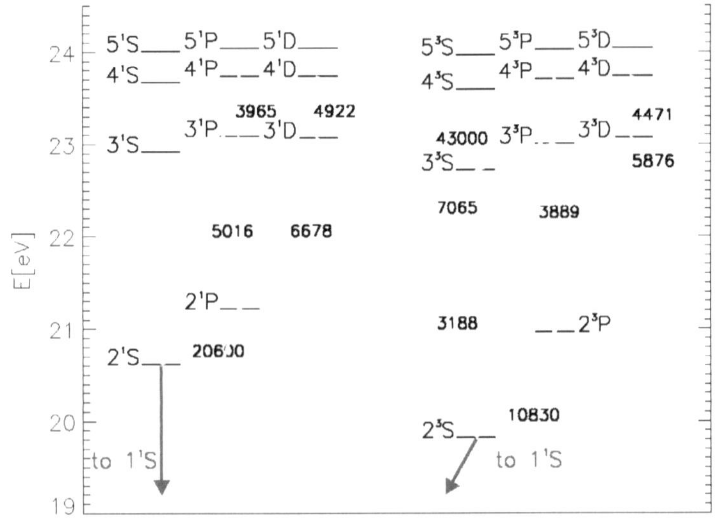

Figure 3. Grothrian diagram. The HeIλ2.058 transition is indicated.

5. FLUXER, an IDL-based vizualization and processing tool for data cubes

FLUXER (see figure 4) is a processing and visualization tool for astronomical data cubes. This tool was created to deal with data delivered by imaging field spectrographs like OSIRIS@Keck or SINFONI@VLT.

The program allows :

- Automatic and manual identification of crazy/dead pixels in data cubes
- Interpolation of these pixels in 1, 2, or 3 dimensions
- Operations on data cubes, like e.g. smoothing each slice of a data cube with a gaussian PSF of median filtering each slice
- Interactive and automatic fitting of emission/absorption lines with various profiles (flux maps, velocity fields, dispersion fields, ...).
- Extraction of spectra from data cubes
- Creation of animated channel maps in gif format (for UNIX only)
- Data cube viewing in 2D with full astrometry (if supplied)

All spectra and flux maps presented here have been extracted from the data cube using FLUXER.

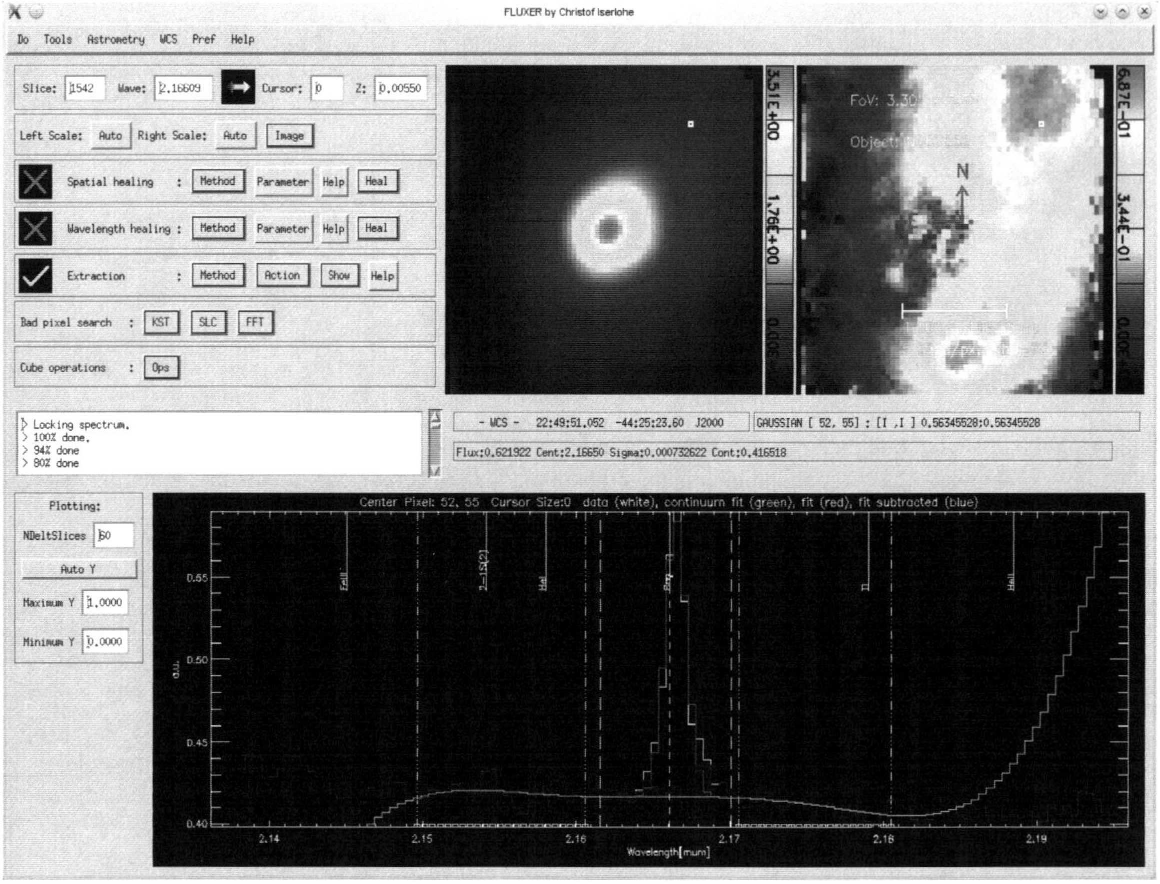

Figure 4. The main GUI of FLUXER. As an example a flux map of Brγ has been extracted from a data cube obtained with SINFONI of NGC7582.

References

[1] Larkin, J., et al. 2006, SPIE, 6269,
[2] Wizinowich, P. L., Acton, D. S., Lai, O., Gathright, J., Lupton, W., & Stomski, P. J. 2000, SPIE, 4007, 2
[3] Bacon, R., et al. 1995, A&AS, 113, 347
[4] Mundell, C. G., Wrobel, J. M., Pedlar, A., & Gallimore, J. F. 2003, ApJ, 583, 192
[5] Kaiser, M. E., Bradley, L. D., II, Hutchings, J. B., et al. 2000, ApJ, 528, 260
[6] Barbosa, F. K. B., Storchi-Bergmann, T., Cid Fernandes, R., Winge, C., & Schmitt, H. 2006, MNRAS, 371, 170
[7] Bentz, M. C., et al. 2006, ApJ, 651, 775
[8] Onken, C. A., et al. 2007, ApJ, 670, 105
[9] Crenshaw, D. M., et al. 2000, AJ, 120, 1731
[10] Das, V., et al. 2005, AJ, 130, 945
[11] Storchi-Bergmann, T., McGregor, P. J., Riffel, R. A., Simões Lopes, R., Beck, T., & Dopita, M. 2009, MNRAS, 394, 1148
[12] Storchi-Bergmann, T., Lopes, R. D. S., McGregor, P. J., et al. 2010, MNRAS, 402, 819
[13] Hutchings, J. B., et al. 1998, ApJL, 492, L115
[14] Hamann, F., Barlow, T. A., Junkkarinen, V., & Burbidge, E. M. 1997, ApJ, 478

Astronomy at High Angular Resolution 2011 IOP Publishing
Journal of Physics: Conference Series **372** (2012) 012015 doi:10.1088/1742-6596/372/1/012015

Unveiling the nucleus of NGC 7172

Semir Smajić[1,2], Sebastian Fischer[1], Jens Zuther[1] and Andreas Eckert[1,2]

[1] I. Physikalisches Institut, Universität zu Köln, Zülpicher Straße 77, 50937 Köln
[2] Max-Planck-Institut für Radioastronomie, Auf dem Hügel 69, 53121 Bonn, Germany

E-mail: smajic@ph1.uni-koeln.de

Abstract. We present the results of near-infrared (NIR) H+K ESO-SINFONI integral field spectroscopy (IFS) of the Seyfert 2 galaxy NGC 7172. The aim is to investigate the central 800pc, concentrating on excitation conditions, morphology and stellar content. The NIR is less influenced by dust extinction than optical light and therefore yields a close undisturbed look at the central region.
In this proceedings we concentrate on emission line measurements in the central 800pc of NGC 7172. The detection of [SiVI] and broad Paα and Brγ components are clear signs of an accreting super-massive black hole (SMBH) which hides behind the prominent dustlane. Temperatures of about 1400K show evidence of hot dust in the nuclear region. Narrow components of Paα and Brγ allow for an extinction measurement. The molecular hydrogen lines, hydrogen recombination lines and [FeII] point out that the excitation of these lines is caused by an AGN. The findings show evidence for nuclear activity located behind the prominent dustlane crossing the central region of the galaxy. The nucleus of NGC 7172 is a Seyfert 1 nucleus either surrounded or hidden by a molecular dust torus. Our observation aids the unified model scheme in its basic proposition, however, we can not conclude the shape of the torus due to our spatial resolution.

1. Introduction

For more than 20 years we try to figure out the mechanisms that lead to the formation of super-massive black holes (SMBH) and how these are fed successfully. The composition and design of the inner kpc around the active galactic nucleus (AGN) is a very important fact since it is in direct 'contact' with the central source. Antonucci [1] proposed the Unified Model (UM) which describes all AGN as being intrinsically the same, but show different characteristics due to the positioning of the central region with respect to the observer. The narrow line region (NLR) extends to several hundred pc and can be seen in both Seyfert types. It harbors high ionization emission lines (e.g., HI recombination lines, [OIII]) which are used to classify a galaxy as active. These lines are supposed to be ionized by high energy photons from the central source. Therefore the higher the ionization potential of the detected line the closer it is situated to the central source (i.e. coronal lines [SiVI]) [2; 3]. The shape of the dusty torus is not exactly known. It can have scales of up to several pc going down to the radius which is dictated by the sublimation temperature of the dust. The most simple model is a donut surrounding the central region, however, a more realistic approach is to assume the donut to be patchy consisting of several separated and coinciding clouds. The composition of the torus is another remark. What makes up the inner parts of this torus and what does it consist of in its outer parts. Is it molecular

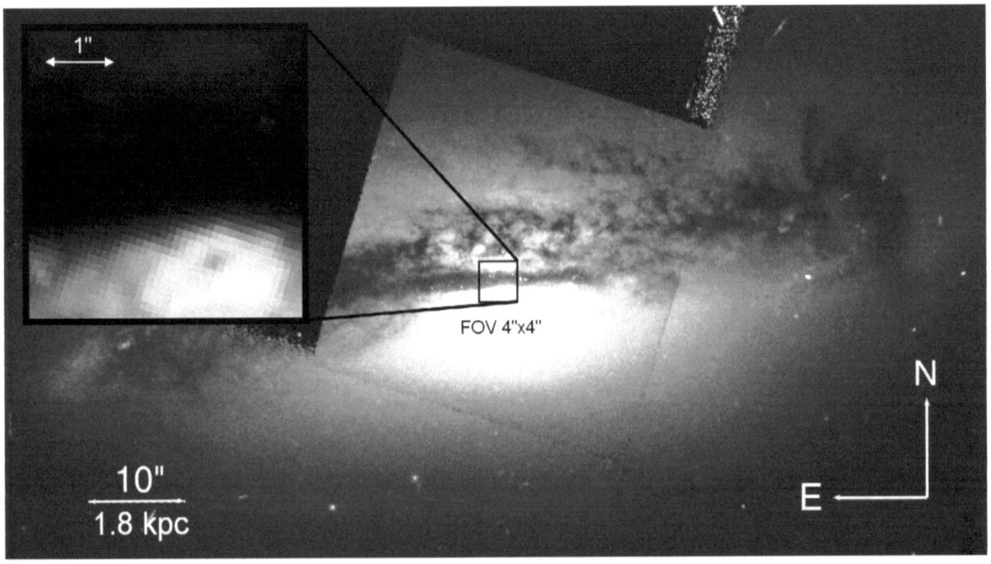

Figure 1. HST F606W image [8]. The dark box shows the FOV of the observation. The brightness scale is linear in the detail image. To be able to show the edges of the galaxy we chose a higher contrast in the big image.

dust or does atomic gas also play a role, especially in the transition phases (inner and outer part of the torus). The broad line region (BLR) are clouds of mainly hydrogen and helium which are close to the central source and show high velocities which can be seen in broad line profiles. This part and the accretion disk of the central black hole cannot be seen directly in Seyfert 2 but can be detected in polarized light due to light refraction and reflection on dust grains. The torus hides the BLR and the rest of the inner part from our view in Seyfert 2's. Therefore the classification of Seyfert galaxies is done in first instance by the emission line ratio $[OIII]/H\beta > 3$ [4] to decide if the ionization of the central source is strong enough and then to divide into Seyfert 1 and 2 galaxies the detectability of a broad $H\alpha$ component is needed. For Seyfert 1 galaxies a broad $H\alpha$ line is detected whereas for a Seyfert 2 no broad lines are detected. This conventions have their roots in the historically preferred optical wavelength regime, however, in the recent years with new detector generations and new telescopes using adaptive optics (AO) systems the near-infrared (NIR) can be investigated even better than the optical. In these years in a few Seyfert 2 galaxies (e.g. NGC 1386, NGC 7582, IRAS 05189-2524) that show no optical broad lines broad NIR emission lines were detected (i.e. Paschen and Bracket series) [5–7].

1.1. NGC 7172

NGC 7172 is a peculiar spiral galaxy S0/a and it is a member of the Hicks Compact Group (HCG) 90. The inclination of the galaxy is almost 90 degrees (edge-on) with a prominent dust lane crossing the nuclear region. The dust lane is visible in the optical as seen in figure 1. NGC 7172 is at a distance of about 36Mpc. It is classified as a Seyfert 2 galaxy [9] with no signs of polarized broad lines [10]. Other observations show a decrease in the flux level in the X-ray since the 80's [11; 12] and a flux short term variability of 30% [11]. A column density of about $10^{23}\mathrm{cm}^{-2}$ was calculated by several authors [11–15]. In the optical only narrow emission lines were detected with deep internal absorption visible in the NaI D absorption [9]. In the NIR with respect to optical observations a spatial offset for the nuclear emission of about 2" was detected [16]. PAH emission at $3.3\mu m$ was not detected in the nuclear region but in the disk which hints at a starburst within the disk but not in the nuclear region [17–19]. In the radio extended emission in the disk was detected [20] and an outflow from the nuclear region to the south-western region can be recognized [21].

1.2. Observation

The galaxy was observed within a sample of Seyfert 2 galaxies. These were chosen to test the UM that got challenged in the last years (e.g., [22], [23]). We used SINFONI, the AO assisted NIR integral field unit (IFU) at the very large telescope (VLT) on mount Paranal in Chile. The nucleus of NGC 7172 was used as a natural guide star (NGS) for the AO. We used the H+K grating with the 3"×3" field of view (FOV). Therefore we got spectra from 1.45 to 2.4 microns with a resolution of about 1500. We used a dithering method to decrease the chance of stacking of dead pixel. This dithering increased the nominal FOV to 4"×4", however, by dithering the FOV by 0.5 arcsec in eight directions separated by 45 degrees we lose integration time on the edges of the FOV. The central region of 2"×2" benefits from the full integration time on the source of 1300 seconds. Another 650 seconds integration time was spent on sky.

Although we had good weather conditions and the AO loop closed we got a spatial resolution of about 0.5 arcsec. The reason for this is that SINFONI uses a wavefront sensor that works in the optical wavelength regime. This is usually not a problem since the NIR and optical point source of a galaxy is the same, the nucleus. As can be seen in figure 2 the brightest point in the optical does not overlap with the brightest point in the NIR. Additionally the brightest source in the optical is not the active galactic nucleus (AGN) but an extended region to the south-west. Hence, the AO correction, which corrects the PSF for a point source, tried to correct something that is not a point source into a point source. Therefore, we did not benefit from the AO in fully which would have given us a near diffraction limited spatial resolution of about 200 marcsec.

2. Emission Line Regions

2.1. Continuum

We compare 2MASS J, H and K images of NGC 7172 [24] with the HST image [8] (see Fig. 1) and our observation. Figure 2 shows the HST image of our FOV overlayed with K-band contours from our observation and showing the centers of J (five pointed star), H (four pointed star) and K (triangle) flux peaks from the 2MASS images. It becomes clear from the 2MASS flux peaks that the flux peak moves from south to north for about 2" as already noted by Sharples [16]. We note that the pixel scale of the 2MASS images have a pixel scale of 1" and a spatial resolution of 2"-3". The pointing accuracy of the 2MASS observation and of our SINFONI observation is about 0.1 arcsec. We use the world coordinate system (WCS) to overplot the data. Furthermore, we overlayed the HST image and the 2MASS images and checked for the locations of two foreground stars. It is also clear that our K-band flux peak (white contours) and the 2MASS K-band flux peak are both behind the dust lane. The dust lane is so opaque that it is still recognizable in J-

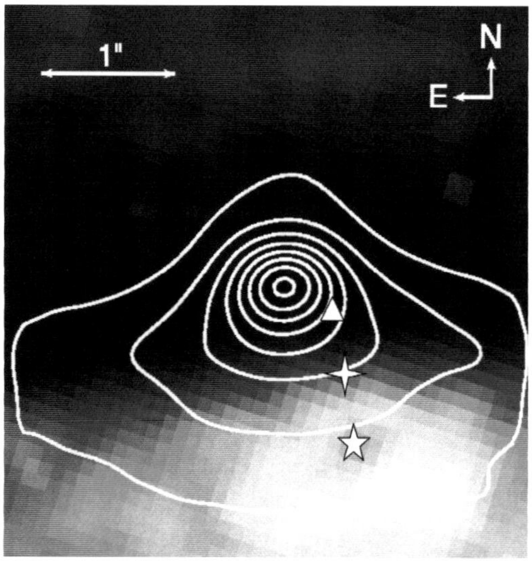

Figure 2. HST F606W image overlayed with SINFONI and 2MASS data.

and H-band and dissolves for the most part eventually in the K-band. Therefore, we conclude that observations in the optical did not point at the actual nucleus since the nucleus is hidden behind the opaque dust lane. Hence, the AO system was not able to actually see the nucleus and therefore was closed on an extended region in the south-west.

2.2. Emission Lines

As already mentioned NGC 7172 was classified as a Seyfert 2 nucleus from optical observations. We detect for the first time broad emission lines in NGC 7172 with broad Paα and Brγ detections (see Fig. 5(a),(b)). Using the flux of Paα we are able to determine the black hole mass [25]. We measure a full width at half maximum (FWHM) for Paα of 5700kms^{-1}s and a luminosity of 9.1×10^{41}ergs^{-1}.

$$M_{BH} = 10^{7.16}\left(\frac{L_{Pa\alpha}}{10^{42}ergs^{-1}}\right)^{0.49}\left(\frac{Pa_{\alpha}}{10^{3}kms^{-1}}\right)^{2}$$

We determine a black hole mass of $M_{BH}=(4.5\pm1.0)\times10^{8}M_{\odot}$. The literature value of $M_{BH}=2.04\times10^{8}M_{\odot}$ [26] affirms our black hole mass.

We detect a strong narrow Paα and a Brγ component (see Fig. 5). Both are detected in the central region with lower luminosity wings in the east-west direction (galactic disk) and both show a flux peak offset to the south-west from the nuclear flux peak. Using the flux of the narrow components of Paα and Brγ we can determine the extinction A_V using the line ratios of the case B recombination [4]. With the formula for A_V [27]

$$A_V = \frac{-2.5\times\log\left(\frac{f_{Br\gamma O}/f_{Pa\alpha O}}{f_{Br\gamma}/f_{Pa\alpha}}\right)}{(a(\lambda_{Br\gamma})+b(\lambda_{Br\gamma})/R_V)-(a(\lambda_{Pa\alpha})+b(\lambda_{Pa\alpha})/R_V)},$$

we determine an average extinction for the central region of A_V 7 with a Paα/Brγ line ratio of about 10.

We also detect the coronal line [SiVI] $\lambda1.963\mu$m (see Fig. 4(a)). Coronal lines have a high ionization potential and therefore are considered as AGN tracer. [SiVI] has an ionization potential of about 150eV, which cannot be produced by mere starbursts. In our observation [SiVI] extends from the nuclear region to the south-west. The flux peak is offset to the south-west with respect to the nucleus, similar to the HI recombination lines Paα and Brγ. Although offset, the flux peaks are situated behind the dust lane. Hence, we can affirm that the nucleus is situated behind the dust lane and the same is true for the NLR.

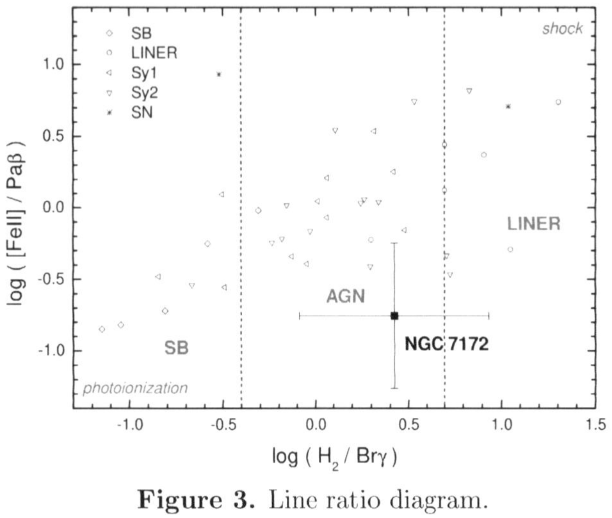

Figure 3. Line ratio diagram.

Two molecular lines were detected. The H_2 $\lambda\lambda1.957\mu$m2.122μm lines (see Fig. 4(a),(b),(c),5(b)). Both peak in the central, nuclear region. No other molecular lines were detected. The forbidden line transition of [FeII] $\lambda1.644\mu$m is detected, however, the CO(7-4) absorption feature is strong and [FeII] is blended by this absorption feature (see Fig. 4(d)).

The line ratio diagram log([FeII]$\lambda1.257\mu$m/Paβ) over log(H_2/Brγ) shows that our data point lies in the AGN regime, which shows combined ionization of shock and photoionization. The data is taken from a circular aperture in the central region with a radius of 5 pixel. The rhombs represent starburst galaxies [28], circles represent LINERs and asterixs supernovae [29], left hand triangles represent Seyfert 1 galaxies [30] and head-long triangles represent Seyfert 2s [31]. The data points are not corrected for extinction. Our data point is slightly lower than other points in that $\log(H_2/Br\gamma)$ regime. But note that [FeII] is heavily biased by the CO absorption feature. Additionally, [FeII] $\lambda 1.257\mu m$ and Paβ were calculated from [FeII] $\lambda 1.644\mu m$ and Paα [4; 32] which are not close by and therefore, extinction plays an important role. Extinction correction and correction for the absorption feature would push our data point further up where the other data points are situated.

3. Conclusion

For the first time broad emission lines are detected in the active galaxy NGC 7172. The HI recombination lines Paα and Brγ are detected with a FWHM of 5700kms^{-1} which is typical for Seyfert 1 galaxies. From these lines we determine a black hole mass of $4.5\times10^8 M_\odot$. The coronal line [SiVI], which is an AGN tracer is detected, too. This confirms our detection of the nucleus behind the dust lane. The molecular hydrogen lines at 1.96μm and 2.12μm are also detected. The [FeII] line at 1.64μm is heavily blended by the stellar absorption feature CO(7-4) and marginally detected in the same region as [SiVI].

Further investigation of stellar composition and gaseous and stellar kinematics is in process. This will help us to determine the recent history of the nuclear region in NGC 7172 and the feeding process of the black hole.

References

[1] Antonucci R 1993 *Annual Review of Astronomy and Astrophysics* **31** 473–521 URL `http://adsabs.harvard.edu/abs/1993ARA%26A..31..473A`

[2] Marconi A, Moorwood A F M, Salvati M and Oliva E 1994 *Astronomy and Astrophysics* **291** 18–28 URL `http://adsabs.harvard.edu/abs/1994A%26A...291...18M`

[3] Murayama T and Taniguchi Y 1998 *The Astrophysical Journal* **497** L9 URL `http://adsabs.harvard.edu/abs/1998ApJ...497L...9M`

[4] Osterbrock D E 1989 *Astrophysics of Gaseous Nebulae and Active Galactic Nuclei* (University Science Books)

[5] Davies R I, Müller Sánchez F, Genzel R, Tacconi L J, Hicks E K S, Friedrich S and Sternberg A 2007 **671** 1388–1412 (*Preprint* `0704.1374`)

[6] Reunanen J, Kotilainen J K and Prieto M A 2002 **331** 154–168 (*Preprint* `arXiv:astro-ph/0111463`)

[7] Reunanen J, Kotilainen J K and Prieto M A 2003 **343** 192–208 (*Preprint* `arXiv:astro-ph/0305100`)

[8] Malkan M A, Gorjian V and Tam R 1998 *The Astrophysical Journal Supplement Series* **117** 25 URL `http://adsabs.harvard.edu/abs/1998ApJS..117...25M`

[9] Veron-Cetty M and Veron P 1986 *Astronomy and Astrophysics Supplement Series* **65** 241–258 URL `http://adsabs.harvard.edu/abs/1986A%26AS...65..241V`

[10] Lumsden S L, Heisler C A, Bailey J A, Hough J H and Young S 2001 *Monthly Notices of the Royal Astronomical Society* **327** 459–474 URL `http://adsabs.harvard.edu/abs/2001MNRAS.327..459L`

[11] Guainazzi M, Matt G, Antonelli L A, Fiore F, Piro L and Ueno S 1998 *Monthly Notices of the Royal Astronomical Society* **298** 824–830 URL `http://adsabs.harvard.edu/abs/1998MNRAS.298..824G`

[12] Dadina M, Bassani L, Cappi M, Palumbo G G C, Piro L, Guainazzi M, Malaguti G, Di Cocco G and Trifoglio M 2001 *Astronomy and Astrophysics* **370** 70–77 URL http://adsabs.harvard.edu/abs/2001A%26A...370...70D

[13] Akylas A, Georgantopoulos I and Comastri A 2001 *Monthly Notices of the Royal Astronomical Society* **324** 521–528 URL http://adsabs.harvard.edu/abs/2001MNRAS.324..521A

[14] Risaliti G 2002 *Astronomy and Astrophysics* **386** 379–398 URL http://adsabs.harvard.edu/abs/2002A%26A...386..379R

[15] Awaki H, Murakami H, Ogawa Y and Leighly K M 2006 *The Astrophysical Journal* **645** 928–939 URL http://adsabs.harvard.edu/abs/2006ApJ...645..928A

[16] Sharples R M, Longmore A J, Hawarden T G and Carter D 1984 *Monthly Notices of the Royal Astronomical Society* **208** 15–24 URL http://adsabs.harvard.edu/abs/1984MNRAS.208...15S

[17] Moorwood A F M 1986 *Astronomy and Astrophysics* **166** 4–12 URL http://adsabs.harvard.edu/abs/1986A%26A...166....4M

[18] Roche P F, Aitken D K, Smith C H and Ward M J 1991 *Monthly Notices of the Royal Astronomical Society* **248** 606–629 URL http://adsabs.harvard.edu/abs/1991MNRAS.248..606R

[19] Imanishi M 2000 *Monthly Notices of the Royal Astronomical Society* **319** 331–336 URL http://adsabs.harvard.edu/abs/2000MNRAS.319..331I

[20] Unger S W, Lawrence A, Wilson A S, Elvis M and Wright A E 1987 *Monthly Notices of the Royal Astronomical Society* **228** 521–531 URL http://adsabs.harvard.edu/abs/1987MNRAS.228..521U

[21] Thean A, Pedlar A, Kukula M J, Baum S A and O'Dea C P 2000 *Monthly Notices of the Royal Astronomical Society* **314** 573–588 URL http://adsabs.harvard.edu/abs/2000MNRAS.314..573T

[22] Wang J and Zhang E 2007 *The Astrophysical Journal* **660** 1072–1092 URL http://adsabs.harvard.edu/abs/2007ApJ...660.1072W

[23] Meléndez M, Kraemer S B, Schmitt H R, Crenshaw D M, Deo R P, Mushotzky R F and Bruhweiler F C 2008 **689** 95–107 (*Preprint* 0808.4154)

[24] Jarrett T H, Chester T, Cutri R, Schneider S E and Huchra J P 2003 *The Astronomical Journal* **125** 525–554 URL http://adsabs.harvard.edu/abs/2003AJ....125..525J

[25] Kim D, Im M and Kim M 2010 *The Astrophysical Journal* **724** 386–399 URL http://adsabs.harvard.edu/abs/2010ApJ...724..386K

[26] Winter L M, Mushotzky R F, Reynolds C S and Tueller J 2009 *The Astrophysical Journal* **690** 1322–1349 URL http://adsabs.harvard.edu/abs/2009ApJ...690.1322W

[27] Cardelli J A, Clayton G C and Mathis J S 1989 *The Astrophysical Journal* **345** 245–256 URL http://adsabs.harvard.edu/abs/1989ApJ...345..245C

[28] Dale D A, Roussel H, Contursi A, Helou G, Dinerstein H L, Hunter D A, Hollenbach D J, Egami E, Matthews K, Murphy T W, Lafon C E and Rubin R H 2004 *The Astrophysical Journal* **601** 813–830 URL http://adsabs.harvard.edu/abs/2004ApJ...601..813D

[29] Larkin J E, Armus L, Knop R A, Soifer B T and Matthews K 1998 *The Astrophysical Journal Supplement Series* **114** 59 URL http://adsabs.harvard.edu/abs/1998ApJS..114...59L

[30] Rodrguez-Ardila A, Pastoriza M G, Viegas S, Sigut T A A and Pradhan A K 2004 *Astronomy and Astrophysics* **425** 457–474 URL http://adsabs.harvard.edu/abs/2004A%26A...425..457R

[31] Rodrguez-Ardila A, Riffel R and Pastoriza M G 2005 *Monthly Notices of the Royal Astronomical Society* **364** 1041–1053 URL http://adsabs.harvard.edu/abs/2005MNRAS.364.1041R

[32] Nussbaumer H and Storey P J 1988 *Astronomy and Astrophysics* **193** 327–333 URL http://adsabs.harvard.edu/abs/1988A%26A...193..327N

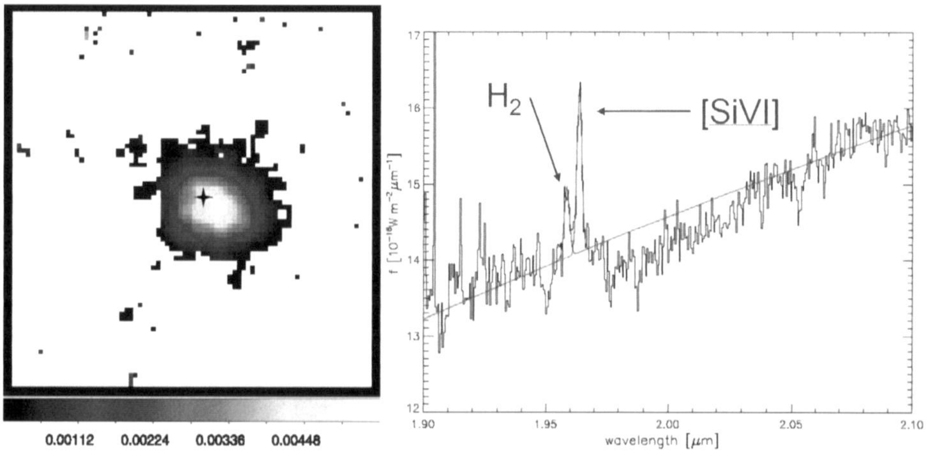

(a) Linemap of [SiVI] with spectrum.

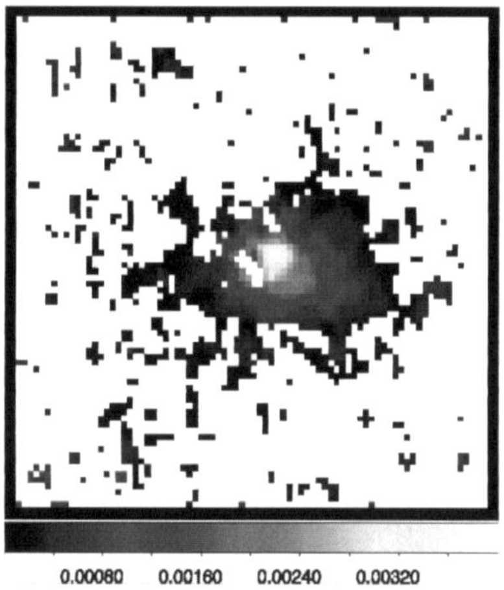

(b) H_2 $\lambda 1.96\mu m$ linemap.

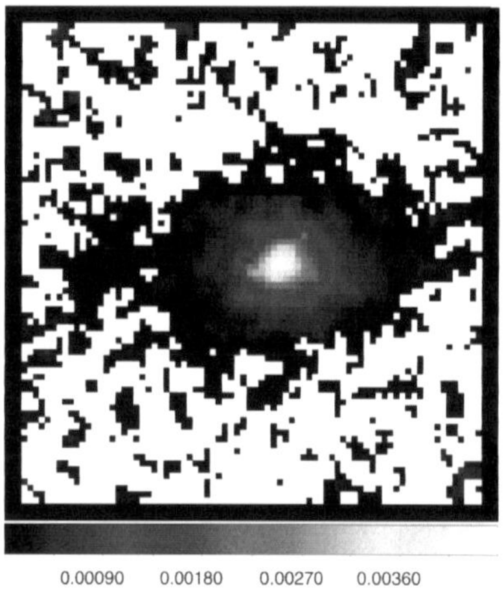

(c) H_2 $\lambda 2.12\mu m$ linemap.

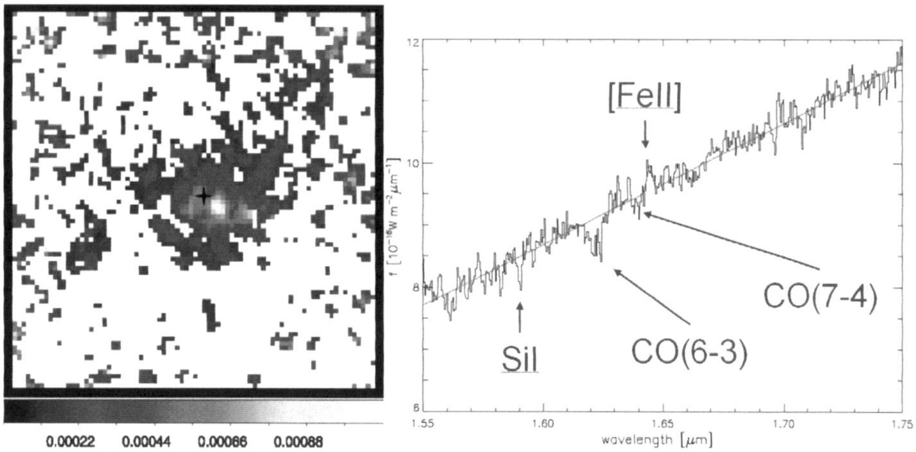

(d) Linemap of [FeII] with spectrum.

Figure 4.

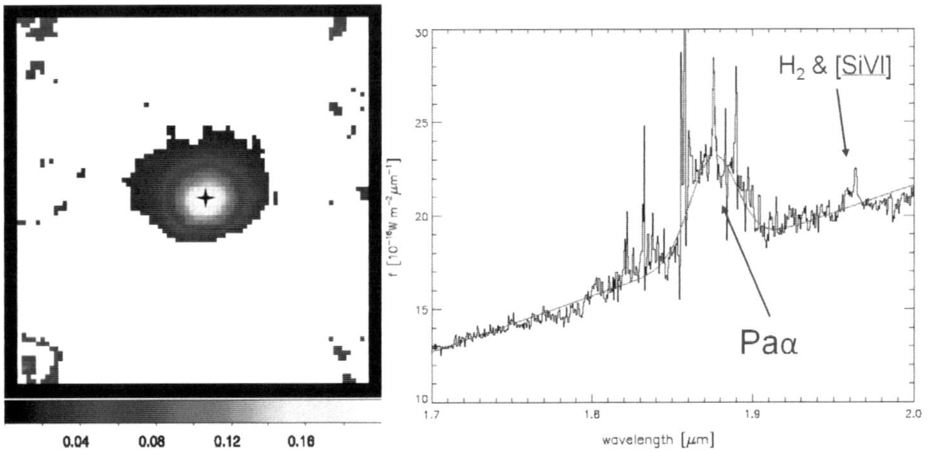

(a) Paα broad component linemap with nuclear spectrum.

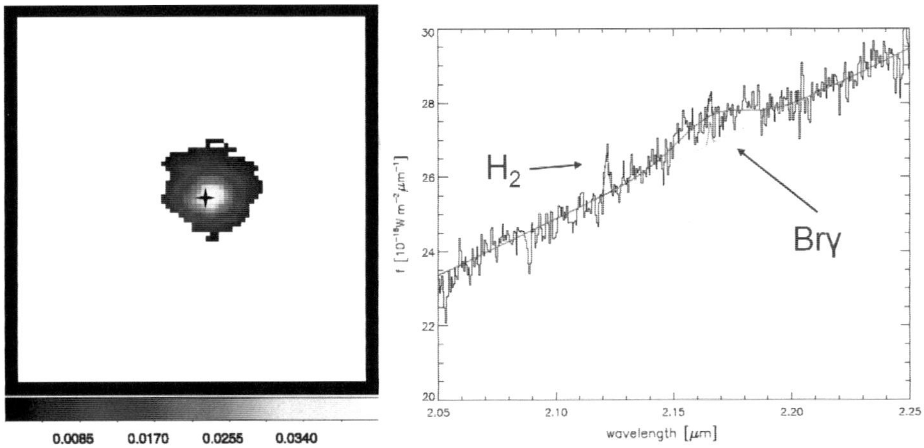

(b) Brγ broad component linemap with nuclear spectrum.

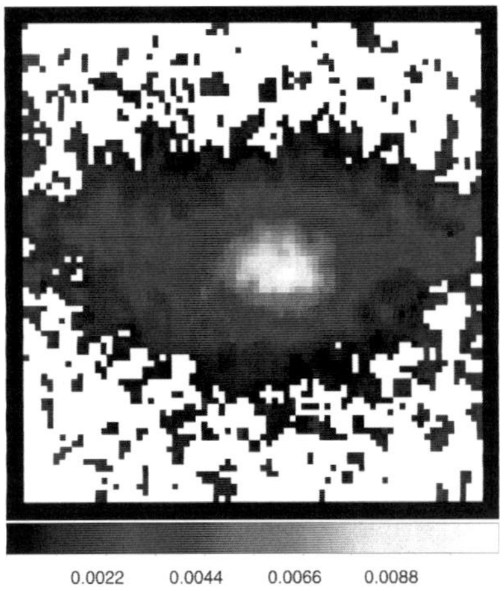

(c) Paα narrow component linemap.

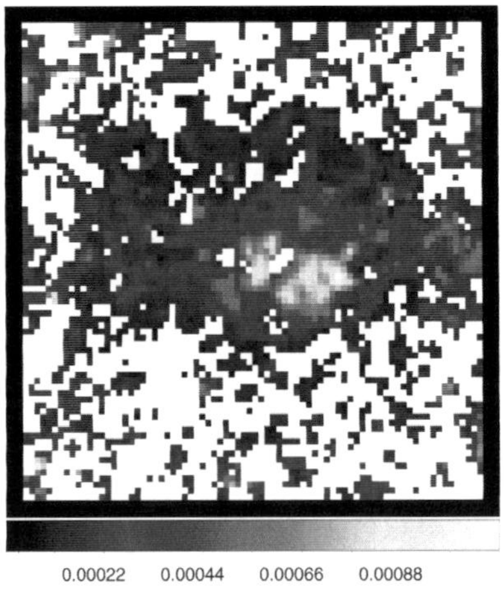

(d) Brγ narrow component linemap.

Figure 5.

Astronomy at High Angular Resolution 2011 IOP Publishing
Journal of Physics: Conference Series **372** (2012) 012016 doi:10.1088/1742-6596/372/1/012016

Measuring the stellar luminosity function and spatial density profile of the inner 0.5 pc of the Milky Way nuclear star cluster

Tuan Do[1], Andrea Ghez[2], Jessica R. Lu[3], Mark R. Morris[2], Sylvana Yelda[2], Gregory D. Martinez[1], Annika H. G. Peter[1], Shelley Wright[4], James Bullock[1], Manoj Kaplinghat[1], K. Matthews[5]

[1]Physics and Astronomy Department, UC Irvine, CA, USA
[2]Physics and Astronomy Department, UCLA, CA, USA
[3]Institute for Astronomy, University of Hawaii, HI, USA
[4]Dunlap Institute, University of Toronto, ON, Canada
[5]Caltech, Pasadena, CA, USA

E-mail: tuan.do@uci.edu

Abstract. We report on measurements of the luminosity function of early (young) and late-type (old) stars in the central 0.5 pc of the Milky Way nuclear star cluster as well as the density profiles of both components. The young (~ 6 Myr) and old stars (> 1 Gyr) in this region provide different physical probes of the environment around a supermassive black hole; the luminosity function of the young stars offers us a way to measure the initial mass function from star formation in an extreme environment, while the density profile of the old stars offers us a probe of the dynamical interaction of a star cluster with a massive black hole. The two stellar populations are separated through a near-infrared spectroscopic survey using the integral-field spectrograph OSIRIS on Keck II behind the laser guide star adaptive optics system. This spectroscopic survey is able to separate early-type (young) and late-type (old) stars with a completeness of 50% at K′ = 15.5. We describe our method of completeness correction using a combination of star planting simulations and Bayesian inference. The completeness corrected luminosity function of the early-type stars contains significantly more young stars at faint magnitudes compared to previous surveys with similar depth. In addition, by using proper motion and radial velocity measurements along with anisotropic spherical Jeans modeling of the cluster, it is possible to measure the spatial density profile of the old stars, which has been difficult to constrain with number counts alone. The most probable model shows that the spatial density profile, $n(r) \propto r^{-\gamma}$, to be shallow with $\gamma = 0.4 \pm 0.2$, which is much flatter than the dynamically relaxed case of $\gamma = 3/2$ to $7/4$, but does rule out a 'hole' in the distribution of old stars. We show, for the first time, that the spatial density profile, the black hole mass, and velocity anisotropy can be fit simultaneously to obtain a black hole mass that is consistent with that derived from individual orbits of stars at distances < 1000 AU from the Galactic center.

1. Introduction

The proximity of the Galactic center offers us the unique opportunity to study the nucleus of a typical spiral galaxy. In particular, the combination of high angular resolution imaging and spectroscopy enabled by ground-based adaptive optics systems have helped to realize this potential with detailed observations of the kinematics and photometry of individual stars within the central parsec. Recent findings include: measurement of the mass of the Milky Way supermassive black hole, Sgr A* [1, 2], and the spectroscopic separation of the young and old stars allowing study of their very different kinematic and spatial distributions [e.g 3, 4]. These observations challenge some long-standing theoretical expectations for a cluster with a massive black hole. In particular, the projected surface density radial profile of the old stars is very flat and inconsistent with the prediction of a steep stellar cusp around massive black holes within dynamically relaxed star clusters [e.g 3, 5–7]. In addition, not only are there numerous young stars ($\sim$ 6 Myr old) in an environment inhospitable to star formation, there may be evidence that the young stars at the Galactic center formed with a very top heavy initial mass function (IMF) compared to the standard IMF [4], and that these stars may be distributed in one or more kinematically distinct structures [8, 9].

These recent findings have motivated us to extend our integral-field spectroscopic survey in Do et al. [3] by a factor of 3 radially out to 0.5 pc from Sgr A* along the major axis of the projected disk plane of the clockwise disk of young stars [e.g. 9]. We obtain spectra for $\sim$ 500 stars with $K' < 15.5$ mag, which allow us to investigate the radial profile and luminosity functions of both the early-type (young) and late-type (old) populations. In Section 2, we discuss our new observations while in Section 3 we review our method for spectral typing. In Section 4, we adapt a method from Bayesian inference with star-planting simulations to spectral type the sources and establish the spectral completeness of our survey. In Section 5, we present the resulting surface number density profiles and K' luminosity functions of the early and late-type stars and in Section 5.2, we discuss the implications of these results for the mass function and origins of the young stars. We also use anisotropic spherical Jeans modeling with three-dimensional velocity measurements for the late-type stars to simultaneously derive the spatial density profile, black hole mass, and velocity anisotropy (Section 6).

2. Observations and data reduction

Near-IR integral-field spectra of the Galactic center were obtained between 2007 and 2010 using the OH-Suppressing Infra-Red Imaging Spectrograph (OSIRIS) in conjunction with the laser guide star adaptive optics (LGS AO) system on the Keck II telescope. The laser guide star was fixed at the center of the field of view, while low-order tip-tilt corrections were made on the R=13.7 mag star, USNO 0600-28577051, which is located $\sim 19''$ from Sgr A*. The primary observations for this work constitute a survey through the narrow-band filter Kn3 (2.121 to 2.220 μm) at a spectral resolution of R = 3800. This includes both observations initially reported in [3] and [10] and new 2010 observations, which extends from $4''$ to $14''$ (0.16 pc to 0.56 pc). We refer to this new extension of our survey as the Galactic Center OSIRIS Wide-field Survey (GCOWS). While our initial work, which covered only a $8'' \times 6''$ region centered on Sgr A*, used OSIRIS's 35 mas plate scale (field of view $1.58'' \times 2.24''$), the new GCOWS observations, which are at larger Galactic center radii where the stellar densities are lower, were obtained with 50 mas plate scale (field of view $2.25'' \times 3.2''$). This provided a good compromise between spatial resolution and field of view in this region. GCOWS covers a region of approximately $10'' \times 7.2''$ east of the survey in [10], along the projected disk plane of the clockwise disk of young stars [9]. Each of the new fields is observed with a six point dither pattern of 900 s per frame, in which the dithered frames have small ($\sim 0''.01$) offsets from one another. The larger plate scale allows us to reach a sensitivity comparable to the previous 35 mas plate scale observations, which have about 9 dithers per field. We also observe skies and an A0V and G2V calibration star after each night's Galactic center observations.

3. Spectral identification

Within the wavelength region (2.121 to 2.220 μm) chosen for this survey, the two prominent features useful for differentiating between the cool late-type giants and the hot OB stars are the Br γ and Na I doublet features. The old stars show strong Na I lines at 2.2062 and 2.2090 μm, but weak Br γ lines at 2.1661 μm. The O and B stars have no Na I lines, and show variable amounts of Br γ absorption depending on spectral type, with Br γ preferentially becoming stronger for B main sequence stars [See 11]. We group the stellar spectra by eye into three groups: 1) early-type, 2) late-type, or 3) unknown, as in Do et al. [3]. More details are given in Do et al. (in prep.). This sample of manually spectral-typed sources is complete to about 50% at $K' = 15.5$, when compared to sources detected from imaging (which are > 96% complete at this magnitude, see Lu et al. in prep.).

4. Completeness correction

In order to determine the luminosity function of the two populations of stars, we need to correct for the incompleteness of our spectral identification. As the luminosity function is an aggregate measurement of all the stars, we can correct for the incompleteness statistically by determining the probability that a given star is early or late-type. In order to assign probabilities for each star, we use a combination of star planting simulations as well as Bayesian inference based on the properties of stars that have been manually spectral-typed. The star planting simulations are used to sample the spatially varying sensitivity to early and late-type stars, which we use in our Bayesian inference to determine the probability of the type of star. While Do et al. (in prep) gives a detailed description, here we give a brief summary of our method:

- An automated method is developed to measure equivalent widths of key lines - in particular, the Br γ and Na I lines.
- We then use a faint ($K' > 14.0$) subset of the sample of manually typed stars to characterize the observed distribution of equivalent widths of the spectral features (Br γ and Na I) seen in the early-type stars compared to the late-type stars.
- The Bayes factor is computed to establish the relative strength of the evidence that an untyped star is either early-type or late-type by using the equivalent width of its spectral features with the distributions in equivalent widths in the previous step as priors.
- In order to translate the Bayes factor into a probability that a star is early or late-type, we plant either early and late-type spectra next to each untyped source. The simulations are repeated 100 times for each type of star. The simulated spectra are extracted and the Bayes factor is computed for all the simulated sources. Using these distributions, we determine our sensitivity for observing early type compared to late-type stars. We find that we are about twice as sensitive to late-type stars as compared to early-type stars at $K' = 15.5$. A half-sample boot-strap is used to determine the errors on these probabilities, which are small compared to Poisson errors.
- In order to obtain the final probabilities, we also include the fact that the two types of stars have different radial surface density profiles. For example, a random star far from Sgr A* would have a lower probability of being an early-type star compared to one that is found close to Sgr A*, since the surface density profile of early-type stars drop rapidly while those of late-type stars is flat [e.g. 3, 7]. This effect is included as a prior on the relative fraction of both types of stars as a function of radial distance from Sgr A*.

Using the above method, we assign a probability for all untyped sources; with these probabilities and the manually typed sources (which have a probability of 1), we have a complete statistical sample of the stars to measure their luminosity function and surface density profiles.

5. Results

5.1. K-luminosity function

We construct the luminosity function of the two populations by star counts in 0.5 magnitude bins. We use all detected stars in our luminosity function, which includes stars that were manually spectral-typed and the rest which have probabilities of being early or late-type. The manually typed population is about 50% complete in the $K' = 15.0 - 15.5$ magnitude bin. The manually typed sources are assigned a probability of 1 while the stars not manually spectral-typed contribute to the luminosity function based on their probabilities. The error bar in each luminosity bin is slightly larger than Poisson errors as they include errors in assigning probabilities (as determined by a half-sample bootstrap of the simulated sources). We also correct for variable extinction using the extinction map from [12] to a uniform A_K value of 2.74. Figure 1 shows the extinction and completeness corrected K′ luminosity functions for both types of stars.

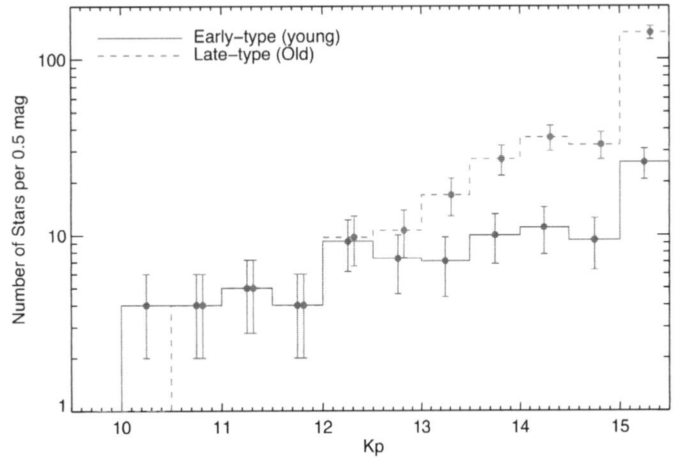

Figure 1. The extinction and completeness corrected K′ luminosity function for the early (blue, solid) and late-type (red, dashed) stars.

5.2. Implications for the mass function of the young stars

It is not trivial to derive an IMF from a luminosity function as there are many variables that can affect the luminosity beyond the distribution of mass. For example, the age, star formation history, and metallicity will all affect the transformation from mass to luminosity. In Lu et al. (in prep), a detailed analysis is performed using the data presented in this paper and a combination of stellar evolution and stellar atmosphere models. Nevertheless, it is illuminating to compare our luminosity function with those determined from previous observations. Compared to a similar study by [4], the luminosity function of the young stars presented here contains significantly more faint B-stars, resulting in a steeper slope for the luminosity function (see Figure 2). This increase in the number of fainter stars should translate to a steeper mass function. The difference in luminosity functions are important to resolve as the observations of [4] are one of only a few that show a significant departure from a standard IMF, with $dN/dm \propto m^{-0.45\pm0.3}$, which is very top-heavy compared to a Salpeter IMF of $dN/dm \propto m^{-2.35}$ [for a review of variations in IMF see 13]. There are two major difference between our study and that of [4]. First, we accounted for incompleteness by using a combination of star planting simulations around each untyped source and Bayesian inference using their spectra and location. In contrast, the authors of [4] planted early-type stars in data cubes and determined the recovery rate by measuring the CO index and applying the fraction recovered as the completeness correction to the young star counts. However, the difference between the number of early-type stars in the faintest bin at $K' = 15.5$ is so large that our result would only be consistent with that of [4] only if we do not apply any completeness correction. It is unclear at the present time how the different methods of completeness correction affect the stars, and more work will have to be done to compare the two methods. The second major difference between the two studies is that the present survey is conducted mainly along the disk plane of the clock-wise disk of young stars, whereas most of the Bartko et al. survey [4] is perpendicular to the disk plane. The differences in luminosity function may be due to differences in the mass function of young stars on the stellar disk and those outside. Explicitly comparing the luminosity function of the on-disk versus off-disk population may help to resolve this issue (see Yelda et al. in prep).

5.3. Surface density profile

We construct the radial surface density profile by averaging the density azimuthally in radial bins centered at Sgr A*. The radial width of the bins grow with radius so that they become larger to accommodate the falling radial density profile of the young stars. The radial surface density profile separated by early and late-type stars corrected for completeness is shown in Figure 3. The surface density profile of the late-type stars is very flat out to the edge of our survey. The completeness corrected surface density profile as a function of projected distance, R, from Sgr A* has a best fit power law, $\Sigma(R) \propto R^{-\Gamma}$, with $\Gamma_L = 0.02 \pm 0.14$ for late-type stars, and $\Gamma_E = 0.87 \pm 0.14$ for early-type stars.

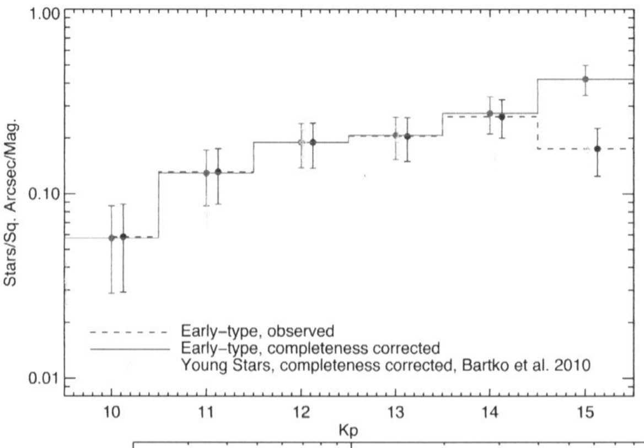

Figure 2. A comparison between the observed (dashed purple) and completeness corrected K' luminosity function of early-type stars (blue) with the completeness corrected K' luminosity function from Bartko et al. [4] (green) for stars located at a projected distance $R > 0.''8$ from Sgr A*. The number counts are normalized between our observations and those from Bartko et al. [4] in the K′ = 11.5 − 12.5 bin in order to compare the relative difference in the slopes. The points for the observed K′ luminosity function are shifted slightly for clarity. We find a significantly greater fraction of faint young stars than in [4].

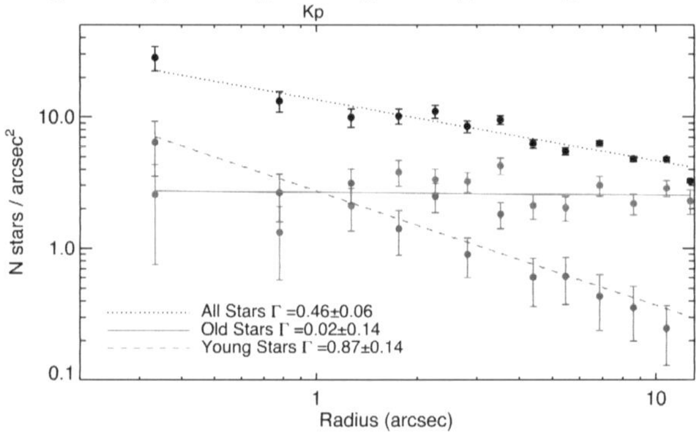

Figure 3. The azimuthally averaged and completeness corrected surface density profile of young stars (blue) and old stars (red). The completeness correction includes our relative sensitivity to both types of stars from star planting simulations, as well as the distance prior, and imaging completeness.

6. Jeans modeling of the nuclear star cluster

While the projected surface density profile of the late-type stars rules out a Bahcall-Wolf cusp, the true physical distribution of stars is difficult to constrain with only positional measurements due to the effects of projection onto the plane of the sky. In order to overcome projection effects, we utilize the kinematic measurements of the old late-type stars to constrain the cluster density profile. Similar to the methodologies described in the literature [14, 15], we use the Jeans equation to relate the velocity dispersion and structure of the cluster to the shape of the gravitational potential. The Galactic center is unique in that we have measurements of three velocity components for each individual star as well as two dimensions in position. Thus, we extend the above methodologies to include the full three dimensional velocity phase space [16].

As mentioned above, in order to relate the velocity and positional measurements, we will use the spherically symmetric Jeans equation that allows for anisotropy:

$$\frac{d(\rho_\star \sigma_r^2)}{dr} + 2\frac{\beta(r)\rho_\star \sigma_r^2}{r} = -G\rho_\star \frac{M(r)}{r^2}, \tag{1}$$

where σ_r is the dispersion in the radial direction (in the sense of spherical coordinates, not the radial velocity). The anisotropy, $\beta(r)$ is parameterized as $\beta(r) = \frac{\beta_0 + \beta_\infty (r/r_\beta)^\eta}{1+(r/r_\beta)^\eta}$, where β_o is the inner anisotropy, β_∞ is the outer anisotropy, r_β is the location of the transition, and η is the sharpness of the transition. The stellar spatial density profile, $\rho_\star(r)$ is defined to be a broken power law:

$$\rho_\star \propto r^{-\gamma}\left(1 + (r/r_b)^\delta\right)^{(\gamma-\alpha)/\delta}, \tag{2}$$

where γ is the inner power law slope, α is the outer slope, r_b is the break radius, and δ is the sharpness of the transition between the two slopes. The mass profile $M(r)$ is defined to be a point source with the mass of the black hole $M(r) = M_{BH}$; we find that including an extended stellar mass component does not significantly any of the resulting cluster parameters or the measured black hole mass, because our observations ($R < 0.5$ pc) are well within the sphere of influence of the black hole ($r_{infl} \approx 2$ pc). The total set of model parameters, including the bulk cluster velocity in cylindrical coordinates $\bar{v}_T$, $\bar{v}_R$, $\bar{v}_z$, and the normalization for the surface density profile A, are thus:

$$\mathcal{M} = \{\bar{v}_T, \bar{v}_R, \bar{v}_z, A, r_b, \alpha, \delta, \gamma, r_\beta, \beta_o, \eta, \beta_\infty, M_{BH}\}. \quad (3)$$

Given this set of model parameters, with the distance to the Galactic center fixed at $R_o = 8$ kpc, we use the Jeans equation to compute the radial dispersion and project it on the sky to compare to the observed velocity dispersions and their covariances. In order to avoid having to bin the data either radially or by velocity, thus losing information, we choose to compute the likelihoods for each source having an observed velocity vector $\mathbf{v}$ and a projected distance from Sgr A* of R, given the set of model parameters $\mathcal{M}$. The probability density function (PDF) for each individual star is defined as:

$$\mathcal{P}(\mathbf{v}, R|\mathcal{M}) \propto \frac{1}{\sqrt{|\mathbf{C}(R)|}} \exp\left[-(\mathbf{v} - \bar{\mathbf{v}}) \cdot \mathbf{C}(R)^{-1} \cdot (\mathbf{v} - \bar{\mathbf{v}})/2\right]. \quad (4)$$

Where $\mathbf{C}(R)$ is the covariance matrix for the velocity moments, including the errors for the three velocity components. The combined likelihood of the whole sample is thus:

$$\mathcal{L}(\mathcal{M}) = \prod_i \mathcal{P}(\mathbf{v_i}, R_i|\mathcal{M}) \quad (5)$$

over the data points $\{\mathbf{v}_i, R_i\}$. The $\mathcal{M}$ parameter space is explored in a Bayesian fashion utilizing a technique based on the MultiNest algorithm [17, 18]. Here, we assume the typical non-informative uniform priors on every parameter considered above.

6.1. Measuring positions and velocities

Since we would like to use the sample of stars with all three dimensions in velocity, we limit the kinematic analysis to only late-type stars with spectroscopy and radial velocity measurements with precision of < 100 km/s. This includes a sample of 248 stars. The proper motions (v_x, v_y) and positions (x, y) are from the results of Yelda et al. [19], which are determined from LGS AO images of the central $20'' \times 20''$ taken with NIRC2 instrument; the positions and velocities are measured in a reference frame defined by radio masers [19]. The line of sight velocity measurements (v_z) are obtained by cross-correlation of the spectra with an M3II stellar template (HD40239) from the SPEX telescope [20]. These radial velocities are then corrected for the local standard of rest.

6.2. Results: spatial density profile, black hole mass, and anisotropy

Given the assumption of spherical symmetry, no rotation, and $R_o = 8$ kpc, we determine the posterior PDFs given the observed positions and velocities. The resulting PDFs for the inner slope, black hole mass, and inner anisotropy is shown in Figure 4. The 1 σ central confidence intervals for the three parameters are: $\gamma = 0.4^{+0.2}_{-0.2}$, $M_{BH} = 4.0^{+0.5}_{-0.4} \times 10^6$ $M_\odot$, and $\beta_o = -0.2^{+0.3}_{-0.3}$. With the kinematic measurements, γ has been much better constrained than with the number counts alone (which was only able to place an upper limit of $\gamma < 1.0$ at the 3 σ level). The kinematic modeling also shows that the black hole mass is consistent with that derived by using the stellar orbit measurement of the star S0-2 [e.g. 1, 2]. Because the data are limited to only 0.5 pc, a few parameters such as the break radius of the density profile and the outer density slope are only weakly constrained. The anisotropy indicates that the cluster is tangentially anisotropic within the radial range of our data set. In Do et al. (in prep.), we will explore these issues further. Here we limit our discussion to the implications of our spatial density measurement.

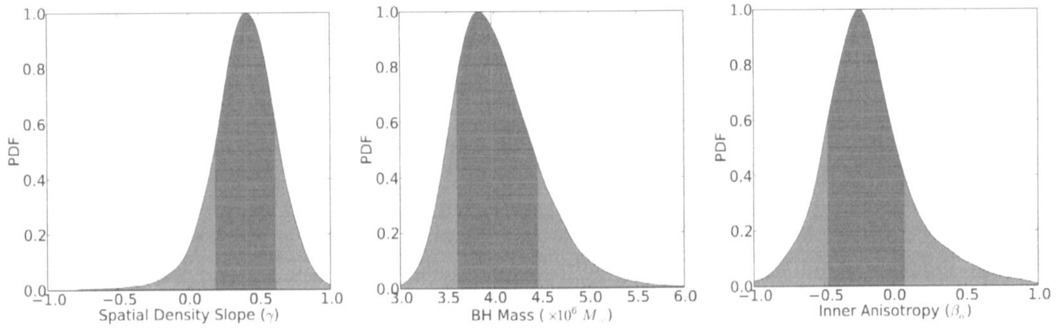

Figure 4. The posterior distribution for the inner slope (left), black hole mass (center), and the inner anisotropy. The line is the location of the central interval with the shaded regions encompassing 68% or 1 σ. The 1 σ central confidence intervals for the three parameters are: $\gamma = 0.4^{+0.2}_{-0.2}$, $M_{BH} = 4.0^{+0.5}_{-0.4} \times 10^6$ $M_\odot$, and $\beta_o = -0.2^{+0.3}_{-0.3}$.

7. Implications for the dynamical history of the MW NSC and black hole mass measurements

Having a measured spatial density profile is an important step in understanding the current dynamical state of the inner regions of the MW nuclear star cluster. Merritt [21] showed that the elimination of highly-bound stellar orbits generically leads to a density profile $\rho \sim r^{-0.5}$ at small radii if the distribution function is isotropic. Tightly bound orbits may be eliminated from the Galactic Center through the inspiral of an intermediate or a second supermassive black hole. A collision-dominated nucleus may also show density slopes as flat as $\gamma = 0.5$ [22]. These predicted density-profile slopes are close to our central value of $\gamma = 0.4$, and $\beta = 0$ is within the 68% credible interval. Thus, our results are consistent with the existence of a major dynamical event in the recent history of the Galactic Center.

Previous attempts at measuring the black hole mass from the star cluster have yielded systematically low black-hole mass estimates compared to the stellar orbit measurements [23–25]. These previous analyses did not always separate the old stars from the young, sometimes enforced isotropy of the stellar orbits, and often fixed the old-star density profile to $\rho \sim r^{-1.8}$ or the entire stellar population to $\rho \sim r^{2.5}$. These effects drive the black-hole mass estimates to be too low (Do et al., in prep.). For example, Schödel et al. [26] showed that by fixing the stellar density profile to the observed old stars, they were able to obtain a black hole mass consistent with the individual stellar orbits using proper motion measurements. For the first time, our results show that given a uniform sample of stars, it is unnecessary to fix either the cluster density profile or the black hole mass to obtain a simultaneous fit for the cluster parameters and the black hole mass through Jeans modeling.

Acknowledgments

A.H.G.P is supported by a Gary McCue Fellowship through the Center for Cosmology at UC Irvine. The infrared data presented herein were obtained at the W. M. Keck Observatory, which is operated as a scientific partnership among the California Institute of Technology, the University of California and the National Aeronautics and Space Administration. The Observatory was made possible by the generous financial support of the W. M. Keck Foundation. The authors wish to recognize and acknowledge the very significant cultural role that the summit of Mauna Kea has always had within the indigenous Hawaiian community. We are most fortunate to have the opportunity to conduct observations from this mountain.

References

[1] Ghez A M, Salim S, Weinberg N N, Lu J R, Do T, Dunn J K, Matthews K, Morris M R, Yelda S, Becklin E E, Kremenek T, Milosavljevic M and Naiman J 2008 ApJ **689** 1044–1062 (*Preprint*

0808.2870)

[2] Gillessen S, Eisenhauer F, Trippe S, Alexander T, Genzel R, Martins F and Ott T 2009 ApJ **692** 1075–1109 (*Preprint* 0810.4674)

[3] Do T, Ghez A M, Morris M R, Lu J R, Matthews K, Yelda S and Larkin J 2009 ApJ **703** 1323–1337 (*Preprint* 0908.0311)

[4] Bartko H, Martins F, Trippe S, Fritz T K, Genzel R, Ott T, Eisenhauer F, Gillessen S, Paumard T, Alexander T, Dodds-Eden K, Gerhard O, Levin Y, Mascetti L, Nayakshin S, Perets H B, Perrin G, Pfuhl O, Reid M J, Rouan D, Zilka M and Sternberg A 2010 ApJ **708** 834–840 (*Preprint* 0908.2177)

[5] Peebles P J E 1972 ApJ **178** 371–376

[6] Bahcall J N and Wolf R A 1976 ApJ **209** 214–232

[7] Buchholz R M, Schödel R and Eckart A 2009 A&A **499** 483–501 (*Preprint* 0903.2135)

[8] Paumard T, Genzel R, Martins F, Nayakshin S, Beloborodov A M, Levin Y, Trippe S, Eisenhauer F, Ott T, Gillessen S, Abuter R, Cuadra J, Alexander T and Sternberg A 2006 ApJ **643** 1011–1035 (*Preprint* arXiv:astro-ph/0601268)

[9] Lu J R, Ghez A M, Hornstein S D, Morris M R, Becklin E E and Matthews K 2009 ApJ **690** 1463–1487 (*Preprint* 0808.3818)

[10] Do T 2010 *Physical Processes in the Vicinity of a Supermassive Black Hole* Ph.D. thesis University of California, Los Angeles

[11] Hanson M M, Conti P S and Rieke M J 1996 ApJS **107** 281–+

[12] Schödel R, Najarro F, Muzic K and Eckart A 2010 A&A **511** A18+

[13] Bastian N, Covey K R and Meyer M R 2010 ARA&A **48** 339–389 (*Preprint* 1001.2965)

[14] Martinez G D, Minor Q E, Bullock J, Kaplinghat M, Simon J D and Geha M 2011 ApJ **738** 55 (*Preprint* 1008.4585)

[15] Martinez G D, Bullock J S, Kaplinghat M, Strigari L E and Trotta R 2009 J. Cosmology Astropart. Phys. **6** 14–+ (*Preprint* 0902.4715)

[16] Strigari L E, Bullock J S and Kaplinghat M 2007 ApJ **657** L1–L4 (*Preprint* arXiv:astro-ph/0701581)

[17] Feroz F, Hobson M P and Bridges M 2009 MNRAS **398** 1601–1614 (*Preprint* 0809.3437)

[18] Skilling J 2004 *American Institute of Physics Conference Series* (*American Institute of Physics Conference Series* vol 735) ed R Fischer, R Preuss, & U V Toussaint pp 395–405

[19] Yelda S, Lu J R, Ghez A M, Clarkson W, Anderson J, Do T and Matthews K 2010 ApJ **725** 331–352 (*Preprint* 1010.0064)

[20] Rayner J T, Cushing M C and Vacca W D 2009 ApJS **185** 289–432 (*Preprint* 0909.0818)

[21] Merritt D 2010 ApJ **718** 739–761 (*Preprint* 0909.1318)

[22] Murphy B W, Cohn H N and Durisen R H 1991 ApJ **370** 60–77

[23] Genzel R, Thatte N, Krabbe A, Kroker H and Tacconi-Garman L E 1996 ApJ **472** 153

[24] Genzel R, Pichon C, Eckart A, Gerhard O E and Ott T 2000 MNRAS **317** 348–374 (*Preprint* arXiv:astro-ph/0001428)

[25] Chakrabarty D and Saha P 2001 AJ **122** 232–241 (*Preprint* arXiv:astro-ph/0103467)

[26] Schödel R, Merritt D and Eckart A 2009 A&A **502** 91–111 (*Preprint* 0902.3892)

Simulations of the Galactic Centre Stellar Discs In a Warped Disc Origin Scenario

A Ulubay-Siddiki[1], **and H Bartko**[2]

[1]Istanbul University, Faculty of Science, Department of Physics, 34134, Vezneciler, Istanbul/Turkey
[2]Max Planck Institute for Extraterrestrial Physics, Giessenbachstr., 85748 Garching/Germany

E-mail: aulubays@istanbul.edu.tr

Abstract. The Galactic Center (GC) hosts a population of young stars some of which seem to form a system of mutually inclined warped discs. While the presence of young stars in the close vicinity of the massive black hole is already problematic, their orbital configuration makes the situation even more puzzling. We present a possible warped disc origin scenario for these stars, which assumes an initially flat accretion disc which develops a warp through Pringle instability, or Bardeen-Petterson Effect. By working out the critical radii and the time scales involved, we argue that disc warping is plausible for GC parameters. We construct time evolution models for such discs considering the discs' self-gravity, and the torques exerted by the surrounding old star cluster. Our simulations suggest that the best agreement for a purely self-gravitating model is obtained for a disc-to-black hole mass ratio of $M_{\rm d}/M_{\rm bh} \sim 0.001$.

1. Introduction

Our Galactic Centre hosts a supermassive black hole (SMBH), SgrA*, of mass $\sim 4 \times 10^6\ M_\odot$ [11, 13, 41, 14]. SgrA* is surrounded by a group of young stars [22, 12, 34, 27, 4, 5] as well as a cluster of old stars [41, 38]. Of the more than 100 young stars observed in the radial range ~ 0.05 pc to ~ 0.5 pc, nearly 60 populate a disc which rotates clockwise (CW) on the plane of sky [12, 25, 34, 27, 28, 5]. Of the rest, 20 stars seem to form an other counter clockwise (CCW) rotating disc which is highly inclined to the CW one [12, 34, 4, 5] (but also see [27]). Ages of these young stars are a few Myr, suggesting that there has recently been an episode of star formation in the Galactic Centre (GC).

The standard recipe of star formation in terms of the collapse of a molecular cloud seems unlikely to work for the GC environment due to the strong tidal field of the massive black hole. However theoretical studies point to the possibility of fragmentation of accretion discs which become self-gravitating beyond a few tenth of parsec [21, 10, 15, 32, 1]. Simulations of the GC show that such discs might form through the infall [6], or collision [19] of molecular clouds. On the other hand, an open question on the geometrical configuration of the discs remain, since a planar accretion disc, such as that assumed in the above mentioned works, most likely leaves behind a planar distribution of stars when it fragments, but recent observations provide evidence for a gradual 60° warp in the CW disc [4]. Therefore even though the stars might have formed in-situ, their current orbital distribution suggests an origin which is more sophisticated than a simple thin accretion disc [28]. Numerical simulations following the time evolution of such discs are also in line with this idea, and show that it is unlikely to perturb the stars to the high inclinations they are observed when they form on a cold flat accretion disc [8].

Warped gaseous discs at similar scales are observed in a few nearby AGN, such as NGC4258, NGC1068 and Circinus [18, 17, 16], which suggests that the GC might be representing a particular stage of a cycle during the lifetime of a typical spiral galaxy nucleus [31].

A planar accretion disc could become warped through various processes such as when torqued by a spinning black hole [3, 2, 26, 30], when exposed to radiation from a central source [35, 36, 37], or through vector resonant relaxation [7, 20] if the self-gravity torques do not dominate over the stochastic torques induced by surrounding stellar cusp.

In this work, we propose a scenario for the formation of a warped disc at the Galactic Centre considering radiation instability, and the Bardeen-Petterson Effect as possible warping mechanisms. We assume that when a supposed active phase of the Galactic Center ends, the disc can cool to fragment and form stars. Thereafter, the stellar disc continues its evolution in the gravitational field of the black hole, its self-gravity, and the surrounding old star cluster. We investigate the conditions for which this scenario could work, and we present results from our simulations of warped discs showing that a low mass model such as as inferred from the observations today fits the data reasonably well. A detailed discussion of our results can be found in [43].

2. Warping the Galactic Centre Disc

In this section, we discuss a plausible scenario for the origin of the warp observed in the orbits of the young stars. After assuming the past existence of an accretion disc around the Galactic Centre black hole, we investigate two well studied mechanisms: radiation pressure instability [35, 36, 37] and Bardeen-Petterson Effect [3] for warping the accretion disc. We further assume that the active phase of the GC ends some time, hence the warped disc can cool to form the stars observed today. Thereafter, the evolution of the disc is governed by gravitational torques only. In light of the recent observations presented in [41, 38], we account for the effects of a slightly flattened surrounding star cluster in addition to the disc's self-gravity.

2.1. Surface Density of the Disc Prior to Fragmentation

In order to determine the plausibly of disc warping prior to star formation, we need to evaluate a few parameters of the gaseous disc, which from the current observations can not be constrained. For those parameters, we mostly make use of the canonical values for AGN discs. However, one parameter, the surface density of the supposed past gaseous disc may be determined by the observations.

When the total number of stars in a stellar system is known, one can determine the total mass in stars by adopting an appropriate initial mass function (IMF), which for the GC discs is shown to be $\xi(M) = \xi_0 M^{-0.45}$ [5]. The total mass in disc stars for the GC reads: $M_\mathrm{s} \sim 5360 M_\odot$, translating into a mass ratio of $M_\mathrm{d}/M_\mathrm{bh} = 0.00134$ when the black hole mass is assumed to be $4 \times 10^6 M_\odot$. Once having evaluated the total mass of the current disc, one can then adopt a star formation efficiency, ϵ_SF, and determine the gaseous disc mass, and translate this into a surface density for which the warping criteria might be evaluated. The observations of the stellar discs indicate a $1/r^{1.4}$ decline of the surface density [5]. Adopting this, we found for the density of the parent molecular disc $\Sigma_{0.1} = 9.52 \times 10^4 \mu_{5360} \epsilon_\mathrm{SF}^{-1} \mathrm{M}_\odot \mathrm{pc}^{-2}$, with $\mu_{5360} = (M_\mathrm{s}/5360 M_\odot)$ [43].

2.2. Radiation Driven Warping of the Galactic Centre Disc

Warping of accretion discs by radiation is studied in detail by many authors [36, 29, 37, 33]. When an optically thick disc exposed to central radiation absorbs and re-emits the incident photons parallel to the disc local normal, it might acquire a warp if it is slightly distorted. The net torque competes with the $r - z$ stress in the disc, the result of which determines whether warping will be pronounced or not. The effect of the $r - z$ stress is to force the disc settle onto a plane on a time scale $t_{\nu_2} = 2r^2/\nu_2$, where ν_2 is the coefficient of vertical viscosity. The warp can grow if the settlement time scale is longer than the precession time induced by radiation. This is given as [37]

$$\frac{12\pi\Sigma r^3\Omega c}{L} \leq \frac{2r^2}{\nu_2}, \tag{1}$$

where L is the luminosity of the central source, c is the speed of light, and α is the Shakura-Sunyaev parameter [39]. For the magnitude of vertical viscosity, we refer to hydrodynamical simulations which give $\eta_\nu = \nu_2/\nu_1 < 3.5/\alpha$, and the radial viscosity coefficient ν_1 can be evaluated using standard equations for accretion discs in steady state [9]. Equation 1 can be rearranged to give the radiation warping critical radius [43]

$$R_{\rm rad} > \frac{2\eta_\nu^2}{\gamma_{\rm crit}^2\epsilon^2}\frac{2GM_{\rm bh}}{c^2} = 9.1 \times 10^{-3}\alpha^{-2}\epsilon_{0.1}^{-2}\ {\rm pc}, \tag{2}$$

where $\gamma_{\rm crit} \simeq 0.32$ [37], and ϵ denotes the radiative efficiency. Thus warping the GC disc at the distance of the young stars (0.03-0.5) pc requires $\alpha \sim 0.3$, or $\alpha \sim 0.1$.

2.3. Bardeen-Petterson Effect and the Galactic Centre Disc

When an accretion disc forms around a rotating (Kerr) black hole, its angular momentum might be misaligned with that of the black hole. The parts of the disc close to the hole experience the so-called Lense-Thirring precession [24] which results from the dragging of inertial frames. Since the viscous timescale at these parts of the disc are shorter than those for the outer parts, the inner disc aligns with the black hole spin when the outer parts are still on their original orbits. This way, the disc develops a warp beyond a critical radius $R_{\rm BP}$. The location of this radius is again determined by the competition between alignment timescale, i.e. viscous time scale in the vertical direction, and precession time scale induced by the Bardeen-Petterson Effect. Using steady state disc equations as given in [9], the precession rate given by [23], and solving for the critical radius $R_{\rm BP}$ we found in [43]

$$R_{\rm BP} = 4.22 \times 10^{-4}\left(\frac{a\epsilon\alpha\mu_{5360}}{\eta_{\rm edd}\epsilon_{\rm SF}}\right)^{1/2.4}\ {\rm pc} \tag{3}$$

for the GC disc where a is the black hole spin parameter, and $\eta_{\rm edd} \equiv L/L_{\rm edd}$, with $L_{\rm edd}$ the Eddington luminosity. This estimate shows that the Bardeen-Petterson radius for the GC is much below the inner edge of the stellar discs. A comparison of the angular momenta of the disc, and the black hole, in addition to the estimate of the alignment time scale suggests that if the disc fragmented to stars after it was warped through Bardeen-Petterson Effect, $\epsilon_{\rm SF} < 1$, and $\alpha < 0.1$ [43].

3. The Model and The Numerical Setup

In the previous sections we argued that for a range of plausible parameters, an initially flat disc around SgrA* could be warped prior to star formation. After that, the disc evolves gravitationally. To simulate the time evolution of the stellar disc, we model a warped disc as a collection of concentric circular rings which are mutually tilted, and are in gravitational interaction with each other. We also account for the torques induced by the surrounding old star cluster. Each ring around the black hole of mass $M_{\rm bh}$ is characterized by it mass m, radius r, and the Euler angles (ψ, θ, ϕ). ψ denotes the location of a star on its ring, where θ, and ϕ are the inclination and the azimuthal angles of the rings. The equations of motion, and the mutual gravity torques for such a system of rings are written as in [42], while for the torque induced by the star cluster, we use equation (2) given in [40].

In our simulations, we use 50 equally spaced rings to model a disc. We adopt two different mass fractions in order to enable our models to be applicable to systems other than the GC. We construct models that involve only the effects of the self-gravity (models SGM), and models which also account for the torques from the surrounding star cluster (models CLM). The indices 1 and 2 in the naming of the models refer to mass fraction of $M_{\rm d}/M_{\rm bh} = 0.00134$ and $M_{\rm d}/M_{\rm bh} = 0.0134$ respectively. For models CLM, the parameter controlling the strength of the cluster torque is set such that it could correspond to a mass of $10^6 M_\odot$ and a flattening of 0.9. The initial warp shapes of all the models is such that the disc has an overall warp of about 60° going from in-to-out, and the azimuths lie on a straight line. We note that it is justified to assume such high amplitude warps when considering both the radiation and Bardeen-Petterson warping as the driving mechanisms [37, 26].

4. Results

The self-gravity and the star cluster torques will have an effect of forcing the disc precess. The temporal evolution of the disc will then be governed by the relative strengths of both torques.

In figures 1 and 2 we show how the warp profiles, i.e. radius-vs-inclination- of the models SGM1 (left) and SGM2 (right) change throughout the simulations. The red lines depict the warp profiles at the start of the simulation, while the blue lines depict the final configuration. The dense black dots show individual ring inclinations at a particular time t. We see that while the warp shape of model SGM1 remains almost unchanged during the evolution, the more massive disc breaks into two pieces. Of these two distinct pieces which precess independently, the inner disc at negative inclinations exhibits an almost planar shape, while the outer one has a warp of about 10°.

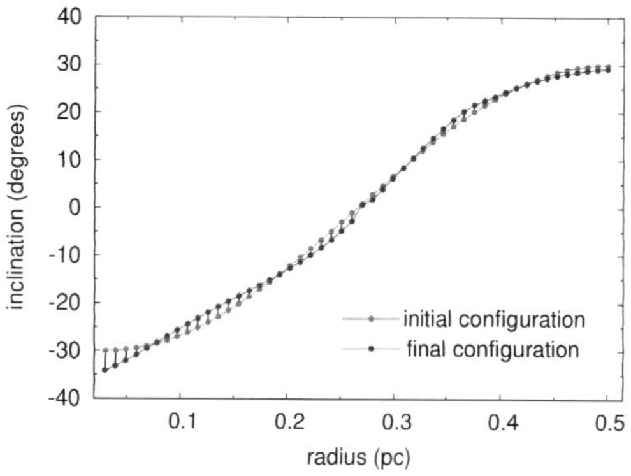

Figure 1. Warp profile of model SGM1 throughout the simulation. The inclination of the disc remains unaltered.

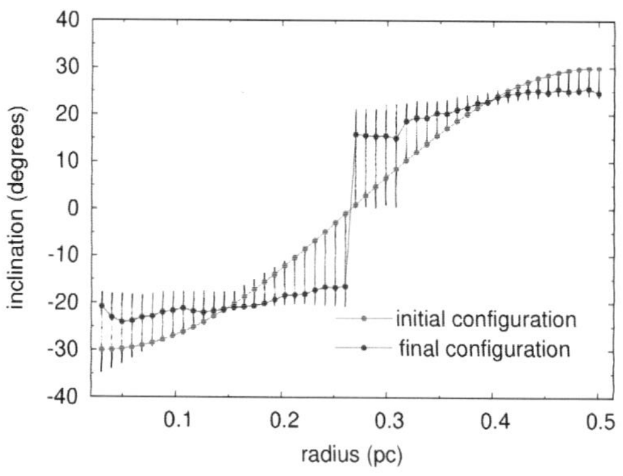

Figure 2. Warp profile of model SGM2 throughout the simulation. The disc breaks into two pieces.

In figures 3 and 4 we show the warp profiles of models CLM1 (left) and CLM2 (right). We see again that the lighter disc changes its inclination only slightly while the more massive one acquires a warp shape where the middle parts of the disc are pushed to more polar orbits, and the disc has a distorted shape.

In the different panels of Fig. 5, we show the xy projections of the 3-dimensional views of the models. The color code depicts the height of the disc (z-axis) above the plane. Fig. 5a shows the projection of the models at the start of the simulation. The rest of the figures are

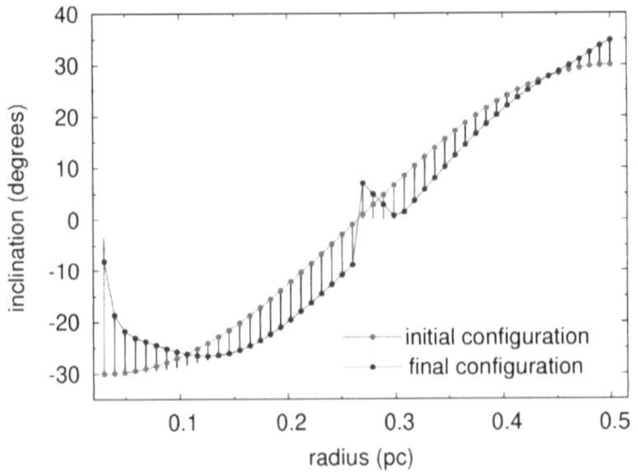

Figure 3. As in Fig. 1 but including the torques from the surrounding star cluster. The disc changes its inclination slightly due to the torques from the star cluster.

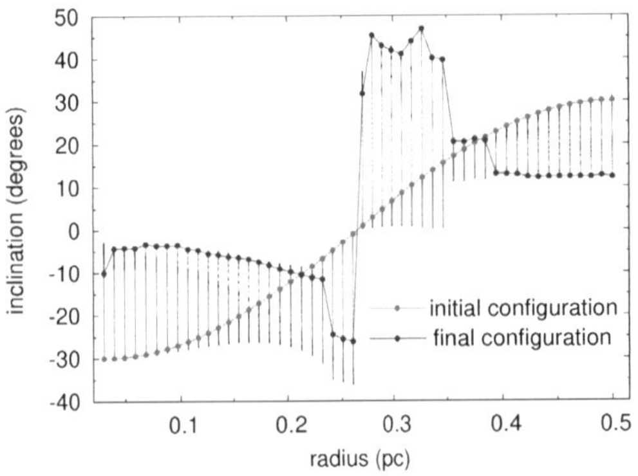

Figure 4. As in Fig. 2 but including the torques from the surrounding star cluster. The disc dissolves due to the torques from the star cluster.

for the final configurations. Fig. 5b is for model SGM1, while the figures 5c, 5d, and 5e are for models SGM2, CLM1, and CLM2 respectively. The precession of the model discs become apparent in these plots, where the self-gravity only model with a low mass fraction is seen to precess the least. The final configuration of this model is very close to its starting configuration. When the disc mass fraction is increased, we see that the coherence between the inner and the outer parts is lost, although these parts remain intact within themselves. When the influence of the surrounding star cluster is taken into account, the model discs precess faster compared to self-gravity only models, which manifests itself with the spiral-like projection of the orbits for the low mass model. When the disc mass is increased, the disc precesses with an even higher rate, and the disc dissolves due to differential precession.

5. Comparison with the Observations

In this section, we will briefly discuss how a comparison of our models with the observations can be made. In our comparison, we use the data presented in [4] which includes 136 stars with measured (x, y) positions, and the Monte Carlo simulated z positions. The respective space velocities are also given. We start by writing the coordinates of a star on its ring, and the components of the velocities. We then generate a random sample of stars on these rings by dicing the positions of the star on the ring, and obtain the phase space distribution of stars for each instant of time t [43].

As a next step, we generate the components of local average angular momenta as a function of average projected distance from the black hole, and rotate the data from the simulations successively until the best fit is obtained. Fig. 6 shows the best fitting warp profile which is obtained for model SGM1. The blue dots represent the data, where shaded are shows the region bound by the error bars. The red triangles depict the model. We see that the gradual change in the disc inclination is successfully produced. When the effects of the star cluster are also taken into account, the observed overall warp shape can not be produced by our simulations. This might mean either that the star cluster is nearly spherical, or that its mass is less than the value we assumed here based on the most recent observations reported in [41, 38]. We also note that, none of our simulations lead to a configuration where parts of a broken disc would appear in counter clockwise rotation on the plane of the sky to mimic the CCW disc.

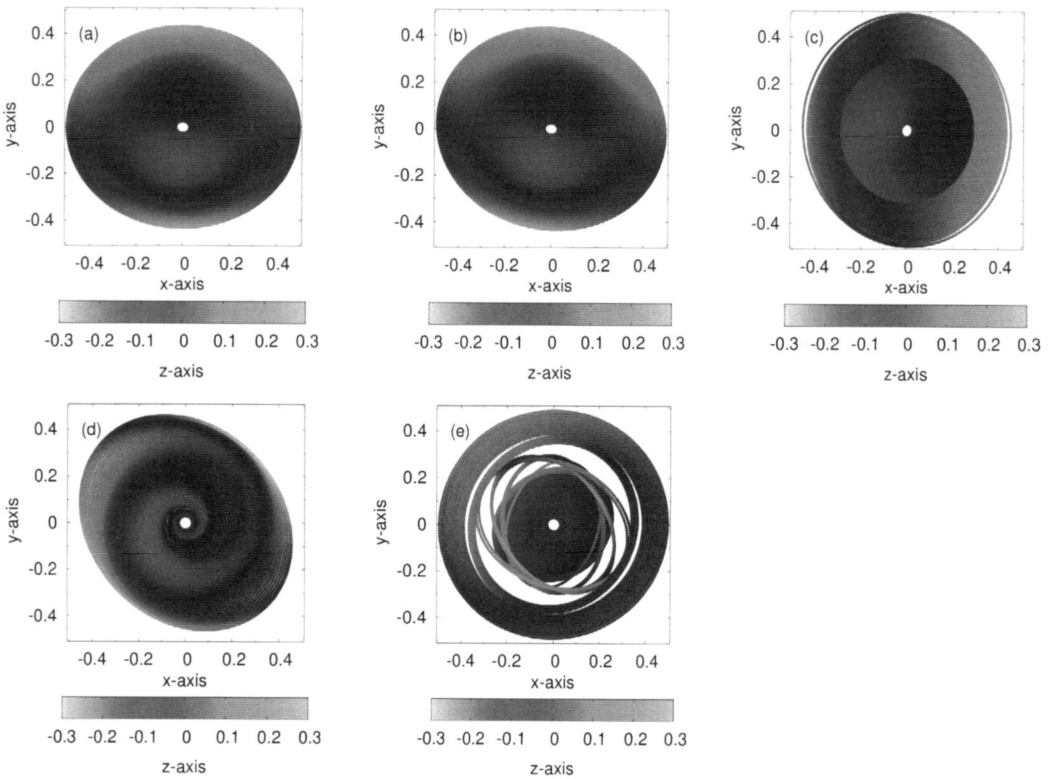

Figure 5. xy projections of the 3-dimensional views of the model discs. (a): for the initial configuration, (b): for model SGM1, (c): for model SGM2, (d): for model CLM1, (e): for model CLM2. The disruption of a massive disc under the influence of the star cluster can be clearly seen in Fig. 5e.

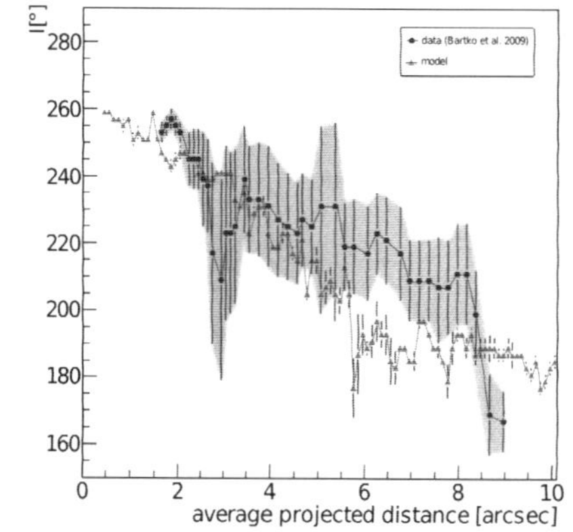

Figure 6. The warp profile of the best fitting model (red triangles) and of the CW disc (blue dots) given in [4]. Figure is from [43].

6. Summary

In this work, we have proposed a new origin scenario for the warped stellar discs of young stars at the Galactic Centre (GC). We assumed that the GC had an active phase during which the star formation was delayed until the past accretion disc was warped. In order to test the plausibility of this idea, we considered two warping mechanisms, the Pringle (radiation) instability, and the Bardeen-Petterson Effect. Using simple arguments, we showed that warping of the GC disc is plausible if the star formation efficiency is high, and the viscosity parameter $\alpha \sim 0.1$.

Keeping in view that the activity in the Galactic Centre has ceased some time ago after which the disc is subject to gravitational torques only, we followed the time evolution of warped discs taking into account the discs' self-gravity in the non-linear regime and the torques exerted by the surrounding old star cluster. In order to let our results be applicable to systems other than the GC, we examined for two different disc-to-black hole mass ratios (mass fractions) the long term temporal behavior of the warped disc models and investigated the interplay between the self-gravity and the cluster torques. We also compared our models with the observations of the warped GC discs. Our findings can be summarized as follows:

i) When the mass fraction is $M_{\rm d}/M_{\rm bh} \approx 0.001$, the overall inclination of the disc is not grossly altered during the simulations. When only the self-gravity of the disc is considered, the disc precesses as a single body throughout its evolution, while inclusion of the torques from the star cluster cause the rings precess apart.

ii) When the disc mass fraction is $M_{\rm d}/M_{\rm bh} \approx 0.01$, the disc breaks into two parts precessing independently in the case of a purely self-gravitating disc, and dissolves due to differential precession when the star cluster torques are taken into account.

iii) In our simulations, we do not find a configuration where parts of a broken disc would counter-rotate on the plane of the sky. Comparing our models with the observations, we see that the low mass model taking into account only self-gravity of the disc agrees better with the data. When the torques from the star cluster are also present, the overall warp shape can not be produced. This disagreement might mean either that the old star cluster at the Galactic Centre is nearly spherical, i.e. does not torque the disc, or that the mass of the cluster within the outer edge of the disc is less than the value assumed here based on the most recent observations. Future observations targeting the old star cluster will therefore help a better understanding of the evolution of the stellar discs.

Acknowledgements

A.U.S. thanks Ortwin Gerhard (MPE) for useful discussions and comments on the manuscript, and acknowledges the support by the Scientific Research Projects Coordination Unit of Istanbul University under project number UDP-16581.

References

[1] Alexander R D, Armitage P J, Cuadra J and Begelman M C 2008 *ApJ* **674** 927
[2] Armitage P J and Natarajan P 1999 *ApJ* **525** 909
[3] Bardeen J M and Petterson J A 1975 *ApJ* **195** L65+
[4] Bartko H *et al* 2009 *ApJ* **697** 1741
[5] Bartko H *et al* 2010 *ApJ* **708** 834
[6] Bonnell I A and Rice W K M 2008 *Science* **321** 1060
[7] Bregman M and Alexander T 2009 *ApJ* **700** L192
[8] Cuadra J, Armitage P J and Alexander R D 2008 *MNRAS* **388** L64
[9] Frank J, King A and Raine D J 2002 *Accretion Power in Astrophysics* by Juhan Frank and Andrew King and Derek Raine p 398 (Cambridge, UK: Cambridge University Press) February 2002
[10] Gammie C F 2001 *ApJ* **553** 174
[11] Genzel R, Pichon C, Eckart A, Gerhard O E and Ott T 2000 *MNRAS* **317** 348
[12] Genzel R *et al* 2003 *ApJ* **594** 812

[13] Ghez A M, Salim S, Hornstein S D, Tanner A, Lu J R, Morris M, Becklin E E and Duchêne G 2005 *ApJ* **620** 744
[14] Gillessen S, Eisenhauer F, Trippe S, Alexander T, Genzel R, Martins F and Ott T 2009 *ApJ* **692** 1075
[15] Goodman J 2003 *MNRAS* **339** 937
[16] Greenhill L J, Booth R S, Ellingsen S P, Herrnstein J R, Jauncey D L, McCulloch P M, Moran J M, Norris R P, Reynolds J E and Tzioumis A K 2003 *ApJ* **590** 162
[17] Greenhill L J and Gwinn C R 1997 *Astrophysics and Space Science* **248** 261
[18] Herrnstein J R, Greenhill L J and Moran J M 1996 *ApJ* **468** L17+
[19] Hobbs A and Nayakshin S 2009 *MNRAS* **394** 191
[20] Kocsis B and Tremaine S 2011 *MNRAS* **412** 187
[21] Kolykhalov P I and Syunyaev R A 1980 *Soviet Astronomy Letters* **6** 357
[22] Krabbe A *et al* 1995 *ApJ* **447** L95+
[23] Kumar S and Pringle J E 1985 *MNRAS* **213** 435
[24] Lense J and Thirring H 1918 *Physikalische Zeitschrift* **19** 156
[25] Levin Y and Beloborodov A M 2003 *ApJ* **590** L33
[26] Lodato G and Pringle J E 2007 *MNRAS* **381** 1287
[27] Lu J R, Ghez A M, Hornstein S D, Morris M, Matthews K, Thompson D J and Becklin E E 2006 *Journal of Physics Conference Series* **54** 279
[28] Lu J R, Ghez A M, Hornstein S D, Morris M R, Becklin E E and Matthews K 2009 *ApJ* **690** 1463
[29] Maloney P R, Begelman M C and Pringle J E 1996 *ApJ* **472** 582
[30] Martin R G 2008 *MNRAS* **387** 830
[31] Milosavljević M and Loeb A 2004 *ApJ* **604** L45
[32] Nayakshin S, Dehnen W, Cuadra J and Genzel R 2006 *MNRAS* **366** 1410
[33] Ogilvie G I and Dubus G 2001 *MNRAS* **320** 485
[34] Paumard T *et al* 2006 *ApJ* **643** 1011
[35] Petterson J A 1977 *ApJ* **216** 827
[36] Pringle J E 1996 *MNRAS* **281** 357
[37] Pringle J E 1997 *MNRAS* **292** 136
[38] Schödel R, Merritt D and Eckart A 2009 *A&A* **502** 91
[39] Shakura N I and Syunyaev R A 1973 *A&A* **24** 337
[40] Sparke L S 1986 *MNRAS* **219** 657
[41] Trippe S, Gillessen S, Gerhard O E *et al* 2008 *A&A* **492** 419
[42] Ulubay-Siddiki A, Gerhard O and Arnaboldi M 2009 *MNRAS* **398** 535
[43] Ulubay-Siddiki A, Bartko H and Gerhard O 2012 *MNRAS* submitted

Astronomy at High Angular Resolution 2011 IOP Publishing
Journal of Physics: Conference Series **372** (2012) 012018 doi:10.1088/1742-6596/372/1/012018

Dynamical evolution of the young stellar disc in the Galactic centre

Ladislav Šubr, Jaroslav Haas

Astronomical Institute, Charles University, V Holešovičkách 2, CZ-180 00 Praha, Czech Republic

E-mail: subr@sirrah.troja.mff.cuni.cz

Abstract. Origin of several hundreds of young stars in within the distance $\lesssim 1\,\mathrm{pc}$ from the Galactic supermassive black hole still represents an open problem of contemporary astrophysics. In this contribution we further investigate the model which assumes their formation in situ via fragmentation of a self-gravitating gaseous disc. We show that currently observed configuration of the system of young stars can be obtained as an outcome of a dynamical evolution of a single, initially very thin stellar disc. Our model assumes the long-term evolution of the stellar disc to be determined by gravitational influence of a distant molecular torus (CND) and mutual resonances of stellar orbits within the disc.

1. Introduction

High resolution observations of the Galactic Centre have brought surprising view of the nuclear star cluster. One of the unexpected discoveries is presence of a numerous population of massive young stars. Current observations distinguish up to ~ 200 of O and B giants, LBVs and Wolf-Rayet stars within a distance $\lesssim 1\,\mathrm{pc}$ form the central supermassive black hole (SMBH), identified as a radio source SgrA*. It has been soon realised (Genzel et al. 2000) that velocity dispersion of the young stars in the Galactic Centre (few tens of them were observed at that times) is strongly anisotropic. Later on, Levin & Beloborodov (2002) have shown that subset of them have coplanar orbits, forming a stellar disc. Nowadays, it is widely accepted that the young stellar disc, so-called clockwise stellar disc (CWS) spans from $\sim 0.04\,\mathrm{pc}$ to $\sim 0.5\,\mathrm{pc}$ and it consists of $\gtrsim 50$ stars. Rest of the stars is on orbits more or less scattered from the CWS plane. Due to lack of reliable knowledge of orbital parameters of individual stars, it is difficult to make strong statements about shape of the structure they form. Some authors (e.g. Genzel et al. 2003, Paumard et al. 2006) suggest presence of two stellar disc, some others (e.g. Bartko et al. 2009) speak about some kind of warped disc(s).

The most surprising fact about these young stars is their location in the vicinity of the SMBH. Standard formation scenario which assumes some kind of (quasi-)spherical collapse through the protostellar phase can hardly work due to strong tidal forces of the SMBH. Well known process of mass segregation that generally transports massive stars towards centres of star clusters acts on much longer time-scales than the life-time of the observed massive stars. Already in the work of Levin & Beloborodov (2002) it has been suggested that the stars may have been formed in situ via fragmentation of a massive gaseous disc. Alternate theories of their origin are based on some transportation mechanism that may have brought them towards the SMBH from more

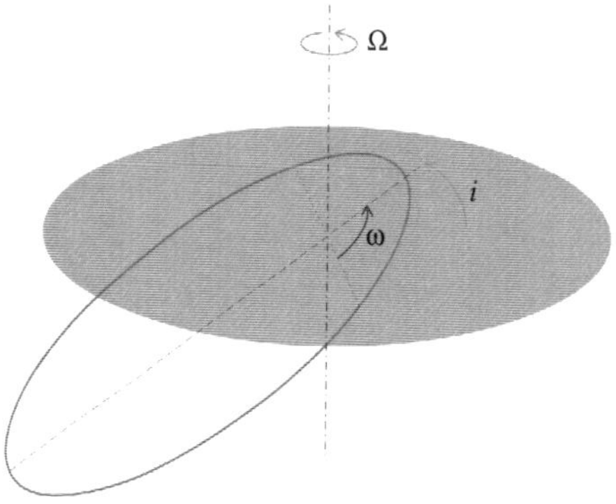

Figure 1. Angles determining orientation of a Keplerian orbit in space.

distant region where their formation is considered to be less tricky. Both classes of models apparently fail to explain some observed properties of the set of young stars in the Galactic Centre. While the in situ formation scenario naturally places stars into the disc-like structure, it has a difficulty to explain the fact that at most only a half of the observed young stars are orbiting the black hole coherently, while the other half is on apparently randomly oriented orbits. Moreover, observational data indicate that orbits of a substantial fraction of the young stars have moderate or large eccentricities ($e \gtrsim 0.5$) while the model more naturally leads to low eccentricities. The main drawbacks of the 'transportation' model is the fact that the parent star cluster would be tidally disrupted already at larger distances from the SMBH, unless it is strongly bound together e.g. via an intermediate-mass black hole. However, observations indicate neither young stars to be present at distances above 1 pc from SgrA*, nor any evidence for an IMBH within the central parsec.

Our work aims to support the hypothesis of star formation in a single thin disc. By means of numerical N-body simulations we show that such a system would evolve into a currently observed state provided it is influenced by a distant massive axisymmetric source of gravity. This setup is justified by an observed gaseous torus (so-called circum-nuclear disc, CND) which orbits the central SMBH at a distance of ~ 1.5 pc.

2. Motion in the perturbed Keplerian potential

In our analysis, we consider gravitational field in the central parsec of the Galaxy to be dominated by the supermassive black hole. Hence, on short time-scales, the stellar motions follow Keplerian ellipses. Their shape is described by semi-major axis, a, and eccentricity, e. Orientation of the orbital plane is determined by two angles – inclination, i, with respect to some reference plane and longitude of the ascending node, Ω (see Fig. 1). One more angle, argument of pericentre, ω, describes orientation of the orbit within its plane and, finally, true anomaly determines position of the body on the orbit.

We further assume that, on longer time-scales, the stellar orbits evolve due to

(i) gravity of a spherical cusp of late-type stars

(ii) gravity of a distant gaseous torus (CND) which is modelled by a massive ring

(iii) mutual torques of stellar orbits from the disc.

2.1. External perturbations

Secular evolution of Keplerian orbits under the influence of a distant massive perturber has been extensively studied in the past. Considering the molecular torus of mass $10^5 M_\odot \lesssim M_{\rm CND} \lesssim$

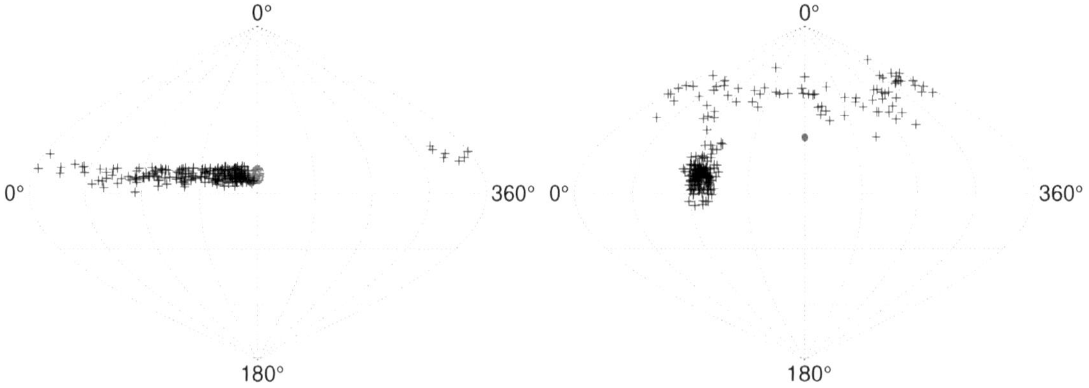

Figure 2. Sinusoidal projection of angular momenta of individual stars. Latitude corresponds to inclination with respect to the CND; longitudinal angle is Ω. Initial angular momenta are confined within the shaded regions with half-opening angle of 5° and 2° in the left and right panels, respectively. Directions of individual orbits after $\sim 7\,\mathrm{Myr}$ of dynamical evolution are marked with crosses. Main constituents of the gravitational field are the central SMBH of mass $4\times10^6 M_\odot$, CND of mass $0.2M_\bullet$ and radius 1.6 pc and a spherical cluster of mass $0.1M_\bullet$ within R_{CND}, which is modelled by smooth potential. In the left panel, stars are treated as test particles and their mean initial inclination with respect to the CND is 80°. In the right panel, all stars have equal masses of $M_\star = 40\,M_\odot$ and the initial mean inclination is 60°. The initial stellar disc has a surface density profile $\propto r^{-2}$ within the interval $\langle 0.04\,\mathrm{pc},\ 0.4\,\mathrm{pc}\rangle$.

$10^6 M_\odot$ and typical radius $R_{\mathrm{CND}} \approx 1.6\,\mathrm{pc}$ as the main perturbation to the dominating Keplerian potential of the central SMBH, we may directly apply well established analysis of a hierarchical triple systems as introduced by Kozai (1962) and Lidov (1962). Note that the secular theory of hierarchical triple systems uses averaging technique for description of potential of the outer body, i.e. its outcomes are directly applicable to our case in which the outer body is replaced with a ring of radius R_{CND} and mass M_{CND}. In general, the star on an orbit with semi-major axis $a \ll R_{\mathrm{CND}}$ would undergo periodic oscillations of inclination and eccentricity, coupled with oscillation or rotation of the argument of pericentre, ω, and monotonic decrease of the longitude of the ascending node, Ω (angles are measured with respect to the plane of the outer body, i.e. plane of the CND in our model). Characteristic time for the secular evolution is then

$$T_{\mathrm{K}} \equiv \frac{M_\bullet}{M_{\mathrm{CND}}} \frac{R_{\mathrm{CND}}^3}{a\sqrt{GM_\bullet a}} \,. \tag{1}$$

In contrary, an extended spherically symmetric source of perturbation which may be represented by the cusp of late-type stars in the Galactic Centre, leads solely to (negative) pericentre advance (i.e. rotation of the angle ω). It has been shown by various authors that combination of both spherical and axisymmetric perturbations may lead to significant damping of 'Kozai' oscillations of eccentricity and inclination. In Šubr & Schovancová (2008) and Šubr et al. (2009) we have shown that under the assumption of massive enough spherical cusp, the only remaining effect of the gravitational influence of the CND upon the orbital planes is monotonic precession around the symmetry axis of the CND with frequency

$$\dot{\Omega} = -\frac{3}{4}\cos i\; a^{3/2}\,\frac{\sqrt{GM_\bullet}}{R_{\mathrm{CND}}^3}\,\frac{M_{\mathrm{CND}}}{M_\bullet}\,\frac{1+\frac{3}{2}e^2}{\sqrt{1-e^2}} \,. \tag{2}$$

Strong dependence of $\dot{\Omega}$ upon the orbit semi-major axis and inclination with respect to the CND (for $i \rightarrow 90°$) led us to formulate a hypothesis that all young stars in the Galactic Centre (except for the S-stars) have been formed in a single gaseous disc. Due to differential precession of individual stellar orbits in the compound gravitational field of the CND and the spherical cusp, the outer and/or less inclined stars have been stripped out of the parent structure. The innermost parts which were less subject to the precession are distinguished as the CWS today. Due to strong dependence of the orbital precession upon the inclination with respect to the CND, this model requires initial inclination of the disc close to 90°. Differential precession than leads to a specific warped structure with angular momenta lying nearly in the plane of the CND (see left panel of Fig. 2)

2.2. Mutual torques of stellar orbits

The above formulated hypothesis of the origin and subsequent orbital evolution of young stars in the Galactic Centre is based on an assumption of negligible mutual interaction of stars which are treated as test particles moving in the external gravitational field. Consequently, their inclination with respect to the CND remains unchanged which leads to a specific pattern of projection of their angular momenta on the sphere. A natural step forward is to consider mutual gravitational interaction of stars from within the disc. In Haas et al. (2011b) we have introduced a set of equations describing secular evolution of two interacting circular orbits in the external field of a distant ring and a spherical cusp. Underlying perturbative Hamiltonian theory is a straightforward variation of the approach of Kozai (1962) and Lidov (1962) mentioned above. We have shown that initially coplanar orbits may interact in two qualitatively different regimes:

(i) In the case of weak interaction, the orbits precess independently around the symmetry axis with frequencies $\dot{\Omega}(a)$ and $\dot{\Omega}(a')$. On time-scale $2\pi/|\dot{\Omega}(a) - \dot{\Omega}(a')|$, they periodically exchange angular momentum which leads to oscillations of their inclinations.

(ii) Strong interaction regime may occur if mass of the stars is large and/or their orbits are close enough. In this case, they also exchange periodically some part of their angular momentum, but the inclination oscillations have generally smaller amplitudes and their precession is locked together.

Fig. 3 shows examples of evolution of two interacting stellar orbits in the external perturbative field of the CND. In two upper panels we plot evolution of inclinations and longitudes of ascending nodes for orbits interacting in the strong regime, while lower panels show weakly interacting system. Difference between the two cases are just masses of the interacting stars, while other parameters are kept unchanged.

We have further generalised our semi-analytical treatment to larger number of interacting stars. It appears that the two basic modes of interaction can be still found in the evolution of the N-body system. In particular, groups of stars interacting within the strong regime keep close values of i and Ω (i.e. orientations of their orbits). On the other hand, several, internally strongly interacting, groups of stars may interact in the weak regime with each other which leads to secular change of their orientation – see Fig. 4 for an example with four stars.

We argue that mutual torques among orbits played an important role in evolution of the young stellar system in the Galactic Centre. The denser inner part of the disc may have survived differential torque from the CND even for moderate initial inclinations. Moreover, due to the weak regime interaction with the outer parts, the surviving inner disc gradually changed its orientation towards higher inclination with respect to the CND which is in accord with present-day mutual orientation of the CND and the CWS. Right panel of Fig. 2 demonstrates this behaviour in terms of projections of angular momenta of the stellar orbits. The system presented here was integrated by means direct N-body integrator NBODY6 with the central SMBH being represented by an external Keplerian potential. The spherical stellar cusp was

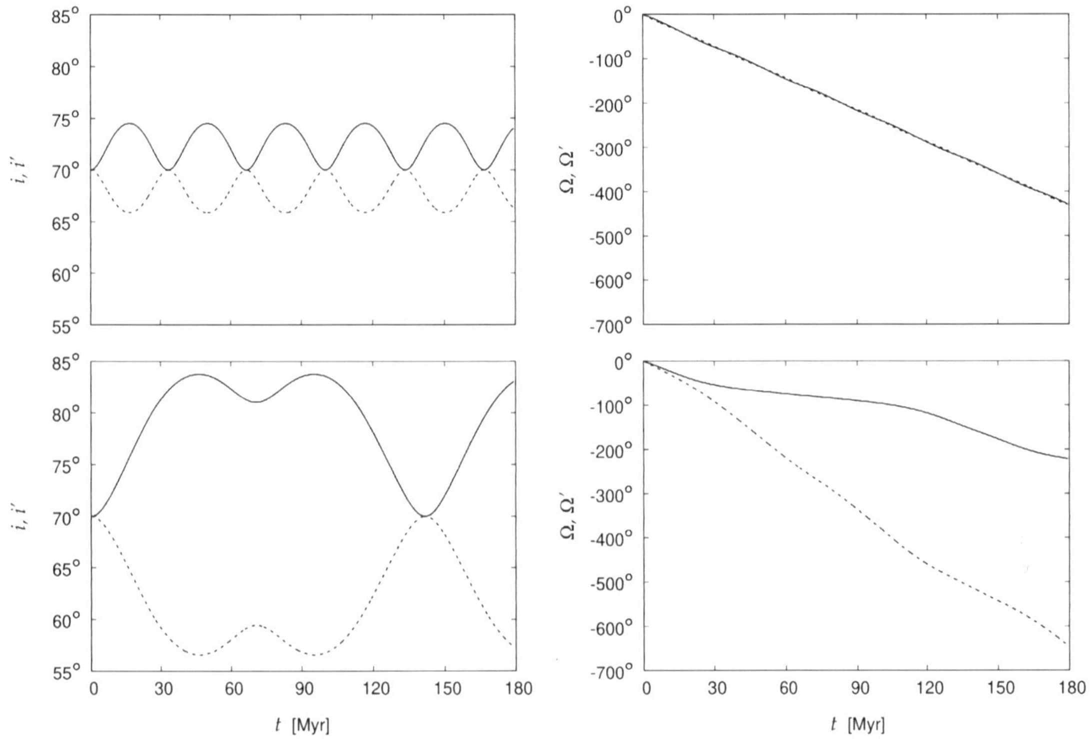

Figure 3. Temporal evolution of inclination and longitude of the ascending node for two mutually interacting stars in a perturbative gravitational field of a massive ring (CND). Upper panels describe to stars in the strong interacting regime (coupled precession) while the bottom ones correspond to weakly interacting orbits. Common parameters are: mass of the central SMBH $M_\bullet = 4 \times 10^6 M_\odot$, $M_{\rm CND} = 0.3\,M_\bullet$ and $R_{\rm CND} = 1.6\,$pc. Semi-major axis of the inner orbit (solid lines) is $a' = 0.04\,R_{\rm CND}$, while the outer one (dashed lines) is at $a = 0.05\,R_{\rm CND}$. Masses of stars are $M = M' = 36\,M_\odot$ and $M = M' = 20\,M_\odot$ in the upper and lower panels respectively.

approximated by an additional external potential corresponding to the Bahcall & Wolf (1976) radial density profile $\rho(r) \propto r^{-7/4}$ with total mass $0.1 M_\bullet$ within $R_{\rm CND}$. In accord with the findings of the perturbative Hamiltonian theory, we modelled the CND by a single massive particle orbiting the SMBH on a circular orbit of radius $R_{\rm CND} = 1.6\,$pc. Finally stellar disc was set up as 200 gravitating particles of equal mass $M_\star = 40\,M_\odot$ which were generated randomly on circular orbits with semi-major axis distribution $n(a) \propto 1/a$ within the interval $\langle 0.04\,\mathrm{pc},\, 0.4\,\mathrm{pc}\rangle$. Half-opening angle of the disc was 2° and it was inclined 60° with respect to the CND. Right panel of Fig. 2 shows both the initial state and a snapshot at $t = 7\,$Myr. Approximately one half of the stars still forms a coherently rotating structure which changed its orientation to nearly perpendicular with the CND. Second half was stripped from the parent structure. Hence, considering mutual gravitational influence of stars from the disc leads to the state compatible with the observations for much wider range of initial conditions than the model in which the stars are treated as test particles (left panel of Fig. 2). In the latter case, the parent disc has to start from orientation close to perpendicular with respect to the CND as its inclination does not change in time. Moreover, disc of test particles is much more vulnerable to destruction through differential precession which is stronger for lower inclinations.

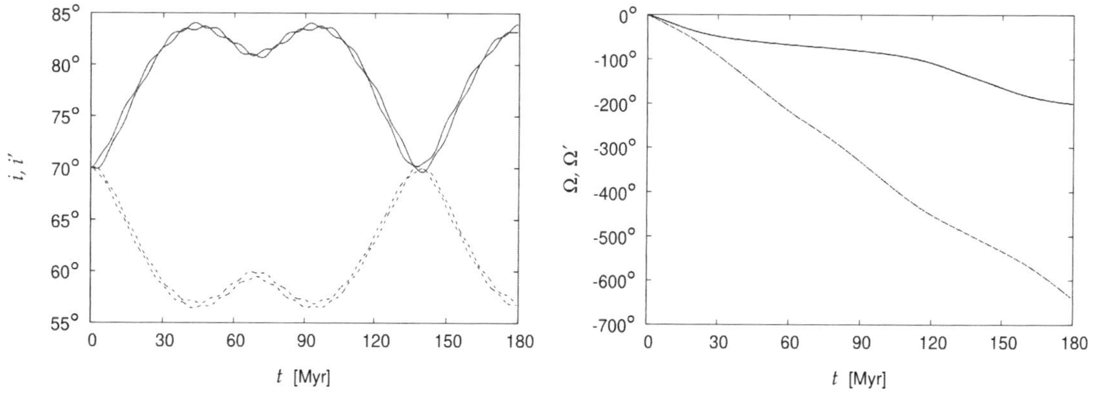

Figure 4. Evolution of elements, i and Ω, determining orientation of orbits of four mutually interacting stars. Parameters of the system are identical to the lower panel of Fig. 3 except for masses of the stars which are all equal to $10\,M_\odot$ and radii of their orbits which are set to $a_1 = 0.0373\,R_{\rm CND}$, $a_2 = 0.0408 R_{\rm CND}$, $a_3 = 0.0478\,R_{\rm CND}$ and $a_4 = 0.0511\,R_{\rm CND}$. Pairs of close stars interact in the strong regime, forming two couples which are mutually in the weak interaction regime.

2.3. Two-body relaxation

The mutual orbit interaction described in the previous section operates on time-scales much longer than the orbital period. In the rare case of random close encounters, it ceases to be valid. Two-body stellar relaxation within the sphere of influence has been already discussed in the literature (Rauch & Tremaine 1996, Hopman & Alexander 2006). A general outcome for this kind of systems is that so called resonant relaxation which leads to change of orbital orientation and angular momentum is more efficient than classical two-body relaxation which, in addition, changes energy of stars. However, this generic statement was formulated under the consideration of spherical symmetry. The situation is different in the case of coherently rotating stellar disc. Our N-body simulations show (see also Haas & Šubr 2012) that energy of the stars from the disc may change significantly. This is definitely due to large stellar densities, exceeding by several orders of magnitude stellar density of the spherical cusp. Fig. 5 shows the effect of two-body relaxation by means of change of eccentricity and semi-major axis distributions for the model of stellar disc evolving under the influence of spherical cusp and the CND. Complementary work of Haas & Šubr (2012) shows that identical result is obtained without both perturbations to the central Keplerian potential, i.e. it is solely a consequence of the intrinsic relaxation of the stellar disc.

3. Conclusions

We have set up a model of stellar disc in the Galactic Centre which assumes its origin in a single thin coherently rotating structure. We argue that such an initial configuration may lead to a state comparable to the currently observed system of young stars in the Galactic Centre. The important agents which are capable to transform the initial state into the final one within ~ 7 Myr are the massive molecular torus, which, on one hand, leads to partial destruction of the parent disc (i.e. it pushes some stars from the disc to randomly oriented orbits on which they are observed now). On the other hand, together with mutual torques of stellar orbits, the CND leads to secular evolution of the orientation of the surviving disc towards nearly perpendicular which is in accord with the observations. The spherical stellar cusp plays an important role in damping the Kozai oscillations induced by the CND, i.e. preventing the stellar disc from

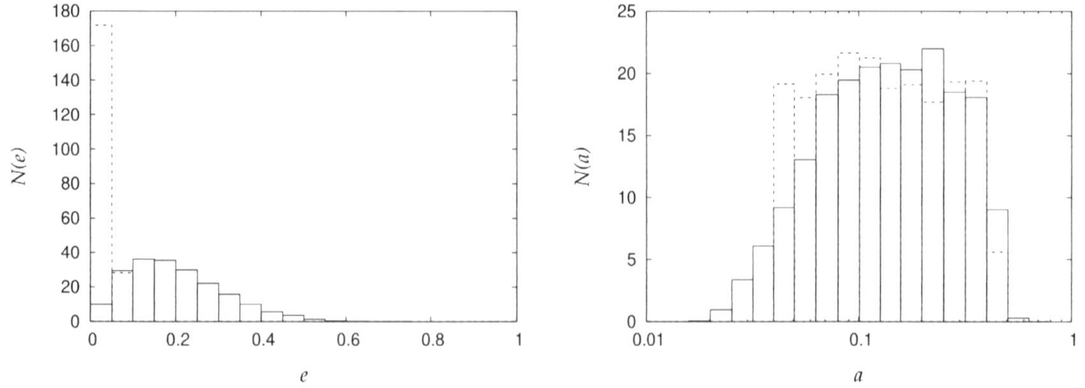

Figure 5. Initial (dashed) and final (solid lines) counts of eccentricities and semi-major axes for the self-gravitating stellar disc evolving under the influence of the CND (corresponds to the model in the right panel of Fig. 2.

complete destruction within its life-time. Two-body relaxation among the stellar disc orbits may lead to modification of the initial semi-major axis distribution and is also capable of pushing some stars to moderate or high eccentricities.

Current orbital state of the young stars in the Galactic Centre is, up to several exceptions, still not known exactly. This uncertainty naturally propagates back to the time of their formation. As a final remark let us note that regardless of the model of their origin, dynamical evolution during the life-time of the young stars has to be taken into account in order to establish correct relation between (current) observations and the initial state. We have shown that realistic external sources of gravity may play crucial role in the dynamical evolution of the young stars. Hence, good knowledge of parameters of the CND and cusp of late-type stars is important for proper answer to the question of the origin of young stars in the Galactic Centre.

Acknowledgments
This work was supported by the Research Program MSM0021620860 of the Czech Ministry of Education.

References
[1] Bahcall J. N., Wolf R. A., 1976, ApJ 209, 214
[2] Bartko H., Martins F., Fritz T. K., Genzel R., Levin Y., 2009, ApJ 697, 1741
[3] Genzel R., Pichon C.,Eckart A., Gerhard O. E., Ott T., 2000, MNRAS 317, 348
[4] Genzel R., Schödel R., Ott T., Eisenhauer F., Hofmann R. et al., 2003, ApJ 594, 812
[5] Haas J., Šubr L., Kroupa P., 2011a, MNRAS 412, 1905
[6] Haas J., Šubr L., Vokrouhlický D., 2011b, MNRAS 416, 1023
[7] Haas J., Šubr L., 2012, JPhCS
[8] Hopman C., Alexander T., 2006, ApJ 645, 1152
[9] Kozai Y., 1962, AJ 67, 591
[10] Levin Y., Beloborodov A. M., 2003, ApJ 590, L33
[11] Lidov M. L., 1962, Planetary and Space Sci. 9, 719
[12] Paumard T., Genzel R., Martins F., Nayakshin S., Beloborodov A. M. et al., 2006, ApJ 643, 1011
[13] Rauch K. P., Tremaine S., 1996, NewA 1, 149
[14] Šubr L., Schovancová J., 2008, JPhCS 131, a2013S
[15] Šubr L., Schovancová J., Kroupa P., 2009, A&A 496, 695

Ices in the Galactic Center?

Jihane Moultaka[1], **Andreas Eckart**[2], **Koralka Mužić**[3]

[1]IRAP, Observatoire Midi-Pyrénées, CNRS, Université Toulouse III, 14 avenue Edouard Belin, 31000 Toulouse

[2]I- Physikalisches Institut, Universität zu Köln, Zülpicher Str. 77, 50937 Köln, Germany

[3]Department of Astronomy and Astrophysics, University of Toronto, 50 ST George Str., Toronto ON M5S 3H4, Canada

E-mail: `jihane.moultaka@irap.omp.eu`

Abstract. We present for the first time, an infrared data-cube of the central parsec of our Galaxy in the 2.8 to 4.2 micron range. This spectral band (the L-band) harbors important signatures of the interstellar and circumstellar medium, since the water ice absorption feature at 3 micron traces the dense medium and the hydrocarbon absorption at 3.4 micron is only observed in the diffuse gas.

Thanks to a calibrator spectrum of the foreground extinction in the L-band derived in a previous paper, we corrected our data-cube for the line of sight extinction. Our observations performed with ISAAC spectrograph at the VLT-ESO telescope suggest that part of the absorption features takes place in the local environment of the Galactic center. This induces the presence of very low temperatures in the central parsec.

1. Introduction

The Galactic Center (GC) is the ideal place where one can study at high angular resolution the direct environment of a central galactic super massive black hole (SMBH). In this work, we present the first results of our study of the half parsec diameter area around the central SMBH SgrA* of our Galaxy (Schödel et al. 2002, Ghez et al. 2003).This study is conducted in spectroscopy in the mid-infrared, from about 2.8 to 5.1 μm.

The present spectral region is of great interest since it harbors prominent absorption features and emission lines that are signatures of the interstellar and circumstellar media (ISM and CSM, respectively). The ISO SWS spectrum of the central region (Lutz et al 1996) shows that the features are actually present in that direction towards SgrA*. Studying the distribution of these features allows to derive valuable information on the physics of the region.

In the L-band domain (from 2.8 to 4.2μm), in addition to the Hydrogen and Helium emission lines that are signatures of the ionized gas and Wolf-Rayet stars, one can distinguish a broad feature at 3μm, due to the presence of water ices in the dense ISM. This feature has been often observed towards Young Stellar Objects (YSOs) (e.g. Ishii et al. 1998). Hydrocarbon absorption features at 3.4μm and 3.48μm, can also be distinguished in this spectral domain. They are generally observed exclusively in the diffuse medium.

In the M-band domain (from 4.4 to 5.1μm), the integrated spectrum of the Galactic center shows also Hydrogen recombination lines as well as two other absorption line complexes attributed to the ^{12}CO and ^{13}CO in gas phase and an absorption line of the ^{12}CO in its solid phase at 4.676μm. Moreover, a broad feature at 4.62μm called XCN, is attributed to CN bonds.

Paradoxally, the central parsec has been very little studied in these bands although the latter harbor important signatures from the ISM and CSM. The only few studies that were made, were done either at low spatial resolution (e.g. Moneti et al. 2001), or at low spectral resolution (e.g. McFadzean et al. 1989). In all cases, only a small number of individual sources were observed (Chiar et al. 2000, 2003).
In general all previous studies resulted in the idea that the previous absorption features arise from molecular clouds along the line of sight towards the Galactic Center.

2. Observations and extinction calibration

Some years ago, we initiated L- and M- band observations of the central parsec of our Galaxy using the capabilities of ISAAC spectrograph located at the UT3-VLT-ESO telescope. We observed a larger number of sources than already studied in these bands, at higher spatial resolution or smaller apertures and in most of the cases, at higher spectral resolution. We mapped the central half parsec in these two bands and finished our program last August 2011. Here we show the first results of our L-band mapping that needed 26 slit positions in order to cover the whole desired central region. This allowed us to build the first L-band data-cube of the central half parsec of our Galaxy.

In a former paper (Moultaka et al. 2004), we used the L-band spectrum of a "CO star" (a late type star showing CO band-heads in its K-band spectrum) in order to calibrate the foreground extinction towards the central parsec at these wavelengths. Since this star is at an almost 13" angular distance from the center (i.e. it is close enough to the studied region but outside the mini-spiral area), it is located at the edge of the SgrA* west region and does not show excess emission in the L-band spectrum. Hence we assumed that it is free from local extinction and that its spectrum is only affected by the line of sight extinction. Thus dividing its spectrum by a blackbody spectrum of 3600 K temperature, we derived the L-band line of sight extinction spectrum (see Fig. 1).

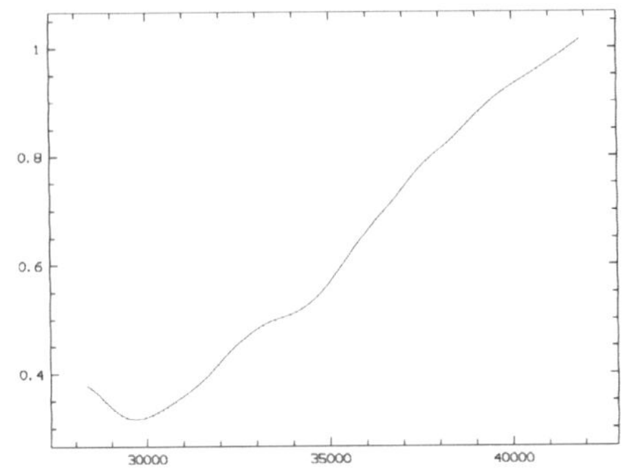

Figure 1. L-band spectrum of the foreground extinction towards the Galactic Center. The x-axis is in Ångström units and the y-axis is the optical depth of the extinction.

Thanks to this calibrator, we derive the first data-cube corrected for the foreground extinction in the L-band (see the integrated L-band maps corrected and non corrected for the line of sight extinction in Figs. 2 and 3, respectively).

Journal of Physics: Conference Series **372** (2012) 012019 doi:10.1088/1742-6596/372/1/012019

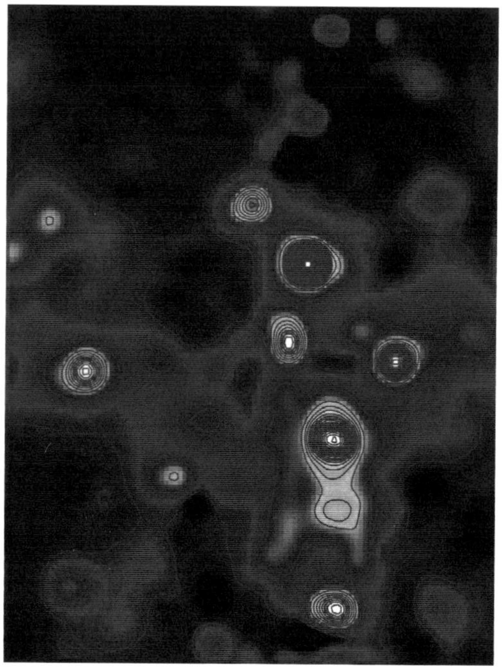

Figure 2. Integrated emission along the L-band spectral region obtained from our L-band data-cube not corrected for the foreground extinction. Contours are also shown for clarity.

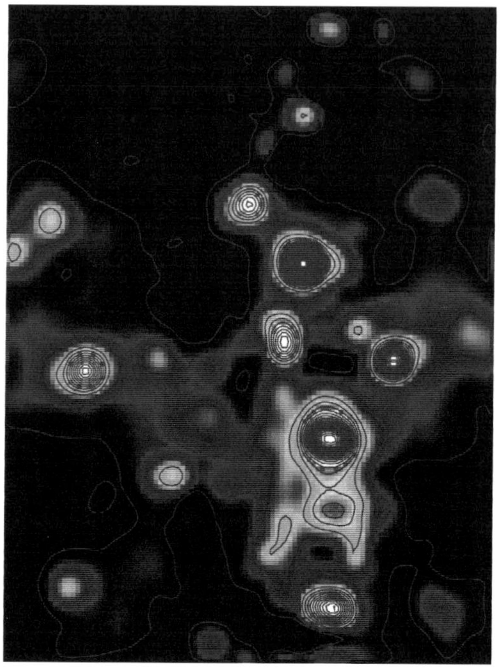

Figure 3. Integrated emission along the L-band spectral region obtained from our L-band data-cube corrected for the foreground extinction. Contours are also shown for clarity.

These maps show that we manage to reproduce the region very well even though ISAAC is not an integral field spectrograph.
In Fig. 4 we overlay our non-corrected map of the L-band integrated emission (in contours) over a NAOS/CONICA (NACO) L-band image of the central parsec. This figure shows again that our mapping is very successful in reproducing the stellar field and that the positions are accurate to within better than a fraction of an acrcsecond.

3. Validation of the line of sight extinction calibration

In our previous work (Moultaka et al. 2004) we derived the intrinsic spectra of a dozen of bright sources in the region. All spectra could be fitted by spectra of blackbodies with temperatures equal to the known temperatures of the sources or of the dust they are embedded in (see Moultaka et al. 2004). This shows that our calibration of the line of sight extinction spectrum is correct at least for the area defined by these sources.

In the present work, use blackbody continua to fit a number of spectra of early-type stars located all around the central half parsec region and outside the mini-spiral. The stars are shown in red in Fig. 5 as well as their spectra and the continua fitted with an appropriate blackbody. All of them were fitted by blackbody spectra of about 26000 K temperature which is in agreement with the temperature of such stars. This result also validates our calibration of the line of sight extinction spectrum.

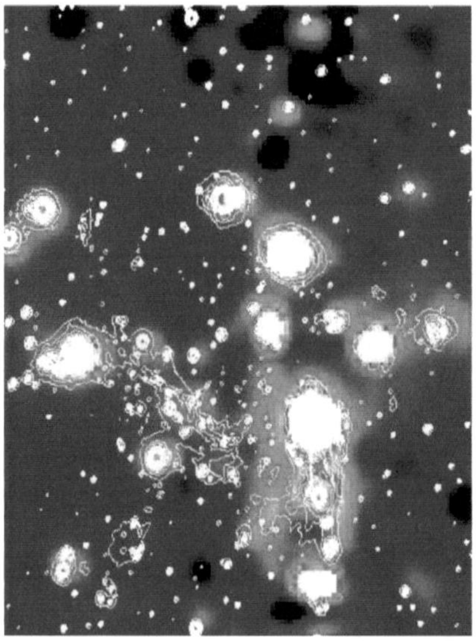

Figure 4. NAOS-CONICA L-band image of the central parsec with contours of our integrated L-band emission image obtained with our data-cube.

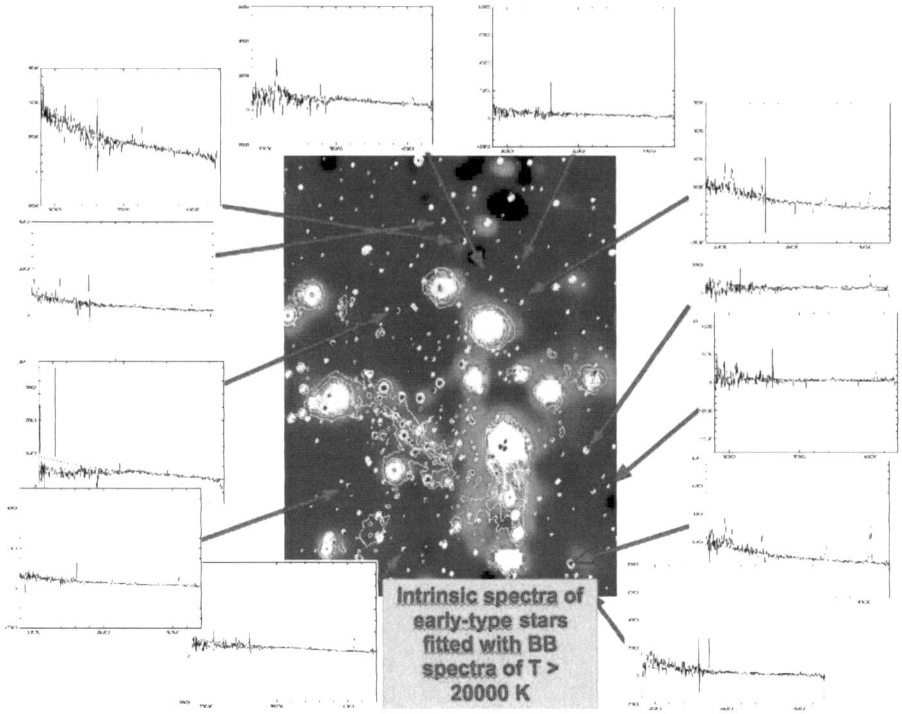

Figure 5. L-band spectra corrected for the foreground absorption of a dozen of early-type stars which locations are shown in red dots and arrows.

4. Results

Given that we validated our foreground extinction spectrum and consequently our extinction corrected data-cube, we are now confident to derive maps of the optical depths of the water ice and hydrocarbon absorptions in our data-cube .

We assume that the continuum of the hydrocarbon absorption can be approximated by a straight line going from 3.32μm to 3.77μm and the absorption of the water ice feature by a

Journal of Physics: Conference Series **372** (2012) 012019 doi:10.1088/1742-6596/372/1/012019

straight line going from 2.84μm to 3.77μm. The optical depths are calculated via the definition $\tau = -ln(\frac{F_{obs}}{F_{cont}})$, where F_{obs} is the integrated observed flux over the absorption feature and F_{cont} is the integrated continuum flux along the spectral feature. The optical depths of the water ice absorption were obtained by subtracting the hydrocarbon absorption. The advantage in using optical depths is that these are independent of the continuum intensity, so one can compare directly all luminosity class objects.

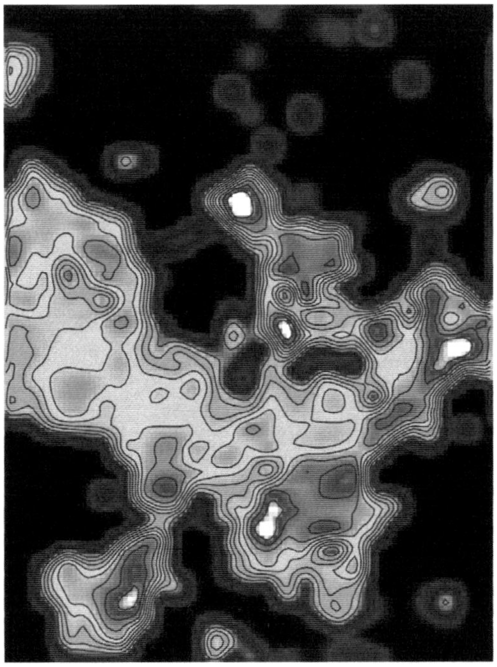

Figure 6. Optical depth map of the water ice absorption feature built with our extinction corrected data-cube. Contours are also shown for clarity.

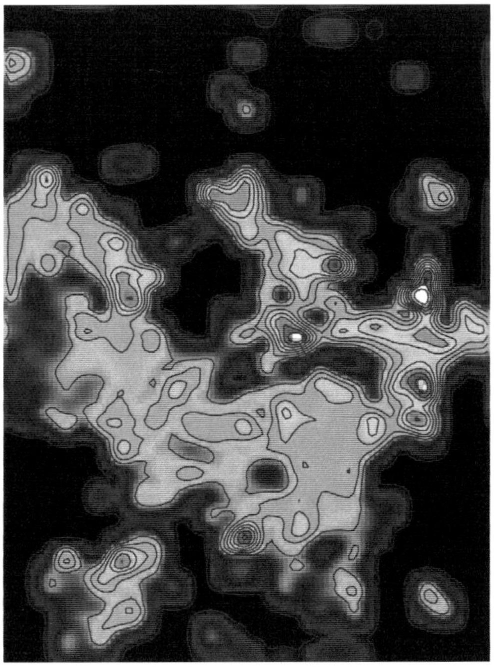

Figure 7. Optical depth map of the hydrocarbon absorption feature at 3.4 μm built with our extinction corrected data-cube. Contours are also shown for clarity.

The corrected and non corrected optical depth maps for foreground extinction are very similar in both cases of the water ice absorption feature and of the hydrocarbon feature. This suggests that we do not over-correct our features for the line of sight extinction and do not overestimate the line of sight absorption since we don't get negative values. In Figs 6 and 7, we show the extinction corrected optical depth maps of the water ice and hydrocarbon absorptions. Our resulting maps show that the features take place locally and can be associated with the neighboring infrared sources.

Moreover, in Figs. 8 and 9 we overlay the K-band extinction map derived by Schödel et al. (2009) on our optical depth maps. We find no correlations between the K-band and the water ice and hydrocarbon extinction maps. This again, stresses the fact that water ice and hydrocarbon absorptions we map are not primarily produced in the foreground material.

The water ice and hydrocarbon absorptions trace the mini-spiral region but also the dusty sources like IRS3, 7 and north of IRS 13. If we compare our maps with figure 5 from Mužić et al. (2007) showing the dusty filaments in the central parsec, we find that the absorption optical

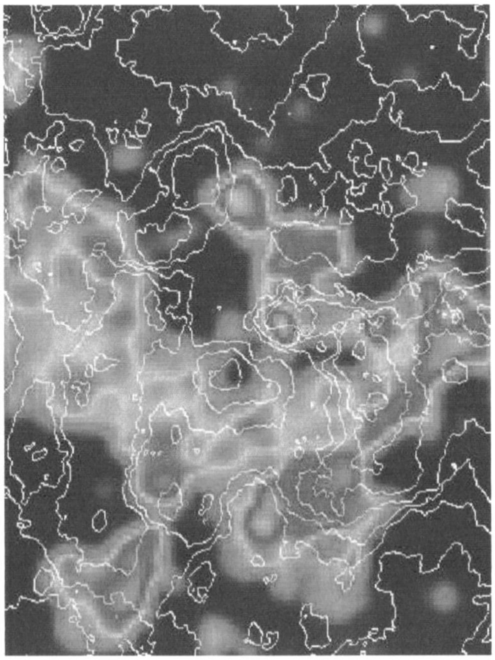

Figure 8. Optical depth map of the water ice absorption with contours of the K-band extinction image from Schödel et al. (2009) overlaid.

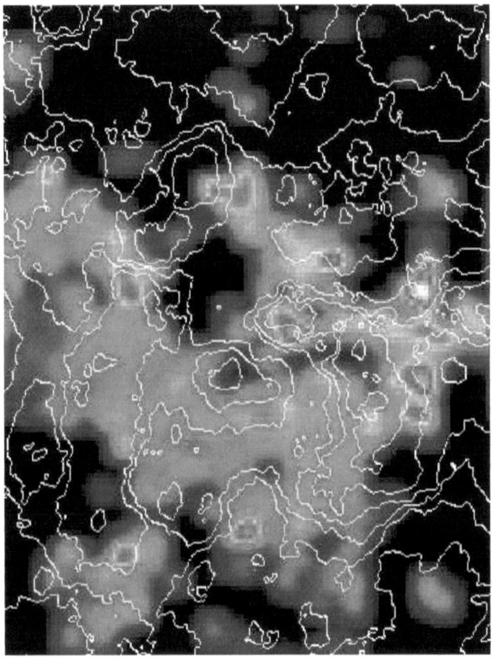

Figure 9. Optical depth map of the hydrocarbon absorption with contours of the K-band extinction image from Schödel et al. (2009) overlaid.

depths peak at the same positions as the dusty filaments tracing well these high density regions.

In Figs. 10 and 11 we show the optical depth maps with contours of the integrated L-band emission overlaid. One can notice in these figures that the minima of the optical depth coincide with the positions of the infrared sources and that peaks of the optical depth are located in the neighboring regions. We interpret these observations as being the resulting shape of mass loosing stars interacting with the local ISM and thereby enhancing their surrounding dust emission (possibly by piling up the local dust).

From the L-band extinction spectrum along the line of sight shown in Fig. 1, we derive an optical depth of about 0.55, this corresponds to an $A_L \sim 0.6$ mag which is consistent with an $A_K \sim 1.52$ mag. This is less than the value obtained by Schödel et al. (2009) of 2.46 mag. We conclude that the absorption features do not account alone for the foreground extinction but an important contribution comes from the continuum as well.

Since there is no direct relation between the optical depths of the features and the optical depth of the overall extinction along the L-band, we build a new optical depth map taking into account the absorption line (i.e. both features) and the continuum contribution within the entire wavelength band. The smoothed version of this map is shown in Fig. 13. We find that the optical depths vary from 0 to 1 in the whole region. This means that A_L varies from 0 to 1.08 mag corresponding to a variation of A_K from 0 to 2.3 mag. This is much higher than the 0.5 mag variation along the line of sight derived by Schödel et al. (2009). This result implies that the absorption features we observe take place locally. The presence of water ices

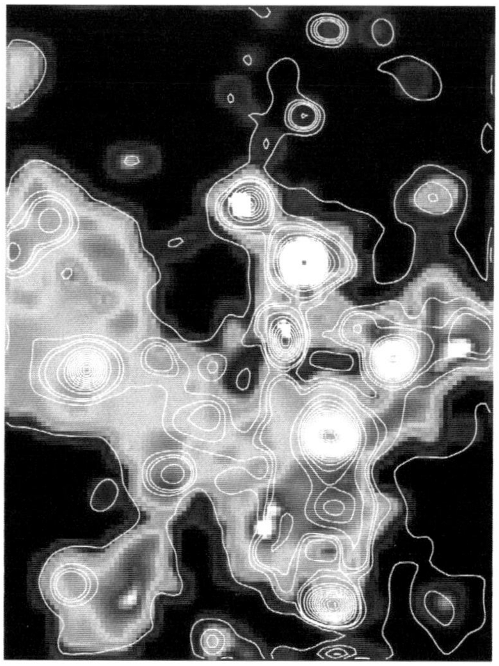

Figure 10. Optical depth map corrected for foreground extinction of the water ice absorption feature with contours of the integrated L-band emission.

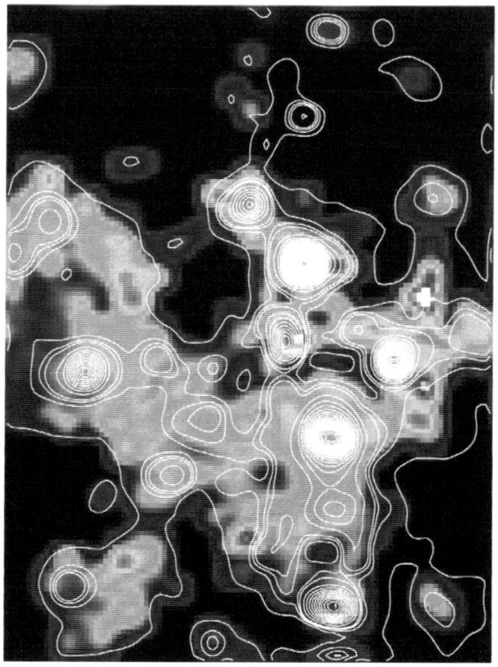

Figure 11. Optical depth map corrected for foreground extinction of the hydrocarbon absorption feature with contours of the integrated L-band emission.

in the Galactic Center implies temperatures of the order of 60K. This is consistent with our finding that (Moultaka et al. 2009) that CO ices are probably present in the region as well. We showed in Moultaka et al. (2009) that, given the complexity of the region with the presence of bow-shock sources, narrow dust filaments and dust embedded YSOs, one cannot discard the presence of high-density pockets and high-optical depths where very low temperatures can also be present. These compact dusty structures can persist and survive while traveling through the central parsec since we showed that the travel time is shorter than the evaporation time of molecular clumps and disks in similarly harsh environments.

Finally, we find a weak correlation between the water ice and the hydrocarbon optical depths with a correlation coefficient of 0.38 (see Figs. 12 and 14). This correlation suggests that the ISM presents itself as a mixture (possibly clumpy) of a dense and a diffuse medium.

References

[1] Chiar, J. E., Tielens, A. G. G. M., Whittet, D. C. B., et al. 2000, ApJ, 537, 749
[2] Chiar, J. E., Adamson, A. J., Pendelton, Y. J., et al. 2002, ApJ, 570, 198
[3] Ghez, A., Duchne, G., Matthews, K., et al. 2003, ApJ, 586, L127
[4] Ishii, M., Nagata, T., Sato, S., et al. 1998, AJ, 116, 868
[5] Lutz, D., Feuchtgruber, H., Genzel, R., et al. 1996, A&A, 315, 269
[6] McFadzean, A. D., Whittet, D. C. B., Bode, M. F., Adamson, A. J., & Longmore, A. J. 1989, MNRAS, 241, 873
[7] Moneti, A., Cernicharo, J. & Pardo, J. R. 2001, ApJ, 549, L203
[8] Moultaka, J., Eckart, A., Schödel, R. 2009, ApJ, 703, 1635

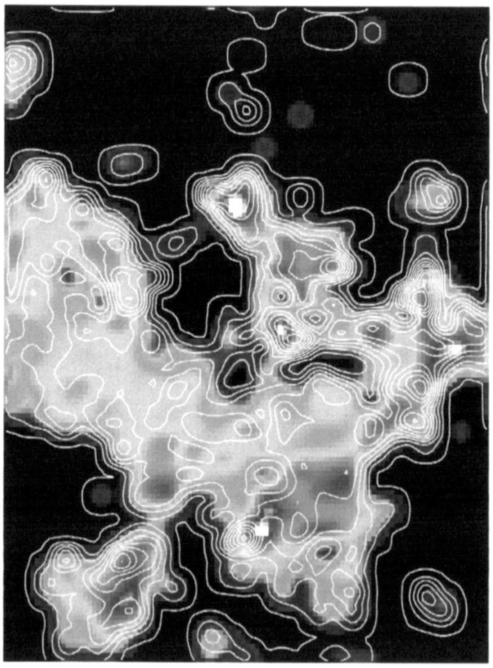

Figure 12. Optical depth map corrected for the foreground extinction of the water ice absorption with contours of the hydrocarbon absorption corrected optical depth map.

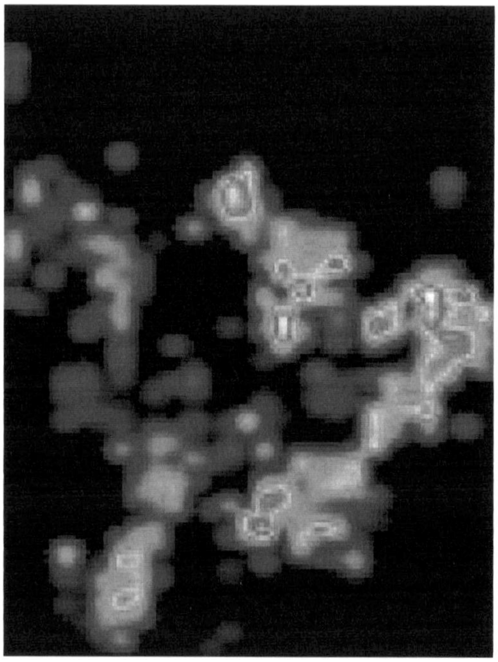

Figure 13. Optical depth map of the extinction along the whole L-band corrected for the foreground extinction.

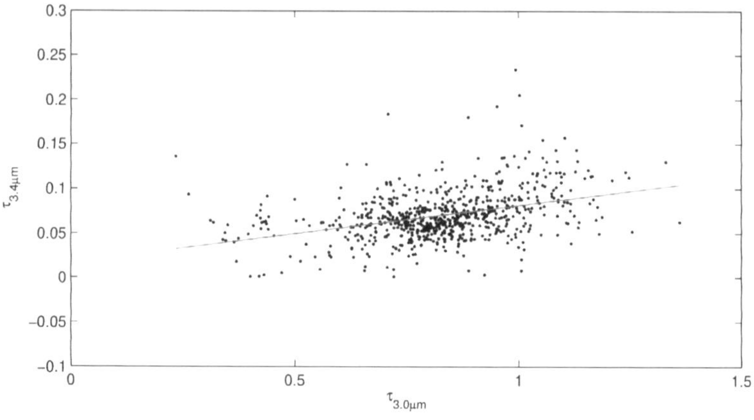

Figure 14. Optical depth of the hydrocarbon versus the ice absorption features. This figure plots the intensities of all pixels of the water ice and hydrocarbon corrected optical depth maps, at which the S/N ratio in the integrated L-band map, is higher than 10. In red, the best linear fit.

[9] Moultaka, J., Eckart, A., Viehmann, T., Mouawad, N., Straubmeier, C., et al. 2004, A&A, 425, 529
[10] Mužić, K., Schödel, R., Eckart, A., Meyer, L., & Zensus, A. 2008, A&A, 482, 173
[11] Schdel, Ott, T., Genzel, R., et al. 2002, Nature, 419, 694

Astronomy at High Angular Resolution 2011 IOP Publishing
Journal of Physics: Conference Series **372** (2012) 012020 doi:10.1088/1742-6596/372/1/012020

Sagittarius A* in the Infrared

Rainer Schödel

Instituto de Astrofísica de Andalucía (CSIC), Glorieta de la Astronomía s/n, 18008 Granada, Spain

E-mail: rainer@iaa.es

Abstract. The central black hole of the Milky Way, Sagittarius A* (SgrA*) is, in terms of Eddington luminosity, the weakest accreting object of its class accessible to detailed observations. It is therefore key to the refinement of theoretical models of radiatively inefficient accretion. Unfortunately, our knowledge of the mean SED of Sgr A* is very limited. Current models rely almost exclusively on cm to mm mean flux measurements and only on upper limits at infrared to soft X-ray wavelengths. Here, we present a new analysis of imaging data of the Galactic center (GC) at 2.2 to 8.6 microns, obtained with NACO and VISIR at the ESO VLT. We used the VISIR burst mode combined with a novel implementation of the holographic image reconstruction algorithm to obtain mid-infrared images with a Strehl ratio $\gg 90\%$ even under conditions of $\sim 2-3$" seeing in the visual. No counterpart of Sgr A* is detected at 8.6 microns. At this wavelength, Sgr A* is located right on top of a dust ridge, which considerably complicates the search for a potential point-source. Based on the available data, it is argued that Sgr A* cannot be detected in the MIR with currently available instruments, not even during flares. At 3.8 and 4.8 μm SgrA* is detected at all times. We measure the time-averaged mean fluxes of Sgr A* at these wavelengths. From the literature there is evidence that SgrA* is also detected at 2.2 μm most of the time. The new measurements of the mean, quiescent emission of SgrA* fill a gap of almost 6 orders of magnitude in its known mean SED and provide novel constraints on accretion/emission models.

1. Introduction

The black hole at the center of the Milky Way, Sagittarius A* (SgrA*), has a mass of about $4 \times 10^6\,M_\odot$ and is located at a distance of 8 kpc. The combined systematic and statistical uncertainties of these values are about 10% for its mass and 5% for its distance [e.g. 1–3]. The electromagnetic emission from SgrA* is powered by accretion, but, as inferred from measurements of Faraday rotation, the actual accretion rate is lower than $2 \times 10^{-7}\,M_\odot$, possibly by up to two orders of magnitude [4]. This low accretion rate is intimately connected to a low efficiency in converting the accreted mass to radiation. Accretion onto SgrA* is generally described with so-called *Radiatively Inefficient Accretion Flow* (RIAF) models, that have efficienies orders of magnitude below the 10% efficienty of a canonical thin-disk accretion flow [5]. Radiating at $\sim 10^{-9}$ Eddington luminosities [6], SgrA* is thus the weakest black hole accessible to detailed measurements and is thus of special interest for the theory of accretion, emission, and outflow in compact objects.

Unfortunately, the Galactic center is located behind $30-50$ magnitudes of visual extinction [7; 8] and characterized by an extreme source density at all wavelengths, which creates great observational difficulties in measuring the mean SED of SgrA*. Although being of fundamental importance for theoretical models, the SED of SgrA* is only known well at radio to millimeter

Journal of Physics: Conference Series **372** (2012) 012020 doi:10.1088/1742-6596/372/1/012020

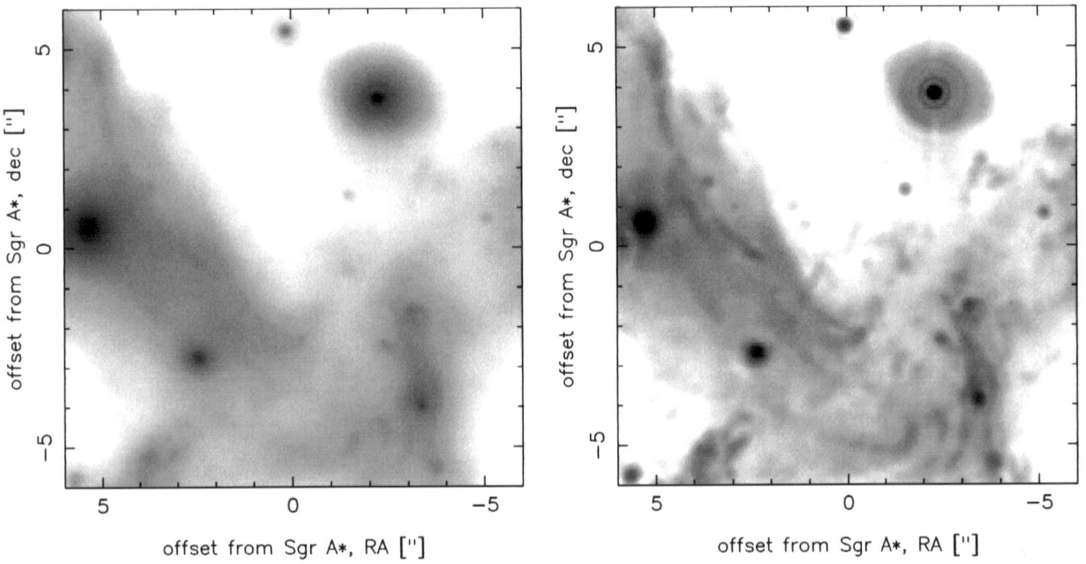

Figure 1. Left: SSA reconstruction of 8.6 μm burst mode data taken with VISIR on 23 May 2007. Right: Holographic reconstruction.

wavelengths. In the X-ray domain, SgrA* is a relatively weak source, that is embedded into extended X-ray emission from the gas within its Bondi radius [9]. At far- to mid-infrared wavelengths, only upper limits on SgrA* could be established so far [e.g. 10]. In the near-infrared (NIR), on the other hand, SgrA* is an extremely variable source, with hour-long flares occurring several times per day. A large number of observing campaigns was focused on these flares in the past years [e.g. 11–15], but the questions whether there is a permanently detectable NIR source at the position of SgrA* and what is its mean flux were neglected. However, it is the mean flux at a given wavelength that provides fundamental constraints for theoretical models.

2. NIR and MIR Imaging of the Galactic center

The Galactic center has been observed numerous times with VISIR/VLT at mid-infrared (MIR) wavelengths, often within the framework of multi-wavelength campaigns on SgrA*. The MIR emission in the central parsec of the GC is dominated by the warm dust in the mini-spiral, which is roughly composed of three streams of (partially) ionized gas and dust [e.g., 16; 17]. Both stars as well as numerous ISM features can be seen in MIR images within 1" of SgrA* [10]. When searching for a MIR counterpart of SgrA* one should therefore use the highest possible spatial resolution. The wavelength-dependence of the FWHM of a point source in seeing-limited observations can be approximated with the Roddier formula: $FWHM \propto \lambda^{-0.2}$, which leads to $FWHM_{8.6\,\mu m} \approx 0.5$" with a seeing of ~0.8". The diffraction limit of the VLT is, however, ~0.25" at 8.6 μm. Therefore, with standard imaging techniques, the diffraction limit at short MIR wavelengths can only be exploited during exceptionally good seeing conditions. This situation has changed with the implementation of the *burst mode* at VISIR, that allows the observer to store individual exposures. The typical exposure time with VISIR is 0.02 s, i.e. much shorter than the coherence time of the atmosphere. Speckle imaging techniques can thus be applied for optimal image reconstruction. Typically, a simple shift-and-add (SSA) algorithm is applied because it is easy to use and implement [e.g. 18]. However, this algorithm is only efficient if the speckle clouds of the sources are dominated by a single speckle, but some of the observations used for this work were obtained under visual seeing of $2-3$", so that this requirement was not

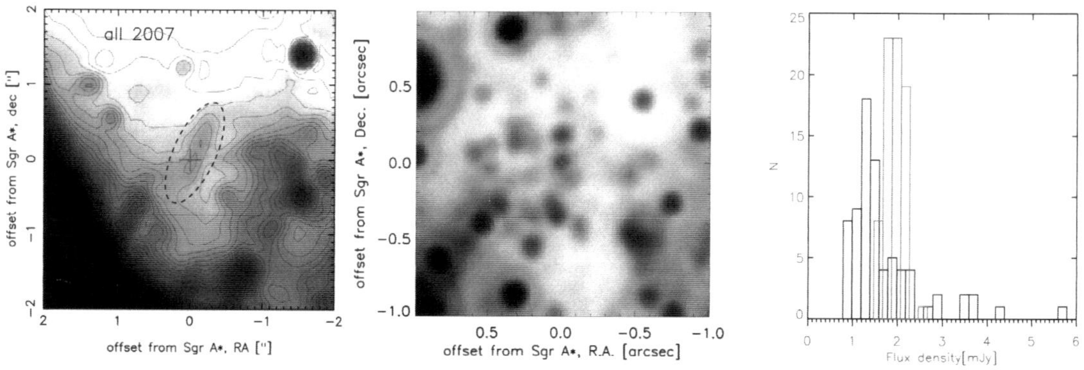

Figure 2. Left: Deep 8.6 μm image obtained from a holographic reconstruction of all 2007 VISIR burst mode data. The *SgrA*-Ridge* is indicated by a dashed ellipse. Middle: Average L'−image from all used NACO L'−data from 2006 - 2009. The red cross in the right and middle panels marks the position of Sgr A*. Right: Histogram of the measured flux densities of the star S2 (red) and of Sgr A* (blue), measured in images of 800 s integration time each.

fulfilled. Therefore, the more complex, but more efficient holography algorithm was used for image reconstruction [19]. The great performance of holography is illustrated in Fig. 1, where an SSA image image from 23 May 2007 is compared to a holographic reconstruction, which has a Strehl ratio $\gg 90\%$.

For the observations in the NIR, the ESO archive was used to access the great amounts of observations of SgrA* with NACO/VLT at L' (3.8 μm), between 2006 and 2009, and at M' (4.8 μm), in 2003, 2004, and 2006. Data reduction was standard. For further details on the data used and on their reduction, see [20].

3. Results

No counterpart of Sgr A* could be detected at 8.6 μm, where Sgr A* lies atop a dust ridge (see Fig. 2). This *Sgr A*-Ridge* limits considerably the point-source sensitivity at the position of Sgr A*. Therefore, to determine an upper limit to the emission from Sgr A* it was first necessary to estimate the background flux density from the ISM at its position. For this purpose, the flux in a circular area centered on Sgr A* was estimated for the image of each observing epoch by interpolation from surrounding pixels. This approximate background was then subtracted and the remnant emission corrected for aperture effects [for a more detailed description of the methodology, see 20]. Four epochs with very low upper limits were excluded from the measurements because they may have been affected by systematic uncertainties due to chopping into bright sources. The weighted mean and error of the remnant flux in the images from the remaining 8 epochs is 7.1 ± 0.9 mJy, leading to a 3σ upper limit of 9.8 mJy for the emission from Sgr A* at 8.6 μm. Assuming en extinction of $A_{8.6} = 2.0 \pm 0.3$ [21], this leads to a de-reddened 3σ upper limit of 84 mJy, which includes the uncertainty of the extinction correction.

In the M'−band, at 4.8 μm, a counterpart of Sgr A* was detected in the images from all epochs. Due to the relatively low resolution at this wavelength ($FWHM \approx 0.135$") , Sgr A* is confused with a dust blob located about 0.094" to the southeast [22; 23]. It therefore appears as an elongated source, which could be approximated well by a combination of two point sources that were fitted simultaneously in order to estimate the emission from Sgr A*. The emission from the dust blob is 1.2 ± 0.3 mJy and the emission from Sgr A* is 1.5 ± 0.3 mJy, where the uncertainty includes the statistical uncertainty from the various measurements as well as a

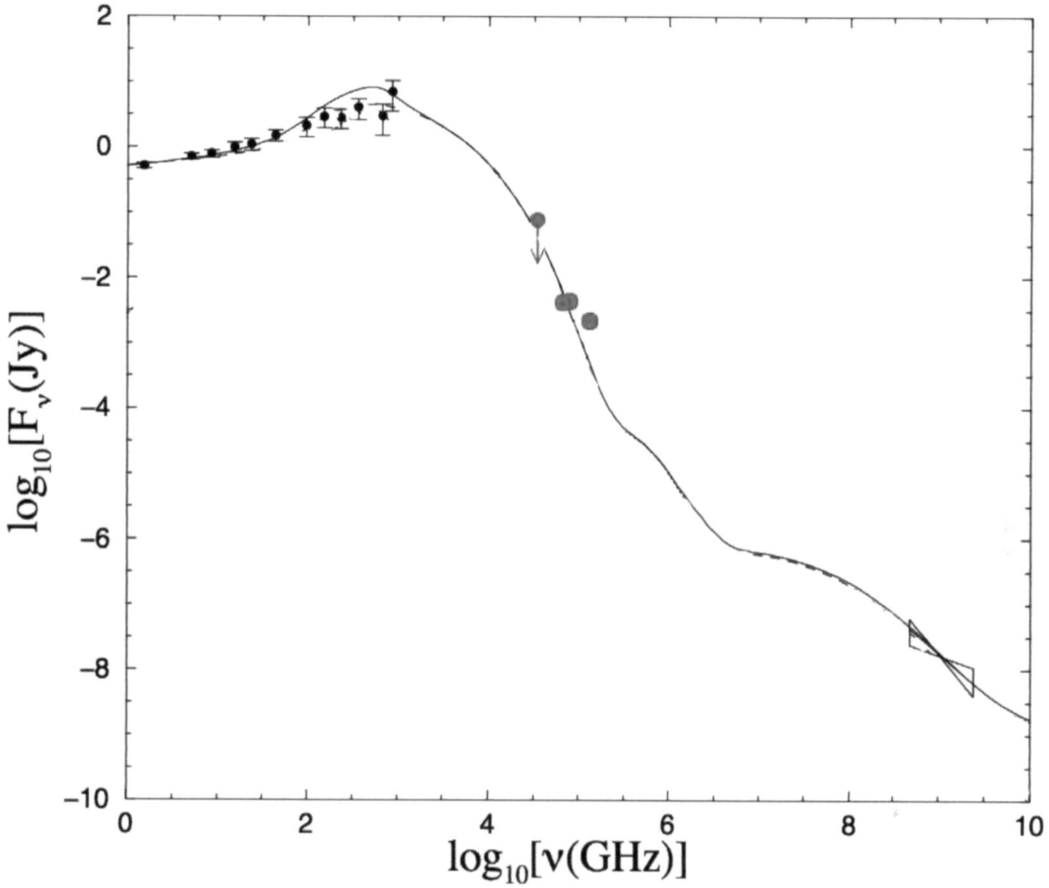

Figure 3. Combined jet-ADAF model for the SED of Sgr A*. See [28] for a description of the model and of the radio and X-ray data. The $3\,\sigma$ upper limit at $8.6\,\mu$m and the mean near-infrared emission from the work described in this article is indicated by red points [20].

possible systematic uncertainty from contamination by stellar sources. The extinction-corrected mean flux from Sgr A* at $4.8\,\mu$m is 3.8 ± 1.3 mJy, taking the uncertainty of the extinction correction into account.

The L'-band, centered at $3.8\,\mu$m, is probably the best band to estimate the mean emission from Sgr A* in the NIR because of an optimal trade-off between relatively high angular resolution, on the one hand, and relatively low contamination from surrounding stars and ISM, on the other hand. Sgr A* was detected in all epochs, with a de-reddened mean flux density of 5.0 ± 0.6 mJy, including the uncertainty from the extinction correction.

In order to examine short timescales, the imaging data at L' were divided into subsets, comprising 800 s of observations. Sgr A* could be detected in all of the resulting images of 800 s integration time. The corresponding histograms of the measured flux densities of Sgr A* and of the nearby star S2 are shown in the right panel of Fig. 2. The width of the histogram for the star S2 indicates the measurement uncertainty. The histogram for Sgr A* shows a broader distribution, with a distinct tail towards high flux densities.

In the $K-$band, at wavelengths around $2.1 - 2.2\,\mu$m, the situation is complicated by the intrinsic faintness of the extremely red source Sgr A* and the extreme stellar confusion at its position [24]. Nevertheless, the available literature indicates that a (highly variable) counterpart of Sgr A* can be detected at all times, with a flux density of $0.5 - 2.5\,\mathrm{mJy}$ [25–27].

4. Discussion

The new infrared measurements of Sgr A* are indicated by red points in Fig. 2, along with a jet-ADAF model of its SED [28]. The new upper limit at $8.6\,\mu\mathrm{m}$ is the tightest one obtained so far at this wavelength. The detection of Sgr A* in the MIR is hindered considerably by the low angular resolution in this wavelength regime, combined with the strong background emission from the ISM near Sgr A*, and the general difficulty of observing in the MIR due to the strong thermal background from sky and telescope. The long integration times necessary to reach the required sensitivity of a few mJy are on the order of 1 hour or longer, which means that any bright flares will be diluted. An analysis of images with about 19 min time resolution indicates that it is probably not possible to detect Sgr A* with currently existing instrumentation, not even during a bright flare [20].

This is the first time that a permanently detectable counterpart of Sgr A* is reported in the NIR. The new data close a previously existing gap of about 6 orders of magnitude in our knowledge of its SED. The measurements of the mean emission from Sgr A* in the NIR agree rather well with the model shown in Fig. 2 and also with a different kind of RIAF model [29], while published models based on relativistic hydrodynamics simulations generally under-predict the newly derived infrared flux densities [20; 30]. It has to be kept in mind, however, that all published models so far are purely based on radio to millimeter and X-ray measurements.

As can be seen in Fig. 2, tighter limits in the MIR can help to constrain models that include a jet component. Finally, the mean flux at $\sim 2.2\,\mu\mathrm{m}$ appears somewhat high. It is rather possible that this is due to contamination by confusion with stars at the position of Sgr A*.

It appears that the L'-band is the best wavelength to monitor the emission from Sgr A* in the NIR. The histogram of flux densities measured in images of 800 s integration time at L' indicates that Sgr A* spends most of the time in a state of low emission with a well defined mean. This state could be termed the *quiescent state* of Sgr A*, with a *flaring state* marked by the tail toward high flux densities [see also 27].

References

[1] Gillessen S, Eisenhauer F, Trippe S, Alexander T, Genzel R, Martins F and Ott T 2009 *ApJ* **692** 1075–1109 (*Preprint* 0810.4674)

[2] Genzel R, Eisenhauer F and Gillessen S 2010 *Reviews of Modern Physics* **82** 3121–3195

[3] Yelda S, Ghez A M, Lu J R, Do T, Clarkson W and Matthews K 2010 *ArXiv e-prints*

[4] Marrone D P, Moran J M, Zhao J H and Rao R 2007 *ApJ* **654** L57–L60

[5] Quataert E 2003 *Astronomische Nachrichten Supplement* **324** 435–443

[6] Markoff S, Bower G C and Falcke H 2007 *MNRAS* **379** 1519–1532

[7] Scoville N Z, Stolovy S R, Rieke M, Christopher M and Yusef-Zadeh F 2003 *ApJ* **594** 294–311

[8] Schödel R, Najarro F, Muzic K and Eckart A 2010 *A&A* **511** A18+

[9] Baganoff F K, Maeda Y, Morris M, Bautz M W, Brandt W N, Cui W, Doty J P, Feigelson E D, Garmire G P, Pravdo S H, Ricker G R and Townsley L K 2003 *ApJ* **591** 891–915

[10] Schödel R, Eckart A, Mužić K, Meyer L, Viehmann T and Bower G C 2007 *A&A* **462** L1–L4

[11] Kunneriath D, Witzel G, Eckart A, Zamaninasab M, Gießübel R, Schödel R, Baganoff F K, Morris M R, Dovčiak M, Duschl W J, García-Marín M, Karas V, König S, Krichbaum T P, Krips M, Lu R, Mauerhan J, Moultaka J, Mužić K, Sabha N, Najarro F, Pott J, Schuster

K F, Sjouwerman L O, Straubmeier C, Thum C, Vogel S N, Teuben P, Weiss A, Wiesemeyer H and Zensus J A 2010 *A&A* **517** A46+

[12] Eckart A, Baganoff F K, Morris M R, Kunneriath D, Zamaninasab M, Witzel G, Schödel R, García-Marín M, Meyer L, Bower G C, Marrone D, Bautz M W, Brandt W N, Garmire G P, Ricker G R, Straubmeier C, Roberts D A, Muzic K, Mauerhan J and Zensus A 2009 *A&A* **500** 935–946

[13] Yusef-Zadeh F, Bushouse H, Wardle M, Heinke C, Roberts D A, Dowell C D, Brunthaler A, Reid M J, Martin C L, Marrone D P, Porquet D, Grosso N, Dodds-Eden K, Bower G C, Wiesemeyer H, Miyazaki A, Pal S, Gillessen S, Goldwurm A, Trap G and Maness H 2009 *ApJ* **706** 348–375

[14] Dodds-Eden K, Porquet D, Trap G, Quataert E, Haubois X, Gillessen S, Grosso N, Pantin E, Falcke H, Rouan D, Genzel R, Hasinger G, Goldwurm A, Yusef-Zadeh F, Clenet Y, Trippe S, Lagage P O, Bartko H, Eisenhauer F, Ott T, Paumard T, Perrin G, Yuan F, Fritz T K and Mascetti L 2009 *ApJ* **698** 676–692

[15] Eckart A, Schödel R, Meyer L, Trippe S, Ott T and Genzel R 2006 *A&A* **455** 1–10

[16] Viehmann T, Eckart A, Schödel R, Pott J and Moultaka J 2006 *ApJ* **642** 861–867

[17] Zhao J H, Morris M R, Goss W M and An T 2009 *ApJ* **699** 186–214

[18] Christou J C 1991 *Experimental Astronomy* **2** 27–56

[19] Schoedel R, Yelda S, Ghez A, Girard J H V, Labadie L, Rebolo R and Perez-Garrido A 2011 *ArXiv e-prints* (*Preprint* `1110.2261`)

[20] Schödel R, Morris M R, Muzic K, Alberdi A, Meyer L, Eckart A and Gezari D Y 2011 *A&A* **532** A83+

[21] Fritz T K, Gillessen S, Dodds-Eden K, Lutz D, Genzel R, Raab W, Ott T, Pfuhl O, Eisenhauer F and Yusef-Zadeh F 2011 *ApJ* **737** 73–+

[22] Eckart A, Baganoff F K, Schödel R, Morris M, Genzel R, Bower G C, Marrone D, Moran J M, Viehmann T, Bautz M W, Brandt W N, Garmire G P, Ott T, Trippe S, Ricker G R, Straubmeier C, Roberts D A, Yusef-Zadeh F, Zhao J H and Rao R 2006 *A&A* **450** 535–555

[23] Ghez A M, Hornstein S D, Lu J R, Bouchez A, Le Mignant D, van Dam M A, Wizinowich P, Matthews K, Morris M, Becklin E E, Campbell R D, Chin J C Y, Hartman S K, Johansson E M, Lafon R E, Stomski P J and Summers D M 2005 *ApJ* **635** 1087–1094

[24] Hornstein S D, Matthews K, Ghez A M, Lu J R, Morris M, Becklin E E, Rafelski M and Baganoff F K 2007 *ApJ* **667** 900–910

[25] Do T, Ghez A M, Morris M R, Yelda S, Meyer L, Lu J R, Hornstein S D and Matthews K 2009 *ApJ* **691** 1021–1034

[26] Meyer L, Do T, Ghez A, Morris M R, Yelda S, Schödel R and Eckart A 2009 *ApJ* **694** L87–L91

[27] Dodds-Eden K, Gillessen S, Fritz T K, Eisenhauer F, Trippe S, Genzel R, Ott T, Bartko H, Pfuhl O, Bower G, Goldwurm A, Porquet D, Trap G and Yusef-Zadeh F 2011 *ApJ* **728** 37–+

[28] Yuan F, Markoff S and Falcke H 2002 *A&A* **383** 854–863

[29] Yuan F, Quataert E and Narayan R 2004 *ApJ* **606** 894–899

[30] Mościbrodzka M, Gammie C F, Dolence J C, Shiokawa H and Leung P K 2009 *ApJ* **706** 497–507

Astronomy at High Angular Resolution 2011 IOP Publishing
Journal of Physics: Conference Series **372** (2012) 012021 doi:10.1088/1742-6596/372/1/012021

Near-infrared polarization in the central parsec of the galactic center

Rainer M Buchholz[1], **Gunther Witzel**[1], **Rainer Schödel**[2], **Andreas Eckart**[1,3], **Markus Bremer**[1], **and Koraljka Muzic**[4]

[1] I. Physikalisches Institut der Universität zu Köln, Zülpicher Strasse 77, 50937 Köln, Germany
[2] Instituto de Astrofísica de Andalucía (CSIC), Glorieta de Astronomía s/n, E-18008 Granada, Spain
[3] Max-Planck-Institut für Radioastronomie, Auf dem Hügel 69, 53121 Bonn, Germany
[4] University of Toronto, Department of Astronomy and Astrophysics, 50 St. George Street, Toronto, ON M5S 3H4

E-mail: buchholz@ph1.uni-koeln.de

Abstract. The GC offers unique opportunities to study stellar and bow-shock polarization effects in a dusty environment. NIR polarimetry of the stellar and bow-shock sources in the central parsec is presented for the first time at NACO resolution (AO assisted, ESO VLT). We present polarization maps of the central 3"x19" and spatially resolved polarimetry of the known extended bow-shock sources in this region. The measured foreground polarization is largely parallel to the Galactic plane, with average values of 5.5% at 15° (Ks-band) and 9.5% at 20° (H-band) in the center of the FOV. These values vary over the field-of-view, and we suggest that this may be caused by local dichroic extinction on dust grains in the Northern Arm of the Minispiral. It was possible to isolate the intrinsic polarization of the two bow-shock sources contained in the sample, IRS 21 and 1W, and both show similar intrinsic polarization degrees of 5.5% respectively 7.8% (Ks) and 6.9% (H, only 1W) at polarization angles coincident with previous MIR findings, both in total and spatially resolved. The spatial polarization pattern of both sources hints at the processes likely responsible for the intrinsic polarization: scattering on and emission from elongated, aligned grains.

1. Introduction

The Galactic Center, located at a distance of ~8 kpc (see e.g. [4]), contains the densest star cluster in the galaxy. Shrouded by more than 30 mag of optical and still ~3 mag of Ks-band extinction [11], it provides the opportunity to study the relation between extinction and polarization. The foreground polarization towards the GC has been measured previously as 4% at 25° at 0.5-1.0" resolution [3, 7]. The data presented here examines this region at much higher resolution and greater depth than ever before, enabling polarization measurements on an unprecedented number of sources in this region, including resolved polarimetry on two known bow-shock sources (IRS 1W and IRS 21, see [12]).

2. Observation and data reduction

The data used here were obtained with the NACO adaptive-optics assisted instrument at the ESO VLT UT4 on Paranal, Chile (programs 073.B-0084(A), H-band, 179.B-0261(A) and 083.B-0031(A), both Ks-band), using a Wollaston prism in conjunction with a half-wave plate. This

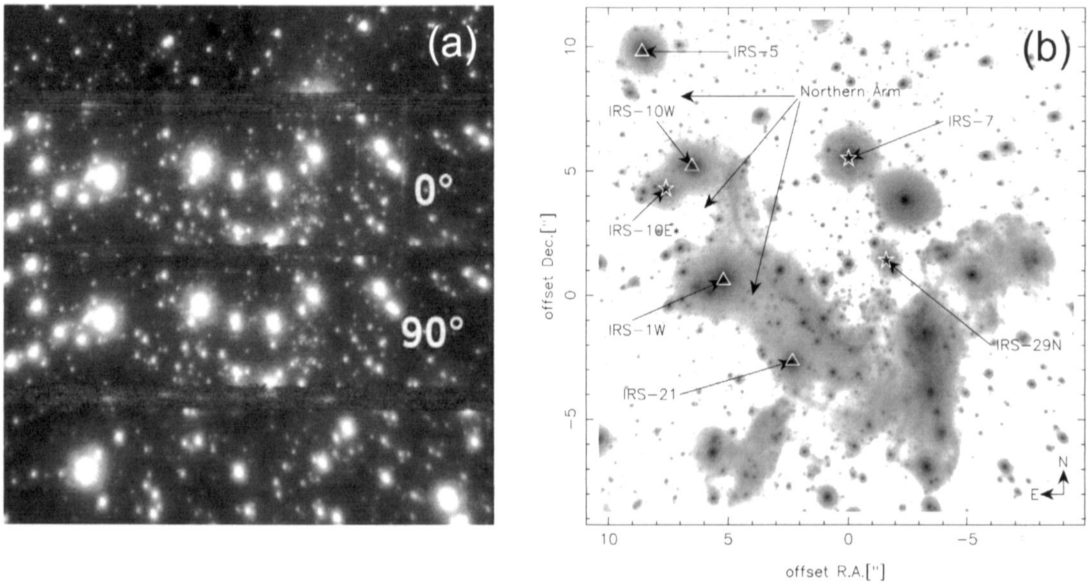

Figure 1. (a): Sample Wollaston FOV (0° and 90° channel. (b): Lp-band image of the central region of the GC. The brightest stellar sources are indicated with stars, while the prominent bow-shocks are marked with triangles.

yielded four images for each dither position, at 0°, 45°, 90°, and 135°. These images were flat-fielded, sky-subtracted, corrected for dead/hot pixels, and combined to a mosaic for each channel.

We applied a photometric method recently presented by [10] in order to achieve the necessary precision: deconvolution-assisted PSF fitting photometry. This method counters the problems of PSF variation over the field-of-view (FOV) due to anisoplanatism, as well as crowding in a densely populated region like the central parsec. We achieved photometric uncertainties of better than 3% (Ks-band) respectively 6% (H-band) for most sources brighter than 16 mag in the Ks-band. For details on the application of this method, please see [2].

For our study of the two bow-shock sources in the FOV, IRS 1W and IRS 21, we applied a Lucy-Richardson deconvolution with a PSF determined from bright sources in the IRS 1 and IRS 16 cluster, followed by aperture photometry (with overlapping apertures of 27 mas covering regions of significant flux).

To account for the instrumental polarization of NACO, we used the model developed by [13] to calibrate the data. The instrumental polarization sets a lower limit for the systematic uncertainties of the polarization parameters, with $dp = 0.5\%$ and $d\theta = 5°$.

The intrinsic polarization of the bow-shock sources was measured by assuming that the foreground polarization can be treated as a simple linear polarizer, which can then be described by a Müller matrix. When this matrix is inverted and applied to the measured Stokes vector of an intrinsically polarized source, this yields the actual intrinsic polarization parameters of that source. The reliability of this method depends on the accuracy of the assumed foreground polarization. This value is estimated based on neighboring, non-extended sources.

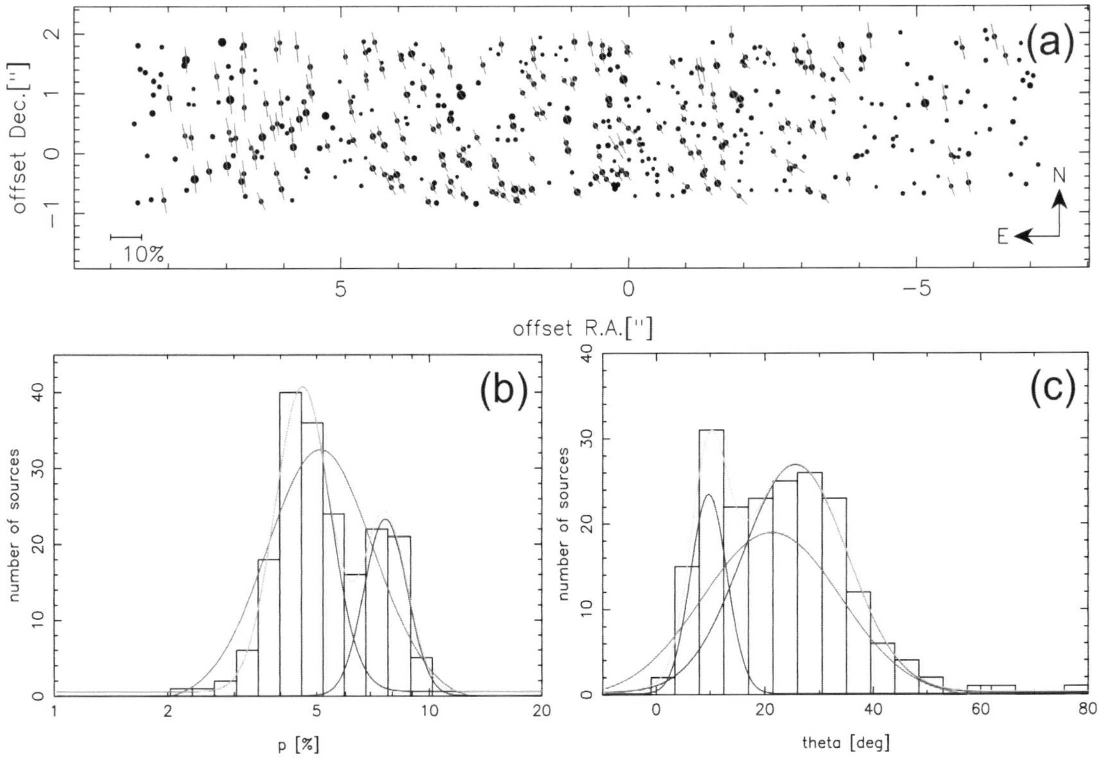

Figure 2. Ks-band polarization in the innermost 3"×19". (a): Polarization map, red lines indicate the polarization (only shown for sources with reliably measured parameters). (b): Polarization degrees, logarithmic plot. The red line indicates a single peak Gaussian fitted to the histogram, while green lines denote two separately fitted Gaussian peaks. The blue line indicates the sum of these two Gaussians. (c): Polarization angles. Lines denote fitted Gaussians as described for (b).

3. Results and discussion

3.1. Ks-band polarization

The polarization parameters of 194 sources brighter than 16 mag were measured in the Ks-band dataset (see Fig.2a). For fainter sources, the photometric uncertainty, insufficient completeness and source crowding start to become too large determine the polarization reliably.

The polarization angles in the central arcseconds mostly follow the orientation of the galactic plane within the uncertainty limits (31.4°, see [8], while angles of ~25-30° are found in this work), while they are slightly steeper (~5-15°) towards the eastern edge of the FOV (see Fig.2a). A small number of sources west of Sgr A* also shows similar steep angles, but there are too few reliable sources there to allow any conclusions.

The distribution of the polarization angles (see Fig.2c) can be fitted with a single Gaussian, peaked at 20° with a FWHM of 30°. Using the FWHM as a measure for the uncertainty (with $\sigma = \frac{FWHM}{2\sqrt{2 \times ln(2)}}$), this yields $\theta = 20° \pm 13°$. Using a fitting function with two Gaussian peaks yields a significantly lower χ^2 (by a factor of 3). The two peaks are fitted at $\theta_1 = 10° \pm 4°$ (FWHM of 10°) respectively $\theta_2 = 26° \pm 8°$ (FWHM of 20°). Considering the uncertainties of the polarization angles (up to 15°), it can be questioned if these two peaks are indeed a real feature, with the distance between the peaks of the order of these typical errors.

The polarization degree also appears to vary over the field, with values of 4-5% in the central region and 8-10% towards the eastern edge (and for some western sources, but with the same

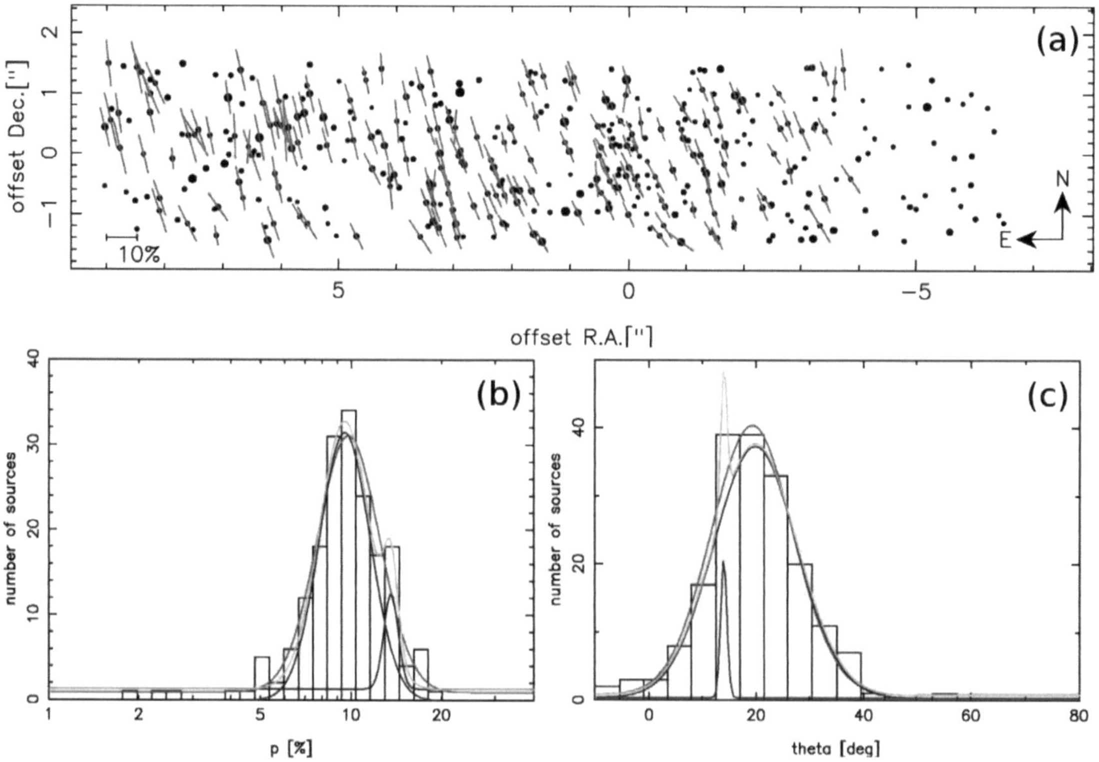

Figure 3. H-band polarization in the innermost 3"×16.5". (a): Polarization map, red lines indicate the polarization (only shown for sources with reliably measured parameters). (b): Polarization degrees, logarithmic plot. The red line indicates a single peak Gaussian fitted to the histogram, while green lines denote two separately fitted Gaussian peaks. The blue line indicates the sum of these two Gaussians. (c): Polarization angles. Lines denote fitted Gaussians as described for (b).

caveat as for the polarization angle). Especially sources in the area around the IRS 1 sources show these higher polarization degrees. The logarithms of the polarization degrees were fitted with a Gaussian (peaked at (5.1 ± 1.7)%, FWHM of 4.0%, see Fig.2b), and just as it was the case for the polarization angles, the fit was quite poor. Repeating the fit with a double Gaussian yielded two peaks at (4.6 ± 0.8)% respectively (7.7 ± 1.2)%, with FWHMs of 1.8% respectively 2.8% and a significantly better χ^2 (by a factor of 8). The relative uncertainties of the polarization degree reach up to 30%, and this limits the confidence in the two fitted peaks. Comparing the two fitted Gaussian distributions for both parameters reveals that a similar number of stars are contained in the 10° and the 7.7% peak (∼25-30%), respectively the 26° and 4.6% peak (∼70-75%). This confirms the general trends found in Fig.2a and indicates that the fitted peaks indeed correspond to a real feature.

3.2. H-band polarization

In the H-band, reliable results could be obtained for 163 sources brighter than 18 mag. The limit of 18 mag corresponds to the limit of 16 mag in the Ks-band, assuming a typical H-Ks of ∼2 mag (with H-$Ks_{intrinsic}$ ∼0 mag and A_H-A_{Ks} ∼2 mag, see e.g. [11]). The lower number of sources with reliable polarization compared to the Ks-band can be attributed to the significantly lower Strehl ratio of the H-band data (0.17 compared to 0.27 in the Ks-band). The polarization angles found here are very similar to those in the Ks-band, but with a more uniform distribution

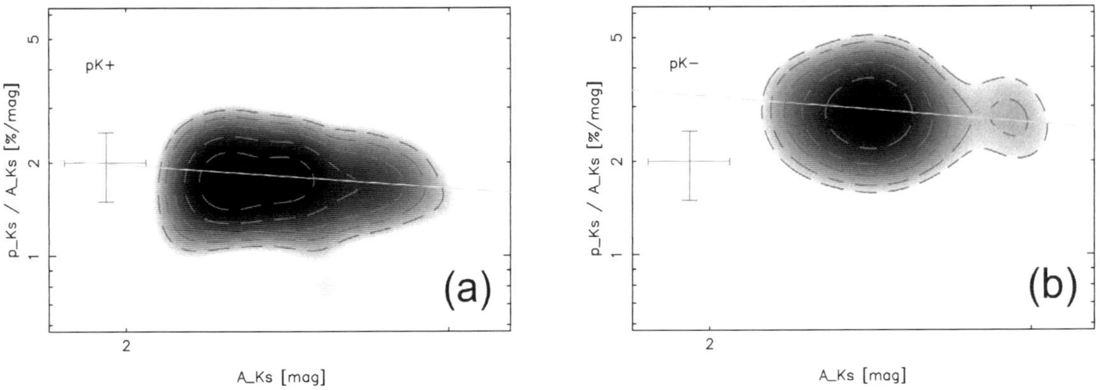

Figure 4. Ks-band polarization efficiency of GC sources, plotted against Ks-band extinction. (a): sources with $p_{Ks} < 6\%$ (pK$^-$). (b): sources with $p_{Ks} > 6\%$ (pK$^+$). There is a significant offset between the two sub-groups.

over the FOV (see Fig.3a). The distribution of the polarization angles can be fitted well with a single Gaussian, peaking at $20° \pm 8°$ (FWHM of 19°, see Fig.3c). Fitting this distribution with a double Gaussian produces a slightly better χ^2, but this can be expected for increasing the number of fitting parameters. A single Gaussian fits the distribution with sufficient accuracy, compared to the poor fit with a single Gaussian function for the Ks-band polarization angles. Typical errors of the polarization angle reach about 12°.
The polarization degree also appears to be quite uniform over the FOV, with typical values of 8-12%. Fitting the logarithms of the polarization degrees with a single Gaussian leads to a peak at $(9.8 \pm 0.7)\%$ (FWHM of 1.7%), satisfyingly matching the data (see Fig.3b). The data was also fitted with two Gaussian peaks for comparison, but as it was the case for the polarization angles, this only improves the fit marginally. The relative uncertainties of the polarization degree are of the order of up to 40%.

3.3. Polarization efficiency

The Ks-band polarization degrees were compared to the extinction map presented by [11]. Fig.4 shows the polarization efficiency $\frac{p_{Ks}}{A_{Ks}}$ plotted against the Ks-band extinction A_{Ks} taken from the extinction map at the location of each source. As it turns out, almost the same distribution of A_{Ks} is found for sources with $p_{Ks} < 6\%$ (pK$^-$) and $p_{Ks} > 6\%$ (pK$^+$), which in turn leads to an offset between the two sub-datasets in polarization efficiency. They are therefore plotted separately in Fig.4. Sources with less than 3% polarization were excluded, since this low value indicates either a foreground source or intrinsic polarization perpendicular to the foreground (e.g. in IRS 1W). In both of these cases, a comparison with an extinction map would not produce meaningful results.
The distributions can be fitted with a power law:

$$\frac{p_\lambda}{A_\lambda} \propto A_{Ks}^{\beta} \tag{1}$$

, yielding power law indices of $\beta_{pK^-} = -0.4 \pm 0.4$ resp. $\beta_{pK^+} = -0.5 \pm 0.7$. Despite the large errors which stem from considerable scatter of the parameters, the pK$^+$ and pK$^-$ values match the relation found by [5], who examined a large number of sources covering a range in optical depth of about a factor of 100. In that study, a power law relation with $\beta \sim -0.25$ was found, and the author proposed a model where the magnetic field along the LOS consists of a constant and a random component, thus leading to different grain alignment in each section along the

Journal of Physics: Conference Series **372** (2012) 012021 doi:10.1088/1742-6596/372/1/012021

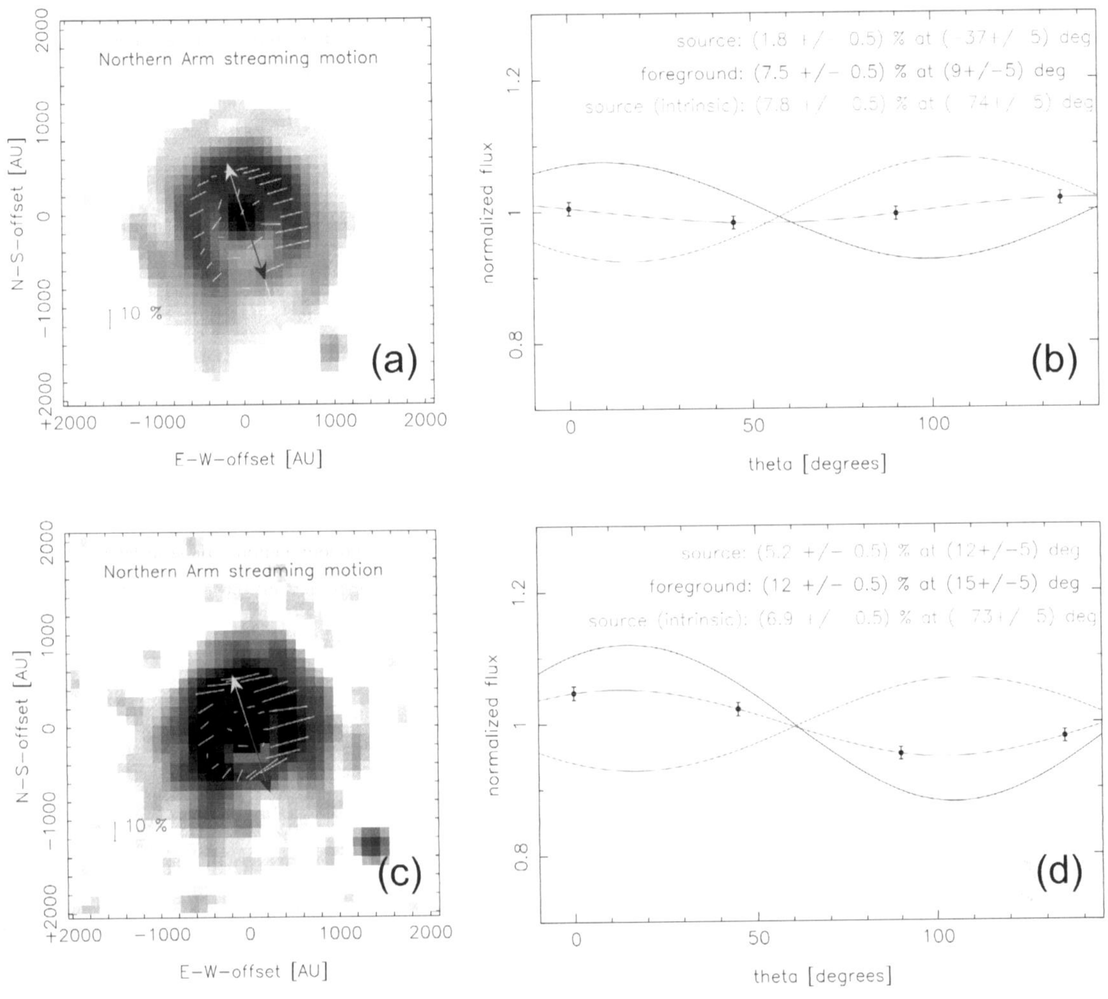

Figure 5. Polarization of the bow-shock source IRS 1W. (a): Ks-band polarization map, with polarization vectors (green lines) and projected streaming/proper motion vectors. (b): total, foreground, and intrinsic polarization (Ks-band). (c): H-band polarization map, see (a). (d): total, foreground, and intrinsic polarization (H-band).

LOS. This reproduces the findings in that study quite well, and it is also consistent with the results presented here within the uncertainties.

Compared to the pK^- values, a significant offset in polarization efficiency is detected for the pK^+ sources, while the underlying power law appears to be very similar. This might indicate that the additional polarization observed for the pK^+-sources is caused by a local contribution, likely by dust in the central parsec itself. Even if this additional local dust only contributes an insignificant amount to the total extinction, it can still lead to such a strong influence on the total polarization, since the local dust can be assumed to be aligned efficiently. A similar dust column density along the LOS might also consist of aligned grains, but the random component would be averaged out and leave only a small net contribution. In order to produce such a deviation along the LOS, a very specific and therefore unlikely dust configuration would be required.

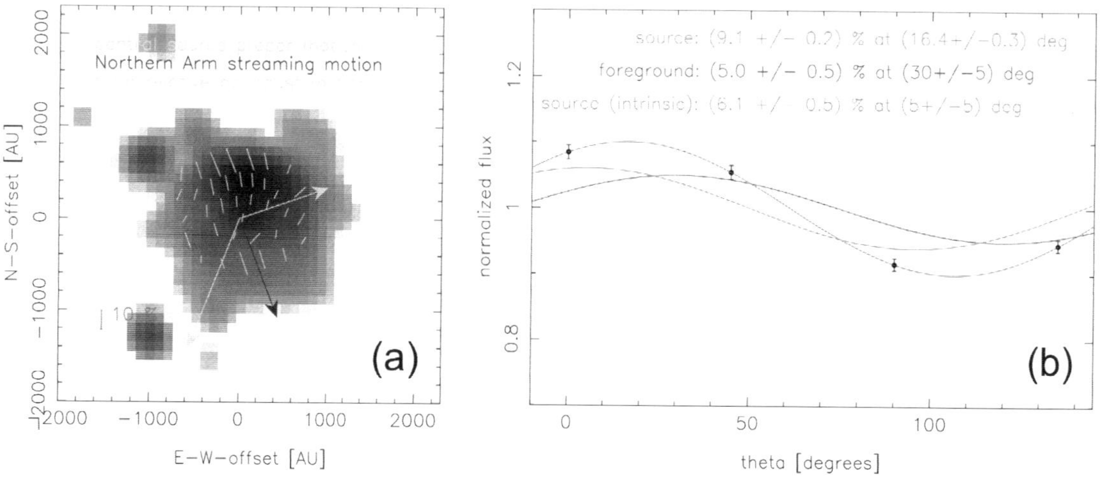

Figure 6. Polarization of the bow-shock source IRS 21. (a): Ks-band polarization map, with polarization vectors (green lines) and projected streaming/proper motion vectors. (b): total, foreground, and intrinsic polarization (Ks-band).

3.4. Intrinsically polarized bow-shock sources

IRS 1W shows the characteristic horseshoe-shape of a bow-shock source (see e.g. [12]), in consistence with the relative velocities of the streaming material of the Northern Arm [6] at the location of the source and the proper motion of IRS 1W itself (velocities plotted in Fig.5, see [9]). The total polarization of IRS 1W was measured as (1.8 ± 0.5) % at (-37 ± 5)° East-of-North in the Ks-band. A higher value of (4.6 ± 2.5)% at $(-85° \pm 8°)$ is given by [7], but this may be an effect of contamination by neighboring sources due to insufficient resolution. If it is assumed that the foreground polarization for this source is the same as for the surrounding objects and a depolarization matrix with the parameters for $p = 7.6\%, \theta = 9.2°$ is applied, this yields a total intrinsic polarization of (7.8 ± 0.5) % at (-75 ± 5)°.

In the H-band, the total polarization of IRS 1W is measured as (5.2 ± 0.5) % at (12 ± 5)° East-of-North. The polarization angle appears typical for a stellar source affected by foreground polarization, but the polarization degree is much lower than the ~12% found for stellar sources in the vicinity. The application of a depolarization matrix with $p = 12\%, \theta = 15°$ leads to an intrinsic polarization of (6.9 ± 0.5) % at (-73 ± 5)°.

Fig.5(a and c) shows resolved polarimetry of IRS1W in both wavelength bands. Polarization degrees of about 10-20 % with very similar polarization angles for regions with significant flux are found. Apparently, the polarization degree is lower by a factor of up to ~3 around the apex compared to the tails (this feature only appears in the Ks-band, while the observed polarization angles and degrees are otherwise very similar in both bands).

IRS 21 is not covered by the 2009, high quality polarimetric dataset, but several data-sets with a rotated FOV are available in the ESO archive that contain this source (the data-set used here has the best Strehl ratio of the available data with this FOV). The total polarization of IRS 21 was measured as (9.1 ± 0.2) % at $(16.4° \pm 0.3°)$, which agrees very well with the value given by [7] (IRS 21 is more isolated than IRS 1W, so lower spatial resolution can still provide accurate results). After applying a depolarization matrix with polarization parameters determined from the surrounding point sources (5% at 30°, determined on only two sources close to IRS 21 and thus not as reliable as the values for IRS 1W), IRS 21 appears to have a total intrinsic polarization of (6.1 ± 0.5) % at (5 ± 5) degrees. Unfortunately no polarimetric H-band data covering IRS 21 are available.

Fig.6a shows the Ks-band polarization of individual regions. Polarization degrees of about 3-8 % are measured here, with less uniform polarization angles than those found for IRS 1W in regions with significant flux. In addition, rather an increase than a decrease of the polarization degree is detected towards the apex.
The intrinsic polarization degree found for this source is slightly lower than that of IRS 1W, while both sources have very different intrinsic polarization angles. This can be explained by the relative motion of both sources to the ambient material and the local magnetic fields (see [9] for the proper motions of both sources, the streaming motion in the Northern Arm was taken from [6], while the magnetic field orientation was measured by [1]). The polarization angles of both sources are perpendicular to the magnetic field lines, but their motion relative to the field differs: IRS 1W moves parallel to the field, which would lead to a *warping* of the field lines around the bow-shock. While the turbulence at the apex tends to randomize grain orientation (and thus reduces the polarization), the concentrated magnetic field in the tails of the bow-shock may be of sufficient strength to align the grains fast enough so that they produce the observed pattern via emission or scattering. In contrast to this, IRS 21 moves perpendicular to the field, so we assume the field lines are compressed in the shock front. This would increase the field strength at the apex and produce a stronger grain alignment (and therefore polarization) there. Multiple scattering and local dichroic extinction may also play a role in both sources (depending on the dust parameters), but the extent of all these effects cannot be determined without thoroughly modeling the conditions in such a bow-shock environment.

4. Conclusions

(i) While the foreground polarization follows the orientation of the Galactic plane on large scales in the H- and Ks-band, local deviations exist and point to the influence of magnetically aligned dust particles in the mini-spiral.

(ii) The polarization efficiency follows a power law that matches previous models of a combination of a constant and a random alignment component along the LOS. The two sub-groups of stronger and weaker polarized sources show an offset in polarization efficiency, which supports the idea that this deviation is a local effect and not caused by a very specific and unlikely alignment of dust clouds along the LOS.

(iii) Both bow-shock sources in the FOV show polarization (in total and resolved) consistent with emission and/or scattering on aligned dust grains. This may indicate that the local fields are strong enough to dominate grain alignment even in the turbulent bow-shock environment.

For a more in-depth analysis of the data presented here, please see [2].

References

[1] Aitken, D, Smith, C, Moore, T, and Roche, P 1998, *MNRAS* **299**, 743
[2] Buchholz, R M, Witzel, G, Schödel, R, Eckart, A, Bremer, M and Muzic, K 2011, *A&A* **534**, A117
[3] Eckart, A, Genzel, R, Hofmann, R, Sams, B J, and Tacconi-Garman, L E 1995, *ApJ* **445**, 23
[4] Gillessen, S, Eisenhauer, F, Trippe, S, Alexander, T, Genzel, R, Martins, F, and Ott, T 2009, *ApJ* **692**, 1075
[5] Jones, T J 1989, *ApJ* **346**, 728
[6] Lacy, J H, Achtermann, J M, and Serabyn, E 1991, *ApJ* **380**, L71
[7] Ott, T, Eckart, A, and Genzel, R 1999, *ApJ* **523**, 248
[8] Reid, M J, and Brunthaler, A 2004 *ApJ* **616**, 872
[9] Schödel, R, Merritt, D, and Eckart, A 2009, *A&A* **502**, 91
[10] Schödel, R 2010a, *A&A* **509**, 58
[11] Schödel, R, Najarro, F, Muzic, K, and Eckart, A 2010b, *A&A* **511**, 18
[12] Tanner, A, Ghez, A M, Morris, M, and Christou, J C 2005, *ApJ* **624**, 742
[13] Witzel, G, Eckart, A, Buchholz, R M, Zamaninasab, M, Lenzen, R, Schödel, R, Araujo, C, Sabha, N, Bremer, M, Karas, V, et al. 2011, *A&A* **525**, 130

Astronomy at High Angular Resolution 2011 IOP Publishing
Journal of Physics: Conference Series **372** (2012) 012022 doi:10.1088/1742-6596/372/1/012022

Flare emission from Sagittarius A*

A. Eckart[1,2], M. García-Marín[1], S. N. Vogel[3], P. Teuben[3], M.R. Morris[4], F. Baganoff[5], J. Dexter[6,7], R. Schödel[8], G. Witzel[1], M. Valencia-S.[1,2], V. Karas[9], D. Kunneriath[9], M. Bremer[1], C. Straubmeier[1], L. Moser[1], N. Sabha[1], R. Buchholz[1], M. Zamaninasab[2], K. Mužić[10], J. Moultaka[11], J. A. Zensus[2]

1) I.Physikalisches Institut, Universität zu Köln, Zülpicher Str.77, 50937 Köln, Germany
2) Max-Planck-Institut für Radioastronomie, Auf dem Hügel 69, 53121 Bonn, Germany,
3) Department of Astronomy, University of Maryland, College Park, MD 20742-2421, USA,
4) University of California, Los Angeles, Department of Physics and Astronomy, USA,
5) Massachusetts Institute of Technology, Kavli Institute for Astrophysics and Space Research, USA,
6) Department of Physics, University of Washington, Seattle, WA 98195-1560, USA ,
7)Theoretical Astrophysics Center and Department of Astronomy, University of California, Berkeley, CA 94720-3411, USA
8) Instituto de Astrofísica de Andalucía (CSIC), Glorieta de la Astronomía s/n, 18008 Granada, Spain,
9) Astronomical Institute AV CR, Prague, Czech Republic,
10) Department of Astronomy & Astrophysics, University of Toronto, 50 St. George Street, Toronto, ON M5S 3H4, Canada
11) LATT, Université de Toulouse, CNRS, 14, Avenue Edouard Belin, 31400 Toulouse, France

E-mail: `eckart@ph1.uni-koeln.de`

Abstract. Based on Bremer et al. (2011) and Eckart et al. (2012) we report on simultaneous observations and modeling of the millimeter, near-infrared, and X-ray flare emission of the source Sagittarius A* (SgrA*) associated with the super-massive ($4\times10^6 M_\odot$) black hole at the Galactic Center. We study physical processes giving rise to the variable emission of SgrA* from the radio to the X-ray domain. To explain the statistics of the observed variability of the (sub-)mm spectrum of SgrA*, we use a sample of simultaneous NIR/X-ray flare peaks and model the flares using a synchrotron and SSC mechanism. The observations reveal flaring activity in all wavelength bands that can be modeled as the signal from adiabatically expanding synchrotron self-Compton (SSC) components. The model parameters suggest that either the adiabatically expanding source components have a bulk motion larger than v_{exp} or the expanding material contributes to a corona or disk, confined to the immediate surroundings of SgrA*. For the bulk of the synchrotron and SSC models, we find synchrotron turnover frequencies in the range 300-400 GHz. For the pure synchrotron models this results in densities of relativistic particles of the order of $10^{6.5}$cm^{-3} and for the SSC models, the median densities are about one order of magnitude higher. However, to obtain a realistic description of the frequency-dependent variability amplitude of SgrA*, models with higher turnover frequencies and even higher densities are required. We discuss the results in the framework of possible deviations from equilibrium between particle and magnetic field energy. We also summarize alternative models to explain the broad-band variability of SgrA*.

1. Introduction

Sagittarius A* (SgrA*) is the closest super-massive black hole (SMBH) and is the prime candidate to study the variability and spectral properties of accreting SMBHs. Since it is at the center of the Milky Way at a distance of about 8 kpc, Doeleman et al. (2008, 2009) and Fish et al. (2011) have recently succeeded in obtaining structural information on event-horizon scales through very long baseline interferometry (VLBI) at 1.3 mm wavelength. SgrA* is strongly variable in the radio and millimeter wavelength regime (Zhao et al. 2003; Mauerhan et al. 2005; Eckart et al. 2008ac; Marrone et al. 2008; Li et al. 2009; Yusef-Zadeh et al. 2008, 2009). It shows an inverted spectrum from the radio to the (sub-)millimeter domain (Falcke et al. 2000) and displays order-of-magnitude flares in the infrared and X-ray domain (Baganoff et al. 2001, 2003; Genzel et al. 2003; Eckart et al. 2004, 2006a, 2010, Ghez et al. 2003, 2004). The sub-mm spectrum of SgrA* itself is rather unexplored owing to the difficulty of separating it from contributions of the surrounding "mini-spiral" and the circum-nuclear disk (CND).

While the so-called sub-millimeter bump is attributed to SgrA* and is thought to be due to relativistic, thermal electrons of a hot, thick, advection-dominated accretion flow (Dexter et al. 2009, 2010; Narayan et al. 1995; Yuan et al. 2003), it is, however, currently unclear whether the radio spectrum is due to contributions from an accretion flow (Yuan et al. 2003), due to a jet (Falcke & Markoff 2000), or a combination of both. The strong variability observed at radio to X-ray emission is most likely due to synchrotron and/or synchrotron self-Compton (SSC) radiation. Part of the radiation may be due to a single hot-spot or a multi-spot model in the mid-plane of the accretion flow or an increased accretion rate (see models by Broderick & Loeb 2006 and Eckart et al. 2006b, 2011, Meyer et al. 2007, Yuan et al. 2008, see also Pecháček et al. 2008, Shcherbakov & Baganoff 2010, Dexter et al. 2010, Zamaninasab et al. 2011). Spiral arm models (e.g. Karas et al. 2007) or jet/jet-base models (Falcke & Markoff 2000, Markoff 2005, Markoff et al. 2007) are also possible. Another result from multi-wavelength observations of SgrA* is that the process of adiabatic expansion of source components may be relevant (Eckart et al. 2006a, 2008abc, 2009, 2010, 2012, Yusef-Zadeh et al. 2006a, 2008, 2009). Here we assume that the sub-mm, NIR and X-ray flux excursions are physically related (However, see a detailed discussion on this in section 5.3 in Eckart et al. 2012). This expansion can explain the observed time lags between the infrared/X-ray and millimeter emission peaks. The fact that one observes different time delays between NIR and sub-mm flares can be understood as a consequence of a spread in source sizes and synchrotron turnover frequencies. The idea of adiabatic expansion is supported by the fact that there is no conclusive observational evidence for sub-mm flares preceeding NIR events and by the fact that it has been detected at radio cm-wavelengths (Yusef-Zadeh et al. 2006ab). Here we summarize results obtained for simultaneous flare emissions in the NIR and X-ray with implications to the observed radio variability. In this summary we include the analysis of the NIR spectral index under the assumption that especially for weak flares synchrotron losses are important. A detailed description of the Synchrotron/SSC analysis is given in Eckart et al. (2012). Throughout the article we use $S_\nu \propto \nu^{-\alpha}$.

2. Observations

The analysis summarized in this paper is based on Bremer et al. (2011) and Eckart et al. (2012) who have studied a sample of NIR H- and Ks-band flares and a sample of all X-ray flares (until 2010) that showed within less than about 10 minutes a near-infrared emission peak. The detailed flare data are listed in Tab.3 by Eckart et al. (2012). In this paper we use a consistent calibration based on the extinction value of A_K=2.46 (Schödel et al. 2010). For the L'-band flares (Dodds-Eden et al. 2009; Trap et al. 2011ab), we used A'_L=1.23 (Schödel et al. 2010) and extrapolated to the Ks-band with a spectral index of +0.7±0.3 (Hornstein et al. 2007, see also Bremer et al. 2011).

3. Interpretation

3.1. The NIR spectral index

Bremer et al. (2011) point out that the broad-band near-infrared spectral index is an essential quantity to investigate the physical mechanism that underlies the observed SgrA* flare emission. The authors present a method to derive the NIR spectral index of SgrA* between the H- and Ks-band from the statistics of the peak flux density of the SgrA* flares. They base their spectral index derivation on an unprecedentedly large time-base of about seven years of monitoring the infrared counterpart of SgrA*. The underlying assumption is that for SgrA* in both NIR bands the same optically thin, dominant emission mechanism is at work and produces similar number distributions of flares. The author could then cross-correlate the peak flare flux histograms and determine the expectation value of the H-Ks-band spectral index during the bright flare phases of SgrA*. (For a detailed analysis of the possible contribution of stars within the central arcseconds to the flux density at (of close to) the position of SgrA* and the statistical analysis of K-band light curves see Sabha et al. 2012 and Witzel et al. 2012).

Using this method, Bremer et al. (2011) can confirm that the spectral index for brighter flares is consistent with α=+0.7 ($S_\nu \propto \nu^{-\alpha}$) which is indicative of optically thin synchrotron radiation. In addition the authors find a tendency for the weaker flares to exhibit a steeper near-infrared spectrum. From this they can conclude that the distribution of spectral indices as a function of peak Ks-band flare flux density can successfully be represented by an exponential cutoff proportional to $exp[-(\nu/\nu_0)^{0.5}]$, were ν_0 is a characteristic cutoff frequency. Such a cutoff is expected if synchrotron losses are at work. If ν_0 varies between the NIR and sub-mm wavelength domain, and if the peak flare sub-mm flux density variation is about one Jansky, then the model explains the observed spectral properties of SgrA* in the NIR. For the model calculations we assume that the intrinsic optically thin spectral index of these synchrotron flares is α=+0.7±0.3. The spectral index cannot be substantially steeper in order not to violate the MIR flux density limit (Schödel et al. 2007, 2011). For a peak flare flux density in the Ks-band of 7-6 mJy, the spectral index of +0.7 results in a ~1.5 mJy flux density difference with respect to the H-band. This corresponds to a $\geq 3\sigma$ quantity for the relative calibration uncertainty of about 10% between the two NIR bands. For the combined ≥30 flares per Ks- and H-band the flux density differences due to the +0.7 spectral index is significant for the bulk of the detected flares.

Fig. 1 shows the relation between the H-Ks-band spectral index and the observed Ks-band peak flux density as obtained for a power law assuming an exponential cutoff due to synchrotron losses. The thick solid black curves show model calculations that cover the expected spectral synchrotron indices and a 1 THz flare strength corresponding to the degree of variability observed at high frequencies. Model α is based on a power law spectral index of α=+0.4 and a flux of 0.071 Jy at 1 THz. As laid out in Eckart et al. (2012) we used for model β a value of α=+0.8 and 0.5 Jy, for model γ values of α=+0.6 and 0.76 Jy, and for model δ values of α=+0.4 and 1.4 Jy. Above a Ks-band flux density of 12 mJy, models α, γ, and δ have flux densities above 8.5 mJy, and model β has a 8.6 μm flux densities above the 22 mJy flux limit (Schödel et al. 2007, Schödel et al. 2011) at which SgrA* has not yet been detected.

In Fig. 1 we also show rectangular boxes that mark the flux densities and NIR spectral index ranges obtained from the broad-band NIR spectroscopy (Gillessen et al. 2006) from the H-/K'-imaging (Hornstein et al. 2007, see also Bremer et al. 2011). The values that correspond to the blue boxes are given in Tab.4 of Bremer et al. (2011). In region A the sub-mm/NIR flares are brighter and have flatter synchrotron spectra. The synchrotron loss turnover frequencies are well above 10^{13}Hz. In region B the sub-mm/NIR flares are weaker and steeper, with synchrotron loss turnover frequencies are below 10^{13}Hz.

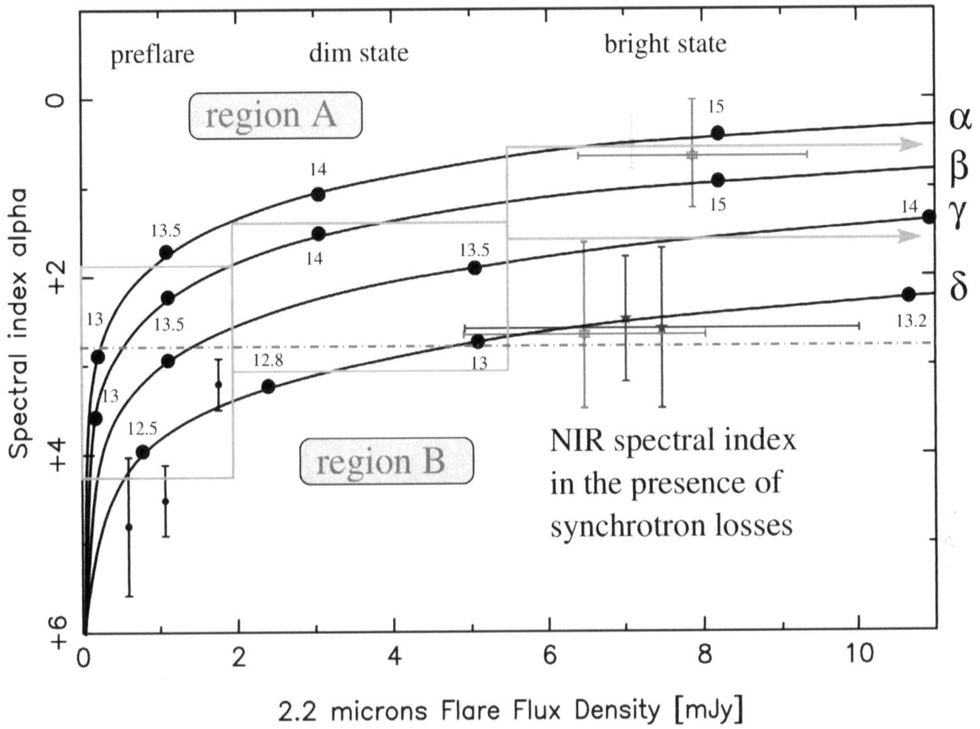

Figure 1. The H-Ks-band spectral index versus the de-reddened Ks-band flux density (Bremer et al. 2011). The black dots are the logarithms of the cutoff frequencies and the black solid lines are theoretical curves α to δ explained in the text. These curves are parameterized and labeled with the logarithm of the cutoff frequency. The boxes and the regions A and B separated by a dashed-dotted line are explained in the text.

3.2. NIR/X-ray flares

In Eckart et al. (2012) we have described physical and statistical aspects of the relation between sub-mm and NIR/X-ray flare emission of SgrA*. For most synchrotron and SSC models, the authors find synchrotron turnover frequencies in the range of 300-400 GHz. For the pure synchrotron models this gives densities of relativistic particles of the order of $10^{6.5}$cm^{-3}. For the pure SSC models, the median densities are about one order of magnitude higher. In order to obtain, however, a realistic description of the frequency-dependent variability amplitude of SgrA*, we require models with substantially higher turnover frequencies and higher densities.

3.2.1. Formalism

Eckart et al. (2012) adopt the formulae previously presented by Marscher (1983, 2009). For the case of the Galactic Center, they find that the SSC X-ray flux density $S_{X,SSC}$ (in μJy), magnetic field B (in G), and column density of relativistic electrons N_0 (in cm^{-3}keV^{-1}) involved in the emission process can be described by

$$S_{X,SSC} = d(\alpha)\ln(\frac{\nu_2}{\nu_m})\theta^{-2(2\alpha+3)}\nu_m^{-(3\alpha+5)}S_m^{2(\alpha+2)}E_X^{-\alpha}\delta^{-2(\alpha+2)} \quad , \tag{1}$$

$$B = 10^{-5}b(\alpha)\theta^4\nu_m^5 S_m^{-2}\delta \quad , \tag{2}$$

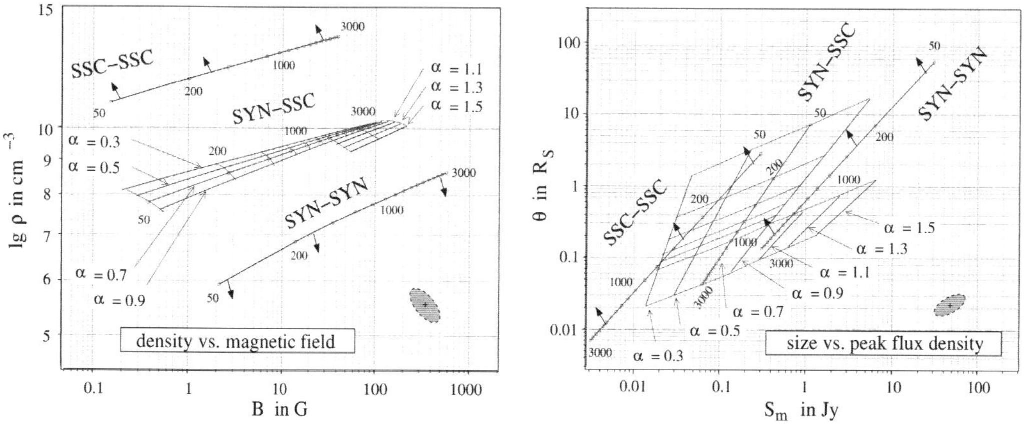

Figure 2. Volume density of relativistic electrons versus magnetic field and source size versus synchrotron peak flux density for the three major source models SYN-SYN, SYN-SSC and SSC-SSC as described in the text and in Eckart et al. (2012). The graphs describe a specific flare observed on 7 July 2004. The ellipses represent uncertainties and the thick arrows indicate the direction into which the SSC-SSC and SYN-SYN model line will move if the synchrotron or self-Compton limit is lowered. The red circles are individual model data points connected with black solid lines. Limits on the density, magnetic field, size and peak flux density S_m result in a spectral distribution of S_m values that can be compared with the distribution of the observed data. Details for the figure are given in the text and by Eckart et al. (2012).

$$N_0 = n(\alpha) D_{Gpc}^{-1} \theta^{-(4\alpha+7)} \nu_m^{-(4\alpha+5)} S_m^{2\alpha+3} \delta^{-2(\alpha+2)} \quad . \tag{3}$$

Here $d(\alpha)$, $b(\alpha)$, and $n(\alpha)$ are dimensionless parameters (see Marscher 1983). For the case that they are all functions of α, D_{Gpc} is the luminosity distance in gigaparsecs and E_X the X-ray photon energy in keV. The underlying relativistic electron distribution is expressed as $N = N_0 \; \exp(-p)$ with an electron power-law index p.

Eckart et al. (2012) then outline that the peak flux S_m, source size θ, magnetic field B, and relativistic particle density N_0 can all be expressed as power laws of the frequency ν:

$$S_m = \kappa_1 \nu_m^{-\alpha} \quad , \tag{4}$$

$$\theta = \kappa_2 \nu_m^{\zeta_1} \quad , \tag{5}$$

$$B = \hat{\rho} \nu_m^{\zeta_2} \quad , \tag{6}$$

and

$$N_0 = \kappa_3 \nu_m^{\zeta_3} \quad , \tag{7}$$

with coefficients and exponents given in their paper.

3.2.2. Three cases of describing the spectra

One can think of three different, basic cases in order to discuss the contributions of a single source emitting synchrotron (SYN) and synchrotron self-Compton (SSC) radiation in the radio, NIR and X-ray domains. The three combinations are abbreviated as SSC-SSC, SYN-SYN and SYN-SSC. The first label SSC-SSC indicates that both the NIR and X-ray fluxes are due to SSC scattered photons and less than 10% are due to a pure SYN contribution. The second label SYN-SYN indicates that a dominant contribution in these two wavelength bands is due to a SYN contribution and again less than 10% is due to a SSC contribution. The third label SYN-SSC indicates that the NIR flux density is mainly due to a SYN contribution and that the predominant amount of X-ray radiation is caused by a SSC contribution. In the two cases of SYN-SYN and SSC-SSC, the spectral indices in the NIR and X-ray domains are the same (unless synchrotron losses are involved). The spectral index can be calculated from the flux densities in the two bands. In Fig. 2 we show the results of model calculations for one particular flare. In the case of SYN-SSC, the graphs are labeled with the value of the optically thin SYN spectral index. The model graphs are only shown as thin solid lines if the NIR and X-ray flux densities are not contaminated by more than 10% by the SSC or pure SYN contributions, respectively. Mixed models in which both contaminating contributions are higher than 10% can be located between the SYN-SYN and SSC-SSC graphs. The models also obey the MIR flux density limit (Schödel et al. 2007, Schödel et al. 2011; see Eckart et al. 2012 for details).

Models in Fig. 2 were calculated for a cutoff frequency range between 50 GHz and 3 THz. The figure shows that the typical ranges of relativistic electron densities for the SYN-SYN, the SYN-SSC, and the SSC-SSC models are around 10^7, 10^9, and 10^{12} cm^{-3}, respectively. In general the relativistic electron density increases with increasing magnetic field strength B, spectral index α, and synchrotron cutoff frequency. From SSC-SSC via SYN-SSC to SYN-SYN models the peak flux density S_m and source size θ both increase. This is also the case for decreasing cutoff frequency ν_m and spectral index α.

For θ<2 R_S, S_m< 2 Jy, and B < 30 G, a larger number of NIR/X-ray flares can be described by SYN-SSC and SSC-SSC models, and cutoff frequencies of a few hundred GHz are preferred. For the rejected SSC-SSC and the SYN-SYN models, the violation of the MIR flux density limit prohibits these models. For the SYN-SSC, models can be rejected on the basis of excessive magnetic field strength B, source size θ, and peak flux density S_m.

We consider the observed frequency-dependent variability amplitude ranges as an important observable that needs to be explained by SgrA* source models. Eckart et al. (2012) list in Tab.5 the upper limits to B, S_m, and θ, the lower limit to $\log(\rho)$, along with the best-fit χ^2 value obtained using the given parameter limits. For the SYN-SSC case, we used the full range of spectral indices to describe the sub-mm flux density variation of SgrA*. With the restricted set of parameters in Tab.5 (Eckart et al. 2012) the models represent the observed frequency-dependent variability amplitudes quite well. As explained in more detail in Eckart et al. (2012) for the SSC-SSC case, the degree of variability at high frequencies is too small to reflect the measurements well. For the SYN-SYN case, the degree of variability at low frequencies is much larger than observed. Also, most models are unable to provide an overall description of the SgrA* variability. Under the assumption of a single source component we can summarize that only the SYN-SSC model is capable in reproducing the observed SgrA* variability.

3.2.3. Equipartition

The analysis by Eckart et al. (2012) has shown that for a realistic description of the observed frequency-dependent variability amplitudes of SgrA* the single source component descriptions require high turnover frequencies (rather a few 100 GHz than a few 10 GHz) and at least densities that are 10 to 100 times the mean density of the larger scale accretion flow. The exact properties of matter close to the SgrA* SMBH are not constrained very well by existing observations. However, the accretion rate and hence the central density may be much higher than the limits derived from Faraday rotation. Also the magnetic field equipartition fraction and the bias field strength in the case of magnetic field reversals are not well constrained (see Marrone et al. 2007 and Igumenshchev et al. 2003 for details). While equipartition in the larger scale SgrA* accretion flow appears to be justified, deviations from equipartition may occur close to the midplane or temporary accretion disk of SgrA*. Such deviations are likely to occur as a

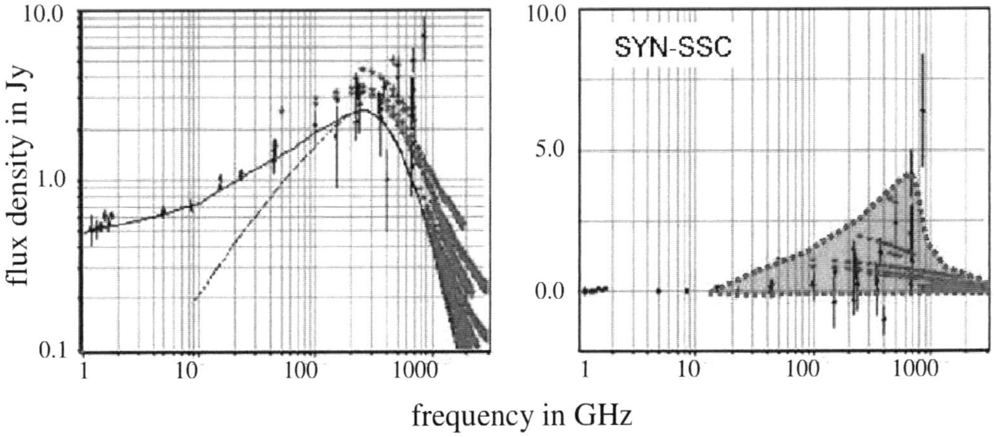

Figure 3. Observed flux densities of Sagittarius A* (Eckart et al. 2012) taken from the literature (blue) compared to a combined model that consists of the fit given by Falcke et al. (2000), Marrone et al. (2008) (black line), and Dexter et al. (2010) (black dashed line). The spectrum of synchrotron self-absorption frequencies for the range of models discussed here is shown in red. We show results for the SYN-SSC models corresponding to the parameters in Tab.5 by Eckart et al. (2012). The SYN-SSC is the preferred model and represents most closely the observed variability of SgrA*. On the left, we show the complete model, on the right we show only the variable flux densities as observed (blue) and modeled (red).

consequence of plasma instabilities in the immediate vicinity of SgrA* as two-stream instabilities of collisionless shocks.

The two-stream instability is known as a common instability in plasma physics. One of the ways to induce it involves a high energetic particle stream injected into a plasma. Another way is by creating conditions in which different plasma species (ions and electrons) have different drift velocities. The energy contained and released from the particles can then lead to plasma wave excitation. The importance of the two-stream instability e.g. in Gamma-Ray Burst Sources (GBR) is highlighted by Medvedev & Loeb (1999). Through the generation of magnetic fields in in a relativistic shock, in this case one expects deviations from equipartition between magnetic field and particle energy by a factor of up to 10^5. The effect of beam-plasma instabilities on accretion disk and the associated flares are discussed in Krishan, Wiita & Ramadurai (2000) and Keppens, Casse & Goedbloed (2002). The importance of acceleration of particles by 1st and 2nd order Fermi processes in the framework of relativistic jets and accretion discs around SMBH is highlighted by e.g. Kowal et al. (2011) and Nishikawa et al (2011).

Particles may interact with each other not through Coulomb collisions but through collective behavior within plasma waves, i.e. collisionless shocks. In these shocks, the transition from pre-shock to post-shock states occurs on a length scale that is much smaller than the mean free path for particle collisions. The Weibel instability (Weibel 1959) represents such a collective response. It can occur both in magnetized and initially non-magnetized low-temperature relativistic flows. In the latter they work apparently best (see details in the review by Bykov & Treumann 2011). Another way to induce them is by having flows parallel to an ambient magnetic field - as e.g. in case of a toroidal magnetic field structure within the mid-plane of an accretion flow or an accretion disk. For initially non-magnetic and for magnetic shocks the average magnetic energy remains at least about 15-20% below the equipartition at the shock. Magnetic fields induced by Weibel instabilities may form vortices, and 3D PIC (particle in cell) simulations indicate that they may result in an extended downstream region with small-scale magnetic fields. In the framework of accretion disks they might then be regarded as single or multiple hot spots.

The case of SgrA:* Following Homan et al. (2006) and Readhead (1994) the intrinsic brightness temperature, T_{int}, is related to the equipartition brightness temperature T_{eq} at energy balance by the expression $T_{int} = \eta^{1/8.5} T_{eq}$, where $\eta = u_p/u_B$ is the ratio of the energy densities of the radiating particles, u_p, to the energy density of the magnetic field, u_B. The exponent of η assumes that the radiating particles follow a power-law Lorentz distribution with index $p = 2.5$ for $N_\gamma d\gamma = K\gamma^p$ and $p = 2\alpha + 1$. This value of p corresponds to a spectral index of $\alpha = 0.75$ ($S \propto \nu^{-\alpha}$) in the optically thin parts of a jet or outflow.

Recent and ongoing VLBI observations at 1.3 mm (Doeleman et al. 2008, 2009; Fish et al. 2011) determined a lower limit to the brightness temperature of the VLBI source SgrA* at 230 GHz of $T_b = 2 \times 10^{10}$ K. This is close to the equipartition brightness temperature of $T_{eq} = 5 \times 10^{10}$ K that has been found by Readhead (1994) for a sample of accreting super-massive black holes. So we may assume that the mm-radiation source SgrA* is close to equipartition in its more quiescent phases. The intrinsic brightness temperature of SgrA* is then limited either by the inverse Compton limit ($T_{int} \leq 10^{12}$ K; Kellermann & Pauliny-Toth 1969, Pauliny-Toth & Kellermann 1966) or by the possibility that the emitting source is near equipartition to begin with so that $T_{int} = T_{eq}$ (Singal & Gopal-Krishna 1985; Singal 1986; Readhead 1994). We measure $T_{obs} = \delta T_{int}$, δ being the relativistic boosting factor and for SgrA* we can assume $\delta \sim 1$ (details in Eckart et al. 2012). In the sub-mm regime SgrA* can easily show flares that are 30% to 100% in excess over the low state (which may be subject to longer variability as well; see Dexter et al. 2010). This already implies significant deviations from equipartition with values for η in the range of 10 to 350. As outlined above, such an imbalance may be the result of an injection or acceleration of particles generated through plasma instabilities close to the SMBH and possibly at the footpoint of a jet or outflow.

3.3. Polarization

The radio/NIR/X-ray flares (e.g. Fig.3) of increased flux last for about 100 min and are usually accompanied by variations in the polarized emission (e.g. Eckart & Genzel 1996, Ghez et al. 1998,2004 , Schödel et al. 2002, Baganoff et al. 2001, 2003, Eckart et al. 2004, 2006b, 2008ab, Nishiyama et al. 2009, Zamaninasab et al. 2010, Kunneriath et al. 2010). This time scale fits well with the mm-wave ~hourly timescales and structure function knees at 30-100 min found on individual occasions by Mauerhan et al. (2005) and Yusef-Zadeh et al. (2011). The mm/sub-mm wavelength linear polarization of SgrA* is just below 10% and variable in magnitude and position angle (PA) on timescales down to a few hours (Bower et al. 1999a). At 227 and 343 GHz Marrone et al. (2007) have determined an almost constant rotation measure of about -5.6$\times 10^{-5}$ rad m^{-1}. The mean intrinsic PA of 167$^o\pm$30° limits with its uncertainties, the accretion rate fluctuations to 25%. The accretion rate is limited to $<2\times 10^{-7}$ M_{solar} yr^{-1} if the magnetic field is near equipartition, ordered, and largely radial, or has values of $<2\times 10^{-9}$ M_{solar} yr^{-1} if there is a sub-equipartition, disordered, or toroidal field. Marrone et al. (2007) also find that the rather constant mean intrinsic PA is probably originating in the sub-millimeter photosphere of SgrA*, rather than arising from rotation measure changes. Circular polarization has also been detected for SgrA* (0.5% at cm-wavelengths), with a rising polarization fraction from 1.4 to 15 GHz (Bower et al. 2002, Bower et al. 1999abc; Sault & Macquart 1999) and further to 345 GHz (1.2% at sub-mm wavelengths; Munoz et al. 2012). This expected circular polarization at mm/sub-mm wavelengths (Beckert & Falcke 2002; Beckert 2003, Tsuboi et al. 2003; Bower et al. 2003ab, 2005) is most likely due to Faraday conversion of linear polarization which is also measured with increasing intensity towards high frequencies (Munoz et al. 2012). Hourly variable sub-mm emission (Marrone et al. 2006, Mauerhan et al. 2005, Miyazaki et al. 2004) and a very compact sub-mm VLBI structure (Doeleman et al. 2008/9) possibly point towards plasma blobs on relativistic orbits close to the event horizon. Hence, simultaneous observations of the variability at mm/sub-mm wavelengths are fundamental for constraining the radiative processes. Contributions to the flare emission from red noise processes are likely to be important (see Do et al. 2009, Eckart et al. 2008a). However, Zamaninasab et al. (2010, 2011) have demonstrated that highly polarized NIR sub-flares (on time scales of $\sim$20 min) are statistically significant compared to the randomly polarized red-noise. In addition to intensity variations due to relativistic boosting, the PA is rotated in the strong gravitational field of the SMBH. In these cases we can therefore estimate spin and spin-vector orientation of the SMBH (i.e. the accretion disk) independently from both simultaneous NIR and mm-polarization data in comparison to disk and jet models that we generate using a relativistic 2D- and 3D-disk codes (Dovciak, Karas, & Yaqoob 2004 and Bursa et al. 2012, this volume; running in Prague and Cologne). This code is capable of handling polarized emission and relativistic radiative transfer. It allows us to describe an orbiting and gradually evolving spot on different trajectories (in a disk or jet) near a rotating black hole including optical depth effects.

3.4. Alternative models
In the SSC-SSC, SYN-SSC, and SYN-SYN models discussed above (see section 3.2.2) a single population of relativistic electrons is responsible for the emission at different wavelengths. One can think of more complex models in which different populations of thermal or relativistic electrons are responsible for the emission. Here we briefly discuss two of these models. In both cases the SYN and SSC contributions have to be carefully tuned such that they contribute predominantly only in one frequency domain and cannot be attributed to the three cases mentioned above.

3.4.1. Double-SYN
If we assume that SgrA* is surrounded by a magnetized corona and - at least temporarily - by an accretion disk then the accreting medium is not uniform, and it may show inhomogeneities due to plasma instabilities operating in the weakly magnetized accretion flow. Such a dilute magnetosphere may contain individual magnetized source components which may lead to flares as they approach the black hole horizon (Galeev, Rosner & Vaiana, 1979, and subsequent works on patchy coronae). These two components may both be the source of synchrotron emission[1]. They may have different sizes, magnetic fields and particle densities and - more importantly - they may exist close to each other and may even be connected causally, e.g. outflows and the possibly back flowing envelopes or cocoons, or they may represent a jet or outflow and its base. In this double-SYN case the more compact or denser component could be responsible for the synchrotron X-ray contribution and the large and more tenuous component could be responsible for the mm/NIR synchrotron contribution.

3.4.2. External Compton scattering
The region responsible for the infrared emission may be different from that of the mm-emission and the mm-emission region may be optically thin for near-IR photons. Then the relativistic electrons in this source component are available to up-scatter near-IR seed photons to X-ray energies (Yusef-Zadeh et al. 2009). Since the scattering source is not identical in this case with the seed photon generating source, one may attribute this mechanism to the external inverse Compton scattering (ICS) mechanism in contrast to the synchrotron self Compton (SSC) excess. In order to allow for a high efficiency for inverse Compton scattering, the density of the scattering electrons and the scattered seed photons must be high. (For non-thermal sources this

[1] We do not see the necessity to propose a double-SSC model, given that substantial synchrotron emission as well as a high density of relativistic electron is required for both components.

is therefore usually best done by the source also responsible for the seed photons i.e. SSC.) Again the individual synchrotron and synchrotron self Compton contributions have to be well tuned such that the pure inverse Compton contribution remains dominant in the X-ray domain. This whole scenario may be complicated even further if one considers that the mm-wavelength emitting cloud may be optically thick and occults the quiescent or flaring emission from SgrA* (Yusef-Zadeh et al. 2010).

However, if one accepts the statement that the mm-wavelength flares are physically associated with the NIR/X-ray flares, then the observed adiabatic expansion velocity of the radio components (typically 0.01 c) and the observed radio/NIR/X-ray flare length of 20 minutes to one hour give a minimum size of the order of 0.3 to 1 R_s (Schwarzschild radii). However this is a typical size of an emitting source component close to the SMBH, such that sources - emitting at different wavelengths at or below this size - have to be treated as a single component (implying SSC rather than ICS) than two separate sources. Alternatively, one is forced to treat the emitting regions as independent regions and treat them in a fully stochastical description. These senarii become rapidly unphysical, since it is difficult to have a source only emitting in a single spectral domain without a significant contribution in neighboring wavelength domains.

3.4.3. Fermion and Boson balls

There is the attempt to explain very massive objects at the centers of galaxies by the concept of Fermion or Boson balls. The mere fact that we observe strong variability from SgrA* may already speak for a rather stabile than a fragile mass concentration at the center. A universal Fermion ball solution for compact galactic nuclei can be excluded by the results from stellar orbits near SgrA* (see references in Eckart et al. 2008abc, 2010). An alternative explanation is that of a massive Boson star (Torres, Capozziello & Lambiase 2000, Lu & Torres 2003 and references therein). This explanation is severely challenged by the general agreement between the measured polarized flare structure and the theoretical predictions (e.g. Eckart et al. 2006b, Zamaninasab et al. 2011). If the indicated signatures of orbiting matter (very close to the event horizon) are considered, then also a stationary Boson star may be excluded as an alternative solution for SgrA* (Eckart et al. 2006b and see also result for the nucleus of MCG-6-30-15 by Lu & Torres 2003). For these objects one expects the orbital periods to be larger than those inferred from modeling the NIR polarized flares. Especially ongoing accretion and star formation in the central cluster results in requirements to keep Fermion or Boson balls stable, which are most likely not met. A similar result is reached by Miller (2006) and Munyaneza & Biermann (2005) who discuss constraints on alternatives to super-massive black holes and their growth of super-massive black holes in galaxies.

3.4.4. Stars, Planets and Asteroids

The models for SgrA* presented above are based on descriptions derived for non-thermal sources associated with both stellar black holes and low- and high-luminosity accreting super-massive black holes in nuclei of external host galaxies. Since the luminosity of SgrA* is rather low compared to other SMBH sources and since it is the closest of such sources that we can study, there are attempts to explain the flare emission of SgrA* via different mechanisms. There are two major recent models: The first one involves a population of asteroids and planets that are disrupted by tidal forces exerted on them by SgrA* (Zubovas, Nayakshin & Markoff 2011). The second uses the coronal radiation of a cusp of spun-up stars to explain the X-ray luminosity of SgrA* (Sazonov, Sunyaev & Revnivtsev 2012). It needs to be investigated if both of these models can explain the full body of observations of SgrA* and if they can be tested by further observations.

4. Summary and Conclusion

Eckart et al. (2012) have shown that the bulk of synchrotron and SSC models applicable to SgrA* have preferred synchrotron turnover frequencies in the range of a few hundred GHz. For the pure synchrotron models, the densities of relativistic particles are of the order of $10^{6.5}\mathrm{cm}^{-3}$. For models involving SSC contributions the median densities are an order of magnitude higher. These values are quite comparable to those quoted for the accretion flow toward SgrA* (e.g. Yuan et al. 2003, 2004, Yusef-Zadeh et al. 2006ab).

For a realistic description of the observed frequency-dependent variability amplitudes of SgrA* higher densities and turnover frequencies are also required (details in Eckart et al. 2012). Higher densities are neither unreasonable nor out of reach. The accretion rate and hence the central density may be much higher than the limits derived from Faraday rotation (Bower et al. 2003, Bower 2003, Marrone 2006a, Marrone et al. 2006abc, and discussion by Marrone et al. 2007). In particular, the magnetic field equipartition fraction as well as the bias field strength in the case of magnetic field reversals - that can be expected for a turbulent flow - are unknown. Variations in the magnetic field structure as well as the field strength with respect to equipartition (Marrone et al. 2007, Igumenshchev et al. 2003) are likely to result in higher densities in the immediate vicinity of the central SMBH. While SgrA* may be close to equipartition, it is quite likely that at least during flux density events associated with very strong X-ray flares and the related NIR and (sub)mm-flux excursions, SgrA* may go through phases in which it is off equipartition.

Currently, it is difficult to decide which of all the models mentioned here describe the dominant contributors to the flare activity of SgrA*, or which combinations of effects are most likely or if one or some of the mechanisms can be excluded entirely. A key appears to be the restrictions imposed onto the models by the frequency dependant flux density variability or polarization properties. Therefore, the simultaneous flux density and polarization monitoring observing campaigns are essential. Any model (or combinations of those) must be able to explain these observations. In addition, it is essential, that high angular resolution observations in the NIR (interferometry with long baselines) and radio domain (mm- and sub-mm-VLBI) are being performed. These observations will help to identify the orbital motion around the SMBH or motion along a jet/outflow.

Acknowledgements: N. Sabha is member of the Bonn Cologne Graduate School (BCGS) for Physics and Astronomy supported by the Deutsche Forschungsgemeinschaft. M. Valencia-S. is member of the International Max-Planck Research School (IMPRS) for Astronomy and Astrophysics at the Universities of Bonn and Cologne supported by the Max Planck Society. F. Baganoff was supported by NASA through Chandra Award Number GO9-0101X and SAO Award Number 2834-MIT-SAO-4018. Part of this work was supported by the COST Action MP0905: Black Holes in a violent Universe and PECS project No. 98040. We are grateful to all members of the NAOS/CONICA, ESO PARANAL, and APEX team. M. García-Marín is supported by the German federal department for education and research (BMBF) under the project number 50OS1101. R. Schödel acknowledges support by the Ramón y Cajal program, by grants AYA2010-17631 and AYA2009-13036 of the Spanish Ministry of Science and Innovation, and by grant P08-TIC-4075 of the Junta de Andalucía. Ongoing CARMA development and operations are supported by the National Science Foundation under a cooperative agreement, and by the CARMA partner universities.

References

[1] Baganoff, F.K., Bautz, M.W., Brandt, W.N., et al. 2001, Nature, 413, 45
[2] Baganoff, F. K., Maeda, Y., Morris, M., et al. 2003, ApJ 591, 891

[3] Beckert, T.; Falcke, H., 2002, A&A 388, 1106
[4] Beckert, T., 2003, ANS 324, 459
[5] Bower, Geoffrey C.; Wright, Melvyn C. H.; Backer, Donald C.; Falcke, Heino, 1999a, ApJ 527, 851
[6] Bower, Geoffrey C.; Falcke, Heino; Backer, Donald C., 1999b, ApJ 523, L29
[7] Bower, Geoffrey C.; Backer, Donald C.; Zhao, Jun-Hui; Goss, Miller; Falcke, Heino, 1999c, ApJ 521, 582
[8] Bower, G.C., Falcke, H., Sault, R.J., Backer, D.C., 2002, ApJ 571, 843
[9] Bower, G.C.; Wright, M.C.H.; Falcke, H.; Backer, D.C.,2003 ApJ 588, 331
[10] Bower, G.C., 2003, Ap&SS 288 69
[11] Broderick, A.E., Loeb, A., 2006, MNRAS 367, 905
[12] Bremer, M., Witzel, G., Eckart, A., Zamaninasab, M., Buchholz, R.M. , Schödel, R., Straubmeier, C., García-Marín, M., Duschl, W., 2011, A&A 532, 26
[13] Bykov, A. M.; Treumann, R. A., 2011, A&ARv 19, 42
[14] Dexter, J.; Agol, E.; Fragile, P.C., 2009, ApJ 703, L142
[15] Dexter, J., Agol, E., Fragile, P. C., McKinney, J. C., 2010, ApJ 717, 2, 1092
[16] Do, T.; Ghez, A.M.; Morris, M.R.; Yelda, S.; Meyer, L.; Lu, J.R.; Hornstein, S.D.; Matthews, K., 2009, ApJ 691, 1021.
[17] Doeleman, S.S.; Fish, V.L.; Broderick, A.E.; Loeb, A.; Rogers, A.E.E., 2009, ApJ, 695, 59
[18] Doeleman, S.S.; Weintroub, J.; Rogers, A.E.E.; et al., 2008, Natur 455, 78
[19] Dodds-Eden, K.; Porquet, D.; Trap, G.; Quataert, E.; et al., 2009, ApJ 698, 676
[20] Dovciak, M.; Karas, V.; Yaqoob, T., 2004, ApJS 153, 205
[21] Eckart, A.; Baganoff, F. K.; Morris, M.; et al., 2004, A&A 427, 1
[22] Eckart, A.; Baganoff, F. K.; Schödel, R.; Morris, M.; Genzel, R.; Bower, G. C.; Marrone, D.; et al. 2006a, A&A 450, 535
[23] Eckart, A.; Schödel, R.; Meyer, L.; Trippe, S.; Ott, T.; Genzel, R., 2006b, A&A 455, 1
[24] Eckart, A.; Baganoff, F. K.; Zamaninasab, M.; Morris, M. R.; Schödel, R.; et al., 2008a, A&A 479, 625
[25] Eckart, A., Schödel, R., Baganoff, F.K., Morris, M., et al. 2008b, JPhCS 131, 2002
[26] Eckart, A.; Garca-Marn; Witzel, G et a, in Proc. of "Identification of dark matter 2008", August 18-22, 2008c, Stockholm, Sweden, p.38,
[27] Eckart, A.; Schödel, R.; Garcia-Marin, M.; Witzel, G.; Weiss, A.; Baganoff, F. K.; Morris, M. R.; Bertram, T.; Dovciak, M.; Duschl, W. J.; et al., 2008d, A&A 492, 337
[28] Eckart, A.; Baganoff, F. K.; Morris, M. R.; Kunneriath, D.; Zamaninasab, M.; Witzel, G.; Schödel, R.; Garcia-Marin, M.; et al., 2009, A&A 500, 935.
[29] Eckart, A.; Garcia-Marin, M.; König, S.; Kunneriath, D.; Muzic, K.; Straubmeier, C.; Witzel, G.; Zamaninasab, M., 2010, Proc. of the 7th International Heidelberg Conference on Dark 2009 . Held 18 - 24 January 2009 in Christchurch, New Zealand. Edited by Hans Volker Klapdor-Kleingrothaus (Heidelberg University, Germany) & Irina V Krivosheina (Radiophysical Research Institute (NIRFI), Russia). Published by World Scientific Publishing Co. Pte. Ltd., 2010. ISBN 9789814293792, pp. 303-319
[30] Eckart, A.; Garca-Marn, M.; Vogel, S. N.; Teuben, P.; Morris, M. R.; Baganoff, F.; Dexter, J.; Schödel, R.; Witzel, G.; Valencia-S., M.;et al., 2012, A&A 537, 52
[31] Eckart, A.; Zamaninasab, M.; Sabha, N.; Garca-Marn, M.; Kunneriath, D.; Straubmeier, C.; Muzic, K.; Witzel, G.; Bremer, M.; Valencia-Schneider, M.; et al., The Galactic Center: a Window to the Nuclear Environment of Disk Galaxies. Proceedings of a workshop held at Shanghai, China on October 19-23, 2009. Edited by Mark R. Morris, Q. Daniel Wang, and Feng Yuan. San Francisco: Astronomical Society of the Pacific, 2011, p. 294
[32] Fish, V.L.; Doeleman, S.S.; Beaudoin, C.; Blundell, R.; Bolin, D.E.; Bower, G.C.; et al., 2011, ApJ 727, L36
[33] Falcke, H.; Markoff, S., 2000, A&A 362, 113
[34] Galeev, A. A.; Rosner, R.; Vaiana, G. S., 1979, ApJ 229, 318
[35] Genzel, R.; Schödel, R.; Ott, T.; Eckart, A.; Alexander, T.; Lacombe, F.; Rouan, D.; Aschenbach, B., Nature 425, 934
[36] Ghez, A. M., Duchéne, G., Matthews, K., et al. 2003, ApJ, 586, L127
[37] Ghez, A.M., Wright, S.A., Matthews, K., et al. 2004, ApJ 601, 159
[38] Gillessen, S.; Eisenhauer, F.; Quataert, E.; et al., 2006, ApJ 640, L163
[39] Homan, D. C.;et al., 2006, ApJ 642, L115
[40] Hornstein, S. D., Matthews, K., Ghez, A. M., Lu, J. R., Morris, M., Becklin, E. E., Rafelski, M., & Baganoff, F. K. 2007, ApJ, 667, 900
[41] Igumenshchev, I.V.; Narayan, R.; Abramowicz, M.A.; 2003, ApJ 592. 1042
[42] Karas, V., Dovciak, M., Eckart, A., Meyer, L., 2007, Proceedings of the Workshop on the Black Holes and Neutron Stars, eds. S. Hledik and Z. Stuchlik, 19-21 September 2007 (Silesian University, Opava), pp. 99-108, arXiv:0709.3836

[43] Keppens, R.; Casse, F.; Goedbloed, J. P., 2002, ApJ 569, L121
[44] Kowal, G.; de Gouveia Dal Pino, E. M.; Lazarian, A., 2011, ApJ 735, 102
[45] Krishan, V.; Wiita, P. J.; Ramadurai, S., 2000, A&A 356, 373
[46] Kunneriath, D.; Witzel, G.; Eckart, A.; Zamaninasab, M.; et al.; et al., 2010, A&A 517, 46
[47] Li, J., Shen, Z.-Q., Miyazaki, A., Huang, L., Sault, R. J., et al., 2009,ApJ 700, L417
[48] Lu, Youjun; Torres, Diego F., 2003, IJMPD 12, 63
[49] Markoff, S., 2005, ApJ 618, L103
[50] Markoff, S.; Bower, G.C.; Falcke, H., 2007, MNRAS 379, 1519
[51] Marrone, D. P. 2006a, PhD thesis, Harvard Univ.
[52] Marrone, D.P.; Moran, J.M.; Zhao, J.-H.; Rao, R., 2006b, ApJ 640, 308
[53] Marrone, D.P.; Moran, J.M.; Zhao, J.-H.; Rao, R., 2006c, JPhCS 54, 354
[54] Marrone, D.P.; Moran, J.M.; Zhao, J.-H.; Rao, R. 2007, ApJ 654, L57
[55] Marrone, D. P.; Baganoff, F. K.; Morris, M.; et al., 2008, ApJ 682, 373
[56] Marscher, A.P. 1983, ApJ, 264, 296
[57] Marscher, A.P. 2009, in Belloni, T. (ed.): The Jet Paradigm - From Microquasars to Quasars, Lect. Notes Phys. 794 (2009), 2009arXiv0909.2576M
[58] Mauerhan et al. 2005, ApJ 623, L25
[59] Medvedev, Mikhail V.; Loeb, Abraham, 1999, ApJ 526, 697
[60] Melia, F. & Falcke, H. 2001, ARA&A 39, 309
[61] Meyer, L.; Schödel, R.; Eckart, A.; Duschl, W. J.; Karas, V.; Dociak, M., 2007, A&A 473, 707
[62] Miller, M.C., 2006, MNRAS 367, L32
[63] Miyazaki, A.; Tsutsumi, T.; Tsuboi, M., 2003, ANS 324, 363
[64] Munyaneza. F & Biermann, P. L., 2005, A&A 436, 805
[65] Munoz, D. J.; Marrone, D. P.; Moran, J. M.; Rao, R., 2012, ApJ 745, 115
[66] Narayan, R.; Yi, I.; Mahadevan, R., 1995, Natur 374, 623
[67] Nishikawa, K.-I. et al., 2011, AdSpR 47, 1434
[68] Nishiyama, S.; Tamura, M.; Hatano, H.; Nagata, T.; Kudo, T.; Ishii, M.; Schödel, R.; Eckart, A., 2009, ApJ 702, L56
[69] Pauliny-Toth, I. I. K.; Kellermann, K. I., 1966, ApJ 146, 634
[70] Kellermann, K. I.; Pauliny-Toth, I. I. K., 1969, ApJ 155L, 71
[71] Pecháček, T., Karas, V., Czerny, B., 2008, A&A 487, 815
[72] Readhead, A. C. S. 1994, ApJ, 426, 51
[73] Sabha, N.; Eckart, A.; Merritt, D.; submitted to A& A, 2012arXiv1203.2625S
[74] Sault, R. J.; Macquart, J.-P., 1999, ApJ526, L85
[75] Sazonov, S.; Sunyaev, R.; Revnivtsev, M., 2012, MNRAS 420, 388
[76] Shcherbakov, R.V.; Baganoff, F.K., 2010, ApJ 716, 504
[77] Schödel, R.; Eckart, A.; Alexander, T.; Merritt, D.; et al., 2007, A&A 469, 125
[78] Schödel, R.; Najarro, F.; Muzic, K.; Eckart, A., 2010, A&A 511, 18,
[79] Schödel, R., Morris, M.R., Muzic, K., Alberdi, A., Meyer, L., Eckart, A., & Gezari, D.Y. 2011, A&A, 532, A83.
[80] Singal, A. K. 2009 ApJ 703, L109
[81] Singal, K. A.; Gopal-Krishna, 1985, MNRAS 215, 383
[82] Torres, D.F., Capoziello, S., Lambiase, G., 2000, Phys. Rev. D 62, 104012
[83] Trap, G.; Weiss, A.; Goldwurm, A.; Dodds-Eden, K.; Terrier, R.; et al., 2011a, ASPC 439, 304
[84] Trap, G.; Goldwurm, A.; Dodds-Eden, K.; Weiss, A.; Terrier, R.; et al., 2011b, A&A 528, 140
[85] Tsuboi, M.; Miyahara, H.; Nomura, R.; Kasuga, T.; Miyazaki, A., 2003, ANS 324, 431
[86] Witzel, G., Eckart, A., Bremer, M., Schödel, R., Shazamanian, B., submitted to ApJ.
[87] Weibel E.S. 1959, Phys Rev Lett 2:83-84.
[88] Yuan, F., Quataert, E., & Narayan, R. 2003, ApJ 598, 301
[89] Yuan, F., Quataert, E., & Narayan, R. 2004, ApJ 606, 894
[90] Yuan, F., Lin, J.; Wu, K.; Ho, Luis C., 2008, MNRAS 395, 2183
[91] Yusef-Zadeh, F., et al., 2006a, ApJ 650, 189
[92] Yusef-Zadeh, F., et al., 2006b, ApJ 644, 198
[93] Yusef-Zadeh, F.; Wardle, M.; Heinke, C.; et al., 2008, ApJ 682, 361
[94] Yusef-Zadeh F.; Bushouse H.; Wardle M. et al. 2009, ApJ, 706, 348
[95] Yusef-Zadeh F., et al. 2011, ApJ 729, 44
[96] Zamaninasab, M.; Eckart, A.; Dovciak, M.; Karas, V.; Schödel, R.; Witzel, G.; Sabha, N.; Garcia-Marin, M.; Kunneriath, D.; Muzic, K.; et al., 2011, MNRAS 413, 322,
[97] Zamaninasab, M.; Eckart, A.; Witzel, G.; Dovciak, M.; Karas, V.; Schödel, R.; Gießübel, R.; Bremer, M.;

Garcia-Marin, M.; Kunneriath, D.; et al., 2010, A&A 510, 3
[98] Zhao, J.-H., Young, K.H., Herrnstein, R.M., Ho, P.T.P., et al., 2003, ApJL, 586, L29.
[99] Zubovas, K.; Nayakshin, S.; Markoff, S., 2012, MNRAS, in press, arXiv1110.6872Z

Astronomy at High Angular Resolution 2011 IOP Publishing
Journal of Physics: Conference Series **372** (2012) 012023 doi:10.1088/1742-6596/372/1/012023

Radiative Models of Sagittarius A* and M87 from Relativistic MHD Simulations

J. Dexter[1], E. Agol[2], P. C. Fragile[3] and J. C. McKinney[4]

[1]Theoretical Astrophysics Center and Department of Astronomy, University of California, Berkeley, CA 94720-3411, USA
[2]Department of Astronomy, University of Washington, Box 351580, Seattle, WA 98195, USA
[3]Department of Physics & Astronomy, College of Charleston, Charleston, SC 29424, USA
[4]Kavli Institute for Particle Astrophysics and Cosmology, Stanford University, Stanford, CA 94305-4060, USA

E-mail: `jdexter@berkeley.edu`

Abstract. Ongoing millimeter VLBI observations with the Event Horizon Telescope allow unprecedented study of the innermost portion of black hole accretion flows. Interpreting the observations requires relativistic, time-dependent physical modeling. We discuss the comparison of radiative transfer calculations from general relativistic MHD simulations of Sagittarius A* and M87 with current and future mm-VLBI observations. This comparison allows estimates of the viewing geometry and physical conditions of the Sgr A* accretion flow. The viewing geometry for M87 is already constrained from observations of its large-scale jet, but, unlike Sgr A*, there is no consensus for its millimeter emission geometry or electron population. Despite this uncertainty, as long as the emission region is compact, robust predictions for the size of its jet launching region can be made. For both sources, the black hole shadow may be detected with future observations including ALMA and/or the LMT, which would constitute the first direct evidence for a black hole event horizon.

1. Introduction

The Galactic center massive black hole candidate, Sagittarius A* (Sgr A*) [1], and the supermassive black hole candidate in the center of M87 are both bright radio sources and are the two largest black holes in angular size ($4-5$ microarcseconds or μas). For both these reasons, they are the two major targets of interest for imaging with ongoing very long baseline interferometry observations at millimeter wavelengths (mm-VLBI), which have already detected time-variable event horizon scale structure in Sgr A* [2, 3]. Data have also been taken for M87, but are as yet unpublished. The Sgr A* data have previously been fit both with simple geometric models (symmetric Gaussian and annulus of constant brightness) and using a radiatively inefficient accretion flow model (RIAF, [4, 5, 6]. While physically motivated, these models are non-relativistic and do not include the magnetic fields that drive turbulence in the accretion flow via the magnetorotational instability (MRI, [7]), causing accretion and leading to the observed synchrotron radiation.

This essential physics can now be captured in general relativistic MHD (GRMHD) simulations of black hole accretion. Simulations have been used to model the millimeter emission in Sgr A* either in the absence of general relativity [8] or in axisymmetry [9, 10, 11]. The millimeter emission in the models arises in the innermost portions of the accretion flow ($r \simeq 5$M, where we use $G = c = 1$ units except where otherwise noted), so that relativistic effects are dominant. In simulations run in axisymmetry, the MRI decays on the local orbital timescale. This leads to unrealistic variability, prevents the disk from reaching quasi-steady state, and can cause the appearance of unphysically hot and bright regions [10]. One group [11] has used 3D GRMHD simulations, but then time-averaged the resulting data. Two groups have compared spectra from 2D GRMHD simulations with the spectrum of M87 [12, 13], but neither considered jet emission or made predictions for mm-VLBI.

Here we discuss radiative models of the mm emission from Sgr A* and M87 for comparison with mm-VLBI constructed by performing relativistic radiative transfer calculations on 3D GRMHD simulation data. These are the first models to self-consistently include variability, and the M87 models are the first predictions for images of a jet launching region from a simulation. The numerical models of Sgr A* are the only ones to have been fit directly to mm-VLBI data to test GRMHD accretion theory and estimate parameters of the black hole and its accretion flow.

2. Radiative Modeling

A radiative model of Sagittarius A* or M87 consists of three components [14]: a dynamical solution for the fluid variables in the accretion flow, a model for the electron distribution function based on the dynamical solution, and an emission model given the electron distribution function. These components give fluid frame emission/absorption coefficients everywhere in spacetime. Observables are calculated from these coefficients in full general relativity via ray tracing [15].

2.1. GRMHD Simulations

The dynamical solution consists of the output of general relativistic MHD simulations of black hole accretion flows. The simulations start from a gas torus in hydrostatic equilibrium in a Kerr spacetime threaded with a weak magnetic field. This configuration quickly develops turbulence driven by the MRI, and the resulting stresses cause angular momentum transport outwards allowing accretion onto the black hole. These simulations are currently limited in several important respects: they only develop a quasi-steady accretion flow over a limited radial domain ($r \lesssim 25$M); and they neglect both the radiation that astronomers observe and the electrons that produce it.

In highly underluminous systems like Sgr A* and M87, ignoring the effects of radiation on the dynamical solution is reasonable (but see [16] and Dibi et al., in prep.). The limited radial domain is also acceptable given the extremely compact size measured by mm-VLBI, and found theoretically for the millimeter emission in Sgr A* and M87.

However, there are also issues with modeling these sources using GRMHD simulations. Their low densities means that the ions and electrons may have significantly different temperatures, with the hot ions providing the fluid pressure captured by the simulation and the cool electrons producing the observed emission. We must then decide how to model the electrons before calculating observables from simulation data (see § 2.2).

In this work, we consider results from four GRMHD simulations (summarized in Table 1 of [17]). All are three dimensional: in axisymmetry, the MRI decays on the local orbital timescale, preventing the accretion flow from reaching quasi-steady state and leading to artificial variability, an inaccurate time-averaged structure, and causing contamination of the spectra by unrealistically hot zones in the "funnel wall" [10]. The simulations are from two codes: `Cosmos++` [18, 19, 20] and a 3D extension of HARM [21, 22, 23, 24, 25]. The set of simulations is limited by the number of existing and available 3D GRMHD simulations of relatively thick ($H/R \simeq 0.2$) black hole accretion flows. The scarcity of simulations and their systematic differences prevent an estimate of the black hole spin. Only the MBD simulation forms a relativistic, magnetically dominated-jet and so this is the only simulation used in the disk/jet modeling of M87. We use the last 2000 − 4000M of simulation data, once transients from the initial conditions have decayed, where 1M is 20s (9hr), 6×10^{11}cm (10^{15}cm), and 5μas (4μas) for Sgr A* (M87).

2.2. Electrons

The millimeter emission in Sgr A* is well described by synchrotron emission from hot, thermal electrons in the innermost portions of a radiatively inefficient accretion flow [26, 4]. A non-thermal tail extending to higher energies may be responsible for the radio and IR emission [4], or it may arise in a jet [27].

Since we focus on the millimeter emission, we model accretion disk components as having entirely thermal electrons. The electron temperature is specified using a constant electron-ion temperature ratio, T_i/T_e [8, 10], which allows the electron temperature to be calculated from the simulation pressure and density. This is the simplest possible prescription for a two-temperature accretion flow from a single fluid simulation.

In M87, it is unclear whether disk or jet emission is dominant at millimeter wavelengths. Extended jet structure is observed at 7mm [28, 29], indicating that the jet emission is at least comparable to that from the disk. Since in the simulations the jet forms self-consistently from an accretion flow, we include both disk and jet components with different assumed electron populations [30].

The ultrarelativistic outflow in the simulation forms in magnetically-dominated region, so to separate the regions for electron modeling we define the jet by $b^2/\rho c^2 > 1$, where b is the magnetic field strength and ρ is the fluid density. The jet electrons are assumed to be entirely non-thermal with a power law distribution. The simulation variables of density and internal energy (pressure) within magnetically-dominated regions are unreliable, since they are set by the artificial numerical floor values used for code stability. We therefore scale the non-thermal particle density with the only reliable simulation variable in the jet region, the magnetic energy density: $n_{\rm nth} \propto b^2$, where $n_{\rm nth}$ is the non-thermal particle density (cf. [31]). Fitting the M87 spectrum requires 5 − 10% of the magnetic energy to be dissipated in non-thermal particles for the simulation considered.

In both cases, the emissivity is assumed to be entirely synchrotron radiation. Inverse Compton scattering is ignored, although it is likely to be very important for M87 at high energies [12, 30]. For the thermal disk electrons, we use an approximate form of the synchrotron emissivity given by Leung et al. (2011) [32]. For the jet power law electron distribution, we use a form of the emissivity that takes into account the low-energy cutoff to the distribution, which is important for the millimeter emission in M87 [14].

2.3. Ray Tracing

Relativistic radiative transfer is performed on the simulation data via ray tracing using the code `grtrans` [14]. Starting from an observer's camera, rays are traced backwards in time toward the black hole assuming they are null geodesics (geometric optics approximation), using the public code `geokerr` [33]. In the region where rays intersect the accretion flow, the radiative transfer equation is solved along the geodesic, which then represents a pixel of the image. This procedure is repeated for many rays to produce an image, and at many time steps of the simulation to produce time-dependent images (movies). Light curves are computed by integrating over the individual images. Repeating the procedure over observed wavelengths gives a time-dependent spectrum.

To calculate fluid properties at each point on a ray, the spacetime coordinates of the geodesic are transformed from Boyer-Lindquist to the regular [34] or modified [35] Kerr-Schild coordinates used in the simulation. Since the accretion flow is dynamic, light travel time delays along the geodesic are taken into account. Data from the sixteen nearest zone centres (eight on the simulation grid over two time steps) are interpolated to each point on the geodesic.

Computing emission and absorption coefficients requires converting simulation fluid variables (pressure/internal energy, mass density, and magnetic field strength) into an electron distribution function in physical units. The black hole mass sets the length and time scales, while the mass of the initial torus provides an independent scale and fixes the accretion rate. The scalings are such that n and b^2 are proportional to the accretion rate.

3. Model Fitting

For Sgr A*, we fit simulated images at 1.3mm to mm-VLBI observations [2, 3] and the total flux at 1.3mm and 0.4mm to observations of the sub-mm spectral index [36]. Details on the statistical techniques used to estimate parameters from the fits can be found in [5, 17]. The free parameters for the Sgr A* models are i) the time-averaged accretion rate, ii) the ratio T_i/T_e, iii) the inclination and iv) sky orientation angles, and v) the simulation.

Generically, the simulated images provide excellent fits to the mm-wavelength observations of Sgr A* (reduced $\chi^2 \lesssim 1$ for many models for each simulation [37, 17]). Probability distributions for the inclination (i) and sky orientation (ξ) angles, the median electron temperature in the region that produces 99% of the millimeter emission (T_e), and the time-averaged accretion rate onto the black hole ($\dot{M}$) including all mm-VLBI data are estimated as follows (all to 90% confidence): $i = 60 \pm 15°$, $\xi = -70°^{+86°}_{-15°}$, $T_e = 6 \pm 2 \times 10^{10}$K and $\dot{M} = 3^{+7}_{-1} \times 10^{-9} M_{\odot}\mathrm{yr}^{-1}$. Including all of the data gives tighter constraints, but consistent estimates, for all parameters as found from the first epoch alone. The viewing angle estimates and probability distributions are in excellent agreement with those from using a RIAF model [5, 6]. The mm-VLBI observations, even with limited array coverage and sensitivity, are meaningfully constraining the relevant parameters, but they cannot yet distinguish between physical accretion flow models (they are, however, beginning to favor physical models over simple geometric ones [6]). The constraints on the accretion rate and electron temperature are consistent with, but independent of, estimates from the observed linear polarization and Faraday rotation measures [38, 39, 40] and spectral fitting [4], respectively.

The first mm-VLBI data for M87 have been obtained [3] but have not yet been published. We fit models to multi-epoch observational data in the optical [41], IR [42, 43], and mm [44], inflating the error bars to account for variability between the non-simultaneous observations and treating data points that may be dominated by dust in the galactic nucleus or large-scale jet emission as upper limits. Since there are roughly the same number of free parameters in the disk/jet models as spectral constraints, we identify fiducial models rather than attempt to quantitatively constrain the parameters. The models can be either disk/jet (DJ1) or jet only (J2), and provide satisfactory descriptions of the emission from the immediate vicinity of the M87 black hole [30].

The viewing geometry of M87, unlike Sgr A*, may be constrained by the Lorentz factor and sky orientation of the large-scale jet, which can reasonably be assumed to correspond to the black hole spin axis. All models assume this favored geometry ($i < 40°$, $-115° < \xi < -75°$).

4. Black Hole Images

The millimeter emission from Sgr A* and M87 arises from the immediate vicinity of the black hole ($r < 5 - 10$M for disk and $r < 2 - 4$M for jet emission). On these scales, relativistic effects tend to dominate the appearance of the images, provided that the emission region is not completely optically thick. The important effects are: i) Doppler beaming from the component of the fluid velocity along the observer's line of sight, which is significant except for face-on viewing, and dominant for edge-on viewing; ii) light bending, which causes the back of the accretion flow to appear above/below the black hole (again except for face-on viewing), and iii) gravitational lensing. Doppler beaming causes significant asymmetry between the sides of the image where the gas is moving towards and away from the observer. The combination of ii)

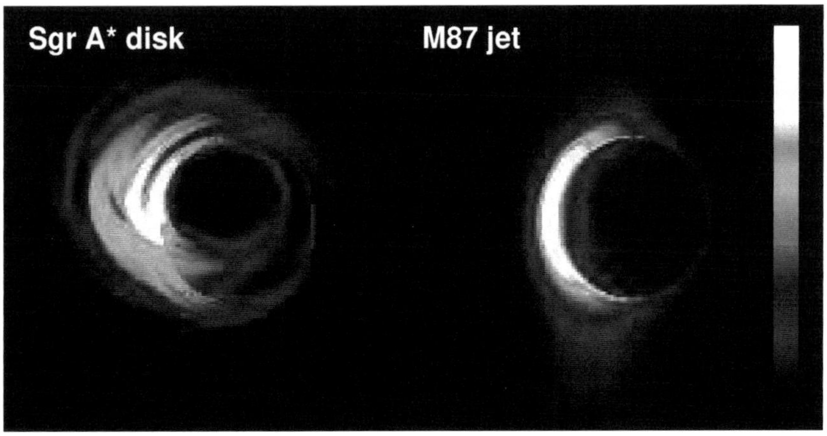

Figure 1. False color fiducial images of an accretion disk model of Sgr A* (left) and a jet image of M87 (right). The left/right asymmetries are caused by Doppler beaming from orbital motion, and the black hole shadow is clearly visible in the center of both images. The intensity scale is linear with a dynamic range of 60, and the panels are each 12 Schwarzschild radii ($\simeq$ 100 microarcseconds) across.

and iii) importantly leads to the black hole "shadow" [45, 46, 17], the transition between bound and unbound photon orbits. For lines of sight inside the shadow, the emission comes from in front of the black hole, where few photon emission angles escape to the observer at infinity. The gravitational lensing in the immediate vicinity often leads to a bright ring at the circular photon orbit, surrounding the shadow. Observing signatures of this shadow would constitute the first direct evidence for a black hole event horizon. The combined effects of asymmetry from Doppler beaming and light bending are typically both important for favored models of M87 and Sgr A*, leading to characteristic crescent image morphologies (an example is shown in the left panel of Figure 1). The shadow is also apparent in the image.

4.1. The Size of M87

For the low inclination of M87, both disk and jet models also lead to crescent images. In the jet case, the emission arises in the *counter-jet*, since the emission from the forward jet is along lines of sight in the shadow and very little of it reaches the observer (see the right panel of Figure 1 for an example). For this reason, we can robustly predict the size of the jet launching region despite significant model uncertainties, as long as the emission region is compact ($r \lesssim 10$M) and not completely opaque ($\tau \lesssim 3$). We also assume the viewing geometry estimated from observations of the large-scale jet.

We predict the size observed by mm-VLBI by interpolating fiducial images to the locations sampled by the current mm-VLBI array [3]. The results are $\simeq 33-44\mu$as, or $4-5$ Schwarzschild radii [30], very similar to existing observations of Sgr A*.

4.2. Black Hole Shadow

Best fit models of Sgr A* and fiducial jet or disk/jet models of M87 all lead to crescent images, where Doppler beaming leads to significant asymmetry but the front and back of the accretion flow lead to significant contrast above and below the black hole. The black hole shadow is then visible on orientations probing the front/back of the accretion flow, while the image appears roughly Gaussian on baselines probing orientations that are significantly affected by

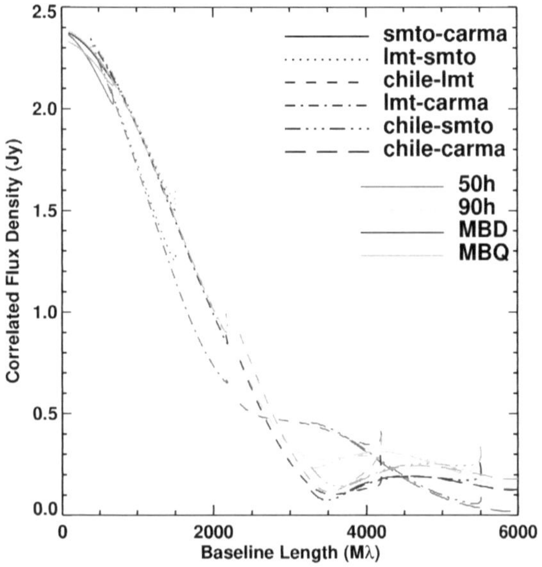

Figure 2. Visibility amplitude (Fourier transform of image intensity) as a function of baseline length for best fit models from [17] at 1.3mm. The telescopes considered are in Arizona (SMTO), California (CARMA), Mexico (LMT) and Chile (APEX/ASTE/ALMA). The BH shadow appears as the local minimum near 3000Mλ on the LMT-Chile and Chile-CARMA baselines.

Doppler beaming. For both sources, the models predict that the black hole shadow is not visible with current telescopes (roughly East-West baselines), but may be observed with North-South baselines including telescopes in Chile (e.g., ALMA) or Mexico (LMT) (see Figure 2). These are also the telescopes with the greatest ability to distinguish between and test the various models. Adding a nearly orthogonal baseline orientation allows a probe of the 2D structure of the image. There is also a signature of the shadow in the closure phase [17], and combining amplitude and phase information probably offers the best chance for its detection.

5. Variability

These radiative models are the first to self-consistently include variability from first principles. The variability in the accretion disk emission is driven by fluctuations in the particle density and magnetic field strength, which in turn are caused by magnetic turbulence driven by the MRI [37]. These fluctuations are highly correlated with those in the accretion rate, although the power law index of the power spectrum differs between the two quantities [47, 17].

6. Discussion

We have discussed the first models of the event horizon scale millimeter-wavelength emission in Sgr A* and M87 from 3D GRMHD simulations. The models are constructed by performing general relativistic radiative transfer on the simulation data via ray tracing, and can then be used to compare to or make predictions for mm-VLBI observations. Even with limited data, the observations are able to constrain the viewing geometry and physical conditions in the Sgr A* accretion flow. Thus far, the uncertainties in the parameter estimates decrease, while the values remain consistent, as more mm-VLBI data become available. For M87, we predict a size of $33-44\mu$as, where within that range the size depends on the total millimeter flux of the source at the time of observation, whether the emission comes from a disk or jet, and on the viewing geometry within the allowed range. In both sources, the variability from the models agrees with observations, especially in Sgr A* where even the morphology of the millimeter flares can be reproduced with no additional free parameters. In both sources, the black hole shadow, a signature of the event horizon, may be accessible to future mm-VLBI observations using ALMA and/or the LMT.

6.1. Uncertainties

The results presented here apply to existing simulations with modest disk scale heights ($H/R \simeq 0.2$) at relatively high black hole spins ($a \simeq 0.5 - 0.9375$) where the black hole spin and accretion disk angular momentum axes are aligned. We expect the accretion flow in low-luminosity sources should be somewhat thicker ($H/R \simeq 1$). The first GRMHD simulations with $H/R \simeq 1$ have recently been carried out (McKinney et al., in preparation), and may be used to study both Sgr A* and M87. These simulations span a wide range of black hole spin and initial magnetic field configuration, and will allow a fairly robust parameter study of these radiative models including estimates of all relevant parameters including the black hole spin.

For the accretion disk component, a constant ratio of T_i/T_e is assumed with no justification. This assumption may be okay, since the millimeter emission region in these sources is so compact that their physical conditions do not vary greatly. However, this approximation should be tested in the future.

The primary uncertainty in the jet modeling is the dissipation physics and jet mass loading, both of which are important for constructing images. When the emission region is compact ($R \lesssim 5$M), relativistic effects tend to shape the images into crescents for a large range of emission geometries. The jet emission region may be extended, however, in which case most of the results for jet images discussed here for 1.3mm do not apply. This is true observationally for M87 at 7mm. We have assumed the non-thermal particle density is proportional to the magnetic energy density, but other sub-grid prescriptions should be tried as well, if only to gauge the range of viable jet image morphologies.

Finally, in geometrically thick sources the accretion flow and black hole spin orientations are likely to remain misaligned. This disk "tilt" has drastic consequences for the structure and evolution of the accretion flow [19, 48, 49] and its observational properties [50]. The conclusions reached here regarding crescent images, the visibility of black hole shadows, and variability driven by MRI turbulence all apparently hold for similar models from simulations with a $15°$ tilt [14]. However, the parameter estimates for Sgr A* discussed above only apply in the aligned case.

Acknowledgments

This work was partially supported by NASA Earth & Space Science Fellowship NNX08AX59H (JD), STScI grant HST-GO-11732.02-A, NSF grant AST 0807385, NASA grant 05-ATP05-96, and NASA Chandra Fellowship PF7-80048 (JCM).

References

[1] Balick B and Brown R L 1974 ApJ **194** 265–270

[2] Doeleman S S, Weintroub J, Rogers A E E, Plambeck R, Freund R, Tilanus R P J, Friberg P, Ziurys L M, Moran J M, Corey B, Young K H, Smythe D L, Titus M, Marrone D P, Cappallo R J, Bock D C J, Bower G C, Chamberlin R, Davis G R, Krichbaum T P, Lamb J, Maness H, Niell A E, Roy A, Strittmatter P, Werthimer D, Whitney A R and Woody D 2008 Nature **455** 78–80 (*Preprint* 0809.2442)

[3] Fish V L, Doeleman S S, Beaudoin C, Blundell R, Bolin D E, Bower G C, Chamberlin R, Freund R, Friberg P, Gurwell M A, Honma M, Inoue M, Krichbaum T P, Lamb J, Marrone D P, Moran J M, Oyama T, Plambeck R, Primiani R, Rogers A E E, Smythe D L, SooHoo J, Strittmatter P, Tilanus R P J, Titus M, Weintroub J, Wright M, Woody D, Young K H and Ziurys L M 2011 ApJ **727** L36+ (*Preprint* 1011.2472)

[4] Yuan F, Quataert E and Narayan R 2003 ApJ **598** 301–312 (*Preprint* arXiv:astro-ph/0304125)

[5] Broderick A E, Fish V L, Doeleman S S and Loeb A 2009 ApJ **697** 45–54 (*Preprint* 0809.4490)

[6] Broderick A E, Fish V L, Doeleman S S and Loeb A 2011 ApJ **735** 110–+ (*Preprint* 1011.2770)

[7] Balbus S A and Hawley J F 1991 ApJ **376** 214–233

[8] Goldston J E, Quataert E and Igumenshchev I V 2005 ApJ **621** 785–792 (*Preprint* arXiv:astro-ph/0411627)

[9] Noble S C, Leung P K, Gammie C F and Book L G 2007 *Class. and Quant. Gravity* **24** 259–+ (*Preprint* arXiv:astro-ph/0701778)

[10] Mościbrodzka M, Gammie C F, Dolence J C, Shiokawa H and Leung P K 2009 ApJ **706** 497–507 (*Preprint* 0909.5431)

[11] Shcherbakov R V, Penna R F and McKinney J C 2010 *ArXiv e-prints* (*Preprint* 1007.4832)
[12] Mościbrodzka M, Gammie C F, Dolence J C and Shiokawa H 2011 ApJ **735** 9–+ (*Preprint* 1104.2042)
[13] Hilburn G and Liang E 2011 *ArXiv e-prints* (*Preprint* 1109.2847)
[14] Dexter J 2011 *Radiative Models of Sagittarius A* and M87 from Relativistic MHD Simulations* Ph.D. thesis University of Washington
[15] Luminet J P 1979 A&A **75** 228–235
[16] Drappeau S, Dibi S, Dexter J, Markoff S and Fragile P C 2011 Self-consistent spectra from GRMHD simulations with radiative cooling: A link to reality for Sgr A *SF2A-2011: Proceedings of the Annual meeting of the French Society of Astronomy and Astrophysics* ed G Alecian, K Belkacem, R Samadi, & D Valls-Gabaud pp 563–565
[17] Dexter J, Agol E, Fragile P C and McKinney J C 2010 ApJ **717** 1092–1104 (*Preprint* 1005.4062)
[18] Anninos P, Fragile P C and Salmonson J D 2005 ApJ **635** 723–740 (*Preprint* arXiv:astro-ph/0509254)
[19] Fragile P C, Blaes O M, Anninos P and Salmonson J D 2007 ApJ **668** 417–429 (*Preprint* 0706.4303)
[20] Fragile P C, Lindner C C, Anninos P and Salmonson J D 2009 ApJ **691** 482–494 (*Preprint* 0809.3819)
[21] Gammie C F, McKinney J C and Tóth G 2003 ApJ **589** 444–457 (*Preprint* arXiv:astro-ph/0301509)
[22] Noble S C, Gammie C F, McKinney J C and Del Zanna L 2006 ApJ **641** 626–637 (*Preprint* arXiv:astro-ph/0512420)
[23] McKinney J C 2006 MNRAS **367** 1797–1807 (*Preprint* arXiv:astro-ph/0601410)
[24] Tchekhovskoy A, McKinney J C and Narayan R 2007 MNRAS **379** 469–497 (*Preprint* 0704.2608)
[25] McKinney J C and Blandford R D 2009 MNRAS **394** L126–L130 (*Preprint* 0812.1060)
[26] Narayan R and Yi I 1994 ApJ **428** L13–L16 (*Preprint* arXiv:astro-ph/9403052)
[27] Falcke H and Markoff S 2000 A&A **362** 113–118 (*Preprint* arXiv:astro-ph/0102186)
[28] Ly C, Walker R C and Wrobel J M 2004 AJ **127** 119–124 (*Preprint* arXiv:astro-ph/0309743)
[29] Walker R C, Ly C, Junor W and Hardee P J 2008 *Journal of Physics Conference Series* **131** 012053–+
[30] Dexter J, McKinney J C and Agol E 2011 *ArXiv e-prints* (*Preprint* 1109.6011)
[31] Broderick A E and McKinney J C 2010 ApJ **725** 750–773 (*Preprint* 1006.5015)
[32] Leung P K, Gammie C F and Noble S C 2011 ApJ **737** 21–+
[33] Dexter J and Agol E 2009 ApJ **696** 1616–1629 (*Preprint* 0903.0620)
[34] Fragile P C and Anninos P 2005 ApJ **623** 347–361 (*Preprint* arXiv:astro-ph/0403356)
[35] McKinney J C 2006 MNRAS **368** 1561–1582 (*Preprint* arXiv:astro-ph/0603045)
[36] Marrone D P 2006 *Submillimeter properties of Sagittarius A*: The polarization and spectrum from 230 to 690 GHz and the submillimeter array polarimeter* Ph.D. thesis AA(Harvard University)
[37] Dexter J, Agol E and Fragile P C 2009 ApJ **703** L142–L146 (*Preprint* 0909.0267)
[38] Agol E 2000 ApJ **538** L121–L124 (*Preprint* arXiv:astro-ph/0005051)
[39] Quataert E and Gruzinov A 2000 ApJ **545** 842–846 (*Preprint* arXiv:astro-ph/0004286)
[40] Marrone D P, Moran J M, Zhao J and Rao R 2007 ApJ **654** L57–L60 (*Preprint* arXiv:astro-ph/0611791)
[41] Sparks W B, Biretta J A and Macchetto F 1996 ApJ **473** 254–+
[42] Perlman E S, Biretta J A, Sparks W B, Macchetto F D and Leahy J P 2001 ApJ **551** 206–222 (*Preprint* arXiv:astro-ph/0012044)
[43] Perlman E S, Mason R E, Packham C, Levenson N A, Elitzur M, Schaefer J J, Imanishi M, Sparks W B and Radomski J 2007 ApJ **663** 808–815 (*Preprint* 0704.1156)
[44] Tan J C, Beuther H, Walter F and Blackman E G 2008 ApJ **689** 775–781
[45] Bardeen J M 1973 Timelike and null geodesics in the Kerr metric. *Black holes (Les astres occlus)* ed DeWitt B S and DeWitt C (New York: Gordon and Breach) p 215
[46] Falcke H, Melia F and Agol E 2000 ApJ **528** L13–L16 (*Preprint* arXiv:astro-ph/9912263)
[47] Noble S C and Krolik J H 2009 ApJ **703** 964–975 (*Preprint* 0907.1655)
[48] Fragile P C and Blaes O M 2008 ApJ **687** 757–766 (*Preprint* 0807.2453)
[49] Fragile P C 2009 ApJ **706** L246–L250 (*Preprint* 0910.5721)
[50] Dexter J and Fragile P C 2011 ApJ **730** 36–+ (*Preprint* 1101.3783)

The Origin of Parsec-Scale Gaseous and Stellar Disks in the Galactic Center and AGNs

F. Yusef-Zadeh
Dept. Physics and Astronomy, Northwestern University, Evanston, IL. 60208
E-mail: zadeh@northwestern.edu

M. Wardle
Department of Physics and Astronomy, Macquarie University, Sydney, NSW 2109, Australia
E-mail: mark.wardle@mq.edu.au

Abstract. The Galactic center stellar disk and the circumnuclear ring provide a unique opportunity to study in detail the dynamics and physical conditions of distant molecular disks in the nuclei of galaxies. One of the key questions is how these disks form so close to their host black holes and under what condition they form stars in a tidally stressed environment. We argue that disk formation around a massive black hole is due to partial accretion of extended molecular clouds that temporarily pass through the central region of the Galaxy. The cancellation of angular momentum of the gravitationally focused gas naturally creates a compact gaseous disk. The disk can potentially become gravitationally unstable and form stars. We apply these ideas to explain the origin of sub-parsec megamaser disks found in the nuclei of Seyfert 2 galaxies. We show that an empirical scaling relation between the mass of the black hole and the size of the disk can be understood in the context of the cloud capture scenario. We conclude that the stellar and gas disks found in our Galactic center act as a bridge to further our understanding of more distant mega-maser disks in the nuclei of Seyfert 2 galaxies.

1. Introduction

Understanding the formation of massive young stars in the immediate vicinity of massive black holes is challenging. Two plausible models for the origin of the disk of massive stars orbiting Sgr A*, the 4×10^6 $M_\odot$ black hole at the Galactic center, have been proposed, namely, in-situ star formation and the dynamical migration of star clusters (see Genzel, Eisenhauer and Gillessen 2010; Levin and Beloborodov 2003; Gerhard 2001; Nayakshin et al. 2007). It has become clear that the in-situ star formation model can account for present observations better than the migration scenario in which a young star cluster undergoes dynamical friction and spirals toward the Galactic center. The question then arises as to what extent disk formation near Sgr A* can provide insight in the formation of sub-parsec gaseous disks orbiting supermassive black holes in AGNs. To address this question, we will first focus on the origin of Galactic center stellar and molecular disks near Sgr A* followed by AGN disks in the context of partial cloud capture by the massive black hole. We then discuss observational signatures such as the scaling relation between the mass of the black hole and the size of the disk as well as the recoil of the black hole in the context of the cloud capture scenario. A more detailed account of the results presented here can be found in Wardle and Yusef-Zadeh (2008, Paper I), Yusef-Zadeh et al. (2009, Paper II) and Wardle and Yusef-Zadeh (2011, Paper III).

2. Galactic Center Young Stellar Disks

Observations of our Galactic center indicate that there are about 100 massive young stars distributed within one or two disks with a sharp inner edge (Paumard et al. 2006; Lu et al. 2010). In the in-situ formation scenario, an interstellar cloud is tidally disrupted and captured by Sgr A* (Nayakshin, Cuadra & Spingel 2007). There are two issues that are of interest in the context of this mode of star formation. One is related to the mechanism in which massive stars are formed in a disk. If the gas disk is significantly massive, it can become gravitationally unstable resulting in the fragmentation and formation of stars. A cooling self-gravitating gas disk around a black hole becomes unstable to fragmentation on a dynamical time scale if Toomre's $Q< 1$. In this picture, the strong tidal force by the black hole plays a dual role. On the one hand, tidal shear acting against self-gravity stretches the cloud in the radial direction, making it more difficult for the cloud to fragment. On the other hand, the gas is tidally squeezed in the vertical direction, parallel to the disk axis, thus, making it easier to resists the gravitational stress by increasing the gas density to the Roche's critical value. As the cooling time is comparable to the dynamical time scale, stars in a hot disk are formed in self-gravitating clumps with a range of eccentricities and inclinations as the disk circularizes (Nayakshin et al. 2007; Wardle and Yusef-Zadeh 2008; Bonnel and Rice 2008; Alig et al. 2011). Simulations of the capture of a cloud suggests that star formation during such an event will occur before the disk is fully circularized and becomes dynamically cold (Mapelli et al. 2008). This is in contrast to a cold self-gravitating disk that forms stars with small eccentricities. Large eccentricities could not easily be built up from initial circular orbits in a cold disk, thus, the hot disk scenario may better explain the observed stellar kinematics

The second issue is related the mechanism in which a cloud loses angular momentum to form a disk so close to Sgr A*. The cloud has no means of getting rid of its angular momentum and it is difficult and rare to capture a compact cloud by a massive black hole without being engulfed. However, nature provides an easier mechanism to bring the gaseous material closer in to the black hole. This is done by the capture of material from an extended cloud passing on opposite sides of the black hole, thus allowing the loss of much of the angular momentum of gravitationally focused gas. The gaseous material with large impact parameters can settle much closer to the black hole, thus higher gas density is achieved. Given the level of asymmetry and inhomogeneity in the spatial distribution of the cloud, the cloud capture event may even account for multiple counter-rotating disks with different inclinations. Figure 1 shows a diagram of a cloud impacting Sgr A*. The gravitational focusing of an incoming molecular cloud is shown in the top panel whereas the outer region of the cloud that is not captured continuous its motion in the direction away from Sgr A*. The inner region of the cloud is carved out, captured and circularized to form a disk.

3. Galactic Center Molecular Disk

Multi-wavelength studies of the Galactic center show that a 2-parsec clumpy molecular ring circles Sgr A* with a rotational velocity of about 100 km s^{-1} (Christopher et al. 2005). The observed CO and HCN (Güsten et al. 1987) emission indicates a total mass of $\sim 10^5$ $M_\odot$. However, HCN observations imply that there is high density material close to the inner edge (Christopher et al. 2005) suggesting that the mass may be closer to 10^6 $M_\odot$. The application of the formation scenario to the circumnuclear ring of gas argues that this disk was recently captured, is currently settling down and is on the verge of forming stars (Paper I, Paper III).

The extent of the ring suggests an initial migrating cloud speed towards the lower end of the 50–100 km s^{-1} range. The Hoyle-Lyttleton radius is about 12 pc which is similar to the outer radius of the molecular ring. On this scale, one expects considerable asymmetry in the cloud material passing by Sgr A* during the capture event, with a corresponding reduction in the net cancellation of angular momentum during the capture process (cf. Bottema & Sanders 1986). At first sight our model suggests that the circumnuclear ring could become unstable to gravitational fragmentation (Paper I), but this assumes that it has kinematically relaxed. The velocity dispersion of the ring is $\sim$30 km s^{-1}, so that it is not unstable unless its mass is 10^6 M$_\odot$.

Methanol and water masers have recently been detected suggesting that star formation is taking place (Paper II). It appears that the circumnuclear ring is still in the process of settling down soon after formation. The ring's orbital time scale at 2 pc is $\sim 10^5$ yr, so this implies that the age of the ring is $\leq 10^6$ yr. If this is so, the remains of the original interloper cloud should lie within $\sim$ 100 pc of Sgr A*. One candidate is the +50 km s^{-1} molecular cloud which extends along the plane from the Galactic center to l$\approx 0.2^0$ and consists of a number of bound cloudlets with a total mass of $\sim 10^6$ M$_\odot$. This cloud is thought to lie about 30 pc behind Sgr A*, consistent with an interaction $\sim 3 \times 10^5$ years ago.

4. AGN Megamaser Disks

Recent VLBA observations have discovered several sub-parsec megamaser disks with high inclinations in the nuclei of Seyfert 2 galaxies (Herrnstein et al. 2008; Kuo et al. 2010). The inner and outer radii are determined by milli-arcsecond observations. The mass of the disks are measured accurately because these disks lie well within the sphere of influence of their host black holes and are not self-gravitating. Figure 2 shows an empirical linear correlation between the disk radius and the black hole mass. There is a well-defined upper envelope to the maser disk radii which scales linearly with black hole mass, given approximately by R$_{\rm max} \sim 0.3$ M$_7$ pc, where the central black hole mass is $M = M_7 \times 10^7$ M$_\odot$.

The remarkable similarity of size scale of the stellar disk surrounding Sgr A* and the megamaser disks suggest that they could be formed the same way. Paper III shows that for plausible estimates of the mass and angular momentum of the captured material, this process naturally reproduces the empirical linear relationship between maser disk size and black hole mass. The capture of clouds with column densities $\leq 10^{23.5}$ cm^{-2} results in a non-self-gravitating disks of the correct scale and sufficient column density to allow X-ray irradiation from the central source to reproduce thin megamaser disks. By contrast, the capture of a cloud with higher column density would instead create a self-gravitating disk giving rise to rapid star formation.

In the context of disk formation, the disk size scales with black hole mass. The reason for such a scaling is as follows. The mass of the disk depends on the Hoyle-Littelton capture radius and the column density of the initial cloud. The size of the disk depends on the capture radius which itself depends on the mass of the black hole and the angular momentum cancellation. Thus, the cloud capture model implies that the radius of the disk depends linearly on the mass of the black hole.

5. Cloud Migration

How do molecular clouds migrate toward Galactic nuclei? The inner 200 pc of the Galaxy is rich in dense molecular clouds, many of which are on eccentric orbits (Bally et al. 1988; Oka et al. 1998; Martin et al. 2004; Jones et al. 2011). In addition to the +50 km s^{-1} molecular cloud and other members of a disk population of molecular clouds distributed within the inner 30 pc of Galactic center. Their non-circular, elongated motion is thought to be induced by the Galaxy's barred potential (see Kim et al. 2011), with dynamical friction aiding migration to the central regions of the Galaxy. The bar potential of the Galaxy can apply a torque to the clouds, thus trigger infall toward the nucleus (see F. Combes in these proceedings). Recent simulations also suggest that gas supplied to galactic centers is controlled by angular momentum transfer from one massive gas clump to another during gravitational encounters (Namekata & Habe 2011).

In the context of a transition from X1 to X2 orbits, the gas clouds lose angular momentum spiraling in toward the central region of the Galaxy. In order to bring gaseous material close in to the inner few pc, the bar must be nested (Namekata et al. 2009).

Some of these inward-moving clouds may interact with their host supermassive black holes. In our own Galaxy, the circumnuclear molecular ring (e.g. Christopher et al. 2005) at a distance of 1.7 pc from Sgr A*, the 50 km s^{-1} molecular cloud, and the circumnuclear rings found on scales of several parsecs from the center of numerous Seyfert galaxies suggest an ample supply of material. The rate of migration of molecular material is estimated to be about $10^4 - 10^5$ $M_\odot$ per $10^6 - 10^7$ years. The age of the stellar disk (Paumard et al. 2006), and the relative youth of the circumnuclear ring (Paper II) is consistent with the age estimate implying that the rate of encounters of massive clouds with Sgr A* is $\sim 10^{-6}$ yr^{-1}. This may have been been ongoing for a significant fraction of the Galaxy's lifetime as the stellar population in the central parsec is consistent with roughly constant star formation over the past 12 Gyr (Maness et al. 2007).

6. Black Hole Recoil

The partial capture of a cloud with incident velocity $V_{initial}$ imparts momentum to the system of black hole with a mass M_{BH} and gaseous disk with a mass M_{Disk}, implying a kick velocity

$$V_{kick} = \frac{M_{Disk} \times V_{initial}}{M_{BH} + M_{Disk}} \qquad (1)$$

to the black hole. Given that black holes like Sgr A* are embedded in their host nuclear cluster, the black hole is expected to move with respect to the center of the cluster and exchanges heat with the cluster due to dynamical friction. The recoil is of the order of few km s^{-1} and is likely to be damped via dynamical friction on the surrounding stars, on a timescale of the stellar crossing time across the sphere of influence, $\sim 10^3$ years. However, the parameter space for the significance of the recoil to the black hole in different nuclear cluster environment has not been explored.

7. Conclusions

In summary, we have investigated the capture of gas clouds by massive black holes via the Hoyle-Lyttleton process. We applied this mechanism to three different evolutionary phases of cloud capture scenario. One is a subset of AGN megamaser disks with high inclinations. There is no apparent star formation in these disks which lie within the sphere of influence of their host black holes. Second application of the cloud capture scenario is to the 2-parsec molecular disk orbiting Sgr A* where the gaseous disk is young. It has been argued that this system is showing early signatures of star formation. Lastly, sub-parsec stellar disk orbiting Sgr A* where massive star formation has already taken place a few million years ago.

The lack of star formation in megamaser disk systems could be due to strong magnetic fields acting against self-gravity of the disk (Johansen and Levin 2008; Gaburov, Johansen and Levin 2011). Alternatively, the low column density may be responsible. Assuming that the tidal shear by the black hole is accounted for, then the mass-to-flux ratio is an important parameter for the comparison of self-gravity and the magnetic field. Thus, the lack of star formation depends on the column density of incoming clouds and the magnetic field of the disks rotating around massive black holes. Future molecular line and magnetic field observations of thermal gas in these disks should be quite useful.

The physical relationship that threads through these different systems is the capture of a cloud by engulfing the massive black hole. This is an effective way to get rid of the angular momentum of the incoming molecular cloud. The engulfing scenario can also be applied when a star approaches the tidal disruption radius of a massive black hole. We speculate that these processes could operate in galaxy mergers, dense stellar cusps around massive black holes, leading to starburst activity, black hole growth and powerful flaring activity in the vicinity of central massive black holes.

We found a scaling relation between the radius of the captured disk and the massive black hole which appears to be supported by the Hoyle-Lyttleton capture process. The outer radius of the captured cloud is understood in terms of a physical truncation of the disk because the scale determined kinematics of cloud capture depends linearly on the black hole mass. The apparent inner radius of the captured disk can be understood if the disk is warped in such a way that the inner disk does not get sufficient exposure to the central source of X-rays, to produce 22 GHz H_2O (Maloney et al. 2002). As for the disk of stars orbiting Sgr A*, the inner radius is 0.03pc. In this case, it is possible that the stars that are formed within the inner 0.03pc are influenced by dynamical processes. In fact, there is a cluster of B-type stars with eccentric orbits distributed isotropically within the inner radius but with a completely different distribution than that of the disk of stars. Alternatively, it is possible that the captured gas within the inner radius is dissipated to high temperature and does not have sufficient density to overcome the tidal stress. This idea could be relevant to the formation of the inner radius of the circumnuclear molecular ring where hot stars of the disk are thought to have ionized the inner 1.7pc of molecular gas and created a cavity of molecular gas in this region.

8. References

[1] Alig et al. 2011, MNRAS, 412, 486
[2] Bally, J. et al. 1988, ApJ, 324, 223
[3] Bonnell, I. A., & Rice, W. K. M. 2008, Science, 321, 1060
[4] Bottema, R. & Sanders, R. H. 1986, A&A, 158, 297
[5] Christopher, M. H., Scoville, N. Z., Stolovy, S. R., & Yun, M. S. 2005, ApJ, 622, 346
[6] Combes, F. 2011, these proceedings
[7] Gaburov, E., Johansen, A. & Levin, Y. 2011, submitted to ApJ
[8] Genzel, R. Eisenhauer, F. & Gillessen, S. 2010, RvMP, 82, 3121
[9] Gerhard, O. 2001, ApJ, 546, L39
[10] Güsten et al. 1987, ApJ, 318, 124
[11] Herrnstein, J. R., Moran, J. M., Greenhill, L. J., & Trotter, A. S. 2005, ApJ, 629, 719
[12] Johansen, A. & Levin, Y. 2008, A&A, 490, 501
[13] Jones, P. A. et al. 2011, MNRAS, in press
[14] Kim, S. S. et al. 2011, ApJ, 735, L11
[15] Kuo, C. Y., et al. 2011, ApJ, 727, 20
[16] Levin, Y. & Beloborodov, A. M. 2003, ApJ, 590, L33
[17] Lu, J. R. et al. 2009, ApJ, 690, 1463
[18] Maloney, P. R. 2002, PASA, 19, 401
[19] Maness, H. et al. 2007, 669, 1024
[20] Mapelli, M., Hayfield, T., Mayer, L. & Wadsley, J. 2008, MNRAS, (arXiv:0805.0185)
[21] Martin, C. L. et al. 2004, ApJS, 150, 239
[22] Namekata, D. et al. 2009, ApJ, 691, 1525
[23] Namekata, D., & Habe, A. 2011, ApJ, 731, 57
[24] Nayakshin, S., Cuadra, J. & Springel, V. 2007, MNRAS, 379, 21
[25] Oka, T. et al. 1998, ApJS, 118, 455
[26] Paumard, T., Genzel, R., Martins, F., Nayakshin, S., Beloborodov, ApJ, 643, 1011
[27] Wardle, M., & Yusef-Zadeh, F. 2008, ApJ, 683, L37 (Paper I)
[28] Wardle, M., & Yusef-Zadeh, F. 2011, ApJ (submitted) (Paper III)
[29] Yusef-Zadeh, F., Braatz, J., Wardle, M. & Roberts, D. 2008, ApJ, 683, L147 (Paper II)

Journal of Physics: Conference Series **372** (2012) 012024 doi:10.1088/1742-6596/372/1/012024

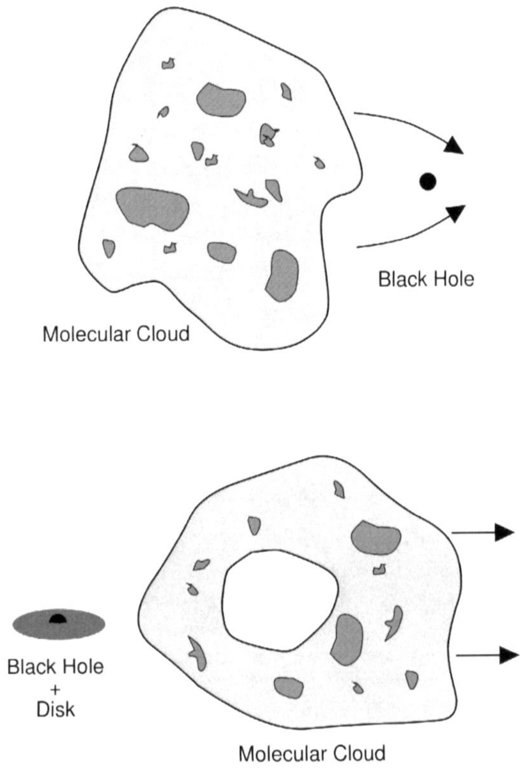

Figure 1. A schematic diagram of a cloud before and after impacting Sgr A*.

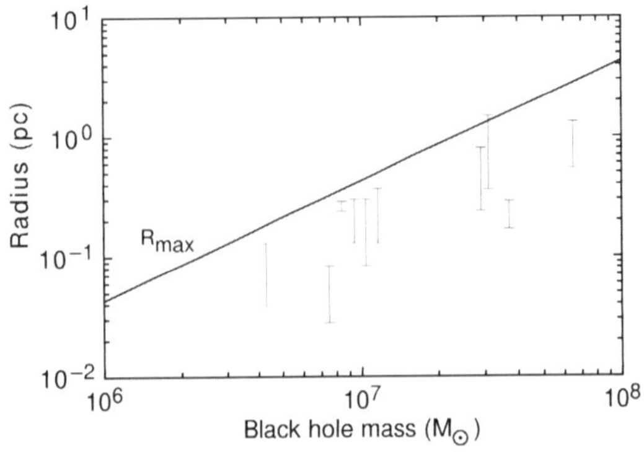

Figure 2. A plot of the black hole mass and the outer and inner radii of AGN mega-maser AGNs plus the Galactic center stellar disk. The upper envelope, labelled "$R_{\rm max}$", is given by $R_{\rm max} = 0.3(M/10^7 {\rm M}_\odot)$ pc.

NaCo/SAM observations of sources at the Galactic Center

J. Sanchez-Bermudez[1], **R. Schödel**[1], **A. Alberdi**[1], **J. U. Pott**[2]
[1] Instituto de Astrofísica de Andalucía (CSIC), C/ Glorieta de la Astronomía S/N, 18008 Granada, Spain
[2] I. Physikalisches Institut, University of Cologne, Zülpicher Str. 77, 50937 Köln, Germany
E-mail: joel@iaa.es, rainer@iaa.es, antxon@iaa.es, pott@ph1.uni-koeln.de

Abstract. Sparse aperture masking (SAM) interferometry combined with Adaptive Optics (AO) is a technique that is uniquely suited to investigate structures near the diffraction limit of large telescopes. The strengths of the technique are a robust calibration of the Point Spread Function (PSF) while maintaining a relatively high dynamic range. We used SAM+AO observations to investigate the circumstellar environment of several bright sources with infrared excess in the central parsec of the Galaxy. For our observations, unstable atmospheric conditions as well as significant residuals after the background subtraction presented serious problems for the standard approach of calibrating SAM data via interspersed observations of reference stars. We circumvented these difficulties by constructing a synthesized calibrator directly from sources within the field-of-view. When observing crowded fields, this novel method can boost the efficiency of SAM observations because it renders interspersed calibrator observations unnecessary. Here, we presented the first NaCo/SAM images reconstructed using this method.

1. Introductory Remarks

The Galactic center (GC) nuclear stellar cluster (NSC), the densest region of the Milky Way ($\rho > 10^6$ $M_\odot$ pc^{-3}) [1; 2], contains the supermassive black hole (BH) SagittariusA* (SgrA*) at its center [3; 4] and exhibits signs of at least two epochs of recent star formation [5; 6]. Theoretical considerations as well as observational evidence suggest that the GC environment favors a top-heavy initial mass function (IMF) [7–9] (see however Do et al., these proceedings). On the order of 100 high-mass stars were created in the most recent ($\approx 4x10^6$ years-old) starburst episode in the central parsec of the GC. Since massive stars are rare in the Galaxy, this makes the Milky Way NSC an important site to study the properties of such objects (e.g. morphology, multiplicity) and their interactions with the interstellar medium (ISM).

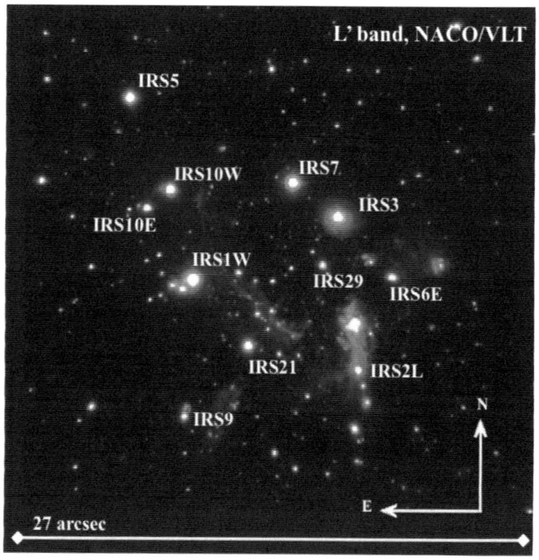

Figure 1: The inner central parsec of the GC in the L'- band with NaCo/VLT. Our targets, the brighter massive stars in the field with known diffuse (resolved and/or unresolved) emission, are identified by their names.

Figure 1 provides an overview of the central parsec of the Galaxy in the form of an L'-band AO image obtained with the NaCo instrument of the Very Large Telescope (VLT). Our targets, the brighter sources ($4 < mag_L < 7$) in the field of view (FOV), are labeled with their names. The principal selection criteria for this sample are that all of the sources (a) exhibit significant mid-infrared excess, (b) are related to recent starburst events (~6 and ~100 Myr ago) and (c) show extended structure like dusty outflows and bow-shocks. Our sources can be divided into three main groups: **(a) Evolved (super-) giants**, like IRS 7, 9, 3; **(b) Massive, post-main sequence stars,** such as IRS 6E, 29NE1 and **(c) Bow-shocks,** as IRS 1W, 21, 10W, and 5.

The importance of a detailed study of such objects with high angular resolution lies in the possibility to understand the role of binarity/multiplicity in the dust formation processes, as well as the circumstellar dust distribution. In this context, Sparse Aperture Masking interferometry + Adaptive Optics is one of the most suitable techniques to address these goals. Nevertheless, because of the difficulties we encountered during the observations (bad atmospheric conditions, difficulty of measuring the sky emission), the analysis of our SAM data requires the development of dedicated software and new reduction techniques to boost the efficiency of the observations. In this work, we present a novel calibration technique, as well as our first reconstructed SAM images of the targets described above.

2. Brief description of the technique

Adaptive Optics (AO) is a powerful technique which is used to overcome the degrading effects (mainly induced by the atmospheric turbulence) in the wavefront (WF) of a specific source, while keeping high sensitivity. Nevertheless, in spite of the great promise of this technique to provide us with diffraction limited images, it still faces a number of problems that make a precise and stable PSF calibration very difficult. Such problems are, for example: (a) The PSF of the AO systems is variable on different timescales and highly sensitive to the observing conditions (e.g. seeing, airmass, etc); (b) the AO performance is constrained by technological limitations of the system (e.g. the number of sensors and actuators that measure and correct the incoming WF); and (c) the WF of the guide star is not the same as the one of the science objects. Therefore, their wavefronts suffer different distortions because they take slightly different paths through the atmosphere (anisoplanatism) [10; 11].

Sparse Aperture Masking (SAM) interferometry is a technique which transforms a single dish telescope into a non-redundant Fizeau interferometer by placing a mask with many holes in the pupil plane of the telescope camera. The main advantage of SAM interferometry is the fact that the non-redundant mask (NRM) removes most of the incoherent noise produced by the atmospheric turbulence. In this way, a very well defined and stable PSF can be obtained. The PSF resembles the diffraction pattern of the mask (see Fig. 2). The nominal angular resolution (θ, which corresponds to the Full-Width Half-Maximum of the beam) achieved by this technique is of the order of $\theta \approx \lambda/2D$. This is a factor of ~2 better than the resolution obtained in standard imaging. Nevertheless, due to the fact that the NRM covers most of the telescope pupil, SAM suffers from low photon efficiency and is limited to bright objects (e.g. in the case of the VLT the magnitudes of the targets range between 4 to 12, depending on the mask used). **A combination of the technologies of SAM and AO offers us a unique tool to recover high fidelity images and, hence, study structures between one and few times the PSF size (e.g. in the case of the VLT, to analyze morphologies in the regime of 50 to 150 mas), with the strengths of the robust PSF calibration provided by SAM, and the higher dynamic range obtained by the AO systems [11].**

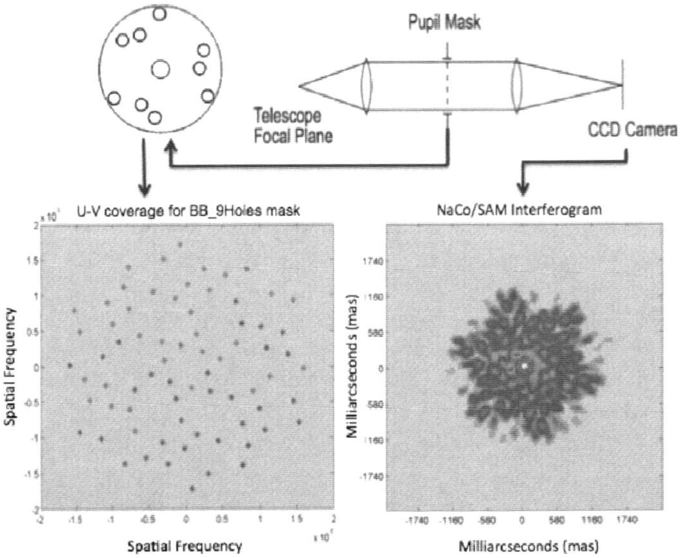

Figure 2: Sparse Aperture Masking layout. In the figure a NRM with nine holes is located in the pupil of the telescope. It is thus transformed into an interferometer with a non-redundant u-v coverage (lower-left gray scheme); note that the number of points sampled in the Fourier space correspond to all baselines formed by all the possible pair of holes combinations plus their complex conjugates (in this case, 36 red dots + 36 blue dots). The lower-right gray scheme shows the SAM interferogram (i. e. interferogram) of a PSF obtained at the camera detector, with a shape that only depends on the geometry of the interferometric array. The upper part of the scheme was adapted from the NaCo manual and Tuthill et al. (2006).

3. NaCo/SAM L-band observations of the Galactic Center

3.1. NaCo/SAM observational setup

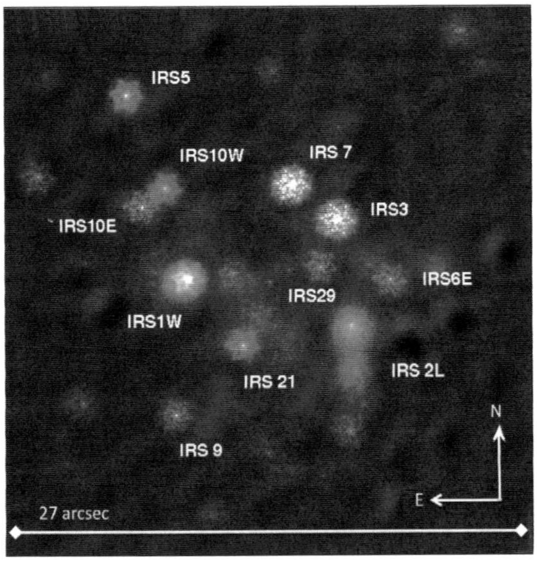

Figure 3: NaCo/SAM FOV. The main targets of our sample show clearly the interference pattern of the NRM overimposed on them and are labeled with their names. Since SAM is a Fizeau type interferometry, it keeps the entire FOV of the camera used.

In June 2010 we performed VLT NaCo/SAM observations of the central parsec of the GC NSC in the L band, using two different filters; (i) L' (3.8 μm) and (ii) NB 3.74 (3.74 μm). The observations in the L' filter were conducted using the *BB_9holes* mask, while the ones in the NB 3.74 filter were taken using the *9_holes* mask. The main reason to use these NRMs is because they are optimized to be utilized with broad band and narrow band filters, respectively. The observations were recorded using the Pupil-Tracking mode of NaCo (which freezes the pupil and makes the FOV rotate in opposite direction to the parallactic angle). The so-called *cube mode* was used, which saves each single data integration time (DIT) frame. The advantage of the cube mode is that it allows the selection of the best frames. Frame selection can become necessary when AO performance is not stable, as it was the case for the observations described here. Our observations consisted of a group of 13 sets of data-cubes composed of 20 - 30 exposures with 5 sec of DIT. Because SAM removes the low spatial frequencies related to the atmospheric noise, chopping techniques were not used to eliminate the near-infrared background radiation. Nevertheless, since the GC NSC is extremely crowded, separate sky observations were taken on a dark cloud a few arcmin to the north-west of Sgra*. IRS7, the brightest target in our FOV, was used as guide star for the the IR WFS of NaCo. Figure 3 shows a single Naco/SAM exposure with our targets identified on it.

3.2. Data Reduction

The reduction process of our NaCo/SAM data-cubes is divided into two principal steps: (i) standard imaging reduction and (ii) interferometric analysis.

3.2.1. Standard Imaging: In this reduction part, as in any other optical-NIR imaging observations, all data-cubes were flat-fielded, bad-pixel corrected and background subtracted. Depending on the observational data set, the sky template for the background subtraction was obtained by calculating either (i) the median of the sky cubes or (ii) through the median of the lower values at each pixel from the dithered images, when no dedicated sky observations close in time were available. As our images were taken using the Pupil Tracking mode of NaCo, all cubes had to be derotated to correctly compute the interferometric observables. All data-cubes were frame selected, discarding bad images through the analysis of the cube flux statistics ($\sim$ 70-90 % of the frames were eliminated). Finally, each one of the individual targets was stored into a sub-cube of 128 x 128 pixels.

3.2.2. Interferometric reduction process: To obtain the interferometric observables (squared visibilities and closure phases) we used Michael Ireland's IDL pipeline. This code calculates the Fourier transform of all the frames in a given data-cube and convolve them with a *matched filter* template, which is the Fourier transform of all baselines formed by the holes in the NRM. Due to the facts that (i) each hole in the NRM has a finite radius, and (ii) that our data cubes are stored into a discrete array of pixels, the normalized flux at every baseline in the Fourier space is spread into a group of pixels, or **splodge**, around the central spatial frequency of the used baseline (see Fig 4). Therefore, every raw (not calibrated) squared visibility is calculated by the squared value of the sum of the normalized flux at a given splodge, divided by the total flux received (i.e. baseline zero). The closure phases are the arguments of the bispectra, i.e. the ensemble of all visibility products for non-redundant baseline triangles. It is important to remark that: the visibility amplitudes give us information about the compactness and extension of the source brightness distribution in all the spatial frequencies sampled by the baselines of our interferometer, while closure phases contains information on the orientation and symmetry of the target structure and are robust to any observational effect (e.g. atmospheric turbulence).

In order to increase the u-v coverage and to get a better sampling of the sources brightness distribution (see Fig. 5) we use the Earth rotation synthesis through the combination of the visibilities obtained by observations of the targets at different parallactic angles. An example of the IRS7 raw visibilities and closure phases is shown in Figure 6.

3.2.3. Calibration of visibilities and closure phases. The standard observing technique for NaCo/SAM is the observation of a science target interwoven with observations of a nearby point-source reference object (i.e. the calibrator). This star should be observed at similar airmass and in the same AO configuration as the science target. Therefore, calibrator sources should be of similar brightness as the targets. Unfortunately, this was not possible for our observations because of the following difficulties:

(a) Rapidly changing observing conditions (clouds, highly variable seeing, etc.) affected the calibrator observations, so that it was hard to obtain data of similar quality on source and calibrators;

(b) The GC is highly extincted. Hence, it is very difficult to find calibrators that are sufficiently bright in the L band (~ 4 mag_L) because the ones that fulfill this condition are too bright in the K-band so that they would saturate the infrared wavefront sensor of NaCo. Therefore, the observed calibrators were systematically fainter than our brighter targets (e.g. IRS7 and IRS 3);

(c) The observations were affected by strong residual patterns that remained in the images after background subtraction. The reason for these residuals are not clear, but they are typical for L-band observations, where the background fluctuates rapidly (see Fig. 7).

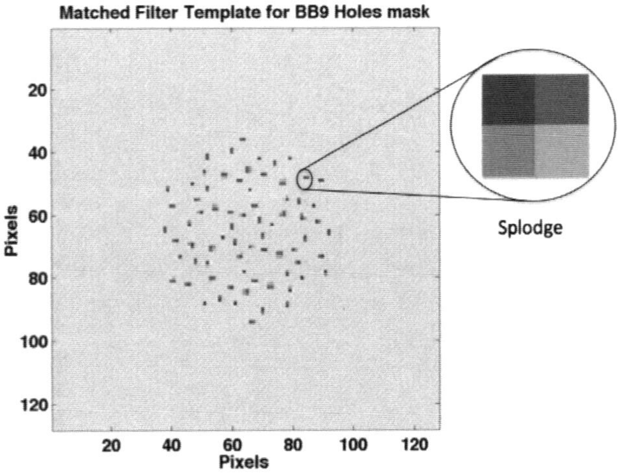

Figure 4: Matched Filter Template. The total flux at every spatial frequency (given by every baseline of the mask) sampled in the matched filter is spread in a series of pixels called "splodges". The total amount of pixels which form a splodge is determined by the size of the mask holes.

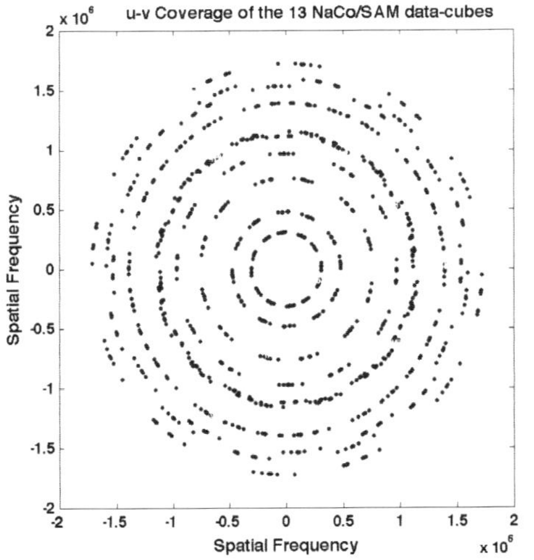

Figure 5: NaCo/SAM total u-v coverage. The total u-v coverage of our interferometer is produced using the Earth Rotation Synthesis through the combination of all data-cubes visibilities. The beam is symmetric in all directions.

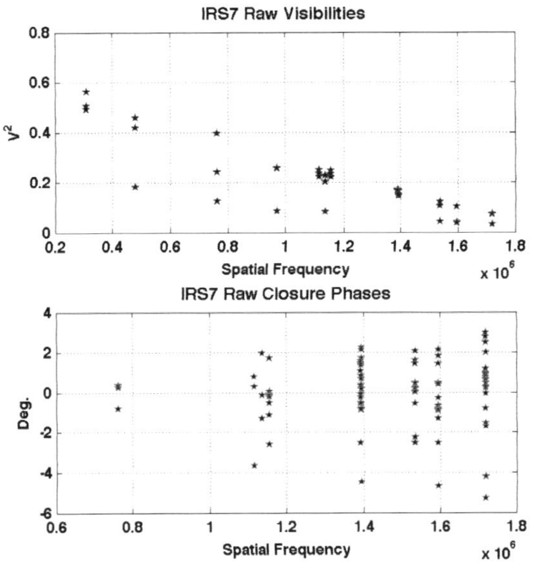

Figure 6: Raw Squared visibilities and Closure phases of IRS7. On the figure we have the uncalibrated 36 visibilities and 84 closure phases of one of the observed IRS7 L-band data-cubes using the *BB_9holes* mask.

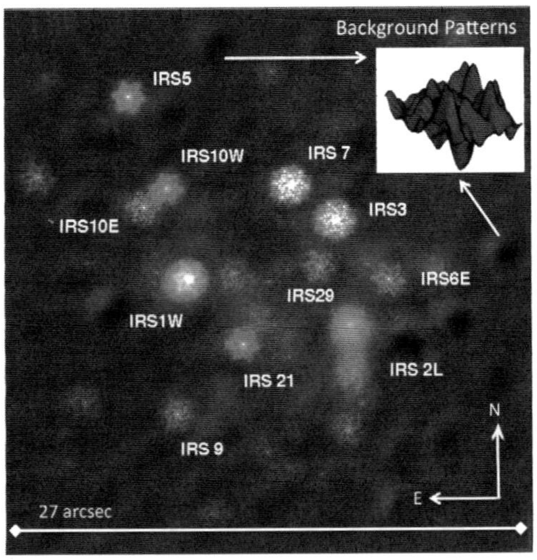

Figure 7: L-band NaCo/SAM background patterns. All frames of our observations exhibit some background patterns spread all over the FOV. The angular scale of the background variations is of the order of the size of the PSF (~ 1").

In order to overcome these problems we developed a calibration technique, which we called the *"Synthetic Calibrator"*. This method takes advantage of the presence of several bright sources in our FOV and creates a PSF by source superposition, using StarFinder [12] routines. The images of stars are extracted from the FOV, cleaned from the contamination of secondary sources, locally background subtracted, centered with sub-pixel accuracy, normalized and combined by a median superposition. The resulting PSF is thus obtained directly from the field, without the need to observe a calibrator source. Some of the advantages are: (a) No interwoven observations of standard calibrators are needed, therefore, the observing conditions are exactly the same for both the calibrator and the target; (b) the calibrators can be as bright as science targets.

Note that, a priori, we do not know the intrinsic source-structure of our targets. This can can be a source of systematic uncertainties when extracting a synthetic calibrator. Nevertheless, the latter effect, as well as the background residuals, are effectively minimized by the median superposition of several sources when creating the synthetic calibrator, as our tests have confirmed. Figure 8 shows and example of the calibrated visibilities and closure phases of two of our targets IRS 7 and IRS 1W obtained with this technique. Of course, the Synthetic Calibrator technique can only be used when several sufficiently bright sources are present within the field-of-view. It may thus be ideal for observations of stars in dense clusters.

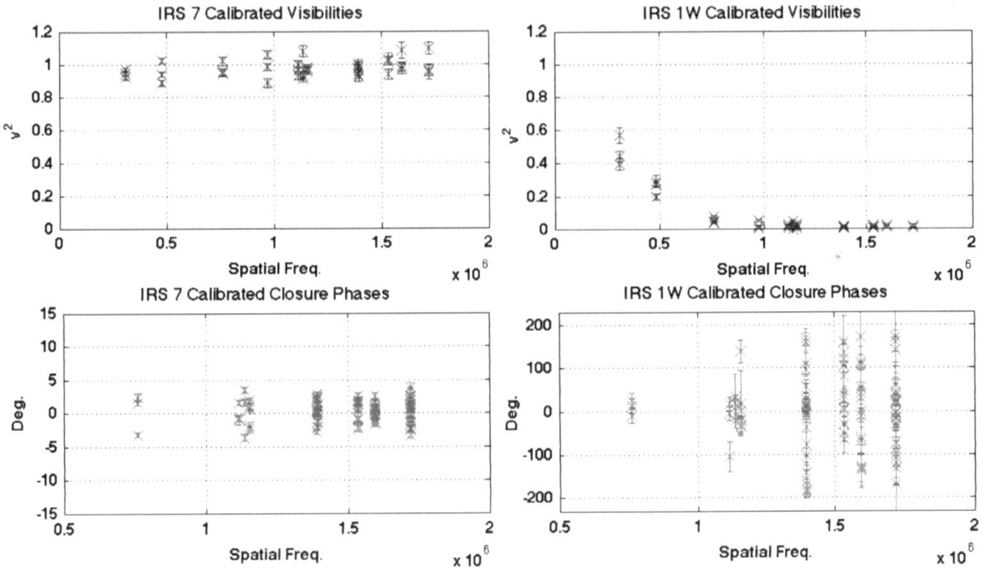

Figure 8: Calibrated Visibilities and Closure Phases of IRS 7 and IRS 1W. Calibrated visibilities of IRS 7 are constant and close to the unit, demonstrating that this source is a point-like object; its closure phases also confirm the point-symmetry of the object (left). On the other hand, the visibilities of IRS 1W suggest the existence of an extended structure larger than the interferometric beam; its closure phases demonstrate the point asymmetry of this source (right).

4. Image reconstruction.

Once the calibrated interferometric observables had been obtained, image reconstruction was done with the BSMEM package [13], which uses a maximum entropy algorithm to reconstruct the interferometric maps. Figure 9 displays a comparison between three of our reconstructed SAM images (IRS 1W, IRS 5 and IRS 10W) in the L-band (upper part) and the AO deconvolved Ks images of the same sources (lower part) obtained with the 10-meter Keck telescope by Tanner et al. (2005) [14]. The resolution achieved with our SAM observations in the L-band (at 3.8 μm, $\theta \approx 60$ mas) is similar to the one obtained by the AO system of Keck telescope in the Ks band (at 2.2 μm, $\theta \approx 45$ mas). This demonstrates that the nominal resolution of the NaCo/SAM technique has been achieved. On the other hand, Figure 10 presents for comparison a set of six L-band

Journal of Physics: Conference Series **372** (2012) 012025 doi:10.1088/1742-6596/372/1/012025

images taken with the VLT. The first column shows our SAM images, the second one displays Lucy-Richardson [15] devoncolved images and the third one corresponds to raw AO images. The resolution and quality of the reconstructed SAM images surpass the raw and deconvolved AO images at the same frequency.

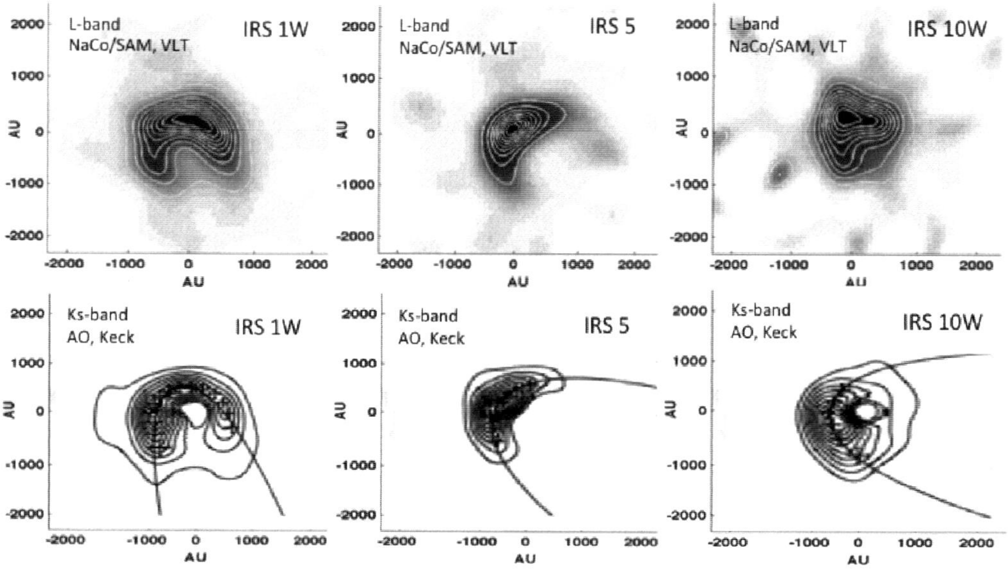

Figure 9: NaCo/SAM vs AO Keck images. At the top, NaCo/SAM images of IRS 1W, IRS 5 and IRS 10W in the L-band; in the bottom, the same sources are presented as deconvolved AO images in the K-band. The angular resolution of both subsets of images is similar, even though there is a difference in wavelength. The K-band images are adapted from Tanner et al (2005). Contours of NaCo/SAM images represent 10, 20, 30 40, 50, 60, 70, 80 and 90 % of the total normalized flux.

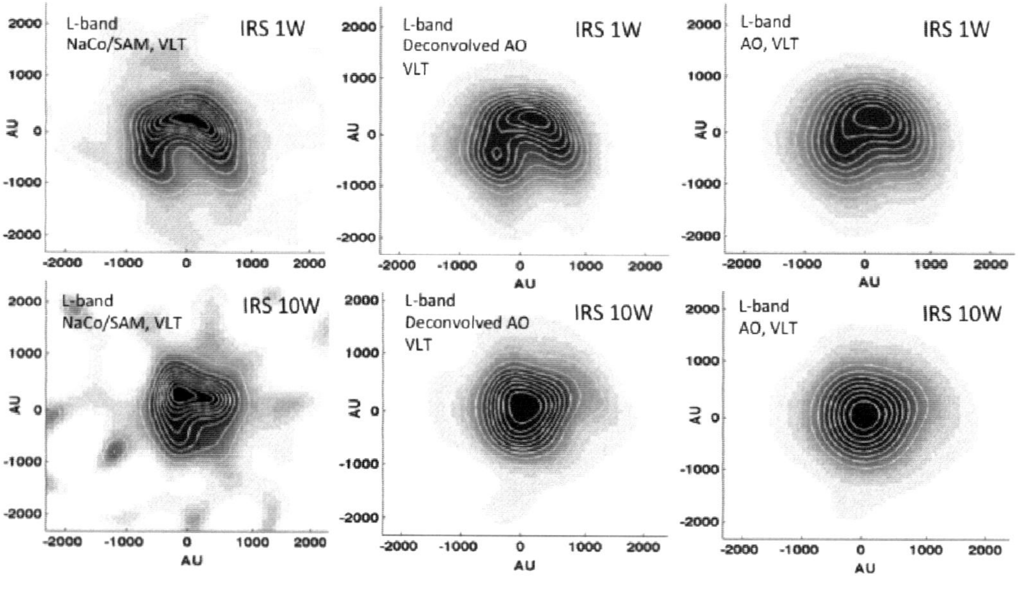

Figure 10: NaCo/SAM vs AO. IRS 1W and IRS 10W are presented in three subset of images: at the left, in the form of NaCo/SAM images; in the middle, as Lucy-Richardson deconvolved images; to the right, as non-deconvolved AO images. Note how the resolution on the NaCo/SAM images overpass the ones obtained by the other two subsets. Contours represent 10, 20, 30 40, 50, 60, 70, 80 and 90 % of the total normalized flux.

5. Summary

We overcome the difficulties of our NaCo/SAM observations (highly variable weather conditions, background instabilities and calibration problems) with the creation of *"The Synthetic Calibrator"* from the median superposition of several sources in the field, which according to our results appears to be a promising new way to calibrate SAM observations of dense stellar fields. The strength of this technique relies on the opportunity to observe multiple targets at the same time in Stellar Clusters, thus increasing the efficiency of the observations by eliminating the necessity of interspersed observations between standard calibrators and targets. From our reconstructed NaCo/SAM images we can conclude that the nominal L-band NaCo/SAM resolution was achieved ($\theta \approx 60$ mas) and that the quality of reconstructed maps quality clearly exceeds Lucy-Richardson deconvolved images. With these results, the scientific analysis of the sources, according to the scientific case described in the introduction, can be addressed.

Acknowledgments

JSB acknowledges support by the "JAE-Pre" programme of the Spanish Consejo Superior de Investigaciones Científicas (CSIC). RS acknowledges support by the Ramón y Cajal programme, by grants AYA2010-17631 and and AYA2009-13036 of the Spanish Ministry of Science and Innovation, and by grant P08-TIC-4075 of the Junta de Andalucía. AA acknowledges support by grant AYA2009-13036 of the Spanish Ministry of Science and Innovation and by grant P08-TIC-4075 of the Junta de Andalucía.

References

[1] Genzel, R and Schödel, R and Ott, T and Eisenhauer, F and Hofmann, R and Lehnert, M and Eckart, A and Alexander, T and Sternberg, A and Lenzen, R and Clénet, Y and Lacombe, F and Rouan, D and Renzini, A and Tacconi-Garman, L E 2003 *ApJ* **594** 812–832

[2] Schödel R, Eckart A, Alexander T, Merritt D, Genzel R, Sternberg A, Meyer L, Kul F, Moultaka J, Ott T and Straubmeier C 2007 *A&A* **469** 125–146

[3] Schödel R, Ott T, Genzel R, Eckart A, Mouawad N and Alexander T 2003 *ApJ* **596** 1015–1034

[4] Ghez A M, Becklin E, Duchjne G, Hornstein S, Morris M, Salim S and Tanner A 2003 *Astronomische Nachrichten Supplement* **324** 527–533

[5] Krabbe A, Genzel R, Eckart A, Najarro F, Lutz D, Cameron M, Kroker H, Tacconi-Garman L E, Thatte N, Weitzel L, Drapatz S, Geballe T, Sternberg A and Kudritzki R 1995 *ApJ* **447** L95

[6] Figer D F 2004 *The Formation and Evolution of Massive Young Star Clusters* (*Astronomical Society of the Pacific Conference Series* vol 322) ed H J G L M Lamers, L J Smith, & A Nota p 49

[7] Morris M 1993 *ApJ* **408** 496–506

[8] Paumard T, Genzel R, Martins F, Nayakshin S, Beloborodov A M, Levin Y, Trippe S, Eisenhauer F, Ott T, Gillessen S, Abuter R, Cuadra J, Alexander T and Sternberg A 2006 *ApJ* **643** 1011–1035

[9] Bartko H, Martins F, Fritz T K, Genzel R, Levin Y, Perets H B, Paumard T, Nayakshin S, Gerhard O, Alexander T, Dodds-Eden K, Eisenhauer F, Gillessen S, Mascetti L, Ott T, Perrin G, Pfuhl O, Reid M J, Rouan D, Sternberg A and Trippe S 2009 *ApJ* **697** 1741–1763

[10] Girard J, Pompei E, Amico P and Lidman C *Very Large Telescope NACO manual* URL `http://www.eso.org/sci/facilities/paranal/instruments/naco/doc/`

[11] Tuthill P, Lloyd J, Ireland M, Martinache F, Monnier J, Woodruff H, ten Brummelaar T, Turner N and Townes C 2006 *Society of Photo-Optical Instrumentation Engineers (SPIE) Conference Series* (*Society of Photo-Optical Instrumentation Engineers (SPIE) Conference Series* vol 6272)

[12] Diolaiti E, Bendinelli O, Bonaccini D, Close L, Currie D and Parmeggiani G 2000 *Astronomical Data Analysis Software and Systems IX* (*Astronomical Society of the Pacific Conference Series* vol 216) ed N Manset, C Veillet, & D Crabtree p 623

[13] Buscher D F 1994 *Very High Angular Resolution Imaging* (*IAU Symposium* vol 158) ed J G Robertson & W J Tango p 91

[14] Tanner A, Ghez A M, Morris M R and Christou J C 2005 *ApJ* **624** 742–750

[15] Lucy L B 1974 *ApJ* **79** 745

The evolution of the Galactic nuclear bulge traced by Cepheid variable stars

Noriyuki Matsunaga
Kiso Observatory, Institute of Astronomy, University of Tokyo, Japan
E-mail: matsunaga@ioa.s.u-tokyo.ac.jp

Abstract. The Galactic nuclear bulge hosts significant populations of young stars, within a few hundred parsecs around the Galactic Centre, unlike the more extended bulge. Recently, we discovered classical Cepheid variable stars (hereinafter Cepheids) in the Galactic nuclear bulge. Their ages are estimated to be $\sim$ 25 Myr based on the period-age relation of Cepheids. They are the first clear evidence of star formation a few tens of Myr ago. In addition, the period distribution of the Cepheids indicates that star formation rate was enhanced around $\sim$ 25 Myr ago compared to between 30 and 70 Myr ago. Such detailed star formation history has never been obtained for central parts of other galaxies. The timescale of the change in star formation seems to be consistent with the value, $\sim$ 20 Myr, suggested for gas inflow into the nuclear bulge. We discuss its implication on the evolution of the nuclear bulge.

1. Introduction

The Galactic Centre region is one of the most interesting places in the Galaxy, with the super-massive blackhole, dense star clusters, massive molecular clouds and so on. It is known that there exist stellar populations with different sizes and different characteristics: the Galactic bulge, the nuclear stellar disk and the nuclear star cluster around the super-massive blackhole [1]. The latter two systems together called the nuclear bulge, within a radius of a few hundred parsecs of the central blackhole, contain young stars of a few Myr old [2], unlike the extended bulge which is dominated by old stars ($\geq$ 10 Gyr, [3]). Conspicuous groups of such young stars are massive stellar clusters: the nuclear star cluster, Arches and Quintuplet [4, 5]. All of these clusters are as young as a few Myr, whereas the star formation history remains unclear for times more than $\sim 10^7$ yr ago.

Figer *et al.* [6] found that a scenario of continuous star formation fits the luminosity function of selected regions towards the nuclear bulge better than scenarios with intensive burst(s) of star formation in the past. On the other hand, Yusef-Zadeh *et al.* [7] estimated star formation rates in the last 10^{5-7} yr based on *Spitzer* photometry of young stellar objects and other methods, and suggested a temporal change in star formation. Although recent works agree that the recent star formation rate of the nuclear bulge has been roughly 0.1 $M_\odot$/yr, further observational efforts are necessary to unveil the population of young stellar objects [8]. Moreover, the estimate of the star formation rate $\sim 10^7$ yr ago obtained by Yusef-Zadeh *et al.* is based on nonthermal radio emission from the nuclear bulge region, whose result has a low time resolution. Some authors claimed that red supergiants and bright red giants may be as young as 10^7 yr [9, 10], but the age estimates for such objects are difficult because of their complicated processes like mass loss [11]. It is necessary to find other population tracers whose ages can be accurately determined.

We here consider Cepheids in order to investigate the star formation a few tens of Myr ago. They are pulsating giants or supergiants and have pulsation periods roughly between 2 and 50 days. It is well known that they follow a period-luminosity relation, and this allows estimating their distances [12]. Another important characteristics of Cepheids is their period-age relation [13], i.e. the younger Cepheids have the longer pulsation periods. We can also estimate their ages from their periods. However, the survey of Cepheids in the Milky Way is far from complete. In Figure 1, positions of currently known Cepheids [14] are indicated by yellow points. Only objects on the near side of the Sun were found, and no Cepheid at distances near the Galactic Centre were known. The main cause for this is the interstellar extinction towards the Galactic plane, which makes optical surveys of Cepheids difficult, and infrared surveys of variable stars are necessary to overcome this difficulty.

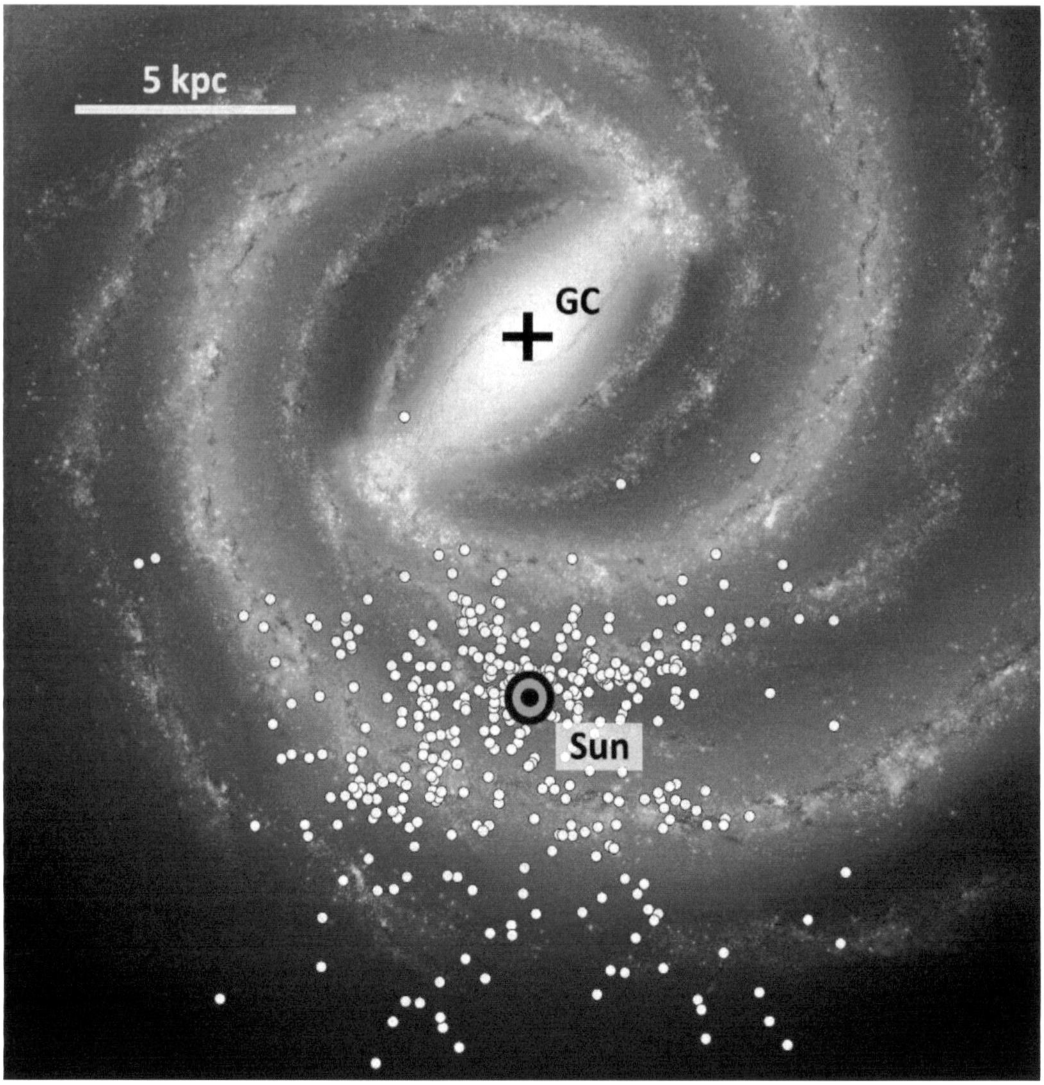

Figure 1. Positions of previously known Cepheids [14], indicated by yellow points, plotted over an artist's impression of the Milky Way (R. Hurt: NASA/JPL-Caltech/SSC) seen from the north Galactic pole.

2. Star formation history of the nuclear bulge traced by Cepheids

We have recently conducted a near-infrared survey of variable stars towards the Galactic Centre using the IRSF telescope and the SIRIUS camera [15] located in South African Astronomical Observatory. In addition to hundreds of Miras [16], we have recently discovered three classical Cepheids [17]. The estimated distances, ~ 8 kpc, agree well with the distance to the Galactic Centre [16, 18]. Also, the Cepheids receive typical foreground extinctions of the objects around the Galactic Centre, ~ 2.5 mag in the K_s band (2.14 μm). These support their membership to the nuclear bulge together with their distribution consistent with the disk-like structure of the nuclear bulge. All the Cepheids we discovered have periods ~ 20 days, which suggests the ages ~ 25 Myr based on the period-age relation [13]. The uncertainty in the age estimate is 20 percent, which is taken from the standard deviation of the period-age relation (see [13]). Therefore, they are the first tracers of a stellar population, in the nuclear bulge, with a few tens of Myr whose ages are accurately determined.

A further important finding from our survey is the lack of Cepheids with $P = 5$–19 days in the nuclear bulge. This is an unexpected result considering period distributions of known Cepheids in the Milky Way and other nearby galaxies [19]. The peak of the period distribution of the MW Cepheids is around 5 days, and objects with longer periods are expected to be less. Considering the period-age relation, the lack of the shorter-period Cepheids in our sample suggests that star formation was less active 30–70 Myr ago compared to 20–30 Myr ago as discussed in [17]. The star formation rate can be estimated from the number of Cepheids assuming the lifetime of Cepheids and the initial mass function. We used the lifetimes from [13] and the initial mass function from [20]. We also assumed that our survey effectively covers roughly 15 per cents of the nuclear bulge to estimate the star formation rate for the entire nuclear bulge. The estimated star formation rate was 0.075 $M_\odot$/yr with a factor of three uncertainty for 20–30 Myr ago. This value is similar to those obtained for the recent few Myr [7, 8]. In contrast, the rate was less than 0.02 $M_\odot$/yr (1 σ upper limit) 30–70 Myr ago.

It is worth while to note that there remain a few uncertainty in our estimates. Especially, the effects of metalicity and the initial mass function are important, and we discussed them in the Supplementary Information of [17]. We concluded that the change in the star formation rate is still significant at the 2 σ level. A similar change was suggested in [7], but our result has a higher time resolution. In addition, the star formation may be also less active between 10–20 Myr. Cepheids in this age range are expected to have periods significantly longer than 20 days, but none of such long-period Cepheids have been found. However, the expected number of such Cepheids is small and we cannot reach a firm conclusion. Massive stars identified in the nuclear bulge are a few Myr old, like the Arches and Quintuplet clusters, but no star aged 10–20 Myr has been confirmed. The star formation history 10–20 Myr ago remains uncertain.

3. Discussion

3.1. Evolution of the nuclear bulge

Here we discuss the implication of the star formation variation on the evolution of the nuclear bulge. It is considered that the star formation rate varies with the gas density, which is known as the Schmidt-Kennicutt relationship [21]. Although it is unclear if such a relation holds for the region close to the Galactic Centre, Yusef-Zadeh *et al.* [7] found that the star formation rate in the nuclear bulge agrees with the rate expected from the relation. Then, the change in the star formation rate occurred during the last 10^7 yr suggests the change in the gas density. The current mass of the gas within the nuclear bulge is estimated to be $\sim 10^7$ $M_\odot$ [1]. Gas needs to be supplied to the nuclear bulge in order to make the star formation over the Hubble time possible. Therefore, the star formation history revealed by tracers with various ages is important to consider the circulation and the consumption of interstellar gas around the Galactic Centre.

How do the star formation and the gas density change in the Galactic nuclear bulge, and what can we learn about the evolution of the Galaxy from the observational results? First of all, the star formation rate obtained for the nuclear bulge is much smaller than those which we find for starburst galaxies. For example, recent observations in the sub-millimetre revealed violent starbursts, $> 100\ M_{\odot}$/yr, at high redshifts [22, 23]. Such starbursts are considered to be driven by mergers [24, 25], and the merger-driven galaxy formation has been important especially in the early Universe. In comparison with those starbursts, the star formation in the Galactic nuclear bulge is (at least currently) smaller by several orders of magnitude. The Galaxy is relatively isolated from other large galaxies. Although dwarf galaxies may have some effects on the structure of the Galaxy [26, 27], their effects are presumably small on phenomena in the central region of the Galaxy. The evolution of the nuclear bulge is driven by a different process from the merger-driven starbursts. It is suggested by several authors that gas is transformed from the inner disk to the nuclear bulge via the bar-like structure [28, 29, 30]. Moreover, Stark *et al.* [31] proposed that the star formation in the nuclear bulge can be cyclic with a timescale of ~ 20 Myr. This scenario is consistent with the change in star formation we found for the nuclear bulge.

This process may be related to evolution of pseudobulges, disk-like stellar components slowly evolved in galaxy centres which are otherwise similar to merger-built classical bulges [32]. The Galactic nuclear bulge seems to be a pseudobulge [33] considering its various characteristics, although its classification and detailed evolution remains to be established [34]. Pseudobulges are considered to be made up of disk gas, and bars play important roles to supply the gas into the central regions of galaxies. In a few cases, star formation in pseudobulges seems to be episodic [35, 36]. The scenario for such evolution of pseudobulges agrees with the observed phenomena, which we discussed above, of the Galactic nuclear bulge.

3.2. Future prospects

Because of its proximity, we can observe in detail individual stars and interstellar matter in the Galactic nuclear bulge. For example, a recent investigation revealed a gas cloud of three times the mass of Earth which is approaching Sgr A* [37]. The Cepheids and other stars with their ages estimated make it possible to trace the star formation history with a high time resolution. Such detailed investigations on the Galactic nuclear bulge provide us with good opportunities to understand the evolution of bulges. In the following, we discuss future prospects of studies on the Cepheids in the nuclear bulge and the related stars.

First, the motion of the Cepheids is important to consider their properties and origins. Spectroscopic and astrometric observations are required to measure their motions within the nuclear bulge. The Arches cluster has been found orbiting around the Galactic Centre at a high velocity [38]. Although the uncertainty in the distance to the Arches makes the interpretation difficult, its motion is consistent with the rotation in the x2 orbit suggested by [39]. A recent simulation demonstrated that the gas transported from the bar to the nuclear bulge may form a ring along the x2 orbit and star formation occur in the ring [30]. It is interesting to see if the nuclear-bulge Cepheids show motions similar to the Arches or not.

Another important clue to the evolution of the bulge may come from chemical abundances of the Cepheids and other stars around. Figure 2 is a schematic diagram to show the distribution of stellar populations around the Galactic Centre. Within the nuclear bulge, $r_{\rm GC} < 0.3$ kpc, a few Myr old massive stars and the ~ 25 Myr old Cepheids are known, but the result of our Cepheid survey suggests the lack of stars aged 30 to 70 Myr. In contrast, no evidence has been found for recent star formations in the outer bulge region, indicated by the gray box at $r_{\rm GC}$ =0.3–3 kpc [3]. Although the exact age distribution is not known, Mira variables and other late-type bright stars represent stellar populations with a wide age range, 1–10 Gyr and possibly younger, in the entire bulge [9, 10, 16]. Furthermore, their properties vary as a function of $r_{\rm GC}$ indicating the age gradient from the outer region to the nuclear bulge [40]. In the disk exist stellar populations with various ages including Cepheids, although only several Cepheids in the inner disk have been found so far [41].

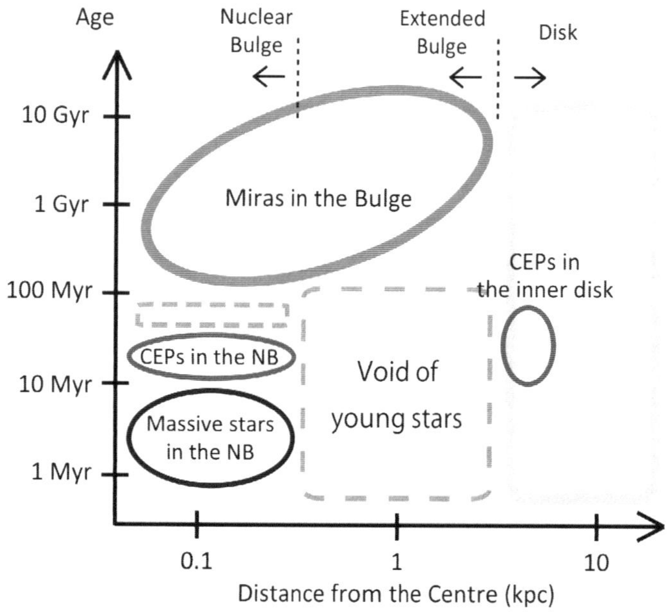

Figure 2. Distribution of stellar populations around the Galactic Centre. Various populations are illustrated according to their ages and distances from the Galactic Centre, and their typical tracers are indicated. CEP and NB stand for Cepheid and the nuclear bulge, respectively. See text for the detail.

It is important to understand how the various populations in Figure 2 have been formed. Spectroscopic measurement of their chemical features is expected to be useful for this purpose. For instance, how are the abundances of the nuclear-bulge Cepheids similar to those of the inner-disk Cepheids? Some of the inner-disk Cepheids are slightly older than those in the nuclear bulge, and they might have been formed of similar gas if a part of the inner-disk gas were provided to the nuclear bulge through the bar. How about the abundances of younger massive stars in the nuclear bulge? Massive stars born $\sim$ 25 Myr ago together with the nuclear-bulge Cepheids might have enriched the gas within the nuclear bulge which later forms the younger population. These discussions are rather speculative, but it would be valuable to compare the chemical features of these tracers.

Finally, more and more monitoring surveys are to be revealing a large number of Cepheids and other variable stars like RR Lyr stars in or around the bulge. Recently, the OGLE-III project reported classical and type II Cepheids towards the bulge but a few degree offset from the Galactic plane [42]. Their current survey OGLE-IV covers a wider region including the Galactic plane.[1] Furthermore, the VVV survey is carrying out a monitor survey in the near-infrared, hence more effective for obscured sightlines, for a wide region of the bulge and the disk (520 deg^2) [43]. Those new samples would be helpful to investigate the evolution of the bulge.

Acknowledgments
We thank the IRSF/SIRIUS team and the staff of South African Astronomical Observatory (SAAO) for their support during our near-IR observations. This work was supported by Grant-

[1] http://ogle.astrouw.edu.pl/

in-Aid for Young Scientists, No. 80580208, from the Japan Society for the Promotion of Science (JSPS).

[1] Launhardt R, Zylka R and Mezger P G 2002 *A&A* **384** 112–139
[2] Serabyn E and Morris M 1996 *Nature* **382** 602–604
[3] Zoccali M, Renzini A, Ortolani S, *et al.* 2003, *A&A* **399** 931–956
[4] Figer D F, Kim S S, Morris M, *et al.* 1999 *ApJ* **525** 750–758
[5] Paumard T, Genzel R, Martins F, *et al.* 2006, *ApJ* **643** 1011–1035
[6] Figer D F, Rich R M, Kim S S, *et al.* 2004, *ApJ* **601** 319–339
[7] Yusef-Zadeh F, Hewitt J W, Arendt R G, *et al.* 2009 *ApJ* **702** 178–225
[8] An D, Ramírez S V, Sellgren K, *et al.* 2011 *ApJ* **736** 133+
[9] van Loon J Th, Gilmore G F, Omont A., *et al.* 2003 *MNRAS* **338** 857–879
[10] Blum R D, Ramírez S V, Sellgren K, *et al.* 2003 *ApJ* **597** 323–346
[11] Gallart C, Zoccali M and Aparicio A 2003 *ARA&A* **43** 387–434
[12] Sandage A and Tammann G A 2006 *ARA&A* **44** 93–140
[13] Bono G, Marconi M, Cassisi S, *et al.* 2005 *ApJ* **621** 966–977
[14] Fernie J D, Evans N R, Beattie B *et al.* 1995 *IBVS* **4148** 1
[15] Nagayama T, Nagashima C, Nakajima Y, *et al.* 2003 *SPIE* **4841** 459–464
[16] Matsunaga N, Kawadu T, Nishiyama S, *et al.* 2009 *MNRAS* **399** 1709–1729
[17] Matsunaga N, Kawadu T, Nishiyama S, *et al.* 2011 *Nature* **477** 188–190
[18] Gillessen S, Eisenhauer F, Trippe S, *et al.* 2009 *ApJ* **692** 1075–1109
[19] Antonello E, Fugazza D and Mantegazza L 2002 *A&A* **388** 477–482
[20] Kroupa P 2001 *MNRAS* **322** 231–246
[21] Kennicutt R C Jr 1998 *ARA&A* **36** 189–232
[22] Hatsukade B, Kohno K, Aretxaga I, *et al.* 2011 *MNRAS* **411** 102–116
[23] Ikarashi S, Kohno K, Aguirre J E, *et al.* 2011 *MNRAS* **415** 3081–3096
[24] Barnes J E and Hernquist L E 1991 *ApJ* **370** L65–L68
[25] Mihos J C and Hernquist L 1996 *ApJ* **464** 641–663
[26] Weinberg M D and Blitz L 2006 *ApJ* **641** L33–L36
[27] Purcell C W, Bullock J S, Tollerud E J, *et al.* 2011 *Nature* **477** 301–303
[28] Binney J, Gerhard O E, Stark A A, *et al.* 1991 *MNRAS* **252** 210–218
[29] Athanassoula E 1992 *MNRAS* **259** 345–365
[30] Kim S, Saitoh T R, Jeon M, *et al.* 2011 *ApJ* **735** L11+
[31] Stark A A, Martin C L, Walsh W M, *et al.* 2004 *ApJ* **614** L41–L44
[32] Kormendy J and Kennicutt R C Jr 2004 *ARA&A* **42** 603–683
[33] Binney J 2009 *IAUS* **254** 145–152
[34] Clarkson and Rich 2009 *Astro10: Science White Papers* #47
[35] Kormendy J and Kennicutt R C Jr 2004 *ARA&A* **42** 603–683
[36] Schnnierer E, Boker T, Emsellem E, *et al.* 2004 *A&A* **462** L27–L30
[37] Gillessen S, Genzel R, Fritz T K, *et al.* 2012 *Nature* **481** 51–54
[38] Stolte A, Ghez A M, Morris M, *et al.* 2008 *ApJ* **675** 1278–1292
[39] Bissantz N, Englmaier P and Gerhard O 2003 *MNRAS* **340** 949–968
[40] Lindqvist M, Habing H J and Winnberg A 1992 *A&A* **259** 118–127
[41] Pedicelli S, Bono G, Lemasle B, *et al.* 2009 *A&A* **504** 81–86
[42] Soszyński I, Udalski A, Pietrukowicz P, *et al.* 2011 *AcA* **61** 285–301
[43] Minniti D, Lucas P W, Emerson J P, *et al.* 2010 *New Astron.* **15** 433–443

Astronomy at High Angular Resolution 2011 IOP Publishing
Journal of Physics: Conference Series **372** (2012) 012027 doi:10.1088/1742-6596/372/1/012027

Disk-Halo interaction: The molecular clouds in the Galactic center region

D. Riquelme[1], J. Martín-Pintado[2], R. Mauersberger[3], M.A. Amo-Baladrón[2], S. Martín[4], L. Bronfman[5]

[1]IRAM. Avenida Divina Pastora 7, local 20 E18012, Granada, Spain
[2] Centro de Astrobiología (CSIC/INTA), Ctra. de Torrejón a Ajalvir km 4, E-28850, Torrejón de Ardoz, Madrid, Spain
[3] Joint ALMA Observatory, Alonso de Córdova 3107, Vitacura, Santiago, Chile
[4] European Southern Observatory, Alonso de Córdova 3107, Vitacura, Casilla 19001, Santiago, Chile
[5] Departamento de Astronomía, Universidad de Chile, Casilla 36-D, Santiago, Chile

E-mail: riquelme@iram.es

Abstract. From a large-scale study of the Galactic center (GC) region in SiO(2 − 1), $HCO^+(1-0)$, and $H^{13}CO^+(1-0)$, we identify shock regions as traced by the enhancement of SiO emission. We selected 9 positions called by us as "interaction regions", because they mark the places where gas in the GC could be interacting with gas coming from higher latitude ("disk-halo interaction") or from larger galactocentric radius. These positions were studied using the $^{12}C/^{13}C$ isotopic ratio to trace gas accretion/ejection. We found a systematically higher $^{12}C/^{13}C$ isotopic ratio (> 40) toward the interaction regions than for the GC "standard" molecular clouds (20 − 25). These high isotopic ratios are consistent with the accretion of the gas from higher galactic latitudes or from larger galactocentric distances. There are two kinetic temperature regimes (one warm at ~ 200 K and one cold at ~ 40 K) for all the positions, except for the positions associated to the giant molecular loops where only the warm component is present. Relative molecular abundances suggest that the heating mechanism in the GC is related to shocks. We mapped one molecular cloud placed at the foot points of the giant molecular loops in 3-mm molecular lines to reveal the morphology, chemical composition and the kinematics of the shocked gas.

1. Introduction

To understand the formation and evolution of galaxies, one must explore their central few hundred parsecs. Observations of selected regions of the center of our Galaxy allow to study in detail the physical processes responsible for gas accretion and ejection, the role of the magnetic fields and galactic winds, in the interaction regions with a very high spatial spatial resolution. We study two kinds of phenomena which are presumably occurring in the GC region, and are believed to be partly responsible for the gas accretion toward the nuclear region of the Galaxy: the barred potential, and the Giant Molecular Loops (GMLs).
[1] showed that the large-scale gas kinematics in the central regions of the Galaxy can be explained by a barred galactic potential with two major families of orbits inside the bar: the X1 orbits parallel to the bar, and the inner X2 orbits, orthogonal to it. The gas falls from X1 to X2 orbits when it self-intersects in the X1-X2 orbits transition. The other phenomenon is

the huge loop structures (GMLs) found by [2] and thought to be formed by a Parker instability. The gas of the loops is thought to flow down along the magnetic field lines and join the Galactic plane, generating shock fronts at the "foot points" of the loops. However, so far there is no clear observational evidence of gas accretion from the X1 orbits and/or from high latitudes in the GMLs.

We performed a large-scale study of the GC region in SiO(2 – 1), $HCO^+(1-0)$, and $H^{13}CO^+(1-0)$ using the 4 m-NANTEN telescope [3]. We found an increased emission in SiO as compared to the HCO^+ line intensity at the foot points of the GMLs and toward the 1.3 complex, which indicates the presence of shocks [4; 5]. From this study we selected 9 positions (see Fig. 1) where the SiO emission shows an enhancement with respect to the HCO^+ emission. The positions are considered in this work as the "interaction regions", because they mark the places where gas in the GC could be interacting with gas coming from higher latitude ("disk-halo interaction") or from larger galactocentric radii, according to the GMLs scenario and the bar potential model, respectively. They were selected including five positions in the GMLs and two positions in the X1-X2 orbits interaction places. The positions in the GMLs were called "halo" to differentiate from the molecular clouds in the Galactic plane. Among them, 3 positions are placed where the gas in the GMLs joins the gas in the disk (foot points of the loops), and 2 positions are placed at the top of the loops. Each of the positions located at the X1-X2 orbits interaction places has two main kinematical components, one associated to the X1 orbits and the other to the X2 orbits. They were called as "disk X1", and "disk X2". Finally, we selected 2 positions close to $b = 0$ (that we called "disk"), which are not associated to neither the GMLs nor the locations of the orbits interaction, and therefore they should trace the typical GC clouds.

2. The $^{12}C/^{13}C$ isotopic ratio

To test if the GMLs scenario is correct, or if the gas is being ejected from the disk in the GC, we studied the $^{12}C/^{13}C$ isotopic ratio to trace gas accretion/ejection. Using the IRAM 30m-telescope, we observed the $J = 1 - 0$ rotational transition of HCO^+, HCN, HNC, and their ^{13}C isotopic substitutions toward the 9 selected positions. While ^{12}C is predicted to be formed in first-generation, massive stars on rapid timescales, ^{13}C is produced primarily via CNO processing of ^{12}C seeds from earlier stellar generations, on a longer timescale. The $^{12}C/^{13}C$ isotopic ratio shows, therefore, the relative degree of primary to secondary processing in stars. Table 1 presents the results. We found a systematically higher $^{12}C/^{13}C$ isotopic ratio (> 40) toward the halo and the X1 orbits than for the GC "standard" molecular clouds (20-25). The results are consistent with a scenario where gas from the halo is accreted to the disk and with the transfer of gas from the outskirt of the disk to the GC through X1 and X2 orbits as suggested by the potential bar scenario [6].

3. The kinetic temperatures towards the interaction regions

We derived the physical conditions of the molecular gas in the interaction regions of the GC. Using the Effelsberg 100m-telescope, we measured the metastable inversion transitions of NH_3 from $(1, 1)$ to $(6, 6)$ in 6 positions that can be observed from this site. Using rotational diagrams and large velocity gradient calculations, we estimated for the first time, the kinetic temperatures of the molecular clouds in these regions (Table 2). The kinetic temperatures are high, not only in the typical GC clouds, but also in the high latitude and high velocity clouds observed in this paper. Therefore, the heating mechanism of the molecular clouds are likely to be a general characteristics of the molecular gas throughout the GC region. We derive two kinetic temperature regimes (one warm at $\sim$ 200 K and one cold at $\sim$ 40 K) for all the positions, except for the halo where only the warm component is present. Using the 30m-telescope, we also observed molecular tracers of different physical processes like SiO, HNCO, CS, $C^{18}O$, and ^{13}CO (shocks, photodissociation, dense gas) to derive the densities, and to reveal the

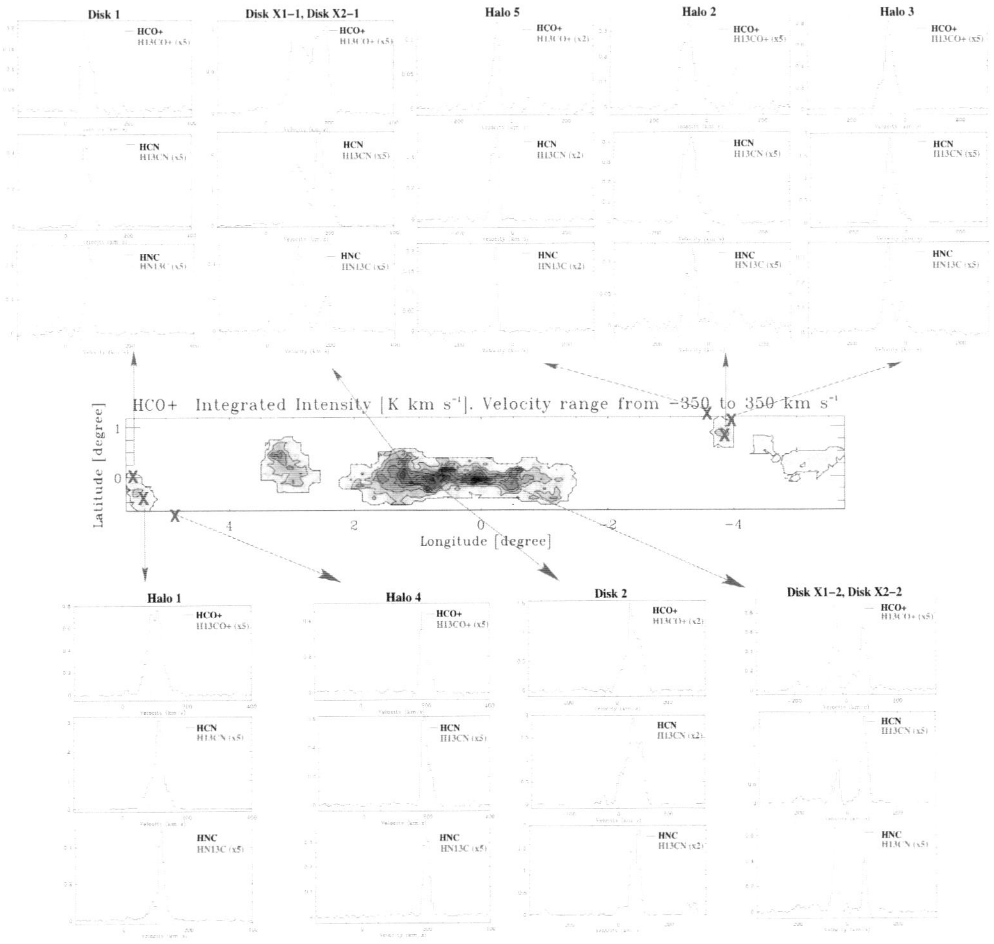

Figure 1. Spectra toward selected positions in the GC in HCO^+, $H^{13}CO^+$, HCN, $H^{13}CN$, HNC, and $HN^{13}C$. The positions are indicated in the HCO^+ integrated intensity map from [3]. The main isotope is plotted in black solid line. The ^{13}C isotope is plotted in red dashed line. The ^{13}C substitution is scaled by a factor of 2 or 5 to allow an easy visualization of the spectra. The factor is indicated in each spectrum. Figure from [6]

heating mechanisms affecting the molecular gas in these regions. From the high gas kinetic temperature, and by the increased SiO abundance, our results support the shock origin of the heating mechanisms in the GC. The high kinetic temperature in the X1 orbits and in the foot point of the GMLs seems to support to the large scale dynamics induced by the bar potential and the GMLs scenario as origins for the shocks.

4. 3 mm mapping of the foot point of the GMLs

We used the Mopra telescope to map one molecular cloud ($M{-}3.8+0.9$) at the foot points of the GMLs in 3mm lines. The maps reveal structures at small scales in the SiO emission, and large differences between the spatial distribution of the SiO and the HCO^+. The SiO emission in the $M{-}3.8+0.9$ cloud has narrow profiles (20 km/s) in comparison with the HCO^+ profiles (50 km/s), thus, shocked gas is dynamically more confined than the HCO^+ emission (Riquelme et al 2012b, in prep.).

Table 1. Intensity ratios from the observed ^{12}C and ^{13}C isotopomers. Table from [6].

Source	Velocity Component LSR [$\mathrm{km\,s^{-1}}$]	Velocity Range [$\mathrm{km\,s^{-1}}$]	$HCO^+/H^{13}CO^+$ ratio of $\int T_A^* dv$	$HCN/H^{13}CN$ ratio of $\int T_A^* dv$	$HNC/HN^{13}C$ ratio of $\int T_A^* dv$
Halo 1	100	[50, 190]	45.5± 5.4	13.5±0.2	≥ 25.6
	87	[50, 97]	≥73.9	25.8±2.0	≥7.5
	117	[97,135]	32 ± 3.7	11.3±0.1	≥37.5
	144	[135,190]	39.1±24.7	16.7±1.8	≥2.7
Halo 2	−62	[−115,−20]	73.1±36.5	14.6 ±1.0	≥15.4
	left wing	[−115,−70]	≥34.4	21.2 ±3.3	≥8
	right wing	[−70, −20]	53.2±26.1	11.6 ±0.9	≥13.6
Halo 3	−60	[−120,−30]	38 ±5.0	13 ±0.3	≥40.1
	left wing	[−120,−80]	54.2 ±37.7	19.2 ±1.9	≥10.4
	central peak	[−80, −30]	35.7 ±4.0	12 ±0.2	48.3 ± 21.8
Halo 4	200	[150, 250]	28.3 ±5.4	11.8 ±0.9	14.9 ±3.1
	peak	[150, 210]	29.2 ±7.5	11.7 ±0.8	17.9 ±3.9
	right wing	[210, 250]	≥10.6	12.4 ±4.0	7.1 ±3.4
Halo 5	−50	[−100,−40]	13.8 ±5.0	6.9 ±0.3	22.8 ±12
Disk X1-1	180	[140,230]	56 ± 6.4	10.8 ±0.1	25.5 ± 7.5
	left wing	[140,180]	57.2 ± 7.5	11.6 ±0.1	≥8
	right wing	[180,230]	54.4 ± 11	9.9 ±0.2	15.7 ± 4.4
Disk X2-1	95	[50,140]	29 ± 1.6	12.1 ± 0.2	22.1 ± 2.7
	left wing	[50,92]	32.4 ± 3.7	13.9 ± 0.4	22.8 ± 4.0
	right wing	[92,140]	27.3 ± 1.7	11.3 ± 0.2	21.7 ± 8.6
Disk X1-2	67	[0,100]	42.1± 8.6	9.4 ± 0.1	9.9 ± 1.3
Disk X2-2	−43	[−80,−20]	21.7 ±1.9	6.6 ±0.1	13.1 ±1.0
Disk 1	66	[35,105]	14.0 ±2.8	13.1 ± 1.0	15.8 ± 3.6
	central peak	[35,70]	16.4 ±4.6	23.1 ± 3.8	≥11.4
	right wing	[70,105]	11.6 ±3.4	7.8 ± 0.7	11.7 ± 2.7
Disk 2	left wing	[0, 43]	≥29.1	9.1 ± 0.2	≥21.9
	55	[43, 97]	4.1 ± 0.1	3.5 ± 0.1	3.7 ± 0.1
	right wing	[97,135]	16.1 ± 2.3	13.8 ± 1.3	≥2.6

We identified four main velocity components: from −140 to −70 $\mathrm{km\,s^{-1}}$, from −70 to −40 $\mathrm{km\,s^{-1}}$, from −40 to −20 $\mathrm{km\,s^{-1}}$, and from −20 to 20 $\mathrm{km\,s^{-1}}$. In these velocity ranges, we identify five main molecular complexes, which are indicated with green boxes in Fig. 2 on the HCN maps. In Fig. 3 we show the integrated intensity ratio of the SiO to HCO^+ in velocity ranges of 10 $\mathrm{km\,s^{-1}}$. While it is clear that both, the complex 3 and 4 are intense in both molecules; HCO^+ dominates towards the complex 2, showing the lowest ratio, and the SiO dominates in complex 1 (which almost does not present HCO^+ emission) and in complex 5. There are differences up to a factor 25 between the region where the HCO^+ dominate in the complex 2, and where the SiO dominates in the complex 5 and in complex 1.

5. Acknowledgments

D.R. and M.A.A-B. were supported by Radionet during the observations. D.R. was supported by DGI grant AYA 2008-06181-C02-02. J.M-P. and S.M. have been partially supported by the Spanish MICINN under grant numbers ESP2007-65812-C02-01 and AYA2010-21697-C05-01. L.B. acknowledges support from CONICYT projects FONDAP 15010003 and Basal PFB-06.

Table 2. Kinetic temperatures derived from LVG calculations.

Source	Cloud number	low temperature T_{kin}	$N(NH_3)$	high temperature T_{kin}	$N(NH_3)$	single temperature from p−NH_3 T_{kin}	$N(NH_3)$
		[K]	cm^{-2}	[K]	cm^{-2}	[K]	cm^{-2}
Halo1	1					> 115	1.20×10^{14}
	2					> 90	$> 1.99 \times 10^{14}$
	3					> 135	6.6×10^{13}
Halo4	1					90	1.99×10^{14}
DiskX1-1	1	38	1.99×10^{14}	$> 300^a$	1.58×10^{14}		
DiskX2-1	1	38	4.47×10^{14}	100	3.3×10^{14}		
DiskX1-2	1	52	1.66×10^{14}	215	1.51×10^{14}		
DiskX2-2	1	28	1.90×10^{14}	95	1.0×10^{14}		
Disk1	1	23	3.98×10^{13}	$> 154^b$	$> 1.58 \times 10^{13}$		
	2	38	6.31×10^{13}	$> 82^b$	$> 4.47 \times 10^{13}$		
Disk2	1	68	1.41×10^{15}	200	1.41×10^{15}		
	2					> 145	2.00×10^{15}
	3	50	1.41×10^{15}	80	1.41×10^{15}		

[a] LVG gives a T_{kin} greater than 300 K which is the value allowed by the collisional rates given by [7].

[b] Due that the values for the (4,4)-(5,5) metastable inversion transitions are upper limits to the actual T_{MB}, the modeled curve in the LVG plot is outside the allowed range, and we give a lower limit to the kinetic temperature using the rotational temperature form the LTE plot.

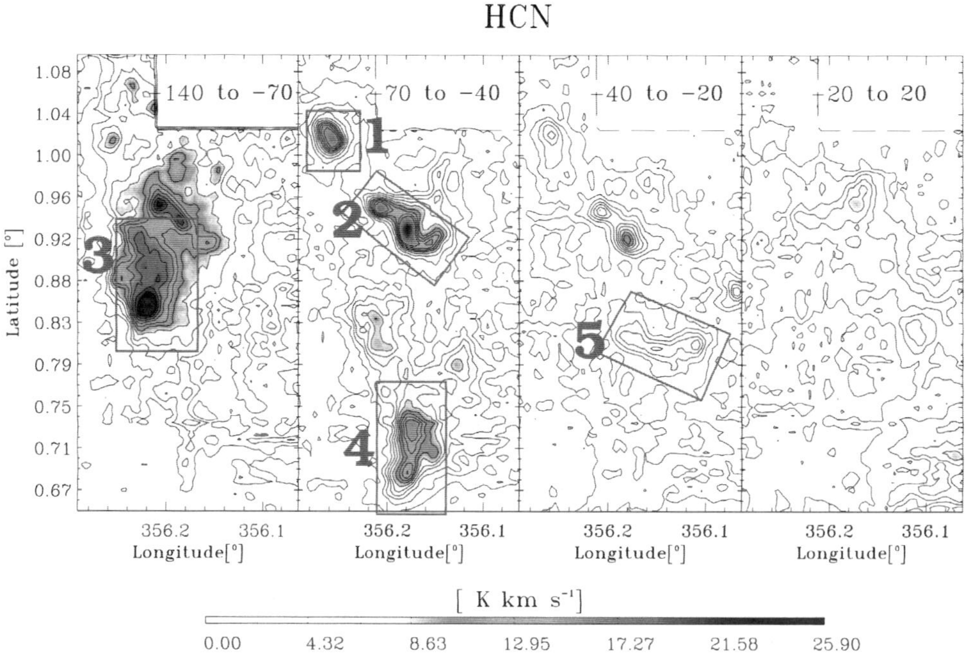

Figure 2. Position of the five molecular complexes discussed in the text.

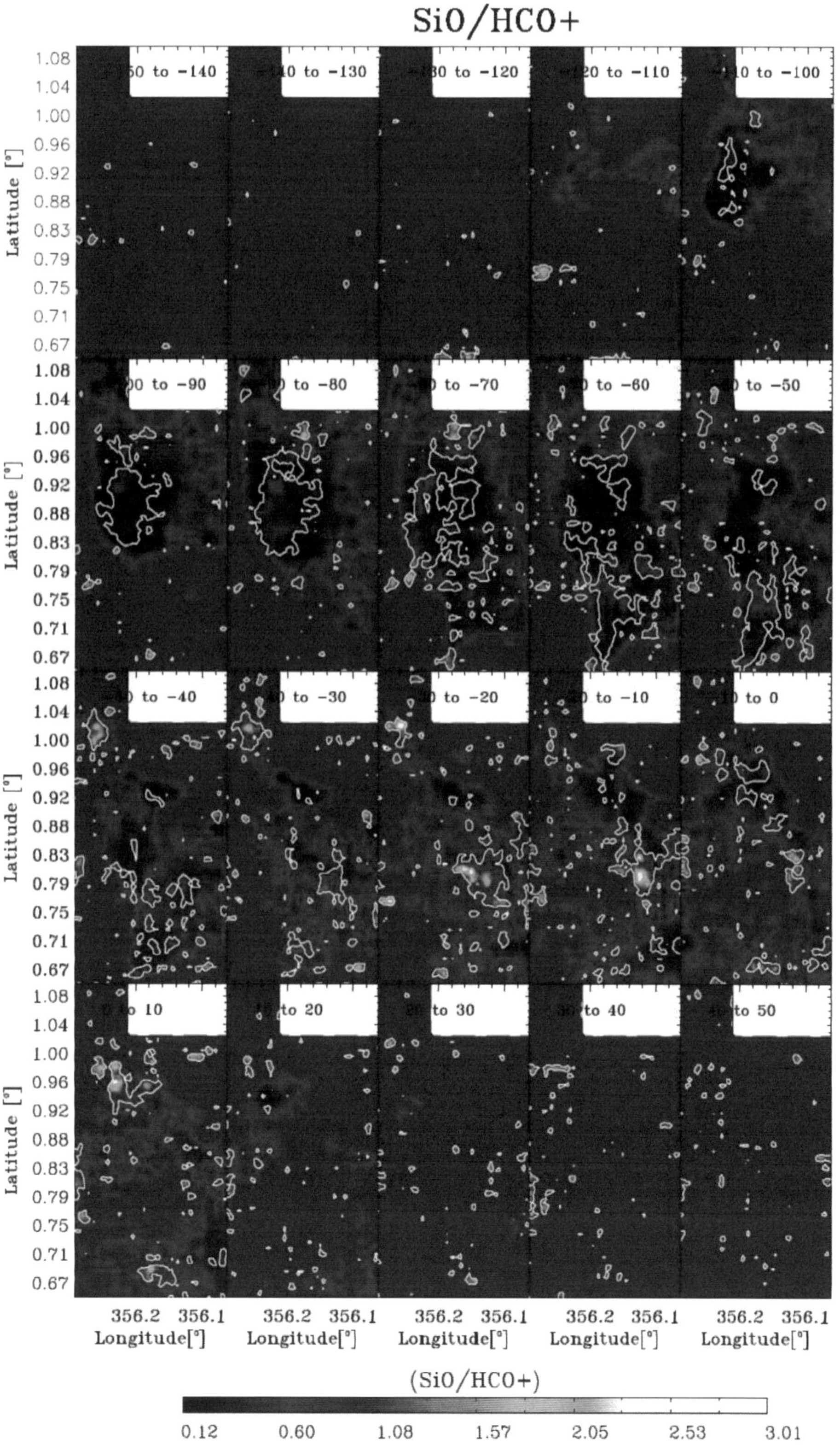

Figure 3. Ratio of the integrated intensity of SiO(2 − 1) to HCO^+, in velocity intervals of 10 $km\,s^{-1}$.

References

[1] Binney J, Gerhard O E, Stark A A, Bally J and Uchida K I 1991 *MNRAS* **252** 210–218

[2] Fukui Y, Yamamoto H, Fujishita M, Kudo N, Torii K, Nozawa S, Takahashi K, Matsumoto R, Machida M, Kawamura A, Yonekura Y, Mizuno N, Onishi T and Mizuno A 2006 *Sci.* **314** 106–109

[3] Riquelme D, Bronfman L, Mauersberger R, May J and Wilson T L 2010 *A&A* **523** A45

[4] Martín-Pintado J, Bachiller R and Fuente A 1992 *A&A* **254** 315

[5] Martín-Pintado J, de Vicente P, Fuente A and Planesas P 1997 *ApJ* **482** L45

[6] Riquelme D, Amo-Baladrón M A, Martín-Pintado J, Mauersberger R, Martín S and Bronfman L 2010 *A&A* **523** A51

[7] Danby G, Flower D R, Valiron P, Schilke P and Walmsley C M 1988 *MNRAS* **235** 229–238

Journal of Physics: Conference Series **372** (2012) 012028 doi:10.1088/1742-6596/372/1/012028

Frame-dragging effects on magnetic fields near a rotating black hole

V. Karas, O. Kopáček and D. Kunneriath

Astronomical Institute, Academy of Sciences, Boční II 1401, CZ-14131 Prague, Czech Republic

E-mail: vladimir.karas@cuni.cz

Abstract. We discuss the role of general relativity frame dragging acting on magnetic field lines near a rotating (Kerr) black hole. Near ergosphere the magnetic structure becomes strongly influenced and magnetic null points can develop. We consider aligned magnetic fields as well as fields inclined with respect to the rotation axis, and the two cases are shown to behave in profoundly different ways. Further, we construct surfaces of equal values of local electric and magnetic intensities, which have not yet been discussed in the full generality of a boosted rotating black hole.

1. Introduction

Various processes can lead to the magnetic topology exhibiting null points, in particular, these can emerge by complex motions of the plasma. In this respect the black hole rotation brings a new situation, as magnetic field lines become twisted in a highly curved spacetime of a rotating black hole, approximately where the ergosphere forms. Here, we examine the resulting structure of the magnetic field, namely, the emergence of critical points in a local frame of a physical observer, resembling the occurrence of X-points.

Previously (Karas & Kopáček, 2009) we considered a special case of a uniform magnetic field in perpendicular orientation with respect to the black hole spin axis, and we demonstrated that magnetic null points can indeed form near a rotating (Kerr) black hole. Here, the embedded magnetic field is allowed a general orientation, i.e., it can be inclined in an arbitrary angle with respect to the rotation axis. The axial symmetry is broken between the gravitational and electromagnetic fields, and this has to result in a truly three-dimensional structure of magnetic field lines along with gravito-magnetically induced electric field. The adopted approach allows us to identify the gravitational effects operating in a magnetically dominated system, where a super-equipartion magnetic field governs the motion of plasma. We further expand the discussion in Karas et al (2012) by constructing surfaces of equal values of local electric and magnetic intensities, even for the case of a general (oblique) orientation of the background magnetic field.

2. Aligned and oblique magnetic fields near black hole horizon

Astrophysical black holes do not support their own intrinsic magnetic field; this has to be generated by external currents and brought down to horizon by accretion. A black hole can also enter a pre-existing magnetic flux tube, and then one asks if the process of magnetic reconnection is influenced by strong gravitational field near horizon. And does the black hole rotation play a significant role in the scenario of this kind?

As we wish to discuss magnetic fields inclined with respect to the spin axis, and we also want to include the fast translatory motion, the following picture appears to be appropriate: Kerr black hole traversing a "magnetic filament", described as an extended (largely one-dimensional) flux tube. Such an idea can be motivated observationally, by highly ordered and elongated arcs (of about 100μG) that are seen in Galactic Center, within a few parsecs from Sagittarius A* compact radio source (Sgr A*; Yusef-Zadeh et al, 1984). They are thought to represent large-scale magnetic flux tubes that are illuminated by synchrotron emission from relativistic electrons (LaRosa et al, 2004; Morris, 2006).

Given a limited resolution that can be achieved with current imaging techniques, the magnetic filaments cannot be traced down to the characteristic size of the black hole.[1] Therefore, the actual mapping of the magnetic structures near black holes is not directly possible. Nonetheless, their role in accelerating the particles is suggestive, especially because they could help us to understand the origin of particle acceleration and the resulting signatures in the electromagnetic signal.

Naturally, in astrophysically realistic solutions the role of non-ideal plasma will need be included. Currently, neither of the frequently discussed limits (i.e., vacuum vs. force-free approximations) are able to account for both the plasma currents as well as the accelerating electric fields. This task will require resistive MHD, which is, however, beyond the scope of this brief paper.

2.1. The effect of black hole rotation and translatory motion

Starting from King et al (1975) and Bičák & Dvořák (1976), the organised electromagnetic test fields surrounding black holes have been discussed and their astrophysical consequences considered in various papers. The case of oblique geometry, however, has been explored only partially (Bičák and Janiš, 1980). This is caused especially by the fact that the off-equatorial fields are lacking any symmetry, and so are more difficult to visualise. Also, qualitatively new effects on the field structure were not expected.

Nevertheless, in Karas & Kopáček (2009) we were able to demonstrate that the asymptotically perpendicular field lines develop a magnetic neutral point in the equatorial plane. This is interesting because such structures of the magnetic field are relevant for processes of electromagnetic acceleration. The magnetic null point emerges in a local physical frame, and could trigger reconnection. However, the asymptotically perpendicular field is a rather exceptional case. Therefore, here we investigate near-horizon magnetic structures in more detail, also for a general orientation of the magnetic field (see also Karas et al 2012).

The main objective of this discussion is to track the location of magnetic null points and to explore a complex three-dimensional configuration inside the ergosphere. We consider the form of magnetic lines together with the induced electric lines for different values of the model parameters: the inclination angle of the asymptotic magnetic field $\theta_{\rm o}$ $(=\arctan(B_{\perp}/B_{\parallel})$, the black hole spin a $(a^2 \leq M)$, and the boost velocity β $(\beta^2 \equiv v_x^2 + v_y^2 + v_z^2 < 1)$. We observe the layers of alternating magnetic orientation to occur also in the general case, i.e., when the black hole rotates and moves with respect to an oblique magnetic field. However, the three-dimensional structure of the field lines is very complicated as they become highly entangled around the null point.

We specify the gravitational field by Kerr metric. Our starting form of the electromagnetic field is a stationary solution of Maxwell's test-field electro-vacuum equations in the given curved

[1] Gravitational radius $r_{\rm g} = GM/c^2 \approx 1.48 \times 10^{13}\,M_8$ cm, where the central black hole mass is expressed in terms of $M_8 \equiv M/(10^8 M_{\odot})$. The velocity of the Keplerian orbital motion of a particle is then $v_{\rm K} \approx 2.1 \times 10^{10} (r/r_{\rm g})^{-1/2}\,{\rm cm\,s^{-1}}$. The corresponding orbital period is $T_{\rm K} \approx 3.1 \times 10^3 (r/r_{\rm g})^{3/2} M_8$ s. Hereafter, we use a dimensionless form of geometrized units, where all quantities are scaled with the black hole mass; M does not appear in the equations explicitly.

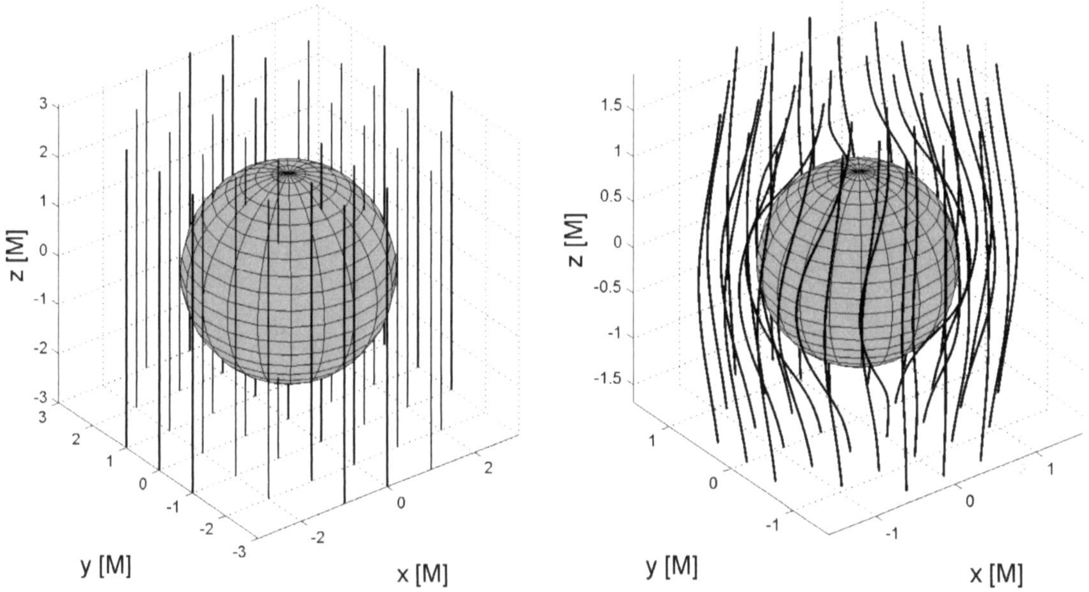

Figure 1. Magnetic lines of a uniform field near a non-rotating black hole. Left: The field lines appear to be perfectly homogeneous (in these suitably defined coordinates), plotted with respect to frame of free-falling observers. Right: magnetic field aligned with the rotation axis of an maximally rotating ($a = M$) Kerr black hole, when they are expelled out of the horizon. (see Kopáček, 2011, for more details).

spacetime.[2] The electromagnetic four-potential $\boldsymbol{A}$ can be then written as superposition of two contributions: $\boldsymbol{A} = B_{\parallel}\,\boldsymbol{a}_{\parallel} + B_{\perp}\,\boldsymbol{a}_{\perp}$, where $B_{\parallel}$ and $B_{\perp}$ define the magnitudes of the two parts.

The aligned field has two non-vanishing components of the normalized electromagnetic four-potential,

$$a_{t\parallel} = B_{\parallel}a\Big[r\Sigma^{-1}\Big(1+\mu^2\Big)-1\Big], \tag{1}$$

$$a_{\phi\parallel} = B_{\parallel}\Big[\tfrac{1}{2}(r^2+a^2)-a^2r\Sigma^{-1}(1+\mu^2)\Big]\sigma^2, \tag{2}$$

where we use Boyer-Lindquist t, r, θ, and ϕ dimension-less spheroidal coordinates ($\mu = \cos\theta$, $\sigma = \sin\theta$). Eqs. (1)–(2) represent an asymptotically homogeneous magnetic field (Wald, 1974).

On the other hand, the perpendicular to axis component of the field is given by (Bičák and Janiš, 1980)

$$a_{t\perp} = B_{\perp}a\Sigma^{-1}\Psi\sigma\mu, \tag{3}$$

$$a_{r\perp} = -B_{\perp}(r-1)\sigma\mu\sin\psi, \tag{4}$$

$$a_{\theta\perp} = -B_{\perp}\Big[(r\sigma^2+\mu^2)a\cos\psi+\Big(r^2\mu^2+(a^2-r)(\mu^2-\sigma^2)\Big)\sin\psi\Big], \tag{5}$$

$$a_{\phi\perp} = -B_{\perp}\Big[\Delta\cos\psi+(r^2+a^2)\Sigma^{-1}\Psi\Big]\sigma\mu, \tag{6}$$

[2] These are astrophysically acceptable approximations which reflect the fact that black holes can only acquire a negligibly small electric charge, while cosmic electromagnetic fields are not strong enough to influence the spacetime metric significantly.

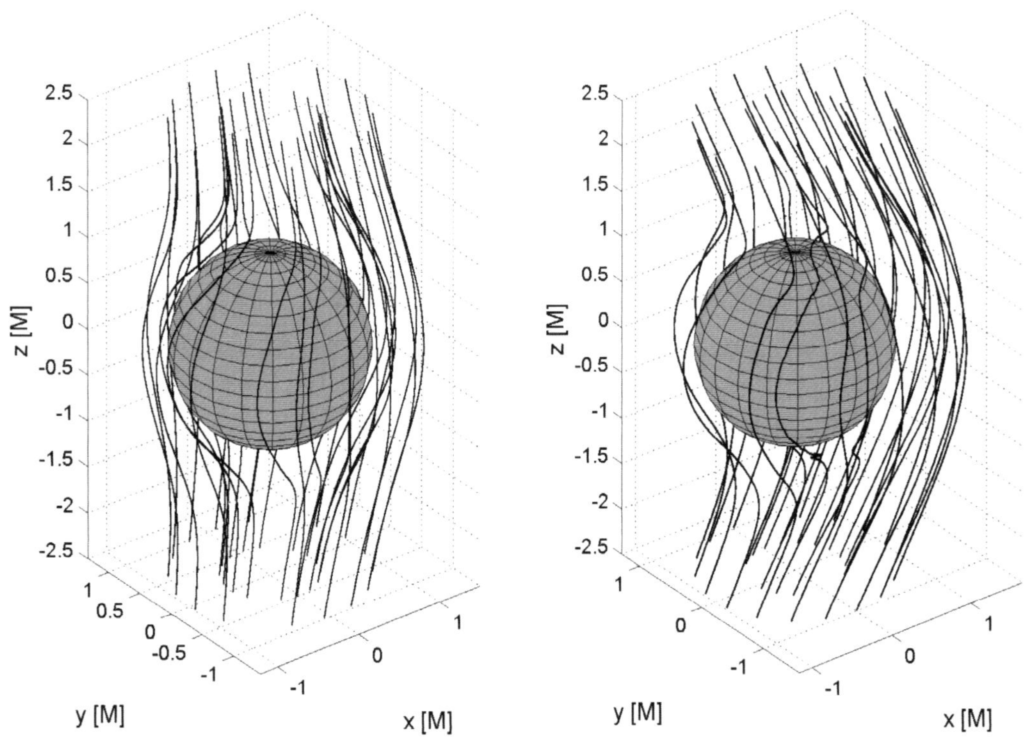

Figure 2. The case of extreme spin, $a = M$, taking into account non-zero velocity of the translatory boost along x-direction: $v_x = 0.1$ (left), $v_x = 0.3$ (right).

with $\Sigma(r, \mu)$ and $\Delta(r)$ being the Kerr metric functions, $\psi \equiv \phi + a\delta^{-1} \ln\left[(r - r_+) / (r - r_-)\right]$, $\Psi = r\cos\psi - a\sin\psi$, $\delta = r_+ - r_-$, and $r_\pm = 1 \pm \sqrt{1 - a^2}$. Knowing the complete set of four-potential components the magnetic field structure is fully determined: $F_{\mu\nu} \equiv A_{[\mu,\nu]}$.

Equations (3)–(6) describe the field lacking the axial symmetry. Such a situation cannot be strictly stationary, however, the alignment time-scale is very long, and so we can neglect the associated energy losses here. The field line structure depends also on the motion of observers determining the field components. In order to construct and discuss our plots, we employ a free-falling physical frame, evaluate the electromagnetic tensor with respect to the local frame, and use these components to draw the field lines.

In absence of the perpendicular component ($B_\perp = 0$), the field is relatively uncomplicated (see figure 1). Although the frame-dragging acts also on the aligned field lines, their shape can be integrated to find the surfaces of constant magnetic flux in a closed (analytical) form. Previously, the aligned fields were explored especially in context of magnetic field expulsion from a maximally rotating black hole. In our notation, the example in Fig. 1 refers to zero velocity of the translatory boost, i.e. $\beta = 0$.

Once a nonzero boost velocity is included, the structure of the aligned field becomes more complicated (figure 2). This is mainly due to an interplay between the boost of the black hole and its rotation acting jointly on the (originally) aligned field. As β increases, the magnetic lines are progressively puffed out of horizon and wound around it (see Kopáček 2011, for more examples and details).

Figure 3. The neighborhood of a magnetic null point of perpendicular magnetic field in the equatorial plane $(x, y) \equiv (r\cos\phi, r\sin\phi)$, i.e., viewed along the rotation axis of an extreme $(a = M)$ black hole. The black circular section in the bottom left corner denotes the horizon, $r = r_+(a)$. The colour scale indicates the magnetic intensity (in arbitrary units).

2.2. Neutral points of the magnetic field

We explore the case of magnetic field with a non-vanishing component inclined with respect to rotation axis ($B_\perp \neq 0$). In fact, Karas & Kopáček (2009) explored a strictly perpendicular case. Confining the magnetic lines in the equatorial plane $\theta = \pi/2$, the nested structure of magnetic layers emerges. These are essential for the magnetic reconnection.

Near-horizon structure of magnetic lines is visualized in figure 3 by using the Line-Integral-Convolution (LIC) method in Matlab. This technique allows us to identify clearly the location of neutral points. It turns out to be particularly useful here with the general orientation of the asymptotic field direction, as the global solution for the field lines is too cumbersome. Further, by introducing $\xi(r; a) \equiv 1 - r_+(a)/r$ as a new radial coordinate, the horizon surface $r = r_+$ collapses into the center and the layered structure of magnetic sheets is seen in more detail.

The fully three-dimensional structure of magnetic lines develops outside the equatorial plane. In figures 4–5 we observe a superposition of two essential effects. Firstly, the X-type structure

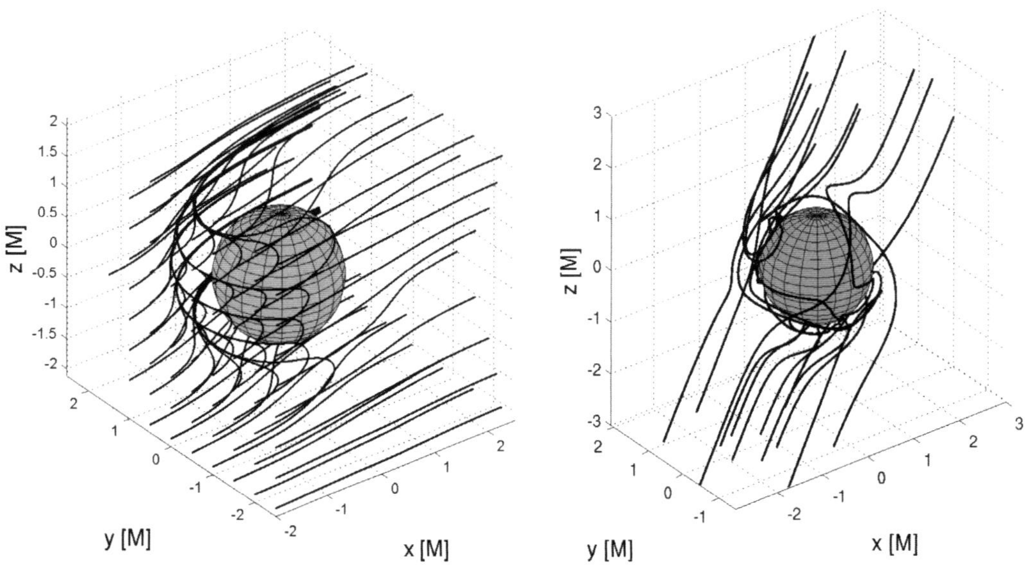

Figure 4. Left: Magnetic field asymptotically perpendicular to the rotation axis (z) of an $a = M$ black hole, centered on the origin. Right: Similar example as on the left side, but with the field in a general direction with respect to rotation axis.

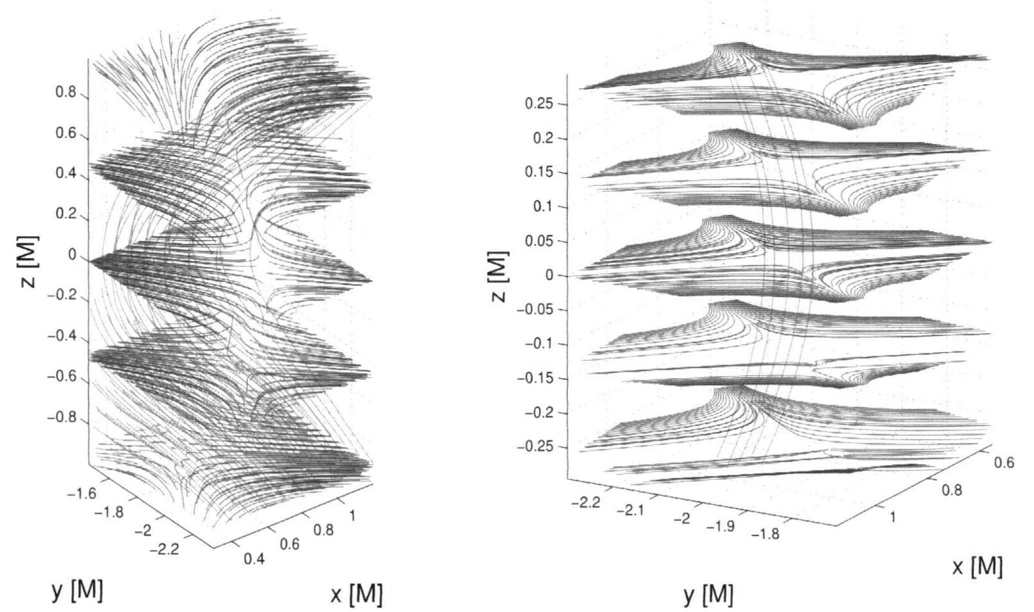

Figure 5. Three-dimensional detail around the equatorial plane. Magnetic lines are plotted in blue color. The equatorial lines reside within the plane $z = 0$, revealing a null point in the origin. The lines adopt some non-zero vertical component outside the equatorial plane. Electric lines are shown in red; they pass through the magnetic null point, crossing the equatorial plane in the vertical direction.

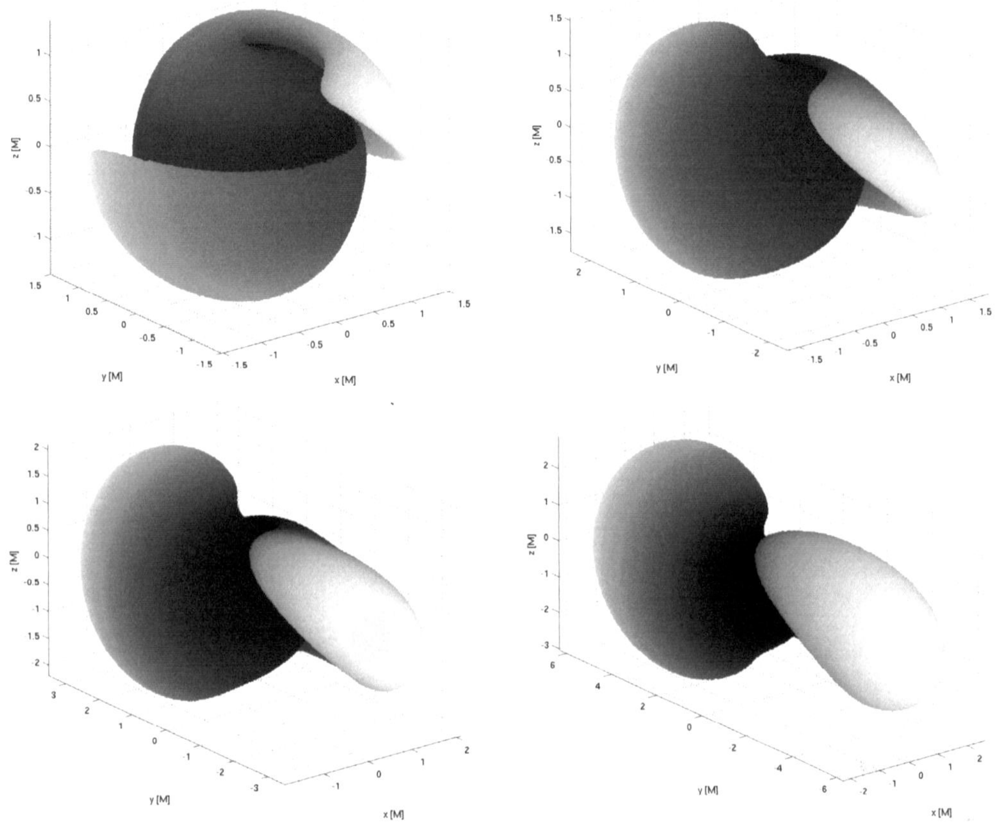

Figure 6. Three-dimensional view of the surfaces of equal values of local electric and magnetic intensities, for the case of an oblique background magnetic field $B_x = B_z$ and the extremal spin $a = M$ of the rotating Kerr black hole. Boost velocity of the linear motion is (from upper left to bottom right panels): $v = 0$, 0.6, 0.8, and 0.9. Different levels of shading of the surface correspond to the distance from origin (in Boyer-Lindquist coordinates), where the black hole is located. Increasing velocity of the translatory motion causes a growing deformation of the surface.

of the magnetic null point persists also outside the equatorial plane. Secondly, however, the magnetic lines acquire a growing vertical component B_z, whereas the electric field passes through the magnetic null point and crosses the equatorial plane vertically. Such a structure suggests that particles can be accelerated by the non-vanishing electric field, and they can stream away from the neutral point.

Finally, another useful way of visualizing the electromagnetic structure is by studying the electromagnetic invariants. These can help us not only to track interesting points of the electromagnetic field but also to identify locations of where charges can be electromagnetically supported or produced via electron-positron pair creation (e.g., Ruffini & Wilson, 1975; Karas & Vokrouhlický, 1991). Recently, Lyutikov (2011) explored the shape of surfaces where the magnitude of the local electric field equates the magnetic field component near a magnetized Schwarzschild black hole. Figure 6 shows these $E = B$ surfaces in our system. Obviously the presence of the black hole as well as its motion and rotation influence the form of these surfaces.

3. Discussion and Conclusions

Rotation is a highly interesting attribute of cosmic black holes. In principle, black holes can be spun up close to extreme $a = M$. In stellar-mass black holes the spin is thought to be chiefly natal (McClintock et al, 2012), whereas supermassive black holes in galactic cores can adjust their angular momentum by accretion, and the outcome of evolution depends largely on the dominant mode of accretion, during their entire life-time. In both cases, the spin is an important characteristic, potentially allowing the efficient acceleration of matter. We showed that it can be also relevant directly for the onset of magnetic reconnection.

We discussed the structure of electromagnetic test-fields and the layered pattern of current sheets that can be induced by the gravito-magnetic action. Neutral points of the magnetic field suggest that magnetic reconnection can occur. The proposed scenario can be astrophysically relevant in circumstances when the black hole rotates and moves across a magnetic flux tube, origin of which is in currents flowing in the cosmic medium far from the black hole. We considered the limit of a magnetically dominated system in which the organised magnetic field dictates the motion to plasma; the opposite limit of a black hole moving through a force-free plasma has been investigated by other authors (recently, Palenzuela et al, 2010; Tchekhovskoy, 2012).

In this paper we considered an idealised situation, starting from an electro-vacuum solution, assuming fast motion and rotation of the black hole, and embedding it in an asymptotically uniform magnetic field. Future simulations should clarify, whether the astrophysically realistic effects of the moving black hole on the surrounding electromagnetic structure in its immediate neighborhood will be similar to those envisaged here. Despite the field-line structure in this paper being induced purely by the action of frame-dragging, the exact choice of the projection tetrad is not very essential for the existence of magnetic layers. The choice of the physical frame does affect the presence and the exact location of the magnetic neutral point, and the associated electric field which accelerates charged particles away from the neutral point. Although the exact location of the neutral point varies with the model parameters, it always occurs very close to the black hole, where the frame-dragging is efficient. Therefore, the point of particle acceleration has to be close to horizon as well.

We thank the ESA PECS grant No. 98040 and the Czech Science Foundation grant No. 205/09/H033 for a continued support.

References

Bičák J and Dvořák L 1976 *GRG* **7** 959
Bičák J and Janiš V 1980 *MNRAS* **212** 899
Karas V and Kopáček O 2009 *CQG* **26** 025004
Karas V, Kopáček O and Kunneriath D 2012 *CQG*, in press (arXiv:1201.0009)
Karas V and Vokrouhlický D 1991 *J Phys I France* **1** 1005
King A R, Lasota J P and Kundt W 1975 *Phys Rev D* **12** 3037
Kopáček O 2011 *Thesis* (Charles University, Prague)
LaRosa T N, Nord M E, Lazia T J W and Kassim N E 2004 *ApJ* **607** 302
Lyutikov M 2011 *Phys Rev D* **83** 064001
McClintock J E et al 2012, see this Volume
Morris M 2006 *J Phys Conf Ser* **54** 1
Palenzuela C, Garrett T, Lehner L and Liebling S L 2010 *Phys Rev D* **82** 044045
Ruffini R and Wilson J R 1975 *Phys Rev D* **12** 2959
Tchekhovskoy A 2012, see this Volume
Wald R M 1974 *Phys Rev D* **10** 1680
Yusef-Zadeh F, Morris M and Chance D 1984 *Nature* **310** 557

Detecting supermassive binary black holes with VLBI - discovery of a ring-structure in 3C454.3

S Britzen[1], M Zamaninasab[1], M Aller[2], H Aller[2], O Kurtanidze[3], S Vercellone[4], G M Richter[5], A Witzel[1], TP Krichbaum[1] and JA Zensus[1]

[1] Max Planck Institute for Radioastronomy, Auf dem Hügel 69, 53121 Bonn, Germany
[2] Department of Astronomy, University of Michigan, Ann Arbor, MI, USA
[3] Abastumani Observatory, Mt. Kanobili, Georgia
[4] INAF - IASF Palermo, Italy
[5] Astrophysikalisches Institut Potsdam, An der Sternwarte 16, 14482 Potsdam, Germany

E-mail: sbritzen@mpifr.de

Abstract. We report the detection of the first VLBI ring-structure around the core of an AGN - around the core of the quasar 3C454.3. This ring-structure starts being visible in VLBI maps around 1996. It expands with an apparent velocity between 0.11±0.01mas/yr and 0.18±0.01mas/yr and dominates the pc-scale structure for at least 14 years. This is the result of a re-analysis of 41 VLBA data sets at six different radio frequencies observed between 1995.57 and 2011.48. We observe a correlation between radio flaring, flux-density variability, a ring-structure and kinematic properties of the jet. Taken together, it is tempting to see a causal connection and to explain all of this geometrically. The kinematic changes as well as the changes in the flaring characteristics might be caused by a change of the angle to the line of sight towards the observer. This behaviour resembles our findings for 0735+178 - with 3C454.3 being the second AGN to reveal kinematic mode changes.
These mode changes could be explained by the presence of a supermassive binary black hole. 3C454.3 had been modelled as a binary black hole before.

1. Introduction

3C 454.3 is a 16th magnitude quasar at a redshift of $z = 0.859$. It belongs to the class of optically violent variable sources. The source has been monitored with telescopes from the radio to the TeV-regime in order to unravel the origin of the high energetic radiation detected with the *FERMI* satellite (Tosti et al. 2008). For some time the source was the brightest gamma-ray emitting AGN in the sky as observed by *FERMI*.
The radio morphology has been studied and monitored within many VLBI observations. In the cm-regime VLBI observations were performed by Moellenbrock et al. (1996). The source has been observed within the VSOP prelaunch survey (Fomalont et al. 2000). It is being monitored within the MOJAVE/2cm survey (Lister et al., 2005; Homan et al. 2006) and has been studied at higher frequencies by Lonsdale et al. (1998), Rantakyro et al. (1998), Lobanov et al. (2000), Marscher et al. (2002), and Jorstad et al. (2010). The connection between mm-continuum variations and the VLBI structure has been investigated by Savolainen et al. (2002).The polarimetry at 43 GHz was studied by Kemball et al. (1996). A detailed analysis of

component motion in the pc-scale jet of this source has been performed by Pauliny-Toth et al. (1987). In particular, the observations and analysis of early VLBI observations performed (e.g., Pauliny-Toth et al. 1987) revealed some pecularities which seem to be unique in this source as opposed to typical "superluminals". These authors report on the detection of structural changes following a flux-density outburst which are unlike those which have been observed in the 'classical' superluminal sources at that time.
Qian et al. (2007) showed that the long-term radio-variability at 4.8 and 8 GHz can be explained with a binary black hole model.

2. Observations

The multi-wavelength flux-density output of this source is peculiar, and so is the VLBI structure. In order to understand the physical origin of the pecularities, we studied in detail the pc-scale morphology and its evolution over roughly 15 years at several radio frequencies. In addition, we collected light-curve information from the radio, the optical, and X-ray wavebands. By studying not only the morphology but also the emission processes we hope to put the pieces together. We expect that all the different properties taken together form a puzzle which - only when as complete as possible - can tell about the phyiscal origin of the many enigmas in this source which at first glance seem to be unrelated.
It turned out that the analysis of the pc-scale jet and multi-wavelength emission reveal even more so far unknown properties of this source.
We (re-)analyzed 32 VLBA observations (between 1995.57 and 2011.48) observed at 15 GHz from the MOJAVE/2cm survey and derived the parameters of the observed VLBI structure. In addition, we investigated VLBA observations obtained at 5.0, 8.3, 15.3, 22.2, 43.1 and 86.2 GHz on May, 19, 2005. Three 43 GHz VLBA observations from March 2001 and May and October 2005 have also been investigated with regard to the spectral properties. We fitted the components of the VLBI structure with Gaussian components. We analyzed the kinematic properties of the pc-scale jet components with regard to the flaring states. We studied the properties of the lightcurves obtained in the radio (4.8, 8.0, 14.5, 22 and 37 GHz), optical (R- and B-band), and X-rays and searched for correlations between the flaring properties and the VLBI structure of the source.

3. Results

We confirm the results concerning the general VLBI structure of 3C 454.3 by Pauliny-Toth et al. (1987). The source shows a core-jet structure with a jet extending about 10 mas from the core with an estimated overall position angle of $\sim$ -50°. Embedded in the jet are several features. Despite the general agreement with earlier studies, the source structure in the epochs analyzed here differs significantly from previous studies when investigated in more detail.

3.1. Detection of a ring-structure

In the time covered by the observations, the structure consists of three substructures: the core (K), the inner jet components (A, B, C), an expanding ring-like structure (R_0, R_1, R_2, R_3, R_4, see Fig. 1) and the outer jet components (Y, Z) at a distance of 8–10 mas from the core. To determine the individual motion of these components, we show in Fig.3 the core separation as a function of time (bottom). Fig.3 (middle) shows the position angle evolution of these components with time. Fig.3 (top) shows the flux-density evolution with time. The different parts of the jet reveal significantly different kinematic properties. Outer jet components separate from the core and are known from earlier observations. Their positions with regard to the position angle are stable. The components building the ring are shown in Fig.2. We detect a ring-like structure around the pc-scale core. To our knowledge, this is the first source for which such a ring-like

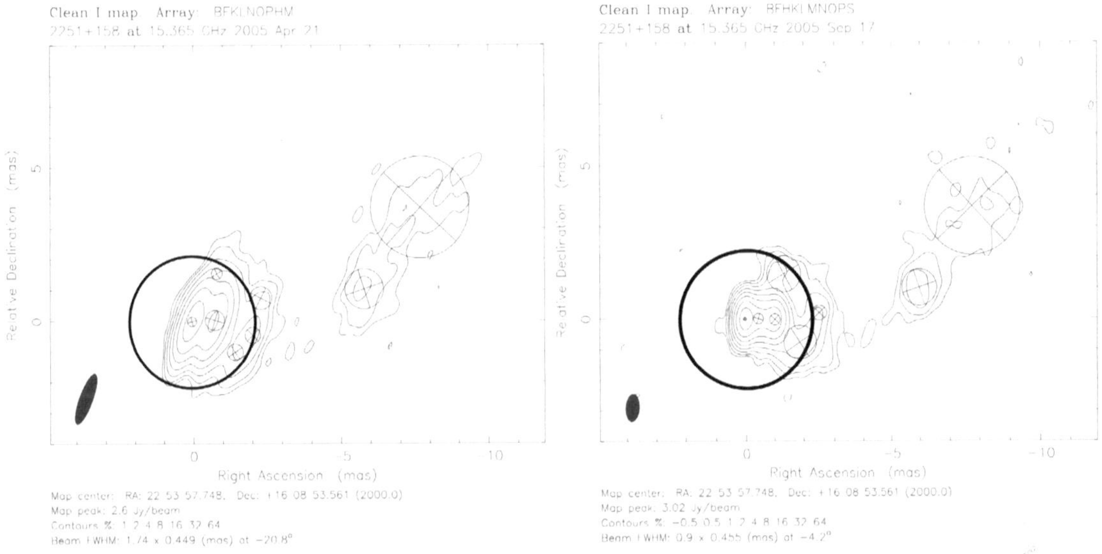

Figure 1. Two maps with the Gaussian modelfit components superimposed. A circle has been drawn to delineate the apparent ring-like structure.

structure has been detected. The ring-like structure expands and increases in radius between March 2003 and 2010. To measure the apparent velocity of this expansion we fitted a linear regression to each ring-component. The estimated velocities range between 0.11±0.02 mas/yr (c) and 0.23±0.03 mas/yr (c). Thus, the ring-like emitting region must have been ejected from the core region before 1996 when it becomes visible in the VLBI maps.

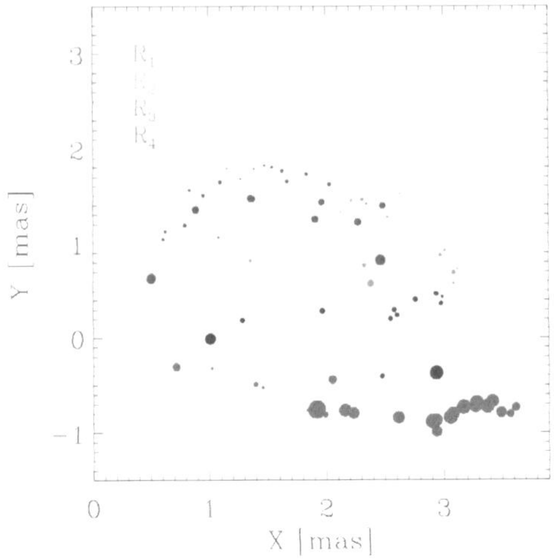

Figure 2. Trajectories of the identified Gaussian components for the extended ring-like structure. The size of each symbol is proportional to the flux density of the component.

3.2. Unusual mode change in jet kinematics

The inner jet region consists of the core (K) and - for the time between 1995 and 2008 - the two components A and B. Between 1995 and 2005 these two components do hardly show any motion with regard to the core (see Fig. 3, left at bottom). In addition, no new component ejections can be detected within this 10 year period. Thus in contrast to typical quasars, this source reveals an unexpected behaviour:

- apparent subluminal motion of two jet components in the vicinity of the core between 1995 and 2005
- no component ejections between 1995 and 2005

The core of 3C 454.3 shows a rapid flaring starting around 2005. At the same time, the 2 inner components (A, B) start moving away from the core with apparent superluminal velocities. After 2008, two new components are detected with VLBI and must have been ejected from the core earlier: component C and D. They separate from the core with apparent superluminal velocities. We thus observe two different kinematic modes for component motion in this source: until 2005 two components remain almost stationary with regard to the core. From 2005 on, the components move with apparent superluminal speeds.

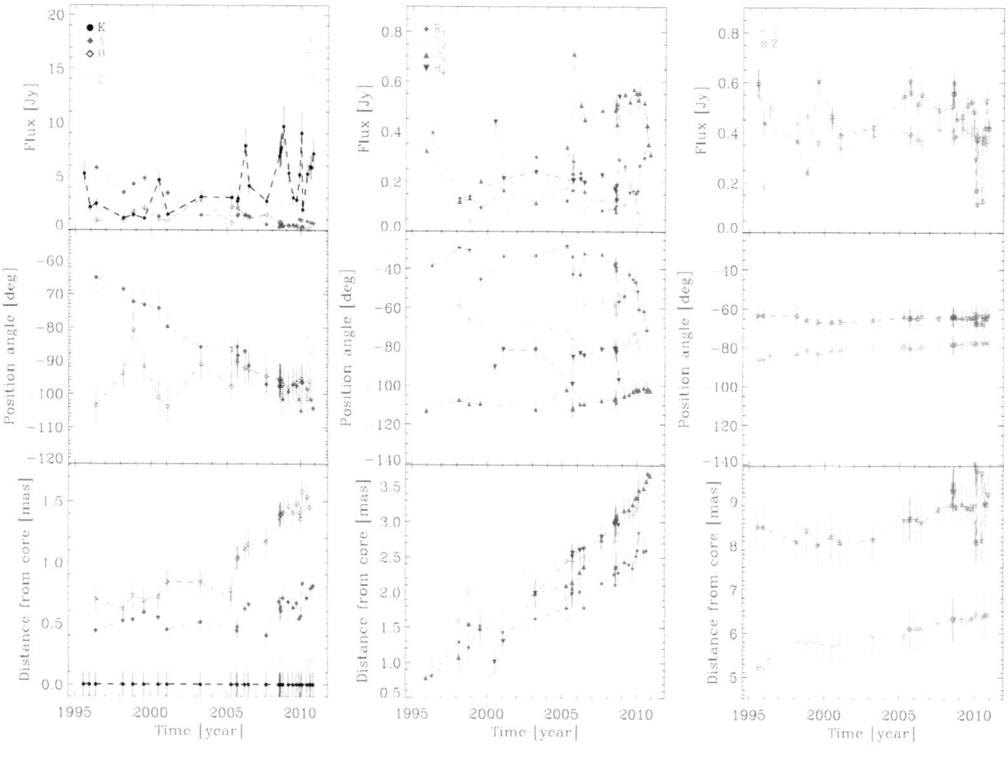

Figure 3. Flux, position angle and core separation as a function of time for the identified components.

4. Flux-density evolution

The radio flux-density of 3C 454.3 is known to be highly variable across the electromagnetic spectrum. In Fig.4 we show the single-dish flux-density evolution as monitored within the

Journal of Physics: Conference Series **372** (2012) 012029 doi:10.1088/1742-6596/372/1/012029

UMRAO monitoring program (4.8, 8.0, 14.5 GHz) and at two additional radio frequencies (22 and 37 GHz). The most prominent variability seems to occur simultaneously at the three frequencies. The first hint for a ring-structure has been seen in VLBI observations at 15 GHz around 1996. From this time onwards the ring-structure expands and becomes more clearly visible.It is likely that this evolving ring-structure could be related with the major radio flare seen in Fig.4 around 1995.

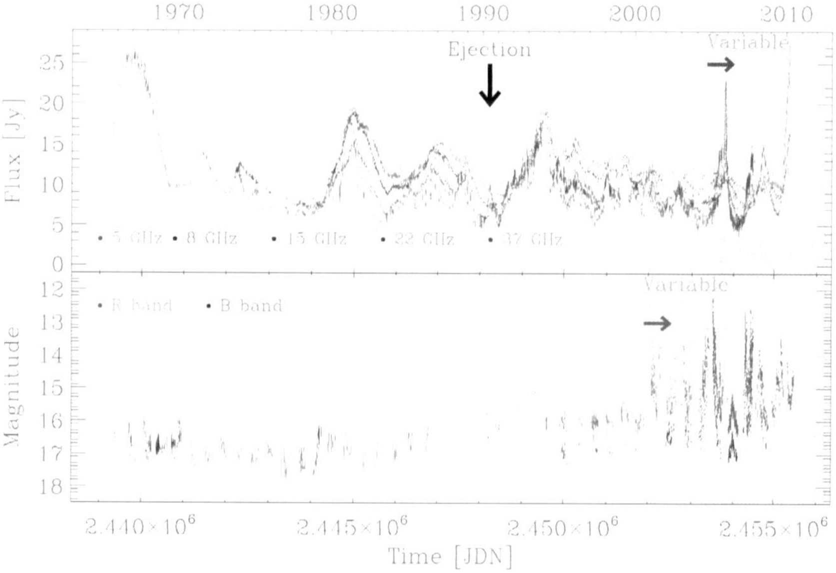

Figure 4. Historical emission behaviour of 3C 454.3 in various radio and optical band. In the top panel, the radio light-curves of 5 (red), 8 (blue), 15 (green), 22 (brown) and 37 GHz (violet) flux densities are presented. In the bottom panel, the R-band light curve (red) starting from 1990 has been complemented with historical B-band data (blue) before that date, shifted by a colour index $B - R = 1.0$, which is appropriate for intermediate-faint states. The grey shaded region indicates the approximate time of the ejection for the ring-like feature.

5. Discussion

The here reported detection of a ring-like structure around the core of 3C454.3 is a novelty in VLBI observations of AGN. The ring-like structure expands and increases in radius between 1996 and 2010 when it seems to become fainter. The estimated velocities range between 0.11±0.02 mas/yr (c) and 0.23±0.03 mas/yr (c). Thus, the emission of this ring might have been ejected from the core region before 1996 and might be correlated with the major flare around 1995.
Based on the spectral properties of this feature it can be ruled out that the object results from a supernova explosion in the central region of 3C454.3. An Einstein ring with a dark matter component can also be ruled out due to the time scales, sizes, and spectral properties of this structure. The remaining explanations of this ring-structure are:

- the angle to the line of sight changes - the jet turns towards us, most likely due to precession. The cause of the precession can be a binary black hole at the centre of 3C454.3. This could lead to a brightening of different regions of jet.

- extended jet region becomes visible due to shock propagation
- a jet from another (second) black hole becomes visible for a short period of time
- a combination of a changing angle to the line of sight and shock propagation

The preferred explanation of the authors is a changing angle to the line of sight due to precession. Since a binary black hole model had already been proposed to explain the observed variability in the radio lightcurves of this source (Qian et al. 2007) the observed ring structure could be the result of changed beaming properties of the source. This scenario could also explain the different variability properties. We would thus observe a precessing jet in this source turning towards us and revealing and allowing a deeper view into the inner core regions. By monitoring the VLBI and lightcurve properties of this source it should be possible to determine the timescales of the precession and deduce the parameters of the binary black hole system.

5.1. Kinematic mode changes in the pc-scale jet

Based on the observations of Pauliny-Toth et al. (1987) the components close to the core remained - after a major radio flare in 1981 at similar core separations for some time -not showing the typical apparent superluminal motion expected for this kind of objects. These structural changes, as reported by Pauliny-Toth et al., do not appear to fit easily into the scheme of relativistic jet models. The structural variations appear to be much more complex ("superluminal brightening"). The observations show a rapid increase in size of the core region and the appearance of some features which are nearly stationary with respect to component 1, while others move with apparent velocities of 20c. The authors conclude that the most likely scenario to explain the observations is that the features in the core region may not represent discrete, physical components, but rather regions where the emission from a relativistic jet is enhanced, for example by shocks. Lind and Blandford (1985) have emphasized that the velocity of shock fronts, which might be observed as a motion of 'components', can be smaller than that of the underlying medium, which determines the Doppler boosting of the emission. In particular, it is possible to obtain stationary emission patterns. Following the outburst, the relativistic fluid excites shocks at progressively larger distances from the core. Near the core, the shock pattern appears stationary. A change in physical conditions at larger distances from the core produces a moving pattern and gives rise to the 'classical' superluminal behaviour of the outer components. We find the same "stationary behavior" for the two inner components for the time after the major radio flare in 1995. Between 1995 and 2005 this stationary behaviour remained and no component ejections were seen. The kinematic behaviour changes drastically for the inner two components from 2005 on. This coincides with the onset of a different kind of radio flaring. From 2005 onwards all components move with apparent superluminal speeds. Both events - both mode changes - occurred together with a significant radio flaring. Not only did the kinematic properties of the jet components change, but the variability properties of the lightcure also changed. The here observed kinematic mode changes compare quite nicely with our findings for the AGN 0735+178 (Britzen et al. 2010). In this source we observed three mode changes in the kinematic properties (from stationary to apparent superluminal motion) within 30 years. These kinematic mode changes correlate with significant changes of the morphology (staircase / straight jet) and correlate as well with flarings observed in the optical light-curve.
We conclude that the physical appearance of the source changed significantly around 1995 and again 2005. Most likely one of the following scenarios can explain the severe changes in the

source:
- the conditions in the jet and/or core changed drastically
- the angle to the line of sight changed and we see now a different part of the jet and/or core region

So far it is unclear, where the true core in this source might be located and whether one of the inner jet components is the core. If the core is not the left-most component then the questions remains what is the nature of the other two bright inner components. It is clear that in this source the radio flaring is not always related to component ejections. This is in accordance with our findings for 0735+178 (Britzen et al. 2010). In addition, the kinematic mode changes between "stationary" and apparent superluminal motion seem to be a more common phenomenon than expected: 0735+178 and 3C454.3 clearly show these mode changes. It would be very important to find out whether these mode changes occur periodically in this source as seems to be the case in 0735+178. This source has been explained within a model of non-ballistic motion plus precession (Gong 2008, Britzen et al. 2010). It is very tempting to conclude that a periodic mode change might be related to a precessing system consisting of two black holes. A series of papers describing the detailed data analysis and the results of the theories applied is in preparation.

5.2. Acknowledgements
We thank J. Attridge, J.-L. Gomez and A. Marscher for providing VLBI data of 3C454.3 from their AGN monitoring programs. Part of this work was supported by the: COST Action MP0905 Black Holes in a Violent Universe. We appreciate the use of the NRAO archive of VLBA data. This research has made use of data from the MOJAVE database that is maintained by the MOJAVE team (Lister et al., 2009). This work made use of the VLBA, which is a facility of the National Science Foundation, operated under cooperative agreement by Associated Universities, Inc. This research has made use of data from the University of Michigan Radio Astronomy Observatory which is supported by a series of grants from the NSF and NASA and by funds from the University of Michigan. This research has made use of the NASA/IPAC Extragalactic Database (NED) which is operated by the Jet Propulsion Laboratory, California Institute of Technology, under contract with the National Aeronautics and Space Administration. The National Radio Astronomy Observatory is a facility of the National Science Foundation operated under cooperative agreement by Associated Universities, Inc. Thiw work is partly supported by the Georgian National Science Foundation through grant GNSF/ST09-521-4-320.

[1] Britzen S, et al. 2010 *A&A* **515** 105
[2] Fomalont E B, Frey S, Paragi Z, Gurvits L I, et al. 2000 *ApJS* **131** 95
[3] Gong B 2008 *MNRAS* **389** 315
[4] Homan D C , Lister M L 2006 *AJ* **131** 1262
[5] Jorstad S G, et al. 2010 *ApJ* **715** 362
[6] Kemball A J, Diamond P J , Pauliny-Toth I I K 1996 *ApJ* **464** 55
[7] Lister M L, Homan D C 2005 *AJ* **130** 1389
[8] Lobanov A P, Krichbaum T P, Graham D A et al. 2000 *A&A* **364** 391
[9] Lonsdale C J, Doeleman S, Phillips R B 1998 *AJ* **16** 8
[10] Lind K R, Blandford R D, 1985 *ApJ* **295** 358-367
[11] Lister M et al. 2009 *AJ* 137 3718
[12] Marscher A P, Jorstad, S G, Mattox, J R, et al. 2002 *ApJ* **577** 85
[13] Moellenbrock G A, Fujisawa K, Preston R A, et al. 1996 *AJ* **111** 217
[14] Pauliny-Toth I I K, Porcas R W, Zensus J A, et al. 1987 *Nature* **328** 778
[15] Qian S J, Kudryavtseva N A, Britzen S, et al. 2007 *ChJAA* **7** 364
[16] Rantakyro, F T, Baath L B, Backer, et al. 1998 *A&AS* **131** 451
[17] Savolainen T, Wiik K, Valtaoja E, et al. 2002 *A&A* **394** 851
[18] Tosti G, Chiang J, Lott, B, et al. 2008 *ATel* 1634

Astronomy at High Angular Resolution 2011 IOP Publishing
Journal of Physics: Conference Series **372** (2012) 012030 doi:10.1088/1742-6596/372/1/012030

Probing the Central Regions of AGN through VHE Gamma-Ray Induced Pair Cascades

Markus Böttcher and Parisa Roustazadeh
Astrophysical Institute, Department of Physics and Astronomy
Ohio University, Athens, OH 45701, USA
E-mail: boettchm@ohio.edu

Abstract. Recent discoveries by VERITAS, H.E.S.S., and MAGIC indicate that very-high-energy (VHE) gamma-rays may be produced in all types of blazars, including radio-loud quasars. In the dense nuclear radiation fields (BLR, dust torus IR) in these objects, VHE gamma-rays may be efficiently absorbed and initiate Compton-supported pair cascades. We have developed a Monte-Carlo code following the full 3-dimensional development of pair cascades in AGN environments. We have shown that even very weak magnetic fields may lead to efficient quasi-isotropization of the cascade emission, escaping predominantly in MeV – GeV gamma-rays. We propose this as a potential contributor to the Fermi gamma-ray flux from several radio galaxies, and present model fits to the Fermi-detected radio galaxies NGC 1275 and Cen A.

1. Introduction

One might wonder what a paper on high-energy (HE: > 100 MeV) and very-high energy (VHE: $E > 100$ GeV) γ-ray emission in AGN is doing at a conference on "The Central Kiloparsec of AGN". The angular resolution of the Large Area Telescope (LAT) on-board the *Fermi Gamma-Ray Space Telescope* and of the current generation of ground-based Atmospheric Cherenkov Telescope (ACT) facilities is of the order of $\sim 0.1^o$ and does therefore not offer the prospect of spatially resolving structures within the central kiloparsec of AGN, neither with current instruments, nor any time in the near future (e.g., with the *Cherenkov Telescope Array*). However, there is evidence that the region of high-energy γ-ray production in AGN is very close to the central engine, possibly within the central ~ 1 pc (although there is also evidence to the contrary, see., e.g., [5]), and the processes of γ-ray emission are sensitive to the details of the radiation environment in this central region of the AGN.

The overwhelming majority of AGN detected as sources of HE and VHE γ-ray emission by *Fermi*-LAT and ground-based ACTs are blazars, which are jet-dominated (radio-loud) AGN with their jets oriented closely aligned with our line of sight. Their broadband spectral energy distributions (SEDs) are dominated by Doppler-enhanced, non-thermal emission originating within the relativistic jet. Blazars are sub-divided into flat-spectrum radio quasars (FSRQs) and BL Lac objects, depending on the presence or absence of broad (EW > 5 Å) emission lines in their optical spectra. While the mechanism responsible for the γ-ray emission in blazars is still a matter of debate (see, e.g., [11] for a recent review of blazar emission models), it is clear that it critically depends on the presence (or absence) of radiation fields from the Broad Line Region (BLR) and/or warm dust in the nuclear region of the AGN. Therefore, γ-ray observations of

AGN are a tool to indirectly study their nuclear radiation environment, possibly on sub-parsec scales.

This paper discusses a method to probe the nuclear environments of radio-loud AGN through the signatures of $\gamma\gamma$ absorption of HE and VHE γ-rays and Compton-supported pair cascades initiated by this process.

2. Intrinsic $\gamma\gamma$ absorption in blazars

While previous generations of ground-based Atmospheric Cherenkov Telescope (ACT) facilities detected almost exclusively high-frequency peaked BL Lac objects (HBLs) as extragalactic sources of VHE γ-rays, in recent years, a number of non-HBL blazars have been detected by the current generation of ACTs, ranging from intermediate BL Lac objects (IBLs; e.g., W Comae [3], 3C 66A [4]) to even flat-spectrum radio quasars (FSRQs; e.g., 3C 279 [7], 4C+21.35 [8]). This suggests that most blazars might be intrinsically emitters of VHE γ-rays. According to AGN unification schemes [26], radio galaxies are the mis-aligned parent population of blazars with the less powerful FR I radio galaxies corresponding to BL Lac objects and FR II radio galaxies being the parent population of radio-loud quasars. Blazars are those objects which are viewed at a small angle with respect to the jet axis. If this unification scheme holds, then, by inference, also radio galaxies may be expected to be intrinsically emitters of VHE γ-rays within a narrow cone around the jet axis.

While there is little evidence for dense radiation environments in the nuclear regions of BL Lac objects — in particular, HBLs —, strong line emission in Flat Spectrum Radio Quasars (FSRQs) as well as the occasional detection of emission lines in the spectra of some BL Lac objects (e.g., [27]) indicates dense nuclear radiation fields in those objects. This is supported by spectral modeling of the SEDs of blazars using leptonic models which prefer scenarios based on external radiation fields as sources for Compton scattering to produce the high-energy radiation in FSRQs, LBLs and also some IBLs (e.g., [15, 11]). If the VHE γ-ray emission is indeed produced in the high-radiation-energy-density environment of the broad line region (BLR) and/or the dust torus of an AGN, it is expected to be strongly attenuated by $\gamma\gamma$ pair production (e.g., [18, 13, 20, 16, 24]).

Aharonian et al. [6] have suggested that such intrinsic $\gamma\gamma$ absorption may be responsible for producing the unexpectedly hard intrinsic (i.e., after correction for $\gamma\gamma$ absorption by the extragalactic background light) VHE γ-ray spectra of some blazars at relatively high redshift. A similar effect has been invoked by Poutanen & Stern [19] to explain the spectral breaks in the *Fermi* spectra of γ-ray blazars.

The ultimate confirmation of the relevance of the $\gamma\gamma$ absorption process in blazars would come from the observation of a $\gamma\gamma$ absorption trough. When caused by a photon field concentrated around a mean photon energy $\epsilon_s = h\nu_s/m_e c^2$, the absorption trough is expected to have a minimum at $\epsilon_\gamma \sim 2/\epsilon_s$. For a radiation field dominated by Lyα emission, this would correspond to a γ-ray energy of $E_\gamma \sim 25\,(1+z)^{-1}$ GeV, and for lower-energy soft photon fields (e.g., thermal IR emission from a dust torus), the absorption trough is expected to be at higher energies. Unfortunately, this places the expected absorption features just at the transition between the energy ranges of *Fermi*-LAT and ground-based VHE γ-ray observatories.

A comprehensive analysis of combined *Fermi*-LAT and ground-based VHE γ-ray spectra of all VHE blazars is currently being performed by G. Şentürk [23]. Figure 1 shows an example of this combined analysis for the LBL S5 0716+714, which may provide evidence for the expected $\gamma\gamma$ absorption feature. The solid line in Figure 1 shows a model of an intrinsic power-law, absorbed through $\gamma\gamma$ pair production by a circumnuclear radiation field dominated by HeII recombination continuum emission. However, it should be noted that the *Fermi*-LAT and MAGIC data shown in Figure 1 (as all data sets included in [23]) are non-simultaneous, and correspond to vastly different integration times between *Fermi*-LAT and the ground-based γ-

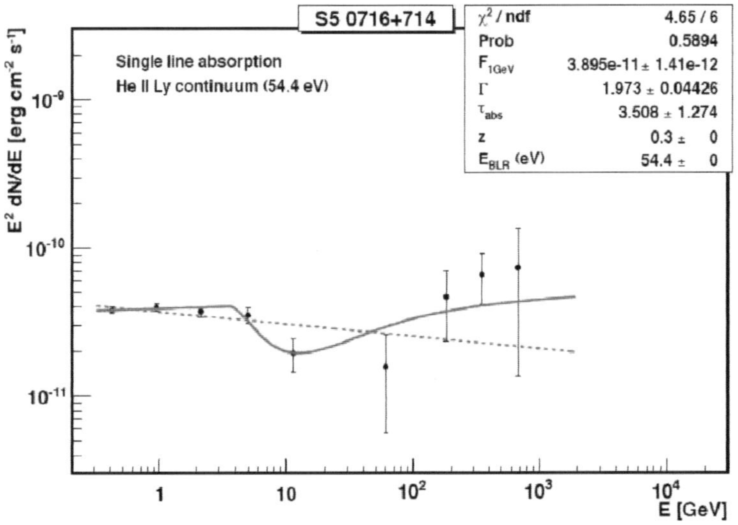

Figure 1. Combined *Fermi*-LAT and MAGIC spectrum of the LBL S5 0716+714. The combined spectrum can be well represented by an intrinsic power-law, absorbed by a photon field dominated by HeII recombination continuum emission (from [23]).

ray observatories. Future, simultaneous data sets, for example from the recent γ-ray flare of BL Lacertae observed both by VERITAS [17] and *Fermi*-LAT [12, 14], may provide conclusive evidence for or against the reality of these absorption features.

3. Compton-Supported Pair Cascades

If the $\gamma\gamma$ absorption process discussed in the previous section does play an important role in AGN environments, it will lead to the development of Compton-supported pair cascades (e.g., [9, 25, 21, 22]). We [21, 22] have developed a Monte-Carlo (MC) code which follows the development of the resulting pair cascades in full 3D geometry. Arbitrary (e.g., line-dominated or thermal) spectra of the circumnuclear radiation field can be simulated, and the code follows the trajectory of the secondary electrons/positrons in a uniform magnetic field of strength B which is mis-aligned with respect to the jet axis by an angle θ_B. This configuration can be seen as a proxy for a helical magnetic field with $\tan\theta_B = B_{\rm tor}/B_{\rm pol}$, where $B_{\rm tor}$ and $B_{\rm pol}$ are the toroidal and poloidal components of the magnetic field. The cascades are Compton supported, which requires the magnetic-field energy density to be substantially lower than the radiation energy density, $u_B \ll u_{\rm ext}$ as otherwise synchrotron losses would quench the cascade development. For more details on the MC code, see Roustazadeh & Böttcher [21, 22]. In those papers, we performed a general parameter study, varying both the strength and orientation of the magnetic field as well as the parameters of the target radiation field.

We aim at a parameter study that is as model-independent as possible, and therefore do not specify the details of the γ-ray production mechanism. It is assumed that HE and VHE γ-rays are produced at $\lesssim pc$ scales from the central engine, and we study the interaction of this emission with external radiation fields. At distances of $R \gg 10^{16}\, M_8$ cm from the central black hole with mass $M = 10^8\, M_8\, M_\odot$, emission from the accretion disk will interact with the γ-rays

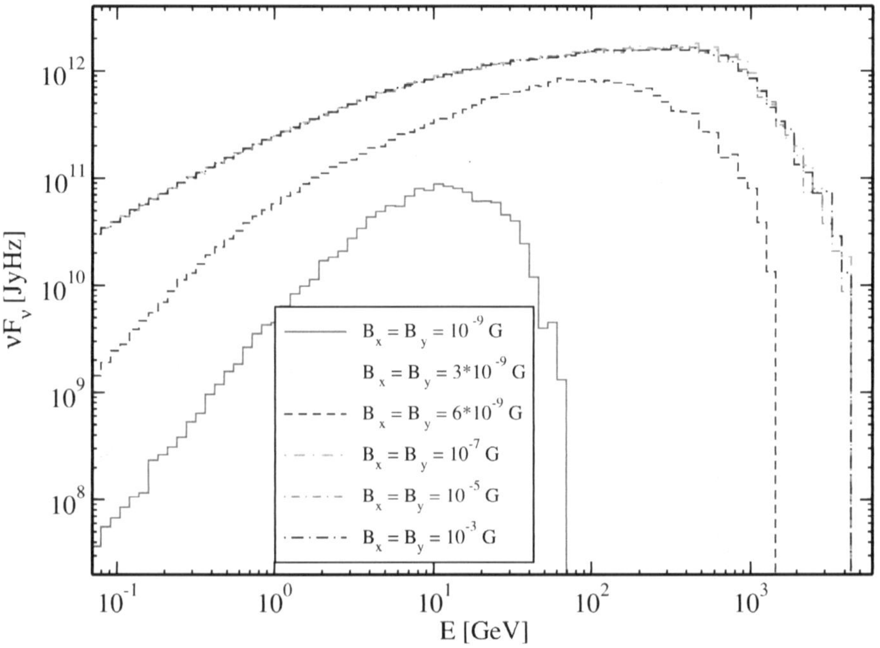

Figure 2. Effect on the off-axis (viewing angle $0.4 \leq \cos\theta_{\rm obs} \leq 0.6$) cascade emission of a varying magnetic field strength in the AGN environment.

(propagating predominantly along the jet axis), under a very unfavorable angle and are therefore not expected to contribute significantly to the $\gamma\gamma$ opacity. This motivates our assumption of isotropic external radiation fields, as appropriate for the BLR radiation field (within the BLR) and IR emission from warm dust in the nuclear region.

One of the salient results of our parameter study is illustrated in Figure 2. It shows that for typical radio-loud AGN parameters even very small ($B_\perp \sim \mu$G) magnetic field components perpendicular to the jet axis in the central ~ 1 pc of the AGN may lead to efficient deflection and quasi-isotropization of the cascade emission in the *Fermi* energy range.

The off-axis cascade emission is generally dominated by a power-law spectrum $\nu F_\nu \propto \nu^{1/2}$ due to Compton emission from radiatively cooling electrons/positrons injected at high energies by the $\gamma\gamma$ pair production process. The spectrum exhibits a high-energy turn-over due to the fact that for the highest-energy particles the Compton cooling length is much shorter than the gyro-radius. They therefore effectively radiate in the forward (blazar) direction. Only cooled particles at lower energy have been deflected enough to contribute to the off-axis emission. At the lowest energies, the effects of particle escape can become important, steepening the spectrum below characteristic Compton energies of particles for which the Compton cooling length becomes longer than the characteristic extent of the target photon field.

4. Application to NGC 1275 and Cen A

Our result that even very weak perpendicular magnetic fields are sufficient to lead to the efficient deflection of the cascades to off-axis directions led us to the idea that this off-axis cascade emission may be observable by *Fermi* in nearby radio galaxies. In Roustazadeh & Böttcher [21], we applied this idea to the radio galaxy NGC 1275 [1]. This galaxy is known to host a low-luminosity AGN, which provides a plausible source of target photons for the development of a pair cascade, assuming that this radio galaxy would appear as a γ-ray bright blazar when

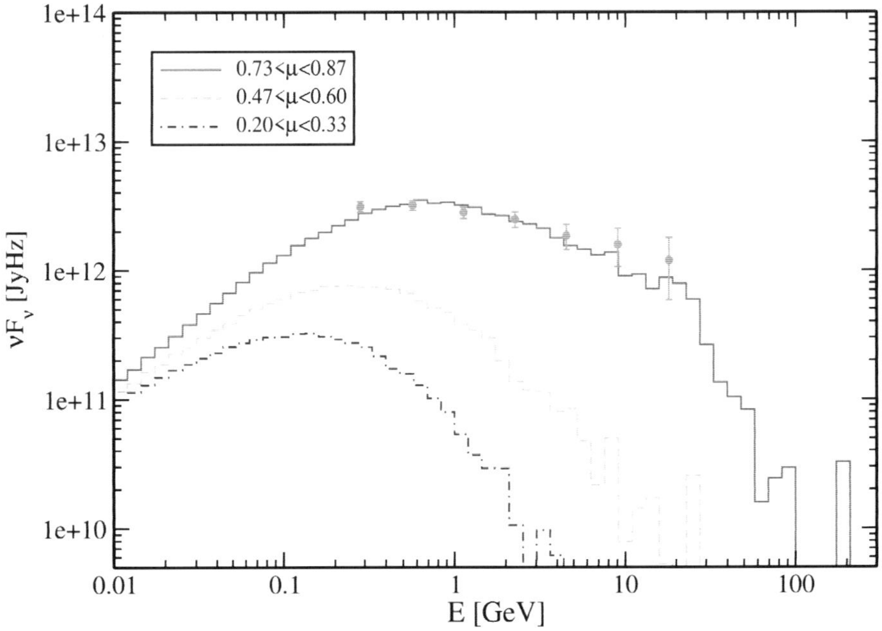

Figure 3. Fit to the *Fermi*-LAT spectrum of the radio galaxy NGC 1275 by Compton emission from a VHE γ-ray induced cascade developing in a Lyα dominated BLR radiation field of a low-luminosity AGN in NGC 1275.

viewed along the jet axis. We found that the expected cascade emission at the well-constrained observing angle toward NGC 1275 allows for a satisfactory fit to the γ-ray flux and spectrum as observed by *Fermi*-LAT (see Figure 3).

As a second test case, we applied this model to the radio galaxy Cen A[2]. In Abdo et al. [2], the radio – γ-ray SED of Cen A was interpreted by several variations of pure off-axis blazar models. Interestingly, all of those required rather extreme parameters — in particular, unusually large magnetic fields, low Doppler factors, and large particle energies. Figure 4 shows our fit to the SED of Cen A using the equilibrium version of the single-zone leptonic jet emission model of Böttcher & Chiang[10]. Apart from the viewing angle of $\theta_{\rm obs} = 70^o$, parameters rather typical of low-frequency-peaked γ-ray blazars were adopted. This fit under-represents the *Fermi*-LAT spectrum. We then added the off-axis cascade emission from the VHE γ-ray emission, which would be visible along the jet axis, in the near-nuclear infrared radiation field of this galaxy. The addition of this cascade component provides an excellent fit to the overall SED of Cen A with plausible parameters.

Acknowledgments

This work was supported by NASA through Fermi Guest Investigator grants NNX09AT81G and NNX10AO49G.

References

[1] A. A. Abdo et al., 2009a, *ApJ*, **699**, 31
[2] A. A. Abdo et al., 2009b, *ApJ*, **700**, 597
[3] V. A. Acciari et al., 2008, *ApJ*, **684**, L74

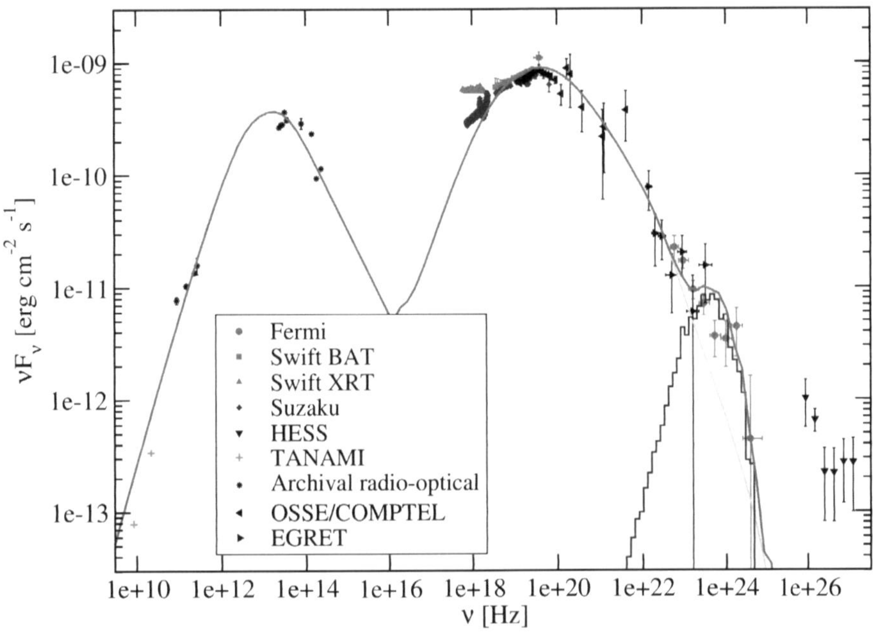

Figure 4. Fit to the spectral energy distribution of Cen A, including a contribution to the *Fermi* γ-ray spectrum from the off-axis emission due to the expected cascade developing in the near-nuclear infrared radiation field.

[4] V. A. Acciari et al., 2009, *ApJ*, **707**, 612
[5] I. Agudo et al., 2011, *ApJ*, **735**, L10
[6] F. A. Aharonian et al., 2008, *MNRAS*, **387**, 1206
[7] J. Albert, et al., 2008, *Science*, **320**, 1752
[8] J. Aleksić, et al., 2011, *ApJ*, **730**, L8
[9] W. Bednarek & J. G. Kirk, 1995, *A&A*, **294**, 366
[10] M. Böttcher & J. Chiang, 2002, *ApJ*, **281**, 127
[11] M. Böttcher, 2010, in proc. *Fermi Meets Jansky*, MPIfR, Bonn, Germany, p. 41
[12] S. Cutini et al., 2011, *ATel* # 3368
[13] A. C. Donea & R. J. Protheroe, 2003, *Astrop. Phys.*, **18**, 337
[14] D. Gasparrini et al., 2011, *ATel* # 3468
[15] G. Ghisellini et al., 1998, *MNRAS*, **301**, 451
[16] H. T. Liu et al., 2008, *ApJ*, **688**, 148
[17] R. Ong et al., 2011, *ATel* # 3459
[18] R. J. Protheroe & P. L. Biermann, 1997, *Astrop. Phys.*, **6**, 293
[19] J. Poutanen & B. Stern, 2010, *ApJ*, **717**, L118
[20] A. Reimer, 2007, it ApJ, **665**, 1023
[21] P. Roustazadeh & M. Böttcher, 2010, *ApJ*, **717**, 468
[22] P. Roustazadeh & M. Böttcher, 2011, *ApJ*, **728**, 134
[23] G. Şentürk et al., 2011, in preparation
[24] J. Sitarek & W. Bednarek, 2008, *MNRAS*, **391**, 624
[25] J. Sitarek & W. Bednarek, 2010, *MNRAS*, **401**, 1983
[26] C. M. Urry, and P. Padovani, 1995, *PASP*, **107**, 803
[27] R. C. Vermeulen et al., 1995, *ApJ*, **452**, L5

The central structure of Broad Absorption Line QSOs: observational characteristics in the cm-mm wavelength domain

G Bruni[1,2,3], K-H Mack[1], D Dallacasa[1,2], F M Montenegro-Montes[4], C R Benn[5], R Carballo[6], J I González-Serrano[7], J Holt[8] and F Jiménez-Luján[3,5,7]

[1]INAF-Istituto di Radioastronomia, via Piero Gobetti, 101, I-40129 Bologna, Italy
[2]Università di Bologna, Dip. di Astronomia, via Ranzani, 1, I-40127 Bologna, Italy
[3]Dpto. de Física Moderna, Univ. de Cantabria, Avda de los Castros s/n, E-39005 Santander, Spain
[4]European Southern Observatory, Alonso de Córdova 3107, Vitacura, Casilla 19001, Santiago, Chile
[5]Isaac Newton Group, Apartado 321, E-38700 Santa Cruz de La Palma, Spain
[6]Dpto. de Matemática Aplicada y Ciencias de la Computación, Univ. de Cantabria, ETS Ingenieros de Caminos, Canales y Puertos, Avda de los Castros s/n, E-39005 Santander, Spain
[7]Instituto de Física de Cantabria (CSIC-Universidad de Cantabria), Avda. de los Castros s/n, E-39005 Santander, Spain
[8]Leiden Observatory, Leiden University, P.O. Box 9513, NL-2300 RA Leiden, The Netherlands

E-mail: `bruni@ira.inaf.it`

Abstract. Accounting for ~20% of the total QSO population, Broad Absorption Line QSOs are still an unsolved problem in the AGN context. They present wide troughs in the UV spectrum, due to material with velocities up to 0.2 c toward the observer. The two models proposed in literature try to explain them as a particular phase of the evolution of QSOs or as normal QSOs, but seen from a particular line of sight.
We built a statistically complete sample of Radio-Loud BAL QSOs, and carried out an observing campaign to piece together the whole spectrum in the cm wavelength domain, and highlight all the possible differences with respect to a comparison sample of Radio-Loud non-BAL QSOs. VLBI observations at high angular resolution have been performed, to study the pc-scale morphology of these objects. Finally, we tried to detect a possible dust component with observations at mm-wavelengths.
Results do not seem to indicate a young age for all BAL QSOs. Instead a variety of orientations and morphologies have been found, constraining the outflows foreseen by the orientation model to have different possible angles with respect to the jet axis.

1. Introduction

Broad Absorption Line (BAL) QSOs are a particular and not yet understood class of AGN: their spectra show wide troughs towards the blue wing of some UV emission lines (Mg II, Al III, Si IV, C IV), due to ionized gas with outflow velocities up to 0.2 c. They account for ~20% of the QSO population. Two main scenarios have been considered to explain their nature: the *orientation* model and the *evolutionary* one. The former, proposed by [6], foresees the BAL outflows to be present in all QSOs, but visible only when intercepting the line of sight of the observer, thus a particular orientation can be supposed for these objects. The latter model ([1, 7]) explains the BAL phenomenon as a particular phase in the evolutionary sequence of QSOs, during which a dusty shell is being expelled from the AGN. In this case a greater amount of dust emission should be present with respect to *normal* QSOs. A third model from Punsly ([11, 12]) proposes that the BAL phenomenon is due to polar winds above the inner jet, thus would be visible only for lines of sight close to the jet axis.

In this framework the radio emission is an additional diagnostic tool to test the proposed models. We selected a sample of Radio-Loud BAL QSOs, cross-correlating the fourth edition of the SDSS Quasar Catalogue ([13]) with the FIRST catalogue ([2]), with a constraint of $S_{1.4\ \mathrm{GHz}} > 30$ mJy for the flux density at 1.4 GHz and $1.7 < z < 4.7$ for the redshift, in order to have sufficient flux density for VLBI studies and to allow the classification as BAL QSO through the CIV and MgII lines in the optical band. The whole sample of 25 objects, together with a comparison sample of 34 non-BAL QSOs, is presented in [4].

2. Radio Observations

In 2009-2010 we carried out an observational campaign to cover the cm-wavelength spectrum of the two BAL and non-BAL QSO samples, using both single dish (100-m Effelsberg) and interferometers (GMRT, VLA) in the frequency range from 240 MHz to 43 GHz. Our target was the investigation of the differences between the two samples, aiming at determining which of the three models is the most probable scenario for BAL QSOs. There are three observables that can give clues about that: (1) the spectral index of the synchrotron spectrum is a statistical indicator of the orientation of the source ([10]), since flatter spectral indices imply lines of sight closer to the jet axis. If a particular orientation was present in BAL QSOs, the spectral index distribution for the sample would be different with respect to the non-BAL sample; (2) the synchrotron peak frequency is related to the age of the source, younger sources having a smaller linear size and a higher peak frequency ([8]). If BAL QSOs were young objects in a particular evolutionary phase, the ratio of Gigahertz-Peaked Spectrum (GPS) sources would be greater than in the non-BAL sample; (3) polarization and Rotation Measure (RM) are indicative of the magnetic field strength and particle density around the central environment of the AGN. Finally, VLBI studies can help in determining the morphology and orientation of this particular class of QSOs.

2.1. Spectra and polarization

We fitted the radio spectra of the sources through χ^2 minimization, with a power law and a parabola in the $logS_\nu - log\nu$ plane, in order to determine whether a synchrotron peak frequency in the GigaHertz range was present. In some cases a low-frequency component was found, likely corresponding to old components, and not indicating a young age for the source. We found 9 BAL QSOs and 8 non-BAL QSOs showing a GPS, and thus a similar fraction of GPS sources in the two samples (36% and 24%), but further observations are needed to confirm their classification, since the angular resolution of the VLA in C configuration only allowed us to give an upper limit of 20 kpc to the linear size in the 8 GHz maps, while the maximum size for a GPS source is 1 kpc. Fits and spectra of the two samples can be found in [4].

The spectral index (defined as $\alpha_{1,2} = [(\log S_2 - \log S_1)/(\log \nu_2 - \log \nu_1)]$, where $\nu_2 > \nu_1$) was calculated for all sources in the frequency interval 4.8-8.4 GHz and 8.4-22 GHz, i.e. well above the peak frequency for all of them. A variety of spectral indices have been found, both flat ($\alpha < 0.5$) and steep ($\alpha > 0.5$), suggesting a wide range of orientations for BAL QSOs, similar to

Journal of Physics: Conference Series **372** (2012) 012031 doi:10.1088/1742-6596/372/1/012031

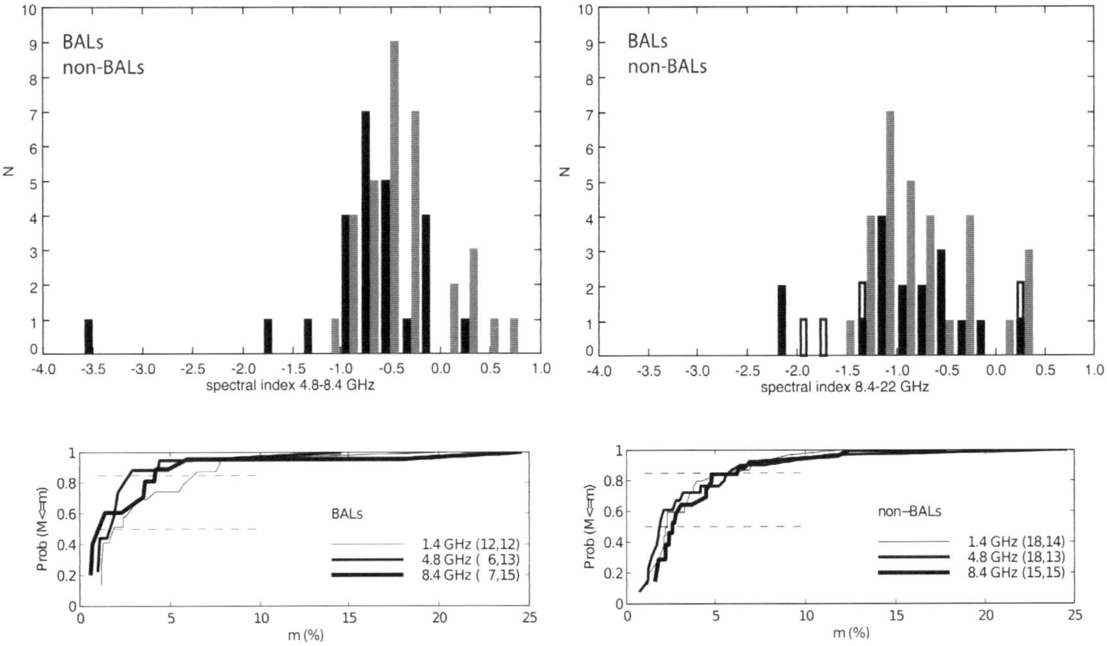

Figure 1. Top panels: distribution of the radio spectral indices ($\alpha_{4.8}^{8.4}$ and $\alpha_{8.4}^{22}$) for the BAL (black) and non-BAL (grey) QSO samples. Upper limits for BAL QSOs are in white. Bottom panels: cumulative distribution of the fractional polarization for the two samples at three frequencies. The pair of numbers in parenthesis indicates the number of detections and the number of upper limits for each frequency. The dashed lines show the 50% percentile (i.e. the median) and the 85% percentile.

the non-BAL QSO sample (see Fig. 1). A Kolmogorov-Smirnov test comparing the two spectral index distributions $\alpha_{4.8}^{8.4}$ and $\alpha_{8.4}^{22}$ for BAL and non-BAL QSOs gives a low significance level (74% and 0.3%) for the two to be different. Moreover, the same test gives a significative evidence that BAL QSOs are not flatter than non-BAL QSOs, in contradiction to the hypothesis from [11, 12].

Measurements or upper limits of the fractional polarization (m) have been obtained for the two samples, and a Kaplan-Meier estimator was used to compare the two, showing that the cumulative distributions are very similar (see Fig. 1). The median value for m is in the range 1.8-2.5% for both samples. We calculated the RMs for all the sources with at least 3 values of m at different frequencies (4 BAL QSOs and 10 non-BAL QSOs): they show similar values despite the small statistics, the only outlier being BAL QSO 1624+37, with the second highest value of RM among extra-galactic sources (RM=-18350 ± 570 rad m^{-2}, [3]).

2.2. Morphology and orientation

In 2009-2010 we obtained also high-resolution images for 11 BAL QSOs in VLBI technique with the EVN (5 GHz) and VLBA (5, 8.4 GHz) arrays ([5]). 9 of 11 sources present a resolved structure (82%), and various morphologies are visible (see Fig. 2). Double (0756+37, 1102+11), core-jet (0044+00, 0816+48, 0849+27, 1304+13) and symmetric (1014+05, 1237+47, 1603+30) structures have been found, so different orientations can be argued. Unresolved sources could be beamed jets towards the observer or extremely young sources, like 1406+34. Projected linear sizes are constrained under 1 kpc: this confirms the GPS classification for 0756+37, 1237+47

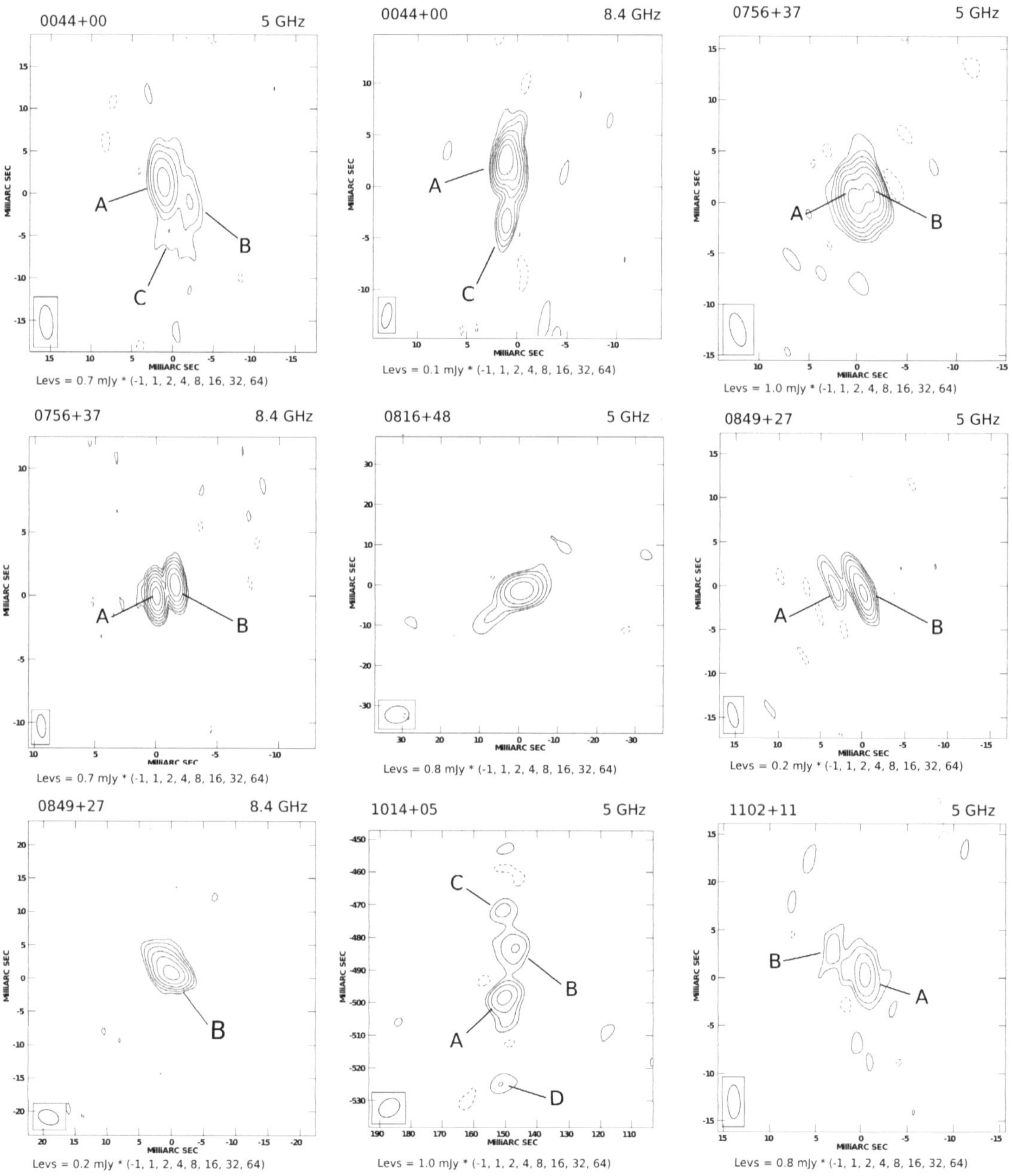

Figure 2. High angular resolution maps of 11 BAL QSOs, obtained with the VLBA and the EVN.

and 1603+30.

Four BAL QSOs were already resolved in VLA maps, showing 1 elongated (1103+11) and 3 core-lobes (0816+48, 0849+27, 1603+30) morphologies, with projected linear sizes between 20 and 200 kpc.

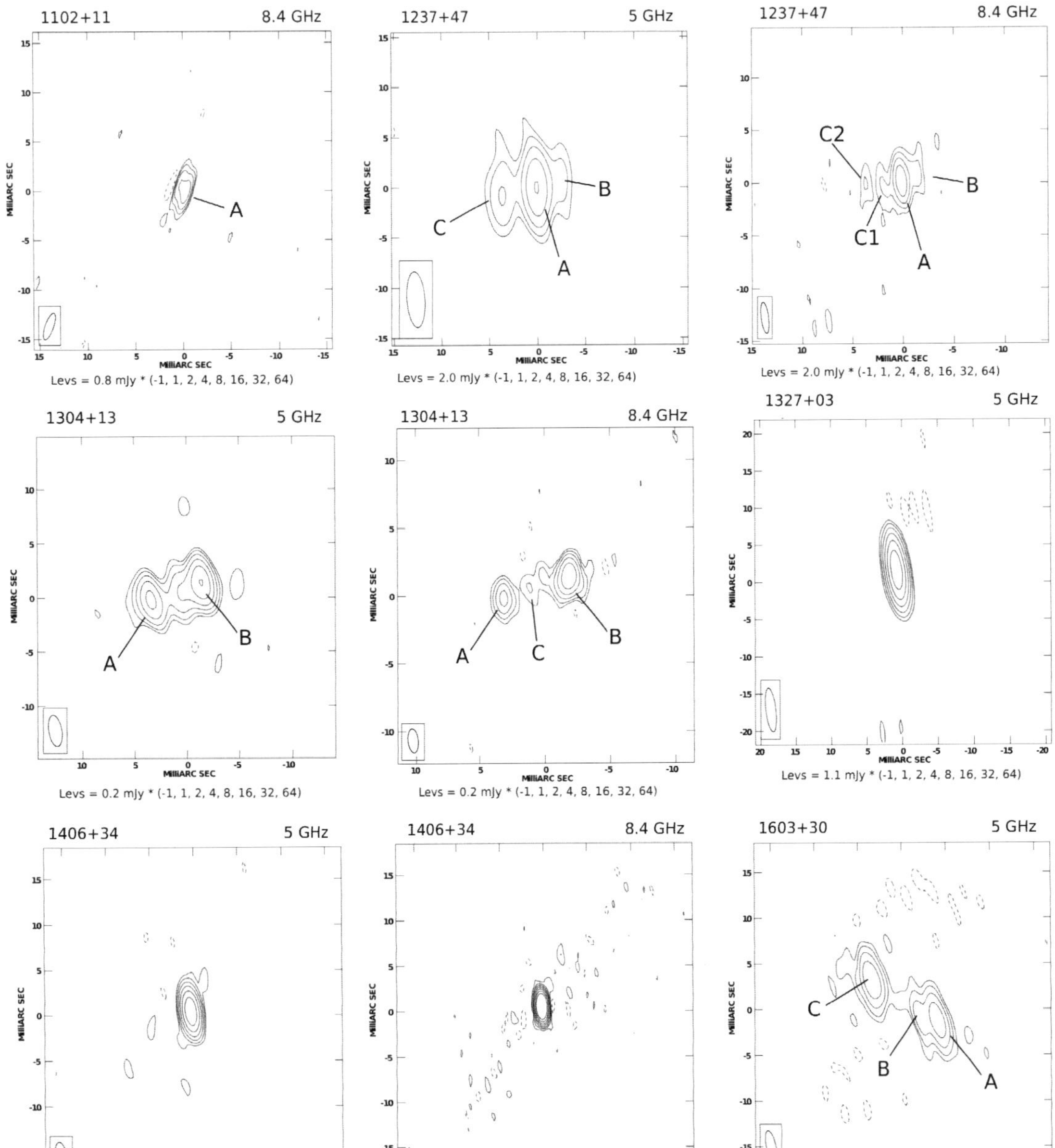

Figure 2. Continued.

3. Dust detection

In the mm-wavelength domain we performed observations with the IRAM-30m single dish (250 GHz) and with APEX (345 and 850 GHz), to detect the grey-body dust emission in the BAL QSO sample. The amount of dust is one of the main differences in the two scenarios proposed to explain the nature of BAL QSOs, since a young source, still expelling its dust cocoon, should present an excess in the mm-band emission (in the redshift range of our sample). The two observations were planned to detect the peak (850 GHz) and the power-law slope (240, 345 GHz) of this emission. We found 1 out of 14 observed sources ($\sim 7\%$) showing a clear dust component not correlated with the synchrotron emission (0756+37, see Fig. 3).

For this study we refer, as a comparison, to the dust emission study published by Omont et al. for QSOs with z$\sim$2 ([9]): they found that $\sim 25\%$ of their sample shows hints of dust at 250 GHz. Since their study has been carried out with the same instrument, and obtaining a similar RMS, we can safely compare with this result, concluding that Radio-Loud BAL QSOs does not seem to present a larger fraction of dust-rich objects with respect to the QSO population.

More observations, already scheduled at APEX, will increase the statistic and thus the reliability of these preliminary results.

4. Conclusions

We can summarize our findings as follows:

- A variety of orientation for BAL QSOs have been found, both from spectral index and from VLBI morphology studies.
- Radio spectra do not seem to indicate a young age, since the fraction of GPS sources is similar to non-BAL QSOs. Moreover, in some cases a low-frequency emission is present, suggesting an old radio component.
- VLBI maps for 11 BAL QSOs show different morphologies and projected linear sizes under 1 kpc. Bigger linear sizes are not excluded for BAL QSOs, since 4 sources were already resolved with the VLA, resulting in projected linear sizes between 20 and 200 kpc.
- Radio-Loud BAL QSOs do not seem to be dust-rich, since preliminary results only show $\sim 7\%$ of the sources with a flux density excess in the mm-band, comparable with previous findings for the QSO population.

Excluding a young age for Radio-Loud BAL QSOs, the most probable scenario seems to be the one proposed by [6], but only if it can account for outflows with a variety of angles with respect to the jet axis, able to justify the different orientations found in radio observations.

Acknowledgments
We are grateful to the AHAR11 LOC for the economical support.
Part of this work was supported by a grant of the Italian Programme for Research of Relevant National Interest (PRIN No. 18/2007, PI: K.-H. Mack) The authors acknowledge financial support from the Spanish Ministerio de Ciencia e Innovación under project AYA2008-06311-C02-02.
This work has benefited from research funding from the European Union's sixth Framework Programme under RadioNet grant agreement no. 227290.
This work has been partially based on observations with the 100-m telescope of the MPIfR (Max-Planck-Institut für Radioastronomie) at Effelsberg.
The National Radio Astronomy Observatory is a facility of the National Science Foundation operated under cooperative agreement by Associated Universities, Inc.
The European VLBI Network is a joint facility of European, Chinese, South African and other radio astronomy institutes funded by their national research councils.
This work has been partially based on observations carried out with the IRAM 30m single-dish. IRAM

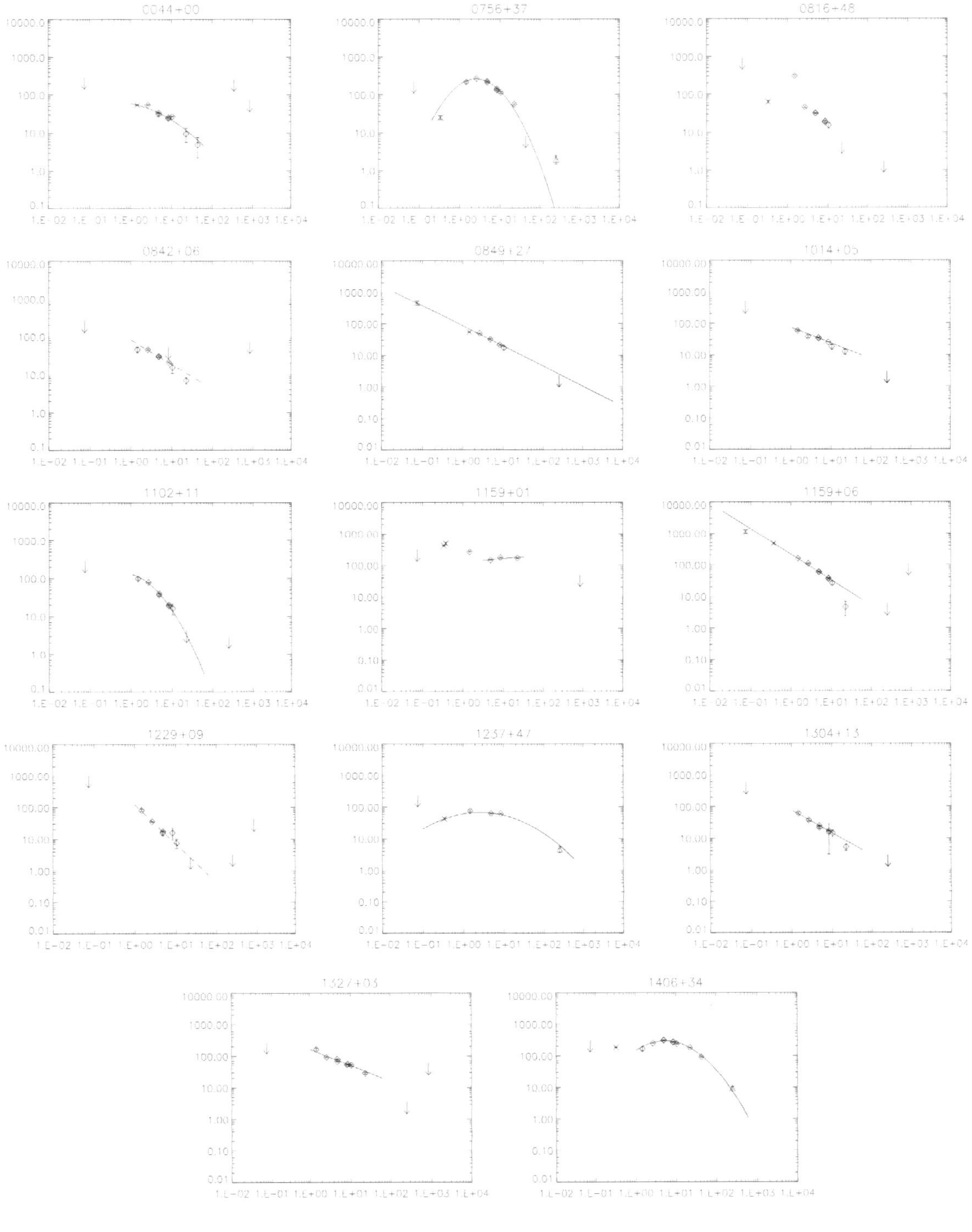

Figure 3. Spectra of 14 BAL QSOs observed in the mm-band (x-axis: GHz; y-axis: mJy). 250 GHz flux densities from IRAM-30m, 345 and 850 GHz from APEX (triangles). Flux densities at lower frequencies are taken from [4] (crosses, $\nu < 1.4$ GHz: literature; rhombi, $\nu \geq 1.4$ GHz: our observational campaign).

is supported by INSU/CNRS (France), MPG (Germany) and IGN (Spain).
This publication is partially based on data acquired with the Atacama Pathfinder Experiment (APEX). APEX is a collaboration between the Max-Planck-Institut fur Radioastronomie, the European Southern Observatory, and the Onsala Space Observatory.
This research has made use of the NASA/IPAC Infrared Science Archive and NASA/IPAC Extragalactic Database (NED) which are both operated by the Jet Propulsion Laboratory, California Institute of Technology, under contract with the National Aeronautics and Space Administration.
Use has been made of the Sloan Digital Sky Survey (SDSS) Archive. The SDSS is managed by the Astrophysical Research Consortium (ARC) for the participating institutions: The University of Chicago, Fermilab, the Institute for Advanced Study, the Japan Participation Group, The John Hopkins University, Los Alamos National Laboratory, the Max-Planck-Institute for Astronomy (MPIA), the Max-Planck-Institute for Astrophysics (MPA), New Mexico State University, University of Pittsburgh, Princeton University, the United States Naval Observatory, and the University of Washington.

References

[1] Briggs F. H., Turnsheck D. A., Wolfe M., 1984, ApJ, 287, 549
[2] Becker R. H., Helfand D. J., White R. L., 1995, ApJ, 450, 559
[3] Benn C.R., Carballo R., Holt J., Vigotti M., González-Serrano J.I., Mack K.-H., Perley R.A., 2005, MNRAS, 360, 1455
[4] Bruni G., Mack K.-H., Salerno E., Montenegro-Montes F.M., Carballo R., Benn C. R., González-Serrano J. I., Holt J. and Jiménez-Luján F., 2012, A&A, in press, (arxiv: 1203.4509)
[5] Bruni G., Mack K.-H., Dallacasa D., 2010, PoS (10th EVN Symposium) 036
[6] Elvis M., 2000, ApJ, 545, 63
[7] Lipari S.L., Terlevich R.J., 2006, MNRAS, 368, 1011
[8] O'Dea C.P, Baum S.A., 1997, AJ, 113, 148
[9] Omont A., Beelen A., Bertoldi F., Cox P., Carilli C.L., Priddey R.S., McMahon R.G., Isaak K.G., 2003, A&A 398, 857
[10] Orr M.J.L., Browne I.W.A., 1982, MNRAS, 200, 1067
[11] Punsly B., 1999a, ApJ, 527, 609
[12] Punsly B., 1999b, ApJ, 527, 624
[13] Schneider D.P. et al, 2007, AJ, 134, 102

Astronomy at High Angular Resolution 2011 IOP Publishing
Journal of Physics: Conference Series **372** (2012) 012032 doi:10.1088/1742-6596/372/1/012032

Slicing the Torus: Obscuring Structures in Quasars

Martin Elvis[1]

[1] Harvard-Smithsonian Center for Astrophysics, 60 Garden St., Cambridge MA02138, USA

E-mail: `elvis@cfa.harvard.edu`

Abstract. Quasars and Active Galactic Nuclei (AGNs) are often obscured by dust and gas. It is normally assumed that the obscuration occurs in an oblate "obscuring torus", that begins at the radius at which the most refractive dust can remain solid. The most famous form of this torus is a donut-shaped region of molecular gas with a large scale-height. While this model is elegant and accounts for many phenomena at once, it does not hold up to detailed tests. Instead the obscuration in AGNs must occur on a wide range of scales and be due to a minimum of three physically distinct absorbers. Slicing the "torus" into these three regions will allow interesting physics of the AGN to be extracted.

1. The Quasar Standard Model

There is a "standard model" for quasars, which was put in place within a decade of the discovery of quasars [1], i.e. by 1973. This standard model consists of three elements: (1) a *supermassive* (10^6-10^9 $M_\odot$) *black hole* (SMBH, [2]), surrounded by (2) an *accretion disk* [3], with (3) a *relativistic jet* [4, 5] emerging perpendicular to the disk and originating at just a few Schwarzchild radii away from the black hole. The elements of this model successfully account for, in turn: (1) the total power output of the quasar, from the gravitational energy released by infall to near the event horizon of the black hole; (2) the maximum temperature of 50,000 K - 100,000 K of the ultraviolet (UV) continuum that dominates the luminosity, from the thermalization of the gravitational energy release in the accretion disk; and (3) the phenomenology of apparent superluminal motion, rapid variability and polarization of the radio emission in those quasars where a jet is pointing almost directly at us (the blazars). This is a pretty good list of successes, and they have held up well against decades of tests.

However, unlike the predictive power of the contemporaneous particle physics standard model [6], the quasar standard model predicts little of the rich phenomenology of quasars: (a) the various 'types' of active galaxies [7, 8]; (b) the maximally hot dust found in quasars, but not in starburst galaxies [9, 10, 11], (c) the strong X-ray emission [12, 13], and (d) all of the many atomic emission and absorption features seen in the spectra of quasars [14], not even the broad ($\sim$1% c) emission lines that led to their initial recognition as exceptional objects [15].

As a response to several of these gaps, a fourth element of the standard model has been commonly accepted since about 1985: a flattened, but large scale-height, *obscuring torus*.

2. The Obscuring Donut Torus

A flattened obscuring torus coaligned with the accretion disk is an appealingly simple addition to the standard model. It is able to explain both the variety of types of AGNs, the existence of maximally hot dust, and the bi-conical morphology often shown by the narrow emission lines of AGNs. This paradigm is known as the Unified Scheme [16, 17, 18, 19, 20]. The Unified Scheme posits obscuration by an optically and geometrically thick "torus", lying between the inner, broad ($\sim 10^4$ km s^{-1}), and outer ($\sim 10^3$ km s^{-1}), narrow, line emitting regions.

The torus is almost universally taken to be a large scale-height (H/R$\sim$1), cold structure, rich in molecular gas and dust, that is co-planar with the accretion disk [18]. As such it resembles a donut with a hole in the center, notably in the famous illustration in Urry & Padovani [20]. This specific form of non-spherical obscuring region has been called the "donut torus" [21]. The "strong" form of the Unified Scheme asserts that our orientation relative to the one jet/disk/torus axis explains *all* of the variety of AGN types.

The strong Unified Scheme cannot be 100% correct as, e.g., the incidence of X-ray obscuration is clearly more common at low luminosities [22, 23, 24, 25], requiring some modification [26].

Nontheless this scheme does explain [16, 27]: (1) the distinction between type 1 (with broad emission lines) and type 2 (no broad emission lines) AGNs, due to optical dust obscuration between the two emission regions; (2) the presence of heavy X-ray obscuration in type 2 AGNs, due mainly to gas; (3) the biconical geometry of the outer narrow line region, due to geometric collimation of the continuum; (4) the finding of polarized broad lines in otherwise purely narrow-lined type 2 AGNs [27, 28, 29], due to scattering off warm electrons above the torus; and (5) the relative space densities of type 1 and type 2 AGNs. All these achievements apply equally to radio-loud AGNs [17], and can be successfully extended to connect the luminosity functions of blazars and radio galaxies [30].

This is a long list of accomplishments, and there can be no doubt that flattened obscuring regions are important in AGNs. However, the elegant reduction of these effects to a single region cannot be sustained for both theoretical and observational reasons.

2.1. Problems with the Donut Torus

There are four theoretical challenges to the donut torus picture of AGN obscuration. While none of them is individually inescapable, together they are a significant challenge. They are:

(i) *The large, H/R$\sim$1, scale height in cold material:* Clearly a cold ($\sim$100 K) medium does not have thermal velocities of greater than the $\sim$1000 km s^{-1} of the narrow emission lines, as would be required to reach H/R$\sim$1 interior to the region emitting those lines. (This assumes that the lines have widths comparable to virial or Keplerian values.) The alternative is that the material is highly clumped and has many clouds on highly inclined orbits. A clumpy torus explains the observed AGN spectral energy distributions (SEDs) better than a continuous medium [31]. A clumpy torus is observationally supported by the emission seen outside of the bicones in NGC 4151, both in optical emission lines (the "rogue clouds" of [32]) and soft X-rays [33]. However these clouds must then collide with one another and collapse on some unclear, but probably short, timescale. A non-static, though possibly steady-state, model is needed (e.g. [34, 35]).

(ii) *The energy and photon deficit problem for the broad emission line region*: Netzer (1987) noted that the observed UV continuum could not be extrapolated in a simple, accretion disk like, way (e.g. [36]) and still provide sufficient photons to ionize the gas producing the broad emission lines. More recent inferred continuum shapes suggest even fewer photons in the extreme UV [37, 38]. Nor does the continuum provide sufficient energy input to power the broad emission lines [39]. Netzer's solution was to locate the broad emission line region directly above the accretion disk. In this way the broad emission line gas sees a more powerful continuum than an observer at a random angle. This explanation works as, above the disk, neither the geometrical cosine θ nor limb darkening [40] diminish the continuum. However, if the torus is co-planar with the disk, and covers the direct view of the disk over 80% of side-on viewing directions, then we can only see the broad emission lines almost perpendicular to the disk, so we also see the same continuum as that gas. The continuum may not be a simple extrapolation of an α-disk spectrum, but could have a second, short-wavelength, peak, e.g. due to blurred Lyman-β and HeII emitted by dense clouds at small radii Lawrence [41], which may solve these problems.

(iii) *The impedence of feedback:* As is now well known, the mass of the stellar bulge and the mass of the SMBH in a galaxy are closely tied together (see, e.g. [42]). Assuming that the two masses are causally linked (but see [43]), some feedback mechanism must keep the two growing in synchrony. There are three ways the power output of a quasar can interact with the host galaxy interstellar medium (ISM): (1) radiation; (2) broad slow winds; (3) narrow relativistic jets. While relativistic jets from their central galaxies are clearly important in clusters of galaxies [44], this mechanism is unlikely to be important in either less massive systems with lower density ISMs, or in the 10 times more numerous radio-quiet quasars. However, both radiation and slow winds will be inhibited by the presence of an exterior donut torus that blocks 80% of the sky as seen from the inner nucleus. This large covering factor increases the efficiency needed to deposit the ~5% of the total radiative quasar power needed to unbind the host galaxy ISM [45, 46] to ~25% of the *escaping* radiation. More efficient scenarios have been proposed however [47].

(iv) *The wide range of obscuring X-ray column density:* The donut torus was invoked to explain AGNs such as NGC 1068, in which the direct central radiation was completely blocked by a Compton thick obscurer [$\tau_{Compton}$ =1 corresponds to N_H=2×10^{24}cm^{-2}]. However, there are many intermediate type AGNs, with a wide range of lower X-ray column densities, and optical reddening values. Values of N_H from ~10^{21}cm^{-2} to >10^{25}cm^{-2} are often observed. This is roughly the difference between a thin sheet of paper and a brick, so that a single physical cause is not required, and may be hard to produce. Usually, in the donut torus scheme, these intermediate column densities are attributed to viewing the donut torus at a grazing angle [48]. In the case of NGC 4151 this may be correct [32]. However, these intermediate column densities are rather common [49]. The extended "atmosphere" of the donut torus must then cover ~35% of the sky seen from the inner nucleus, which is a substantial change to the picture. A clumpy wind can produce the observed N_H distribution [31].

One recurring *observational* problem is that the broad line widths seem to be dominated by orbital rotation [50, 51, 52]. That requires that most broad emission line regions are seen at a large angle to the rotation axis, but the donut torus would prevent this.

These problems can be eased if one or more of the assumptions of the strong Unified Scheme are relaxed: if the ratio of type 2:type 1 AGNs is smaller than had been thought, if the torus is not co-planar, if the X-ray and optical (or equivalently the gas and dust) obscurers are not tightly connected, or if the obscuring region can be sliced into several layers. I will discuss this last possibility next.

3. Slicing the Torus

In the years since the donut torus was proposed there has been much detailed observational work on each of the features that the scheme was invented to explain. One consequence is accumulating evidence that obscuration in AGNs occurs on both smaller and larger scales than that of the hot dust limited radius of the donut torus alone. Here I slice the obscuration into three scales (Figure 1).

3.1. Smaller Scale Obscuration

Large amplitude changes in absorbing X-ray column density have been seen on briefer and briefer time intervals. There are now numerous examples of large changes (ΔN_H >10^{23-24}cm^{-2}) in a few

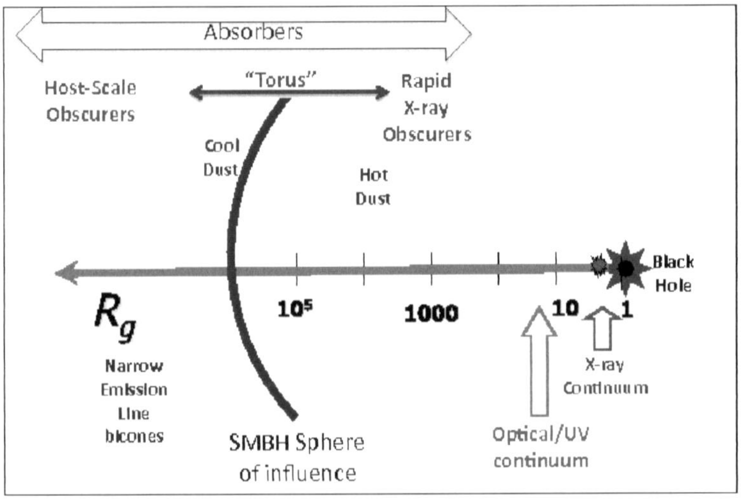

Figure 1. Scales of obscuration in quasars and AGNs. The horizontal scale is in log(R_g).

days or hours (e.g. [53, 54]). The most dramatic example is NGC 1365, in which the X-ray source underwent a total eclipse by a Compton-thick cloud within two days, and emerged within another two days [55]. For any velocity comparable to the Keplerian velocity for the eclipsing clouds, these events all imply that they lie at about the same distance as the broad emission line gas ($r < 10^{3-4}\ R_S$) and have comparably high densities ($n_e > 10^{9-10}\mathrm{cm}^{-3}$). The obvious conclusion is that the eclipsers are discrete clouds of broad emission line gas. Further investigations seem to be revealing unexpected features in these clouds with surprising implications [56, 57].

For our purpose here, these eclipses show that obscuration, at least by gas, occurs well within the dust sublimation radius. The observed low dust-to-gas ratios, from comparisons of optical reddening with X-ray absorption [58, 59], may be explained by a mix of inner, dust-free, and outer, dusty, absorbers [60].

As the UV continuum source has ∼10 times the radius and so ∼100 times the area of the X-ray continuum, it is less likely that similar large amplitude eclipses could be seen at UV wavelengths, which makes it hard to tell if the clouds are dusty, as suggested by Czerny & Hryniewicz [61]. (In principle transiting exoplanet techniques could look for these partial eclipses.)

3.2. Larger Scale Obscuration

Since 1980 there have been many papers demonstrating a clear connection between the obscuration of the AGN and the inclination angle of the host galaxy [62, 16, 63, 64, 65, 66, 67, 68]. Evidently there is some obscuring region dominated by the host galaxy dynamics, and so outside the SMBH sphere of influence, though the radial scale of this obscurer is uncertain. Moreover, as the radio jets of these galaxies lie at random angles to the host galaxy disk plane, within a very wide cone [69, 70], the obscurer, the jet and the accretion disk cannot all share the same axis.

Hubble imaging of type 1 and type 2 hosts also clearly shows that there are many more dust lanes crossing the nuclei of type 2 AGNs than of type 1s on a several 100 pc scale [22]. This obscuration lies beyond the region producing the infrared emission from these AGNs. As these type 2 hosts are of later morphological type than the type 1s, they undermine the strong Unified Scheme.

A different argument against the donut torus, coming from properties on similarly large scales of up to ~1 kpc, is based on the bi-conical regions marked out by the narrow emission lines. A series of papers using Hubble STIS long slit spectroscopy [71, 72] has found that these bi-cones are not made up of pre-existing ISM illuminated by the AGN, as in the donut torus model, but by outflowing hollow cones with wide opening angles. Hence the bi-cone shapes are often not formed by a collimated incident radiation pattern, but are matter-bounded, i.e. limited by the shape of the wind coming from the nucleus. Moreover, the wide angles derived for the wind bi-cones (~50° [73]) imply that any torus covers a more limited solid angle than in the donut torus model. (These bi-cones also show feedback in action, [74]).

Larger scale obscuration may be due to matter streaming toward the accretion disk, as observed in H_2 emission [75, 76], and predicted in some galaxy merger models [77]. Improved imaging in molecular lines in the mm- and sub-mm bands, primarily of CO, have begun to image obscuring structures at the ~100 pc radius scale. In the prototypical type 2 AGN, NGC 1068 (for which the donut torus was largely originally created [27]), a Compton thick warped disk with an inner radius of ~70 pc curls up to just cross our line of sight to the nucleus [78]. This molecular disk appears to be disconnected from any smaller structure stretching down to the dust sublimation radius. ALMA will revolutionize this field.

3.3. Hot Dust Scale Obscuration

Nonetheless, a substantial fraction of the AGN nuclear power is absorbed and re-radiated by hot dust in most AGNs [79, 80]. The peak of the AGN dust emission lies in the 10 μm to 20 μm range, implying temperatures of order 200 K, and reaches shortward to ~1 μ, implying T~1800 K, comparable to the maximum dust sublimation temperature. For the simple black body case [24] this region spans a factor of ~1000 in radius, with most of the emitting area, and presumably most of the mass, at the larger radii.

Indeed, direct mid-IR imaging at 12μm with VLT [81, 77, 81], shows that the mid-IR emission becomes tightly correlated with the AGN continuum only within ~600 R_{sub}, comparable with the expected scale. More extended 12μm emission dilutes the correlation and so presumably arises from star formation.

The dust emitting radius [$R_{dust}(T)$] can be expressed in terms of the Eddington ratio of the accretion rate onto the black hole ($L/0.1L_{Edd} = \lambda_{0.1}$), and the Schwarzchild radius (R_g): $R_{dust}(T) = 1.5 \times 10^6 \lambda_{0.1}\ L_{44}^{0.5}\ T_{200K}^{-2.8} R_g$ [24][1]. Radii of about $10^6 R_g$ apply to the 200 K dust, while the hottest dust, at T~1800 K, has R(hot)~3000 R_g.

It is instructive to compare these dust emitting radii with size of the SMBH sphere of influence, $R_{BHinfl} = G\ M_{BH}/\sigma^2$ [42], so that $R_{BHinfl} = 10^6 \sigma_{200}\ R_g$. The main AGN dust emission thus spans the region in which the host galaxy potential takes over from the black hole potential (Figure 1). We might expect interesting things to happen as this boundary is crossed. A stable H/R~1 torus is unlikely to be continuous across this region.

The most discussed alternative to explain this dust emission is a dusty wind off the accretion disk. The wind may be driven by magneto-centrifugal forces [83, 84, 21], by external irradiation of the disk making it unbound [35], or by star formation in the gravitationally unstable part of the disk [85]. A steady state accretion model [34] can also sustain a large scale height torus, but only at locally super-Eddington accretion rates.

A warped disk [79] can have a large covering factor without needing a large scale height, and so is an alternative to a planar torus. This form has gained new relevance as the means by which a black hole is fed and spun up has been investigated in more detail. Individual accretion events with random angular momentum vectors will naturally produce a low spin black hole [86]. Lawrence & Elvis [24] point out that a warped disk will result, and will produce a type 1

[1] Following [82], for τ=3, and a typical dust grain size $a = 0.05\mu m$. Here L_{44} is the AGN UV continuum luminosity in units of 10^{44} erg s^{-1}, and T_{200K} is the dust temperature in units of 200 K.

or type 2 appearance depending on orientation, similar to that from a donut torus, but with observable differences. Small angle warps are seen in some AGN megamasers on parsec scales [87], and are implied in the Galactic Center [88].

A simple warped disk, with no rotation of the line of nodes, and random accretion directions, predicts a 1:1 type 2:type 1 ratio [24]. These authors then show that this ratio is consistent with observations, if LINERS are excluded, for AGN samples selected by reasonably isotropic indicators (radio, [OIII], mid-IR fluxes). The exception is samples selected in X-rays, which show a higher fraction of X-ray obscured objects. This suggests an extra, dust-free, absorber of X-rays, probably at smaller radii. These absorbers could be the broad emission line clouds noted above (§3.1).

A 1:1 ratio, though, also eases several of the objections to the donut torus. The resulting wider opening angles ease the photon and energy budget deficits, allow wide angle biconical winds to escape, and so greater feedback, and require a more physically reasonable smaller scale height torus. The solution is still open.

Another prediction of the warped disk model is that the radio jets and the bi-cones will not generally be aligned. Several anecdotal examples suggest that this is the case [24], but a general survey is needed. The canonical Hubble snapshot survey [70] needs to be repeated to greater depth, perhaps with AO on 8-meter class telescopes. Infrared interferometers are now beginning to resolve this region as well as, or better than, Hubble [89, 90, 91], so we may be able to see warped AGN dust disks directly in a few years. Eventually, imaging reverberation mapping will let us image the broad line region motions too, in all three spatial and velocity components [92].

4. Conclusions

A complete physical model for quasars is developing, but has some way to go: The atomic emission and absorption features all have promising explanations in terms of winds or failed winds subject to radiation pressure [57]. Although a great deal is know about the behavior of the X-ray source, physically it is normally ascribed to a "hot corona" which is not understood. I anticipate that accurate measurements of the temperature of the X-ray emission by NuSTAR [93] will begin to unpick this knot.

The origin of the AGN powered hot dust and of the various types of AGN is becoming more refined, but also, perhaps, more interesting and tractable. The view of AGN obscuration as being due to a single donut torus does not capture the complexity now realized to exist within AGNs. While acknowledging that the illustration from Urry & Padovani is graphic and immediate, I would urge that it should not be our default picture, as this form tends to become imprinted, and is then hard to get beyond.

In the emerging, more complex, picture - as in the donut torus model - the various types of AGNs continue to be explained by obscuration. However, at least three separate obscuring regions must exist in quasars and AGNs: (1) an inner, probably dust-free, region producing rapid X-ray eclipses, most likely due to broad emission line clouds; (2) an intermediate scale dusty region that reprocesses much of the AGN luminosity into thermal dust emission, including the hottest dust. This region spans the boundary between the black hole- and host-dominated gravitational potentials, and is perhaps in the form of a warped disk; (3) outer regions on 0.1-1 kpc scales, connected to the host disk and/or dust lanes.

By separating out the effects of these different obscurers we are starting to learn much more about the inner structure of AGNs, and their feeding on sub-kiloparsec scales .

Acknowledgments

The author would like to the thank his many collaborators in this area over the years, especially Andy Lawrence and Guido Risaliti, Bozena Czerny, and Gordon Richards, and is grateful to the scientific organizing committee for the opportunity to present this work at AHAR 2011.

References

[1] Schmidt M 1963 *Nature* **197** 1040
[2] Lynden-Bell D 1969 *Nature* **223** 690–694
[3] Shakura N I and Sunyaev R A 1973 *Astronomy & Astrophysics* **24** 337–355
[4] Rees M J 1966 *Nature* **211** 468–470
[5] Blandford R D and Rees M J 1974 *Monthly Notices of the Royal Astronomical Society* **169** 395–415
[6] Weinberg S 1967 *Physical Review Letters* **19** 1264–1266
[7] Khachikian E Y and Weedman D W 1974 *Astrophysical Journal* **192** 581–589
[8] Osterbrock D E 1981 *Astrophysical Journal* **249** 462–470
[9] Rieke G H 1978 *Astrophysical Journal* **226** 550–558
[10] Elvis M and al 1994 *Astrophysical Journalap* **95** 1–68
[11] Glikman E, Helfand D J and White R L 2006 *Astrophysical Journal* **640** 579–591
[12] Elvis M, Maccacaro T, Wilson A S, Ward M J, Penston M V, Fosbury R A E and Perola G C 1978 *Monthly Notices of the Royal Astronomical Society* **183** 129–157
[13] Tananbaum H and al 1979 *Astrophysical Journal Letters* **234** L9–L13
[14] Peterson B M 1997 *An Introduction to Active Galactic Nuclei* (Cambridge, Cambridge University Press)
[15] Seyfert C K 1943 *Astrophysical Journal* **97** 28
[16] Lawrence A and Elvis M 1982 *Astrophysical Journal* **256** 410–426
[17] Barthel P D 1989 *Astrophysical Journal* **336** 606–611
[18] Krolik J H and Begelman M C 1988 *Astrophysical Journal* **329** 702–711
[19] Antonucci R 1993 *Annual Reviews of Astronomy and Astrophysics* **31** 473–521
[20] Urry C M and Padovani P 1995 *Publications of the Astronomical Society of the Pacific* **107** 803
[21] Elitzur M and Shlosman I 2006 *Astrophysical Journal Letters* **648** L101–L104
[22] Malkan M A, Gorjian V and Tam R 1998 *Astrophysical Journals* **117** 25
[23] Polletta M, Weedman D, Hönig S, Lonsdale C J, Smith H E and Houck J 2008 *Astrophysical Journal* **675** 960–984
[24] Lawrence A and Elvis M 2010 *Astrophysical Journal* **714** 561–570
[25] Burlon D, Ajello M, Greiner J, Comastri A, Merloni A and Gehrels N 2011 *Astrophysical Journal* **728** 58
[26] Lawrence A 1991 *Monthly Notices of the Royal Astronomical Society* **252** 586–592
[27] Antonucci R R J and Miller J S 1985 *Astrophysical Journal* **297** 621–632
[28] Miller J S and Goodrich R W 1990 *Astrophysical Journal* **355** 456–467
[29] Tran H D 1995 *Astrophysical Journal* **440** 597
[30] Urry C M, Padovani P and Stickel M 1991 *Astrophysical Journal* **382** 501–507
[31] Nenkova M, Sirocky M M, Nikutta R, Ivezić Ž and Elitzur M 2008 *Astrophysical Journal* **685** 160–180
[32] Das V and al 2005 *Astronomical Journal* **130** 945–956
[33] Wang J and al 2011 *Astrophysical Journal* **742** 23
[34] Vollmer B, Beckert T and Duschl W J 2004 *Astronomy & Astrophysics* **413** 949–957
[35] Thompson T A, Quataert E and Murray N 2005 *Astrophysical Journal* **630** 167–185
[36] Mathews W G and Ferland G J 1987 *Astrophysical Journal* **323** 456–467
[37] Laor A, Fiore F, Elvis M, Wilkes B J and McDowell J C 1997 *Astrophysical Journal* **477** 93
[38] Zheng W, Kriss G A, Telfer R C, Grimes J P and Davidsen A F 1997 *Astrophysical Journal* **475** 469
[39] Binette L, Fosbury R A and Parker D 1993 *Publications of the Astronomical Society of the Pacific* **105** 1150–1163
[40] Sunyaev R A and Titarchuk L G 1985 *Astronomy & Astrophysics* **143** 374–388
[41] Lawrence A 2011 *ArXiv e-prints* (*Preprint* 1110.0854)
[42] Ferrarese L and Ford H 2005 *Space Science Reviews* **116** 523–624
[43] Jahnke K and Macciò A V 2011 *Astrophysical Journal* **734** 92
[44] McNamara B R and Nulsen P E J 2007 *Annual Reviews of Astronomy and Astrophysics* **45** 117–175
[45] Di Matteo T, Springel V and Hernquist L 2005 *Nature* **433** 604–607
[46] Hopkins P F, Hernquist L, Cox T J, Di Matteo T, Robertson B and Springel V 2006 *Astrophysical Journal* **163** 1–49
[47] Hopkins P F and Elvis M 2010 *Monthly Notices of the Royal Astronomical Society* **401** 7–14
[48] Osterbrock D E and Shaw R A 1988 *Astrophysical Journal* **327** 89–98
[49] Risaliti G, Maiolino R and Salvati M 1999 *Astrophysical Journal* **522** 157–164
[50] Wills B J and Browne I W A 1986 *Astrophysical Journal* **302** 56–63
[51] Smith J E, Robinson A, Young S, Axon D J and Corbett E A 2005 *Monthly Notices of the Royal Astronomical Society* **359** 846–864
[52] Young S, Axon D J, Robinson A, Hough J H and Smith J E 2007 *Nature* **450** 74–76
[53] Elvis M, Risaliti G, Nicastro F, Miller J M, Fiore F and Puccetti S 2004 *Astrophysical Journal Letters* **615**

L25–L28
[54] Puccetti S, Fiore F, Risaliti G, Capalbi M, Elvis M and Nicastro F 2007 *Monthly Notices of the Royal Astronomical Society* **377** 607–616
[55] Risaliti G, Elvis M, Fabbiano G, Baldi A, Zezas A and Salvati M 2007 *Astrophysical Journal* **659** L111–L114
[56] Maiolino R and al 2010 *Astronomy & Astrophysics* **517** A47
[57] Elvis M 2012 *ArXiv e-prints* (*Preprint* 1201.3520)
[58] Maccacaro T, Perola G C and Elvis M 1982 *Astrophysical Journal* **257** 47–55
[59] Maiolino R, Marconi A, Salvati M, Risaliti G, Severgnini P, Oliva E, La Franca F and Vanzi L 2001 *Astronomy & Astrophysics* **365** 28–36
[60] Risaliti G, Elvis M and Nicastro F 2002 *Astrophysical Journal* **571** 234–246
[61] Czerny B and Hryniewicz K 2011 *Astronomy & Astrophysics* **525** L8
[62] Keel W C 1980 *Astronomical Journal* **85** 198–203
[63] de Zotti G and Gaskell C M 1985 *Astronomy & Astrophysics* **147** 1–12
[64] Thompson I B and Martin P G 1988 *Astrophysical Journal* **330** 121–129
[65] Kirhakos S D and Steiner J E 1990 *Astronomical Journal* **99** 1722–1739
[66] Kirhakos S D and Steiner J E 1990 *Astronomical Journal* **99** 1435–1441
[67] Simcoe R, McLeod K K, Schachter J and Elvis M 1997 *Astrophysical Journal* **489** 615
[68] Rigby J R, Rieke G H, Donley J L, Alonso-Herrero A and Pérez-González P G 2006 *Astrophysical Journal* **645** 115–133
[69] Kinney A L, Schmitt H R, Clarke C J, Pringle J E, Ulvestad J S and Antonucci R R J 2000 *Astrophysical Journal* **537** 152–177
[70] Schmitt H R, Donley J L, Antonucci R R J, Hutchings J B and Kinney A L 2003 *Astrophysical Journals* **148** 327–352
[71] Crenshaw D M and Kraemer S B 2000 *Astrophysical Journal Letters* **532** L101–L104
[72] Fischer T C, Crenshaw D M, Kraemer S B, Schmitt H R, Mushotsky R F and Dunn J P 2011 *Astrophysical Journal* **727** 71
[73] Fischer T C, Crenshaw D M, Kraemer S B, Schmitt H R and Trippe M L 2010 *Astronomical Journal* **140** 577–583
[74] Wang J, Fabbiano G, Elvis M, Risaliti G, Karovska M, Zezas A, Mundell C G, Dumas G and Schinnerer E 2011 *Astrophysical Journal* **742** 23
[75] Storchi-Bergmann T, Lopes R D S, McGregor P J, Riffel R A, Beck T and Martini P 2010 *Monthly Notices of the Royal Astronomical Society* **402** 819–835
[76] Riffel R A and Storchi-Bergmann T 2011 *Monthly Notices of the Royal Astronomical Society* **417** 2752–2769
[77] Horst H, Duschl W J, Gandhi P and Smette A 2009 *Astronomy & Astrophysics* **495** 137–146
[78] Schinnerer E, Eckart A, Tacconi L J, Genzel R and Downes D 2000 *Astrophysical Journal* **533** 850–868
[79] Sanders D B, Phinney E S, Neugebauer G, Soifer B T and Matthews K 1989 *Astrophysical Journal* **347** 29–51
[80] Elvis M, Wilkes B J, McDowell J C, Green R F, Bechtold J, Willner S P, Oey M S, Polomski E and Cutri R 1994 *Astrophysical Journals* **95** 1–68
[81] Horst H, Gandhi P, Smette A and Duschl W J 2008 *Astronomy & Astrophysics* **479** 389–396
[82] Barvainis R 1987 *Astrophysical Journal* **320** 537–544
[83] Blandford R D and Payne D G 1982 *Monthly Notices of the Royal Astronomical Society* **199** 883–903
[84] Konigl A and Kartje J F 1994 *Astrophysical Journal* **434** 446–467
[85] Collin S and Zahn J P 1999 *Astronomy & Astrophysics* **344** 433–449
[86] Volonteri M, Sikora M and Lasota J P 2007 *Astrophysical Journal* **667** 704–713
[87] Greenhill L J, Booth R S, Ellingsen S P, Herrnstein J R, Jauncey D L, McCulloch P M, Moran J M, Norris R P, Reynolds J E and Tzioumis A K 2003 *Astrophysical Journal* **590** 162–173
[88] Wardle M and Yusef-Zadeh F 2008 *Astrophysical Journal Letters* **683** L37–L40
[89] Tristram K R W and al 2009 *Astronomy & Astrophysics* **502** 67–84
[90] Kishimoto M, Hönig S F, Antonucci R, Barvainis R, Kotani T, Tristram K R W, Weigelt G and Levin K 2011 *Astronomy & Astrophysics* **527** A121 (*Preprint* 1012.5359)
[91] Burtscher L, Jaffe W, Raban D, Meisenheimer K, Tristram K R W and Röttgering H 2009 *Astrophysical Journal Letters* **705** L53–L57
[92] Elvis M and Karovska M 2002 *Astrophysical Journal Letters* **581** L67–L70
[93] Harrison F A and al 2010 *Society of Photo-Optical Instrumentation Engineers (SPIE) Conference Series* (*Society of Photo-Optical Instrumentation Engineers (SPIE) Conference Series* vol 7732)

Astronomy at High Angular Resolution 2011 IOP Publishing
Journal of Physics: Conference Series **372** (2012) 012033 doi:10.1088/1742-6596/372/1/012033

Probing the innermost dusty structure in AGN with mid-IR and near-IR interferometers

M. Kishimoto[1], S. F. Hönig[2], R. Antonucci[2], R. Barvainis[3], T. Kotani[4], F. Millour[5], K. R. W. Tristram[1], G. Weigelt[1]

[1] Max-Planck-Institut für Radioastronomie, Auf dem Hügel 69, 53121 Bonn, Germany
[2] Physics Department, University of California, Santa Barbara, CA 93106, USA.
[3] National Science Foundation, 4301 Wilson Boulevard, Arlington, VA 22230, USA.
[4] ISAS, JAXA, 3-1-1 Yoshinodai, Sagamihara, Kanagawa, 229-8510 Japan.
[5] Observatoire de la Côte d Azur, Departement FIZEAU, Boulevard de l'Observatoire, BP 4229, 06304, Nice Cedex 4, France

E-mail: mk@mpifr-bonn.mpg.de

Abstract. With mid-IR and near-IR long-baseline interferometers, we are now mapping the radial distribution of the dusty accreting material in AGNs at sub-pc scales. We currently focus on Type 1 AGNs, where the innermost region is unobscured and its intrinsic structure can be studied directly. As a first systematic study of Type 1s, we obtained mid-/near-IR data for small samples over ~3–4 orders of magnitudes in UV luminosity L of the central engine. Here we effectively trace the structure by observing dust grains that are radiatively heated by the central engine. Consistent with a naive expectation for such dust grains, the dust sublimation radius $R_{\rm in}$ is in fact empirically known to be scaling with $L^{1/2}$ from the near-IR reverberation measurements, and this is also supported by our near-IR interferometry. Utilizing this empirical relationship, we normalize the radial extent by $R_{\rm in}$ and eliminate the simple $L^{1/2}$ scaling for a direct comparison over the samples. We then find that, in the mid-IR, the overall size in units of $R_{\rm in}$ seems to become more compact in higher luminosity sources. More specifically, the mid-IR brightness distribution is rather well described by a power-law, and this power-law becomes steeper in higher luminosity objects. The near-IR flux does not seem to be a simple inward extrapolation of the mid-IR power-law component toward shorter wavelengths, but it rather comes from a little distinct brightness concentration at the inner rim region of the dust distribution. Its structure is not well constrained yet, but there is tentative evidence that this inner near-IR-emitting structure has a steeper radial distribution in jet-launching objects. All these should be scrutinized with further observations.

1. Introduction

Using long-baseline interferometers in both the mid-IR and near-IR, we are probing the innermost dusty structure in the putative torus in active galactic nuclei (AGNs) at unprecedented spatial resolutions in the IR. One of the first goals of our study has been to directly map the *radial distribution* of the accreting material. We currently focus on Type 1 AGNs, where our line of sight is thought to be more or less face-on and the innermost region can directly be studied without significant effects from inclination and obscuration. We aim to summarize below the status of our systematic Type 1 study in both the mid-IR and near-IR.

2. Removing the simple $L^{1/2}$ dependency

Let us start with addressing a simple, but naive, luminosity scaling that should be carefully considered in our study. The mid- and near-IR radiation in AGNs is generally thought to be the thermal emission from dust grains that are radiatively heated by the UV/optical emission of the central engine. Effectively, we are using these dust grains to trace the inner structure. When the temperature of the observed dust grains is simply determined by the central engine's *direct* illumination and the dust thermal re-emission, the radial distance of the dust grains at a given temperature will be proportional to $L^{1/2}$. Here we assume that the incident spectral shape does not change with the UV luminosity L, which we discuss more below. Then our observations at a fixed wavelength would find the size of the structure to also scale with $L^{1/2}$, since they would be sensing the dust grains of the same temperature range.

This simple $L^{1/2}$ scaling may not realize due to various quantities and properties potentially changing with luminosity. However, at least for the innermost dust sublimation region, this seems to hold – the dust sublimation radius $R_{\rm in}$ as probed by the near-IR reverberation measurements is known to be proportional to $L^{1/2}$ [1]. Here the relation is found for optical luminosity, rather than the more relevant UV luminosity, implying the same UV–optical spectral shape over the range of luminosities at least approximately. Therefore, we assume a generic spectral shape and define the UV luminosity L as a scaled optical luminosity. This innermost $L^{1/2}$ scaling is in fact supported also by the results from our near-IR interferometry, although there is a distinction between our interferometric size measurements and the reverberation measurements, as we discuss in Sect.4. Based on this empirical relationship, which matches the naive expectation, it would be physically insightful if we normalize the radial extent in our study by the dust sublimation radius $R_{\rm in}$ given by the $L^{1/2}$ fit to the reverberation measurements. This will remove the simple $L^{1/2}$ scaling and let us investigate what is left behind it.

Each interferometric measurement gives a visibility at a certain spatial frequency, or its reciprocal, spatial wavelength, which roughly corresponds to the spatial scale probed by that particular configuration of the interferometer (see more in Sect.3). We normalize this probed spatial scale by $R_{\rm in}$ for each observed object (in this case, spatial frequency will have the units of cycles per $R_{\rm in}$). Differences in distance, as well as in luminosity, will be folded into the different angular sizes of $R_{\rm in}$, where visibility curves as a function of the spatial wavelength in units of $R_{\rm in}$ for different objects can be compared uniformly [2]. The normalization gives a physical meaning to the probed spatial scale, which makes the interferometric observables more interpretable physically and directly. This is quite important since our *uv* converage is normally poor and we still have to deal with direct observables in Fourier space (i.e. visibility), rather than quantities converted to those in image space.

We note that the dust sublimation radii from the reverberation measurements are systematically smaller by a factor of $\sim$3 [3] than that expected for a representative property of dust grains in the Galactic interstellar medium for each given L [4]. We have specifically discussed this issue in [3], including the case for large grains as well as anisotropic radiation and absorption. The case for large grains in the nuclear region has long been suggested on different grounds (e.g. [5; 6]), while the effect of anisotropy is extensively studied recently by [7].

3. Results: mid-IR

Figure 1*a* is such a uniform comparison plot. It shows the visibility amplitudes measured with MIDI/VLTI for 6 Type 1 objects at the rest-frame wavelength range of 10–12 μm as a function of normalized spatial frequency in log scale (or normalized spatial wavelength; see upper axis). Figure 1*b* is for the reader who prefers to see the same plot in linear scale.

First, *if* the structure scales only with $L^{1/2}$, i.e. if we see intrinsically the same distribution (only with the radius-temperature scaling discussed above), the visibility curves should look all the same in the normalized units. However, this is *not* the case. Visibilities at a given normalized

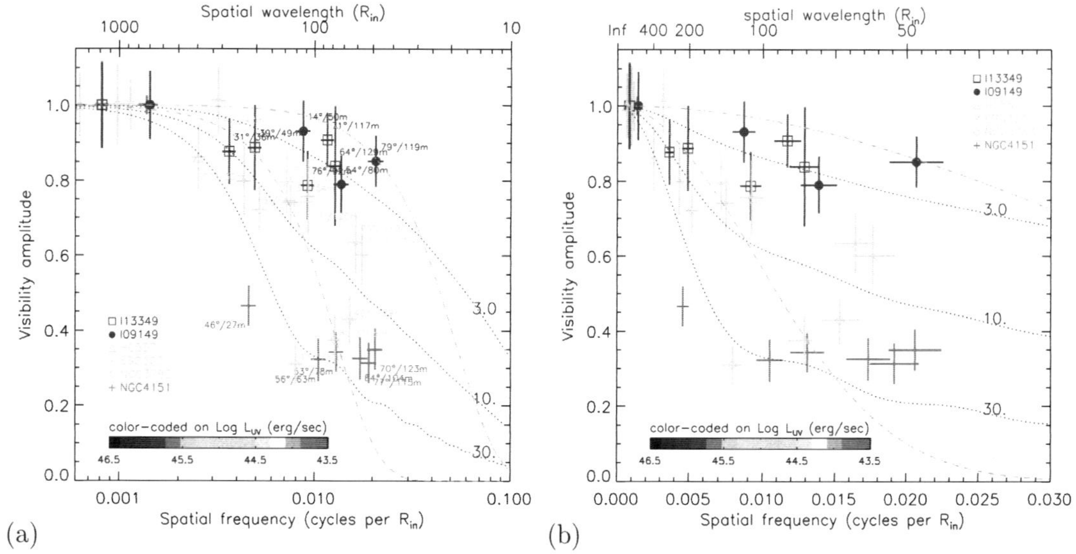

Figure 1. (a) Observed visibility amplitudes for 6 Type 1 AGNs, averaged over 10 to 12 μm in the rest frame of each object, shown as a function of normalized spatial frequency (or its reciprocal, normalized spatial wavelength shown in the upper axis). The data are plotted using different symbols for different objects (see legend), and color-coded using the UV luminosity of the central engine in each object (see color bar), labeled with the position angle and length of the projected baseline. Dotted curves are the visibility functions for power-law brightness distributions with designated half-light radii in units of $R_{\rm in}$. Dashed and dot-dashed gray curves are for Gaussians with HWHM 20 and 5 $R_{\rm in}$, respectively. (b) The same plot as (a) but with x-axis in linear scale. Both panels are from [8].

spatial frequency, or at a given spatial scale probed, are systematically different: within the small sample, which nevertheless spans over ~2.5 orders of magnitudes in UV luminosity, the overall mid-IR emission size in higher luminosity objects (shown in bluer colors) seems more compact than that in lower luminosity ones (redder colors), when seen in units of $R_{\rm in}$.

Here we need to be careful when we talk about the size. In Fig.1*a*, we also have plotted visibility curves of Gaussians with HWHM of 20 and 5 $R_{\rm in}$. In this log-scale plot, the visibility curve for a Gaussian with another HWHM will simply be shifted in the horizontal direction with no change in shape, with the shift amount proportional to the HWHM ratio. This conveniently tells us which visibility data points probe approximately which spatial scales[1], but this also immediately tells us that a single Gaussian seems quite inadequate for describing the mid-IR visibility curve for each object. This is also evident in Fig.1*b* where none of the objects seem to follow Gaussian-like visibility curves.

Rather, the observed visibility curves show that the mid-IR flux originates from a much wider range of spatial scales in each object. The dotted curves in both panels of Fig.1 are for power-law brightness distributions with different steepnesses, where the steepness is characterized as a half-light radius, within which the half of the total integrated light at a given wavelength is contained (with the inner cut-off at radius $R_{\rm in}$; the outer radius is set to $100R_{\rm in}$ here). Each object seems to follow a power-law brightness distribution of a different steepness, with higher luminosity objects looking steeper, while lower luminosity ones being more extended. Furthermore, since the radial

[1] For instance, the HWHM=$20R_{\rm in}$ Gaussian gives visibility V=0.5 at spatial wavelength Λ=4.5$\times$HWHM=$90R_{\rm in}$ (more exactly, 4.5 means $\pi/\ln 2$), and conversely, an observation at Λ=$90R_{\rm in}$ would roughly probe a spatial scale of radius $\sim R_{\rm V0.5} \equiv \Lambda/4.5$=$20R_{\rm in}$. This quantity $R_{\rm V0.5}$ would be called a half-visibility radius [8].

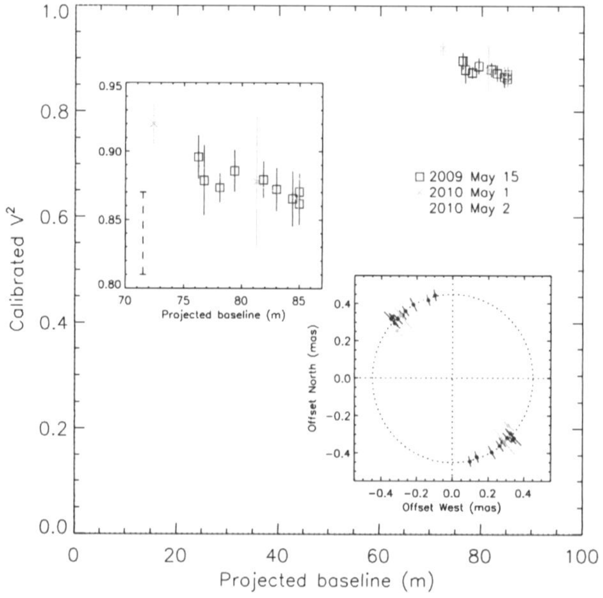

Figure 2. Squared visibility V^2 of NGC4151 as a function of projected baseline lengths observed in two epochs (see legend) from [10; 12]. The possible, perhaps conservative, systematic uncertainty in the V^2 calibration of up to ~0.03 is shown as dashed bar in the left inset (see sect.3.2 of [12]). In the right inset, ring radii corresponding to each data point are plotted along the position angle of each projected baseline. The dotted line in each panel corresponds to the best-fit thin-ring model with a radius of 0.45 mas for our May 2009 data.

temperature run of the heated (illuminated) dust grains as a function of the normalized radius is expected to be roughly the same for these face-on targets, the steepness of the brightness distribution would correspond to that of the surface density of these heated dust grains. Thus we can infer that the higher luminosity objects have steeper dust density distributions.

4. Results: near-IR

In the near-IR, the first measurement of the K-band (2.2 μm) visibility was made for the brightest Type 1 AGN NGC4151 with the Keck interferometer (KI), which showed squared visibility $V^2 \sim 0.84$ at a projected baseline b of 83 m [9]. The result was recently confirmed by us [10] and also by [11]. Furthermore, we measured the K-band visibility for in total 8 Type 1 AGNs [10; 12]. All of the targets showed high visibilities $V^2 \sim 0.8 - 0.9$ at $b \sim 80$ m. We interpret this as an indication of the partial resolution of the dust sublimation region. First, we have marginally detected a decrease of visibility over increasing projected baselines for NGC4151 (Fig.2) [10; 12], which would be robust evidence that we are resolving a structure. Secondly, from this visibility curve, we derived a thin-ring radius for NGC4151, and also for the other AGNs from the observed visibilities. These interferometric ring radii turned out to be quite comparable to the dust sublimation radii from the near-IR reverberation measurements (Fig.3a) [10; 12], which strongly suggests that we are resolving the dust sublimation region.

A further close look at the ring radii and reverberation radii shows that the former are slightly and systematically larger than the latter on average. This is not surprising since the reverberation measurements are known to be most sensitive to the smallest responding radii [13], meaning that they are most probably probing a radius very close to the inner boundary of the dust distribution (thus adequately adopted as the dust sublimation radius). On the other hand, the interferometric ring radii are rather brightness-weighted, effective radii of the overall near-IR brightness distribution. Therefore, we can postulate that the ratio of the ring radius to the reverberation radius $R_{\rm in}$, i.e. ring radius in units of $R_{\rm in}$, can approximately describe the steepness of the structure, in the sense that the ratio is closer to unity when the structure is steeper, while much larger than unity when it is more extended (Fig.3a).

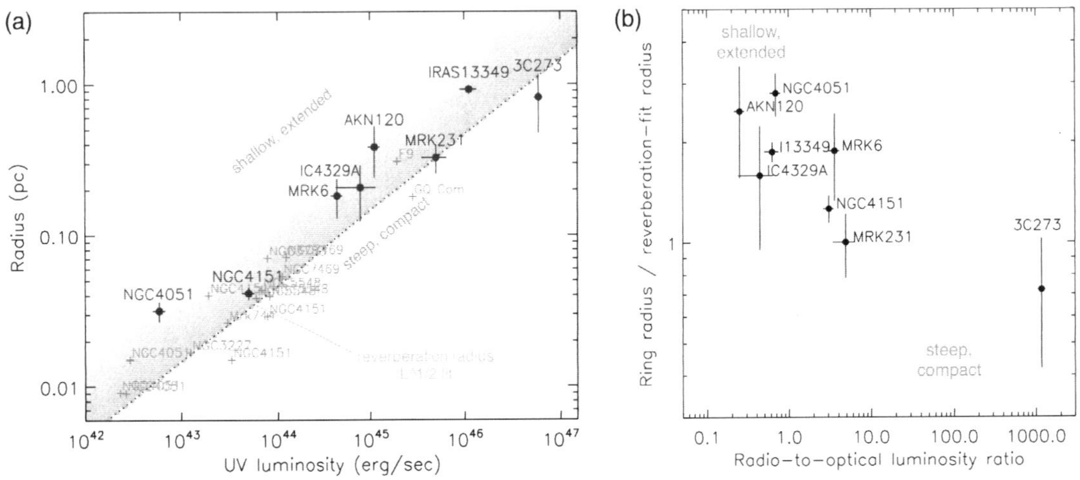

Figure 3. (a) Interferometric thin-ring radii at 2.2 μm for 8 Type 1 AGNs (filled circles [10; 12]) as a function of UV luminosity L. The near-IR reverberation radii are shown as gray plus signs, and the dotted line is the $L^{1/2}$ fit ([1]), which defines the dust sublimation radius $R_{\rm in}$ we adopt here. The ratio of the ring radius to $R_{\rm in}$ would indicate the steepness of the brightness distribution, and thus the darker and lighter shades correspond to steeper and shallower structures, respectively. (b) The ring radii in units of $R_{\rm in}$ are plotted against radio (5 GHz) to optical luminosity ratio [12].

This description of course has a limitation. As the power-law becomes more shallow and extended, the single ring radius description becomes more inadequate, as we have seen in the mid-IR visibility curves[2]. Here, the underlying assumption is that the near-IR brightness distribution is steep enough to be approximately described by a single ring radius, independent of baseline lengths (though still at relatively low spatial frequencies). Currently, the near-IR structure is not well constrained yet, and this should be followed up with multiple, longer baseline observations.

Under this approximate description of the inner radial structure, do we see any relationship between the inner radial structure and the central engine? In fact, there is tentative evidence that jet-launching objects tend to have a steeper structure (Fig.3*b*) [12]. However, the statistical significance is not high and this should be investigated further with a larger sample.

5. Correlated flux as the spectrum of an unresolved structure

We discussed above the mid-IR and near-IR emission regions separately. Obviously, we do want to know whether these two are nicely interconnected. Can we think these two are coming from a single component? The answer does not seem to be affirmative, or at least not in a very obvious way.

Certainly we do see that the structure looks more and more compact toward shorter wavelengths. Figure 4 shows the data for NGC4151, and some of their aspects are shared by other objects. Figure 4*a* shows visibilities both in the mid- and near-IR, where the color of each data point represents the observing wavelength as indicated by the color bar at the top. Toward shorter wavelengths, the overall visibility becomes larger quite quickly, and thus the corresponding structure becomes smaller. However, when we look into the total flux spectrum, shown in black in Figure 4*b*, it can be easily seen that if we try to extrapolate the mid-IR power-law component toward shorter wavelengths, its color is too red to consistently explain

[2] Note that we do not differentiate a ring from a Gaussian since they have essentially the same visibility curve at low spatial frequencies, though a ring model is more physically motivated here.

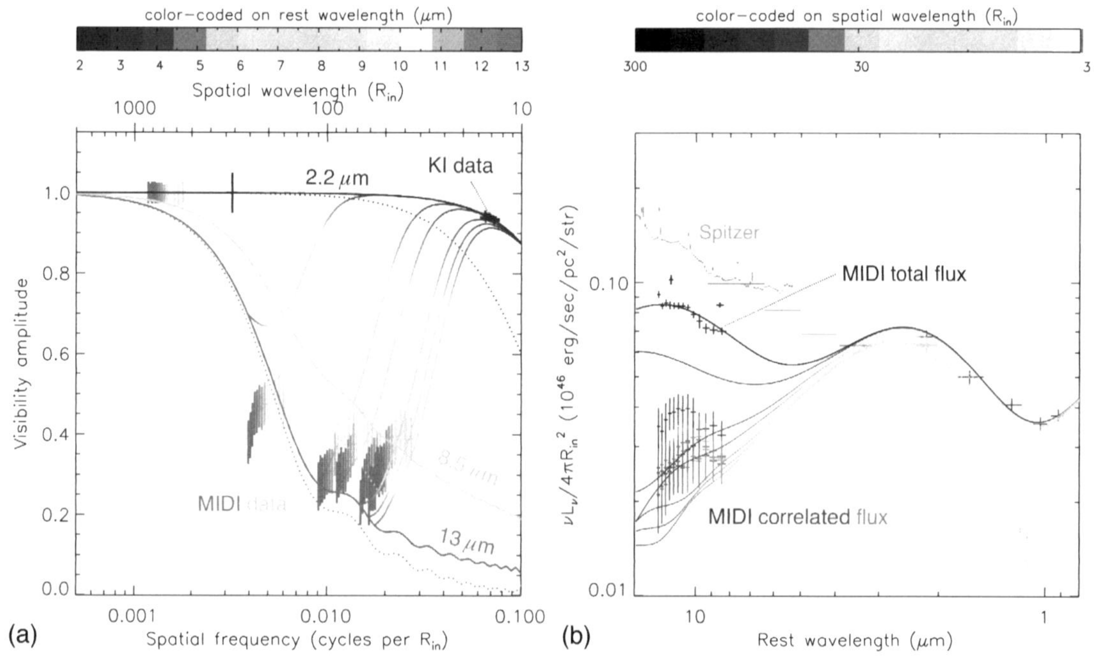

Figure 4. The mid-/near-IR interferometric and photometric data set for NGC4151 [8]. (a) Observed visibilities as color-coded in observing wavelengths as indicated by the color bar at the top. The visibilities from a temperature-/density-gradient model plus an inner 1400K ring are shown in solid curves for 13, 8.5, and 2.2 μm, while dotted curves are the visibilities only for the former power-law component. (b) The total flux spectrum in the mid-IR with MIDI and host-galaxy-subtracted total flux spectrum in the near-IR are shown in black. The two mid-IR data points slightly above the MIDI spectrum are VISIR measurements with a slightly larger aperture, and gray data points are the Spitzer measurements with even larger apertures. The correlated flux given by the MIDI and KI data are shown beneath the total fluxes, color-coded to the corresponding spatial wavelength (reciprocal of spatial frequency) in units of $R_{\rm in}$ (see the color bar at the top). The model curves for each MIDI baseline configuration are also shown in the same color-coding. See [8] for more details.

the near-IR part of the spectrum. Instead, the near-IR part makes up a somewhat distinct component on top of the mid-IR power-law component[3]. As we have seen in the previous section, the near-IR size scale is quite close to the dust sublimation radius $R_{\rm in}$, and the near-IR total flux spectrum shows that its color temperature is around 1400 K, quite generically for other objects as well (e.g. $\sim$1–2 μm SED fit in Fig.2 of [10], and Fig.3 of [3]). From the measured size and the absolute flux level of this near-IR component, its brightness temperature is estimated to be quite close to the color temperature (see Fig.11 of [8]), meaning that the surface filling factor is quite close to unity and the radiation is almost like that of a black body. Thus the near-IR flux seems to originate from an optically-thick, bright rim region of the dust grains at around $R_{\rm in}$ and at the sublimation temperature of $\sim$1400 K.

Figure 4*b* also shows the correlated flux spectra at progressively longer baseline with different colors (see the color bar, coded for spatial wavelength). For the visibilities at relatively low spatial frequencies in the first lobe (i.e. before the first null) and in the case of a roughly centro-symmetric object (which we would expect for Type 1s), the correlated flux approximately shows the spectrum of the part of the source unresolved by the interferometer. The correlated flux at

[3] Historically, the bump around 3 μm in the SED of QSOs has been known for decades [14].

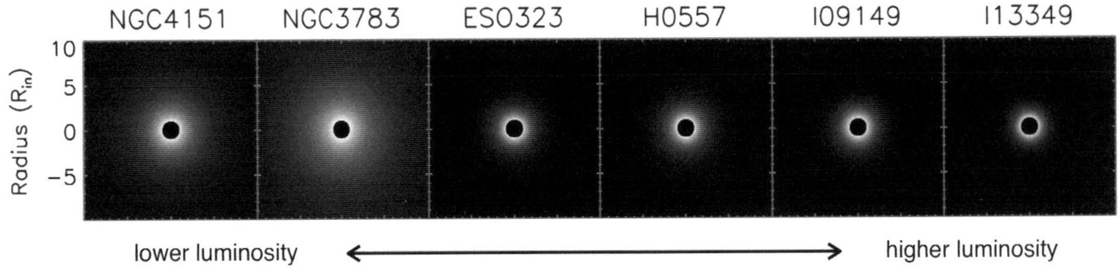

Figure 5. Model images of the power-law component at 13 μm in the rest frame of each object, shown in linear scale, using the best-fit parameters for the temperature and surface density indices [8], with spatial scales shown in units of $R_{\rm in}$. Higher luminosity objects are put toward the right.

longer baselines excludes the flux from the outer region, leaving the spectrum only for the inner core part of the source. Since we also see that the visibility curves look relatively flat at long baselines (Fig.4*a*), they probably show the flux fraction of an almost unresolved source, with the corresponding correlated flux showing the spectrum of this unresolved core. The correlated flux spectrum at long baselines in Fig.4*b* then indicates that the spectrum of this unresolved core does not seem to be as blue as the mid-IR tail of the near-IR-emitting hot rim component we inferred above — it rather suggests a core at a much lower temperature. In this sense, this lower temperature core seems to co-exist with the near-IR hot bright rim at the spatial scale of $R_{\rm in}$.

This hybrid description seems also consistent with the mid-IR and near-IR data for the other 5 targets, having a difference in the steepness of the mid-IR brightness distribution [8].

6. Temperature/density gradient parameterization

All analyses above involve *almost no modeling* – we can essentially infer all these conclusions from the direct observables. However, we can also make our arguments a little more quantitative by introducing a very simple, physical parameterization. To describe the face-on structure physically, we simply use power-law distributions of temperature and surface density, as we have already seen evidence for the distribution resembling a power law. In addition, we approximate the near-IR rim emission as a ring with a certain effective radius. The accretion disk component, which seems to become important only at wavelengths $\lesssim 1$ μm, is assumed to be unresolved and have the near-IR spectral shape of $f_\nu \propto \nu^{+1/3}$, which is supported both theoretically and observationally [15; 16].

This parameterization can describe the whole data sets at least approximately. Figure 4 includes the fitted model curves for NGC4151, and those for the whole sample are found in Figs.1 and 2 of [8]. Using the best-fit parameters, we can construct the model images which visually illustrate what we have discussed. The panels of Figure 5 are such images at 13 μm in the rest frame of each object, with spatial scales in units of $R_{\rm in}$, showing that the structure becomes steeper for higher luminosity objects (to the right). We can also show this structural change back in unnormalized, physical spatial scales, e.g. in pc, by looking at the change of characteristic radii at different wavelengths. Figure 6 shows the mid-IR half-light radii in pc for the 6 mid-IR targets at the rest-frame 13 and 8.5 μm derived from the model fits above, together with the thin-ring radii at 2.2 μm for the 8 near-IR targets (with two overlapping targets). If the surface density structure were the same over the sample, the ratios of the mid-IR to near-IR size would just be the same over the sample. However, since mid-IR radii in pc increase with luminosity L *much slower* than $L^{1/2}$ (specific values are given in Fig.6), the difference between radii at different wavelengths becomes smaller toward higher luminosities, meaning that the

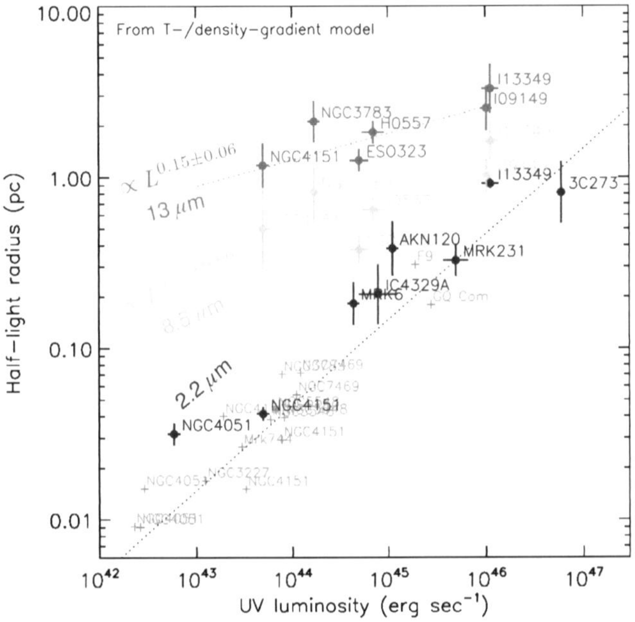

Figure 6. The mid-IR half-light radii $R_{1/2}$ in pc from best-fit models at rest-frame 13 and 8.5 μm (red and green, respectively), plotted against UV luminosity. Thinring radii in pc at 2.2 μm are plotted in purple for the KI-observed sample. The near-IR reverberation radii ([1] and references therein) are indicated as gray plus signs with the $L^{1/2}$ fit shown as a black dotted line (which is our definition of $R_{\rm in}$). The mid-IR part of this Figure is more model-dependent than, but essentially consistent with, that of Fig.8 in [8]. The fits to $R_{1/2}$ are shown in red and green dotted lines with quoted slopes, where the normalizations are 2.6±0.5 and 1.2±0.2 pc at $L \equiv 6\nu L_\nu(5500\text{Å})=10^{46}$ erg/sec for 13 and 8.5 μm, respectively.

radial density structure becomes steeper at higher luminosities.

Note that Figure 6 is essentially the same as Fig.8 of [8] which shows the mid-IR half-light radii measured with a simple geometrical power-law brightness at each wavelength. The half-light radii from the temperature-/density-gradient model are certainly more model-dependent, but utilize more information for each object including the total flux spectrum. The results are essentially consistent within the errors.

The structural steepening could potentially mean that, since the accretion disk luminosity is directly linked to the mass accretion rate, the radial density distribution might become steeper in higher accretion rate objects (here we refer to the absolute rate, not normalized by the Eddington rate). We now even have quantitative estimations for the radial surface density index of the heated dust distribution, which ranges from $\sim r^0$ to $\sim r^{-1}$ for the luminosities covered. We plan to further investigate correlations with different quantities such as the Eddington ratio using a larger sample in order to advance our physical understanding.

7. Conclusions

We have summarized the results of our first systematic study of Type 1 AGNs with IR interferometry, where we attempted to map the radial structure of the inner dusty accreting material. For the warm mid-IR power-law-like component, we seem to see steeper density distributions in objects with higher luminosities, or higher mass accretion rates. For the near-IR compact component in the dust rim region, we see a tentative correlation of steeper structures with more radio-loud central engines. We plan to further scrutinize all the results with a larger sample.

References

[1] Suganuma M, Yoshii Y, Kobayashi Y, Minezaki T, Enya K, Tomita H, Aoki T, Koshida S and Peterson B A 2006 ApJ **639** 46–63 (*Preprint* `arXiv:astro-ph/0511697`)

[2] Kishimoto M, Hönig S F, Tristram K R W and Weigelt G 2009 A&A **493** L57–L60 (*Preprint* `0812.1964`)

[3] Kishimoto M, Hönig S F, Beckert T and Weigelt G 2007 A&A **476** 713–721 (*Preprint* arXiv:0709.0431)

[4] Barvainis R 1987 ApJ **320** 537–544

[5] Maiolino R, Marconi A and Oliva E 2001 A&A **365** 37–48 (*Preprint* arXiv:astro-ph/0010066)

[6] Gaskell C M, Goosmann R W, Antonucci R R J and Whysong D H 2004 ApJ **616** 147–156 (*Preprint* arXiv:astro-ph/0309595)

[7] Kawaguchi T and Mori M 2010 ApJ **724** L183–L187 (*Preprint* 1010.5799)

[8] Kishimoto M, Hönig S F, Antonucci R, Millour F, Tristram K R W and Weigelt G 2011 A&A **536** A78 (*Preprint* 1110.4290)

[9] Swain M, Vasisht G, Akeson R, Monnier J, Millan-Gabet R, Serabyn E, Creech-Eakman M, van Belle G, Beletic J, Beichman C, Boden A, Booth A, Colavita M, Gathright J, Hrynevych M, Koresko C, Le Mignant D, Ligon R, Mennesson B, Neyman C, Sargent A, Shao M, Thompson R, Unwin S and Wizinowich P 2003 ApJ **596** L163–L166 (*Preprint* arXiv:astro-ph/0308513)

[10] Kishimoto M, Hönig S F, Antonucci R, Kotani T, Barvainis R, Tristram K R W and Weigelt G 2009 A&A **507** L57–60 (*Preprint* 0911.0666)

[11] Pott J, Malkan M A, Elitzur M, Ghez A M, Herbst T M, Schödel R and Woillez J 2010 ApJ **715** 736–742 (*Preprint* 1003.4757)

[12] Kishimoto M, Hönig S F, Antonucci R, Barvainis R, Kotani T, Tristram K R W, Weigelt G and Levin K 2011 A&A **527** A121 (*Preprint* 1012.5359)

[13] Koratkar A P and Gaskell C M 1991 ApJS **75** 719–750

[14] Neugebauer G, Oke J B, Becklin E E and Matthews K 1979 ApJ **230** 79–94

[15] Koratkar A and Blaes O 1999 PASP **111** 1–30

[16] Kishimoto M, Antonucci R, Blaes O, Lawrence A, Boisson C, Albrecht M and Leipski C 2008 Nature **454** 492–494 (*Preprint* arXiv:0807.3703)

Journal of Physics: Conference Series **372** (2012) 012034 doi:10.1088/1742-6596/372/1/012034

Nuclear mid-infrared properties of nearby low-luminosity AGN

D Asmus[1], S F Hönig[2], P Gandhi[3], A Smette[4] and W J Duschl[1,5]

[1] Institut für Theoretische Physik und Astrophysik, Christian- Albrechts-Universität zu Kiel, Leibnizstr. 15, 24098 Kiel Germany

[2] Department of Physics, University of California in Santa Barbara, Broida Hall, Santa Barbara, CA 93106-9530, USA

[3] Institute of Space and Astronautical Science (ISAS), Japan Aerospace Exploration Agency, 3-1-1 Yoshinodai, chuo-ku, Sagamihara, Kanagawa 252-5210, Japan

[4] European Southern Observatory, Casilla 19001, Santiago 19, Chile

[5] Steward Observatory, The University of Arizona, 933 N. Cherry Ave, Tucson, AZ 85721, USA

E-mail: asmus@astrophysik.uni-kiel.de

Abstract. We present ground-based high-spatial resolution mid-infrared (MIR) observations of 20 nearby low-luminosity AGN (LLAGN) with VLT/VISIR and the preliminary analysis of a new sample of 10 low-luminosity Seyferts observed with Gemini/Michelle. LLAGN are of great interest because these objects are the most common among active galaxies, especially in the nearby universe. Studying them in great detail makes it possible to investigate the AGN evolution over cosmic timescale. Indeed, many LLAGN likely represent the final stage of an AGN's lifetime. We show that even at low luminosities and accretion rates nuclear unresolved MIR emission is present in most objects. Compared to lower spatial resolution *Spitzer*/IRS spectra, the high-resolution MIR photometry exhibits significantly lower fluxes and different PAH emission feature properties in many cases. By using scaled *Spitzer*/IRS spectra of typical starburst galaxies, we show that the star formation contribution to the 12 μm emission is minor in the central parsecs of most LLAGN. Therefore, the observed MIR emission in the VISIR and Michelle data is most likely emitted by the AGN itself, which, for higher luminosity AGN, is interpreted as thermal emission from a dusty torus. Furthermore, the 12 μm emission of the LLAGN is strongly correlated with the absorption corrected 2-10 keV luminosity and the MIR–X-ray correlation found previously for AGN is extended to a range from 10^{40} to 10^{45} erg/s. This correlation is independent of the object type, and in particular the low-luminosity Seyferts observed with Michelle fall exactly on the power-law fit valid for brighter AGN. In addition, no dependency of the MIR–X-ray ratio on the accretion rate is found. These results are consistent with the unification model being applicable even in the probed low-luminosity regime.

1. Introduction

One of the major discoveries of the last two decades in extragalactic astronomy was that the structure and evolution of the central kiloparsec region of galactic nuclei are strongly connected to the properties and evolution of the central super-massive black hole (SMBH) that presumably resides in the center of each galaxy. In particular, the growth of the SMBH seems to be related to major galaxy mergers leading to various phases of nuclear activity (see contribution by Duschl et al.). These active galactic nuclei (AGN) can be described by the unification model of AGN [1] that contains the SMBH in the nucleus, surrounded by an accretion disk and, further outward, a dusty obscuring torus-like structure. This torus presumably blocks the direct view onto the central region depending on the viewing angle of the observer, therefore explaining the different AGN sub-classes observed. However, for a more accurate description, various modifications to this standard model may arguably be necessary (e.g., as suggested in the contributions by Elvis et al. and Ramos Almeida et al.). In addition, in the low-luminosity regime various changes in the structure of the AGN might occur. This is indicated by a different spectral energy distribution (SED) observed in these low-luminosity AGN (LLAGN)[2], theoretically explained by a change in the accretion disk structure from an optically-thick and geometrically-thin accretion disk [3] to an optically-thin and geometrically-thick accretion flow (e.g., advection-dominated accretion flow; ADAF [4]) at low accretion rates (Eddington ratio $\eta_{\mathrm{Edd}} \lesssim 10^{-2}$). At even lower rates ($\eta_{\mathrm{Edd}} \lesssim 10^{-5}$), a jet of outflowing plasma along the rotation axis of the accretion disk might even dominate the whole SED [5]. Finally, theoretical work predicts that the dusty obscuring torus will disappear at low luminosities/accretion rates [6, 7]. Observational indications for its disappearance have been found mainly for radio-loud LLAGN [8], while remaining unproven for the LLAGN in general (but see also contribution by Fernández Ontiveros et al.).

Searches for the putative torus are most promising in the mid-infrared (MIR), where the thermal emission of the dust is peaking. In particular, ground-based high-spatial resolutions MIR observations yield valuable information about this probably compact ($\lesssim 1\,\mathrm{pc}$) structure. It was found that the observed MIR emission of local AGN can be best explained by a torus consisting of many clouds [9, 10] (and also contribution by Ramos Almeida et al.). Although MIR interferometry provides an excellent possibility to investigate and resolve the torus in nearby AGN (see contributions by Burtscher et al., Kishimoto et al., and Tristram et al.), current its sensitivity is not sufficient to detect most LLAGN.

Even though lacking the high-spatial resolution, early MIR observations led to the discovery of a strong correlation between the nuclear MIR and X-ray emission [11, 12]. This MIR–X-ray correlation can in principle be understood by a close connection between the ultraviolet and X-ray emission originating in the inner accretion disk region. While the hard X-ray emission can penetrate the obscuring torus almost unabsorbed as long as the column density remains in the Compton-thin regime, the ultraviolet is absorbed by the dust and re-emitted in the MIR as thermal radiation. Note that this explanation is independent of the torus orientation, because the MIR emission is nearly isotropic. Therefore hard X-rays are also a good direct tracer of the accretion activity and the current intrinsic power output of an AGN. By using high-spatial resolution MIR VLT/VISIR imaging, we could recently demonstrate for a sample of ~ 40 local AGN that the MIR–X-ray correlation is indeed very strong, in particular when the hard 2-10 keV X-ray emission is corrected for absorption [13]. The present work represents a direct extension of our previous work into the low-luminosity regime and is driven by the main goal of probing the existence of a torus-like obscuring structure in LLAGN.

However, because of the large number of possible MIR emitters in this luminosity regime (Fig.1), only a first step will be presented: high-spatial resolution Gemini/Michelle and VLT/VISIR MIR imaging of LLAGN. We can investigate the validity of the MIR–X-ray correlation for LLAGN and constrain possible contamination by circum-nuclear star formation affecting the MIR photometry. Star formation can significantly contribute or even dominate the emission at low luminosities. For a detailed description of the VLT/VISIR data we refer to [14], while comparable observations are also presented in the contribution by Mason et al.

2. Methods

In order to separate the nuclear MIR emission in LLAGN from surrounding emission regions, high angular resolution data is necessary. Only this guarantees that non-nuclear contamination is mostly avoided, as demonstrated, e.g., in the contribution by Prieto et al. Therefore, we used

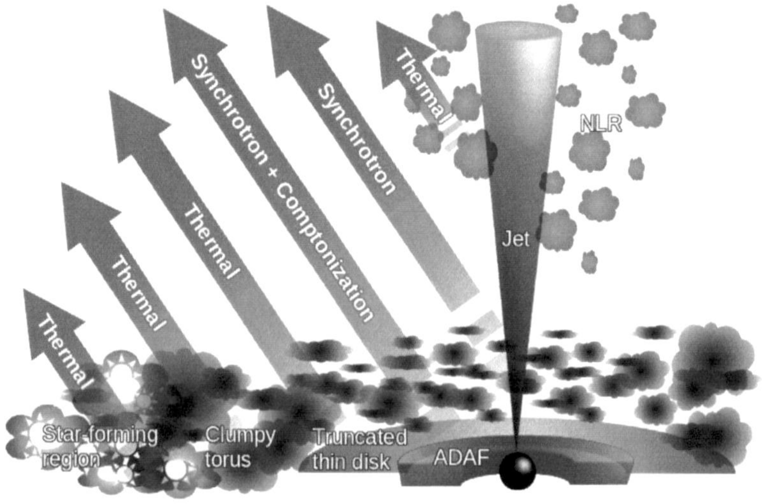

Figure 1. Schematic view of the central structure of an LLAGN. The red arrows indicate the MIR emission of the various components and its nature.

VLT/VISIR and Gemini/Michelle imaging in narrow-band filters to probe the 12 μm continuum and polycyclic aromatic hydrocarbon (PAH) at 11.3 μm for two samples of nearby LLAGN. These were selected to have absorption-corrected X-ray fluxes $F_{2-10\mathrm{keV}} > 10^{-12}$erg/s/cm^2, taken from the literature, mainly [15] and [16]. The X-ray flux limit corresponds to an estimated 12 μm flux of 10 mJy, assuming the MIR–X-ray correlation from [13], and represents roughly the lowest detectable flux with the ground-based MIR instruments that we used. The sample consists of LLAGN with a mean X-ray luminosity $\langle \log L_{2-10\mathrm{keV}}/\mathrm{erg\,s^{-1}} \rangle = 40.4$ at a mean redshift of $\langle z \rangle = 0.005$, corresponding to a distance of $\langle D \rangle = 19$ Mpc and a resolved physical scale of $\langle r_0 \rangle = 40$ pc. The southern sample of 18 LLAGN was observed in 2009, between April and September, with VLT/VISIR in the PAH2 ($11.25 \pm 0.59\,\mu$m) and NeIIref1 ($12.27 \pm 0.18\,\mu$m) filters, and is described in more detail in [14]. These LLAGN have mixed properties, e.g., optical classification and are not complete in any sense. The northern sample consists of 10 low-luminosity Seyferts drawn from the Palomar LLAGN survey [17] and were observed with Gemini/Michelle in February 2010 in the Si-5 ($11.6 \pm 0.55\,\mu$m) and Si-6 ($12.5 \pm 0.6\,\mu$m) filters. All observations were performed in the standard chopping and nodding mode. The data were reduced with the observatory delivered pipeline packages and photometric fluxes were measured with custom IDL routines [9].

3. Results

We detected 7 out of 18 southern and all 10 northern LLAGN with all detected objects being visible in both filters. The high number of non-detected LLAGN can be explained by the high uncertainties in the X-ray luminosities that were used for the sample selection. Indeed, many of these have been revised to much lower values in more recent X-ray observations (further discussed in [14]). Therefore we included 9 additional LLAGN with VISIR observations taken from the literature. All detected LLAGN appear point-like in the MIR at an angular resolution of $\sim 0.4''$, and no extended or off-nuclear emission is visible in the central $4''$ around these nuclei, similar to the findings for most of the brighter AGN [18]. The corresponding VISIR images can be found in [14] and a detailed analysis of the Michelle data will be presented in a future work.

3.1. Comparison to Spitzer – star formation contribution

Because *Spitzer*/IRS spectra (angular resolution $\sim 4''$) are widely used throughout the literature to infer AGN properties, a comparison with the Michelle and VISIR photometry is of great interest. In Fig. 2 we show a selection of objects from the VISIR sample. While the fluxes agree well for some objects, e.g., NGC 1566, the majority of objects ($\sim 60\%$) show significantly lower fluxes in VISIR than in *Spitzer* data (e.g, NGC 1097, NGC 4261, NGC 4579). In particular, many LLAGN do not possess PAH 11.3 μm emission on a 0.4″ scale (e.g. NGC 7213). However, NGC 1097 and NGC 1566 most likely exhibit nuclear PAH 11.3 μm emission as indicated by the comparably high PAH2 filter fluxes. We attribute the missing flux to extended diffuse emission regions that surround the nucleus on a 4″ scale. These have a low surface brightness and, therefore, are over-resolved and not visible in the VISIR and Michelle images while being included in the *Spitzer* beam. In general, there is no evidence for either any correlation between the VISIR and *Spitzer* fluxes, or any trend with the optical classification.

PAH emission is usually attributed to star formation and hence routinely used as its tracer. Furthermore, a tight correlation between the PAH 11.3 μm emission line flux and the 12 μm continuum is present in local starburst galaxies [14]. Reversely we use the PAH emission to estimate the amount of MIR continuum emission due to star formation in the innermost 0.4″ of the observed LLAGN. For this purpose we construct a starburst template SED by normalizing the IRS spectra from [19] by their PAH 11.3 μm flux. This template SED can in turn be applied to the individual LLAGN by scaling with the PAH 11.3 μm emission occurring in those objects. With the data at hand, this PAH emission can only be constrained by using either the IRS spectrum or the filter containing this feature (i.e. PAH2 for VISIR), depending on which provides the lower upper-limit. Hence, the scaled starburst template SEDs (orange-filled curves in Fig. 2) in turn represent upper limits for the star formation related MIR emission. For the majority of LLAGN ($\sim 70\%$), we can constrain the 12 μm continuum flux contamination through star formation to be less than 50% on a 0.4″ scale, while for the remaining 30% the contamination can possibly be up to 100%.

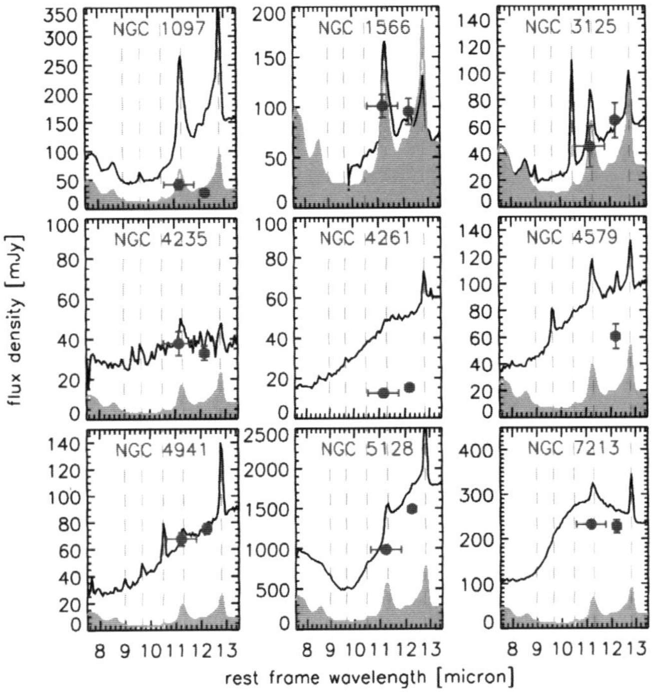

Figure 2. Comparison of VLT/VISIR photometry (red symbols) and *Spitzer* IRS spectra (black line) for LLAGN. Horizontal error bars correspond to the filter pass band. Commonly observed emission lines are indicated by the dotted lines. In addition, the scaled starburst template SED (orange-filled curves) correspond to the derived upper limit for the starburst contribution to the VISIR fluxes.

3.2. MIR–X-ray correlation

As mentioned in Section 1, a strong correlation exists between the absorption-corrected hard X-ray and the observed 12 μm emission for local AGN. Fig. 3 shows the luminosities at these wavelengths for the LLAGN and the sample of local AGN from [13]. The LLAGN tightly follow the same correlation and extend it down to luminosities of $\sim 10^{40}$ erg/s. However, the observed LINERs might exhibit a minor systematic offset (~ 0.3 dex) towards higher MIR or lower X-ray luminosities. Nevertheless, the numbers are too low for any statistically significant difference, and in general, the correlation remains independent of the optical classification within the measurement uncertainties. A power-law fit using `fitexy` [20] yields:

$$\log\left(\frac{\lambda L_\lambda(12\,\mu\mathrm{m})}{10^{43}\,\mathrm{erg\,s^{-1}}}\right) = (0.41 \pm 0.03) + (1.12 \pm 0.04)\log\left(\frac{L_{2-10\mathrm{keV}}}{10^{43}\,\mathrm{erg\,s^{-1}}}\right), \quad (1)$$

which is similar to previous results [13].

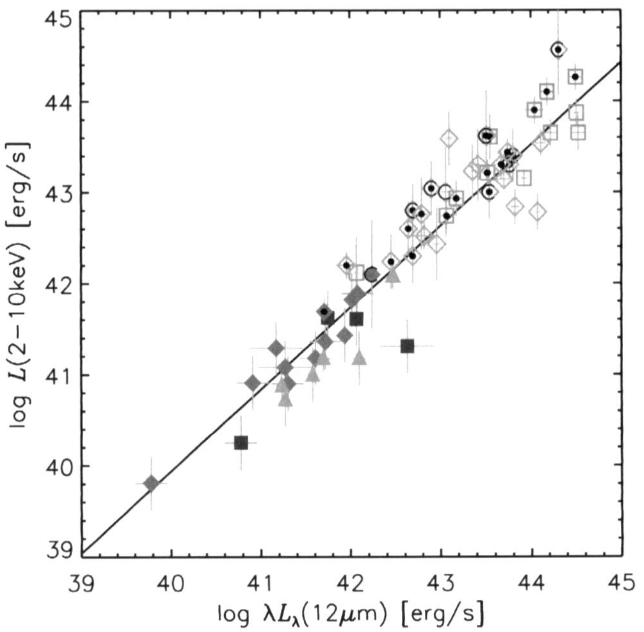

Figure 3. Absorption-corrected hard X-ray luminosities vs. the nuclear MIR luminosities for LLAGN (preliminary; filled symbols) and the AGN from [13] (empty symbols); blue squares: type 1 Seyferts (type 1.5 or lower); red diamonds: type 2 Seyferts; green triangles: LINERs; objects highlighted with central black-filled circles: "well-resolved" AGN from [13] ; solid line: power-law fit to all LLAGN and AGN.

3.3. Dependency on accretion rate

One of the fundamental parameters determining the AGN structure is assumed to be the accretion rate, which can be estimated by the Eddington ratio $\eta_{\mathrm{Edd}} = \log L_{\mathrm{bol}}/L_{\mathrm{Edd}}$, with L_{bol} being the bolometric luminosity of the AGN, $L_{\mathrm{Edd}} = 1.26 \times 10^{38}(M_{\mathrm{BH}}/M_\odot)$ erg/s its Eddington luminosity depending on M_{BH}, the mass of the black hole. We estimate the black hole masses of the LLAGN mainly by using the M_{BH}-σ correlation [21] with the stellar velocity dispersions σ taken from the literature (mainly [22] and the Hyperleda database [23]). Furthermore, the bolometric luminosity is simply assumed as $L_{\mathrm{bol}} = 10L_{2-10\mathrm{keV}}$, which seems reasonable for LLAGN [2].

The Eddington ratios of all observed LLAGN and the AGN from [13] extend over a large range of 5 orders of magnitude, $10^{-5} \lesssim \eta_{\mathrm{Edd}} \lesssim 1$. The MIR–X-ray luminosity ratio is constant, $\log(\lambda L_\lambda(12\,\mu\mathrm{m})/L_{2-10\mathrm{keV}}) = 0.31 \pm 0.36$, over the whole range of η_{Edd}, and therefore independent of the accretion rate for all observed AGN.

4. Discussion & conclusions

The high-angular resolution MIR imaging of $\sim$ 30 nearby LLAGN showed point-like nuclear emission at $\sim$ 50 pc scale in all detected objects. No correlation to larger aperture data like Spitzer is evident from a comparison of VISIR photometry and IRS spectra. Therefore one has to be careful when deriving AGN properties from these larger aperture data. In many cases the observed fluxes are significantly lower and spectral features, as the PAH 11.3 μm emission feature are absent on smaller scales. Furthermore, by using a starburst template SED, we showed that only $\lesssim 30\%$ of the high-resolution data may still be dominated by MIR emission from star formation occurring on $\lesssim$ 50 pc scale. The majority ($\gtrsim 70\%$) of LLAGN are, however, dominated by AGN emission on this scale, while its exact origin – torus, jet or accretion disk – remains undetermined at this point. Furthermore, the MIR–X-ray correlation is valid for all AGN observed at high-angular resolution in the MIR over the whole probed luminosity range from 10^{40} to $\sim 10^{45}$ erg/s. Owing to the fact that no dependence of either optical classification, or radio-loudness, or host morphology was found, this correlation might put strong constraints on the universal AGN structure. Interestingly, none of the theoretically predicted changes in the accretion structure with decreasing accretion rate, e.g., onset of ADAF or the jet-dominance, seem to have any imprint on the MIR–X-ray luminosity ratio. However, the sample is likely incomplete (selection based on availability of 2-10 keV data) and possibly biased against highly obscured (Compton-thick) objects. This along with the small total sample size prevents us from reaching any final conclusions on the AGN structure and nature of the MIR emitter in the low-luminosity regime. At this point we can only conclude that the results agree with those for brighter AGN, for which the MIR emission is well explained by a dusty clumpy torus. Therefore, such a torus might also exist in LLAGN, i.e. the standard unification model of AGN might still hold in this regime, in agreement with [16, 24]. Furthermore, these results imply that in LINERs, the emission is AGN-dominated in at least central tens of parsecs (compare [25]). Finally, the MIR–X-ray correlation represents a powerful observational tool to convert between both wavelength bands for any AGN. In order to settle the question about the nature of the MIR emission, SED modeling with particular focus on the high-resolution infrared data seems most promising (see also contributions from Prieto et al. and Fernández Ontiveros et al.).

References

[1] Antonucci R 1993 *ARA&A* **31** 473–521
[2] Ho L C 2008 *ARA&A* **46** 475–539
[3] Shakura N I and Sunyaev R A 1973 *A&A* **24** 337–355
[4] Narayan R and Yi I 1994 *ApJ* **428** L13–L16
[5] Falcke H and Markoff S 2000 *A&A* **362** 113–118
[6] Hönig S F and Beckert T 2007 *MNRAS* **380** 1172–1176
[7] Elitzur M and Shlosman I 2006 *ApJ* **648** L101–L104
[8] Donato D, Sambruna R M and Gliozzi M 2004 *ApJ* **617** 915–929
[9] Hönig S F, Kishimoto M, Gandhi P, Smette A, Asmus D, Duschl W, Polletta M and Weigelt G 2010 *A&A* **515** 23
[10] Alonso-Herrero A, Ramos Almeida C, Mason R, Asensio Ramos A, Roche P F, Levenson N A, Elitzur M, Packham C, Rodríguez Espinosa J M, Young S, Díaz-Santos T and Pérez-García A M 2011 *ApJ* **736** 82
[11] Krabbe A, Böker T and Maiolino R 2001 *ApJ* **557** 626–636
[12] Lutz D, Maiolino R, Spoon H W W and Moorwood A F M 2004 *A&A* **418** 465–473
[13] Gandhi P, Horst H, Smette A, Hönig S, Comastri A, Gilli R, Vignali C and Duschl W 2009 *A&A* **502** 457–472
[14] Asmus D, Gandhi P, Smette A, Hönig S F and Duschl W J 2011 *A&A* **536** 36
[15] González-Martín O, Masegosa J, Márquez I, Guainazzi M and Jiménez-Bailón E 2009 *A&A* **506** 1107–1121
[16] Panessa F, Bassani L, Cappi M, Dadina M, Barcons X, Carrera F J, Ho L C and Iwasawa K 2006 *A&A* **455** 173–185
[17] Ho L C, Filippenko A V and Sargent W L W 1997 *ApJSS* **112** 315
[18] Horst H, Duschl W J, Gandhi P and Smette A 2009 *A&A* **495** 137–146
[19] Brandl B R, Bernard-Salas J, Spoon H W W, Devost D, Sloan G C, Guilles S, Wu Y, Houck J R, Weedman

D W, Armus L, Appleton P N, Soifer B T, Charmandaris V, Hao L, Higdon J A M and Herter T L 2006 *ApJ* **653** 1129–1144

[20] Press W H, Teukolsky S A, Vetterling W T and Flannery B P 1992 *Numerical recipes in FORTRAN. The art of scientific computing*

[21] Gültekin K, Richstone D O, Gebhardt K, Lauer T R, Tremaine S, Aller M C, Bender R, Dressler A, Faber S M, Filippenko A V, Green R, Ho L C, Kormendy J, Magorrian J, Pinkney J and Siopis C 2009 *ApJ* **698** 198–221

[22] Ho L C, Greene J E, Filippenko A V and Sargent W L W 2009 *ApJSS* **183** 1–16

[23] Paturel G, Petit C, Prugniel P, Theureau G, Rousseau J, Brouty M, Dubois P and Cambrésy L 2003 *A&A* **412** 45–55

[24] Maoz D 2007 *MNRAS* **377** 1696–1710

[25] Yan R and Blanton M R 2011 *ArXiv e-prints* (*Preprint* 1109.1280)

The complexity of parsec-scaled dusty tori in AGN

K. R. W. Tristram[1], M. Schartmann[2,3], L. Burtscher[4,5], K. Meisenheimer[4], W. Jaffe[6], M. Kishimoto[1], S. F. Hönig[7], G. Weigelt[1]

[1] Max-Planck-Institut für Radioastronomie, Auf dem Hügel 69, 53121 Bonn, Germany
[2] Max-Planck-Institut für extraterrestrische Physik, Gießenbachstr., 85748 Garching, Germany
[3] Universitäts-Sternwarte München, Scheinerstraße 1, 81679 München, Germany
[4] Max-Planck-Institut für Astronomie, Königstuhl 17, 69117 Heidelberg, Germany
[6] Leiden Observatory, Leiden University, Niels-Bohr-Weg 2, 2300 CA Leiden, The Netherlands
[7] Physics Department, University of California, Santa Barbara, CA 93106, USA

E-mail: `tristram@mpifr-bonn.mpg.de`

Abstract. Warm gas and dust surround the innermost regions of active galactic nuclei (AGN). They provide the material for accretion onto the super-massive black hole and they are held responsible for the orientation-dependent obscuration of the central engine. The AGN-heated dust distributions turn out to be very compact with sizes on scales of about a parsec in the mid-infrared. Only infrared interferometry currently provides the necessary angular resolution to directly study the physical properties of this dust. Size estimates for the dust distributions derived from interferometric observations can be used to construct a size–luminosity relation for the dust distributions. The large scatter about this relation suggests significant differences between the dust tori in the individual galaxies, even for nuclei of the same class of objects and with similar luminosities. This questions the simple picture of the same dusty doughnut in all AGN. The Circinus galaxy is the closest Seyfert 2 galaxy. Because its mid-infrared emission is well resolved interferometrically, it is a prime target for detailed studies of its nuclear dust distribution. An extensive new interferometric data set was obtained for this galaxy. It shows that the dust emission comes from a very dense, disk-like structure which is surrounded by a geometrically thick, similarly warm dust distribution as well as significant amounts of warm dust within the ionisation cone.

1. Introduction

Molecular gas and dust play an important role in the appearance and physics of active galactic nuclei (AGN). A significant portion of the energy released by the accretion disk is intercepted by the surrounding dusty material and re-emitted at infrared wavelengths. The direct connection between the emission from the accretion disk and the reprocessed emission from the torus is reflected by the tight correlation between the X-ray and mid-infrared luminosities of AGN [1, 2, 3]. Several lines of evidence (ionisation cones, collimated outflows, broad emission lines in polarised light, etc.) suggest that the molecular and dusty material is located in a geometrically thick, toroidal distribution, the so-called *dusty torus*. The viewing-angle dependent obscuration of such a distribution – or, in general, the supposed axisymmetric geometry of AGN – is

[5] Present address: Max-Planck-Institut für extraterrestrische Physik, Gießenbachstraße, 85748 Garching, Germany

made responsible for many of the different observed properties of AGN, most importantly the differences between type 1 (torus face-on) and type 2 (torus edge-on) sources. There are, however, certainly other parameters apart from the viewing angle which influence the observed properties of AGN (see for example the contribution by C. Ramos Almeida).

Most information about the distribution of the nuclear gas and dust has been obtained indirectly, as even the largest single-dish telescopes do not provide enough angular resolution to resolve the nuclear material. Only in the last years it has become possible to spatially resolve the thermal emission from the AGN-heated dust using infrared interferometry. While interferometry in the near-infrared probes the hot ($T \sim 1400\,\mathrm{K}$) dust close to sublimation at the inner rim of the torus (c.f. the contribution by M. Kishimoto), mid-infrared interferometry explores the warm ($T \sim 300\,\mathrm{K}$) body of the dust distribution. The following will focus on studies of AGN with the MID-infrared Interferometric instrument (MIDI) at the Very Large Telescope Interferometer (VLTI). MIDI is currently the only instrument capable of resolving the emission from the warm dust component in AGN.

2. Observations of AGN with MIDI

Several observing campaigns of AGN have been carried out with MIDI. First, the brightest four active galaxies were studied individually, mainly using science demonstration time (SDT) or guaranteed time observations (GTO): NGC 1068 [4, 5], the Circinus galaxy [6], Centaurus A (NGC 5128) [7] and NGC 4151 [8]. In addition to these four sources, seven further sources were observed using the remaining GTO in a snapshot survey [9]. The goal was to test the feasibility of observations for these fainter sources at the sensitivity limit of MIDI and to obtain first size estimates of their emission regions. After the successful completion of the feasibility tests, a large observing programme containing 13 sources was initiated with the aim to investigate a statistical sample of AGN (L. Burtscher et al. in prep.). A sample of six type 1 sources studied with MIDI has been presented in [10]. A few other sources have been observed successfully, among them NGC 3783 [11, 10] and NGC 424 (S. F. Hönig et al. in prep.). Here, the results obtained from the snapshot survey as well as the results obtained from new measurements for the Circinus galaxy will be summarised.

A list of all the galaxies from GTO observations and the snapshot survey is shown in table 1. Both type 1 and type 2 AGN as well as a radio galaxy and a quasar are among these sources. The last column lists a short, verbal result from the interferometric observations for the respective galaxy. These results range from more or less unresolved sources (Mrk 1239) to very well resolved sources, where a disk component can be distinguished from a more extended, diffuse structure (NGC 1068 and the Circinus galaxy).

3. The size–luminosity relation in the mid-infrared

3.1. Building the size–luminosity relation

From the visibilities V measured by the interferometer, simple size estimates for the mid-infrared emission regions in these galaxies can be derived by assuming a Gaussian brightness distribution. Although this is certainly a very simplifying assumption and the true brightness distributions of AGN tori are significantly more complex, a Gaussian distribution is the simplest and most general initial guess for the true brightness distribution. Furthermore, it does not rely on any other assumption (e.g. about the location of the inner or outer edge of the torus as is needed for power-law distributions) and its half width at half maximum $HWHM$ is related to the visibility V, the projected baseline length BL and wavelength λ of the interferometric observation by a simple, analytical relation:

$$HWHM = \frac{\lambda}{BL} \cdot \frac{1}{\pi}\sqrt{-\ln 2 \cdot \ln V}. \qquad (1)$$

Table 1. List of galaxies from GTO observations and the snapshot survey successfully observed with MIDI. The columns are (1) the name, (2) the right ascension (3) the declination, (4) the type and (5) the total mid-infrared flux of the galaxy at 12 μm as well as (6) a short, verbal summary of the results from the interferometric measurements.

Galaxy name (1)	RA (2)	DEC (3)	type (4)	F_{MIR} [Jy] (5)	result (6)
NGC 1068	02 42 41	−00 00 48	Sy 2	16.5	disk + elongated component
NGC 1365	03 33 36	−36 08 26	Sy 1.8	0.4	partially resolved
IRAS 05189-2524	05 21 01	−25 21 45	Sy 2	0.4	faint detection (resolved?)
MCG-05-23-016	09 47 40	−30 56 56	Sy 2	0.6	well resolved
Mrk 1239	09 52 19	−01 36 44	Sy 1.5	0.4	unresolved
NGC 3783	11 39 02	−37 44 19	Sy 1	0.6	partially resolved
NGC 4151	12 10 33	+39 24 21	Sy 1.5	1.2	well resolved
3C 273	12 29 07	+02 03 09	QSO	0.3	possibly resolved
Centaurus A	13 25 28	−43 01 09	FR I	1.3	core + extended component
IC 4329A	13 49 19	−30 18 34	Sy 1.2	1.0	slightly resolved
Circinus	14 13 10	−65 20 21	Sy 2	10.2	disk + elongated component
NGC 7469	23 03 16	+08 52 26	Sy 1.2	0.6	well resolved

This method yields viable size estimates with errors of up to a factor of 4 as long as $0.2 \lesssim V \lesssim 0.9$ (for a more detailed discussion of the accuracy see [12]).

In figure 1 the Gaussian *HWHM* is plotted as a function of the hard X-ray luminosity L_X, which is considered as a proxy for the intrinsic luminosity of the AGN when corrected for foreground and intrinsic absorption. The X-ray fluxes were taken from the first 22 months of data of the hard X-ray survey (14 − 195 keV) with the Burst Alert Telescope (BAT) on the Swift satellite [13] and corrected for absorption. These measurements were obtained before or contemporaneous with the interferometric measurements. For the only source not in the BAT sample, Mrk 1239, the luminosity from [14] was used after correcting it from the soft to the hard X-ray band.

The data shown in figure 1 is still consistent with the naively expected relation $r \sim L^{0.5}$, i.e. that the size of the dust distribution scales with the luminosity of the accretion disk: $r = (0.76 \pm 0.11) \cdot 10^{-18}\,(\mathrm{L/W})^{0.5}\,\mathrm{pc}$. This is in contrast to the much slower dependency than $L^{0.5}$ found for type 1 objects in [10]. However, we also see a tendency for type 1 objects with high luminosities to lie below the relation and a lot depends on the highest luminosity object 3C 273. Certainly, this figure has to be filled with more measurements of a larger sample of sources and the uncertainties as well as selection effects have to be reduced. This is one of the goals of the MIDI AGN Large Programme.

3.2. Comparison to models

To better understand the implications of the interferometrically measured sizes of the dust distributions, the size–luminosity relation is compared to the predictions from two models for clumpy tori of AGN: (1) the clumpy torus model from [16, 17] and (2) the hydrodynamical torus model from [18]. To obtain torus sizes which can be directly compared to the measured size estimates, interferometric observations of the brightness distributions of the two models are simulated and then the sizes are calculated according to equation 1. The range of sizes predicted by the two models are shown by the thick red and orange bars in figure 1. The sizes from the

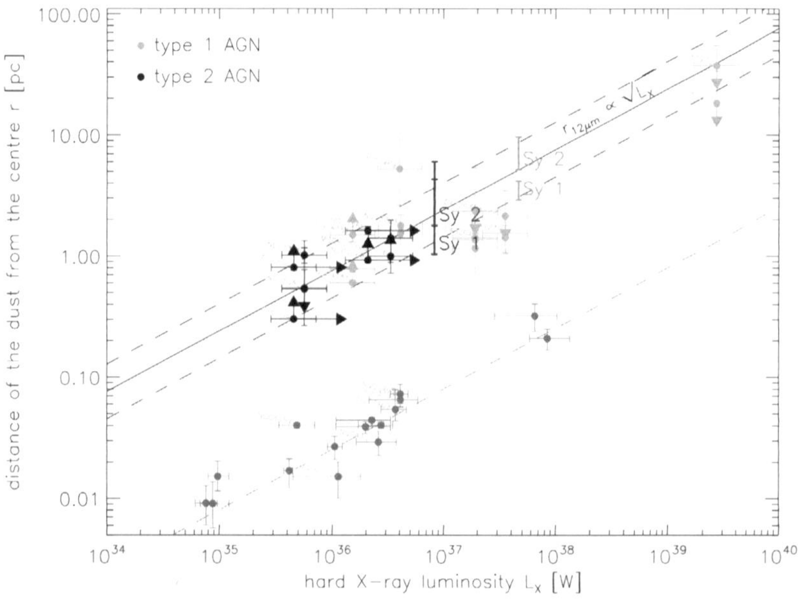

Figure 1. Size of the mid-infrared emitter as a function of the X-ray luminosity, taken as a proxy for the intrinsic AGN luminosity. Type 1 objects are plotted in light blue, type 2 objects in dark blue. Upper and lower limits on the size estimates are marked by arrows. The fitted relation $r = p \cdot L^{0.5}$ is delineated by the black continuous line, the scatter in the relation by the dashed lines. The red dashed line shows the inner radius of the torus derived from reverberation measurements and near-infrared interferometry. The individual measurements of near-infrared radii are plotted as red points and were adapted from figure 30 in [15].

models are in general consistent with the observed sizes at a given luminosity. The models themselves cannot be used to test whether the size–luminosity relation is correct in itself, as they are explicitly assumed to scale with $L^{0.5}$. More interesting is however that both models show a clear offset between the sizes of type 1 (more compact) and type 2 (more extended) tori. Such an offset is *not* observed in this data and the dust distributions in individual galaxies of a single object class appear more different than the differences between the classes. This might be interpreted in the way that actually not so much the inclination but other torus properties are responsible for the type 1/2 differences. This is also suggested by the fitting of torus models to the infrared SEDs of AGN (see contribution by C. Ramos Almeida). In any case this is further evidence that not all AGN have the same dust distribution – even not in nuclei with the same luminosity and in the same object class.

4. The dust distribution in the Circinus galaxy

Being the second brightest AGN in the mid-infrared (c.f. table 1) and relatively nearby, at a distance of only 4 Mpc, the Circinus galaxy is one of the prime targets for studying the nuclear dust distribution in detail with infrared interferometry. Observations in the optical, X-rays, and radio indicate that the Circinus galaxy has a Seyfert 2 nucleus. As a typical type 2 AGN, the galaxy shows broad emission lines in polarised light [19] and an ionisation cone [20, 21, 22]. The mid-infrared spectrum shows a deep silicate absorption feature, caused both by dust lanes within Circinus (which is inclined by $i = 65°$) as well as by the nuclear dust itself. First results

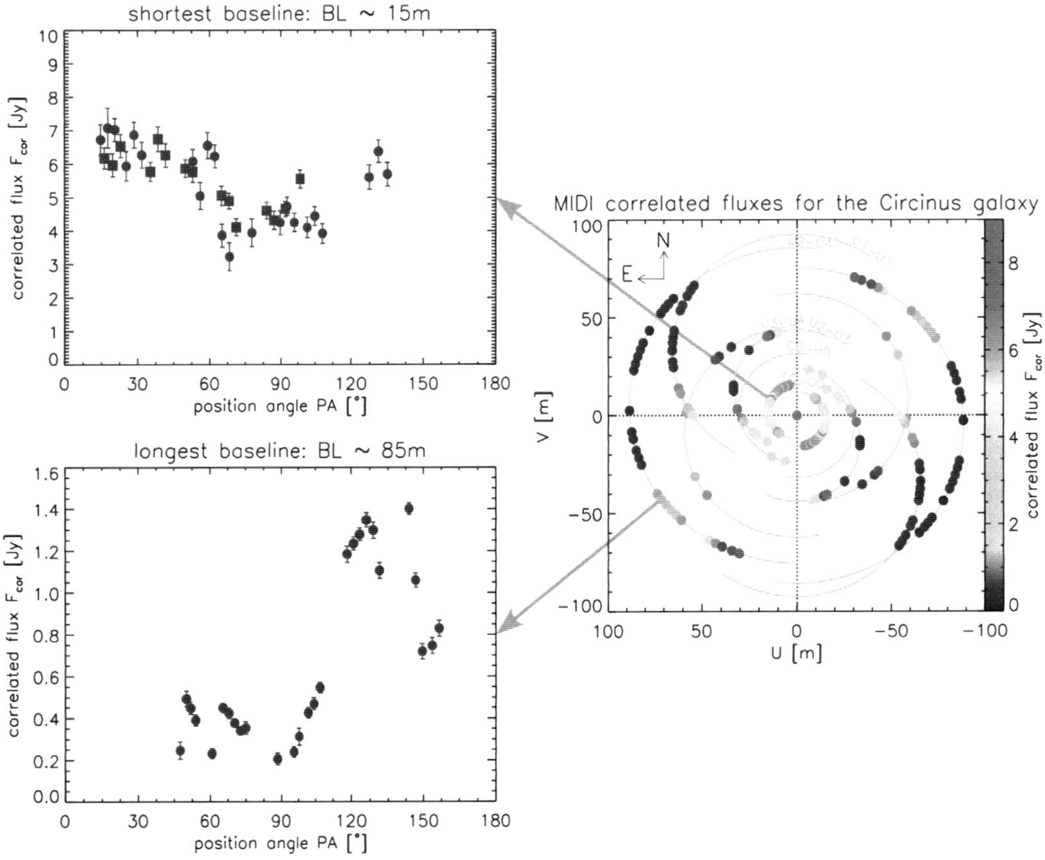

Figure 2. Coverage of the UV plane for the Circinus galaxy achieved by observations with MIDI (right panel). The position in the UV plane corresponds to the angle and the length of the projected baseline of the two telescopes, when seen from the centre. The UV points are colour-coded according to their interferometrically measured correlated fluxes at $12\,\mu$m. On the left hand side, these correlated fluxes are shown as a function of the position angle for two baselines with essentially constant baseline lengths (i.e. resolution power of the interferometer): on the top left for the shortest baseline (E0-G0), probing the largest spatial scales, and on the bottom left for the longest baseline (UT2-UT4), probing the smallest structures.

obtained with MIDI have shown that the emission is resolved into a disk-like component plus an extended emission region [6]. Since then, extensive new observations have been carried out and the first results of these are presented here.

4.1. A direct analysis of the MIDI data

The current coverage of the UV plane achieved by MIDI is shown in the right panel of figure 2. With more than 100 measurements, a respectable UV coverage has been achieved for optical interferometry. However, the data are very inhomogeneous in instrument settings and observing technique, due to the large spread of time over which the data has been obtained. Furthermore, there are some inconsistencies in the data, which make a quick interpretation of the data difficult. Nevertheless, important properties of the dust emission can be directly inferred from the interferometric measurements without the need for any modelling.

Table 2. Best fit parameters for the two component black body Gaussian model for the dust distribution in the nucleus of the Circinus galaxy.

parameter	disk component		extended component	
size of major axis	$\Delta_1 =$	18 mas	$\Delta_2 =$	99 mas
axis ratio	$r_1 =$	0.50	$r_2 =$	0.56
position angle	$\alpha_1 =$	61°	$\alpha_2 =$	94°
silicate depth	$\tau_1 =$	1.4	$\tau_2 =$	2.1
temperature	$T_1 =$	295 K	$T_2 =$	358 K
surface filling factor	$f_1 =$	1.00	$f_2 =$	0.17

On the shortest baseline, essentially only the position angle changed while the baseline length remained more or less the same. This means the same, large spatial scales were probed at different position angles. In the top left panel of figure 2, the correlated flux at 12 μm for this baseline is plotted as a function of the position angle. The correlated flux has a broad minimum at $PA \sim 100°$, which means the emission is resolved most along this position angle, i.e. it is extended most in this direction. This is consistent with the result of deep mid-infrared imaging with 8 m class telescopes, where an elongation along $PA \sim 100°$, i.e. along the southern edge of the ionisation cone, is seen [23].

The same can be done for the UT2-UT4 baseline, which is a very long baseline and hence probes very small spatial structures. In the lower left panel of figure 2, the correlated flux from this longer baseline is plotted as a function of position angle in the same way as for the shortest baseline. A very strong dependency of the correlated flux on the position angle is evident, with a clear maximum at $PA \sim 140°$. The mid-infrared emission hence has a minimum extension along this direction. In fact, the observed dependency of the correlated flux on the position angle is consistent with a relatively thin and highly inclined disk.

4.2. Simple two component model fit

To obtain a more accurate idea about the brightness distribution, the interferometric data has to be modelled. The most unambiguous result which is also physically motivated and does not rely on any further assumptions can be achieved using a two component model: The two components are elongated Gaussian distributions emitting black body or grey body radiation. Furthermore, the two components are located behind an absorption screen giving rise to the silicate absorption feature. The model thus has a total of 12 free parameters, and their values for the best fit are shown in table 2. In figure 3, the two model components are sketched: The inner component has a size of 20 mas, i.e. 0.4 pc at the distance of Circinus. It is very elongated and has a temperature of about 300 K. The second component has a size of about 100 mas, that is 2 pc, and is elongated along $PA \sim 94°$. It has a slightly higher temperature of about 360 K. Most of the total flux of the source is located in the extended component.

We interpret the two Gaussian emitters as signs for a geometrically thick distribution, "torus", of warm dust surrounding a very dense, disk-like structure as well as significant amounts of dust within the ionisation cone. The smaller and denser disk is seen at a high inclination. It is optically thick and it partly exhibits the silicate band in emission. The disk matches H_2O maser emitters tracing a mildly warped, thin disk [24] in orientation and size. The disk is surrounded by the larger, less dense and most likely clumpy or filamentary torus component which gives rise to strong silicate absorption. The elongation of this component indicates that there are significant amounts of dust within the ionisation cone. Since the dust in the ionisation cone is

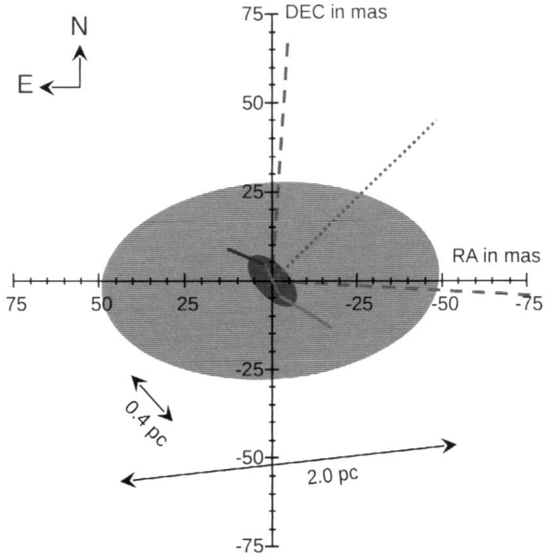

Figure 3. Sketch of simple two component model fit to the data for the Circinus galaxy: a highly elongated warm emission region with a temperature of $\sim$ 300 K (brown) is surrounded by an elongated and slightly warmer emission region (orange). In the centre of the sketch, the location of the H_2O maser emitters in a disk from [24] is overplotted: the blue line to the north-east traces the receding masers and the red line to the south-west the approaching masers. The dashed blue lines trace the edge of the observed ionisation cone seen in [OIII] [21], the dotted line is the cone axis.

roughly at the same temperature as the dust of the disk, this dust must be directly exposed to the radiation from the accretion disk to be heated to such high temperatures at these distances.

4.3. Comparison with NGC 1068

It is interesting to compare these results for the Circinus galaxy to those for NGC 1068, which hosts the brightest active nucleus in the mid-infrared and which is the second Seyfert 2 nucleus where the dust distribution has been studied in detail [4, 5]. There are a few similarities between the two nuclei: Both tori have a highly-inclined dust disk which is surrounded by an extended, geometrically thick dust distribution. And in both nuclei, the inner disk component roughly matches a H_2O maser disk in orientation and size. However, there are also a few significant differences: While in Circinus the torus is nicely aligned perpendicular to the ionisation cone and outflow as expected in the standard picture, this is not the case in NGC 1068. While in NGC 1068 the dust disk, which is located closest to the accretion disk, has a significantly higher temperature than the surrounding dust, there are no indications for hot dust in the Circinus nucleus from our data. And finally, while in the Circinus nucleus the depth of the silicate feature decreases towards the nucleus, the opposite is the case in NGC 1068: here the silicate feature gets deeper towards the smaller disk component. Although it is still not clear what the reasons for these differences are, it is clear that the dust distributions in these two Seyfert 2 galaxies must be quite different.

Conclusions

These proceedings should have convinced the reader of the following three points: (1) Optical interferometry is currently the only way to *directly* study the nuclear dust distributions in AGN – and it is a very successful tool to do so. (2) The warm, AGN-heated dust distributions in AGN are compact with sizes of only a few parsecs. (3) There are significant differences between the dust distributions in individual galaxies, even for objects of the same class and with similar luminosities. These differences are probably larger than the differences between the distributions in type 1 and type 2 objects. This will certainly have impact on our understanding of how accretion onto the super-massive black hole works on parsec scales.

References

[1] Horst H, Gandhi P, Smette A and Duschl W J 2008 *A&A* **479** 389–396 (*Preprint* arXiv:0711.3734)
[2] Gandhi P, Horst H, Smette A, Hönig S, Comastri A, Gilli R, Vignali C and Duschl W 2009 *A&A* **502** 457–472 (*Preprint* 0902.2777)
[3] Levenson N A, Radomski J T, Packham C, Mason R E, Schaefer J J and Telesco C M 2009 *ApJ* **703** 390–398 (*Preprint* 0908.0036)
[4] Jaffe W, Meisenheimer K, Röttgering H J A, Leinert C, Richichi A, Chesneau O, Fraix-Burnet D, Glazenborg-Kluttig A, Granato G L, Graser U, Heijligers B, Köhler R, Malbet F, Miley G K, Paresce F, Pel J W, Perrin G, Przygodda F, Schoeller M, Sol H, Waters L B F M, Weigelt G, Woillez J and de Zeeuw P T 2004 *Nature* **429** 47–49
[5] Raban D, Jaffe W, Röttgering H, Meisenheimer K and Tristram K R W 2009 *MNRAS* **394** 1325–1337 (*Preprint* 0901.1306)
[6] Tristram K R W, Meisenheimer K, Jaffe W, Schartmann M, Rix H W, Leinert C, Morel S, Wittkowski M, Röttgering H, Perrin G, Lopez B, Raban D, Cotton W D, Graser U, Paresce F and Henning T 2007 *A&A* **474** 837–850 (*Preprint* arXiv:0709.0209)
[7] Meisenheimer K, Tristram K R W, Jaffe W, Israel F, Neumayer N, Raban D, Röttgering H, Cotton W D, Graser U, Henning T, Leinert C, Lopez B, Perrin G and Prieto A 2007 *A&A* **471** 453–465 (*Preprint* arXiv:0707.0177)
[8] Burtscher L, Jaffe W, Raban D, Meisenheimer K, Tristram K R W and Röttgering H 2009 *ApJ* **705** L53–L57 (*Preprint* 0909.5191)
[9] Tristram K R W, Raban D, Meisenheimer K, Jaffe W, Röttgering H, Burtscher L, Cotton W D, Graser U, Henning T, Leinert C, Lopez B, Morel S, Perrin G and Wittkowski M 2009 *A&A* **502** 67–84 (*Preprint* 0903.4892)
[10] Kishimoto M, Hönig S F, Antonucci R, Millour F, Tristram K R W and Weigelt G 2011 *A&A* **536** A78 (*Preprint* 1110.4290)
[11] Beckert T, Driebe T, Hönig S F and Weigelt G 2008 *A&A* **486** L17–L20 (*Preprint* arXiv:0806.0531)
[12] Tristram K R W and Schartmann M 2011 *A&A* **531** A99 (*Preprint* 1105.4875)
[13] Tueller J, Baumgartner W H, Markwardt C B, Skinner G K, Mushotzky R F, Ajello M, Barthelmy S, Beardmore A, Brandt W N, Burrows D, Chincarini G, Campana S, Cummings J, Cusumano G, Evans P, Fenimore E, Gehrels N, Godet O, Grupe D, Holland S, Kennea J, Krimm H A, Koss M, Moretti A, Mukai K, Osborne J P, Okajima T, Pagani C, Page K, Palmer D, Parsons A, Schneider D P, Sakamoto T, Sambruna R, Sato G, Stamatikos M, Stroh M, Ukwata T and Winter L 2010 *ApJS* **186** 378–405 (*Preprint* 0903.3037)
[14] Grupe D, Mathur S and Komossa S 2004 *AJ* **127** 3161–3167 (*Preprint* arXiv:astro-ph/0402268)
[15] Suganuma M, Yoshii Y, Kobayashi Y, Minezaki T, Enya K, Tomita H, Aoki T, Koshida S and Peterson B A 2006 *ApJ* **639** 46–63 (*Preprint* arXiv:astro-ph/0511697)
[16] Hönig S F, Beckert T, Ohnaka K and Weigelt G 2006 *A&A* **452** 459–471 (*Preprint* astro-ph/0602494)
[17] Hönig S F and Kishimoto M 2010 *A&A* **523** A27 (*Preprint* 0909.4539)
[18] Schartmann M, Meisenheimer K, Klahr H, Camenzind M, Wolf S and Henning T 2009 *MNRAS* **393** 759–773 (*Preprint* 0811.1425)
[19] Oliva E, Marconi A, Cimatti A and Alighieri S D S 1998 *A&A* **329** L21–L24 (*Preprint* astro-ph/9711054)
[20] Veilleux S and Bland-Hawthorn J 1997 *ApJ* **479** L105 (*Preprint* astro-ph/9703040)
[21] Wilson A S, Shopbell P L, Simpson C, Storchi-Bergmann T, Barbosa F K B and Ward M J 2000 *AJ* **120** 1325–1341 (*Preprint* astro-ph/0006147)
[22] Prieto M A, Meisenheimer K, Marco O, Reunanen J, Contini M, Clenet Y, Davies R I, Gratadour D, Henning T, Klaas U, Kotilainen J, Leinert C, Lutz D, Rouan D and Thatte N 2004 *ApJ* **614** 135–141 (*Preprint* astro-ph/0406620)
[23] Packham C, Radomski J T, Roche P F, Aitken D K, Perlman E, Alonso-Herrero A, Colina L and Telesco C M 2005 *ApJ* **618** L17–L20
[24] Greenhill L J, Booth R S, Ellingsen S P, Herrnstein J R, Jauncey D L, McCulloch P M, Moran J M, Norris R P, Reynolds J E and Tzioumis A K 2003 *ApJ* **590** 162–173 (*Preprint* astro-ph/0302533)

Astronomy at High Angular Resolution 2011 IOP Publishing
Journal of Physics: Conference Series **372** (2012) 012036 doi:10.1088/1742-6596/372/1/012036

The Nuclear Infrared Emission of Low-Luminosity AGN

R E Mason[1], E Lopez-Rodriguez[2], C Packham[2], A Alonso-Herrero[3,11], N A Levenson[4], J Radomski[4], C Ramos Almeida[5,12], L Colina[6], M Elitzur[7], I Arextaga[8], P F Roche[9], N Oi[10]

[1] Gemini Observatory, Northern Operations Center, 670 N. A'ohoku Place, Hilo, HI 96720, USA

[2] University of Florida, Department of Astronomy, 211 Bryant Space Science Center, P.O. Box 112055, Gainesville, FL 32611, USA

[3] Instituto de Física de Cantabria, CSIC-UC, Avenida de los Castros s/n, 39005 Santander, Spain

[4] Gemini Observatory, Southern Operations Center, c/o AURA, Casilla 603, La Serena, Chile

[5] University of Sheffield, Department of Physics & Astronomy, S3 7RH, UK

[6] Departamento de Astrofísica, Centro de Astrobiología (CSIC/INTA), Instituto Nacional de Técnica Aeroespacial, Crta de Torrejón a Ajalvir, km 4, 28850 Torrejón de Ardoz, Madrid Spain

[7] Department of Physics and Astronomy, University of Kentucky, Lexington, KY 40506, USA

[8] Instituto Nacional de Astrofísica, Óptica y Electrónica (INAOE), Aptdo. Postal 51 y 216, 72000 Puebla, Mexico

[9] Astrophysics, Department of Physics, University of Oxford, DWB, Keble Road, Oxford OX1 3RH, UK

[10] Department of Astronomy, School of Science, Graduate University for Advanced Studies (SOKENDAI), Mitaka, Tokyo 181-8588, Japan

E-mail: `rmason@gemini.edu`

Abstract.
We have obtained high-resolution mid-infrared (MIR) imaging, nuclear spectral energy distributions (SEDs) and archival Spitzer spectra for 22 low-luminosity active galactic nuclei (LLAGN; $L_{bol} < 5 \times 10^{42}$ erg s^{-1}). Infrared (IR) observations may advance our understanding of the accretion flows in LLAGN, the fate of the obscuring torus at low accretion rates, and, perhaps, the star formation histories of these objects. However, while comprehensively studied in higher-luminosity Seyferts and quasars, the nuclear IR properties of LLAGN have not yet been well-determined. In these proceedings we summarise the results for the LLAGN at the relatively high-luminosity, high-Eddington ratio end of the sample. Strong, compact nuclear sources are visible in the MIR images of these objects, with luminosities consistent with or slightly in execss of that predicted by the standard MIR/X-ray relation. Their broadband nuclear SEDs are diverse; some resemble typical Seyfert nuclei, while others possess less of a well-defined MIR "dust bump". Strong silicate emission is present in many of these objects. We speculate that this, together with high ratios of silicate strength to hydrogen column density, could suggest optically thin dust and low dust-to-gas ratios, in accordance with model predictions that LLAGN do not host a Seyfert-like obscuring torus.

[11] Augusto González Linares Senior Research Fellow

[12] Current affiliations: Instituto de Astrofísica de Canarias (IAC), C/Vía Láctea, s/n, E-38205, La Laguna, Tenerife, Spain; Departamento de Astrofísica, Universidad de La Laguna, E-38205, La Laguna, Tenerife, Spain.

1. Introduction

We have been investigating the nuclear infrared (IR) properties of LLAGN, loosely defined as a low-ionisation nuclear emission region (LINER) or Seyfert galaxy with $L_{\rm bol}$ below about 10^{42} erg s^{-1}. The IR continuum emission from an LLAGN may contain contributions from a variety of processes: thermal emission from a dusty torus surrounding the central supermassive black hole (SMBH), synchrotron emission from a jet, and thermal emission from a truncated accretion disk. Dust in a narrow-line region (NLR) or associated with the surrounding stellar population may also contribute. Silicate emission or absorption features at 10 and 18 μm contain information about dust geometry and heating, and the suite of IR polycyclic aromatic hydrocarbon (PAH) bands may trace star formation in the nuclear environment. The IR wavelength regime therefore contains much information that could potentially advance our understanding of the nature and lifecycle of LLAGN. However, we do not yet have a good overview of the *nuclear* IR properties of these objects.

Studies of the spectral properties, luminosities, fuel supply etc. of LLAGN suggest that the standard optically thick and geometrically thin accretion disk is truncated, and replaced by a geometrically thick, radiatively inefficient accretion flow (RIAF) interior to the disk truncation radius [1, 2, 3]. The truncation of the thin disk should shift its thermal emission peak towards longer wavelengths and indeed, [4] present models in which the bulk of the nuclear luminosity at $\sim$1–10 μm can come from the truncated disk. [4] note, though, that the available, large-aperture IR data cannot constrain the presence or properties of a truncated disk in the objects modeled. IR photometry, at higher angular resolution than is generally available in the literature, may therefore aid our understanding of the accretion physics of LLAGN.

In higher-luminosity AGN, the difference between types 1 and 2 is explained at least to first order by the presence of a toroidal cloud of dust and gas obscuring the AGN from certain viewing directions while permitting a direct view from others. Some models of the torus explain its existence through inflows of gas from larger scales [5, 6]. During periods when little material is reaching the centre of the galaxy, the torus may become thin and transparent [7]. Conversely, the torus may be part of a dusty, outflowing wind [8]. The disk wind model of [9] predicts that below $L_{\rm bol} \sim 10^{42}$ erg s^{-1}, accretion onto the black hole can no longer sustain the outflow necessary to obscure the nucleus. In either case, low-luminosity AGN may show little nuclear obscuration and dust emission.

Observationally, there are indications that LLAGN do tend to have unobscured nuclei. For instance, nuclear UV point sources have been detected in LINERs of both types 1 and 2 [10]. A decline in absorbing column density has also been observed in X-ray studies of LLAGN [11]. On the other hand, the detection of broad Hα in polarised light in some LINERs [12] suggests that these objects host dust-obscured AGN. X-ray studies of single objects show that some LLAGN do have substantial absorbing columns, and the fraction of such objects may be significant [13]. [14] present average spectra of type 1 and 2 LINERs which suggest an extra hot dust component in the type 1s relative to the type 2s, analogous to "conventional" Seyferts. The role of dust in LLAGN, and in particular whether there exists a luminosity, accretion rate, or other property at which the torus ceases to exist, remains unclear. A search for the IR signatures of the torus – thermal emission and silicate emission and absorption features – promises a better understanding of these issues.

Until recently, high-resolution imaging at $\lambda \sim 10$ μm existed only for a handful of relatively bright, well-known objects. More recently, [15] have presented ground-based MIR imaging of a number of LLAGN, with several new detections (see also contributions from Asmus and Fernandez-Ontiveros in this volume). NIR imaging is also available in the literature, but it has rarely been considered in the context of the multi-wavelength emission of LLAGN. Published LLAGN SED compilations contain little or no high-resolution IR data [16, 17].

To establish the overall nuclear IR properties of LLAGN, we have acquired high-resolution, Gemini MIR imaging of 20 IR-faint nuclei with $L_{bol} < 5 \times 10^{42}$ erg s^{-1}. We combine these data with published measurements in the IR and at other wavelengths to produce nuclear spectral energy distributions (SEDs) for these objects. In terms of high spatial resolution IR data, these SEDs are by far the most detailed yet available. Finally, we also analyse archival Spitzer low-resolution spectroscopy for the 18/22 galaxies with available data. In these proceedings we briefly describe the morphological, spectral and broadband characteristics of the 12 galaxies with relatively high Eddington ratios (log $L_{bol}/L_{Edd} > -4.6$; cf median log $L_{bol}/L_{Edd} \sim -6$ for the LINERs and log $L_{bol}/L_{Edd} \sim -4$ for the Seyferts in the Palomar galaxy sample, [2]). Results for the full sample are presented in a forthcoming paper (Mason et al., submitted).

2. Morphology and Silicate Features

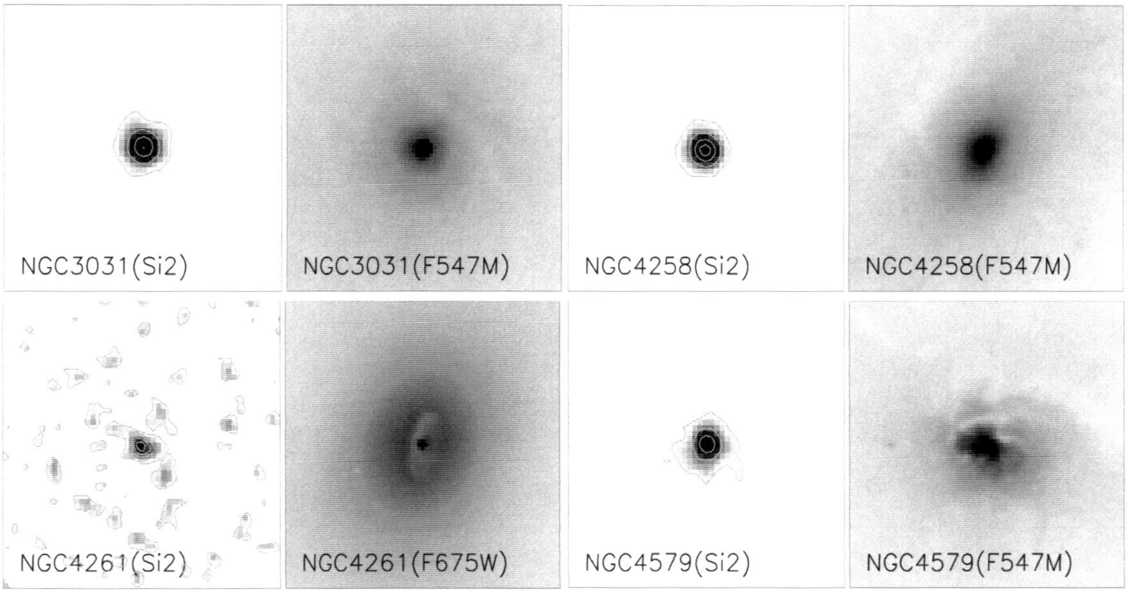

Figure 1. Images of four of the LLAGN at 8.8 μm, along with HST optical images. N is up and E left on the images, and $5'' \times 5''$ regions are shown.

Example 8.8 μm images of four of the LLAGN are presented in Figure 1. Unlike the LLAGN of lower Eddington ratio in the sample, the images of these relatively high-L_{bol}/L_{Edd} objects tend to be dominated by compact nuclei. In higher-luminosity AGN, such as QSOs and "conventional" Seyferts, the torus manifests itself as a bright, pointlike MIR source. To investigate whether the compact MIR nuclei in these LLAGN are evidence of torus emission, we use the well-known MIR/X-ray relation [18, 19] to predict the amount of MIR emission expected from a dusty torus reprocessing high-energy emission from an active nucleus. We fit the MIR and X-ray luminosities of the combined AGN samples of [18, 19] and extrapolate the fit to the luminosities of the LLAGN.

A MIR deficit in the MIR/X-ray plot would be consistent with suggestions that a Seyfert-like torus does not exist in the LLAGN. However, the high-L_{bol}/L_{Edd} LLAGN lie close to the MIR/X-ray relation or exhibit excess MIR emission. In most cases the MIR excess is only significant at the 1-2σ level, but it is noteworthy that only two of the 12 high-L_{bol}/L_{Edd} objects lie to the MIR deficit side of the fit. Similar results are reported by [15]; while finding that

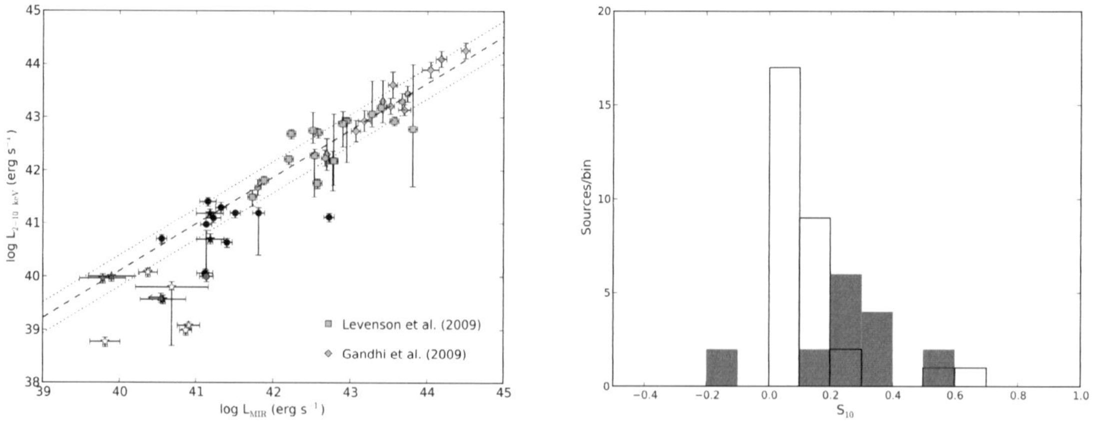

Figure 2. Left: X-ray vs MIR luminosities for the LLAGN, as well as the Seyfert and QSO samples of [18, 19]. Solid black symbols represent the LLAGN discussed in this paper. Right: Strength of the 10 μm silicate feature in the LLAGN with log $L_{bol}/L_{Edd} > -4.6$ (grey) and in the type 1-1.5 Seyferts of [20]. Positive values of S_{10} indicate emission. For clarity, NGC 7479 from the LLAGN sample and UGC 5101 from the Seyfert 1 sample, which have deep absorption features (S_{10} = -2.19 and -1.52 respectively), are not shown.

the MIR/X-ray relation is formally unchanged down to L$\sim 10^{41}$erg s^{-1}, their LLAGN are offset from the relation by about 0.3 dex to the MIR excess side.

[21] find that, even in a sample of Seyfert galaxies with pointlike nuclei in high-resolution MIR imaging, objects with FWHM > 560 r_{sub} (the dust sublimation radius) are systematically offset from the MIR/X-ray relation for "well-resolved" AGN. They attribute this to contamination from nuclear MIR sources other than the torus. All of the LLAGN in this sample have FWHM > 560 r_{sub}, so a contribution from unresolved, non-torus emission sources – nuclear star formation, synchrotron radiation, truncated accretion disk, etc. – is possible. We return to this issue in §4.

9/12 of these LLAGN have been observed with Spitzer's InfraRed Spectrograph. Strikingly, all but two of them exhibit silicate emission bands, in many cases very strong. Figure 2 compares the strength of the 10 μm silicate feature in the LLAGN with the Seyfert 1-1.5 sample of [20]. The LLAGN sample is neither complete nor unbiased, and Figure 2 is not intended to indicate that the distributions of S_{10} differ systematically between the various AGN types. However, it does demonstrate that the 10 μm silicate features found in these particular AGN tend to be comparable to the stronger emission features in the Thompson et al. Seyfert 1 sample. As our sample selection was partly based on the existence of previous, large-aperture photometry, objects with strong silicate emission features could in principle have been more likely to be included in this study. As the flux increases by less than a factor of two across the 10 μm feature, though, it is not clear that this would be a significant effect. In §4 we suggest that the strong silicate emission features may represent a particular stage in the evolution of the torus.

3. Spectral Energy Distributions

Example high-resolution radio – X-ray SEDs of the high-L_{bol}/L_{Edd} LLAGN are shown in Figure 3. The SEDs are quite diverse. If they have one distinguishing feature, it is that many of them are more radio-loud than the average Seyfert galaxy. This is expected from previous work showing that radio loudness increases with decreasing Eddington ratio (e.g. [22]).

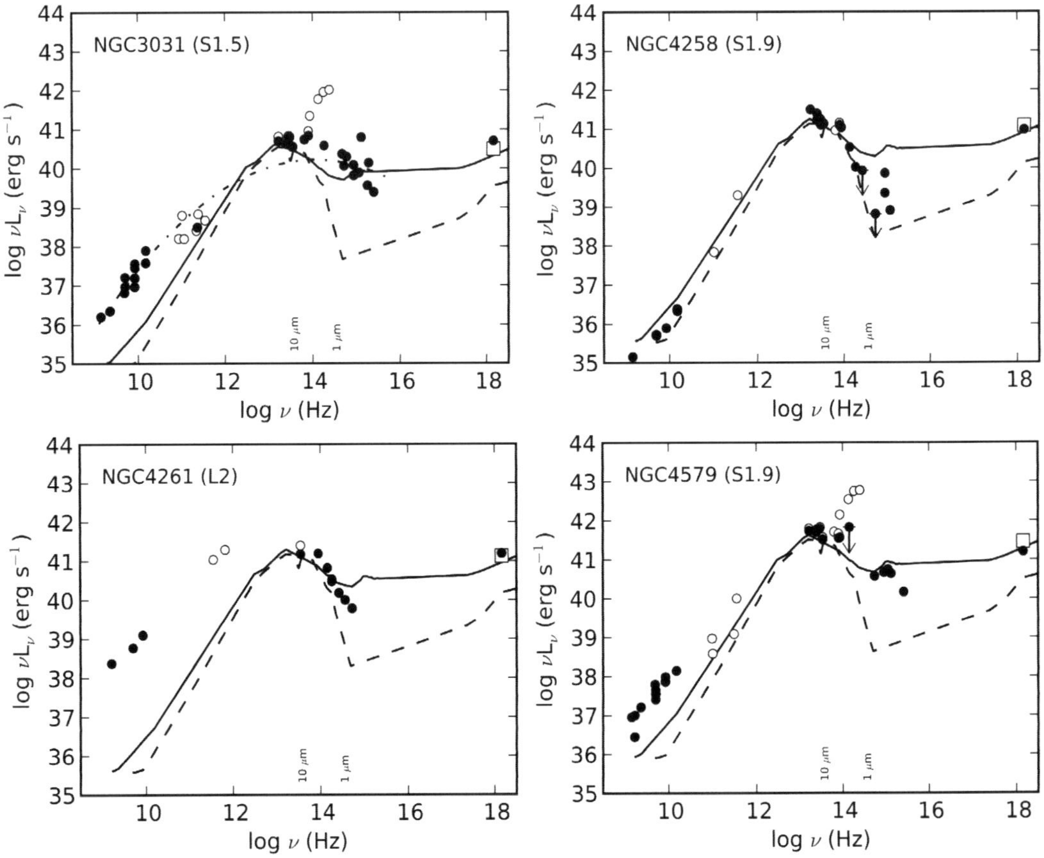

Figure 3. Example SEDs of the LLAGN. Solid points indicate high-resolution data representing the AGN emission. The solid and dashed lines denote the mean type 1 and 2 Seyfert SEDs of [23], respectively.

Some of the SEDs – such as those of NGC 4258, NGC 4261 and NGC 4579 – show a well-defined, Seyfert-like peak at MIR wavelengths. The MIR - NIR/optical spectral slopes in these objects are within the range bracketed by the Seyfert 1 and 2 templates of [23]. Amongst these objects, nuclei of the same AGN type do not necessarily have the same spectral slope; for instance, the type 1.9 Seyferts NGC 4258 and NGC 4579. However, if intermediate-type LLAGN have similar nuclei to more luminous intermediate-type objects, then a variety of spectral shapes is to be expected; [24] find a wide range of spectral indices in Seyferts of types 1.8 and 1.9 in the CfA sample. Judged solely by their SEDs, there is little indication that these objects are anything but "scaled-down" Seyferts, complete with a MIR "dust bump", albeit with extra radio emission.

A few of the galaxies – NGC 3031, for example – have rather flat MIR – NIR/optical SEDs compared to the Seyfert templates, and a less well-defined MIR peak. These are also characteristics of the radio-loud, low-Eddington ratio objects in the sample, whose IR emission we find to be dominated by the jets (Mason et al., submitted). In some objects, such as NGC 1097, the "excess" NIR/optical emission is probably related to a young, massive nuclear star cluster [25, 26].

Overall, many of the high-Eddington ratio LLAGN have SEDs that broadly resemble those of conventional Seyferts, although often with enhanced radio emission. In some cases, though, the MIR – NIR/optical SED slopes are flatter than expected for Seyferts, which hints at possible stellar and/or jet contributions in the IR. Other sources, such as a truncated accretion disk, may also influence the IR properties of the SEDs. Detailed modeling of the SEDs with jet, RIAF, thin disk and torus components is beyond the scope of the present work, but in §4 we suggest a scenario which is consistent with the morphological, spectral and broadband SED characteristics of these LLAGN.

4. Discussion

The new, high-resolution IR imaging and SEDs presented here represent the first step towards establishing the nuclear IR properties of a significant number of LLAGN. At the median distance of the galaxies studied, 16.8 Mpc, the AGN is well isolated in objects with log $L_{2-10\,\mathrm{keV}} > 40.5$ erg s^{-1} (log $L_{\mathrm{bol}} > 41.8$ erg s^{-1}). Deeper observations would allow a cleaner detection of the pointlike central engine in less luminous or more distant objects, but the cost in observing time would be large with current facilities.

The log $L_{\mathrm{bol}}/L_{\mathrm{Edd}} > -4.6$ nuclei discussed in these proceedings, which lie at the high luminosity end of the sample, tend to have prominent nuclear point sources in the MIR. They often have strong silicate emission features compared to those typically observed in Seyfert 1 nuclei, and their MIR luminosities are consistent with or slightly in excess of those predicted by the standard MIR/X-ray relation. The nuclear broadband SEDs of these objects are rather mixed. Some are essentially indistinguishable from those of "conventional" Seyferts, many have excess radio emission compared to higher-luminosity Seyferts, and a few have unusually flat MIR - NIR/optical slopes.

Considered in isolation, these characteristics offer no compelling evidence that the torus is absent in these LLAGN or that it differs from that observed in higher-luminosity Seyferts. However, we do find some indications that the IR emission may not arise in the standard, Seyfert-like torus of the AGN unified model. In Figure 4, we plot the strength of the 10 μm silicate feature against HI column density, for both the Seyfert galaxies of [27] and the high-$L_{\mathrm{bol}}/L_{\mathrm{Edd}}$ LLAGN. In the Seyferts of [27], S_{10} and N_{H} are loosely correlated. The scatter presumably reflects the fact that the observed S_{10} is likely a complicated function of the precise arrangement of clouds in the torus [28], and may also indicate a contribution from absorption in the host galaxy [29, 30, 31]. Many of the high-Eddington ratio LLAGN lie on the upper envelope of the Seyfert points, with relatively high values of S_{10} per unit N_H. One possible explanation is that the dust-to-gas ratio is lower in these particular objects than in most Seyfert galaxies. This may be expected if the torus is an optically thick region of an accretion disk wind that becomes less powerful at low accretion rates [32], when the amount of material reaching the dust sublimation radius and able to form dust grains is reduced. In this case we may expect either a torus with fewer clouds and a higher probability of observing a hot, directly-illuminated cloud face, or simply optically thin dust emission. Both of these configurations would cause relatively strong silicate emission features, consistent with our finding that S_{10} in many of these high-Eddington ratio LLAGN is strong compared to that in typical type 1 Seyferts (Figure 2).

A contribution to the observed silicate features from circumstellar shells in the host galaxy is also possible [33, 34]. As the luminosity of the central engine diminishes, the relative strength of features arising in the surrounding stellar population will increase. This could also account for changes in S_{10}/N_{H} among LLAGN.

The high-Eddington ratio LLAGN have at least as much MIR emission as predicted by the MIR/X-ray relation for Seyferts and quasars. If the torus in these LLAGN does indeed contain less dust, the IR continuum emission must be produced by some other mechanism. A contribution from synchrotron radiation is likely, and may explain the unusually flat MIR

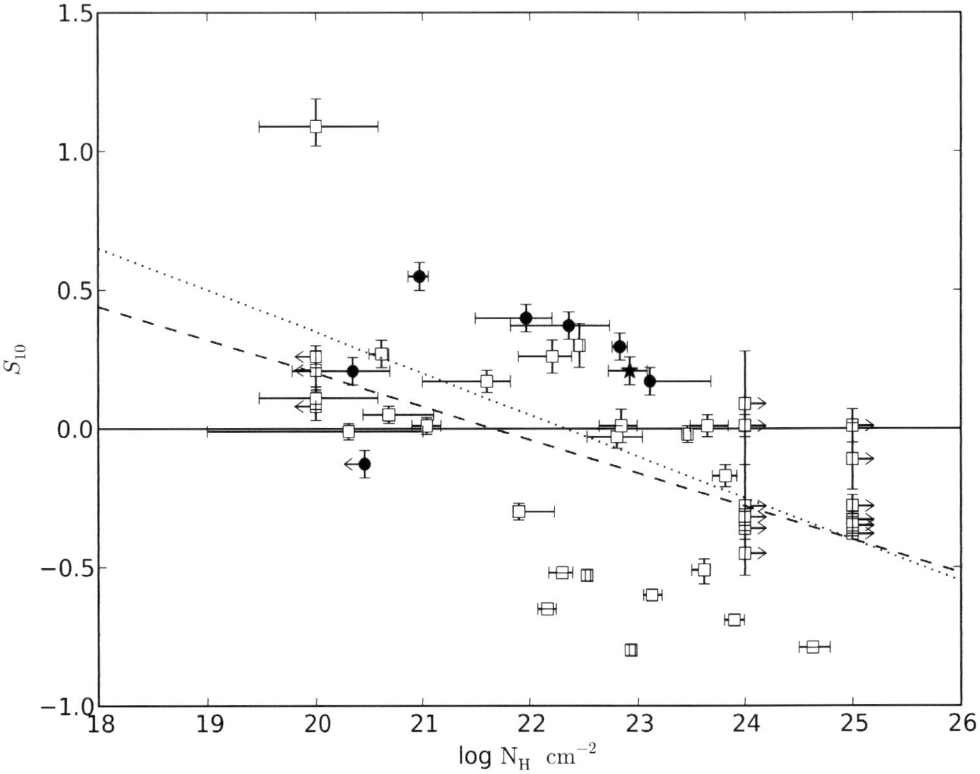

Figure 4. Strength of the 10 μm silicate feature vs X-ray column density for the high-Eddington ratio LLAGN (black). Square symbols indicate the Seyfert galaxies of [27]. The dashed line shows Shi et al.'s fits to their Seyferts, the dotted line their fit to their whole sample (mostly comprised of Seyferts and various classes of quasar). Positive values of S_{10} indicate emission, negative values absorption. As in Figure 2, NGC 7479, which has a deep absorption feature ($S_{10} = -2.19$), is omitted for clarity.

-NIR/optical slopes observed in some of the SEDs. Another possibility is that some of the IR emission comes from a truncated accretion disk. In the models of [4], the disk emission peaks between 1 – 10 μm and can account for essentially all the luminosity at these wavelengths. Emission associated with nuclear star clusters, such as that known to exist in NGC 1097 [25] may also play a rôle. However, [15] find that star formation contributes <30% of the small-scale 12 μm flux in most LLAGN of comparable luminosity to those studied in this paper. Detailed modelling of the LLAGN, taking advantage of the new constraints from the well-sampled IR SEDs presented here, will provide more insight into the processes responsible for the IR emission of LLAGN and the role of the torus in these objects.

Acknowledgments
This paper is based on observations obtained at the Gemini Observatory, which is operated by the Association of Universities for Research in Astronomy, Inc., under a cooperative agreement with the NSF on behalf of the Gemini partnership: the National Science Foundation (United States), the Science and Technology Facilities Council (United Kingdom), the National Research Council (Canada), CONICYT (Chile), the Australian Research Council (Australia), Ministério

da Ciência e Tecnologia (Brazil) and SECYT (Argentina). A.A.-H. and L. C. acknowledge support from the Spanish Plan Nacional de Astronomía y Astrofísica under grants AYA2009-05705-E and AYA2010-21161-C02-1.

[1] Yuan F 2007 Advection-dominated Accretion: From Sgr A* to Other Low-luminosity AGNs *The Central Engine of Active Galactic Nuclei* (*Astronomical Society of the Pacific Conference Series* vol 373) ed L C Ho & J-W Wang pp 95–+
[2] Ho L C 2009 *ApJ* **699** 626–637
[3] Trump J R, Impey C D, Kelly B C, Civano F, Gabor J M, Diamond-Stanic A M, Merloni A, Urry C M, Hao H, Jahnke K, Nagao T, Taniguchi Y, Koekemoer A M, Lanzuisi G, Liu C, Mainieri V, Salvato M and Scoville N Z 2011 *ApJ* **733** 60
[4] Nemmen R, Storchi-Bergmann T and Eracleous M 2011 *ArXiv e-prints 1112.4640*
[5] Wada K, Papadopoulos P P and Spaans M 2009 *ApJ* **702** 63–74
[6] Schartmann M, Burkert A, Krause M, Camenzind M, Meisenheimer K and Davies R I 2010 *MNRAS* **403** 1801–1811
[7] Vollmer B, Beckert T and Davies R I 2008 *A&A* **491** 441–453
[8] Konigl A and Kartje J F 1994 *ApJ* **434** 446–467
[9] Elitzur M and Shlosman I 2006 *ApJl* **648** L101–L104
[10] Maoz D, Nagar N M, Falcke H and Wilson A S 2005 *ApJ* **625** 699–715
[11] Zhang W M, Soria R, Zhang S N, Swartz D A and Liu J F 2009 *ApJ* **699** 281–297
[12] Barth A J, Filippenko A V and Moran E C 1999 *ApJ* **525** 673–684
[13] González-Martín O, Masegosa J, Márquez I and Guainazzi M 2009 *ApJ* **704** 1570–1585
[14] Sturm E, Rupke D, Contursi A, Kim D C, Lutz D, Netzer H, Veilleux S, Genzel R, Lehnert M, Tacconi L J, Maoz D, Mazzarella J, Lord S, Sanders D and Sternberg A 2006 *ApJl* **653** L13–L16
[15] Asmus D, Gandhi P, Smette A, Hönig S F and Duschl W J 2011 *A&A* **536** A36
[16] Ho L C 1999 *ApJ* **516** 672–682
[17] Eracleous M, Hwang J A and Flohic H M L G 2010 *ApJS* **187** 135–148
[18] Gandhi P, Horst H, Smette A, Hönig S, Comastri A, Gilli R, Vignali C and Duschl W 2009 *A&A* **502** 457–472
[19] Levenson N A, Radomski J T, Packham C, Mason R E, Schaefer J J and Telesco C M 2009 *ApJ* **703** 390–398
[20] Thompson G D, Levenson N A, Uddin S A and Sirocky M M 2009 *ApJ* **697** 182–193
[21] Horst H, Gandhi P, Smette A and Duschl W J 2008 *A&A* **479** 389–396
[22] Sikora M, Stawarz Ł and Lasota J P 2007 *ApJ* **658** 815–828
[23] Prieto M A, Reunanen J, Tristram K R W, Neumayer N, Fernandez-Ontiveros J A, Orienti M and Meisenheimer K 2010 *MNRAS* **402** 724–744
[24] Alonso-Herrero A, Quillen A C, Rieke G H, Ivanov V D and Efstathiou A 2003 *AJ* **126** 81–100
[25] Storchi-Bergmann T, Nemmen R S, Spinelli P F, Eracleous M, Wilson A S, Filippenko A V and Livio M 2005 *ApJl* **624** L13–L16
[26] Mason R E, Levenson N A, Packham C, Elitzur M, Radomski J, Petric A O and Wright G S 2007 *ApJ* **659** 241–249
[27] Shi Y, Rieke G H, Hines D C, Gorjian V, Werner M W, Cleary K, Low F J, Smith P S and Bouwman J 2006 *ApJ* **653** 127–136
[28] Hönig S F and Kishimoto M 2010 *A&A* **523** A27
[29] Roche P F, Packham C, Aitken D K and Mason R E 2007 *MNRAS* **375** 99–104
[30] Deo R P, Richards G T, Crenshaw D M and Kraemer S B 2009 *ApJ* **705** 14–31
[31] Alonso-Herrero A, Ramos Almeida C, Mason R, Asensio Ramos A, Roche P F, Levenson N A, Elitzur M, Packham C, Rodríguez Espinosa J M, Young S, Díaz-Santos T and Pérez-García A M 2011 *ApJ* **736** 82
[32] Elitzur M and Ho L C 2009 *ApJl* **701** L91–L94
[33] Bressan A, Panuzzo P, Buson L, Clemens M, Granato G L, Rampazzo R, Silva L, Valdes J R, Vega O and Danese L 2006 *ApJl* **639** L55–L58
[34] Buson L, Bressan A, Panuzzo P, Rampazzo R, Valdés J R, Clemens M, Marino A, Chavez M, Granato G L and Silva L 2009 *ApJ* **705** 356–360

Astronomy at High Angular Resolution 2011 IOP Publishing
Journal of Physics: Conference Series **372** (2012) 012037 doi:10.1088/1742-6596/372/1/012037

The dust emission of high-redshift quasars

C Leipski[1], K Meisenheimer[1]

[1] Max-Planck Institut für Astronomie (MPIA), Königstuhl 17, D -69117 Heidelberg, Germany

E-mail: `leipski@mpia-hd.mpg.de`

Abstract.
The detection of powerful near-infrared emission in high redshift ($z > 5$) quasars demonstrates that very hot dust is present close to the active nucleus also in the very early universe. A number of high-redshift objects even show significant excess emission in the rest frame NIR over more local AGN spectral energy distribution (SED) templates. In order to test if this is a result of the very high luminosities or redshifts, we construct mean SEDs from the latest SDSS quasar catalogue in combination with MIR data from the WISE preliminary data release for several redshift and luminosity bins. Comparing these mean SEDs with a large sample of $z > 5$ quasars we could not identify any significant trends of the NIR spectral slope with luminosity or redshift in the regime $2.5 < z \lesssim 6$ and $10^{45} < \nu L_\nu(1350\text{Å}) \lesssim 10^{47}$ erg/s. In addition to the NIR regime, our combined *Herschel* and *Spitzer* photometry provides full infrared SED coverage of the same sample of $z > 5$ quasars. These observations reveal strong FIR emission ($L_{\rm FIR} \gtrsim 10^{13}\,L_\odot$) in seven objects, possibly indicating star-formation rates of several thousand solar masses per year. The FIR excess emission has unusally high temperatures ($T \sim 65$ K) which is in contrast to the temperature typically expected from studies at lower redshift ($T \sim 45$ K). These objects are currently being investigated in more detail.

1. Introduction

The presence of dust seems to be a ubiquitous property of galaxies throughout the observable universe. Even the most distant quasars at $z \sim 6$ show evidence for copious amounts of dust (e.g. Bertoldi et al. 2003; Wang et al. 2008; Leipski et al. 2010). This indicates the rapid metal enrichment of the interstellar medium within the first billion years after the big bang. Assuming that the observed far-infrared (FIR) emission of these objects is powered by star-formation, the luminosies imply star formation rates of up to a few thousand solar masses per year, possibly indicating the rapid formation of early galactic bulges.

For a more comprehensive picture of the dust emission at high redshifts we are currently analyzing PACS (100+160 μm) and SPIRE (250+350+500 μm) photometry of 71 quasars at $z > 5$ that have been obtained as part of our *Herschel* key project "The Dusty Young Universe". Complementary *Spitzer* data at shorter wavelengths were secured by our group to enable the study of the full optical through infrared SED of a large sample of quasars in the early universe.

In the course of this project we have now identified a number of high-z ($z > 5$) quasars with considerable FIR emission. They are clearly detected in our PACS and SPIRE observations and were previously unknown to have large infrared luminosities (Fig. 1). In combination with our existing *Spitzer* photometry, the *Herschel* data allow us study the full SED of these objects in the rest frame wavelength range 0.5 – 80 μm, which – most importantly – also covers the FIR peak of the SED.

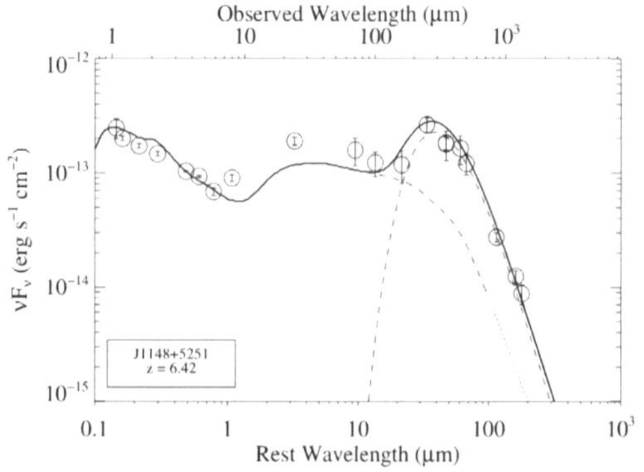

Figure 1. SED for the $z = 6.42$ quasar SDSS J1148+5251 including new *Herschel* data as well as information from the literature. *Left*: The quasar SED was fitted by a linear combination of the SDSS quasar template (Richards et al. 2006) and a modified black body (representing dust emission from star formation). Despite a reasonable match in the rest frame UV/optical and in the FIR, there exist a clear discrepancy between $\sim 1 - 10\,\mu$m.

2. The signatures of hot dust in the rest frame near infrared

Most powerful AGN are strong sources of infrared radiation. This emission originates from dust heated by the accretion disk to various temperatures depending on its distance from the nucleus and its spatial distribution. In such a configuration, the sublimation temperature of the dust (~ 1500 K) sets the inner boundary of the dust distribution in AGN. The dust at these temperatures predominately re-emits the absorbed energy in the rest frame near infrared. Thus, the spectral energy distribution in this regime provides us with a tool to study the AGN dust only a few parsec out from the core.

Using *Herschel* data and archival information, we noticed that common AGN templates (like the SDSS quasar template, Richards et al. 2006) do not match the amounts of hot dust emission between $\sim$1–10 μm in the UV/optical through sub-mm SED of the $z = 6.42$ quasar SDSS J1148+5251 (Leipski et al. 2010; Fig. 1). Interestingly, in the wavelength range where the discrepancy is most apparent ($\sim 1 - 5\,\mu$m), the SED is particularly sensitive to the radial distribution of the innermost dust (however, this is not the only parameter that can influence the shape of the SED in this region; see Schartmann et al. 2008, Hönig & Kishimoto 2010).

The observed mismatch might be due to SDSS J1148+5251 being exceptional with a very high redshift ($z = 6.42$) and having extreme luminosity ($L_{bol} \sim 10^{47}$ erg/s). It is conceivable to assume that the dust distribution might be sensitive to the total luminosity of the heating source (i.e. the accretion disk) or even the evolutionary state of the system (i.e. high-z vs. low-z).

The SDSS template is largely dominated by quasars around a redshift of $z \sim 2$. Only few objects at $z > 3$ are included. Furthermore, while all objects have SDSS and *Spitzer* IRAC data available, many objects lack NIR photometry (which was in these cases determined from scaling the Elvis et al. (1994) quasar template to the nearest observed data point). However, for a robust comparison with our $z > 5$ quasar SEDs – in particular with an emphasis on slope in the hot dust component – we need a sample that provides appropriate numbers of high-redshift quasars with observational data that cover the typical "1 μm inflection" in the quasar SEDs and sample the onset of the hot dust emission. In addition, the indidviual SEDs used to construct the SDSS template have been corrected for host galaxy contributions (see Richards et al. (2006) for details) which has the largest effect (~ 35 % host contribution) in the rest frame NIR where the assumed elliptical host SED peaks.

In order to build template SEDs that cover higher redshifts at various luninosities, we here use as a starting point the DR7 SDSS quasar sample presented by Shen et al. (2011). These authors also provide their quasar table matched with the preliminary data release from the WISE (Wright et al. 2010) MIR survey. This results in ~36000 quasars with SDSS optical phtometry and WISE MIR detections in three bands (3.4, 4.6, 12 μm; we here do not consider the 22 μm band in which many objects are not detected). We further limited the objects to redshifts greater than 2.5 so that the WISE photometry samples the hot dust part of the rest-frame quasar SED. From the remaining sources we then created mean SEDs in a fairly straight forward manner by first de-redshifting the sources into the rest frame, then interpolating them linearly in log space onto a common wavelength grid, normalzing them at 1 μm (rest frame) and eventually calculating the mean for various subsamples. We then compared these mean SEDs with our sample (Fig. 2). Specifically, we searched for trends with increasing redshift (at fixed luminosity) or with luminosity (at fixed redshift). A few important findings are:

- We see quite a large spread in SED shapes for our sample as well as in the comparison samples after normalization, which results in substantial error bars (here: mean absolute deviation). Such a spread of quasar SEDs was also seen by e.g. Elvis et al. 1994 and Richards et al. 2006 in their studies.
- No trend (within the errors) is observed when increasing the luminosity over 2 orders of magnitude ($\sim 10^{45}$–10^{47}) while keeping the redshift bin fixed ($2.5 < z < 3$).
- Equally no trend (within the errors) can be found with redshift between $2.5 < z < 4$ when using only sources at comparable luminosity to the high-z objects.

This implies that the hot dust properties do not change significantly with either redshift or luminosity for the parameter space considered here. In fact, the diversity of individual quasar SEDs appears much greater than any differences between the sample mean values.

We stress, however, that this study and the results presented here have to be considered preliminary. We are currently looking into expanding the observational data base for the quasars by including NIR photometery from the UKIDSS survey. This will allow us to expand this study to even lower redshifts while still appropriately sampling the hot-dust part of the SED. Also, no correction for host galaxy contributions has been performed at this point which could be especially important for the lower luminosity objects. Such a correction could potentially influence the spectral shape in the NIR where the stellar population of an elliptical galaxy of old or intermediate age is brightest.

In fact, in a recent MIR interferometric study, Kishimoto et al. (2011) report a steepening of the radial dust distribution in the innermost few parsec with increasing luminosity. In the SEDs this would translate into a steeper NIR spectral slope for objects at higher luminosity which indicates that a carefull accounting of the host galaxy contribution may prove to be important for the less luminous sources.

3. Far-infrared emission powered by star formation

About 30% of the known luminous $z \sim 6$ quasars are detected at 250 GHz and/or in CO (e.g. Bertoldi et al. 2003; Wang et al. 2008). Such studies suggest that most of the rest-frame far-infrared (FIR) emission comes from massive star formation, possibly indicating the formation of early galactic bulges. Thus, these objects signify an important stage in the connection between the build-up of stellar mass and black hole growth.

However, for the vast majority of high-redshift objects we lack full FIR/sub-mm spectral energy distributions (SEDs). Far-infraed luminosities (L_{FIR}) and dust masses (M_{dust}) are commonly determined using single photometric measurements (typically at 250 GHz) and applying standard values for the dust temperature as determined from lower redshift objects. But the questions remains how far these assumptions are appropriate for the high-redshift objects.

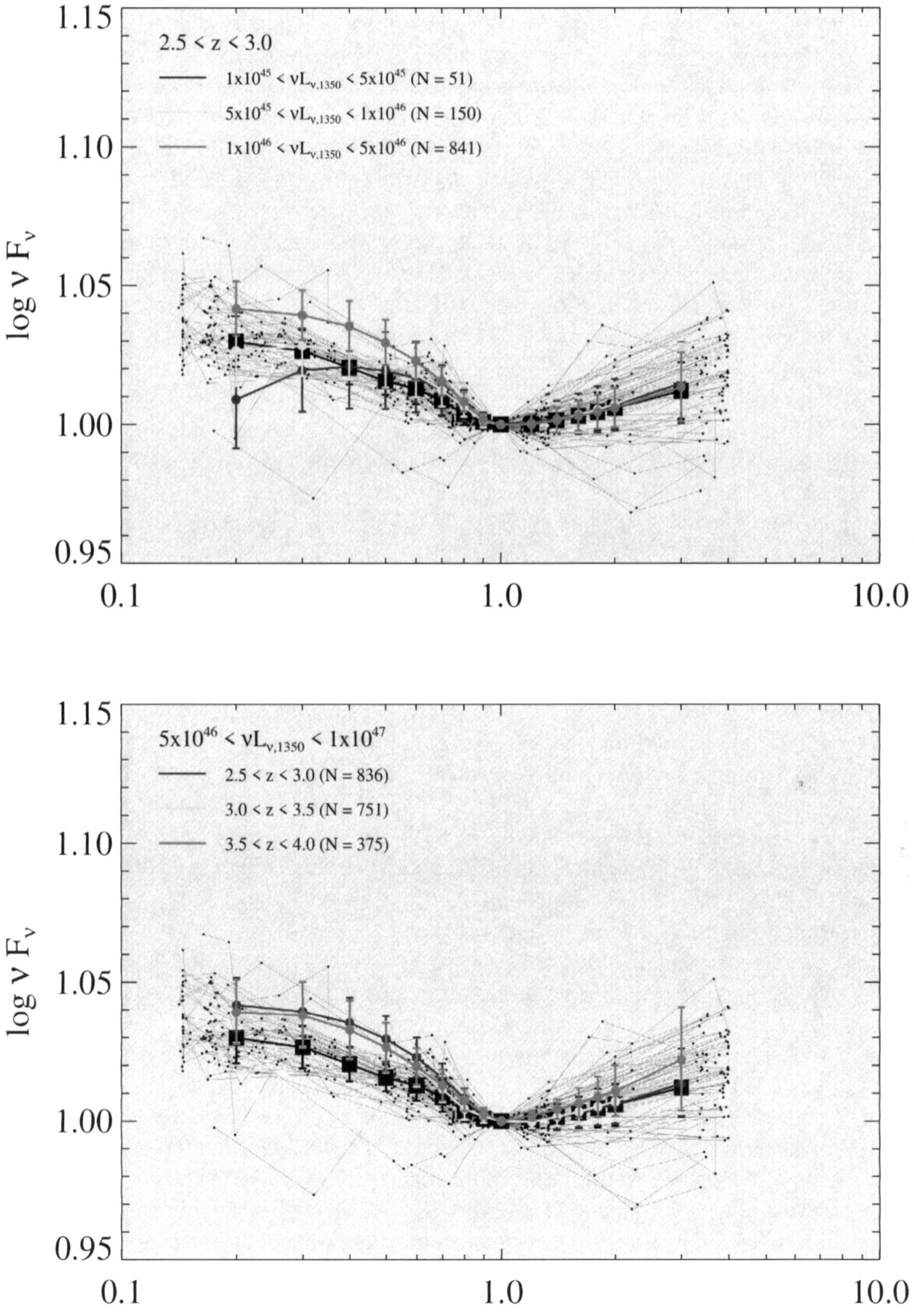

Figure 2. Comparison of normalized SEDs. In both panels the black points indicate the observed photometry for our $z > 5$ sample (here 68 individual sources are shown). The black squares represents a mean SED for our sample created in a similar fashion as for the comparison SEDs. *Top*: Our sample compared with mean SEDs for various luminosity intervals but a common redshift regime $2.5 < z < 3.0$. We give the luminosity intervals and number of objects used to create a specific SED in the plot. *Bottom*: Same as above, but now for a fixed luminosity, and creating the mean SEDs for various redshift intervals.

From our new photometry providing full SED information in the rest frame wavelength range $0.5 - 80\,\mu$m, we have discovered seven quasars at $z > 5$ with considerable FIR excess emission, including SDSS J1148+5251 for which strong FIR and mm emission has been reported previously (e.g. Bertoldi et al. 2003, Beelen et al. 2006). In order to extract additional information from the SEDs, we fitted the observed photometry with a linear combination of a power-law in the UV/optical/NIR (emission from the accretion disk), a clumpy torus model (AGN heated dust from Hönig & Kishimoto (2010)), and a modified black body (starburst-powered excess dust emission). The emmissivity value β for this last component was fixed to a value of 1.6 to be comparable to previous studies at high z. In all cases the additional FIR component was required to describe the observed SEDs at the longest wavelengths (see Fig. 3 for a few examples). These SED fits also allow us to determine quantative parameters: first, the scaling of the torus dust emission can be converted into the bolometric luminosity required to power the observed emission (Hönig & Kishimoto 2010). For the sources considered here we find $L_{bol} = 1 - 5 \times 10^{47}$ erg/s using this approach. These bolometric luminosities translate into dust sublimation radii of $\sim 3 - 5$ pc. We also see that in the combined fits, torus models with steep radial dust distributions are preferred, indicating that most of the dust resides close to the nucleus. Integrating only the FIR excess component between 8 and $1000\,\mu$m we determine luminosities of $\gtrsim 10^{13}\,L_{\odot}$. If we identify this component as being powered by star formation then we observe star formation rates of the order of a few times 10^3 solar masses per year in these early quasar host galaxies. For SDSS J1148+5251 the star-formation rate we determine from the SED fits is consistent with the results from a [CII] imaging study (Walter et al. 2009). These spatially resolved observations also show that the bulk of the detected star formation takes place in the innermost 1.5 kpc of the host galaxy.

One interesting fact related to the FIR excess emission is that the temperature of this component in all sources is fairly high ($T \sim 65$ K), much higher than the $T \sim 45$ K typically found for lower redshift objects (e.g. Beelen et al. 2006). This latter value is commonly used to extrapolate from the 250 GHz measurements in high-redshift quasars (e.g. Wang et al. 2008). Since the determination of dust masses is quite sensitive to the temperature of the dust, the measured difference in FIR temperatures could potentially have an impact on the dust masses derived for the high-redshift quasars.

Our multi-wavelength imaging allows us to check for possible confusion with other (lower redshift) FIR bright sources which could contribute in the large SPIRE beams and mimick the higher FIR dust temperatures. No such confusion problems could readily be identified, at least on scales of a few arc seconds. We might also have to consider quasar heating of the dust as a possible soure to raise the temperature of FIR emitting dust. The very luminous AGN puts out copious amounts of radiation which in principle might be able to heat dust in the host galaxy to the temperatures we observe. In such a case, only some fraction of the FIR luminosity will actually be powered by star formation and in our calculations above we would overestimate the star-formation rates in these objects. A more detailed analysis of these highly interesting sources is currently underway (Leipski et al. 2012, in preparation) and follow-up strategies have been initiated.

4. Summary

We present first results from our *Herschel* and *Spitzer* observations of $z > 5$ quasars. We find that existing quasar templates do not match the amount of hot dust emission observed in almost all high-z quasars. Using the latest SDSS quasar catalogue including data from the WISE MIR survey we do not detect any significant trends in the shape of the NIR SED when considering mean SEDs at various redshifts and luminosities. Our full SED coverage in the infrared reveals considerable FIR excess emission in seven objects. If powered by star formation, the FIR luminosities determined from SED fitting indicate star-formation rates of thousands of

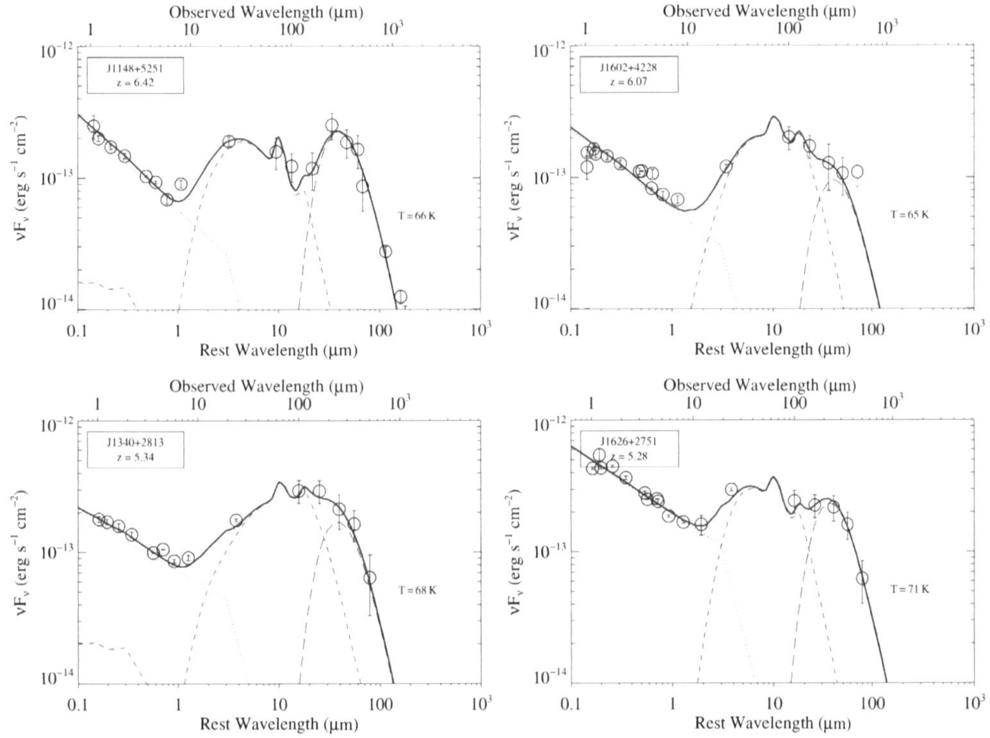

Figure 3. Full optical through infrared SEDs of four sources showing FIR excess emission with high temperatures. In order to disentangle AGN and starburst powered emission, the photometry has been fitted by a linear combination of three components: a power-law in the UV/optical/NIR (dotted line), a clumpy torus model (dashed line), and a modified black body (long-dashed line). The blue solid line shows the total fit. For all sources shown here, the temperature of the FIR component is T $\sim$ 65 K. Note also that compared to Fig. 1 the AGN dominated MIR dust emission in SDSS J1148+5251 is much better described by this model, as is the overlap region of the AGN and the starburst "bump".

solar masses per year. Furthermore, the FIR emission in these objects shows unusually high temperatures ($\sim$ 65 K as opposed to 45 K at lower z). The analysis on these aspects is ongoing.

Acknowledgments
We would like to thank M. Schartmann and S. Hönig for providing their respective torus models as well as for their valuable insight into the modeling process.

References
Beelen A, Cox P, Benford D J et al. 2006, *Astrophysical Journal*, **642**, 694
Bertoldi F, Carilli C L, Cox P et al. 2003, *Astronomy & Astrophysics*, **406**, L55
Elvis M, Wilkes B J, McDowell J C et al. 1994, *Astrophysical Journal Supplement Series*, **95**, 1
Hönig S and Kishimoto M 2010, *Astronomy & Astrophysics*, **523**, 27
Kishimoto M, Hönig S F, Antonucci R et al. 2011, *Astronomy & Astrophysics*, **536**, A78
Leipski C, Meisenheimer K, Klaas U et al. 2010, *Astronomy & Astrophysics*, **518**, L34
Richards G T, Lacy M, Storrie-Lombardi L J et al. 2006, *Astrophysical Journal Supplement Series*, **166**, 470
Walter F, Riechers D, Cox P et al. 2009, *Nature*, **457**, 699
Wang R, Carilli C L, Wagg J et al. 2008, *Astrophysical Journal*, **687**, 848
Wright E L, Eisenhardt P R M, Mainzer A K, et al. 2010, *Astronomical Journal*, **140**, 1868

Journal of Physics: Conference Series **372** (2012) 012038 doi:10.1088/1742-6596/372/1/012038

Multiple-Line Study of NGC 1068: Hot Molecular Gas Caused by Jet-Gas Interaction in the Central 100pc?

Melanie Krips

IRAM, Domaine Universitaire, 300 Rue de las Piscine, 38400 Saint Martin d'Héres, France

E-mail: krips@iram.fr

Abstract. A multiple molecular line and line transition study is presented for the circumnuclear disk (CND) of the proto-typical Seyfert galaxy NGC 1068. A detailed analysis of the kinematics and excitation conditions of the molecular gas, as traced by ^{12}CO, ^{13}CO, HCN and HCO$^+$, suggests that part of the molecular gas in the CND is shocked, expanding and heated to high kinetic temperatures most likely as a consequence of an interaction between the radio jet and the CND. We further find support for an X-ray altered chemistry of the molecular gas in the CND based on the significantly elevated abundance of HCN when compared to star-forming, starbursting or quiescent gas regions.

1. Introduction

The central 100 pc of active galaxies present fascinating laboratories to study not only the feeding mechanisms of their activities, wether in form of an accreting supermassive black hole (SMBH) or a starburst (SB), but also the feedback of these activities onto their environments. While the understanding of feeding processes enforces detailed analyses of the dynamics of the material being accreted, the disclosure of the feedback mechanisms necessitates profound knowledge of the chemical composition and physical properties of the surrounding material, i.e., mainly the molecular gas. The kinematics of the molecular gas can be best studied by ^{12}CO(J=1–0) that traces the general characteristics of the molecular gas. Concerning the chemistry of the gas, a multitude of different molecular species, their different transitions and isotopes are necessary to understand the effect of different feedback processes, whether they are mechanical nature, i.e., via shocks, winds and/or jets, or radiative nature, i.e., via X-ray, ultra-violet or cosmic radiation.

A significant set of active galaxies exist whose ^{12}CO emission has been observed at high angular and spectral resolution (e.g., NUGA series of papers: [1], [2], [3], [4], [5], [6], [7], [8]; see also [9], [10]) allowing to draw first conclusions on accretion mechanisms in active galaxies. Similarly high amounts of multiple line data sets do not yet exist in order to study the feedback in the centers of active galaxies; only a few case studies have been conducted so far (e.g., M82, NGC 253, IC342, and NGC 1068; see for instance: [11], [12], [13], [14]). In these studies, starburst galaxies are dominating as they are rich in molecules and hence possess bright and thus easily detectable molecular line emission. Active galaxies dominated by accretion onto a SMBH show less bright molecular emission and are hence more difficult to be detected in several molecular species and transitions at high angular and spatial resolution so that only the brightest and closest objects among them are being studied so far.

NGC 1068 most likely represents the most prominant nearby galaxy dominated by its SMBH. It is classified as a Seyfert type-2 galaxy and has been the posterchild for the viewing angle unification theory [15] since a long time due to the discovery of an obscured Seyfert type-1 nucleus in its center [16]. NGC 1068 was also the first galaxy to show significantly enhanced HCN emission with respect to the ^{12}CO emission in its center as opposed to SB/star-forming galaxies such as M82 (e.g., [17], [18]). It has been already suggested early on that the enhanced HCN emission is due to an increased abundance of HCN that is caused by the strong X-ray radiation of the SMBH creating a giant X-ray dominated region (XDR) there (e.g., [17], [18], [19]). XDRs are defined in a similar way to the photon-dominated regions (PDRs) in SB galaxies (as in M82, e.g. [20]) except that the chemistry altering radiation are X-rays there and not UV-rays. After the discovery of such strong central HCN emission in NGC 1068, more Seyfert/LINER galaxies have been found to exhibit enhanced HCN emission, such as M51 and NGC 6951 for instance (e.g., [21], [22], [10], [23], [24]). A multi-transition study carried out with IRAM 30m telescope of the HCN and HCO^+ emission in a series of nearby active galaxies, including SB and SMBH dominated galaxies, support the idea of enhanced HCN abundances in the centers of SMBH when compared to the SB dominated galaxies in the sample [24]. Also, indications for increased kinetic temperatures were found for the SMBH galaxies in this study (e.g., NGC 1068 and M51) which in turn could lead to a hypo-excitation of the ^{12}CO(J=1–0) emission and hence to high HCN(J=1–0)/^{12}CO(J=1–0) line ratios falsely simulating enhanced HCN emission.

Together with the detection of other species such as SiO and CN in the CND of NGC 1068 (see [25]), a quite complicated puzzle started to form for the chemistry, dynamics and excitation condititons of the molecular gas. This article discusses the interferometric observations of several ^{12}CO, ^{13}CO, HCN and HCO^+ line transitions at high angular and spectral resolution (see also [14]) in order to shed more light onto the complex nature of the molecular gas in the CND of this source.

2. The Molecular Gas Emission in NGC 1068's CND

The low-J transitions (J=1–0 and J=2–1) of the ^{12}CO, ^{13}CO, HCN and HCO^+ line emission in the CND of NGC 1068 have been observed with the IRAM Plateau de Bure Interferometer (PdBI) while the higher-J transitions (J=2–1 and higher) with the Submillimeter array (SMA). A detailed description of the observations can be found in [14]. The velocity integrated line emission of each observed molecule is shown in Fig.1. All molecules are more or less detected along the ring-like structure of the CND in NGC 1068 that splits up into a western and eastern knot. The weaker molecules mostly show up in the northen "bridge" part of the CND. The J=1–0 transition of HCN and HCO^+ is not plotted here as they will be published later by Usero et al.. However, their integrated intensities have been used to calculate the line ratios for HCN and HCO^+.

2.1. Dynamics of the Molecular Gas

At first sight, the kinematics of the molecular gas within the CND seems to follow normal Keplerian rotation (see grey line profiles of the molecules in Fig.2 and Fig.5 in [14]). However, by looking in more detail into the dynamics of the different molecules, i.e., separating the eastern and western knot, one finds that redshifted and blueshifted emission is found towards both knots (see also Fig. 3) indicating a much more complex underlying kinematics. Standard Keplerian rotation leads to well spatially separated red- and blueshifted emission. [9] argued early on that the CND in NGC 1068 might be warped. Standard (symmetric) tilted ring models, however, cannot explain such an overlap of red- and blueshifted emission. Elliptical orbits, or more precisely non-circular motions, could be one possible explanation. However, given that the

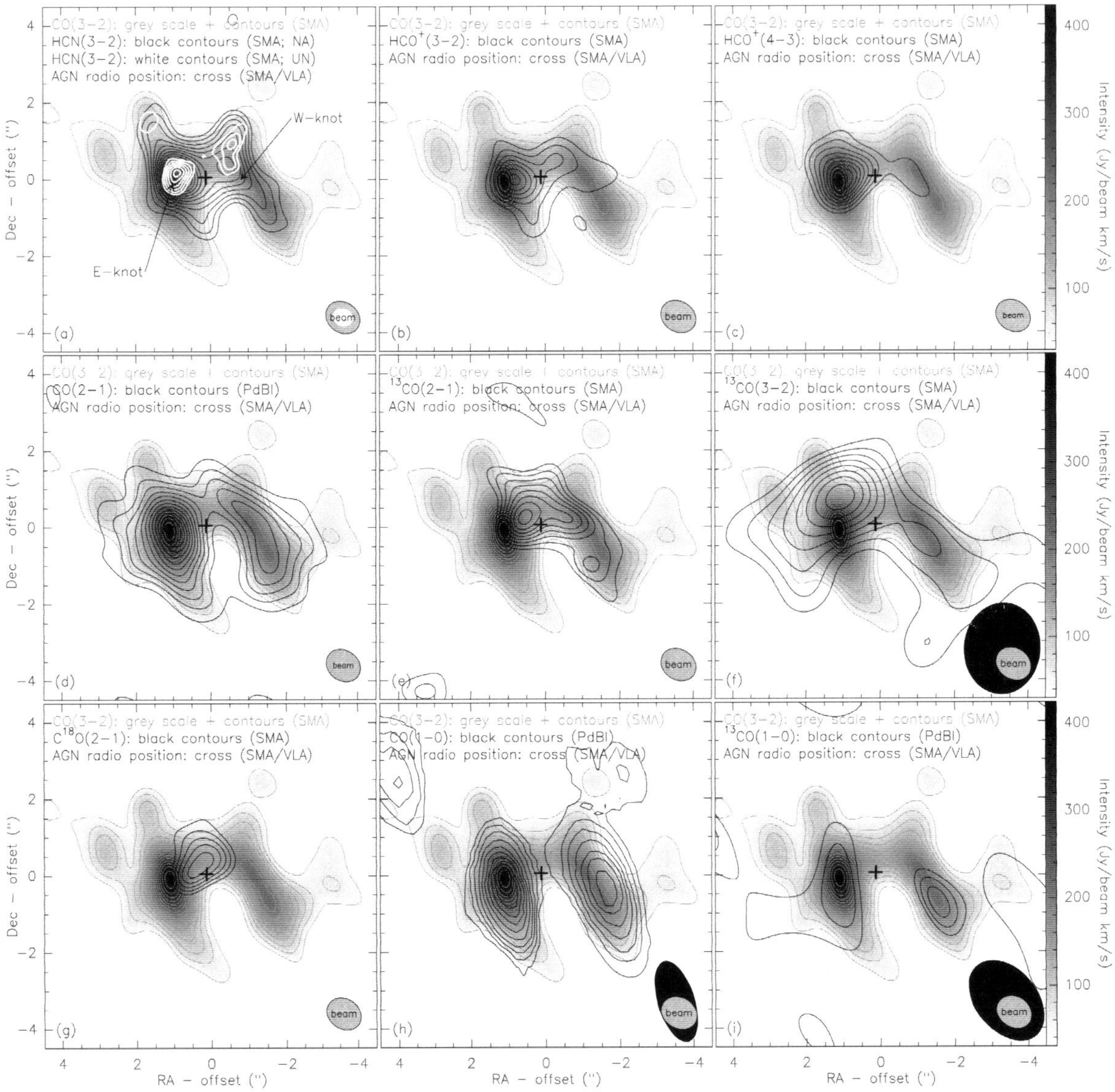

Figure 1. Velocity integrated line emission of ^{12}CO(J=3–2) (*a-g*), HCN(J=3–2) (*a*), HCO$^+$(J=3–2) (*b*), HCO$^+$(J=4–3) (*c*), ^{12}CO(J=2–1) (*d*), ^{13}CO(J=2–1) (*e*), ^{13}CO(J=3–2) (*f*), C^{18}O(J=2–1) (*g*), ^{12}CO(J=1–0) (*h*) and ^{13}CO(J=1–0) (*i*) in NGC 1068, observed with the SMA and the IRAM PdBI (taken from [14]). Contour levles are: ^{12}CO(J=3–2) – from 10σ by 6σ with 1σ=4.8 Jy km s^{-1}; HCN(J=3–2) – from 3σ by 1σ with 1σ=2.6 Jy km s^{-1}; HCO$^+$(J=3–2) – from 2σ by 1σ with 1σ=3.4 Jy km s^{-1}; HCO$^+$(J=4–3) – from 3σ by 1σ with 1σ=2.4 Jy km s^{-1}; ^{12}CO(J=2–1) – from 5σ by 5σ with 1σ=1.2 Jy km s^{-1}; ^{13}CO(J=2–1) – from 3σ by 1σ with 1σ=0.9 Jy km s^{-1}; ^{13}CO(J=3–2) – from 5σ by 1σ with 1σ=3.5 Jy km s^{-1}; C^{18}O(J=2–1) – from 3σ by 1σ with 1σ=0.9 Jy km s^{-1}; ^{12}CO(J=1–0) – from 5σ by 3σ with 1σ=0.4 Jy km s^{-1}; ^{13}CO(J=1–0) – from 1σ by 1σ with 1σ=0.1 Jy km s^{-1}.

intercepting blue- and redshifted parts of the emission are well aligned with the axis of the jet (compare Fig.4) we propose an alternative scenario. Assuming and interaction of the jet with

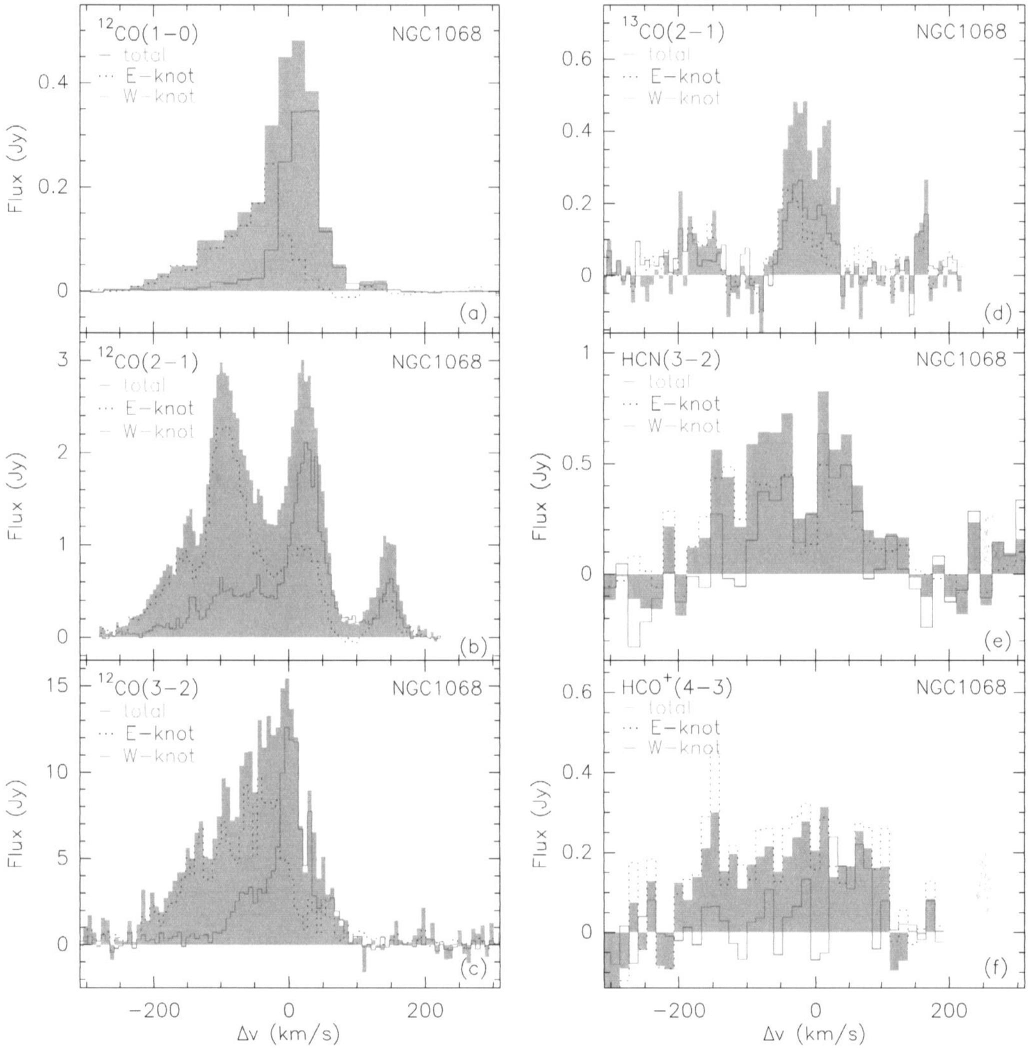

Figure 2. Spatially integrated spectrum of the ^{12}CO(J=1–0), taken from [9], ^{12}CO(J=2–1) and ^{12}CO(J=3–2) (*left column*) and HCN(J=3–2) and HCO^{+}(J=4–3) emission over the E-knot (*dotted blue*) and W-knot (*solid red*) component of NGC 1068.

the CND so that the jet has to "digg" its way through this larger scale disk, parts of the gas can be easily shocked and hence blown outward of the disk. The dynamics will therefore be a mix of outwards moving gas and Keplerian rotating gas with the latter being the dominating mechanism. A toymodel of this hypothesis is shown in Fig. 5 and a more sophisticated dynamical model can be found in [14]. The possibility of the jet hitting parts of the disk in NGC 1068 has been also found by other studies (e.g., [25], [26], [27], [28]). Also in M51, that exhibits similar characteristics to NGC 1068, indications have been found for an interaction of the jet with the surrounding CND (see [22], [10]). However, we cannot entirely exclude at this point that the molecular gas is moving inwards and not outwards although the nature of the emission points more toward an outward motion.

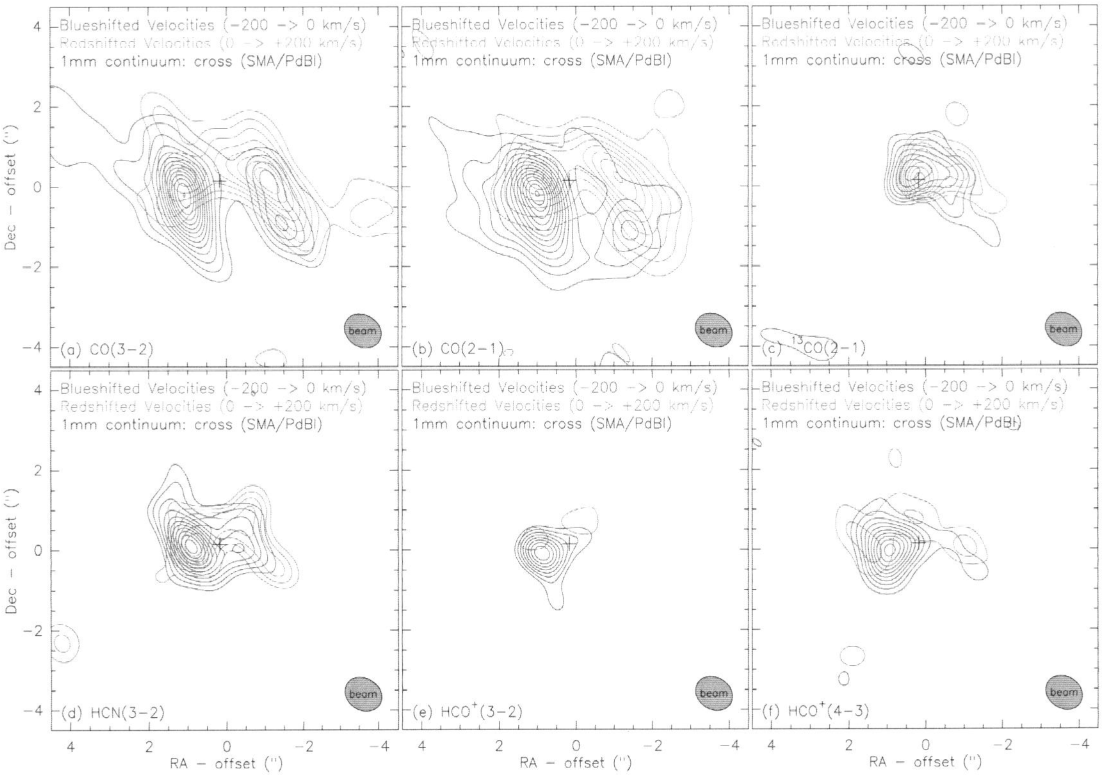

Figure 3. Red- and blueshifted emission of NGC 1068. Contours are in steps of a) from 7σ by 5σ with 1σ=2.0 Jy km s^{-1}; b) from 5σ by 5σ with 1σ=0.9 Jy km s^{-1}; c) from 4σ by 1σ with 1σ=0.5 Jy km s^{-1}; d) from 3σ by 1σ with 1σ=1.9 Jy km s^{-1}; e) from 3σ by 1σ with 1σ=2.3 Jy km s^{-1}; f) from 2σ by 1σ with 1σ=1.4 Jy km s^{-1}.

2.2. Excitation and Chemical Conditions of the Molecular Gas

Given all the different molecular species and transitions obtained in this study a significant set of independent line ratios can be derived in order to form the basis for a detailed simulation of the excitation and chemical conditions of the molecular gas (for the different line ratio maps and values please see [14]). In this study, we used the RADEX code to conduct a large velocity gradient (LVG) modelling (for more details on the analysis see [14]; the code was developed by [29]). The results of the modelling (see Fig.6-7) support previous findings by [24] that the molecular gas is hot in the bridge of the CND in NGC 1068, assumed to be the interaction zone of the jet and the disk. As can be seen, the HCN(J=3–2) emission follows nicely that of the H_2 emission that further indicates an expansion of the gas. Furthermore, we find HCN abundances that are significantly elevated when compared to star-forming/starbursting or quiescent regions. This leads to the conclusion that the excitation and chemical conditions are influenced by a complex combination of different mechanisms including shocks and strong X-ray radiation from the SMBH. Indeed, significant X-ray radiation has been found within the center of NGC 1068 that extends toward the CND (see Fig.12 in [25], and [30]).

3. Conclusions

In this article we discussed a multi-species, multi-transition study of the molecular emission in the CND of NGC 1068 at high spatial resolution. Supporting previous results, we find strong

Journal of Physics: Conference Series **372** (2012) 012038 doi:10.1088/1742-6596/372/1/012038

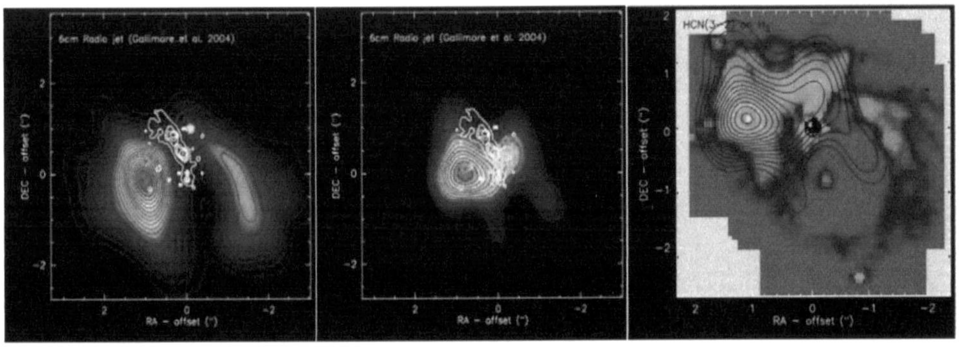

Figure 4. Comparison of the ^{12}CO(J=2–1) and HCN(J=3–2) emission with the radio jet as observed by [28] and the H_2 emission as observed by [26].

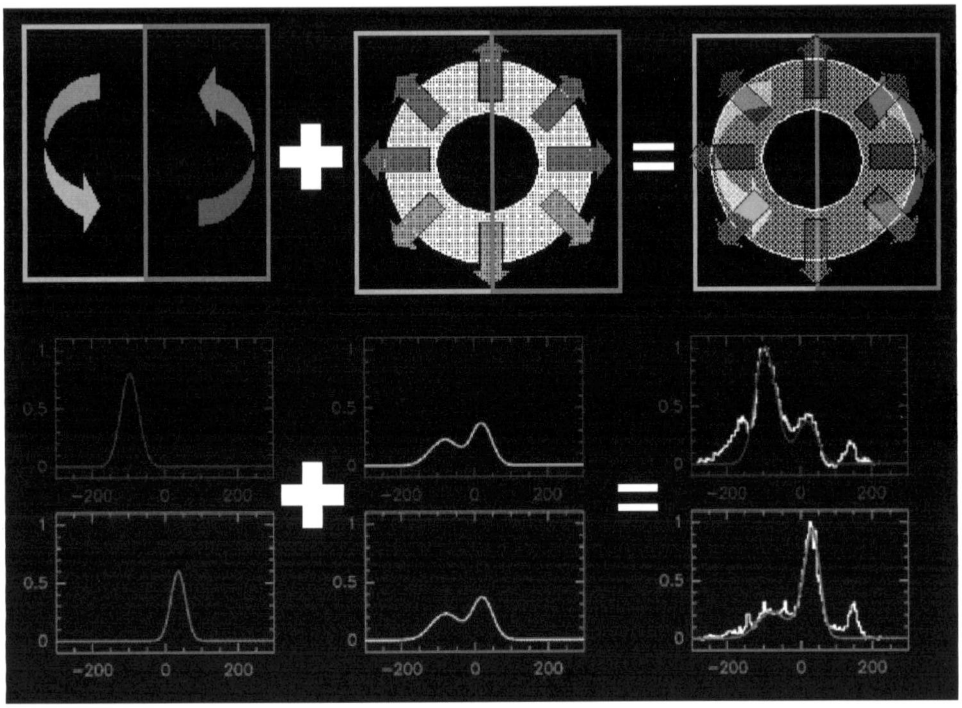

Figure 5. Toymodel to explain the merging of a normal disk rotation with an expanding gas component on top.

indications of an interaction between the jet and CND in NGC 1068 that does not only lead to shocked gas that seems to be blown outwards of the CND but also to high kinetic temperatures within the bridge (northern part) of the CND. Furthermore, modelling of the excitation and chemical conditions of the molecular gas within the CND of NGC 1068 suggests an enhancement of the HCN abundance with respect to starforming/SB and quiescent galaxies. This is thought be a consequence of the strong and large X-ray radiation field from the SMBH that penetrates deep into the CND of NGC 1068 and alters the chemistry of the molecular gas there.

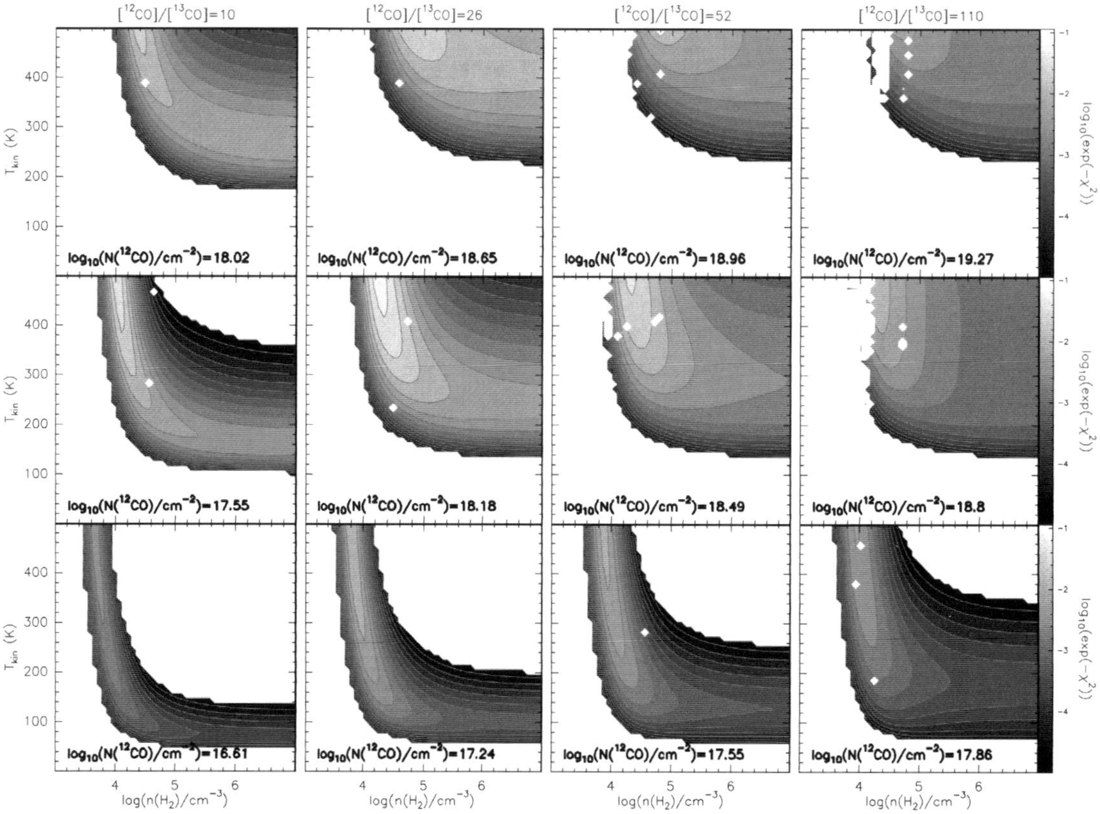

Figure 6. χ^2-fit results obtained from the RADEX simulations of the excitation conditions of the molecular gas. Shown are four different [^{12}CO]/[^{13}CO] abundance ratios (=10,26,52,110) for three different ^{12}CO column densities respectively. The middle panel shows the best χ^2-fit found for each abundance ratio.

3.1. Acknowledgments

The Submillimeter Array is a joint project between the Smithsonian Astrophysical Observatory and the Academia Sinica Institute of Astronomy and Astrophysics and is funded by the Smithsonian Institution and the Academia Sinica. This article is based on observations carried out with the IRAM Plateau de Bure Interferometer. IRAM is supported by INSU/CNRS (France), MPG (Germany) and IGN (Spain).

[1] García-Burillo, S.; Combes, F.; Schinnerer, E.; Boone, F.; Hunt, L. K. 2005, A&A, 441, 1011
[2] Combes, F.; et al., 2004, A&A, 414, 857
[3] Krips, M., et al., 2005, A&A, 442, 479
[4] Boone, F., et al., 2007, A&A, 471, 113
[5] Hunt, L.K., et al., 2008, A&A, 482, 133
[6] Casasola, V.; Combes, F.; García-Burillo, S.;
[7] Hunt, L. K.; Léon, S.; Baker, A. J., 2008, A&A, 490, 61
[8] van der Laan, T.P.R., 2011, A&A, 529, 49
[9] Schinnerer, E., Eckart, A., Tacconi, L.J., Genzel, R. & Downes, D., 2000, ApJ, 533, 850
[10] Matsushita, S.; Muller, S.; Lim, J., 2007, A&A, 468, 49
[11] Aladro, R.; Martín, S.; Martín-Pintado, J.; Mauersberger, R.; Henkel, C.; Ocana Flaquer, B.; Amo-Baladrón, M. A., 2011, A&A, 535, 84
[12] Martín, S.; Mauersberger, R.; Martín-Pintado, J.; Henkel, C.; García-Burillo, S., 2006, ApJS, 164, 450
[13] Meier, D.S.; Turner, J.L., 2005, ApJ, 618, 259
[14] Krips, M., et al., 2011, ApJ, 736, 37

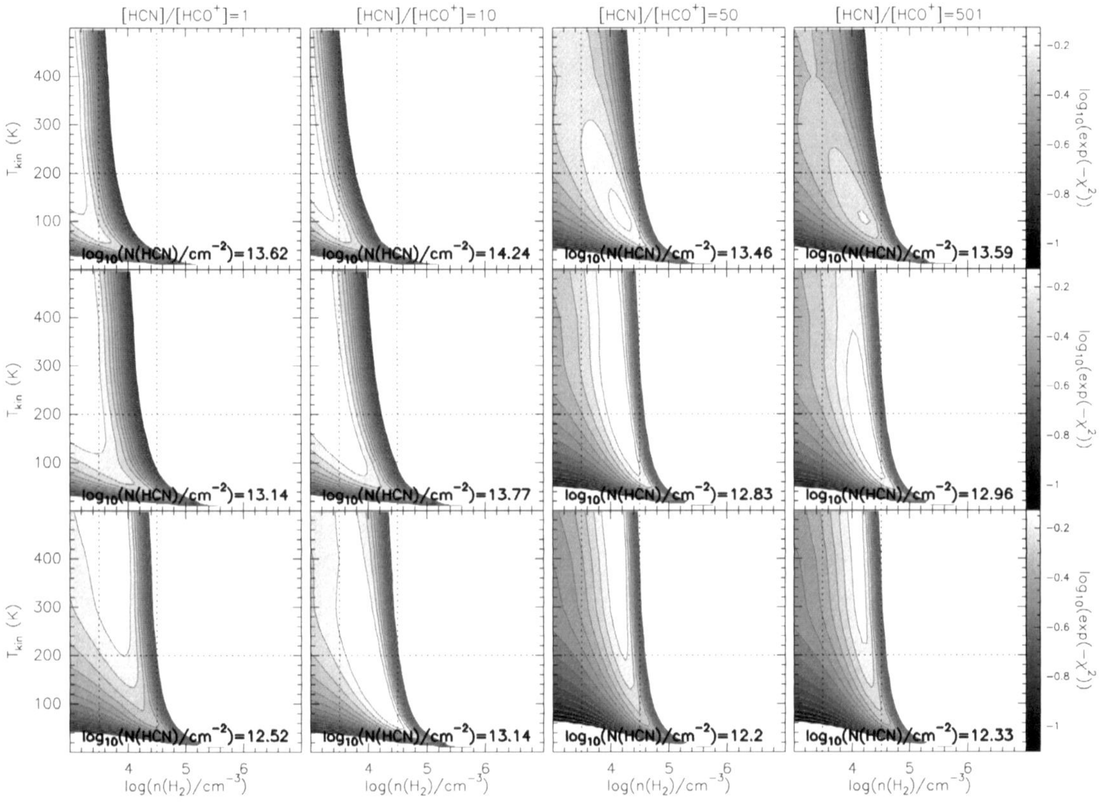

Figure 7. χ^2-fit results obtained from the RADEX simulations of the excitation conditions of the molecular gas. Shown are four different [HCN]/[HCO^+] abundance ratios around the standard galactic value of [HCN]/[HCO^+]$\simeq$10 for three different HCN column densities respectively. The middle panel shows the best χ^2-fit found for each abundance ratio.

[15] Krolik, J. H., & Kallman, T. R. 1987, ApJ, 320, L5
[16] Antonucci, R. R. J., & Miller, J. S. 1985, ApJ, 297, 621
[17] Sternberg, A., Genzel, R., & Tacconi, L. 1994, ApJ, 436, L131
[18] Tacconi, L. J., Genzel, R., Blietz, M., Cameron, M., Harris, A. I., Madden, S., 1994, ApJ, 426, L77
[19] Usero, A., García-Burillo, S., Fuente, A., Martín-Pintado, J., & Rodríguez-Fernàndez, N. J. 2004, A&A, 419, 897
[20] Fuente, A., García-Burillo, S., Gerin, M., Teyssier, D., Usero, A., Rizzo, J. R., de Vicente, P., 2005, ApJ, 619, L155
[21] Kohno, K., Nakanishi, K., Tosaki, T., Muraoka, K., Miura, R., Ezawa, H., Kawabe, R., 2008, Ap&SS, 313, 279
[22] Matsushita, S., et al., 2004, ApJ, 616, L55
[23] Krips, M., et al., 2007, A&A, L468
[24] Krips, M., Neri, R., García-Burillo, S., Martín, S., Combes, F., Graciá-Carpio, J., Eckart, A., 2008, ApJ, 677, 262
[25] García-Burillo, S., et al., 2010, A&A, 519, 2
[26] Müller-Sànchez, F., Davies, R. I., Genzel, R., Tacconi, L. Jz., Eisenhauer, F., Hicks, E. K. S., Friedrich, S., Sternberg, A., 2009, ApJ, 691, 749
[27] Bock, J. J., et al., 2000, AJ, 120, 2904
[28] Gallimore, J.F., Baum, S.A., O'Dea, C.P., 2004, ApJ, 613, 794
[29] Van der Tak, F.F.S., Black, J.H., Schöier, F.L., Jansen, D.J., van Dishoeck, E.F. 2007, A&A 468, 627
[30] Young, A.J., Wilson, A.S., & Shopbell, P.L., 2001, ApJ, 556, 6

Chemical complexity in NGC 1068

R. Aladro[1], S. Viti[1], D. Riquelme[2], S. Martín[3], R. Mauersberger[4], J. Martín-Pintado[5], E. Bayet[6]

[1] Department of Physics & Astronomy, University College London, Gower Street, London WC1E 6BT, UK.
[2] Instituto de Radioastronomía Milimétrica, Avda. Divina Pastora, 7, Local 20, E-18012 Granada, Spain.
[3] European Southern Observatory, Avda. Alonso de Córdova 3107, Vitacura, Casilla 19001, Santiago 19, Chile.
[4] Joint ALMA Observatory, Avda. Alonso de Córdova 3107, Vitacura, Santiago, Chile.
[5] Centro de Astrobiología (CSIC-INTA), Ctra. de Torrejón Ajalvir km 4, E-28850 Torrejón de Ardoz, Madrid, Spain.
[6] Sub-Department of Astrophysics, University of Oxford, Denys Wilkinson Building, Keble Road, Oxford OX1 3RH, UK.

E-mail: `r.aladro@ucl.ac.uk`

Abstract.
We aimed to study the chemistry of the circumnuclear molecular gas of NGC 1068, and to compare it with those of the starburst galaxies M 82 and NGC 253. Using the IRAM-30 m telescope, we observed the inner 2 kpc of NGC 1068 between 86.2 GHz and 115.6 GHz. We identified 35 spectral features, corresponding to 24 different molecular species. Among them, HC_3N, SO, N_2H^+, CH_3CN, NS, ^{13}CN, and $HN^{13}C$ are detected for the first time in NGC 1068. Assuming local thermodynamic equilibrium (LTE), we calculated the column densities of the detected molecules, as well as the upper limits to the column densities of some undetected species. The comparison among the chemistries of NGC 1068, M 82, and NGC 253, suggests that, apart from X-rays, shocks also determine the chemistry of NGC 1068. We propose the column density ratio between CH_3CCH and HC_3N as a prime indicator of the imprints of starburst and AGN environments in the circumnuclear interstellar medium. This ratio is, at least, 64 times larger in M 82 than in NGC 1068, and, at least, 4 times larger in NGC 253 than in NGC, 1068. Finally, we used the UCL_CHEM and UCL_PDR chemical codes to constrain the origin of the species, as well as to test the influence of UV radiation fields and cosmic rays on the observed abundances.

1. Introduction

The proximity of NGC 1068 (D=14.4 Mpc, corresponding to a spatial scale on $1'' = 72$ pc, [24]) as well as its Seyfert 2 activity turn this galaxy into the best target to study the chemical composition of molecular clouds in the vicinity of an AGN. The molecular material close to the central engines of AGN is thought to be heavily pervaded by X-ray radiation originating in the nuclear accretion disk [18]. This strong radiation field has been claimed to be responsible for the observed abundance ratios of some species, such as HCO^+/HCN, or HNCO/CS, which appear to be clearly different between AGN and starburst environments [14, 16, 20]. Observations of NGC 1068 in a number of other molecules, such as SiO, HOC^+, and CN, have revealed a peculiar chemistry interpreted as the result of a giant X-ray dominated region (XDR) in its

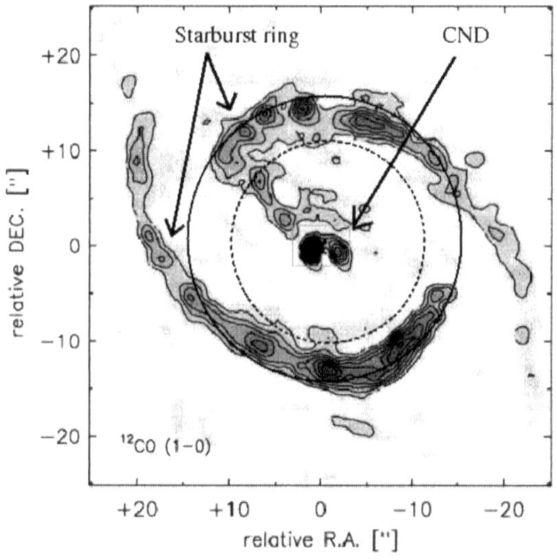

Figure 1. Left: ^{12}CO map of the central region of NGC 1068 taken from [23]. The beam sizes at the lowest and highest frequencies of the survey are plotted with a solid and dotted circles respectively. The starburst ring and circumnuclear disk are indicated.

nucleus, where shocks could be also playing a role [25, 10, 17]. However, an overall chemical study of NGC 1068 has not been done so far due to a lack of unbiased observations of a large number of molecules. Here we present a molecular inventory of this archetypal AGN, which is similar to those already obtained for the starburst galaxies NGC 253 [19] and M 82 [2].

2. Observations and data analysis

The observations were carried out with the IRAM-30 m telescope (Pico Veleta Observatory, Spain) between October 2009 and July 2010. We observed the circumnuclear disk of NGC1068 at a nominal position α_{2000} =02:42:40.9, δ_{2000}=-00:00:46.0. The frequencies range observed is between 86.2 GHz and 115.6 GHz. We used the band E0 of the EMIR receiver and the WILMA autocorrelator. This receiver-backend configuration allowed us to cover 8 GHz simultaneously in the vertical and horizontal polarizations, and led to a channel width spacing of $7-9\,\mathrm{km\,s^{-1}}$. The data were calibrated using the standard dual load method. The observations were done wobbling the secondary mirror with a switching frequency of 0.5 Hz and a beam throw of $\pm 220''$ in azimuth. We checked the pointing accuracy every hour towards the nearby bright continuum sources 2251+158 and 0113-118. The pointing corrections were always better than $4''$. The focus was also checked at the beginning of each run and during sunsets. The rejection of the image sideband was better than 10 dB along the band. The beam sizes ranged from $21''$ to $29''$. This implies that the observations at the lowest frequencies could have picked up some emission from the starburst ring, while the highest frequencies only detected the CND emission (see Fig. 1). We assume that, in the worse case scenario, the starburst ring contribution is lower than 30% [25].

The observed spectra were converted from antenna temperatures (T_A^*) to main beam temperatures ($T_{\rm MB}$) using the relation $T_{\rm MB} = (F_{\rm eff}/B_{\rm eff})\,T_A^*$, where $F_{\rm eff}$ is the forward efficiency of the telescope, which values were between 0.94 and 0.95, and $B_{\rm eff}$ is the main beam efficiency, ranging from 0.77 to 0.81. Linear baselines were subtracted in all cases. The rms achieved is ≤ 2 mK across the whole survey. The data were also corrected by the beam dilution effect as

$T_B = [(\theta_s^2 + \theta_b^2)\,/\,\theta_s^2]\,T_{MB}$, where T_B is the source averaged brightness temperature, θ_s is the source size and θ_b is the beam size. Based on NGC 1068 interferometric observations of ^{12}CO, HCN, and ^{13}CO [12, 23], we assumed an average source size of 4″ for all the molecular species.

3. Molecules in NGC 1068

We detected a total of 35 spectral features towards the AGN of NGC1068. There are no new extragalactic molecular detections. However, seven species or isotopologues are detected for the first time in this galaxy, namely HC_3N, SO, N_2H^+, CH_3CN, NS, ^{13}CN, and $HN^{13}C$. The line identification largely follows the method explained in [2]. We fitted Gaussian profiles to all the detected species. The resulting Gaussian parameters, as well as figures showing the data, and more specific comments about each molecule, will be presented in Aladro et al in preparation.

Assuming local thermodynamic equilibrium conditions and optically thin emission, we used rotation diagrams to determine the column densities (N_{mol}) of the detected species [11]. However, some molecules, such as CO, HCN or HCO^+, seem to be moderately optically thick in NGC 1068, as will be discussed in Aladro et al. in preparation. Therefore, their column densities are likely to be higher. On the other hand, in those cases where we detected only one transition of a given molecule, we assumed a rotational temperature (T_{rot}) of 20 ± 10 K. The T_{rot} value might have an important impact in the column density determination only in extreme cases, such as $T_{rot} < 5$ K or $T_{rot} > 100$ K. Table 1 lists the N_{mol} obtained for each molecule, as well as the 3σ upper limits to the column densities of some other undetected molecules in NGC 1068.

We found that the CO species and their carbon and oxygen isotopologues, ^{13}CO and $C^{18}O$, are the most abundant molecules. On the contrary, HOC^+ is the less abundant of the detected species in our survey. We took from the literature some molecular transitions that lie outside our 3 mm frequency range (see Table 1 for references). Few molecules show two different components in the Boltzmann diagrams. This is the case of ^{12}CO, ^{13}CO and CS. One of such components belongs to a colder gas with higher column densities, while the other one shows warmer rotational temperatures and lower column densities. This latter gas component is likely to arise from the inner regions of the molecular clouds.

4. On the origin of the molecular emission

We used the UCL_CHEM code [26, 27] as well as the UCL_PDR code [6] to model the molecular emission of the detected species shown in Table 1. UCL_CHEM is a depth- and time-dependent chemical model that consists of two separated phases. In phase I, the free-fall collapse of a molecular cloud with atomic composition is simulated. The hydrogen density, initially 10^2 cm^{-3}, is increased until it reaches a final density (free parameter), while the temperature remains low and constant, at 10 K. During this phase, atoms and molecules deplete on to the grains, and surface reactions, including hydrogenation, take place. Phase II simulates the chemistry of the molecular cloud *after* a source of radiation is switched on. We computed the temperatures at different visual extinctions using the UCL_PDR code. During Phase II we assume instantaneous evaporation of the molecular ices. Therefore, only gas-phase reactions take place during this second phase. We have considered 205 different molecules, of which 51 are surface species. 2345 reactions, taken from the UMIST database (http://www.udfa.net, [28]), are involved in the calculations.

Table 2 shows the values of the main parameters of UCL_CHEM. We used the NGC 1068 metallicity (1.056, [29]), and the standard initial element abundance ratios of C, O, N, S, He, Mg, and Si, compiled by [3]. We run a set of four models varying the radiation field and the cosmic-ray ionization rate, in order to sample a variety of conditions that may co-exist in the AGN. We used a number of assumptions when running our models. First, we used a constant density for Phase II. However, molecules are expected to trace environments of different densities, probably ranging from $\sim 10^3$ cm^{-3} to $\sim 10^6$ cm^{-3}. Therefore, our models do not aim to simulate density

Table 1. Column densities of the observed molecules, and 3σ upper limits to the column densities of some undetected species in our survey.

Molecule	N_{mol} [cm^{-2}]	Molecule	N_{mol} [cm^{-2}]	Molecule	N_{mol} [cm^{-2}]
^{12}CO [a]	$(4.0 \pm 0.3) \times 10^{18}$	HCO	$(3.8 \pm 2.7) \times 10^{14}$	H_2CO	$\leq 1.1 \times 10^{16}$
	$(1.2 \pm 0.3) \times 10^{18}$	SO [N]	$(3.2 \pm 3.3) \times 10^{14}$	OCS	$\leq 5.4 \times 10^{14}$
^{13}CO [a]	$(3.3 \pm 0.3) \times 10^{17}$	NS [N]	$(3.2 \pm 1.6) \times 10^{14}$	CH_2NH	$\leq 5.4 \times 10^{14}$
	$(7.8 \pm 2.8) \times 10^{16}$	^{13}CN [N]	$(3.2 \pm 1.6) \times 10^{14}$	CH_3CCH	$\leq 4.5 \times 10^{14}$
$C^{18}O$ [b]	$(1.0 \pm 0.1) \times 10^{17}$	HNCO	$(2.8 \pm 0.1) \times 10^{14}$	SO_2	$\leq 3.6 \times 10^{14}$
C_2H	$(1.5 \pm 0.7) \times 10^{16}$	N_2H^+ [N]	$(1.6 \pm 0.8) \times 10^{14}$	c-C_3H	$\leq 2.9 \times 10^{14}$
CN	$(8.1 \pm 3.9) \times 10^{15}$	$C^{34}S$ [b]	$(1.6 \pm 0.6) \times 10^{14}$	H_2CS	$\leq 2.6 \times 10^{14}$
HCN	$(2.7 \pm 1.3) \times 10^{15}$	$H^{13}CN$	$(1.2 \pm 0.6) \times 10^{14}$	C_2S	$\leq 8.9 \times 10^{13}$
CH_3OH	$(2.3 \pm 0.1) \times 10^{15}$	$H^{13}CN$	$(1.2 \pm 0.6) \times 10^{14}$	$HOCO^+$	$\leq 7.0 \times 10^{13}$
CS [c]	$(7.6 \pm 2.2) \times 10^{14}$	SiO	$(9.1 \pm 4.5) \times 10^{13}$	^{13}CS	$\leq 3.6 \times 10^{13}$
	$(7.0 \pm 4.2) \times 10^{13}$	$H^{13}CO^+$	$(4.2 \pm 2.0) \times 10^{13}$	NH_2CN	$\leq 2.8 \times 10^{13}$
HC_3N [N]	$(6.0 \pm 4.0) \times 10^{14}$	CH_3CN [N]	$(3.9 \pm 0.1) \times 10^{13}$		
HCO^+ [e]	$(5.3 \pm 0.2) \times 10^{14}$	$HN^{13}C$ [N]	$(3.2 \pm 5.0) \times 10^{13}$		
HNC [d]	$(5.2 \pm 0.1) \times 10^{14}$	HOC+	$(1.5 \pm 1.0) \times 10^{13}$		

[N]: New detection in NGC 1068.[a] Using extra detections by [13]. [b] Using extra detections by [20]. [c] Using extra detections by [5]. [d] Using extra detections by [22]. [e] Using extra detections by [16]. To calculate the upper limits of the NGC 1068 column densities, we assumed for all the cases a $T_{rot} = 20 \pm 10$ K and a width of 230 km s^{-1}.

gradients within molecular clouds. Second, we took a visual extinction of $A_v = 2$ mag as an example of a photon-dominated region (PDR), and a $A_v = 10$ mag as an example of a molecular cloud with a dense core. Third, in some cases molecular abundances vary with time. This mainly occurs in the dense cores of the simulated molecular clouds, where species do not reach a steady-state. We thus averaged the abundances over a range of time from 10^3 to 10^7years. However, some molecules may have shorter or larger life cycles, and therefore their abundances may differ.

As can be seen in Table 2, the four models differ in the values of the radiation fields (G_0 = 1 and 1000 in Harbing units) and the cosmic-ray ionization rate (1.3×10^{-17} s^{-1} and 1.3×10^{-15} s^{-1}). Model *a* aims to simulate the physical conditions of hot cores, which have both low external radiation fields and cosmic ray ionization rates; in model *b* we increased the radiation field by a factor of 1000 with respect to model *a* in order to see the effects of strong UV fields in the chemistry of the molecular clouds; on the contrary, model *c* has low UV field, but a high cosmic ray rate; finally, model *d* has high values of both parameters. Table 3 shows whether, according to each model, the molecules are tracing dense gas regions, PDRs (i.e. the external zones of the molecular clouds), or both. Some species seem to not be strongly affected by the variations of UV and cosmic ray fields. Examples are C_2H and CN, which always trace PDRs; or CH_3OH, CS, SO, HNCO and SiO, which always trace dense gas regions. On the other hand, other molecules such as HCO^+, HNC or N_2H^+, trace dense gas regions or PDRs depending on the physical conditions.

5. Molecular abundances in AGN and starburst galaxies

We aimed to compare the molecular abundances of NGC 1068 with those of the starburst galaxies M 82 and NGC 253. These are the three extragalactic objects whose chemical compositions are better known so far. NGC 253 and M 82 were also targets of unbiased molecular line surveys

Table 2. Main parameters of the UCL_CHEM models

Parameter	Model a	Model b	Model c	Model d
Initial H density (phase I)	$10^2\,cm^{-3}$	$10^2\,cm^{-3}$	$10^2\,cm^{-3}$	$10^2\,cm^{-3}$
Final H density (phases I and II)	$10^6\,cm^{-3}$	$10^6\,cm^{-3}$	$10^6\,cm^{-3}$	$10^6\,cm^{-3}$
Temperature (phase I)	10 K	10 K	10 K	10 K
Temperature (phase II)	300 K	300 K	300 K	300 K
Visual extinction	2 & 10 mag	2 & 10 mag	2 & 10 mag	2 & 10 mag
Gas-to-dust mass ratio	100	100	100	100
External UV radiation intensity	1 Habing	1000 Habing	1 Habing	1000 Habing
Cosmic-ray ionization rate	$1.3\times10^{-17}\,s^{-1}$	$1.3\times10^{-17}\,s^{-1}$	$1.3\times10^{-15}\,s^{-1}$	$1.3\times10^{-15}\,s^{-1}$

Table 3. Regions traced by each molecule according to the four models presented in Table 2.

Molecule	Model a		Model b		Model c		Model d	
	PDR	Dense Gas	PDR	Dense gas	PDR	Dense gas	PDR	Dense gas
CO	✓	✓		✓	✓	✓		✓
C_2H	✓		✓		✓		✓	
CN	✓		✓		✓		✓	
HCN		✓		✓	✓	✓		✓
CH_3OH		✓		✓		✓		✓
CS		✓		✓		✓		✓
HC_3N		✓	✓		✓		✓	✓
HCO^+	✓		✓	✓	✓			✓
HNC	✓	✓		✓	✓	✓		✓
HCO	✓		✓		✓	✓	✓	
SO		✓		✓		✓		✓
NS		✓	✓			✓	✓	✓
HNCO		✓		✓		✓		✓
N_2H^+	✓			✓	✓			✓
SiO		✓		✓		✓		✓
CH_3CN		✓		✓	✓			✓
HOC^+	✓	✓	✓	✓	✓	✓	✓	✓
CH_3CCH		✓		✓		✓		✓

Ticks show whether each molecules trace PDRs/CRDRs, dense gas, or both. Boldface ticks indicate when the abundances given by the models match the observed abundances within one order of magnitude.

in the mm range, with 25 and 18 molecular species detected in their respective nuclear regions [19, 2]. The starburst in the center of M 82 is old, with an average stellar population age of $\sim$ 15 Myr [15]. The nuclear starburst, together with a high supernova rate, create strong UV fields, and thus PDRs dominate the nucleus of M 82. On the other hand, NGC 253 is claimed to be in an early stage of the starburst evolution [20, 1], with younger stellar populations in its nucleus ($\sim$ 6 Myr [7]) where the PDRs, although present, do not drive the heating of molecular clouds. Instead, the nucleus of NGC 253 is dominated by large-scale shocks between molecular complexes. A detailed comparison of the M 82 and NGC 253 chemistries done by [2] showed that the ISM in both galaxies have different composition, and that molecular abundances can be used to distinguish the physical processes that dominate galaxy nuclei.

Figure 2 shows the molecular abundances in the central regions of M 82, NGC 253, and

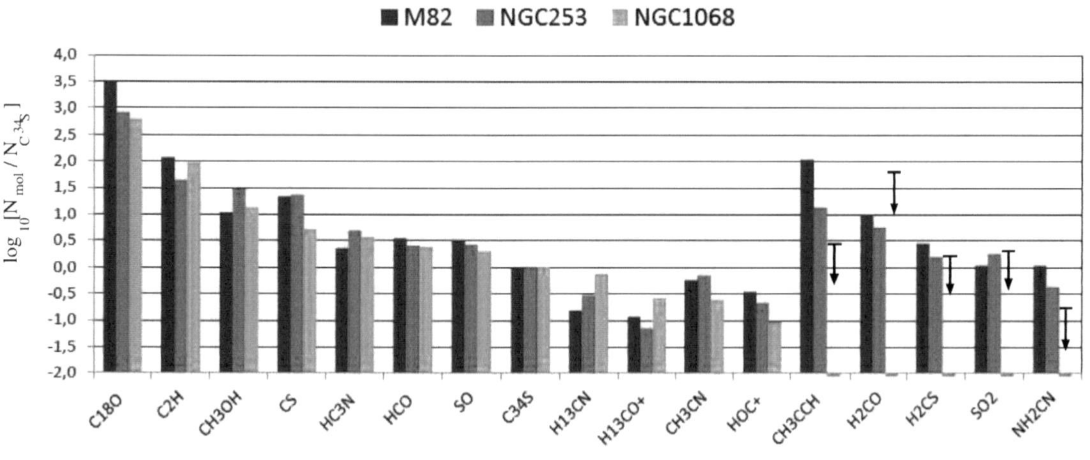

Figure 2. Fractional abundances (with respect to $C^{34}S$) of several molecules in the center of the galaxies M 82 and NGC 253 (data taken from [2] and references therein), and NGC 1068. The vertical arrows show the upper limits of the fractional abundances to the undetected molecules in NGC 1068.

NGC 1068. The molecular ratios were calculated with respect to $C^{34}S$ because, apart from being optically thinner than the main sulfur isotopologue, it does not seem to be significantly affected by the type of nuclear activity in galaxies [20]. Furthermore, $C^{34}S$ has similar column densities in the three galaxies (few times 10^{14} cm^{-3}, [2]). We note that while M 82 and NGC 253 are located at approximately the same distance, $\sim$ 3 Mpc [8], NGC 1068 lies nearly five times further, at 14.4 Mpc [24]. While the comparison of the column densities in these galaxies would be biased by the distance, this should not affect the comparison of the fractional abundances. On the other hand, because of the different distances the IRAM-30m beam covered the $1.5 - 2.0$ central kiloparsecs of the AGN, while for the starburst galaxies, the beam covered the inner 360 pc.

Among all the species compared, only $H^{13}CN$ and $H^{13}CO^+$ show higher abundances with respect to $C^{34}S$ in NGC 1068 than M 82 and NGC 253. On the other hand, C_2H, CH_3OH, HC_3N, HCO, and SO show similar abundances in all three galaxies (within a factor three). A third group is composed by those molecules that are clearly more abundant in M 82 and NGC 253 than in NGC 1068. From this group it can be drawn the clearest chemical differentiation between starbursts and AGN activities. Among the molecules detected in the three galaxies, $CS/C^{34}S$ is almost 5 times higher in NGC 253 than in NGC 1068. Similarly, $CH_3CN/C^{34}S$ shows a difference of a factor 3 between NGC 253 and NGC 1068, while $HOC^+/C^{34}S$ is a factor 4 more abundant in M 82 than in NGC 1068. It is remarkable that these three molecular ratios are very similar in M 82 and NGC 253 (see Fig. 2). Moreover, some of the biggest chemical differences are given by those molecules that were clearly detected in both starburst galaxies, but not in NGC 1068, such as CH_3CCH, H_2CO, H_2CS, SO_2 and NH_2CN. Among them, $CH_3CCH/C^{34}S$ is the ratio showing the highest contrast, being > 39 times higher in M 82 than in NGC 1068 and > 5 times higher in NGC 253 than in NGC 1068. Therefore, we computed other fractional abundances of CH_3CCH with respect to the molecules having similar abundances in the three galaxies. We found that the most striking differences among the AGN and the starburst galaxies are given by R_{CH_3CCH/HC_3N} and R_{CH_3CCH/C_2H}. CH_3CCH and HC_3N are claimed to be tracers of dense gas in galaxies [1], while C_2H might be tracing both PDRs and dense gas regions [4]. To avoid comparisons of molecules that might be arising from different regions, we select [CH_3CCH/HC_3N] as the best ratio to chemically differentiate between AGN and starburst galaxies. We found that $R_{CH_3CCH/HC_3N} > 64$ when comparing M 82 with NGC 1068, and $R_{CH_3CCH/HC_3N} > 4$ when comparing NGC 253 with NGC 1068. We note that a factor of, at least, 64 is much larger than the $R_{HCO^+/HCN}$ commonly used to differentiate between AGN and starburst environments [14, 16]. On the other hand, our upper limit to the column density of CH_3CCH in NGC 1068 is not very tight, so the molecular ratios could be even larger. The values found for R_{CH_3CCH/HC_3N} indicate that the dense ISM in the NGC 1068 nucleus resembles more NGC 253 than M 82. This points at the existence of shocks (rather than PDRs) accompanying the strong X-ray fields present in the central regions of NGC 1068.

6. Conclusions

We carried out a molecular line survey with the IRAM 30-m telescope towards the active galactic nucleus of NGC 1068. We covered the frequencies between 86.2 GHz and 115.6 GHz, detecting a total of 35 spectral line features. Seven species, or their carbon isotopologues, are detected for the first time in this galaxy, namely HC_3N, SO, N_2H^+, CH_3CN, NS, ^{13}CN, and $HN^{13}C$.

Some molecules, such as CH_3CCH, c-C_3H_2, OCS, SO_2, NS, $HOCO^+$, and NH_2CN, are not detected in NGC 1068. We compared the molecular abundances of the NGC 1068 nucleus (claimed to be a giant XDR, e.g. [10]), with those of M,82 (claimed to be a giant PDR, [9]), and NGC 253 (claimed to be dominated by low-velocity shocks, [19]). We found that the ratio R_{CH_3CCH/HC_3N} shows the largest differences among the three galaxies. Specifically, R_{CH_3CCH/HC_3N} is at least 64 times larger in M 82 than in NGC 1068, and at least 4 times larger in NGC 253 than in NGC 1068. Therefore, this ratio can be used to disentangle between AGN and starburst physical environments. Furthermore, R_{CH_3CCH/HC_3N} indicates that UV fields are not probably strong enough in the inner 2 kpc of NGC 1068. Instead, shocks, could be playing a secondary role, after X-rays.

We used the time-dependent chemical (UCL_CHEM) and PDR (UCL_PDR) models to simulate the chemistry in molecular clouds under several physical conditions that might co-exist in AGN. We note, however, that our aim was not to simulate the particular environment in the center of NGC 1068, but to study the influence of UV and cosmic ray fields in the abundances of a large number of molecules. We found that some species seem to always trace photon-dominated regions and/or cosmic ray-dominated regions independently of the strength of the UV and cosmic ray fields (such as C_2H and CN), while other species always trace dense gas regions (e.g. CH_3OH, CS, SO, HNCO, and SiO). Finally, some molecules might be arising from the inner and/or outer layers of the molecular clouds depending on the strength of the external radiation fields (examples are HCO^+, HNC and N_2H^+).

References

[1] Aladro, R., Martín-Pintado, J., Martín, S., Mauersberger, R., & Bayet, E. 2011a, A&A, 525, A89
[2] Aladro, R., Martín, S., Martín-Pintado, J., et al. 2011b, A&A, 535, A84
[3] Bayet, E., Viti, S., Williams, D. A., & Rawlings, J. M. C. 2008, ApJ, 676, 978
[4] Bayet, E., Viti, S., Williams, D. A., Rawlings, J. M. C., & Bell, T. 2009a, ApJ, 696, 1466
[5] Bayet, E., Aladro, R., Martín, S., Viti, S., & Martín-Pintado, J. 2009b, ApJ, 707, 126
[6] Bell, T. A., Roueff, E., Viti, S., & Williams, D. A., 2006, MNRAS, 371, 1865
[7] Fernández-Ontiveros, J. A., Prieto, M. A., & Acosta-Pulido, J. A., 2009, MNRAS, 392, L16
[8] Freedman, W. L., et al. 1994, ApJ, 427, 628
[9] Fuente, A., García-Burillo, S., Gerin, M., et al. 2005, ApJL, 619, L155
[10] García-Burillo, S., et al. 2010, A&A, 519, A2
[11] Goldsmith, P. F., & Langer, W. D. 1999, ApJ, 517, 209
[12] Helfer, T. T., & Blitz, L. 1995, ApJ, 450, 90
[13] Israel, F. P. 2009, A&A, 493, 525
[14] Kohno, K., Matsushita, S., Vila-Vilaró, B., et al. 2001, The Central Kiloparsec of Starbursts and AGN: The La Palma Connection, 249, 672

[15] Konstantopoulos, I. S., Bastian, N., Smith, L. J., Westmoquette, M. S., Trancho, G., & Gallagher, J. S. 2009, ApJ, 701, 1015
[16] Krips, M., Neri, R., García-Burillo, S., et al. 2008, ApJ, 677, 262
[17] Krips, M., Martín, S., Eckart, A., et al. 2011, ApJ, 736, 37
[18] Maloney, P. R., Hollenbach, D. J., & Tielens, A. G. G. M. 1996, ApJ, 466, 561
[19] Martín, S., Martín-Pintado, J., & Mauersberger, R. 2006, A&A, 450, L13
[20] Martín, S., Martín-Pintado, J., & Mauersberger, R. 2009, ApJ, 694, 610
[21] Myers, S. T., & Scoville, N. Z. 1987, ApJl, 312, L39
[22] Pérez-Beaupuits, J. P., Spaans, M., van der Tak, F. F. S., et al. 2009, A&A, 503, 459
[23] Schinnerer, E., Eckart, A., Tacconi, L. J., Genzel, R., & Downes, D. 2000, ApJ, 533, 850
[24] Tully, R. B. 1988, Nearby Galaxies Catalog, Cambridge University Press
[25] Usero, A., García-Burillo, S., Fuente, A., Martín-Pintado, J., & Rodríguez-Fernández, N. J. 2004, A&A, 419, 897
[26] Viti, S., & Williams, D. A. 1999, MNRAS, 305, 755
[27] Viti, S., Collings, M. P., Dever, J. W., McCoustra, M. R. S., & Williams, D. A. 2004, MNRAS, 354, 1141
[28] Woodall, J., Agúndez, M., Markwick-Kemper, A. J., & Millar, T. J. 2007, A&A, 466, 1197
[29] Zaritsky, D., Kennicutt, R. C., Jr., & Huchra, J. P. 1994, ApJ, 420, 87

Astronomy at High Angular Resolution 2011 IOP Publishing
Journal of Physics: Conference Series **372** (2012) 012040 doi:10.1088/1742-6596/372/1/012040

General Relativistic Modeling of Magnetized Jets from Accreting Black Holes

Alexander Tchekhovskoy[1], Jonathan C. McKinney[2] and Ramesh Narayan[3]

[1]Center for Theoretical Science, Jadwin Hall, Princeton University, Princeton, NJ 08544, USA; Princeton Center for Theoretical Science Fellow

[2]Kavli Institute for Particle Astrophysics and Cosmology, Stanford University, P.O. Box 20450, MS 29, Stanford, CA 94309, USA

[3]Institute for Theory and Computation, Harvard-Smithsonian Center for Astrophysics, 60 Garden Street, MS 51, Cambridge, MA 02138, USA

E-mail: `atchekho@princeton.edu`

Abstract. Recent advances in general relativistic magnetohydrodynamic modeling of jets offer unprecedented insights into the inner workings of accreting black holes that power the jets in active galactic nuclei (AGN) and other accretion systems. I will present the results of recent studies that determine spin-dependence of jet power and discuss the implications for the AGN radio loud/quiet dichotomy and recent observations of high jet power in a number of AGN.

1. Introduction

Relativistic jets are one of the most spectacular manifestations of black hole (BH) accretion. Jet-producing accretion systems span 9 decades in central BH mass: from stellar-mass BHs in black hole binaries (BHBs) and gamma-ray bursts (GRBs) to supermassive BHs in active galactic nuclei (AGN). If a single physical mechanism is responsible for producing jets throughout the BH mass spectrum, it must be robust and scale invariant. Magnetic fields are a promising agent for jet production because they are abundant in astrophysical plasmas and because the properties of magnetically-powered jets scale trivially with BH mass [1–5].

How are jets magnetically launched? Figure 1 shows a cartoon depiction of this. Consider a vertical magnetic field line attached on one end to a perfectly conducting sphere, which represents the central compact object, and on the other end to a perfectly conducting "ceiling" which represents the ambient medium (panel a). As the sphere is spinning, after N turns the initially vertical field line develops N toroidal field loops (panel b). This magnetic spring pushes against the ceiling due to the pressure of the toroidal field. As more toroidal loops form and the toroidal field becomes stronger, the spring pushes away the ceiling and accelerates any plasma attached to it along the rotation axis, forming a jet (panels c and d in Figure 1, see the caption for details). In the case when the central body is a black hole, which does not have a surface, the rotation of space-time causes the rotation of the field lines, and jets form in a similar fashion via a process referred to as Blandford-Znajek mechanism (BZ, hereafter) [1].

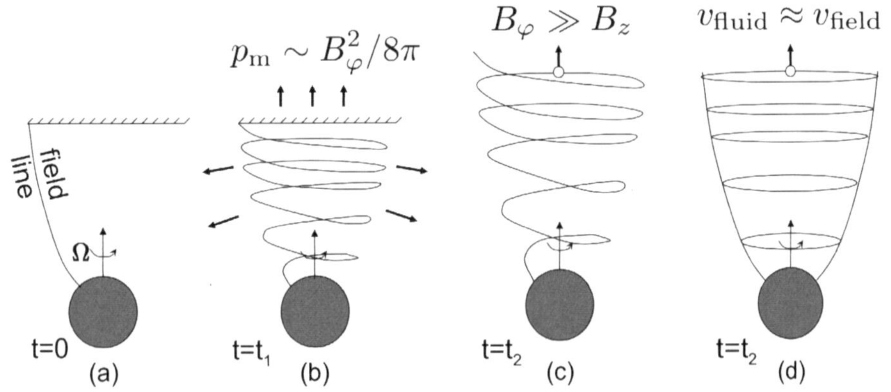

Figure 1. Illustration of jet formation by magnetic fields. (a) Consider a purely poloidal (i.e., $B_\varphi = 0$) field line attached on one end to a stationary "ceiling" (which represents the ambient medium and is shown with hashed horizontal line) and on the other end to a perfectly conducting sphere (which represents the central BH or disk and is shown with grey filled circle) rotating at an angular frequency Ω. (b) After N rotations, at time $t = t_1$, the initially purely poloidal field line develops N toroidal loops. This magnetic spring pushes against the "ceiling" with an effective pressure $p_m \sim B_\varphi^2/8\pi$ due to the toroidal field, B_φ. As time goes on, more toroidal loops form, and the toroidal field becomes stronger. (c) At some later time, $t = t_2$, the pressure becomes so large that the spring pushes away the "ceiling" and accelerates the plasma attached to it along the rotation axis, forming a jet. Asymptotically far from the center, the toroidal field is the dominant field component and determines the dynamics of the jet. (d) It is convenient to think of the jet as a collection of toroidal field loops that slide down the poloidal field lines and accelerate along the jet under the action of their own pressure gradient and hoop stress. The rotation of the sphere continuously twists the poloidal field into new toroidal loops at a rate that, in steady state, balances the rate at which the loops move downstream.

2. Radio-Loud/Quiet Dichotomy of Active Galactic Nuclei

Despite decades of research, we still do not understand what determines jet power in accretion systems or how to accurately infer jet powers observationally [6–11]. There are a number of unresolved puzzles. For the same nuclear B-band luminosity, radio-loud (RL) AGN have a factor of 10^3 higher total (core plus extended) radio luminosity than radio-quiet (RQ) AGN. Figure 3 in [12] clearly shows that these two flavors of AGN follow two well-separated tracks on the radio-loudness diagram, which shows the radio-loudness parameter, $\mathcal{R}$, the ratio of radio to optical luminosities at specified frequencies vs. the Eddington ratio, λ. This dichotomy is especially pronounced at low $\lambda \lesssim 0.01$ when the accretion systems are thought to be be in a radiatively inefficient state of accretion, with a geometrically thick accretion disk. For higher λ-values, accretion systems can undergo spectral state transitions (as seen in BHBs, [13]), which might cause differences in $\mathcal{R}$. Therefore, we limit ourselves to low-λ systems.

If total (core plus extended) radio luminosity is a tracer of jet power, could the dichotomy be due to the differences in the spin of central BHs that power the relativistic jets in RL and RQ AGN [14]? Characteristic values of BH spin in RQ AGN, which lie predominantly at the centers of spiral galaxies, can be quite low, $|a| \lesssim 0.3$. This can be understood if the orientation of angular momentum accreted by the central BHs changes randomly between accretion events [15]. On the other hand, the spin of central BHs in RL AGN, which are hosted predominantly by elliptical galaxies, can be much higher, $a \sim 1$ [15]. Can this difference in BH spin lead to a factor of 10^3 dichotomy?

In the presence of large-scale magnetic fields, rotating BHs produce outflows via the BZ mechanism, which extracts BH rotational energy at a rate,

$$P_{\rm BZ} = \frac{\kappa}{4\pi c}\Phi_{\rm BH}^2\frac{a^2}{16r_g^2} \qquad \text{(standard BZ formula, low-spin limit, } a^2 \ll 1\text{)}, \tag{1}$$

where $\kappa \approx 0.05$ weakly depends on magnetic field geometry, $\Phi_{\rm BH}$ is the magnetic flux through the BH horizon, and $r_g = GM/c^2$ is BH gravitational radius [1]. This low-spin approximation, which we refer to as the standard BZ formula, remains accurate up to $a \lesssim 0.5$ [16, 17], which can be seen in Figure 2. Clearly, the $\propto a^2$ scaling does not appear steep enough to explain the 10^3 dichotomy: BH power varies by a factor of ~ 10 if a varies from 0.3 to 1. Can the power dependence becomes steeper as $a \to 1$? Figure 2 illustrates that an expansion in the powers of BH angular frequency, $\Omega_{\rm H} = ac/2r_{\rm H}$,

$$P_{\rm BZ} = \frac{\kappa}{4\pi c}\Phi_{\rm BH}^2\Omega_{\rm H}^2 f(\Omega_{\rm H}) \qquad \text{(BZ6 formula, accurate for all values of } a\text{)}, \tag{2}$$

remains accurate up to $a \lesssim 0.95$ for $f = 1$ (referred to as the BZ2 formula), where $r_{\rm H} = r_g[1 + (1-a^2)^{1/2}]$ is BH horizon radius. Equation (2) with $f(\Omega_{\rm H}) \approx 1 + 1.38(\Omega_{\rm H}r_g/c)^2 - 9.2(\Omega_{\rm H}r_g/c)^4$ remains accurate for all spins [18], and this gives us our BZ6 formula. Figure 2 shows that while at small spin the BZ6 formula (2) and the standard BZ formula (1) agree, as $a \to 1$ the BZ6 formula gives about 3 times more power. However, the spin dependence is still not steep enough to explain the dichotomy: BH power varies by a factor of ~ 30 if a varies from 0.3 to 1.

So far we considered razor-thin accretion disks, which is an implicit assumption behind the above expressions for power, eqs. (1)–(2). However, disks in low-luminosity AGN are thick, with disk angular thickness as large as $H/R \sim 1$ (including disk body and magnetized corona). Such a thick disk mass-loads equatorial field lines on the BH, so that they become part of a sub-relativistic disk wind, and only the magnetic flux in the highly magnetized funnel above and below the disk contributes to the jet. If the total magnetic flux, $\Phi_{\rm BH}$, through the BH is held constant, this disk occultation effect can cause a much steeper spin-dependence of jet power, $P_{\rm jet} \propto \Omega_{\rm H}^4$, as is seen in Fig. 2. Then, variation of power by a factor of 10^3 is possible, which can explain the radio loud/quiet dichotomy of AGN [18].

3. What Sets Jet Power in Black Hole Accretion Systems?

We have shown that BH power is directly proportional to the square of BH magnetic flux, $\Phi_{\rm BH}$, and the square of BH angular frequency, $\Omega_{\rm H}$ (see eq. 2), with small corrections beyond $a \gtrsim 0.95$. In Nature, $\Phi_{\rm BH}$ is a free parameter, whose value is poorly observationally constrained. Clearly, we have $\Phi_{\rm BH}^2 \propto \dot{M}c^2$. But what sets the dimensionless ratio, $\phi_{\rm BH}^2 = \Phi_{\rm BH}^2/\dot{M}r_g^2c$, which characterizes the degree of inner disk magnetization and controls energy extraction from the BH [19–24]? Using $\phi_{\rm BH}$, we define BZ efficiency as BZ6 power (eq. 2) normalized by $\dot{M}c^2$:

$$\eta_{\rm BZ} = \frac{\langle P_{\rm BZ}\rangle}{\langle \dot{M}\rangle c^2} \times 100\% = \frac{\kappa}{4\pi}\langle\phi_{\rm BH}^2\rangle\left(\frac{\Omega_{\rm H}r_g}{c}\right)^2 f(\Omega_{\rm H}) \times 100\%, \tag{3}$$

where $\langle ... \rangle$ is a time-average. Previous GRMHD simulations found $\eta_{\rm BZ} \lesssim 20\%$, even for nearly maximally spinning BHs [25–28]. Are larger values of $\eta_{\rm BZ}$ possible? This is an especially important question since observations suggest that some AGN produce jets and winds at a high efficiency, $\eta = (P_{\rm jet} + P_{\rm wind})/\langle \dot{M}c^2\rangle \times 100\% \gtrsim 100\%$ [6–10], where $P_{\rm jet}$ and $P_{\rm wind}$ are jet and wind powers, respectively. Are BHs capable of powering such highly efficient outflows? We tested this with global time-dependent GRMHD accretion disk-jet simulations for different values of BH spin. As is standard, we initialized the simulations with an equilibrium hydrodynamic torus around a spinning BH and inserted a weak, purely poloidal ($B_\varphi = 0$) magnetic field loop into the

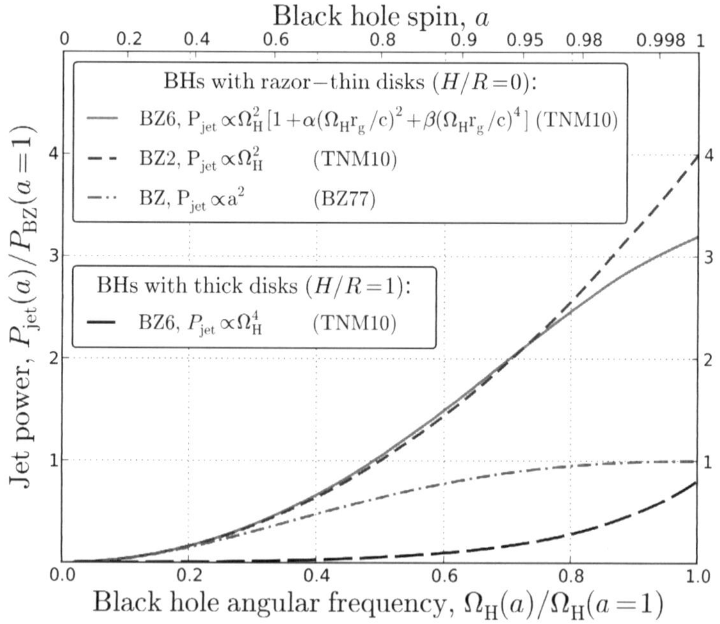

Figure 2. Comparison of various approximations for jet power, $P_{\rm jet}$, versus BH angular frequency, $\Omega_{\rm H}$ (lower x-axis), and BH spin, a (upper x-axis). All powers are normalized to the maximum achievable power in the standard low-spin BZ approximation, $P_{\rm jet} \propto a^2$, which is shown with dot-dashed green line. The standard BZ approximation remains accurate only for moderate values of spin, $a \lesssim 0.5$, and for maximally-spinning BHs ($a = 1$) it under-predicts the true jet power by a factor of ≈ 3. This can be seen by comparing to the BZ6 approximation, $P_{\rm jet} \propto \Omega_{\rm H}^2\left[1+\alpha(\Omega_{\rm H}r_g/c)^2+\beta(\Omega_{\rm H}r_g/c)^4\right]$, which is shown with red solid line and which is uniformly accurate for all values of a (the constant factors, $\alpha \approx 1.38$ and $\beta \approx -9.2$, are obtained in [18]). BZ2 approximation, $P_{\rm jet} \propto \Omega_{\rm H}^2$, shown with blue short dashed line, is accurate up to $a \lesssim 0.95$, beyond which it requires a modest correction. If a BH is surrounded by a thick accretion disk, the disk intercepts BH power that is emitted into a meridional band occulted by the disk. This reduces jet footprint and power relative to the full BH power for razor-thin disks given above and shown with red solid line (see also [18]), provided that BH magnetic flux is held constant. The resulting BZ6 power for thick disks, with characteristic angular thickness $H/R = 1$ expected in low luminosity AGN, is shown with long dashed black curve. Disk occultation effect on jet power is most pronounced at low spins: it causes a substantial steepening of jet power dependence on spin and can explain a factor of 10^3 radio loud/quiet dichotomy of AGN.

torus. Clearly, jet efficiency (eq. 3) depends on the magnetic flux, and time-dependent numerical simulations show that the larger the large-scale vertical magnetic flux in the initial torus, the more efficient the jets [25, 29–31]. To maximize jet efficiency, we populated the torus with a much larger magnetic flux than in previous work. In fact, our torus contained more magnetic flux than the inner disk can push into the BH. The outcome for BH spin $a = 0.99$ is shown in Figure 3. The magnetorotational instability (MRI, [32]) brings gas and magnetic flux to the BH, and an accretion disk of angular density scale height, $h/r \approx 0.3$, forms. Figure 3(d),(e) shows that both $\phi_{\rm BH}$ and η increase until $t \approx 6000 r_g/c$, beyond which they saturate and oscillate around the mean, with time-average $\eta \approx 140\%$. At this time, the BH is saturated with magnetic flux, and the magnetic field on the hole is so strong that it obstructs the accretion and leads to a magnetically-arrested disk, MAD [19, 23, 24, 33–37]. In this state both BH dimensionless magnetic flux, $\phi_{\rm BH}$, and outflow efficiency, η, are maximum, so it is not surprising that we find

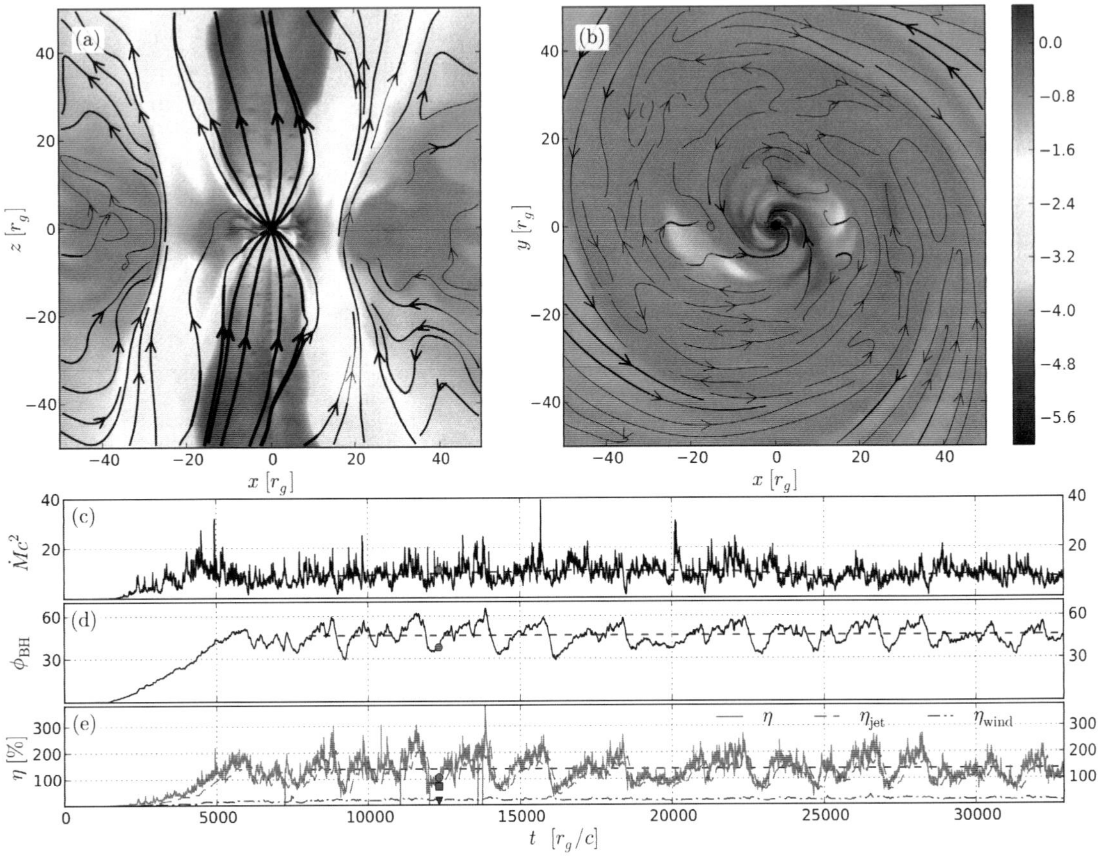

Figure 3. A snapshot and time-dependence of a magnetically-arrested accretion disk [19]. A movie is available at http://youtu.be/nRGCNaWST5Q . The top panels, (a) and (b), show the vertical ($x - z$) and horizontal ($x - y$) slices through the accretion flow. Color shows the logarithm of density: red shows high and blue shows low values (see color bar). The black circle shows a BH ($a = 0.99$) and black lines show field lines. The bottom 3 panels show, from top to bottom, mass accretion rate, $\dot{M}$ (panel c), dimensionless flux through the BH, $\phi_{\rm BH}$ (panel d), and outflow efficiencies of the whole outflow, η (red solid line), the jet, $\eta_{\rm jet}$ (green dashed line), and the wind, $\eta_{\rm wind}$ (blue dash-dotted line) (panel e). Since $\eta_{\rm jet}$ and $\eta_{\rm wind}$ are measured at $r = 100r_g$, they lag by $\Delta t \approx 100r_g/c$ relative to η, which is measured at $r = r_{\rm H}$. Colored symbols show values at the snapshot time, $t \approx 12305r_g/c$. At $t = 0$, the accretion flow contains a large amount of large-scale magnetic flux. Accretion brings mass and flux to the hole, and $\phi_{\rm BH}$ increases until it reaches a maximum value at around $t \approx 6000$ time units. At this time the BH is saturated with magnetic flux and produces as much power as possible. However, the accretion flow brings in even more flux, which impedes the accretion and leads to a magnetically arrested disk. Some of the flux escapes from the BH via magnetic interchange and flux eruptions, two of which are seen in panels (a) and (b), which frees up room for new flux. This process continues in a quasi-periodic fashion, and on average the BH produces outflows at $\eta \approx 140\%$ efficiency, i.e., outflows carry more energy than the entire rest-mass energy supplied by accretion.

much higher values of η than previously reported [19]. In fact, $\eta > 100\%$, therefore jets and winds carry more energy than the entire rest-mass supplied by the accretion, and this unambiguously shows that *net* energy is extracted from the accreting BH. This is the first demonstration of *net* energy extraction from a BH in a realistic astrophysical scenario. Thicker MADs ($h/r \approx 0.6$)

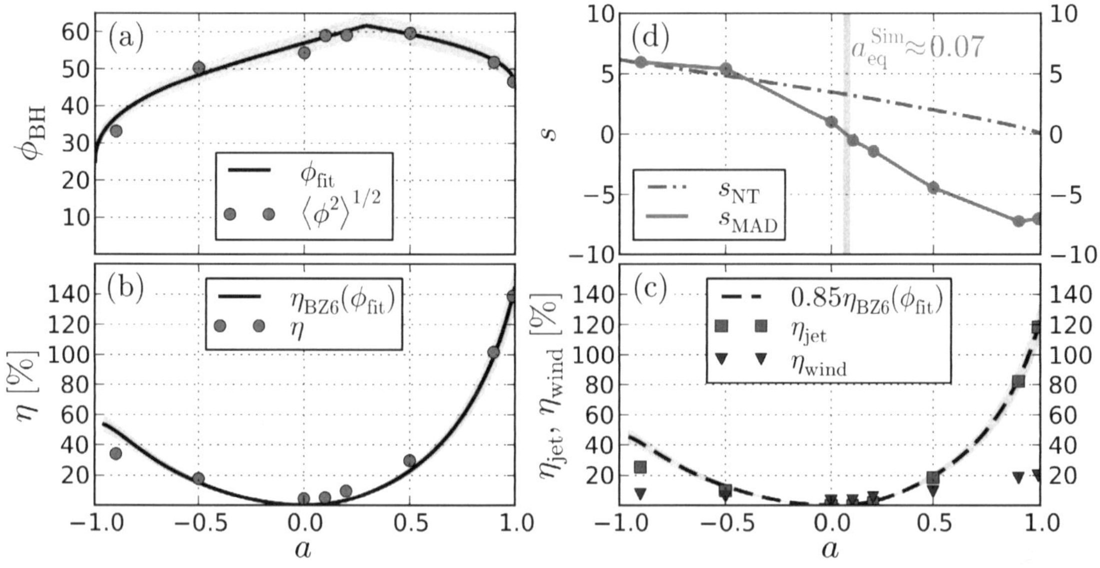

Figure 4. Spin-dependence of various quantities for MADs with $h/r \approx 0.3$. (a) Dimensionless BH spin, ϕ_{BH}: simulation results (red dots), by-eye fit, ϕ_{fit}, comprised of two linear segments in a ϕ_{BH}–Ω_H plane (black line), and 5% uncertainty on the fit (blue band). (b) Energy outflow efficiency, η: simulation results (red dots), BZ6 efficiency (black line, eq. 2) assuming $\phi_{BH}(a) = \phi_{fit}(a)$, and 10% uncertainty (blue band). (c) Jet efficiency (green squares) and wind efficiency (inverted blue triangles). Dashed line shows 85% of the above BZ6 efficiency (a good estimate of jet power for prograde BHs [23]), and blue color shows a 10% uncertainty band. (d) BH spin-up parameter, s, for a thin Novikov-Thorne disk [38] (green dash-dotted line) and for the simulations (red dots). Whereas for thin disks the equilibrium value of BH spin is $a_{eq}^{NT} = 1$, for our simulations it is much lower, is $a_{eq}^{Sim} \approx 0.07$ (vertical red band).

produce outflows at an even higher efficiency, $\eta \simeq 300\%$ [24].

Importantly, in the MAD state η is independent of the initial amount of magnetic flux in the accretion flow, i.e., η depends only on BH spin, a, and disk density angular thickness, h/r [23]. This allows us to reliably study spin-dependence of various quantities, shown in Figure 4. Dimensionless BH flux, ϕ_{BH}, shows $\sim 50\%$ variation with spin (see panel (a)). Prograde BHs ($a > 0$) are a few times more efficient than retrograde BHs ($a < 0$) for the same value of $|a|$ (see panel (b)). Quite the opposite—10 times larger efficiency of retrograde BHs than of prograde BHs—is predicted by the "gap" model [39], which assumes that zero magnetic flux populates the "gap" between the BH horizon and the innermost stable circular orbit (ISCO). We find that this interpretation of the ISCO—as the surface inside of which all magnetic flux nearly freely falls into the BH—does not hold: interchange instability [40] and/or flux eruptions (see Fig. 3) populate the ISCO interior with flux. In fact, the thinner the disk, the larger the fraction of BH magnetic flux that resides in the region between the BH horizon and the ISCO [23, 24].

Panel (c) shows the division of total outflow efficiency into highly magnetized jet and weakly magnetized wind components, with efficiencies η_{jet} and η_{wind}, respectively. Since jets are BH spin-powered (eq. 2), for $a = 0$ jet efficiency vanishes, but winds still derive their power from an accretion disk via a Blandford-Payne–type mechanism [41]. The larger the spin, the more efficient jets and winds. For rapidly spinning BHs most of the energy—about 85% for prograde BHs—is carried by relativistic jets. Importantly, even rather slowly spinning BHs, with $a \lesssim 0.5$, produce prominent BH spin-powered jets. This is in agreement with the recent evidence that jets in BHBs are powered by BH rotation over a wide range of BH spin, $0.1 \lesssim a \lesssim 0.9$ [42].

Do our highly efficient jets affect the spin of central BHs? Figure 4(d) shows spin-dependence of BH spin-up parameter, $s = M/\dot{M} \times da/dt$ [43]. For standard geometrically thin accretion disks, $s > 0$ (see Figure 4(d) and [38]), and the equilibrium spin is $a_{\rm eq}^{\rm NT} = 1$ [44] (we neglect photon capture by the BH, which would limit the spin to $a \approx 0.998$ [45]). Thick accretion flows in time-dependent numerical simulations [29, 43, 46] and semi-analytic studies [47] typically have $a_{\rm eq} \sim 0.9$ (however, see [48]). Figure 4(d) shows that the equilibrium value of spin for our MADs (with $h/r \approx 0.3$) is much smaller, $a_{\rm eq}^{\rm Sim} \approx 0.07$, due to large BH spin-down torques by our powerful jets. Since jet efficiency is $\sim 10^3$ times lower at this value of spin than for rapidly spinning BHs, such equilibrium spin systems are possible candidates for RQ AGN.

4. Conclusions

We confirm that the standard BZ power formula, $P_{\rm jet} \propto a^2$ (eq. 1), remains accurate only for $a \lesssim 0.5$ and present a higher order BZ6 approximation (eq. 2) that is accurate for all values of spin (see Figure 2). We find that if the BH magnetic flux is held constant, the presence of a thick accretion disk in low-luminosity AGN leads to a steep dependence of BH jet power on spin, $P_{\rm jet} \propto \Omega_{\rm H}^4$. This steep dependence allows to explain a factor of 10^3 radio loud/quiet dichotomy of AGN via having two galaxy populations different by the spin of central BHs.

We carried out a series of time-dependent global GRMHD simulations of BH accretion that contain a large amount of magnetic flux. We show that the accumulation of magnetic flux around the center can lead to the formation of a strong centrally-concentrated magnetic field that saturates the BH, obstructs the accretion, and leads to a magnetically-arrested disk (MAD, [19, 23, 24, 33–37]). We show that in this state the outflow efficiency, η, depends only on the BH spin, a, and the angular density thickness of the accretion flow, h/r, and is independent of the initial amount of magnetic flux in the disk. Since BH magnetic flux is as large as possible, we expect MAD systems to achieve the maximum η for a given BH spin and disk thickness. Indeed, we find highly efficient outflows, with $\eta \gtrsim 100\%$, which suggests that MADs might explain observations of AGN with apparent $\eta \sim \text{few} \times 100\%$ [6–11]. We determine the spin-dependence of jet, wind, and total efficiencies in our simulations (Figure 4) and expect that this information can be useful for calibration of semi-analytic jet power models, for studies seeking to infer BH spin from the observed jet efficiency, and for tests of general relativity [11, 49–52].

AT was supported by the Princeton Center for Theoretical Science fellowship. AT and RN were supported in part by NSF grant AST-1041590 and NASA grant NNX11AE16G. We acknowledge NSF support via TeraGrid resources: NICS Kraken and Nautilus, where simulations were carried out and data were analyzed, and NCSA MSS and TACC Ranch, where data were backed up, under grant numbers TG-AST100040 (AT), TG-AST080025N (JCM), TG-AST080026N (RN).

References

[1] Blandford R D and Znajek R L 1977 *MNRAS* **179** 433–456
[2] Chiueh T, Li Z Y and Begelman M C 1991 *ApJ* **377** 462–466
[3] Heinz S and Sunyaev R A 2003 *MNRAS* **343** L59–L64
[4] Tchekhovskoy A, McKinney J C and Narayan R 2008 *MNRAS* **388** 551–572
[5] Tchekhovskoy A, McKinney J C and Narayan R 2009 *ApJ* **699** 1789–1808
[6] Rawlings S and Saunders R 1991 *Nature* **349** 138–140
[7] Ghisellini G *et al.* 2010 *MNRAS* **402** 497–518
[8] Fernandes C A C *et al.* 2011 *MNRAS* **411** 1909–1916
[9] McNamara B R, Rohanizadegan M and Nulsen P E J 2011 *ApJ* **727** 39
[10] Punsly B 2011 *ApJ* **728** L17

[11] Martínez-Sansigre A and Rawlings S 2011 *MNRAS* **414** 1937–1964
[12] Sikora M, Stawarz Ł and Lasota J P 2007 *ApJ* **658** 815–828
[13] Fender R P, Belloni T M and Gallo E 2004 *MNRAS* **355** 1105–1118
[14] Wilson A S and Colbert E J M 1995 *ApJ* **438** 62–71
[15] Moderski R and Sikora M 1996 *A&AS* **120** C591
[16] Komissarov S S 2001 *MNRAS* **326** L41–L44
[17] Tanabe K and Nagataki S 2008 *Phys. Rev. D* **78** 024004
[18] Tchekhovskoy A, Narayan R and McKinney J C 2010 *ApJ* **711** 50–63
[19] Tchekhovskoy A, Narayan R and McKinney J C 2011 *MNRAS* **418** L79–L83
[20] Gammie C F, Narayan R and Blandford R 1999 *ApJ* **516** 177–186
[21] Komissarov S S and Barkov M V 2009 *MNRAS* **397** 1153–1168
[22] Penna R F *et al.* 2010 *MNRAS* **408** 752–782
[23] Tchekhovskoy A and McKinney J C 2012 *arXiv:1201.4385*
[24] McKinney J C, Tchekhovskoy A and Blandford R 2012 *arXiv:1201.4163*
[25] McKinney J C 2005 *ApJ* **630** L5–L8
[26] De Villiers J P, Hawley J F, Krolik J H and Hirose S 2005 *ApJ* **620** 878–888
[27] Hawley J F and Krolik J H 2006 *ApJ* **641** 103–116
[28] Barkov M V and Baushev A N 2011 *New Ast.* **16** 46–56
[29] McKinney J C and Gammie C F 2004 *ApJ* **611** 977–995
[30] Beckwith K, Hawley J F and Krolik J H 2008 *ApJ* **678** 1180–1199
[31] McKinney J C and Blandford R D 2009 *MNRAS* **394** L126–L130
[32] Balbus S A and Hawley J F 1991 *ApJ* **376** 214–233
[33] Bisnovatyi-Kogan G S and Ruzmaikin A A 1974 *Ap&SS* **28** 45–59
[34] Bisnovatyi-Kogan G S and Ruzmaikin A A 1976 *Ap&SS* **42** 401–424
[35] Igumenshchev I, Narayan R and Abramowicz M 2003 *ApJ* **592** 1042–1059
[36] Narayan R, Igumenshchev I V and Abramowicz M A 2003 *PASJ* **55** L69–L72
[37] Igumenshchev I V 2008 *ApJ* **677** 317–326
[38] Novikov I D and Thorne K S 1973 *In Black Holes-Les Astres Occlus, ed. C. De Witt & B. S. De Witt* (New York: Gordon & Breach)
[39] Garofalo D 2009 *ApJ* **699** 400–408
[40] Stehle R and Spruit H C 2001 *MNRAS* **323** 587–600
[41] Blandford R D and Payne D G 1982 *MNRAS* **199** 883–903
[42] Narayan R and McClintock J E 2012 *MNRAS* **419** L69–L73
[43] Gammie C F, Shapiro S L and McKinney J C 2004 *ApJ* **602** 312–319
[44] Bardeen J M 1970 *Nature* **226** 64–65
[45] Thorne K S 1974 *ApJ* **191** 507–520
[46] Krolik J H, Hawley J F and Hirose S 2005 *ApJ* **622** 1008–1023
[47] Benson A J and Babul A 2009 *MNRAS* **397** 1302–1313
[48] Moderski R, Sikora M and Lasota J 1998 *MNRAS* **301** 142–148
[49] Daly R A 2011 *MNRAS* **414** 1253–1262
[50] Buliga S D *et al.* 2011 *Astrophysics* **54** 548–552
[51] Lei W H and Zhang B 2011 *ApJ* **740** L27
[52] Bambi C 2012 *arXiv:1201.1638*

Dynamical processes in galaxy centers

Francoise Combes
Observatoire de Paris, LERMA, CNRS, 61 Av de l'Observatoire, 75014, Paris
E-mail: francoise.combes@obspm.fr

Abstract. How does the gas get in nuclear regions to fuel black holes? How efficient is the feedback? The different processes to cause rapid gas inflow (or outflow) in galaxy centers are reviewed. Non axisymmetries can be created or maintained by internal disk instabilities, or galaxy interactions. Simulations and observations tell us that the fueling is a chaotic and intermittent process, with different scenarios and time-scales, according to the various radial scales across a galaxy.

1. Introduction

Black hole growth in the center of galaxies appears to be closely related to the growth of bulges, either through star formation or mergers (Magorrian et al 1998, Gültekin et al 2009). Analysis of the BH demography today shows that $\sim 25\%$ of the black holes mass is contained in late-type galaxies; the most massive BH have grown faster, at higher redshifts, and the BH mass growth through mergers is not dominant (Shankar et al 2004). Most of the BH growth occurs in type 2 AGN, obscured by dust, which supports the association with star formation.

The dynamical processes able to produce non-axisymmetries in galaxy disks and consequently gravity torques on the gas, driving inflows towards the center, are multiple, and depend strongly on the scale considered. At kiloparsecs scale, stellar bars are the most robust and efficient instabilities, they can be re-inforced or triggered by galaxy interactions. Below one kiloparsec, their action can be continued through secondary, nuclear bars, decoupling and interacting non-linearly with the primary bars. At 10-100pc scale, nuclear disks may become unstable through $m = 1$ modes, lopsidedness, warps and bending. Below that, the clumpiness of the interstellar medium plays a large role so that dynamical friction, and also viscous torques due to turbulence could take over.

In recent years, it has been realised that feedback phenomena, due to star formation or the AGN itself, can produce substantial gas outflows, self-regulating the inflow and the fueling. In the following, all these processes are discussed.

2. Bar gravity torques

When a stellar bar has grown in a galaxy disk, it triggers spiral arms in the gas, which tends to follow periodic orbits but dissipates energy at crossing flows. The gas response and the stellar bar are not in phase, which results in gravity torques exerted on the gas, negative inside corotation, and positive outside (Garcia-Burillo et al. 2005, 2009). The torques are efficient to drive the gas from corotation to the center, or down to a ring at the inner Lindblad resonance (ILR), which is then the site of a starburst.

The amplitude of the torques and their efficiency can be computed from the observations: the gravitational potential is derived from the near-infrared image, and the tangential forces are computed on the gas distribution, allowing to draw the torque maps, sometimes showing negative torques down to the center (e.g. Casasola et al 2008). According to the bar strength, the gas can lose 10-30% of its angular momentum in one rotation, i.e. the inflow time-scale is of the order of 300 Myr. Note that the stellar potential can also exert torques on the stars themselves, but this is much weaker, with time-scales of 4 Gyr (e.g. Foyle et al 2010).

During the gas inflow inside corotation, the gas transfers its angular momentum to the stars (the gas exerts a reciprocal torque on the bar), and this weakens or destroys the bar, which is a wave with negative angular momentum (Bournaud & Combes 2002). When external gas accretion is taken into account, a new bar instability occurs in a disk replenished in gas, and a cycle alternating bar formation and AGN fueling is triggered (Bournaud et al 2005).

The fact that torques change sign at each resonance prevents the gas to inflow in only one step, all across the disk. Clearly, several dynamical processes have to contribute at the various scales. Inside corotation, there are frequently ILRs, and the gas piles up in rings about these radii. When the mass concentration is sufficient in the center, the frequency of orbital precession ($\Omega - \kappa/2$) is large enough that a nuclear bar can decouple from the primary one. There is interaction between the two bars through resonances, for instance the corotation of the nuclear bar is the ILR of the primary bar. The secondary bar can then continue the gas inflow towards the very center. This double inflow has been seen in the type 2 AGN NGC2782, and has been reproduced in fitted simulations (Hunt et al 2008). It can be shown that the actual AGN fueling is a very short phase, i.e. when torques are negative down to the very center. If the gas continues to inflow inside 200pc during almost 1 Gyr, it inflows inside 20pc only during 100 Myr.

3. Instabilities m=1

Lopsidedness may occur at large scale in the outer parts of galaxies, but also in the very center (see e.g. Jog & Combes 2009). A central mass concentration favors the latter, since orbits tend to be keplerian. In a thin self-gravitating disk, the $m = 1$ resonant orbits have a very rapid differential precession rate $\Omega - \kappa$, while in the presence of a massive black hole, the potential becomes spherical and $\Omega = \kappa$. In these nearly keplerian disks, slow $m = 1$ modes can develop, an example being the conspicuous stellar disk around the black hole in M31. This mode develops with a strong self-gravity, and the action of the indirect potential, due to the off-centring of the central BH mass (Bacon et al 2001, Tremaine 2001).

When gas is present in the disk, $m = 1$ asymmetry torques allow the gas to lose angular momentum, and flow to the center. This lopsidedness is frequent at small scales in nearby galaxies, where spatial resolution is sufficient (our Milky Way is an example), and larger samples have been studied at large scales. Over thousands of galaxies, a correlation has been found between the lopsidedness, BH growth and the youth of stellar populations in the center (Reichard et al 2009).

Recent simulations by Hopkins et al (2011) illustrate all these dynamical processes, that occur at decreasing scales. After a cascade of $m = 2$ and $m = 1$ perturbations, the gas is focussing towards the center, and forms a thick disk of size 1-10pc around the black hole. Zoomed simulations allow to go further, and study the fate of the gas that has piled up in the nucleus, with a gas fraction up to 90%. Column densities can then reach 10^{22}-10^{25} cm^{-2}, the gas is unstable to formation of clumps, but also warps, twists and bending. At this stage, dynamical friction of Giant Molecular Clouds (GMC) of M= 10^6 $M_\odot$ against the bulge can occur in a time-scale t$\sim$ 80Myr $(r/100pc)^2$, at a distance r from the center, and varies in 1/M.

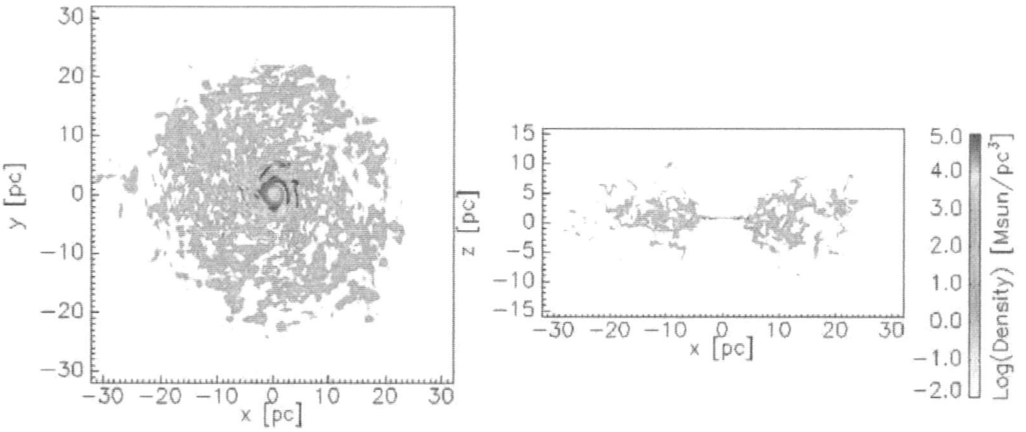

Figure 1. The molecular gas distribution in the galaxy center, in the X-Y plane (left) and in a perpendicular slice (X-Z), in a torus simulation including H_2 formation by Wada et al (2009).

4. Formation of the torus

The resulting dense, compact and thick disk around the BH could be identified as the molecular torus, often invoked to explain type 2 AGN. The warp and bending instabilities can be maintained through continuous gas accretion, and this explains why the torus and outer disk in galaxies are frequently misaligned. The morphology and thickness of the torus obtained depends strongly on the physics assumed, and in particular on the velocity dispersion, the heating or the star formation feedback adopted, in general at a scale below the resolution of the simulations.

Both warm thin and cold thick, fluffly, disks/torus could co-exist, according to the multi-phase simulations by Schartmann et al (2009). Under the influence of stellar winds, from a nuclear stellar cluster, a nuclear disk of radius 10pc forms with cold filaments and a tenuous hot interstellar medium.

Thinner molecular disks are obtained by Wada et al (2009), through introducing the formation of molecular hydrogen (H_2), and consistent chemistry, in presence of an UV flux of G=10 to 100 in Habing units. They simulate a disk of 30pc radius, heated by supernovae feedback, with a finest resolution of 0.125pc. The H_2 fraction at equilibrium is 0.4. The disk is extremely thin in the central 5pc, but then flares at a radius of 10pc, as shown in Figure 1.

The influence of stellar feedback is of course crucial on the morphology of the disks/torus. Hopkins & Quataert (2010) have introduced an equation of state of the gas, to take into account the subgrid physics, and in particular the consequence of star formation on the dipersion and turbulence of the gas. Exploring the whole SN efficiency, the simulations can reproduce the physical properties (density, dispersion) observed for various objects, from quiescent disks to perturbed ULIRGs.

They show through several successive re-simulations how the inflow rate has a chaotic evolution at all scales (cf Figure 2). The accretion is episodic, and is much more contrasted in a merger, than during secular evolution.

5. Quasar feedback

In the standard model of galaxy formation, it is quite difficult to reproduce the galaxy mass function, both at the small mass and large mass ends of the spectrum. For dwarf galaxies, supernovae feedback has long been invoked to eject baryons out of the shallow potential wells, but this does not work for massive galaxies. For the latter, AGN feedback seems appropriate, given the BH-bulge mass correlation, although no global observation has yet confirmed its

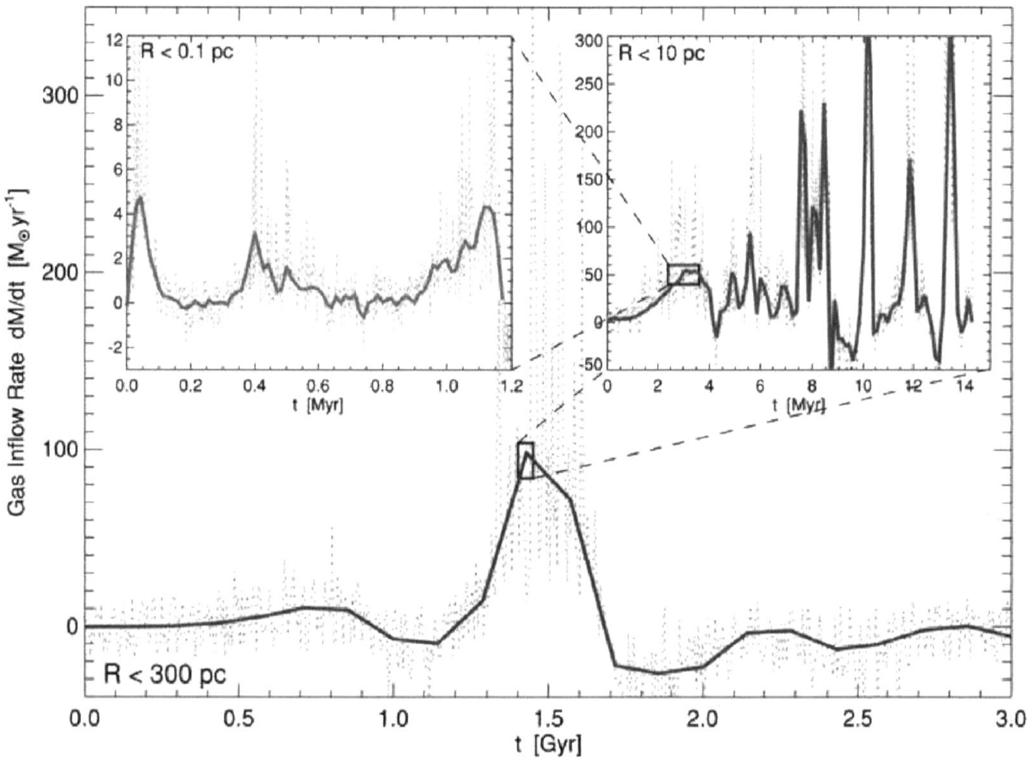

Figure 2. The inflow rate as a function of time, during the merger of two galaxies, and in particular the coalescence of the two nuclei. The main simulation (bottom) shows the accretion inside 300pc, and is resimulated twice at higher spatial and time resolution (top right inflow inside 10 pc and top left inside 0.1 pc). The dash line is the instantaneous rate, smoothed to 5 local dynamical times in the full line, from Hopkins & Quataert (2010).

efficiency. Simulations (e.g. Di Matteo et al 2005) have shown how in galaxy mergers the energy released by the AGN quenches both star formation and AGN growth. Evidence of regulating or quenching star formation is observed in special cases of central galaxies in clusters, where radio jets moderate cooling flows.

Recently, more evidence has been obtained of gas outflows in AGN/starburst galaxies. Many examples of OH and H_2O line absorptions have been detected with the Herschel satellite, a prototype being Mrk 231 (Fischer et al 2010). In the same object, wide wings ($\sim$ 800 km/s) in the CO emission have been detected with the IRAM interferometer (Feruglio et al 2010). The ejected gas amounts to 700 $M_\odot$/yr, more than 3 times the star formation rate of 200 $M_\odot$/yr. These neutral flows correspond to massive ejections, while previoulsy many more ionized gas flows had been detected, but with less mass involved. For example, the [Fe II] emission line in Mrk 1157 has been observed with a kinematic corresponding to outflow along the radio jet, which liberates the iron from dust grains in shocks. The ejected ionized gas amounts to 6 $M_\odot$/yr (Riffel & Storchi-Bergmann 2011).

Selecting optical AGN observed in the H_2 rotational lines with Spitzer, we have recently found several cases of broad wings, implying molecular gas outflows. In 4C12.50 for example (cf Figure 3), there is an H_2 component flowing at 640km/s, and corresponding to an outflow rate of 130 $M_\odot$/yr, much lower than the star formation rate, however (Dasyra & Combes 2011).

Feedback might be invoked also to reduce AGN activity to a rapid phase in galaxy mergers.

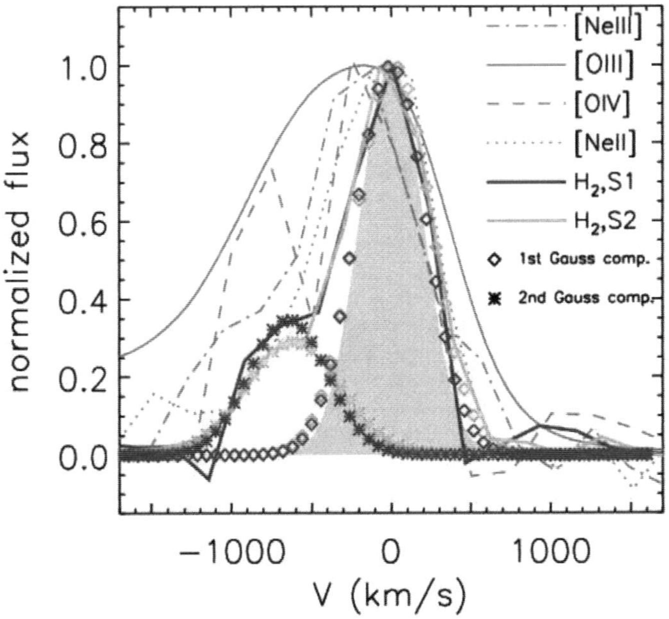

Figure 3. Profiles of several lines, and in particular H_2 S1 and S2, towards the quasar/starburst 4C12.50, showing broad line wings, implying massive gas outflow, from Dasyra & Combes (2011).

Indeed, given the frequency of galaxy mergers, the presence of a massive BH in each galaxy center, and the gas inflow in galaxy interactions, binary AGN are predicted to be more frequent than observed. Recently Green et al (2010) have studied a rare example of binary quasar in a merger at z=0.44, and suggest that many others could be obscured by dust. Locally, binary AGN are very rare: NGC6240 has been shown to host two active nuclei through obscured X-ray emission and two strong neutral Fe Kα lines (Komossa et al 2003). The same evidence has been found for Arp 299 (Ballo et al 2004).

6. Statistics - time-scales, 10-100pc fueling

Among a sample of 16 nearby active galaxies with low-luminosity (Seyfert or Liners), which was observed at high resolution at multi-wavelengths, and in the molecular component with the IRAM interferometer, it is possible to derive preliminary statistics about their fueling state. From their torque maps, the current gas (in/out)flows have been derived (see also Garcia-Burillo, this meeting). Only ~35% (or 6 out of 16 galaxies of the NUGA sample) have negative torques in the center, observed at a resolution scale of 1"~50-100pc. Most of the remaining galaxies show positive torques towards the center, maintaining the gas in a nuclear ring.

This must imply that fueling phases are short, a few 10^7 yrs, possibly due to feedback. This order of magnitude is compatible to what is found in N-body simulations through bar fueling, as presented above, and also to the rarity of binary AGN. With so short fueling phases, it is difficult to identify the trigger, bars or nuclear bars might have weaken when the gas finally reaches the center. This explains the difficulty to find correlations between AGN activity and a given dynamical process. Star formation is also fueled by the same torques, is frequently associated to AGN activity, and is characterised by longer time-scales, therefore correlations are easier to detect with their dynamical trigger.

7. Conclusion

Dynamical processes depend essentially on the radial scale in galaxies. Primary bars are essential to drive gas from 10kpc to R $\sim$ 100pc. Then nuclear bars take over from 100pc to 10pc. These bars could be part of the internal secular evolution, maintained by external gas accretion, or could be triggered by galaxy interactions. Non-axisymmetries of the $m = 1$ type, off-centring and lopsidedness, may play a role, in the outer parts (gas accretion and tides), or in the very center, when the potential is nearly keplerian, close to the black hole. In most cases, there are embedded structures at the various scales to fuel nuclear starbursts and AGN. At scales $\sim$ 1-10pc, when the gas fraction is still high, violent clump instabilities drive turbulence, and consequent viscosity and torques. Dynamical friction, warps and bending instabilities will drive gas to the nucleus, form a thick disk/torus and produce chaotic and intermittent fueling. When the gas is consumed, the nuclear stellar disk remains such that in M31.

Given the various dynamical instabilities of warping and bending, alignment between small scale and large scale disks is not expected. The fueling is certainly moderated by feedback processes, from star formation or the AGN itself. Their efficiency is still poorly known, although gas outflows are present, sometimes with huge amounts, up to three times the star formation rate.

References

[1] Bacon R., Emsellem E., Combes F. et al.: 2001, A&A 371, 409
[2] Ballo, L., Braito, V., Della Ceca, R. et al.: 2004, ApJ 600, 634
[3] Bournaud, F., Combes, F.: 2002, A&A, 392, 83
[4] Bournaud, F., Combes, F., Semelin, B.: 2005, MNRAS, 364, L18
[5] Casasola, V., Combes, F., Garcia-Burillo, S. et al.: 2008, A&A 490, 61
[6] Dasyra, K., Combes, F.: 2011, A&A 533, L10
[7] Di Matteo T., Springel, V., Hernquist, L.: 2005 Nature, 433, 604
[8] Feruglio, C., Maiolino, R., Piconcelli, E. et al.: 2010, A&A 518, L155
[9] Fischer, J., Sturm, E., Gonzalez-Alfonso, E. et al.: 2010, A&A 518, L41
[10] Foyle, K., Rix, H-W., Zibetti, S.: 2010, MNRAS 407, 163
[11] Garcia-Burillo, S., Combes, F., Schinnerer, E. et al.: 2005, A&A 441, 1011
[12] Garcia-Burillo, S., Fernandez-Garcia S., Combes, F. et al.: 2009, A&A 496, 85
[13] Green, P. J., Myers, A. D., Barkhouse, W. A. et al.: 2010, ApJ 710, 1578
[14] Gültekin, K., Richstone, D. O., Gebhardt, K. et al.: 2009, ApJ 698, 198
[15] Hopkins P., Hayward, C. C., Narayanan, D., Hernquist, L.: 2011, MNRAS, arXiv1108.3086
[16] Hopkins P., Quataert E.: 2011, MNRAS 415, 1027
[17] Hopkins P., Quataert E.: 2010, MNRAS 407, 1529
[18] Hunt L., Combes F., Garcia-Burillo, S. et al.: 2008, A&A 482, 133
[19] Jog C., Combes F.: 2009, Physics Report 471, 75
[20] Komossa, S., Burwitz, V., Hasinger, G. et al.: 2003, ApJ 582, L15
[21] Magorrian J., Tremaine S., Richstone D. et al.: 1998, AJ 115, 2285
[22] Reichard T.A., Heckman, T. M., Rudnick, G. et al.: 2009, ApJ 691, 1005
[23] Riffel R. A., Storchi-Bergmann, T.: 2011 MNRAS 417, 2752
[24] Schartmann, M., Meisenheimer, K., Klahr, H. et al.: 2009 MNRAS 393, 759
[25] Shankar, F., Salucci, P., Granato, G.L. et al.: 2004 MNRAS 354, 1020
[26] Tremaine S.: 2001 AJ 121, 1776
[27] Wada K., Papadopoulos, P. P., Spaans, M.: 2009, ApJ 702, 63

Astronomy at High Angular Resolution 2011 IOP Publishing
Journal of Physics: Conference Series **372** (2012) 012042 doi:10.1088/1742-6596/372/1/012042

Resolving Gas Chemistry at Arcsecond Scales in Nearby Spiral Nuclei

David S Meier[1], Jean L Turner[2] and Eva Schinnerer[3]

[1] New Mexico Institute of Mining and Technology, 802 Leroy Place, Socorro, NM 87801
[2] Department of Physics and Astronomy, UCLA, Los Angeles, CA 90095–1562
[3] Max-Planck-Institut für Astronomie, Königstuhl 17, D-69117 Heidelberg, Germany

E-mail: dmeier@nmt.edu

Abstract. Astrochemistry is rapidly becoming one of the most powerful tools to study the structure and evolution of the central kiloparsec in spiral nuclei. Imagining the distribution of quiescent ion-molecule, photon-dominated region (PDR) and shock chemistry permits the triggers of nuclear starbursts to be identified and the bursts subsequent feedback to be constrained. Two new chemical methods for identifying the evolutionary phase of the starbursts in IC 342, NGC 6946 and Maffei 2 are discussed. The first method is to use the HCO^+/N_2H^+ ratio to constrain the degree of penetration of dense clumps by UV radiation. The second determines the evolutionary phase by mapping the amount and physical conditions of the densest molecular component via multi-transition HC_3N observations. With the full capabilities of radio facilities such as the VLA and ALMA, probing the changing gas chemistry on sub arcsecond scales in external galaxies will very soon be routine.

1. Introduction

Centers of galaxies are characterized by short dynamical times, strong tidal forces, nuclear bars and outflows, both ionized and molecular. This makes for a tumultuous interstellar medium (ISM). Systems that harbor active galactic nuclei (AGN) or are in the midst of episodes of intense star formation have ISMs that are further altered by strong UV and X-ray radiation fields. These in turn influence the physical and chemical conditions of the ISM and therefore the future evolution of star formation, AGN feeding and nuclear dynamics. With recent advances in millimeter interferometry it is now possible to follow this evolution directly in the molecular component at scales smaller than an individual giant molecular cloud (GMC).

2. Astrochemistry in an extragalactic context

Astrochemistry provides a direct probe of the physical state of the ISM. In Galactic studies much attention has been focussed on the chemistry present at small scales in the vicinity of localized areas of activity, such as young stellar objects and their outflows, hot/dark cores, individual massive star/supernovae (SNe) molecular cloud interfaces, circumstellar envelopes and comets [1]. Given the difficulties of surveying a number of faint molecular lines over large angular sizes, less work has been done on the large scale chemistry that dominates a galaxy's volume [2-4. Observations of chemistry in nearby galaxies have the sensitivity, resolution and field of view to permit the study of the chemistry across scales of hundreds of parsecs.

From a theoretical point of view, away from extreme environments, interstellar chemistry is fairly well formulated [5-8]. The low average densities and temperatures make most neutral-neutral reactions inefficient throughout the large volumes averaged over in extragalactic observations. Allowing for some approximation, we can characterize much of astrochemistry on these scales as the result of grain or ion catalyzed molecule formation. Ions induce dipolar attractive forces in neutrals. The attraction between the ion and neutral drives a rapid nearly temperature-independent chemistry. In quiescent clouds, cosmic ray ionization of H_2, leading immediately to H_3^+, is the main catalytic ion. Molecules formed primarily in this quiescent 'ion-molecule' chemistry include CO, HCN, HNC, N_2H^+ and OH.

When intense UV radiation fields are present C^+ takes over the role as the dominant catalytic ion. In theses regions, so-called photon-dominated regions (PDRs), molecules formed from C^+ + H_2, including CH, C_2H, CN and c-C_3H_2 are particularly important [9].

Certain large organic molecules, such as CH_3OH, have no known gas-phase formation pathway capable of explaining the abundances observed in both the Galactic and extragalactic ISM. High abundances of these species are observed in cool grains' ice mantles [10]; where the grain surfaces catalyze reactions between atoms, neutral radicals and hydrogen [11-13]. Liberating these mantles can readily explain the observed gas-phase abundances. However, in all but the most extreme starbursts, liberation of mantles by star formation heating (hot cores) occur in regions too compact to be detected with current telescopes. Liberation mechanisms need to operate coherently on scales greater than 10 pc to explain observed abundances in nearby galaxies. The prime mechanism for liberating these large organic species appears to be shocked molecular gas in bars and spirals or starburst winds [14-16].

3. Spatially resolved chemical maps in spiral nuclei

By imaging representative species from the separate categories of quiescent ion-molecule, PDR and shock chemistries it is possible to deconstruct the dynamical structure and evolution of starbursts in external galaxies. Figure 1 shows five representative species, including at least one from each of the three above chemistries, mapped across the central 0.5 kpc of the nearby starbursts, IC 342, Maffei 2 and NGC 6946, along with CO and millimeter continuum. The IC 342 data was imaged with the OVRO interferometer at ~6" resolution (80 pc) [16], the Maffei 2 data with the BIMA interferometer at ~8" (110 pc) [17] and the NGC 6946 data with the Plateau de Bure Interferometer at ~2" resolution (50 pc).

The most striking feature of the maps is how much the morphologies differ amongst the tracer molecules, and from CO and the 3 mm continuum (star formation intensity). For example, C_2H and CH_3OH are nearly anti-correlated in all three nuclei. Therefore the PDR and shock chemistries are distributed very differently. Comparing C_2H with millimeter continuum in Figure 1 demonstrates that PDRs are more intimately associated with star formation than regions emitting in CH_3OH. This is further evidence that CH_3OH does not trace ice mantles evaporated by star formation. Towards IC 342, CH_3OH spans the entire molecular arms, with some emphasis towards the outer ends of the bar/spiral. In Maffei 2 and NGC 6946 the outer ends of the (nuclear) bar are even more strongly pronounced. In [16] a consistent picture for the mutual evolution of star formation and gas chemistry is laid out. Shocks in the outer bar arms cause dissipation of energy and angular momentum. Gas flows inward. Just before it reaches the intersection of the arms with the central ring, gas collides with existing ring gas, piling up and triggering starbursts. Inertia causes the star formation and remaining residual gas to continue inward, settling onto the small nuclear ring. The starbursts just outside the ring-intersection are young and the ring star formation is expected to somewhat more evolved. With little modification this chemical picture carries over to the other two nuclei.

No starbursts are perfectly coeval. It is expected that individual star forming regions in the nuclear of spirals manifest a spread in burst phase. The burst begins with a 'proto-starburst' (in

Journal of Physics: Conference Series **372** (2012) 012042 doi:10.1088/1742-6596/372/1/012042

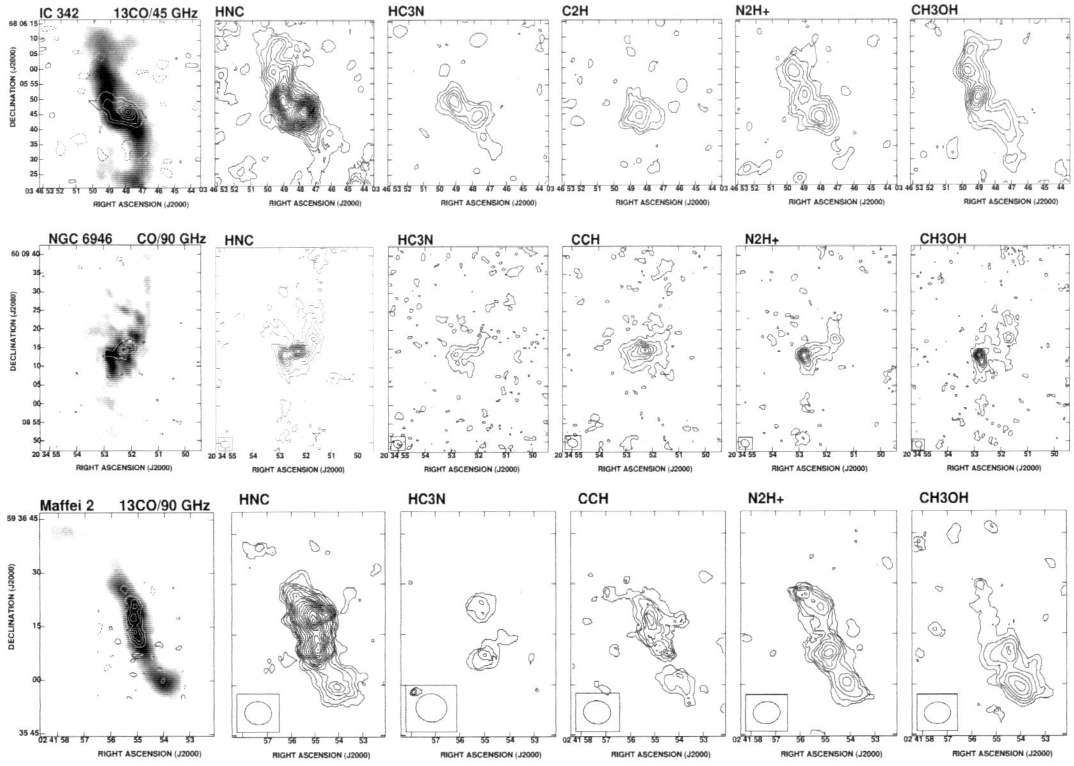

Figure 1. The integrated intensities of six molecular transitions towards the nuclei of the nearby starbursts, IC 342 [16], Maffei 2 [17] and NGC 6946. For each nucleus the leftmost plane shows the molecular gas distribution traced by ^{13}CO(1–0) [except for NGC 6946 which is CO(1–0)] in grayscale. For IC 342 and Maffei 2, the grayscale ranges from 1 - 10 Jy beam^{-1} km s^{-1}, while for NGC 6946 it ranges from 1.5 - 15 Jy beam^{-1} km s^{-1}. The contours are 7 or 3 mm continuum which traces star formation. Contours are 1.0, 0.25 and 1.2 mJy beam^{-1} for IC 342, NGC 6946 and Maffei 2, respectively. The remaining planes are HNC(1–0), HC_3N(10–9), C_2H(1–0; 3/2-1/2), N_2H^+(1–0) and $CH_3OH(2_k - 1_k)$ as labeled. For IC 342 the contours and resolution are the same as in Figure 2 of [16]. For NGC 6946 contours are 0.125, 0.11, 0.15, 0.125 and 0.125 Jy beam^{-1} km s^{-1} for the five trace species (left to right). Beamsizes for these maps are $\sim 2.''5 \times 2''$. For Maffei 2 contours are 1.0, 0.60, 1.1, 0.75 and 1.5 Jy beam^{-1} km s^{-1}. Beamsizes for these maps are $\sim 8'' \times 7''$ except HC_3N which is $\sim 10'' \times 8.''5$.

analogy with a protostar) having high columns of cool, dense gas with faint, perhaps optically thick radio continuum emission. The burst matures, large numbers of massive stars form, bathing the surrounding dense clumps in dissociative UV radiation, creating dense PDRs, an expanding HII region and bright thermal radio continuum. As an individual burst ages the dense molecular gas is consumed or dispersed, HII regions fade and the radio continuum spectrum steepens with the generation of synchrotron emission after the onset of SNe. For an instantaneous burst all three phases typically last $\sim 10^7$ yrs or less. Short, but longer than chemical equilibrium timescales, so we expect to be able discriminate them with chemistry.

Averaged over the nucleus it is quite difficult to separate such phases but with the high angular resolution molecular spectroscopy now available, evidence of this evolution is now being observed in the most nearby systems. This is particularly true in the three nuclei discussed here. All three systems have organized kinematics consistent with inflow driven along the nuclear bar arms followed by star formation triggered at the arm-inner ring intersection regions. Therefore a connection between the location of gas and star formation relative to the arm-inner ring intersection and its age is expected. This connection suggests that changing starburst phases should be morphologically separated and their chemical signatures resolvable.

4. Localized changes in dense gas chemistry: the HCO^+/N_2H^+ ratio

C_2H has a modest critical density and so is more sensitive to PDR gas at low to moderate densities. In IC 342, C_2H tracers the cloud interfaces oriented towards the evolved nuclear cluster not the embedded, young bursts [16]. This tendency to avoid the dense clumps is more pronounced in the outskirts of Maffei 2, while in NGC 6946 we do not have the spatial resolution to tell for sure. In fact in Maffei 2 and to a lesser extent IC 342, C_2H critical densities are such that it the presumably modest density molecular outflows along the minor axis of the nuclear disk are sampled. To investigate whether the gas chemistry in a particular starburst has transitioned from the 'proto-burst' phase to the phase when the PDRs permeate the bulk of the dense cores, we need a high density probe that is PDR sensitive.

We propose that the $HCO^+(1–0)/N_2H^+(1–0)$ integrated intensity ratio could be just such a probe [17]. In quiescent gas, proton transfer reactions of H_3^+ with CO and N_2 form HCO^+ and N_2H^+ respectively:

$$H_3^+ \ + \ CO \ \rightarrow \ HCO^+ \ + \ H_2. \tag{1}$$

$$H_3^+ \ + \ N_2 \ \rightarrow \ N_2H^+ \ + \ H_2, \tag{2}$$

However, in PDR gas only HCO^+ maintains a relatively rapid formation route:

$$C^+ + OH \rightarrow CO^+ + H \tag{3}$$

$$H_2 + CO^+ \rightarrow HCO^+ + H. \tag{4}$$

No corresponding route exists for N_2H^+ in PDRs [18]. So toward the PDR regions associated with the starbursts, HCO^+ can maintain appreciable abundances while N_2H^+ is readily destroyed. Therefore, for irradiated gas we expect a precipitous drop in N_2H^+ intensity that is not matched in HCO^+. Since both HCO^+ and N_2H^+ have high critical densities this ratio is more strongly sensitive to the properties of the dense clumps than is C_2H. Hence we predict that as a starburst strengthens or evolves we should see a dramatic, localized increase in the HCO^+/N_2H^+ ratio.

Consistent with the fact that in IC 342 PDRs appear primarily as surface features, the strong decrease of N_2H^+ relative to HCO^+ is not seen [16]. But the above predicted behavior is observed in Maffei 2 and NGC 6946 (Figure 2). For Maffei 2, the starbursts at the ring intersection region exhibit slightly elevated HCO^+/N_2H^+ ratios but not as extreme as found toward the burst on the ring.

5. Dense gas evolution: towards high angular resolution chemistry

The HCO^+/N_2H^+ ratio suggests that gas chemistry evolves with starburst strength and evolutionary phase. Unlike Maffei 2 and NGC 6946, N_2H^+ remains bright towards both arm-ring intersection starbursts in IC 342. This hints that the gas at these locations remain relatively unprocessed and possibly quite young. To pursue this further we investigate the excitation of very dense molecular component in IC 342 at much higher spatial resolutions ($\sim$25 pc).

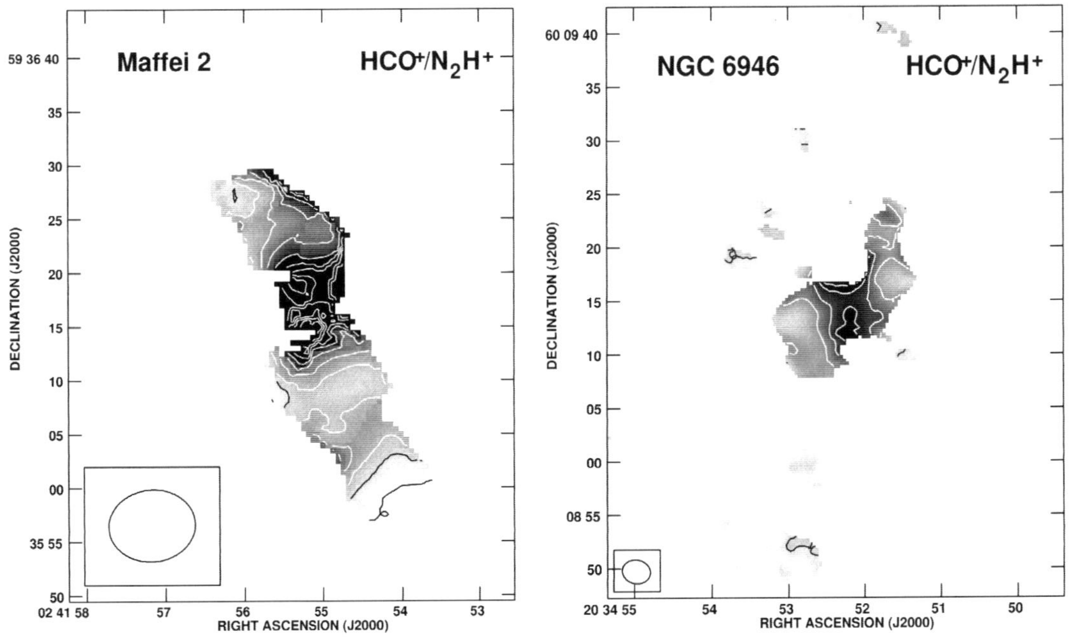

Figure 2. The $HCO^+(1\text{–}0)/N_2H^+(1\text{–}0)$ integrated intensity ratio toward Maffei 2 [17] and NGC 6946. The grayscale ranges from 1 - 6 for Maffei 2 and 1 - 9 for NGC 6946, with dark being large values. Contours are in steps of 1 for Maffei 2 and steps of 2 for NGC 6946. The beam size of each map is displayed in the bottom left corner of each image.

On 100 pc scales HC_3N was the molecule found to be morphologically most closely connected to the starburst [16]. Therefore studies of this molecule are excellent for characterizing the physical conditions of the gas out of which the starburst directly forms. Not only do we improve the spatial resolution but we map multiple transitions to constrain gas densities, excitation temperatures and thermal pressure.

Figure 3 displays the J =5–4 transition of HC_3N imaged in the center of IC 342 with the Very Large Array at 45 GHz [19]. At this resolution, the cloud towards the main starburst, marked by the bright 45 GHz continuum source in the grayscale, breaks up into a small collections of barely detectable sub-GMC clumps. By contrast, the cloud at the corresponding dynamical location of the other side of the ring remains an extremely bright, coherent emission source. Total fluxes were measured over a 2" aperture in J=10–9 and J=16–15 (from the Plateau de Bure interferometer) towards the two arm-ring intersection regions [19]. Figure 3 also shows the corresponding spectral line energy distributions for the three transitions.

The amount and excitation of the dense gas differs signification at the two locations. Toward the main starburst total fluxes are surprisingly low but excitation is not ($T_{ex} \simeq 40$ K). The low absolute brightness together with high excitation implies that the total column of dense gas towards the burst is small. The location of the strongest star formation in the nucleus of IC 342 is *not* a source of abundant dense gas. On the eastern side of the ring, where the star formation indicators are less intense, dense gas columns are high, but excitation remains low ($T_{ex} < 15$ K), implying large amounts of dense gas.

We generate a multi-component, large velocity gradient model that can simultaneously explain the excitation of six low energy transitions of the CO isotopologues [20,21] and the three

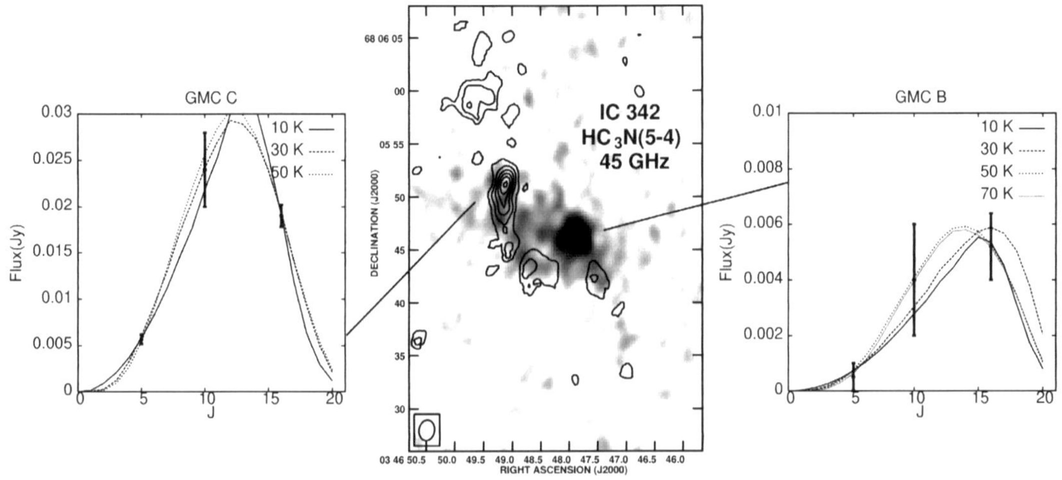

Figure 3. The HC_3N LVG modeling of the GMCs in IC 342 [19]. The central image is a zoom in on the $HC_3N(5–4)$ transition at $< 2''$ resolution overlaid on the adjacent continuum. Spectral line energy distributions from the dense 'proto-starburst' region on the left and the dominant starburst to the right are displayed. The three transitions modeled are J=5–4, J=10–9 and J=16-15. Large-velocity gradient radiative transfer models of flux versus J_u state are shown for kinetic temperatures ranging from 10 K to 70 K. $X(HC_3N)/dv/dr = 10^{-9}$ is assumed for the displayed models.

transitions of HC_3N. For an adopted abundance of HC_3N, solutions are obtained for kinetic temperatures ranging from 5 - 100 K. A subset of these solutions are displayed with different lines in Figure 3. The degeneracy between high kinetic temperatures and high densities means that it is not possible to separately constrain both precisely. However the gas thermal pressure is tightly constrained, with $n_{H2}T_k$ being $\simeq 6 \times 10^6$ cm^{-3} K. From the size and ionization rate of the associated starburst nebula [22], we estimate the thermal pressure of the HII region to be the same within uncertainties.

Toward the main IR/radio continuum source, temperatures and densities of the dense gas are high enough to be in pressure equilibrium with the HII region. The eastern region has lower thermal pressure and large total dense gas masses (few $\times 10^6$ $M_\odot$), despite the much weaker instantaneous star formation. This dichotomy is interesting in the context of the well established global dense gas versus star formation rates relationships [23]. If such simple relations apply universally we would expect the behavior of the dense component of the two star forming regions to be reversed. We interpret this breakdown as a manifestation of the evolution of the starburst across the nucleus. The currently dominant starburst near the bright IR/mm continuum source (Figure 3) has reached an evolutionary phase where it has consumed, or dispersed much of its natal dense gas. The dense gas that remains there is remnant detritus embedded in, and in pressure equilibrium with, the expanding HII region. The younger non-dispersed location appears to be a 'proto-starburst' just beginning an active phase.

6. Conclusions

Astrochemistry is rapidly becoming one of the most powerful tools to study the structure and evolution of the central kiloparsec in galactic nuclei. Here we have used it to map the locations of dense, quiescent ion-molecule, PDR and shock-liberated grain mantle chemistries across the nuclei of three nearby spirals. The chemistry in these nuclei are observed to be quiet variable. Here we present two new methods for watching dense molecular gas properties evolve with starburst phase — the degree of PDR penetration by comparison of C_2H versus the HCO^+/N_2H^+ ratio and the evolution of dense gas clouds near the starburst through HC_3N modeling of the gas density at arcsecond scales. The high spatial resolution and wide simultaneous spectral bandwidths available with the upgraded Karl G. Jansky Very Large Array and the newly arrived Atacama Large Millimeter/Submilleter Array implies that these two diagnostics are only the beginning of what should become possible in the next decade.

Acknowledgments

DSM acknowledges support from NSF grants AST-0506669 and AST-1009620. Additional support for this work is provided by NSF grant AST-0506469 to JLT.

[1] van Dishoeck E F and Blake G A 1998 *ARA&A* **36** 317
[2] Ungerects H, Bergin E A, Goldsmith P F, Irvine W M, Schloerb F P and Snell R L 1997 *ApJ* **482** 245
[3] Bergin E A, Ungerects H, Goldsmith P F, Snell R L, Irvine W M and Schloerb F P 1997 *ApJ* **482** 267
[4] Jones P A et al. 2011 *Preprint* astro-ph/1110.1421
[5] Herbst E and Klemperer W 1973 *ApJ* **185** 505
[6] Watson W D 1973 *ApJL* **183** 17
[7] Prasad S S and Huntress W T Jr. 1980 *ApJS* **43** 1
[8] Lee H.-H, Bettens R P A and Herbst E 1996 *A&AS* **119** 111
[9] Sternberg A and Dalgarno A *ApJS* **99** 565
[10] Gibb E L, Whittet D C B, Boogert A C A and Tielens A G G M 2004 *ApJS* **151** 35
[11] Mllar T J, Herbst E and Charnley SB 1991 *ApJ* **369** 147
[12] Hasegawa T I, Herbst E and Leung, C M 1992 *ApJS* **82** 167
[13] Garrod R T, Widicus-Weaver S L and Herbst E *ApJ* **682** 283
[14] Usero A, Garcia-Burillo S, Martin-Pintado J, Fuente A and Neri R 2006 *A&A* **448** 457
[15] Garcia-Burillo S, Martin-Pintado J, Fuente A and Neri R 2001 *ApJL* **563** 27
[16] Meier D S and Turner J L 2005 *ApJ* **618** 259
[17] Meier D S and Turner J L 2011 *ApJ* submitted
[18] Turner B E 1995 *ApJ* **449** 635
[19] Meier D S, Turner J L and Schinnerer E 2011 *AJ* **142** 32
[20] Meier D S, Turner J L and Hurt R L 2000 *ApJ* **531** 200
[21] Meier D S and Turner JL 2001 *ApJ* **551** 687
[22] Tsai C.-W, Turner J L, Beck S C, Crosthwaite L P, Ho P T P and Meier D S 2006 *AJ* **132** 2383
[23] Gao Y and Solomon P M 2004 *ApJ* **606** 271

10 pc Scale Circumnuclear Molecular Gas Imaging of Nearby AGNs

Satoki Matsushita

Institute of Astronomy and Astrophysics, Academia Sinica, P.O. Box 23-141, Taipei 10617, Taiwan, R.O.C.

Joint ALMA Observatory, Alonso de Córdova 3107, Vitacura 763 0355, Santiago, Chile

E-mail: `satoki@asiaa.sinica.edu.tw`

Abstract. We present the images and kinematics of circumnuclear molecular gas from 100 pc scale down to 10 pc scale in nearby active galactic nuclei (AGNs) using the Submillimeter Array (SMA) and the Plateau de Bure Interferometer (PdBI). We have observed several nearby galaxies that host AGNs, such as the nearest radio galaxy Centaurus A (NGC 5128), the Seyfert 2 galaxy M51 (NGC 5194), the Seyfert 2 galaxy NGC 1068, the Seyfert 1 galaxy NGC 1097, and the Seyfert 2 / starburst composite galaxy NGC 4945, in CO lines to see whether the molecular gas distribution, kinematics, and physical conditions at 10 – 100 pc scale follows the AGN unified model or not. In 100 pc scale, most of the circumnuclear molecular gas shows smooth velocity gradient, suggesting a regular rotating feature, and also shows abnormal line ratios, suggesting the existence of active sources to make the circumnuclear molecular gas dense and/or warm conditions or abnormal chemical compositions. In 10 pc scale, on the other hand, the molecular gas kinematics shows various characteristics, some shows very disturbed kinematics such as a jet-entrained feature in the galaxies that have jets, but some still shows regular rotation feature in a galaxy that does not have obvious jets. These results indicate that the kinematics and physical/chemical conditions of the circumnuclear molecular gas at the scale less than 100 pc is highly affected by the AGN activities, and at this scale, there is no clear evidence of any unified feature seen in the circumnuclear molecular gas.

1. Introduction

It was well known for a long time that there are two types of active galactic nuclei (AGNs) at the centers of galaxies based on optical spectra: Type 1 AGNs show broad permitted lines as well as narrower forbidden lines, and type 2 show relatively narrower permitted and forbidden lines [1]. In 1985, Antonucci and Miller [2] published a result of a polarization observation toward the nucleus of the type 2 AGN host galaxy NGC 1068, showing that broad permitted lines have been observed in the polarized (i.e., scattered) light, suggesting that the type 1 AGN is hidden behind the dense obscuring material in this type 2 nucleus. This result has led to the AGN unification model, which is, the broad line region is located close to a supermassive black hole of a galaxy surrounded by a dense obscuring torus, and it will be type 1 if one sees from the pole of the torus, and will be type 2 if one sees through the torus [3, 4].

2. Circumnuclear Molecular Gas Distribution at 100 pc Scale

After the suggestion of the AGN unification model, intense searches of the obscuring torus have been started, especially using millimeter-wave interferometers. Jackson *et al* [5] and Kohno *et*

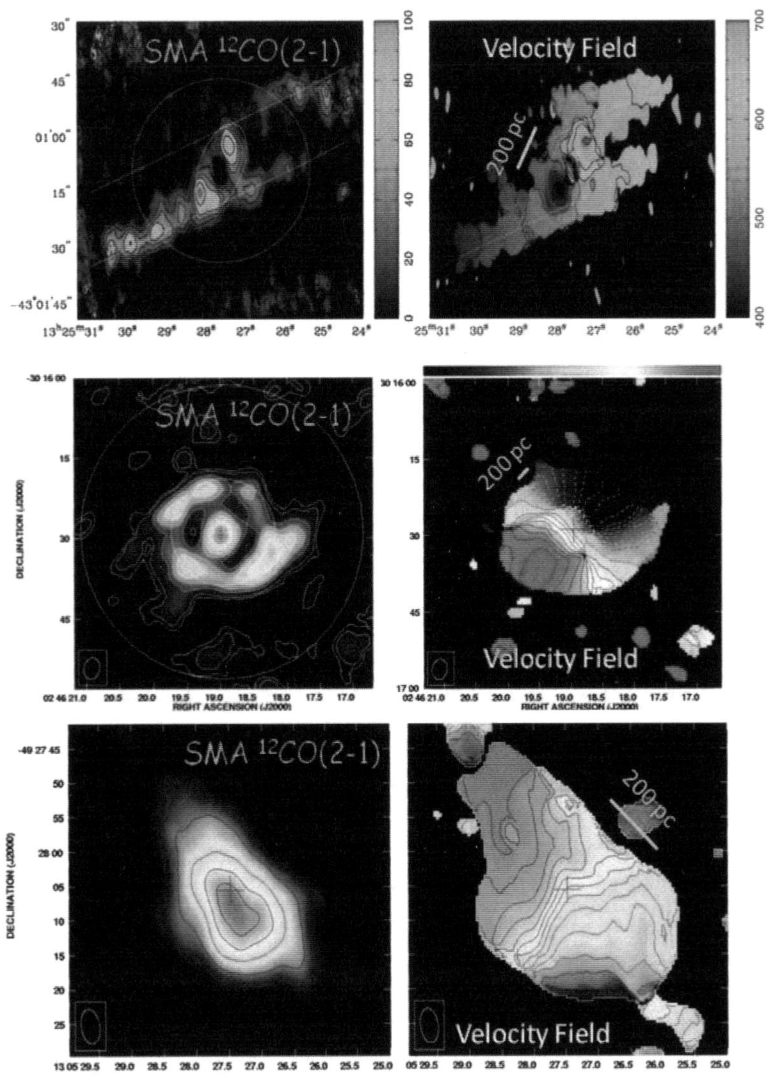

Figure 1. CO(2-1) integrated intensity (moment 0; left) and intensity-weighted velocity field (moment 1; right) maps of *Above:* Centaurus A [11], *Middle:* NGC 1097 [9], and *Bottom:* NGC 4945 [10]. All the images were obtained with the SMA. The linear scale of 200 pc is shown as a bar in the velocity field map of each galaxy.

al [6] observed nearby Seyfert 2 galaxies NGC 1068 and M51 (NGC 5194), respectively, with the HCN(1-0) line at a spatial resolution of several arcseconds, and successfully imaged the dense gas around the AGNs with the velocity gradients almost perpendicular to the radio jets. These results can be explained as dense gas disks or tori with the radii of a few hundred pc, rotating around the AGNs perpendicular to the radio jets. Furthermore, the column density of the observed molecular gas of a few $\times 10^{23}$ cm^{-2} [6] was consistent with the absorption column density measured in the X-ray observations of $(4.2 \pm 1.5) \times 10^{23}$ cm^{-2} [7]. These results matched very well with the AGN unification model.

Since the Submillimeter Array (SMA) [8] has started scientific observations, we have started to observe nearby AGNs in the higher transition (J=2-1 and/or 3-2) CO lines at the spatial resolutions of around a few hundred pc. We observed the galactic centers (central a few kpc) of the Seyfert 1 galaxy NGC 1097 [9], the Seyfert 2 / starburst composite galaxy NGC 4945 [10], and the nearest radio galaxy Centaurus A (NGC 5128) [11] with the CO(2-1) line, and the obtained images are shown in Figure 1. All these galaxies have central molecular gas concentrations with

Journal of Physics: Conference Series **372** (2012) 012043 doi:10.1088/1742-6596/372/1/012043

the scale of several hundred pc (note that Centaurus A has strong absorption features at the galactic center, so that the image shows a hole at the nucleus [12]), and the velocity fields exhibit a smooth velocity gradient, namely rigid rotation feature, around the nucleus (note that NGC 4945 is an edge-on galaxy with a bar, so that the velocity field around the nucleus is affected by the non-circular motion [13]). These velocity features again suggest rotating gas disks or tori with the radii of a few hundred pc, similar as those observed in NGC 1068 and M51 mentioned above. Since the kinematics of these galaxies shows the rigid rotation feature, the gravitational fields of all these galaxies are dominated by the bulge, not the supermassive black holes in the AGNs.

On the other hand, we also obtained higher-J CO lines toward the central few kpc region of NGC 1068 [14], NGC 1097 [9, 15], and M51 [16], and compared with the previously observed lower-J CO line images (Figure 2). Central molecular condensations can be seen in all the images (in whatever transitions) presented here, but the intensities are stronger in higher-J lines than lower-J lines, indicating that the circumnuclear molecular gas has either higher densities and/or higher temperature conditions than, or abnormal chemical conditions compared to the molecular gas at outer radii. Lines ratios at outer radii, namely at the spiral arm regions or at the starburst ring, have similar values as those regions in other galaxies, so that this abnormal line ratios are obviously affected by the existence of the AGNs.

These results suggest that the AGN activities *do not* affect the kinematics of the circumnuclear molecular gas at 100 pc scale, but *do* affect the physical/chemical conditions of the circumnuclear molecular gas.

3. Circumnuclear Molecular Gas Distribution at 10 pc Scale

In our SMA M51 CO(3-2) data presented above, we noticed that the kinematics of the circumnuclear molecular gas deviates from the circular rotation as suggested before [16], but could not study in detail due to the limited (large) spatial resolution. We therefore observed the circumnuclear molecular gas of M51 with $\sim$ 10 pc spatial resolution to spatially resolve the gas kinematics around the Seyfert 2 nucleus in this galaxy. We used the new-A (newly extended) configuration (maximum baseline length = 760 m) of the IRAM Plateau de Bure Interferometer (PdBI), and observed with the CO(2-1) line, since the CO(2-1) line flux is higher than that of the CO(1-0) line, and it is possible to obtain higher spatial resolution images.

Figure 3 shows the newly obtained circumnuclear CO(2-1) image of M51 [21]. The spatial resolution reached to $0.41'' \times 0.31''$, which corresponds to the linear scale of 14 pc $\times$ 11 pc at the distance of M51 (7.1 Mpc [22]). Figure 3(a) is the integrated intensity map of the CO(2-1) data. As you can see, there are mainly two molecular gas clouds at the eastern and western sides of the nucleus, perpendicular to the radio jets. The cloud at the western side is elongated along north-south direction, almost parallel to the radio jets, with the projected distance to the nucleus of only $\sim$ 30 pc. The one at the eastern side only has one strong peak with the size of about 30 pc ($\sim 1''$), roughly the size of the typical Giant Molecular Cloud (GMC) in our Galaxy. There is a weak emission extending toward the nucleus from this cloud. The column density of the molecular gas in front of the nucleus calculated from this weak emission is very small, only 6.2×10^{21} cm^{-2}, much smaller than the column density of atomic gas derived from the recent X-ray absorption measurements of 5.6×10^{24} cm^{-2} [24]. The missing flux due to our interferometric observations cannot explain this difference; this is because the amount of the detected flux has to be at the order of only 0.1% of the total flux for explaining the difference with the missing flux, but our observation obviously detected much more than this flux level. This suggests that the absorbing material in front of the X-ray source (i.e., AGN) is much smaller than our beam size of $\sim$ 10 pc.

Figure 3(b) is the intensity-weighted velocity field map of the CO(2-1) data. The eastern cloud exhibits the velocity gradient from west (blueshift) to east (redshift). This direction of

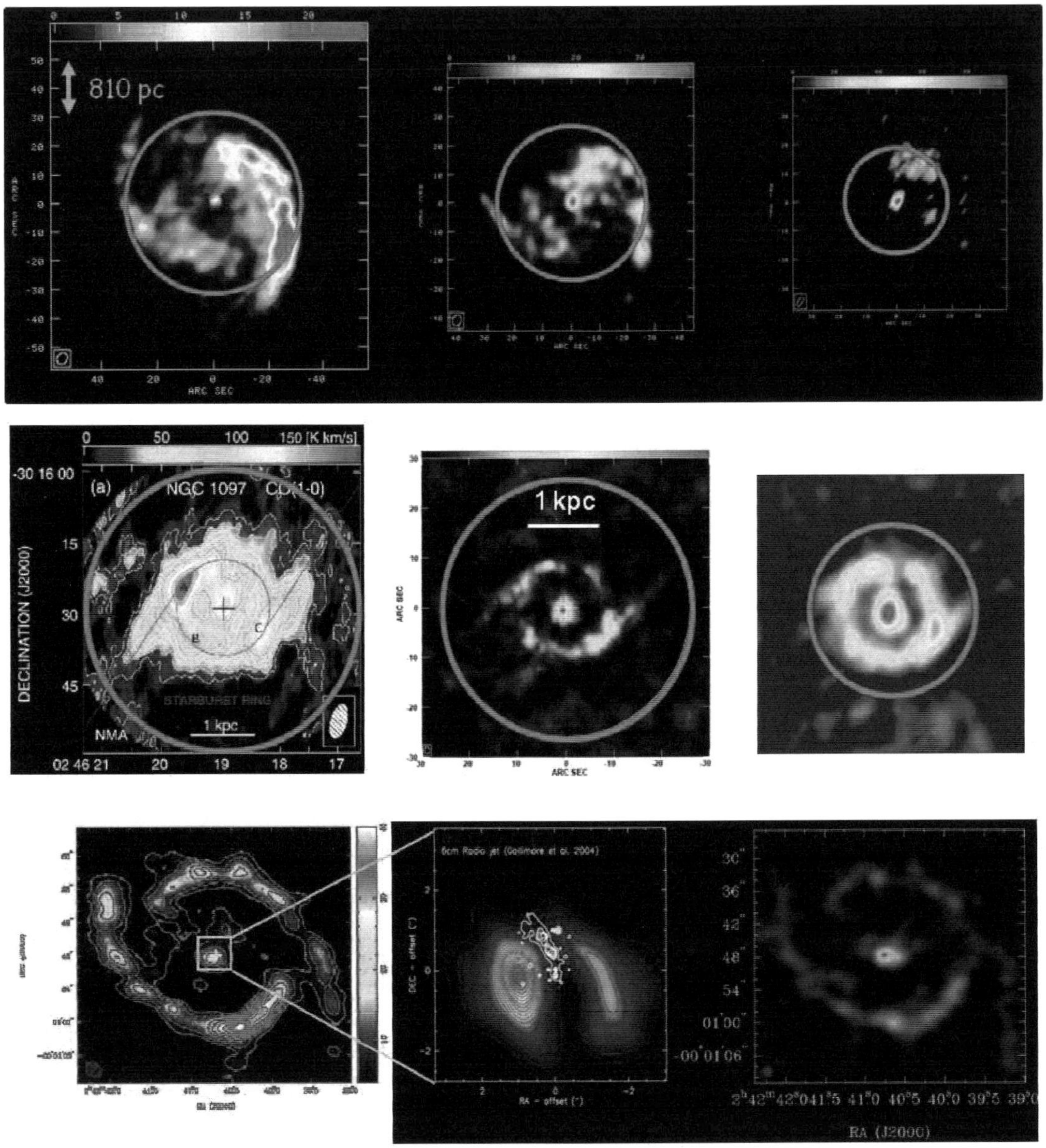

Figure 2. Integrated intensity maps of *Left:* CO(1-0), *Middle:* CO(2-1), and *Right:* CO(3-2) toward *Top:* the Seyfert 2 galaxy M51 (NGC 5194), *Middle:* the Seyfert 1 galaxy NGC 1097, and *Bottom:* the Seyfert 2 galaxy NGC 1068. Scales are matched for each galaxy, except for the CO(2-1) image of NGC 1068. The circles in the M51 and NGC 1097 images show the half power beam width of the primary antenna. M51 CO(1-0) is from Sakamoto *et al* (1999) [17], M51 CO(3-2) from Matsushita *et al* (2004) [16], NGC 1097 CO(1-0) from Kohno *et al* (2003) [18], NGC 1097 CO(2-1) from Hsieh *et al* (2008) [9], NGC 1097 CO(3-2) from Hsieh *et al* (2011) [15], NGC 1068 CO(1-0) and CO(3-2) from Tsai *et al* (2012) [14], and NGC 1068 CO(2-1) from Krips *et al* (2011) [19] (in this image, the 6 cm radio jet contours are also overlaid [20]).

the velocity gradient is totally opposite sense with that observed in the past with $\sim$ 10 times larger spatial resolution [6]. On the other hand, the western cloud exhibits the velocity gradient

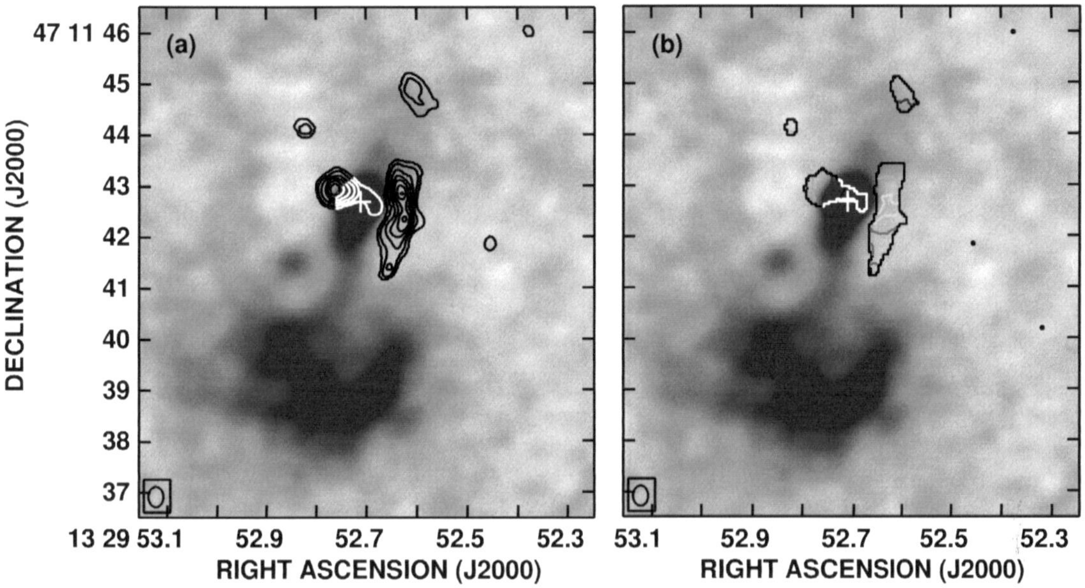

Figure 3. CO(2-1) *(a)* integrated intensity (moment 0) and *(b)* intensity-weighted velocity field (moment 1) maps of M51 at $0.41'' \times 0.31''$ (14 pc × 11 pc) resolution [21] overlaid on the VLA 6 cm radio continuum greyscale image [23]. The cross in each figure marks the position of the AGN.

from north (blueshift) to south (redshift), namely the velocity gradient along the radio jets. The amount of the velocity gradient of the western cloud is 2.2 ± 0.3 km s^{-1} pc^{-1}, which is consistent with that of the ionized gas along the radio jets [25]. These distribution and the kinematics of the western cloud together with the kinematics of the ionized gas along the radio jets suggest that the western cloud is entrained by the radio jets [21]. Note that if we smooth our CO(2-1) data to ten times larger spatial resolution, the kinematics is consistent with that of the past observations [6], since the average velocities of the eastern and the western clouds are consistent with the blueshifted and the redshifted velocities, respectively, of the past observations.

In summary, our $\sim$ 10 pc spatial resolution CO(2-1) results show that there is no clear evidence of rotating disk or torus around the Seyfert 2 nucleus of M51 as suggested in the past, but exhibit that there are highly disturbed molecular gas features very close to the nucleus, possibly affected by the AGN activities. Furthermore, the obscuring material of this type 2 AGN has the size much smaller than our beam size of $\sim$ 10 pc.

Such disturbed circumnuclear gas near AGNs is also recently suggested in NGC 1068 [19]: Recent multiple molecular line observations toward the nucleus of NGC 1068 at a spatial resolution of $\sim$ 30 pc with the SMA and the PdBI revealed that the circumnuclear molecular gas kinematics is better explained by a rotation with an outflowing motion, rather than by a warped disk as suggested in the past [26]. The radius of the rotating disk is estimated to be around 60 pc. The molecular outflow is suggested to be flowing along the radio jets and the ionization cone with the velocity gradient of 3 km s^{-1} pc^{-1}, similar to that observed in the jet-entrained molecular gas in M51 (see above).

On the other hand, a circumnuclear molecular gas disk without any disturbance has also been observed: Recent CO(2-1) observation toward NGC 4945 at a spatial resolution of $\sim$ 20 pc with the SMA has displayed a circumnuclear molecular gas disk with no disturbance in the kinematics (i.e., rigid rotation feature in the position-velocity diagram) [27]. The radius of the rotating disk is around 20 pc, which is closer to the AGN than the distance of the circumnuclear clouds of M51 as mentioned above. Note that NGC 4945 does not have any obvious jet from its Seyfert 2 nucleus, in contrast to the nuclei of M51 and NGC 1068 that exhibit radio and/or ionized gas jets.

4. Discussion

We only have three observations of the circumnuclear molecular gas in $\sim$ 10 pc spatial resolution, but all the samples show different features. The circumnuclear gas of M51 does not show any evidence of clear rotation feature, but show totally disturbed kinematics due to the radio jets. That of NGC 1068 exhibits rotation feature with outflowing gas along the radio jets, but in the case of NGC 4945, which does not have obvious jets from the AGN, only rotation feature can be seen. These results suggest that the jet activity from the AGN affects the circumnuclear molecular gas around $\sim$ 10 pc scale, which could not be seen at $\sim$ 100 pc scale. This suggests that there is no clear unified feature in the circumnuclear molecular gas around AGNs at the scale of 10 pc. Together with the line ratio anormaly, which means the physical conditions or chemical composition anormaly, in the circumnuclear molecular gas as mentioned above, the circumnuclear molecular gas close to AGNs is suggested to be highly affected by the AGN activities.

This also suggests that the star formation near AGNs can be highly affected by the AGN activities; for instance, AGNs with strong jet activities can have less star formation in the circumnuclear (within 100 pc) regions, since too large turbulence induced by the jet activities. Due to very limited sample we have, it is difficult to say anything about this kind of possibilities, but if we increase samples (with ALMA for example), we may be possible to provide some hints for the AGN-starburst connection.

Acknowledgments

We thank all the colleagues who support this SMA Seyfert Survey program, including Jeremy Lim, Dinh-Van-Trung, Sebastien Muller, Daniel Espada, Pei-Ying Hsieh, Richard C.-Y. Chou, Meng-Chun Tsai, Ya-Lin Wu, Nagisa Oi, Kotaro Kohno, Kazushi Sakamoto, Frederic Boone, and Melanie Krips. This work is supported by the National Science Council (NSC) of Taiwan, NSC 97-2112-M-001-021-MY3 and NSC 100-2112-M-001-006-MY3. The Submillimeter Array is a joint project between the Smithsonian Astrophysical Observatory and the Academia Sinica Institute of Astronomy and Astrophysics and is funded by the Smithsonian Institution and the Academia Sinica.

References

[1] Weedman D W 1977 *Ann. Rev. Astron. Astrophys.* **15** 69
[2] Antonucci R R J and Miller J S 1985 *Astrophys. J.* **297** 621
[3] Antonucci R 1993 *Ann. Rev. Astron. Astrophys.* **31** 473
[4] Urry C M and Padovani P 1995 *Publ. Astron. Soc. Pacific* **107** 803
[5] Jackson J M, Paglione T A D, Ishizuki S and Nguyen-Q-Rieu 1993 *Astrophys. J. Lett.* **418**, L13
[6] Kohno K, Kawabe R, Tosaki T and Okumura S K 1996 *Astrophys. J. Lett.* **461** L29
[7] Makishima K, Ohashi T, Kondo H, Palumbo G G C and Trinchieri G 1990 *Astrophys. J.* **365** 159
[8] Ho P T P, Moran J M and Lo F 2004 *Astrophys. J. Lett.* **616** L1
[9] Hsieh P-Y, Matsushita S, Lim J, Kohno K and Sawada-Satoh S 2008 *Astrophys. J.* **683** 70
[10] Chou R C Y, Peck A B, Lim J, Matsushita S, Muller S, Sawada-Satoh S, Dinh-V-Trung, Boone F and Henkel C 2007 *Astrophys. J.* **670** 116
[11] Espada D, *et al* 2009 *Astrophys. J.* **695** 116
[12] Espada D, *et al* 2010 *Astrophys. J.* **720** 666
[13] Lin, L-H, Taam R E, Yen D C C, Muller S and Lim J 2011 *Astrophys. J.* **731** 15
[14] Tsai, M, Hwang C-Y, Matsushita S, Baker A J and Espada D 2012 *Astrophys. J.* **746** 129
[15] Hsieh P-Y, Matsushita S, Liu G, Ho P T P, Oi N and Wu Y-L 2011 *Astrophys. J.* **736** 129

[16] Matsushita S, *et al* 2004 *Astrophys. J. Lett.* **616** L55
[17] Sakamoto K, Okumura S K, Ishizuki S and Scoville N Z 1999 *Astrophys. J. Suppl.* **124** 403
[18] Kohno K, Ishizuki S, Matsushita S, Vila-Vilaró B and Kawabe R 2003 *Publ. Astron. Soc. Japan 55* L1
[19] Krips M, *et al* 2011 *Astrophys. J.* **736** 37
[20] Gallimore J F, Baum S A and O'Dea C P 2004 *Astrophys. J.* **613** 794
[21] Matsushita S, Muller S and Lim J 2007 *Astron. Astrophys.* **468** L49
[22] Takáts K and Vinkó J 2006, *Mon. Not. Royal Astron. Soc.* **372** 1735
[23] Crane P C and van der Hulst J M 1992 *Astron. J.* **103** 1146
[24] Fukazawa Y, Iyomoto N, Kubota A, Matsumoto Y and Makishima K 2001 *Astron. Astrophys.* **374** 73
[25] Bradley L D, Kaiser M E and Baan W A 2004 *Astrophys. J.* **603** 463
[26] Schinnerer E, Eckart A, Tacconi L J, Genzel R and Downes D 2000 *Astrophys. J.* **533** 850
[27] Lim J, Muller S, Boone F, Matsushita S and Dinh-V-Trung 2011 *Astrophys. J.* submitted

Are circumnuclear (starburst) rings important in the evolution of the central kiloparsec?

T. P. R. van der Laan
MPIA, Königstuhl 17, D69117 Heidelberg, Germany
E-mail: vdlaan@mpia.de

Abstract. Circumnuclear starburst rings are effective barriers against gas inflow. The large 'pile-up' of gas leads to distinct star forming events in the rings of NGC 5248 and NGC 6951. In our work, the rings are distinctly seen as the locations where the bulk of the central kiloparsec star formation in the last 2 Gyr has taken place. This in turn has direct implications for the large scale bars that have driven the formation of the circumnuclear rings in these galaxies. They have to be at least as old. The system of large-scale bar and circumnuclear ring has therefore ample time to stop gas inflow on the scale of the last 100 pc. The stars formed in the ring will help build the (pseudo-)bulge and eventually alter the gravitational potential.

1. Introduction

Gas inflow to the nuclear regions of a galaxy is an important factor in the secular evolution of said galaxy. Not in the least since gas inflow is a necessary factor in fueling AGN. However, between the 1 kpc scale and the final parsec are still many mechanisms, such as asymmetries in the gravitational potential, that can hinder or induce gas inflow. Large scale stellar bars are the prime component in driving gas at the 100's of parsec scales. This was also discussed by Combes (this series).

Bar perturbations in the gravitational potential of the galactic disk induce gravitational torques on the material in the disk. Gas, being a collisional entity, will experience angular momentum changes due to these gravitational torques. Between the co-rotation (CR) radius and inner Lindblad resonance (ILR) radius of the large scale bar gas will loose angular momentum and flow inward. These non-circular motions can be seen at the dust and gas lanes along the leading edges of bars. The location of the resonances is set by the bar pattern speed and the rotation of the galaxy disk. At the ILR the gas gets stalled in its inward motion and settles again on circular orbits, unless another mechanism is present that exerts a new pull on the gas. The 'pile-up' of gas at the ILR is more commonly known as a circumnuclear ring.

CO observations of circumnuclear rings in various nearby galaxies have shown that the rings contain large reservoirs of molecular gas (10^8-10^9 $M_\odot$). Such an amount of molecular gas on settled orbits must lead to strong star forming events. Indeed, many circumnuclear rings are very visible in Hα (e.g. Mazzuca et al. 2008), indicating current star formation. To understand what the influence of star formation in the circumnuclear rings on the secular evolution of their host galaxies can be, we need to investigate the amount and manner of the star formation in the circumnuclear rings.

In my contribution I focused on the star formation in the circumnuclear rings of NGC 5248 and NGC 6951. These two galaxies both have circumnuclear rings driven by a large scale bar. Both galaxies are nearby (15.6 and 24.1 Mpc, resp.) and host well visible circumnuclear rings in CO and Hα (Jogee et

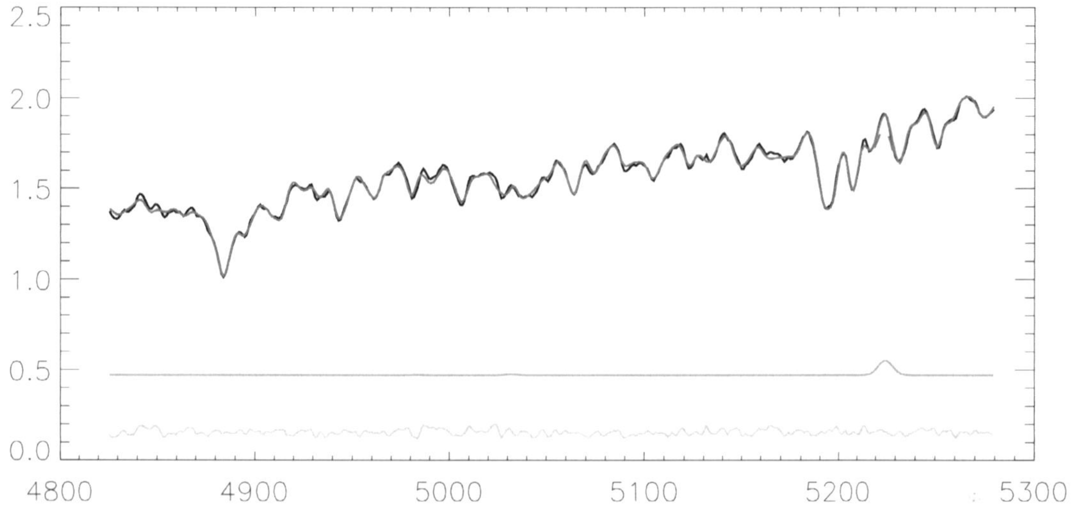

Figure 1. Example of the PPXF/GANDALF fitting of a single SAURON spaxel. The observed spectrum is given in black, the spectral fit in red, the residual in green and the fitted emission line spectrum in blue (offset from zero). Flux is in arbitrary units. The wavelength is in [Å] and is in the observed frame.

al. 2002, Maoz et al. 2001, Van der Laan et al. 2011, Van der Laan in prep. (a) & (b)) as well as in other tracers.

2. Star formation in the ring

In galaxies where circumnuclear rings are driven by a large scale bar the 'amount' of star formation is directly related to the life time of the bar.

The 'manner' of star formation may also be dependent on the large scale bar. The dust/gas lanes of the bar are connected to the circumnuclear ring at two distinct positions. Two modes of star formation have been proposed that deal with this fact. They are called the 'popcorn' and 'pearls-on-a-string' model, respectively. In the popcorn scenario, first put forth by Elmegreen (1994) gas density in the circumnuclear ring builds up across the ring until the gas density is high enough that the gas fragments and starts forming stars. Several of these star formation events might occur over time and without pattern within the ring, hence the designation 'popcorn'. The 'pearls-on-a-string' model by comparison proposes an azimuthal age gradient to the star formation in the circumnuclear ring. Since the large scale bar connects to the circumnuclear ring via the gas and dust spiral arms at its leading edge, the gas density at the ring contact points must therefore be preferred locations for gas overdensities. In the 'pearls-on-a-string' scenario these overdensities would lead to star formation occurring primarily at those locations. As the system rotates over time, an azimuthal stellar age gradient in the circumnuclear ring will be created (Böker et al. 2008).time scale of the gas must be much shorter than viscous time scale.

3. Full view on the ring's stellar population

We wish to obtain a full view of the stellar content in the rings of NGC 5248 and NGC 6951, in order to understand the star formation. We do not wish to constrain ourselves solely on star clusters, but turn to integral field spectroscopy to observe the full stellar content of the ring.

Both galaxies were observed with the SAURON instrument as part of a larger campaign (Dumas et al.

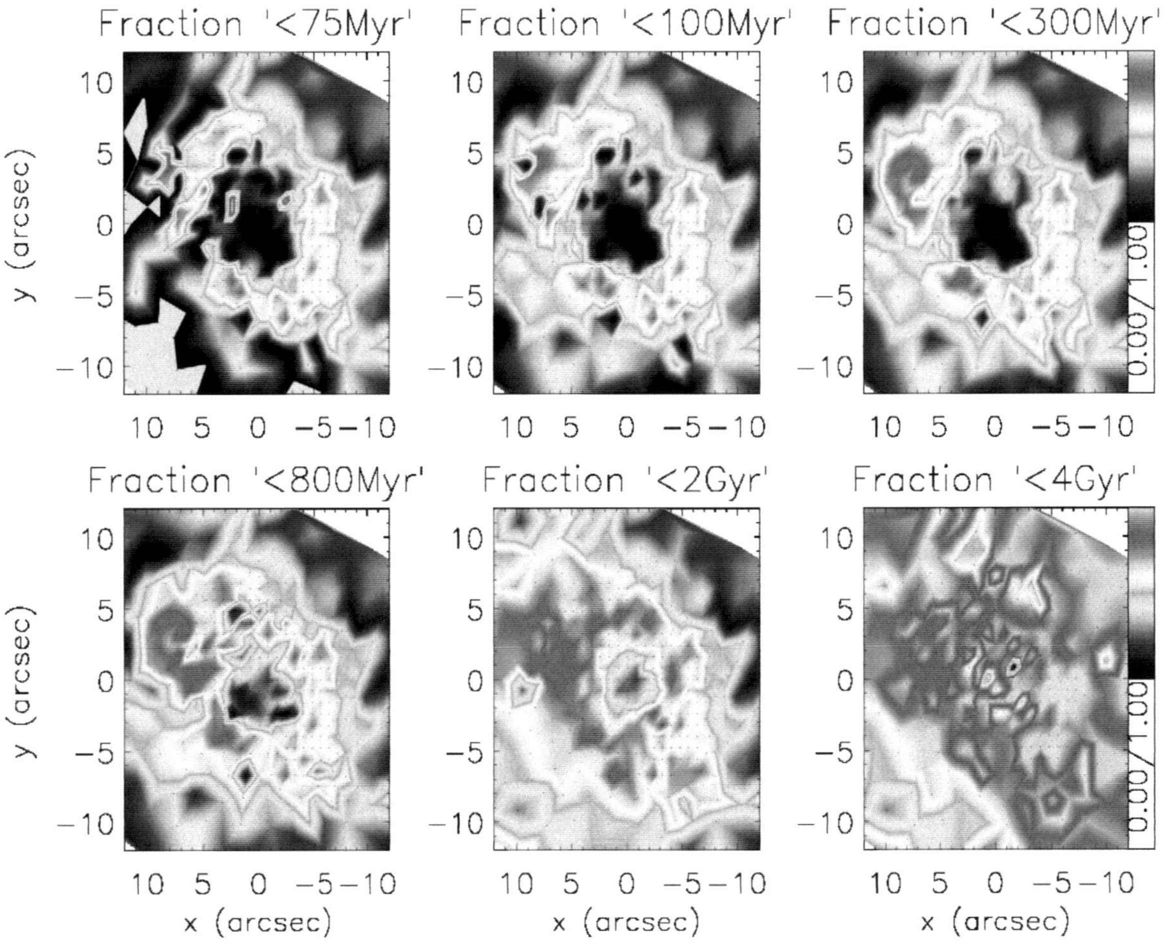

Figure 2. Spatial distribution of the fraction of stellar mass younger than the indicated age. From the top-left to bottom-right this age is 100 Myr, 200 Myr, 400 Myr, 800 Myr, 1 Gyr, 2 Gyr.

2007) between 2001 and 2004. Individual exposures with the SAURON instrument cover a field of view of about 41" × 33", with spaxels of $0\farcs94$, a spectral resolution of 4.2 Å and sampling of 1.15 Å spectra extending from 4780 Å up to about 5350Å. After calibration, individual exposures were recentered (using a direct image of the galaxy as a reference) and merged, by projecting each data cube on a common grid with $0\farcs8^2$ spaxels. The rings in both galaxies are fully included within the field of view and well resolved at this angular resolution.

The main strength of IFU data is that we are not limited to *only* the star clusters, but obtain light from all stars in the ring, also old(er) populations if they are present. Also by utilizing the full available spectral range, we are able to overcome a larger extinction range, which interfere with photometric observations and obscure older star clusters.

We assume that the total stellar content emitting in each spaxel can be split into a combination of single-stellar population (SSP) fractions. We obtained SSP spectra from MILES (Vazdekis et al. 2010). We assumed a Kroupa IMF and instantaneous formation and match the spectral resolution of the model to our data. The SSP spectra span lifetimes from 65 Myr to 15Gyr, with time steps that are selected to follow stellar evolution. The strongest drivers of the fit will be the Hβ and Mgβ absorption lines, which increase and decrease with age, respectively.

We use the spectral fitting codes PPXF (Cappellari & Emsellem 2004) and GANDALF (Sarzi et al.

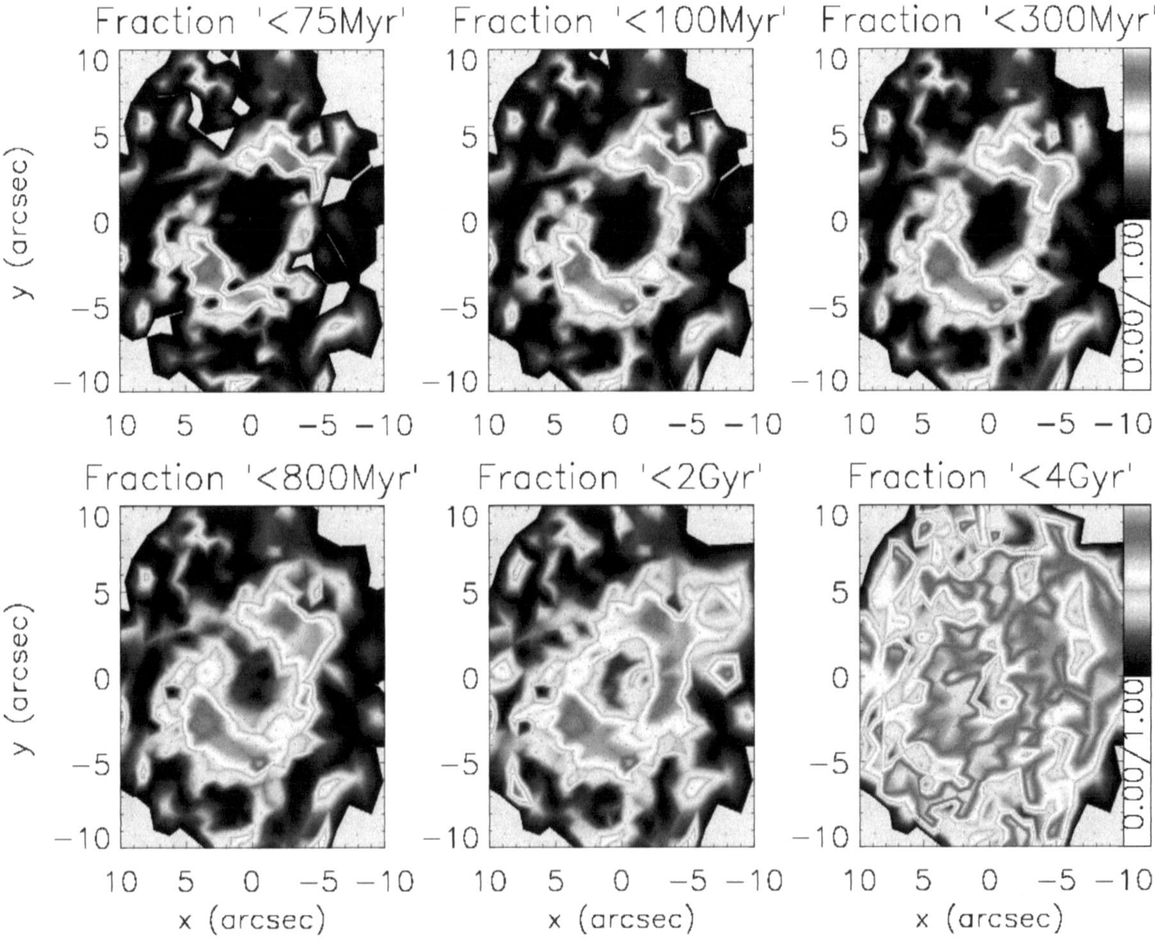

Figure 3. Spatial distribution of the fraction of stellar mass younger than the indicated age. From the top-left to bottom-right this age is 100 Myr, 200 Myr, 400 Myr, 800 Myr, 1 Gyr, 2 Gyr.

2006). GANDALF enables us to correctly deal with the emission lines from the gas that is present in the ring. Among the strongest emission lines we expect is also the Hβ emission line. If we do not account for it, gaseous Hβ emission could fill the stellar absorption line and we would desensitize our results against young stellar populations. We further fit for the [OIII] doublet at 5000Å and [NI] at 5200Å . A representative example of the fitting of a single spaxel is shown in Figure 1.

In Figure 2 and 3, the preliminary results for NGC 5248 and NGC 6951 are shown. From the fitting we obtain the fraction of flux in each SSP. We can convert this into fraction of mass per SSP by using the known M/L ratio for each SSP. We have obtained the presented results by summing over age.

In both galaxies the circumnuclear ring is immediately visible in the younger age bins. In NGC 5248 the ring has a near uniform azimuthal fraction of young stars. In NGC 6951 the highest < 75 Myr mass fractions are seem at two locations that are close to the bar contact points mentioned previously. As we include more age bins the stellar mass fraction in the rings increases, while the stellar mass fraction in- and outside the ring stays low. This can be seen up to 800 Myr/2 Gyr in NGC 5248 and 2 Gyr in NGC 6951. By 4 Gyr the fraction of 'young' stars is no longer distinct between ring and the rest of the central region.

4. Implications for the evolution of the central kiloparsec

Our results show that the circumnuclear rings have existed as distinct star formation events in the central regions of these galaxies for approximately 2 Gyrs. This has several direct implications.life time of the large scale bar that drives the resonances.

- The rings in both NGC 5248 and NGC 6951 are visible in 'fraction of young stars' out to an age of 2 Gyr. Therefore, for the last 2 Gyr gas has been driven to these locations, where it encountered a barrier to further inflow, i.e. a resonance. This means that the large scale bar that leads to the resonance must have been in place for at least 2 Gyr.
- The width of the young stars fraction in NGC 6951 is similar to the CO distribution as presented in Van der Laan et al. 2011. Therefore, the location of the resonances in this system cannot have significantly changed in the last 2 Gyr. Which also means that the large scale bar can't have significantly changed strength in that time. For NGC 5248 this is less clear, as the CO distribution shows a spiral pattern (Van der Laan et al., in prep (a)).
- At 2 Gyr the stellar mass fraction in the rings are close to 1 in both galaxies. This implies that the star formation was so numerous as to outshine any previous stellar populations that were present at this location before the circumnuclear ring was formed.
- The 'manner' of star formation in circumnuclear rings remains ambiguous. In NGC 5248 the youngest stars are distributed nearly uniform throughout the ring, which corresponds to the 'popcorn' model. In contrast, in NGC 6951 the youngest stars are located at 2 positions related to the contact points with the large scale bar, which would be the 'pearls-on-a-string' model.

5. Conclusions

Circumnuclear rings form effective barriers against inflow. Since they are large reservoirs of molecular gas (10^8-$10^9\,M_\odot$), they must be strongly star forming. We investigate what the full influence of circumnuclear rings on the secular evolution of the galaxy is, in particular in the central kiloparsec, by looking at the amount and manner of the star formation in the circumnuclear rings of NGC 5248 and NGC 6951. By using the SAURON integral field spectrograph we are able to probe the full stellar content of the rings.

We find that the rings are distinct star formation regions in the central kiloparsecs of these galaxies, where all, or at least the vast majority, of all star formation in the last 2 Gyr has taken place. This places direct implications on the life time of the large scale bar that drives the resonances in this system; it has to be at least as old to explain these results.

With these life times for the large scale bar and circumnuclear ring and the amount of stellar mass added to the ring in the last 2 Gyr, we conclude that circumnuclear rings are an important mechanism to alter the stellar distribution in the central regions and lead to the (further) formation of the (pseudo-)bulge.

References

[1] Böker, T., Falcón-Barroso, J., Schinnerer, E., Knapen, J. H., & Ryder, S. 2008, , 135, 479
[2] Cappellari, M., & Emsellem, E. ,2004, , 116, 138
[3] Combes, F. 2011, arXiv:1111.4770
[4] Dumas, G., Mundell, C. G., Emsellem, E., & Nagar, N. M. 2007, , 379, 1249
[5] Elmegreen, B. G. 1994, , 425, L73
[6] Jogee, S., Shlosman, I., Laine, S., et al. 2002, , 575, 156
[7] Mazzuca, L. M., Knapen, J. H., Veilleux, S., & Regan, M. W. 2008, , 174, 337
[8] Maoz, D., Barth, A. J., Ho, L. C., Sternberg, A., & Filippenko, A. V. 2001, , 121, 3048
[9] Sarzi, M., Falcón-Barroso, J., Davies, R. L., et al. 2006, , 366, 1151
[10] van der Laan, T. P. R., Schinnerer, E., Boone, F., et al. 2011, , 529, A45

[11] van der Laan, T. P. R., et al., in prep. (a)
[12] van der Laan, T. P. R., et al., in prep. (b)
[13] Vazdekis, A., Sánchez-Blázquez, P., Falcón-Barroso, J., et al. 2010, , 404, 1639

Astronomy at High Angular Resolution 2011 IOP Publishing
Journal of Physics: Conference Series **372** (2012) 012045 doi:10.1088/1742-6596/372/1/012045

Central regions of LIRGs: rings, hidden starbursts, Supernovae and star clusters

Petri Väisänen[1], Andres Escala[2], Erkki Kankare[3], Jari Kotilainen[4], Seppo Mattila[3], Vinesh Rajpaul[5], Zara Randriamanakoto[1,5], Juha Reunanen[3], Stuart Ryder[6], Albert Zijlstra[7]

[1]South African Astronomical Observatory, P.O. Box 9 Observatory 7935, South Africa,
[2] Departamento de Astronomia, Universidad de Chile, Casilla 36-D, Santiago, Chile
[3] Tuorla Observatory, University of Turku, FI-21500 Piikkiö, Finland
[4] FINCA, University of Turku, Väisäläntie 20, FI-21500 Piikkiö, Finland
[5] Department of Astronomy, University of Cape Town, Rondebosch 7701, South Africa
[6] Anglo-Australian Observatory, Epping, NSW 1710, Australia
[7]Jodrell Bank Centre for Astrophysics, University of Manchester, Manchester M13 9PL, UK

E-mail: petri@saao.ac.za

Abstract. We study star formation (SF) in very active environments, in luminous IR galaxies, which are often interacting. A variety of phenomena are detected, such as central starbursts, circumnuclear SF, obscured SNe tracing the history of recent SF, massive super star clusters, and sites of strong off-nuclear SF. All of these can be ultimately used to define the sequence of triggering and propagation of star-formation and interplay with nuclear activity in the lives of gas rich galaxy interactions and mergers. In this paper we present analysis of high-spatial resolution integral field spectroscopy of central regions of two interacting LIRGs. We detect a nuclear 3.3 μm PAH ring around the core of NGC 1614 with thermal-IR IFU observations. The ring's characteristics and relation to the strong star-forming ring detected in recombination lines are presented, as well as a scenario of an outward expanding starburst likely initiated with a (minor) companion detected within a tidal feature. We then present NIR IFU observations of IRAS 19115-2124, aka the Bird, which is an intriguing triple encounter. The third component is a minor one, but, nevertheless, is the source of 3/4 of the SFR of the whole system. Gas inflows and outflows are detected in their nuclei locations. Finally, we briefly report on our on-going NIR adaptive optics imaging survey of several dozen LIRGs. We have detected highly obscured core-collapse SNe in the central kpc, and discuss the statistics of "missing SNe" due to dust extinction. We are also determining the characteristics of hundreds of super star clusters in and around the core regions of LIRGs, as a function of host-galaxy properties.

1. Introduction

Luminous Infrared Galaxies, LIRGs, defined as galaxies with $10^{11} > L_{IR} > 10^{12}$, combine very strong star-formation (SF) with various levels of nuclear activity. They are usually more diverse objects regarding morphology and mode of SF than ULIRGs ($L_{IR} > 10^{12}$), which in the Local Universe tend to nearly always be remnants of major mergers with singularly centralised SF. Some LIRGs also have centralised SF, but in many cases a significant part of the SF is "extended" (e.g. Alonso-Herrero et al. 2006; Rodríguez-Zaurín et al. 2011). It may thus be that local LIRGs might physically represent more closely what high-redshift SF looks like, where a diversity of modes of strong SF are present (e.g. Rodighiero et al. 2011).

"Extended SF" is often defined as merely a fraction of MIR or hydrogen recombination line emission coming from outside a given nuclear region, and as such it could mean either truly diffuse SF, or rather one or more widely off-nuclear, but clearly localised starbursts. This latter scenario is playing out for example in the well-studied closest gas-rich merger system, the Antennae, where the SF is dominated by an overlap region between the main nuclei. Studying such cases in detail, and disentangling various kinematic components, requires high spatial resolution, which has not often been available except for a handful of local galaxies, in addition to spectroscopic data.

We are conducting an ongoing large survey of LIRGs and ULIRGs using adaptive optics NIR imaging mainly with VLT/NACO and also Gemini/ALTAIR/NIRI. Both instruments deliver images with a spatial resolution of ≈ 0.1". The data are very well complemented with existing HST optical data for many of the targets, and there is also an on-going optical spectroscopic campaign with SALT/RSS. The galaxies are at distances ranging from 40 to 180 Mpc, span the range from starbursts to LIRGs, and are at various stages of merging, interaction, or isolation. Some galaxies have been studied in more detail already – here we will present analysis of two such objects, and will finish with a brief discussion of some initial results from the larger sample.

2. NGC 1614: PAH rings and a minor companion

NGC 1614 is a LIRG at a distance of 64 Mpc ($z = 0.016$). Optically it is classified as barred and interacting (Fig. 1), and as a borderline case between a LINER and HII-dominated galaxy. It has also been classified as a composite starburst + AGN. Previous studies have identified a strong nuclear star forming ring of approximately 600 pc diameter (Kotilainen et al 2001; Alonso-Herrero et al. 2001). We observed the central 1.5 kpc region with UKIRT/UIST integral field unit in the L-band (Väisänen et al. 2012). This wavelength region has been shown to be ideal for disentangling AGN signatures from starbursts, using both the continuum slope and the 3.3 and 3.4 μm PAH band features (e.g. Imanishi & Dudley 2000; Risaliti et al. 2006). The method has not, however, been much used with both spatial and spectroscopic information simultaneously, i.e. with IFU, due to the rarity of suitable instrumentation.

We detect a strong central continuum source which is slightly resolved ($\sim$ 80 pc; Fig. 1 lower left, Fig. 2), and a clear signature of a PAH 3.3 μm ring around the nucleus (Fig. 3). Parameterising the PAH flux and the L-band continuum following Risaliti et al. (2006) we find no evidence at all for an AGN, obscured or otherwise: the continuum is flat, the PAH feature is strong even within the central 30 pc resolution element, and there is no 3.4 μm absorption (Fig. 2).

The PAH ring is especially clear when inspected in terms of its equivalent width (EW) and more or less coincides with, but also extends well outside of, the star-formation ring detected with recombination-lines and radio continuum. Figure 3 shows the PAH distribution compared to the SF ring measured from archival NICMOS Paα data. We note that the PAH flux extends to the core regions, where Paα is not seen, while the PAH EW strength simultaneously extends quite far outside of the star formation ring. Using optical and NIR colour information of the core, and the PAH EW strengths in the various regions, and comparing to previous studies linking PAH strengths with ages of SF regions (Tacconi-Garman et al. 2005; Díaz-Santos et al. 2008), we find the central kpc of NGC 1614 to be consistent with the following scenario: the core has experienced a starburst some 20 to 100 Myr ago, has no more ongoing SF and much of the initial ISM has evacuated the region. The ring at 300 pc radius is the site of strongest current SF, and the age of the SF is continuously decreasing as one moves outwards from the core to mid-ring, judging by the change of the PAH-to-Paα ratio. This ratio grows again outward of the ring, but there the level of star-formation is likely (still) too low to be detected in recombination lines, while the strong PAH EW might instead be a signal of the very youngest SF (2-4 Myr time scale; Beirão et al 2009). It thus appears that NGC 1614 is experiencing an outward propagating

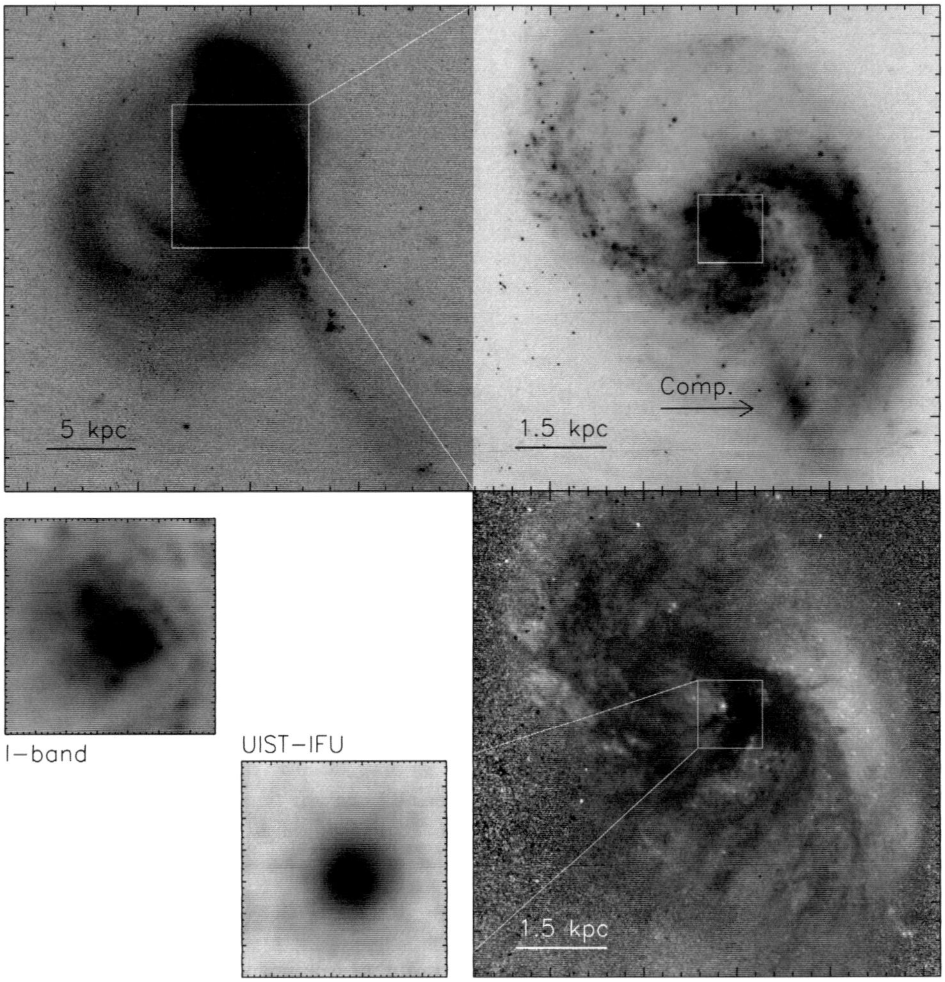

Figure 1. Archival ACS/HST images of NGC 1614. North is up and East left in each image, the tick-marks are 5" in the top-left image and 1" in the right side panels. *Top-left*: an *I*-band (FW814) image showing the outer structures of NGC 1614 at 20 kpc scales. The white rectangle shows the area of the same image zoomed and re-scaled at *top-right*, showing the more regular spiral pattern. The higher surface density structure 9" SW of the nucleus is pointed out, which is the likely remnant of a companion galaxy, and possibly the trigger for the starburst core of the whole system. The *bottom-right* panel shows the ACS/HST $B-I$ colour map of the same region as the panel above it, where darker areas mean redder colours, i.e. higher extinction; the main nucleus lies behind a dust feature. The innermost 1 kpc area is marked with the white rectangles in both right-hand-side panels and is shown zoomed-in at *bottom-left*. The left one is the *I*-band image, and at right we show our UIST *L*-band continuum image, equivalent of Fig. 2., right panel. The bright very blue point source $\sim$ 1.5" NE of the nucleus visible in the I-band and in the colour-map is a likely young super star cluster.

starburst, the detected PAH ring highlighting the location where the SF is just propagating into the "new" molecular gas clouds.

Moreover, we detect a likely remnant of the companion galaxy and likely trigger of the LIRG

Journal of Physics: Conference Series **372** (2012) 012045 doi:10.1088/1742-6596/372/1/012045

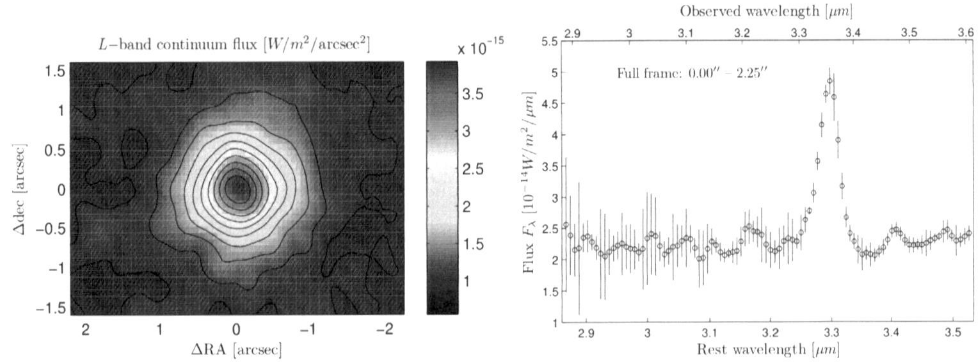

Figure 2. *Left:* *L*-band continuum map of the NGC 1614 nuclear regions from the IFU observations showing the strong marginally resolved nuclear source. *Right:* L-band spectrum of the central ~ 700 pc area showing a flat continuum and an EW ≈ 70 nm PAH feature, typical in starbursts. The spectrum is similar within 0.3" of the core as well.

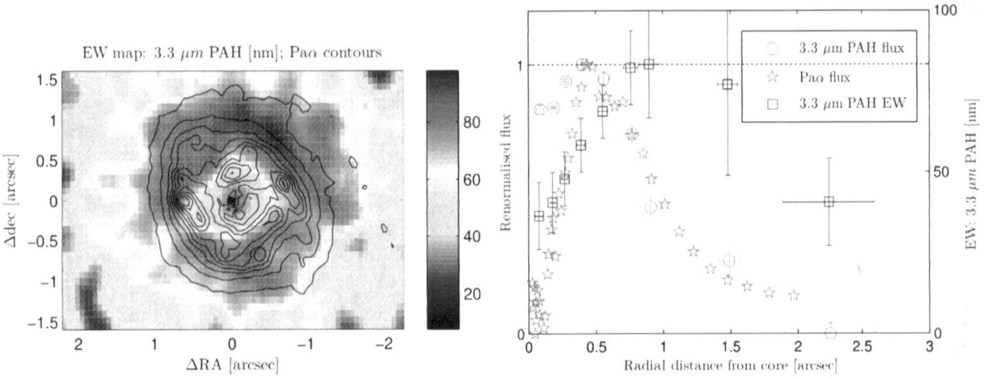

Figure 3. *Left:* The NGC 1614 3.3 μm PAH emission EW shows a wide ring around the nucleus; the contours show Paα emission*Right:* The EW of the 3.3 PAH (blue squares), the PAH feature flux (green circles), and Paα flux (red stars) shown as a function of radius. Each profile is normalised to unity, though the actual values of the 3.3 PAH EW can be read from the right-side axis. PAH EW clearly peaks outward of the main strong SF ring as defined by recombination line strengths.

phenomenon in the linear tidal tail extending SSW of the nucleus, ~9" away. Comparing to dynamical models (Johansson et al. 2009), this feature is consistent with a remnant of a smaller spiral galaxy which is now falling towards the main component for the second time, and has likely lost much of its mass, and perhaps is the source of the whole linear tidal tail. The timescales would put the first passage somewhere at 50 Myr in the past, consistent with the estimates of the age of the central starburst.

3. A triple merger with a hidden companion

Väisänen et al. (2008) presented NIR AO-imaging and optical spectroscopy of the system IRAS19115-2124. The galaxy, which we dubbed the Bird, was expected to consist of a pair-interaction based on optical HST imaging. The VLT/NACO imaging and SALT spectroscopy, however, suggested a third component. (U)LIRGs are complicated dusty objects, and optical data alone, even of high spatial resolution, often can miss whole nuclei and major components of the interaction (Fig. 4.; see also Haan et al. 2011). Moreover, there was evidence that this

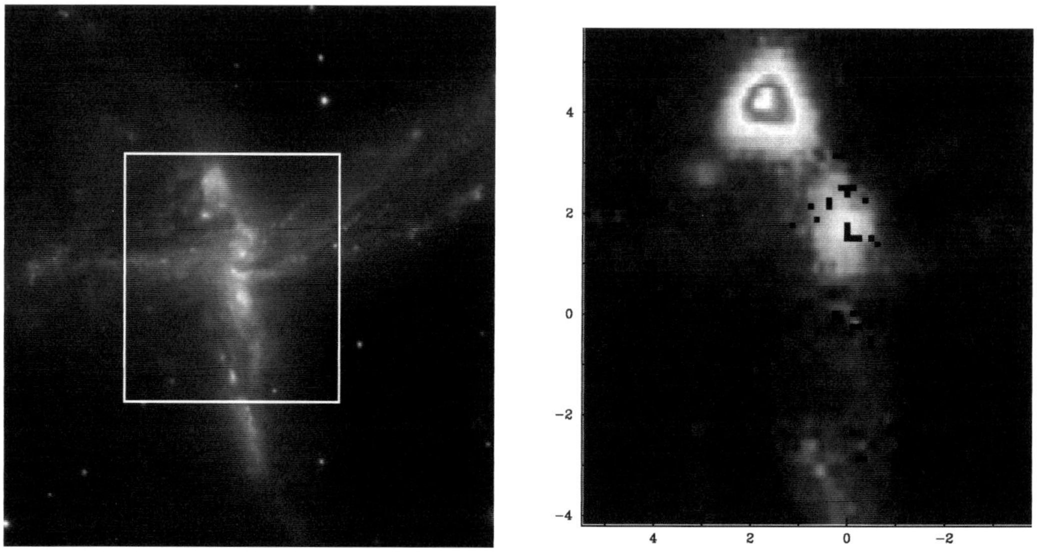

Figure 4. *Left*: A BIK-three colour image of the Bird, where optical bands are from HST and the K-band is our VLT/NACO AO-data. Two major nuclei are seen in the centre. While the optical HST images show spectacular tidal tails extending both vertically and horizontally, the larger Southern nucleus ("Body") is virtually invisible in the HST B-band image, and also the third component ("Head" of the Bird) North-East of the two central nuclei is also totally inconspicuous. The white box shows the region observed with VLT/SINFONI. *Right*: The SINFONI Paβ map of the Bird. The current SF of the Bird is dominated by the small and irregular "Head" component, which has an estimated mass of 1/10 of the two other components. There is some SF also happening in the smaller of the major nuclei ("Heart"), which still resembles a spiral galaxy in the K-band images regardless of the fairly advanced interaction. The tick marks are in arcsec, and 1 arcsec corresponds to approximately 1 kpc.

third, least massive and irregular component dominates the current star-formation output of the whole system. A similar case was recently presented by Inami et al. (2010). This is in contrast to the widely held picture that tidal interactions are expected to drive large quantities of gas into the *central regions* of the interaction resulting in nuclear starbursts. We have now observed the Bird system in more detail using MIR imaging in the 11 μm band with VLT/VISIR and JK IFU observations with VLT/SINFONI.

Figure 4., right panel, shows the Paβ flux over the Bird system, which clearly locates the current SF into the "Head" component. About 3/4 of the current SFR originates from this minor component. While the analysis of the whole dataset is still ongoing, several physical characteristics are immediately obvious from the SINFONI IFU spectra. The Brγ EW is an order of magnitude larger in the "Head" than in the "Body", and mid-way in the "Heart", indicating youngest SF in the Head. Interestingly the [FeII]/Paβ ratio in the Head component is larger in its outer parts compared to its inner regions, suggesting the current SF is more compact than the somewhat older SF of few 10s of Myr on the outer rim. Ratios of molecular hydrogen to hydrogen recombination lines show LINER-like values over the Body component while being pure starburst-like elsewhere.

Figure 5 shows the 11.8 μm contours from VISIR data, corresponding to 11.2 μm PAH-feature at z=0.049, overlaid on the K-band NACO data. The strongest PAH emission is coming from the "spiral galaxy" at the Heart of the Bird. While there is significant PAH flux from the Head as well, it does appear that the flux is depleted there, possibly by the destruction of PAH carriers or e.g. by dilution of hotter dust continuum. The Spitzer/IRAC 3.6 – 4.5 μm colour of

Journal of Physics: Conference Series **372** (2012) 012045 doi:10.1088/1742-6596/372/1/012045

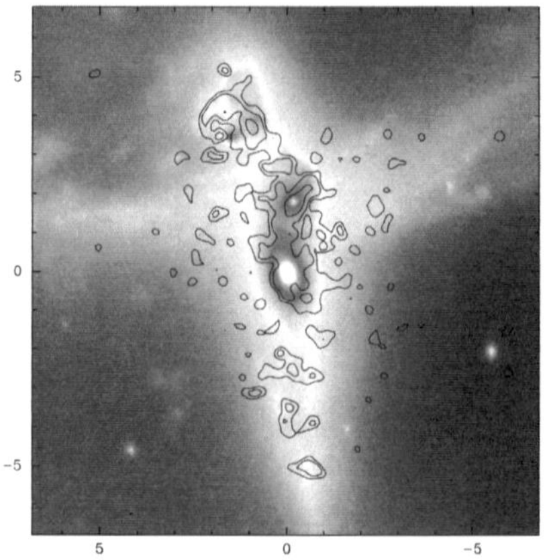

Figure 5. VLT/VISIR 11.8 μm contours on the K-band image. There appears to be a depletion of PAH strength at the location of the hottest component, the Head of the Bird.

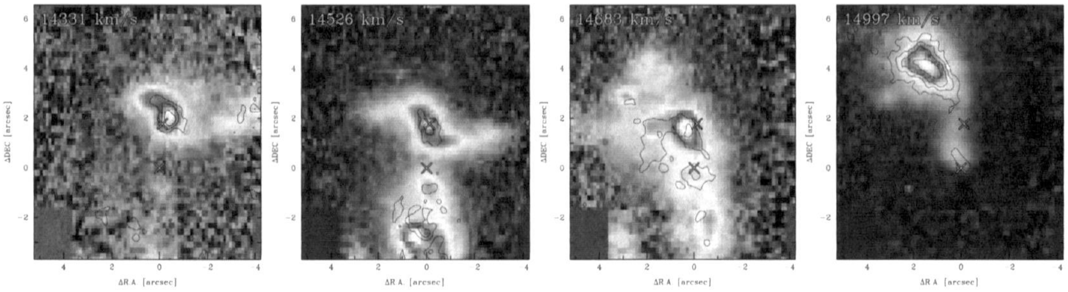

Figure 6. Paα maps at selected velocities from our SINFONI data, with the corresponding H_2 emission shown as contours. The latter follow the Paα distribution very well over the Head component, but the distributions are more varied over the main nuclei. The positions of the two main K-band nuclei seen in Fig. 5 above, are marked with crosses.

the Head, in fact, is very red, typical to dwarf galaxies (e.g. Smith et al. 2009).

Some example velocity slices extracted from the Paα and H_2 lines are shown in Fig. 6 mainly to show the quality of the data and the main kinematic components. The velocity of the Head component is much higher than the others, and actually extends to 15200 km/s. When examining the full data-cube, it also appears that there are ionised gas *outflows* from both the two main galaxy nuclei – cool gas outflows were already detected by the optical SALT spectroscopy in NaD absorption. In addition, there are also indications of *inflows* of gas at the Body and Head components.

4. SNe in the nuclei of LIRGs

SF dominated (U)LIRGs are expected to hide in their central regions large numbers of undetected core-collapse supernovae (CCSNe), i.e. stars more massive than $\sim$8 $M_\odot$ exploding at the end of their (short) lives. Such SNe cannot be detected at optical wavelengths, even in the local universe, because of severe ($A_V > 10$ mag) dust extinction. Recently, Horiuchi et al. (2011) indicated that the measured cosmic CCSN rate is a factor $\sim$2 smaller than that predicted from the measured cosmic SFR, suggesting there is a "SN rate problem". Therefore, an accurate correction for the number of SNe lost in starbursts, LIRGs and ULIRGs will be crucial for the present and future CCSN surveys at high-z.

Our SN search using Gemini/ALTAIR/NIRI has so far produced the detections of 6 SNe. The most recent ones, SNe 2010cu and 2011hi, both occurred in the same host galaxy IC 883 and were located at only $\sim$200-400 pc projected galacto-centric distances. Both the SNe seem to

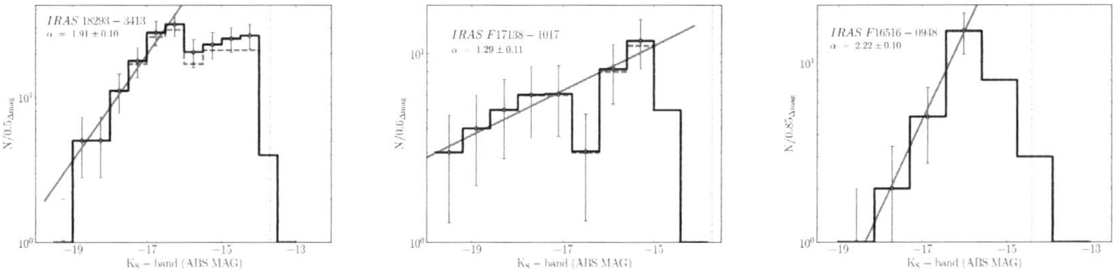

Figure 7. Examples of K-band luminosity functions of three galaxies. A clear break in the power-law slope is seen on the left, while the other two examples show shallow ($\alpha \sim -1.3$) and steep $\alpha \sim -2.2$slopes, respectively, with a possible two-component power-law in the distribution on the right as well.

suffer from relatively low host galaxy extinction ($A_V < 1$ and < 7, respectively; Kankare et al. 2012) suggesting they are not deeply buried in the nuclear regions of IC 883. Despite the rather low extinctions such SNe occurring close to the LIRG nuclei have remained undiscovered by the current SN searches and will therefore contribute to the "SN rate problem". Moreover, some of the other SNe we have discovered have had significantly higher extinctions (e.g. Kankare et al. 2008).

Based on the IR-luminosities for our sample of 7 LIRGs in the multi-epoch Gemini observations, the intrinsic number of CCSNe exploding over the program is about 15 CCSNe. We expect about one third of these to be detectable depending on the assumed distribution of extinctions and concentration of SNe within the innermost nuclear regions. Although the number statistics are arguably still low and the nature of (U)LIRGs at higher redshifts not precisely known, it already appears clear that a significant fraction of CCSNe at intermediate and high-z remains undetected in these systems. Therefore, AO-assisted NIR observations provide an important window to detect and study SNe in the obscured nuclear regions of the local counterparts of the galaxies dominating the SF at higher-z.

5. Super Star Clusters

Young and massive super star clusters (SSCs) are found whenever very active star formation is going on, such as that in interacting LIRGs (e.g. Whitmore et al.1995). We are studying the luminosity and mass functions of SSCs as a function of SFR and type of host galaxy, interaction stage, etc., in an effort to constrain the cluster mass function (CMF), study the formation, evolution, and disruption of SSCs, which all have relevance also to star formation laws in general (Escala & Larson 2008). As a first step, we have studied the K-band LFs of the sample of galaxies from our Gemini observations. According to theoretical expectations and many optical studies of nearby star-forming galaxies, SSC LFs are well fitted by a single power-law index of $\alpha \sim -2$ in a $\Phi(L)dL \propto L^{-\alpha}dL$ distribution. We, in contrast, find a surprising variety of slopes, ranging from much shallower ones of $\alpha \sim -1.3$, to more or less the "nominal" -2 case. In addition, intriguingly, most of the galaxies in the initial sample have clear break-points in their SSC LF slope, where the bright section is significantly steeper than the faint magnitude ranges, or we find distributions which might be better fit with e.g. Gaussian distributions. Fig. 7 shows three examples of the LFs, one with a clear break-point, and others with the opposite ends of the range of slopes, with $\alpha \sim -1.3$ and -2.2, respectively, as fitted to the bright part of the LF. These effects persist through rigorous photometric, blending, and detection/completeness analyses and simulations in the complex background conditions of the target galaxies.

A range of LF slopes and/or breaks in the slopes could indicate mass-dependent disruption of the SSCs, or environment and/or host SFR dependent cluster CMFs (e.g. Bastian 2008;

Chandar et al. 2010; Larsen 2009). Most previous determinations are from less luminous IR-galaxies, while recently relatively shallow slopes for SSC LFs have been found in blue compact dwarfs (e.g. Adamo et al. 2011 and references therein). Our analysis is still ongoing, and will be presented in Randriamanakoto et al. (2012, in preparation), and the full sample of 40 galaxies will also be added to the SSC analysis to get a handle on how universal CFLs are, and how they depend on host galaxy properties.

6. Summary

We are studying a sample of several dozen strongly star-forming galaxies, mostly LIRGs, using NIR adaptive optics imaging, and optical spectroscopy, and IFU observations for selected targets. Adaptive optics and near- and mid-IR observations are often necessary to uncover how many nuclei are involved in the interactions, and to characterise the components, and how star formation is triggered and is propagating in the systems. We find the L-band and in particular the 3.3 μm PAH emission to trace well various levels of star formation the a LIRG nucleus. We also find hundreds of super star clusters, and detect a population of dust-obscured core-collapse SNe in our LIRGs.

[1] Adamo, A. et al., 2011, *MNRAS*, 415, 2388
[2] Alonso-Herrero, A., 2001, *ApJ*, 546, 952
[3] Alonso-Herrero, A., 2006, *ApJ*, 650, 835
[4] Bastian, N., 2008, *MNRAS*, 390, 759
[5] Beirão, P. et al. 2009, *ApJ*, 693, 1650
[6] Chandar, R., et al. 2010, *ApJ*, 719, 966
[7] Díaz-Santos, T. et al. 2008, *ApJ*, 685, 211
[8] Escala, A. & Larson, R.B., 2008, *ApJ*, 685, L31
[9] Haan, S., et al., 2011, *AJ*, 141, 100
[10] Imanishi, M. & Dudley, C. C. 2000, *ApJ*, 545, 701
[11] Inami, H., et al., 2010, *AJ*, 140, 63
[12] Horiuchi et al., 2011, *ApJ*, 738, 154
[13] Johansson, P.H. et al. 2009, *ApJ*, 707, L184
[14] Kankare, E., et al., 2012, *ApJ*, 744, L19
[15] Kankare, E., et al., 2008, *ApJ*, 689, L97
[16] Kotilainen, J., et al., 2001, *A&A*, 366, 439
[17] Larsen, S.S. 2002, *AJ*, 124, 1393
[18] Larsen, S.S. 2009, *A&A*, 494, 539
[19] Risaliti G., et al., 2006, *MNRAS*, 365, 303
[20] Rodighiero, G., et al., 2011, *ApJ*, 739, L40
[21] Rodríguez-Zaurín, J., et al., 2011, *A&A*, 527, 60
[22] Smith, B., et al. 2009, *AJ*, 138, 130
[23] Tacconi-Garman, L.E. 2005, *A&A*, 432, 91
[24] Väisänen, P. et al. 2008, *MNRAS*, 384, 886
[25] Väisänen, P. et al. 2012, *MNRAS*, in press (arXiv:1111.2753)
[26] Whitmore, B. C., et al. 1999, *AJ*, 118, 1551

Do stellar winds play a decisive role in feeding AGN?

R. Davies, L. Burtscher, K. Dodds-Eden, G. Orban de Xivry
Max-Planck-Institut für extraterrestrische Physik,
Postfach 1312, Giessenbachstr., 85741 Garching, Germany
E-mail: davies@mpe.mpg.de

Abstract. While the existence of a starburst-AGN connection is undisputed, there is no consensus on what the connection is. In this contribution, we begin by noting that the mechanisms which drive gas inwards in disk galaxies are generally inefficient at removing angular momentum, leading to stalled inflows. Thus, a tiered series of such processes is required to bring gas to the smallest scales, each of which on its own may not correlate with the presence of an AGN. Similarly, each may be associated with a starburst event, making it important to discriminate between 'circumnuclear' and 'nuclear' star formation. In this contribution, we show that stellar feedback on scales of tens of parsecs plays a critical role in first hindering and then helping accretion. We argue that it is only after the initial turbulent phases of a starburst that gas from slow stellar winds can accrete efficiently to smaller scales. This would imply that the properties of the obscuring torus are directly coupled to star formation and that the torus must be a complex dynamical entity. We finish by remarking on other contexts where similar processes appear to be at work.

1. Introduction

The starburst-AGN connection has been a long disputed topic. A large part of the reason is that any conclusion will depend on the spatial and temporal scales being considered, as well as the luminosities of the AGN and the properties of their host galaxies. The other important issue is whether one is assessing a 'casual' or 'causal' relation (both of which play a role in our understanding of galaxy evolution and the starburst-AGN connection). Casual connections – in which star formation and an AGN are independent products of another phenomenon – abound. Causal connections – in which either the star formation or AGN has a direct impact on the other – are more elusive, and form the rationale for the work discussed here. In this contribution we focus on local galaxies, and hence on Seyferts and Low Luminosity AGN (LLAGN). In addition, we consider primarily AGN with spiral hosts, since local Seyferts typically reside in disk galaxies (although the situation changes if one includes LINERs; Ho 2008). Our aim is to touch on a number of issues related to the starburst-AGN connection in such galaxies, starting with the largest scales ($\gtrsim 1$ kpc) in Section 2, moving in towards the 0.1–1 kpc scale circumnuclear region in Section 3, and then the central tens of parsecs in Sections 4 and 5. We argue that on these small scales there is indeed a causal relation between star formation and the AGN – with stellar outflows from a young post-starburst first hindering and then helping gas inflow towards the central supermassive black hole (BH).

2. Are large scales important?

It is easy to forget that the requirements for fuelling star formation (i.e. cold, dense gas) and a central BH (i.e. low angular momentum gas) are different. That star formation requires cold, dense gas has recently been beautifully demonstrated by Lada et al. (2010). They showed that the number of young stellar objects in galactic clouds is not related to the total cloud mass, but to the mass above a threshold K-band extinction of $A_K = 0.8$ mag. One can argue that this should lead to an anti-correlation between star formation and AGN: only if inflow is stalled (ultimately resulting in a reduced accretion rate onto the central BH) will the gas pile up enough to form the cold dense cores that lead to star formation.

In contrast, Jogee (2006) emphasized that even gas located at a radius of a few hundred parsecs has to lose 99.99% of its angular momentum before it reaches the central BH. This means that large scale structures in a galaxy can only be related to AGN activity if they remove sufficient angular momentum. A galaxy interaction is such a process, and leads to the infall of a huge gas reservoir to small spatial scales on a short timescale. It is for this reason that the most luminous AGN tend to be associated with mergers. However, it is also important to realise that this infall is unlikely to feed the BH(s) directly, but only bring the gas down to scales of order 100 pc. From there other mechanisms, perhaps related to disk processes as suggested by the nested simulations of Hopkins & Quataert (2010), will drive the gas to smaller radii. For Seyferts in disk galaxies, the efficiency of angular momentum removal is the reason why there is at best only a marginal relation between the presence of large scale bars and AGN (Laine et al., 2002; Laurikainen et al., 2004). Bars are inefficient because they only drive modest inflows, and what there is always stalls at a ring on scales larger than 100 pc (Regan & Teuben, 2004). In these cases, it is processes acting on smaller scales that will predominantly regulate the rate at which gas can reach the central regions.

It is in this context that Narrow Line Seyfert 1 (NLS1) galaxies are so unusual. They provide an example of how, in a specific subset of AGN, the large scale host galaxy properties play a prominent role in the properties of the central BH. Combining several analyses from the literature, Orban de Xivry et al. (2011) showed that most NLS1s are associated not only with strong bars, but also with circumnuclear grand design dust spirals. In these objects, it seems likely that circumnuclear spiral pattern is a standing wave driven by the bar. Indeed, this is the way that such spirals are produced in hydrodynamical simulations (Maciejewski, 2004). Thus, the AGN activity is associated with the bar, because the bar stimulates a circumnuclear spiral to drive gas to smaller scales than it otherwise could. Thus the strong bars in NLS1s can be considered efficient at removing angular momentum. Orban de Xivry et al. (2011) also measured the Sérsic index of the bulge luminosity profiles of NLS1s, together with a sample of Broad Line Seyfert 1s. They found that while the indices for the BLS1s covered the full range expected for classical and pseudobulges, the NLS1s were all clustered at the low end. The implication is that these galaxies have pure pseudobulges, and have therefore grown and evolved primarily through secular processes. The authors speculated that in this case, because the BHs have grown through accretion of disk material, they should be spinning rapidly. This would make them radiatively efficient, which in turn slows their growth. If this scenario is confirmed, it could provide a natural explanation for the high BH spins observed in the majority of NLS1s via the 6.4 keV Fe Kα line, as well as the typically low masses of BHs in all galaxies with pure pseudobulges.

3. Circumnuclear Inflow

It has been known now for a long time that the circumnuclear region (scales of 100–1000 pc) plays a crucial role in regulating gas inflow, but the complexity of its dynamics has only begun to be appreciated over the last decade (Wada, 2004). Observationally, the first clues came from imaging surveys with *HST* of dust structures, which are presumed to be associated with gas and hence the inflow mechanisms. In matched samples of active and inactive galaxies, Martini et al. (2003) found a variety of dust structures but no correlation with the presence of an AGN. The first hints of such a relation were reported by Simões Lopes et al. (2007) who, by separating AGN according to host galaxy Hubble type, showed that while circumnuclear dust structures are necessary for AGN activity, they are not sufficient, and that other factors must also be important.

Taking a different approach, the NUGA team looked at the molecular gas directly using 1–2″ resolution CO data. They calculated the gravitational torques, to assess whether these could drive the gas into the central ~ 100 pc. Intriguingly, of the 7 galaxies analysed by Haan et al. (2009), only 4 had a significant global negative torque driving a net inflow through the circumnuclear region; and Garciá-Burillo (2011) reported that of the 25 objects in the full study, only 1/3 revealed torques that could drive gas into the central 100 pc. It is possible that this perplexing result can be understood if one takes account of the AGN luminosity, which is typically a hidden parameter. The NUGA sample was designed to span a range of AGN types which means that a significant number are LINERs. And it contains a large number of LLAGN: none of the objects in Haan et al. (2009) appears among the AGN detected at 14–195 keV in the *Swift* BAT 58-month catalogue (Baumgartner et al., 2010); and the 2–10 keV luminosities (where available) are all $< 10^{40}\,\mathrm{erg\,s^{-1}}$. We speculate that the very modest inflow needed to fuel a LLAGN – less than $10^{-3}\,M_{\odot}\,\mathrm{yr}^{-1}$ – is so small that it does not require, and perhaps may even tend to preclude, a coherent dynamical process in the circumnuclear region. In contrast, such processes may typically help supply the larger 0.01–0.1 $M_{\odot}\,\mathrm{yr}^{-1}$ inflow rates of more luminous Seyferts. We argue that to advance our understanding of the role of the circumnuclear region in fuelling AGN, it is necessary to assess the mechanisms and inflow rates on these scales as a function of AGN luminosity.

None of the work described above directly uses the observed kinematics. Such analyses can be difficult and require high signal-to-noise data that traces kinematics at high spatial resolution across a full 2D field. This is because the velocity fields are usually dominated by rotation, and it is only the deviations from this – the residuals – that contain information about the inflow. Haan et al. (2009) performed a harmonic analysis on the CO kinematics for their sample, a technique that was also used by van de Venn & Fathi (2010) on the ionised gas kinematics in NGC 1097 (and is also the basis of kinemetry (Krajnović et al., 2006) which was used in the analysis of the SAURON sample of early type galaxies). An alternative is to fit a disk model to the circular motions and assess the residuals, a technique employed by a number of authors (e.g. Storchi-Bergmann et al., 2007; Riffel et al., 2008; Davies et al., 2009; Schnorr Müller et al., 2011, amongst others). While this can be effective at revealing non-circular motions, interpreting them robustly can be hard. For example, Davies et al. (2009) have pointed out that the line-of-sight kinematic signature of inflow along a 2-arm spiral is a 1-arm kinematic spiral (superimposed on the global rotation) – making it challenging to relate the kinematics to inflow along spiral arms in a specific way. While the case studies performed so far have demonstrated the potential of such work, systematic studies using integral field spectroscopy (e.g. Hicks et al. in prep) or millimetre line interferometry to probe the kinematics and their residuals will be needed to assess the variety of mechanisms that may be at work in the circumnuclear region, and how they relate to AGN activity in different luminosity regimes.

4. Young Nuclear Starbursts

The mechanisms discussed in the preceeding sections are inefficient at removing angular momentum. As a result, the inflows they drive are stalled in the central regions, leading to intense periodic bursts of star formation. The key issue we want to address here is whether these young nuclear starbursts have an impact on AGN fuelling or is simply a serendipitous by-product of a stalled inflow. One can only begin to answer this by first clarifying what we mean by 'nuclear' and 'young'.

Historically, all starbursts that occur in the central kiloparsecs of a galaxy have been called 'nuclear'. Now that we can spatially resolve this region, such nomenclature is inappropriate, and it is instead crucial to discriminate between 'circumnuclear' meaning 0.1–2 kpc and 'nuclear' meaning $\lesssim$50 pc (and in the future, even smaller scales). The separation of these scales, although to some extent arbitrary, is based on both observational and theoretical work. Comerón et al. (2010) presented a survey of 113 star-forming rings in the central regions of 107 galaxies, finding that such rings occur in 20% of all disk galaxies. The radii of these rings spans 0.1–2 kpc, matching the range in the hydrodynamical simulations of Regan & Teuben (2003). These authors found that rings migrate inwards from $\sim$ 1 kpc as they accumulate gas, and can reach sizes as small as $\sim$ 0.2 kpc. It is a small step to realise that the frequency of rings should be greater if one includes those that are not currently forming stars (i.e. without Hα emission). Indeed, molecular rings are commonly found in interferometric millimetre observations such as those from the PdBI – e.g. at a radius of 140 pc (1.7″) in NGC 3227 (Davies et al., 2006) and at 500 pc (5″) in NGC 6951 (Krips et al., 2007). And spatially resolved stellar population synthesis has revealed intermediate age ($\sim$ 0.5 Gyr) rings at radii of 600 pc (1.5″) in Mkn 1066 (Riffel et al., 2010) and 300 pc (1″) in Mkn 1157 (Riffel et al., 2011).

That rings are indeed common, and may be rather small, must be borne in mind when interpreting observations where spatial resolution is insufficient to disentangle 'circumnuclear' and 'nuclear'. One example of this is the analysis of the 3.3 μm PAH emission feature (which traces star formation) in type 1 and 2 Seyferts by Imanishi & Wada (2004). At the typical redshift of their sample, the spectroscopic aperture covered $\sim$ 0.8 kpc. Thus most of the PAH detections in Seyfert 2s and as many as half of those in Seyfert 1s may be attributable to circumnuclear star formation. Such a hypothesis would have to be verified by higher resolution observations, but it would imply that the star formation results from stalled inflows, at radii too large to have a direct causal impact on BH accretion.

The second issue here is to ask what we mean by 'young' and whether such populations are associated with AGN. The largest survey to address this made use of stellar population synthesis for the central $\sim$ 200 pc of 65 Seyfert 2 (and 14 other) galaxies (Cid Fernandes et al., 2004). By fitting the absorption features in optical spectra with a library of model templates, the authors derived the star formation history of their targets. Most of the sample showed evidence for a 20–60% contribution from what could be either a young (<25 Myr) population or a featureless continuum. But the higher the contribution of this component, the stronger were the weak broad emission lines, suggesting that it was associated with the AGN. The authors noted that while in some objects there may be a young dusty starburst, its contribution to the optical continuum is $\lesssim$ 20%. Their conclusion was simply that "1/3 to 1/2 of Seyfert 2s have experienced significant star formation in the recent past" (specifically, 40% of their galaxies have a mean age in the central $\sim$ 200 pc less than 300 Myr). Riffel et al. (2009) found a similar result from near infrared spectral synthesis of the central $\sim$ 300 pc of 24 Seyferts. In this case, a significant contribution from either the young stellar population or featureless continuum was found in 50% of the Seyfert 2s and most of the Seyfert 1s. This is what one would qualitatively expect if the component was associated with the AGN. Thus while a young (< 50 Myr) population may be present in some objects, it is not possible to quantify definitively its scale or location. In an analysis of STIS spectra of 22 Seyfert 2s (with an aperture radius of $\sim$ 30 pc, thus avoiding issues associated with circumnuclear rings), Sarzi et al. (2007) found that nearly half of the objects required a young starburst or featureless continuum, and most of the rest had only old populations. They argued that the young starburst interpretation of this component contradicted the dearth of intermediate age populations, and that instead its strong correlation with high excitation lines argues for its association with the AGN.

There seems to be some disagreement in the studies above about whether AGN are associated with intermediate age (0.1–1 Gyr) stars; and it seems that some, but probably not all, do host such a population. But a common conclusion appears to be that AGN are not obviously associated with young (< 50 Myr) starbursts in the nuclear (< 50 pc) region.

5. The Role of Stellar Outflows

Using adaptive optics to achieve resolutions of order ~ 10 pc, Davies et al. (2007) estimated the characteristic age of the most recent burst of star formation in a number of AGN and compared this to the accretion rate. They found ages of < 50 Myr were associated with lower Eddington ratios, and the accretion rate was only high for ages > 50–100 Myr. They suggested that this was because at very young ages, the outflows from stars (OB winds and type II supernovae) injected so much turbulence into the surrounding inter-stellar medium (ISM), that any gas would tend to be ejected rather than accreted. However, for a short burst of star formation, this phase would end after 50–100 Myr, after which the slower outflows from AGB stars, which would remain gravitationally bound to the cluster, would be available to accrete to smaller scales. Based on these arguments, Vollmer et al. (2008) put together a scenario in which the star formation and torus properties would be time-dependent, driven by (the stochastic nature of) changes in gas flow from larger scales. At about the same time, 3D hydrodynamical simulations, subsequently scaled to match a typical Seyfert, were yielding similar conclusions about stellar feedback (Schartmann et al., 2009, 2010). These showed that for a stellar cluster of $10^8\,M_\odot$ in the tens of parsecs around a $10^7\,M_\odot$ BH, outflows with speeds greater than about $200\,\mathrm{km\,s^{-1}}$ would tend to disrupt the ISM and drive gas out; but slower outflows could be accreted in streamers to smaller scales on a timescale between the end of the type II supernova phase and before type Ia supernova begin to appear. In this particular case, scaled to match NGC 1068, the streamers fed a dense and compact ($10^6\,M_\odot$, 0.5–1 pc) turbulent disk. The size of this disk is set by the 'angular momentum barrier' and depends on the kinematic properties of the surrounding stellar cluster. Remarkably, it matched the size of the maser disk in this object and also the size of the compact mid-infrared continuum source detected with VLTI. Although this is so far only one example, it provides a tantalising suggestion that we may be able to directly link the properties of a star forming cluster on scales of 10–50 pc with those of more compact structures on scales of 0.1–1 pc. It is reasonable to assume that these structures are related to the putative obscuring torus. This leads immediately to two important conclusions: (i) that the torus is complex, comprising structures on at least 2 different scales that may be responsible for different observational aspects (see also Elitzur, 2008; Elvis, 2012); (ii) the torus is a dynamical entity, the time-dependent properties of which are related to star formation in its local environment, and ultimately to gas inflow from larger scales.

It is important to realise that other models are possible. The analytical model of Kawakatu & Wada (2008) and detailed 3D hydrodynamical simulations of Wada et al. (2009) suggest that inflow and accretion onto the BH should increase with the turbulence in the ISM, and hence occur simultaneously to the star formation. Hobbs et al. (2011) suggest that supernovae should be able to drive enough of the ambient ISM inwards in a 'ballistic' mode to fuel an AGN. All these scenarios rely on the role played by stellar outflows in driving gas to smaller scales. The difference between them is whether gas inflow is synchronous with, or subsequent to, the star formation episode. Further testing with detailed observations and simulations for more objects will be needed to discriminate between them. It is, however, notable that the same physical processes as proposed by Davies et al. (2007) and Schartmann et al. (2009, 2010) have been invoked to explain similar situations in a number of other environments. These include the inner 0.5 pc of the Galactic Center, where simulations have shown that fast stellar winds, and the angular momentum of their stars from which they come, are inhibiting accretion to smaller scales (Cuadra et al., 2006, 2008). In the central 2 pc of M 31, Chang et al. (2007) have argued that mass loss from the older red stars in the eccentric stellar disk (nuclei P1 and P2 in the usual notation) can be driven to smaller scales where it is able to produce a compact gravitationally unstable gas disk of $\sim 10^5\,M_\odot$ every 0.1–1 Gyr. It can thus fuel the star formation that has been observed in the smaller (0.2 pc scale) younger (200 Myr) cluster P3. It is also believed that the multiple stellar populations of globular clusters may arise from secondary generations of stars being formed from the outflows of earlier generations. D'Ercole et al. (2008) have shown that there is a specific window around an age of 40 Myr – between the phases of type II and type Ia supernovae – when this can happen.

6. Conclusions

The main points raised in this contribution can be summarised as follows:

- Global trends between AGN and large scale phenomena are masked by the variety and number of mechanisms required to bring gas to small scales; but relations may be found in specific subsets of objects (e.g. NLS1s). The AGN luminosity – as a proxy for accretion rate – may be an important (hidden) parameter when searching for such relations, since different phenomena will lead to different gas inflow rates and efficiencies.
- Each time a gas inflow is stalled, one may expect star formation to be triggered. It is therefore important to discriminate between 'nuclear' star formation on scales of < 50 pc which may directly influence further gas inflow, and serendipitous 'circumnuclear' star formation on scales of 0.1–2 kpc that does not obviously do so.
- There is no overwhelming evidence for active young (< 50 Myr) starbursts in the tens of parsecs around AGN. In contrast, it appears there is a delay of $\sim$ 100 Myr between a starburst and the onset of accretion to an AGN. This may be because it is only during phases when slow stellar winds predominate that gas can be accreted to small scales. Such a scenario points to a direct link between the properties of star-forming clusters on 10–50 pc scales and the compact structures observed on 0.1–1 pc scales. It would also imply that the torus is a complex dynamical entity, the properties of which are coupled to nuclear star formation.
- In answer to the question posed in the title of this contribution, it is now becoming clear that stellar winds *do* play a decisive role in feeding Seyferts. But it is also clear that other processes are also important.

References

Baumgartner W., et al., 2010, ApJS submitted
Chang P., Murray-Clay R., Chiang E., Quataert E., 2007. ApJ, 668, 236
Cid Fernandes R., et al., 2004. MNRAS, 355, 273
Comerón S., et al. 2010. MNRAS, 402, 2462
Cuadra J., Nayakshin S., Springel V., Di Matteo T., 2006. MNRAS, 366, 358
Cuadra J., Nayakshin S., Martins F., 2008. MNRAS, 383, 458
Davies R., 2006. ApJ, 646, 754
Davies R., et al. 2007. ApJ, 671, 1388
Davies R. et al., 2009. ApJ, 702, 114
D'Ercole A., Vesperini E., Antona F., McMillar S., Recchi S., 2008. MNRAS, 391, 825
Elitzur M., 2008. NewAR, 52, 274
Elvis 2012, in *The Central Kiloparsec in Galactic Nuclei*, held Aug/Sept 2011, Bad Honnef, Germany (arXiv:1201.5101)
Garciá-Burillo S., 2011. in *The Central Kiloparsec in Galactic Nuclei*, held Aug/Sept 2011, Bad Honnef, Germany.
Haan S., et al. 2009. ApJ, 92, 1623
Ho L., 2008. ARA&A, 46, 475
Hobbs A., Nayakshin S., Power C., King A., 2011. MNRAS, 413, 2633
Hopkins P., Quataert E., 2010. MNRAS, 407, 1529
Imanishi M., Wada K., 2004. ApJ, 617, 214

Jogee S., 2006. in *Physics of Active Galactic Nuclei at all Scales*, ed. Alloin D., Johnson R., Lira P., LNP, 693, 143
Kawakatu N., Wada K., 2008. ApJ, 681, 73
Krajnović D., Cappellari M., de Zeeuw T., Copin Y., 2006. MNRAS 366, 787
Krips M., et al. 2007. A&A, 468, L63
Lada C., Lombardi M., Alves J., 2010. ApJ, 724, 687
Laine S., Shlosman I., Knapen J., Peletier R., 2002. ApJ, 597, 97
Laurikainen E., Salo H., Buta R., 2004. ApJ, 607, 103
Maciejewski W., 2004. MNRAS, 354, 892
Martini P., Regan M., Mulchaey J., Pogge R., 2003. ApJ, 589, 774
Orban de Xivry G., et al., 2011. MNRAS, 417, 2721
Regan M., Teuben P., 2003. ApJ, 582, 723
Regan M., Teuben P., 2004. ApJ, 600, 595
Riffel R.A., et al. 2008. MNRAS, 385, 1129
Riffel R., Pastoriza M., Rodríguez-Ardila A., Bonatto C., 2009. MNRAS, 400, 273
Riffel R.A., Storchi-Bergmann T., Riffel R., Pastoriza M. 2010. ApJ, 713, 469
Riffel R., Riffel R.A., Ferrari F., Storchi-Bergmann T., 2011. MNRAS, 416, 493
Sarzi M., Shields J., Pogge R., Martini P., 2007. in *Stellar Populations as Building Blocks of Galaxies*, ed. Vazdekis A., Peletier R., Proc. IAUS, 241, 489
Schartmann M., Meisenheimer K., Klahr H., Camenzind M., Wolf S., Henning T., 2009. MNRAS, 393, 759
Schartmann M., Burkert A., Krause M., Camenzind M., Meisenheimer K., Davies R., 2010. MNRAS, 403, 1801
Schnorr Müller A., et al., 2011. MNRAS, 413, 149
Simões Lopes R., Storchi-Bergmann T., de Fátima Saraiva M., Martini P., 2007. ApJ, 655, 718
Storchi-Bergmann T., et al., 2007. ApJ, 670, 959
Van de Venn G., Fathi K., 2010. ApJ, 723, 767
Vollmer B., Beckert T., Davies R., 2008. A&A, 491, 441
Wada K., 2004. in *Carnegie Observatories Astrophysics Series, Vol. 1: Coevolution of Black Holes and Galaxies*, ed. Ho L. (Cambridge: CUP)
Wada K., Papadopoulos P., Spaans M., 2009. ApJ, 702, 63

Linking the Supermassive Black Hole Growth with the Megamaser Emission

Anca Constantin
James Madison University, Harrisonburg, VA 22807
E-mail: `constaax@jmu.edu`

Abstract. High-resolution observations of the central few 100 pc of the galactic nuclear environments remain prohibitive for large statistical samples, which are crucial for tracing the links between central black hole formation, galaxy formation and AGN activity over cosmic time. With this contribution, we present novel ways of connecting the physics of black hole accretion with its immediate environs via a new quantitative evaluation of the degree to which the strength and spatial configuration of the water maser emission is associated with the nuclear nebular galactic activity. We discuss possible evolutionary/causal connections between these two types of emission, together with criteria that could dramatically enhance our search for mega-maser systems in nearby galactic centers. These are timely results given the interest in combining high-resolution observations with extremely large optical telescopes and large arrays that start to conquer the sub-millimeter window.

1. Introduction

Mega-masers are remarkable natural phenomena. Mapping of the H_2O maser emission with $L_{H_2O} \geq 10L_\odot$ in a disk configuration has provided the first direct evidence in an active galactic nucleus (AGN) for the existence of a thin Keplerian accretion disk with turbulence [15], highly compelling evidence for the existence of a massive black hole (BH) [28, 27], as well as a cosmic distance determination of extremely high precision [19, 20]. To date, such observations and associated calculations have been performed on only less than a handful such systems [31, 27]. Identifying and studying a large sample of the mega-masers in this particularly rare disk configuration is of crucial relevance for our progress in understanding the accretion processes in galaxy centers, and more importantly, for constraining the proposed cosmological models and the nature of Dark Energy. Unfortunately, good candidates are extremely few; the success rate in maser detection remains a mere $\sim 3-5\%$, out of which only $\sim 40\%$ appear to originate in disks [3, 4, 5, 16, 26, 6].

To efficiently search for new powerful mega-maser disks, we need a good understanding of the special physical characteristics that facilitate their production in galaxy centers. There is some evidence that mega-masers may be associated with the molecular disk or torus that surrounds and (partially) obscures an actively accreting massive black hole ($M_{\rm bh} \sim 10^7 M_{\rm sun}$) harbored by a galactic nucleus, i.e., an AGN [28, 5]. However, mega-masers are also detected in face-on, unobscured disks or elliptical galaxies where no molecular torus is expected, and even along jets protruding long distances from the nucleus [12, 13]. So, either masers are not always produced in association with an AGN torus, or these particular cases are unrecognized AGN. Indeed, for the majority of the nearby active galaxies, an AGN classification remains ambiguous; in these sources, an actively accreting black hole might be either obscured by large amounts of dust in the surrounding star-forming regions, or simply hidden in a mix of other ionization sources (shocks, turbulence, etc.) or host galaxy light [21]. Thus, if mega-maser production is genuinely related to black hole accretion, the true detection rate of disk mega-maser emission remains unknown as long as the AGN census is incomplete. To shed light on these matters, we need to investigate all possible connections between the H_2O emission and their host properties, for a wide variety of galactic nuclear activity. We only now have the number statistics necessary to perform such a study.

We introduce here an in-depth multi-parameter characterization of the optical properties of the largest available sample of 146 galaxy centers hosting H_2O masers together with a ~ 20 times larger control sample of galaxies in which maser detection failed. This effort complements the very recent work by Zhu et al. [36], where they investigate the optical properties of the maser host galaxies and find that the maser detection rate increases for higher optical luminosity, larger velocity dispersion, higher extinction and higher [O III] $\lambda 5007$ line luminosity. Our analysis adds a novel comparative investigation of the optical host and nuclear features separated per spectral types that correlate with the strength and type of nuclear galactic activity. If, as previously suggested [9, 10, 32, 33], there is an evolutionary sequence in which galaxies transform from star-forming via AGN to quiescence, accretion onto the central BH should happen as early as the star formation (or H II) phase, and should be active in Transition Objects (or Ts). Thus, if maser activity is related to accretion, we should look for (mega)masers in H II's and Ts as well. Restricting our searches to Seyferts (Ss) and Low Ionization Nuclear Emission Regions (or LINERs; Ls), as it appears to be the case with current efforts, could overlook a good portion, or even the lion's share of mega-masers.

2. The Sample Galaxies & Parameters Involved

The Mega-maser Cosmology Project (MCP; [31, 6]) makes publicly available the entire sample of galaxies detected in H_2O by May 2011. The data is presented in the form of an atlas that comprises sky positions, recession velocities, the maser spectra, and the corresponding discovery reference, for 146 such systems. The MCP atlas also includes the list of all > 3300 galaxies that have been surveyed for H_2O maser emission, along with their Green Bank Telescope (GBT) spectra, recession velocities, the sensitivity of each observation, and the corresponding source brightness temperature. These data provide for the first time a sufficiently large sample of maser galaxies and a control sample of non-detections whose properties can be compared at a statistically significant level.

Optical spectral and photometric measurements for these objects are obtained mainly from wide sky surveys; for the purpose of the analysis presented here, we exploited the nearby galaxy sample of the seventh data release of the SDSS [1]. Measurements of the host and nuclear nebular emission of the SDSS galaxies are drawn from the catalog built by the MPA/JHU collaboration [7]. For these objects, the emission-line component is measured after it is separated from the host stellar emission, and subtracted from the total galaxy spectrum based on fits of stellar population synthesis templates. Black Hole masses ($M_{\rm BH}$) and corresponding accretion rates dM/dt (measured via the Eddington ratio $L/L_{\rm Edd}$) are obtained based on estimates of the stellar velocity dispersion (σ^*), that are available via careful spectral fitting of the host stellar light component in galaxy spectra. To relate the maser and the central BH accretion activity to the host properties, we employ stellar masses and the D_n4000 break index 3850-3950Åand 4000-4100Å; Balogh et al. 1999) or Hδ_A Balmer absorption-line index as proxies for the age of the associated stellar population, as calculated and presented by Kauffman et al. [22, 23].

Cross-matches of the maser and control galaxy lists with these SDSS-based catalogs uncover 79 maser and ~ 1600 control galaxies with photometric information, among which high quality spectroscopic measurements ($> 2\sigma$ confidence) are available for 46 and 1181 maser and non-maser systems respectively. We present here only a comparison between the spectroscopic characteristics. It may be important to note that considering only the spectroscopic samples the 3% maser detection rate remains in place.

3. A statistical comparison of the properties of maser and non-maser galaxies

Optical emission line ratios have often proven to be useful in identifying the dominant ionization mechanism or in delineating differences between various types of galaxy nuclei [2, 35, 25]. A comparison of the emission-line activity in maser and non-maser galaxies is presented in Figures 1 and 2 where we also show how the distributions of the line ratios compare for the two samples. To classify the galaxies into Seyferts (Ss), LINERs (Ls), Transition objects (Ts), and H II galaxies, we adopt the criteria of Kewley et al. [25].

With the exception of the [O III/Hβ] ratio, which is clearly larger for the maser sample, there is no obvious statistical difference between the maser and non-maser (control) galaxies. The difference in [O III/Hβ] could simply be a consequence of the fact that the bulk of the masers have been detected in surveys of Seyferts, whose definition requires such high ratios; however, when only Seyfert maser and control galaxies are compared, the difference is still statistically significant ($KS_{prob} = 0.053$), implying that it reflects an intrinsic relation to maser activity.

The maser detection rate among Seyferts is high ($\sim 8\%$), however, it is unclear whether this is because the disk-maser emission is genuinely connected to this type of nuclear activity (i.e., the BH accretion process) or is simply a consequence of the same survey bias. The maser disks are usually mega-masers, but not all mega-masers are disks. The disks have only been found among the Seyferts, however, mega-masers appear to cover the whole spectral spectrum. The maser detection rate is clearly non-zero for non-AGN types, in fact, there is an equal probability to find masers among Ts as in Ls ($\sim$1.6%), and twice as likely to find them in H IIs ($\sim 3\%$).

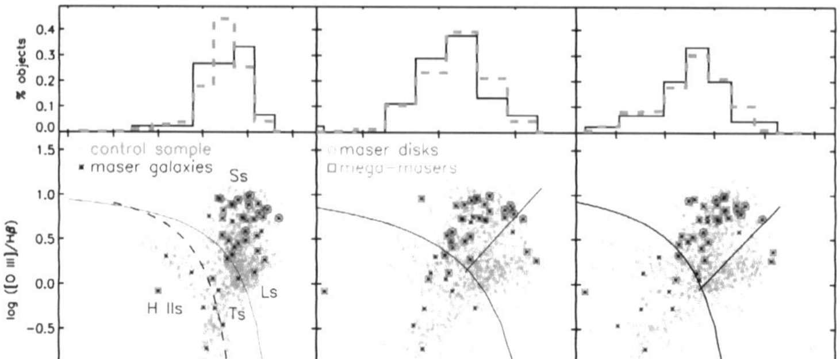

Figure 1. Line-diagnostic diagram for all maser and control galaxies, with a comparison of histograms of the involved line ratios. The non-maser galaxies are plotted with grey symbols, the masers in black, the mega-masers ($L_{H_2O} \geq 10L_{sun}$) are shown as magenta squares while the disks among them are the blue circles. The lines in the bottom panels are semi-empirical fits to the distributions of objects in these diagrams.

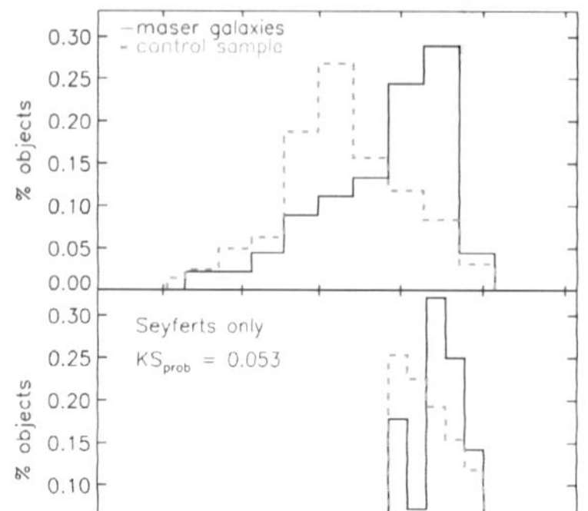

Figure 2. Comparison of the distributions of [O III]/Hβ line ratios for maser and control galaxies. [O III]/Hβ is clearly higher in maser hosts.

3.1. The host characteristics that differentiate between maser and non-maser galaxies

Our comparisons of overall distributions of a variety of optical host properties of maser and non-maser galaxies support previous findings that: 1) maser galaxies show statistically higher levels of reddening and extinction, as measured by the Hα/Hβ Balmer decrements; only the maser galaxies show values of Hα/Hβ > 10; 2) maser galaxies are more massive (higher stellar mass M^*), more luminous (in M_r, not shown here), and harbor more massive BHs in their centers (as shown by the comparison in σ^*). We also show here for the first time that there is a possible connection between the maser activity and the age of the stellar population associated with their host galactic nuclei; the maser galaxies exhibit younger stellar populations (weaker D_n4000, suggesting an order of magnitude difference in age, on average). Figure 3 presents these results.

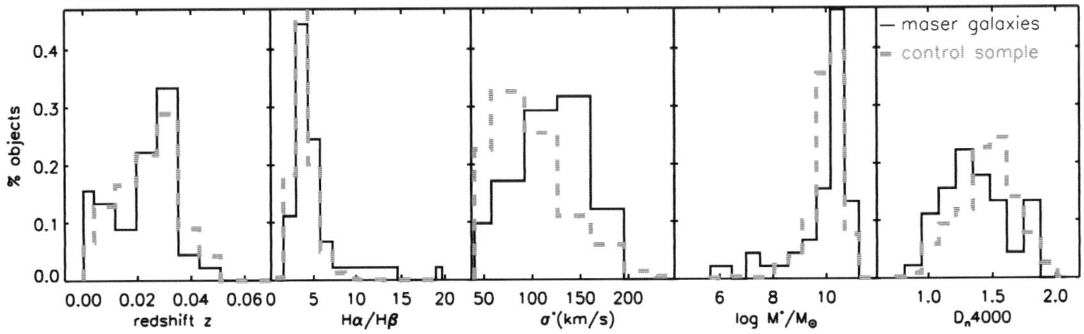

Figure 3. Distributions of redshifts z, Balmer Decrements $H\alpha/H\beta$, stellar velocity dispersions σ^*, host stellar mass M^*, and D_n4000 break index for maser galaxies (black lines) and the control sample (dashed gray lines).

Given the apparent bias of the current maser surveys toward Seyfert galaxies, interpreting these global trends might benefit from looking into how these properties compare when maser and non-maser galaxies are separated per spectral types, as defined based on the line-diagnostic diagrams (Figure 1). Figure 4 illustrates the individual measurements of redshift, Balmer decrements, stellar masses, and D_n4000 parameters, for maser and non-maser galaxies, separated by their respective spectral types H II, T, S, and L. This comparison reveals some features worth mentioning: i) H II's with masers are only at very low z, and only if they are dwarfish, i.e., with very low stellar masses; ii) Ts associated with maser emission are more nearby, less massive and clearly younger in their stellar populations than their counterparts without maser activity; iii) interestingly, Ss with and without masers don't separate much from each other in these host properties; iv) it is only the high-z Ls with the youngest stellar populations that host masers, which are all mega-masers. Interestingly, the mega-maser disks are not among the galaxies with the youngest stellar populations, or among those with extremely large Balmer decrements. It is probably surprising that the only H II hosting mega-maser emission has the highest Balmer decrement in the whole maser galaxy sample; nevertheless, this is not a mega-maser disk.

3.2. The nuclear features that differentiate between maser and non-maser galaxies

A very similar type of comparison for measures that pertain more directly to the process of BH growth in these systems is presented in Figure 5. Here we show individual measurements in 1) stellar velocity dispersion σ^*, as a measure of the BH mass; 2) [O III] line luminosity, L[OIII], which is corrected for reddening and extinction based on a $\tau \propto \lambda^{-0.7}$ attenuation law [8] and the corresponding Balmer decrements; 3) a measure of the Eddington ratio $L/L_{\rm edd}$, where $L = L_{\rm bol} = 600 \times L$[OIII] was used for the bolometric correction [18, 24] and $L_{\rm edd}$ is the Eddington luminosity corresponding to the $M_{\rm BH}$ values obtained via $M_{\rm BH} - \sigma^*$ relation of

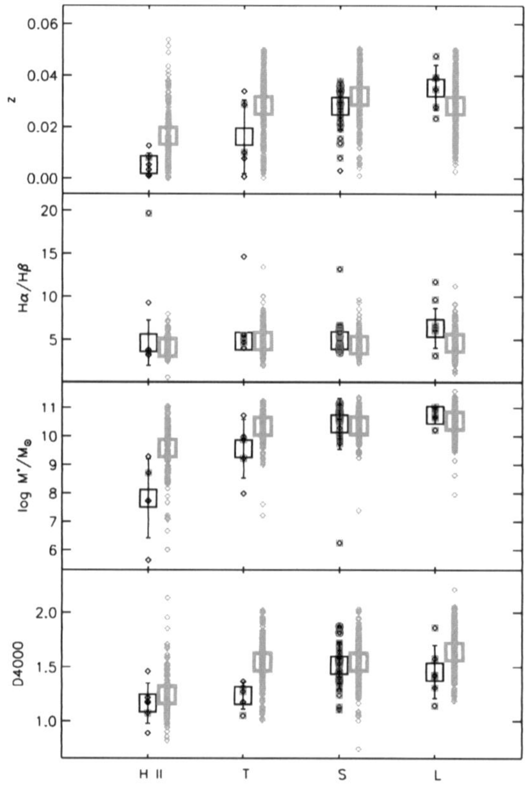

Figure 4. Host properties compared as a function of the different optical spectral types H II, T, S, and L. Symbol types are as in Figure 1. Large open squares indicate averages per spectral type.

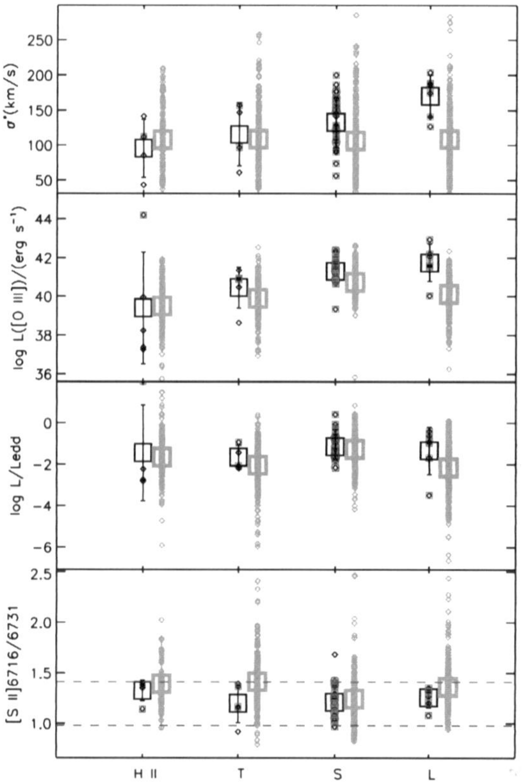

Figure 5. Nuclear properties compared as a function of the different optical spectral types H II, T, S, and L. Symbol types are as in Figure 1. Large open squares indicate averages per spectral type.

Graham et al. [14]; and 4) the [S II] λ 6716/λ6731 line ratio as a proxy for the density n_e of the line-emitting gas [30], which may also be viewed as a gauge of the availability of fuel for accretion and/or masing activity.

We find here that the previously claimed trends for higher BH masses and L[OIII] in maser galaxies [36] are correct not only for the Seyfert hosts, but for galaxies of all spectral types. While it seems especially true that among Ls, maser activity is detected only in systems with the highest L[OIII] and the most massive BHs, Seyferts with and without masers exhibit a relatively weak statistical trend in this direction. Note also that galaxies with very massive BHs ($\sigma^* > 200$ km/s) do not host masers; physical conditions in the vicinity of such systems may not be stable enough to sustain strong mega-maser emission.

Additionally, our comparison, reveals some new and potentially very useful constraints that could be posed to future maser surveys, particularly when searching for mega-maser disks:

1) the combination of $M_{\rm BH}$ and L[OIII] used to calculate $L/L_{\rm edd}$ exposes a potentially narrow range in the accretion rate of nuclei associated with maser emission, which on average is higher

than in the control sample, for all the different spectral types compared here; it seems that the maser activity, and especially the mega-maser disk-like emission, is associated to accretion onto BHs characterized by $logL/L_{\rm edd} \sim -2$. Incidentally, this value also appears to correspond to a change in the accretion flow geometry or mode, from efficient (i.e., Shakura-Sunyaev standard accretion disk/corona, or quasar-like, [34]) to inefficient (i.e., the radiatively inefficient, or advection dominated accretion flow; RIAF/ADAF; e.g., [29]) with decreasing $L/L_{\rm edd}$ [10, 17]. It may thus be possible that the disk mega-maser emission is related to a certain (short) phase in the galaxy-AGN evolution, reflected by this particular change.

2) the comparison of the [S II] λ 6716/λ6731 line ratio unfolds a potential "goldilocks range" in n_e for the maser galaxies, spanning only a range of $\sim 100-300$ cm^{-3} (corresponding to a range in [S II] λ 6716/λ6731 = 1.4 − 1). The galaxies with the highest and lowest n_e in the surveyed sample do not host maser activity at the levels that can be currently detected. Because these measurements of n_e reflect physical conditions of regions located way outside the masing areas, this parameter relates to the maser detection rate only indirectly; nevertheless, this connection is not less essential than that between L[OIII] and the maser activity, which suffers from similar caveats.

4. Summary

We presented here some results of the first multi-parameter comparative analysis of the largest sample available of SDSS-based galaxies with and without maser activity in their centers. Besides confirming previously claimed tendency for high L[OIII] in maser galaxies, we found new criteria for more efficient targeting of maser galaxies, and more importantly, of hosts of mega-maser disks, based on their optical spectral properties. Our study suggests that mega-masers could be found in emission-line galaxies that are not necessarily classified spectroscopically as Seyferts, and that more appropriate survey designs should consider, at least to start with, a cross-match of narrow ranges in their $L/L_{\rm edd}$ values and the [S II] λ 6716/λ6731 line ratios, as described above. Considering a survey of all galaxies included in this study with $-2 < logL/L_{\rm edd} < 0$ and [S II] λ 6716/λ6731 = 1 − 1.4, the maser detection rate is boosted by a factor of 4, to $\sim$ 12%, while the maser disk detection rate increases to > 25%. More details on this analysis, along with a comprehensive Principal Analysis that encompasses a wider variety of parameters than those included in this discussion will be available in Constantin et al [11].

Support for this research has been provided by the Thomas F. Jeffress and Kate Miller Jeffress Memorial Trust, and by the Research Corporation for Science Advancement via a Single Investigator Cottrell College Science Award.

References

[1] Abazajian, K.N., Strauss, M.A., et al., 2009, *ApJS*, **182**, 543
[2] Baldwin, J. A., Phillips, M. M., & Terlevich, R. 1981, *PASP*, **93**, 5
[3] Braatz, J. A., Wilson, A. S., & Henkel, C. 1996, *ApJS*, **106**, 51
[4] Braatz, J. A., Wilson, A. S., & Henkel, C. 1997, *ApJS*, **110**, 321
[5] Braatz, J. A., Henkel, C., Greenhill, L., Moran, J. & Wilson, A., 2004, *ApJ*, **617**, L29
[6] Braatz, J. A., Reid, M. J., Humphreys, E. M. L., Henkel, C., Condon, J. J., & Lo, K. Y., 2010, *ApJ*, **718**, 657
[7] Brinchmann, J., Charlot, S., Heckman, T. M., Kauffmann, G., Tremonti, C., & White, S. D. M. 2004, arXiv:astro-ph/0406220
[8] Charlot, S., & Fall, S. M. 2000, *ApJ*, **539**, 718
[9] Constantin, A., Hoyle, F., & Vogeley, M. S., 2008, *ApJ*, **673**, 715
[10] Constantin et al. 2009, *ApJ*, **705**, 1336
[11] Constantin et al. 2012, in preparation.
[12] Falcke, H., Nagar, N.M., Wilson, A. S., & Ulvestad, J.S., 2000, *ApJ*, **542**, 197
[13] Ferruit, P., Wilson, A. S. & Mulchaey, J., 2000, *ApJ*, **128**, 139

[14] Graham, A.W., Onken, C.A., Athanassoula, E., & Combes, F., 2011, *MNRAS*, **412**, 2211
[15] Greenhill, L.J., Henkel, C., Becker, R., Wilson, T.L. & Wouterloot, J.G.A., 1995, *A&A*, **304**, 21
[16] Greenhill, L. J., Kondratko, P. T., Lovell, J. E. J., Kuiper, T. B. H., Moran, J. M., Jauncey, D. L., & Baines, G. P. 2003, *ApJ*, **582**, L11
[17] Gu, M., & Cao, X. 2009, *MNRAS*, **399**, 349
[18] Heckman, T. M., Kauffmann, G., Brinchmann, J., Charlot, S., Tremonti, C., & White, S. D. M. 2004, *ApJ*, **613**, 109
[19] Herrnstein, J. et al., 1999, *Nature*, **400**, 539
[20] Herrnstein, J., et al. 2005, *ApJ*, **629**, 719
[21] Ho, L.C., 2008, *ARA&A*, **46**, 475
[22] Kauffmann, G., et al. 2003, *MNRAS*, **341**, 33
[23] Kauffmann, G., et al. 2004, *MNRAS*, **353**, 713
[24] Kauffmann, G., & Heckman, T. M. 2009, *MNRAS*, **397**, 135
[25] Kewley, L.J., Groves, B., Kauffmann, G., & Heckman, T., 2006, *MNRAS*, **372**, 961
[26] Kondratko, P. T. et al. 2006, *ApJ*, **638**, 100
[27] Kuo, C. Y., Braatz, J. A., Condon, J. J., Impellizzeri, C. M. V., Lo, K. Y.; Zaw, I., Schenker, M., Henkel, C., Reid, M. J., & Greene, J. E., 2011, *ApJ*, **727**, 20
[28] Miyoshi, M., Moran J.M., Herrnstein J., Greenhill L., Nakai N., et al.1995, *Nature*, **373**, 127
[29] Narayan, R., & Yi, I. 1994, *ApJ*, **428**, 13
[30] Osterbrock, D. E. 1989, *Astrophysics of Gaseous Nebulae and Active Galactic Nuclei* (Mill Valley: University Science Books)
[31] Reid, M. J., Braatz, J. A., Condon, J. J., Greenhill, L. J., Henkel, C., & Lo, K. Y. 2009, *ApJ*, **695**, 287
[32] Schawinski, K., et al. 2007, *MNRAS*, **382**, 1415
[33] Schawinski et al. 2010, *ApJ*, **714**, 108
[34] Shakura, N. I., & Syunyaev, R. A. 1973, *A&A*, **24**, 337
[35] Veilleux, S., & Osterbrock, D. E. 1987, *ApJS*, **63**, 295
[36] Zhu, G., Zaw, I., Blanton, M. R., & Greenhill, L. J., 2011, *ApJ*, **742**, 73

Discovery of an Intermediate Mass Black Hole at the center of the starburst/Seyfert composite galaxy IRAS 01072+4954

M Valencia-S.[1,2], A Eckart[1,2], J Zuther[1], S Fischer[1], S Smajic[1,2], C Iserlohe[1], M García-Marín[1], L Moser[1], M Bremer[1] and M Vitale[1,2]

[1] I.Physikalisches Institut, Universität zu Köln, Zülpicher Str.77, 50937 Köln, Germany
[2] Max-Planck-Institut für Radioastronomie, Auf dem Hügel 69, 53121 Bonn, Germany

E-mail: `mvalencias@ph1.uni-koeln.de`

Abstract. The starburst/Seyfert composite galaxy IRAS 01072+4954 ($z = 0.0236$) is an enigmatic source that combines a Seyfert 1-like X-ray emission with a starburst optical spectrum that lacks broad line emission. We performed High Angular Resolution observations of the central kiloparsec of this galaxy in the near-infrared. Combining our data with 2MASS images of the whole galaxy, we obtain and model the surface brightness profile. We find indications for the presence of an elongated bar-like structure in both data sets. We also model the line of sight velocity distribution of the stars in the bulge. The derived photometrical and kinematical parameters of the bulge are used to evaluate the black hole mass through scaling relations. We find that all reliable estimations of the black hole mass are consistent with the presence of an intermediate mass black hole of $M_{\rm BH} \lesssim 10^5\,{\rm M}_\odot$.

1. Introduction

Tight correlations between the mass of supermassive black holes and the properties of their host galaxies [1, 2, 3, 4, 5] suggest a co-evolution triggered by merging events. At the lower end of the black hole mass range, $M_{\rm BH} < 10^7\,{\rm M}_\odot$, where dwarf galaxies and bulges of late type spirals – among other sources – are located, such correlations are less clear [6, 7, 8, 9, 10, 11]. For example, pseudobulges, barred galaxies and Narrow Line Seyfert 1s (NLSy1s) appear to lie below the $M_{\rm BH} - \sigma_*$ relation established for higher mass systems [12, 13, 14], which has been interpreted as a sign of evolution triggered by secular processes and/or not fully grown systems [15, 16, 17, 18]. Intermediate mass black holes at the center of galaxies (IMBHs; $M_{\rm BH} \lesssim 10^6\,{\rm M}_\odot$) constitute ideal cases to trace the behavior of these correlations at the low-mass end. The characteristics of this population are of particular interest to study galaxy evolution and to put constraints on models of primordial black hole seed formation [19, 20, 21].

During the past decade, IMBHs and their hosts have been studied in the optical – mainly from *SDSS* data, but recently also with images taken with WFPC2 on *HST* – and in the X-rays [22, 23, 24, 25, 14]. We make use of the high angular and spectral resolution that can be achieved in the near-infrared (NIR) when assisted by adaptive optics systems, to discover an IMBH candidate at the center of the starburst/Seyfert composite galaxy IRAS 01072+4954 ($z = 0.0236$). In the optical, this source displays only narrow emission lines and is classified as a transition object in the Baldwin-Phillips-Terlevich diagnostic diagram [26]. This means that the main ionizing source of the emitting species could be star formation, an active nucleus (Seyfert 2) or both. However, in the X-ray domain, it shows an emission typical of Seyfert 1 AGN: steep power-law photon-spectral index ($\Gamma = 2.1$) and very low hydrogen column density ($N_H < 0.04 \times 10^{22}\,\mathrm{cm}^{-2}$). Long- and short-term X-ray flux variations were also detected [27].

In our analysis of its nuclear NIR emission ($r \approx 75$ pc), we find that the main reason for the ambiguous classification is the faintness of the AGN [28]. The low bolometric luminosity, $L_{\mathrm{bol}} \simeq 2.5 \times 10^{42}\,\mathrm{erg\,s^{-1}}$, combined with a very low M_{BH} produces broad emission lines that are very faint and narrow, and therefore not detectable in the optical spectrum taken by Moran et al. [26] from the central $(1 \times 2)\,\mathrm{kpc}^2$ region. The predicted flux and width of the Hα broad component are $F \simeq 5 \times 10^{-14}\,\mathrm{erg\,s^{-1}\,cm^{-2}}$ and $FWHM \sim (400 - 600)\,\mathrm{km\,s^{-1}}$ (determined using the expressions from [29]; see [28] for details). Nevertheless, this mini-AGN seems to accrete at a high rate, $L_{\mathrm{bol}}/L_{\mathrm{Edd}} \simeq 0.2$, which makes it comparable to other NLSy1s (e.g., [13]).

In this paper we focus on the black hole mass determination using K-band integral field observations of the bulge of this galaxy. In the following Sect. 2, we briefly present the data. In Sect. 3, we show the construction and fitting of the surface brightness profile and discuss the probable presence of an inner structure – disk or bar. The fitting of the line-of-sight velocity distribution of the stars in the bulge is presented in Sect. 4. Finally, in Sect. 5 we use those photometric and kinematical measurements to estimate the black hole mass via scaling relations and discuss the results. Conclusions are presented in Sect. 6.

2. The observations

The central $\sim 1\,\mathrm{kpc}^2$ of IRAS 01072+4954 was observed on October 2008 with the Near-Infrared Integral Field Spectrometer NIFS [30] mounted on the "Frederick C. Gillett" Gemini North telescope on Mauna Kea, Hawaii. The adaptative optics (AO) module ALTAIR was used in Laser Guide Star mode. NIFS has a $3'' \times 3''$ field of view and provides high spatial and spectral resolution simultaneously. The observations cover the H- and K-bands, although in this paper we refer only to the K-band data. The spatial resolution achieved with the AO in the K-band was $0.15''$, which corresponds to 72 pc on source. The instrumental spectral resolution is $57\,\mathrm{km\,s^{-1}}$. The data was reduced using the GEMINI IRAF package (released Version 2.14, of September 15, 2008) and standard IRAF tasks. Absolute flux calibration and extinction corrections were also applied to the data cubes.

3. K-band photometry

3.1. Surface brightness profile

The 2MASS [31] provides NIR images that cover the whole galaxy. However, in the K-band, the source is very faint – at the center only $\sim 4 - 5$ times brighter than the background and has a small radius,$\sim 12''$, so that a proper 2D photometric modeling is not possible. We also notice that the bulge size is about the same as the 2MASS K-band point-spread function (PSF), meaning that it is not resolved in the 2MASS image. Therefore, we combine the 2MASS with our NIFS data to construct a 1D surface brightness profile (Fig. 1). We model the bulge as a Sersic profile with effective radius r_e and index n. The disk is described by an exponential profile with scale radius h. A Gaussian with the width of the NIFS PSF represents the emission from the AGN and the hot dust at the center. The photometric parameters resulting from the best fit of these three components are: $r_e = (159.2 \pm 17.8)$ pc, $n = 1.17 \pm 0.08$ and $h = (1.47 \pm 0.53)$ kpc with central surface brightness $I_0 = (74.6 \pm 2.0)\,\mathrm{L_{\odot,K}\,pc^{-2}}$.

The presented model is only a first approximation to the description of the surface brightness. There are mainly two reasons to believe that the bulge contribution found in this way is overestimated. First, the two data sets do not properly overlap with each other. Assuming that the NIFS data represent reasonably well the bulge of the galaxy, and that the emission from $r \gtrsim 2.5''$ of the 2MASS image corresponds to the disk, we convolve the Sersic model to

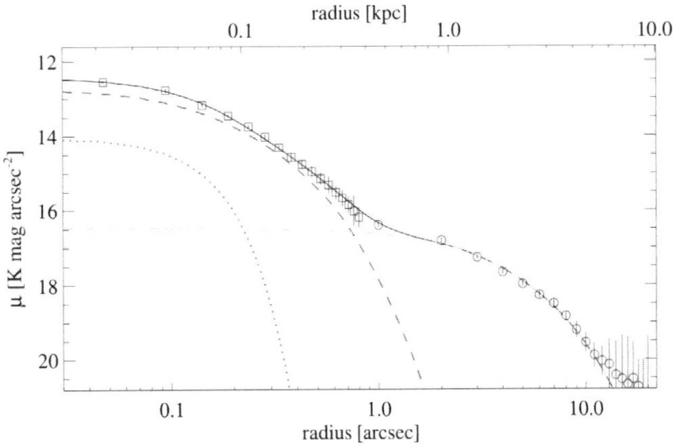

Figure 1. K-band brightness profile of IRAS 01072+4954 obtained by the combination of the NIFS (squares) and 2MASS (circles) data. The dashed red line corresponds to a Sersic profile fitted to the bulge, and the dash-dotted green line to an exponential function that fits the disk component. The dotted magenta line represents the AGN + hot dust contribution, while the solid blue line show the overall fit.

the resolution of the NIFS data and the exponential function to the one of the 2MASS image. However, these components are not independent of each other and the disk contribution at the center might be underestimated. The second one is that IRAS 01072+4954 is a barred galaxy, and although the bar does not seem to dominate the emission at the center – as we show in the following – , its contribution has not been taken into account in the axially symmetric averaged 1D profile.

3.2. IRAS 01072+4954 bar: Photometrical evidence

At first glance, the NIR surface brightness of IRAS 01072+4954 in the 2MASS and in the NIFS data look smooth and uniform. In order to recognize the presence of sub-structures in the disk or in the bulge, we use the unsharp masking method. It essentially consists on subtracting a smoothed image of the galaxy from the original one, adjusting the smoothing function to strengthen the compact structures.

Figure 2 shows in the second panel the result of this process applied to the $15'' \times 15''$ 2MASS K-band image (first panel). The subtracted image was convolved with a 2D Gaussian with a width of $\sigma = 1.5''$. An elongated structure, that resembles a bar, emerges at the center. To evaluate how significant this elongation is, we compare its shape with the 2MASS PSF (image at the bottom right corner on the second panel). Fitting elliptical 2D Gaussians to both shapes, we obtain axis ratios (a/b) of 1.00 for the PSF, and 1.68 with orientation P.A. $= 105°$ for the central structure. This seems to indicate that the shape is intrinsic to the sub-structure. However, the sizes are comparable, implying that the bar-like feature is not resolved in the 2MASS image.

When the same procedure is applied to the $3'' \times 3''$ NIFS image, the AGN and hot dust emission at the center dominate the residuals. Their contribution corresponds to the PSF, which is assumed to be well described by the image of a star close to the field of view. To remove this emission, the PSF is subtracted from the K-band continuum image, in the maximum amount possible such that the residual image is still smooth at the center. Then, the unsharp masking procedure is applied convolving the image with a Gaussian of $\sigma \approx 0.07''$. The second panel in figure 3 shows the results. In the first panel the K-band continuum image, created by taking a median of the NIFS data cube over the wavelength range, is shown. The unsharp masking method reveals an elongated structure. The faint and more extended emission at the center probably corresponds to residuals of the PSF subtraction. When fitting the elongated feature with a 2D Gaussian, we find that the axis ratio $a/b = 1.74$ and the orientation P.A. $= 97°$ are comparable to the values found previously with the 2MASS data. This allows us to suspect that both residuals are related to the same sub-structure. In this case, the sensitivity of the observations is not high enough to obtain a better map of it, and for the same reason, cannot be

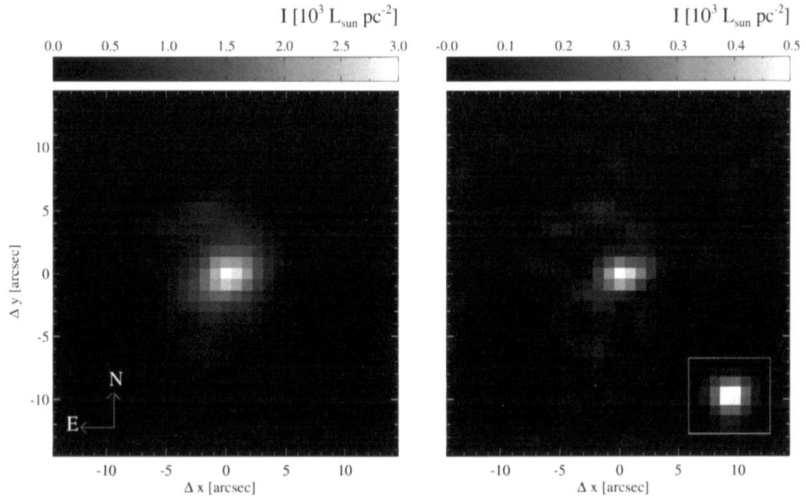

Figure 2. *Left panel:* 15″ × 15″ K-band surface brightness distribution of IRAS 01072+4954 from the 2MASS image. *Right panel:* Result of the unsharp masking procedure applied to the image. A star in the field that represents the 2MASS PSF is shown at the bottom right.

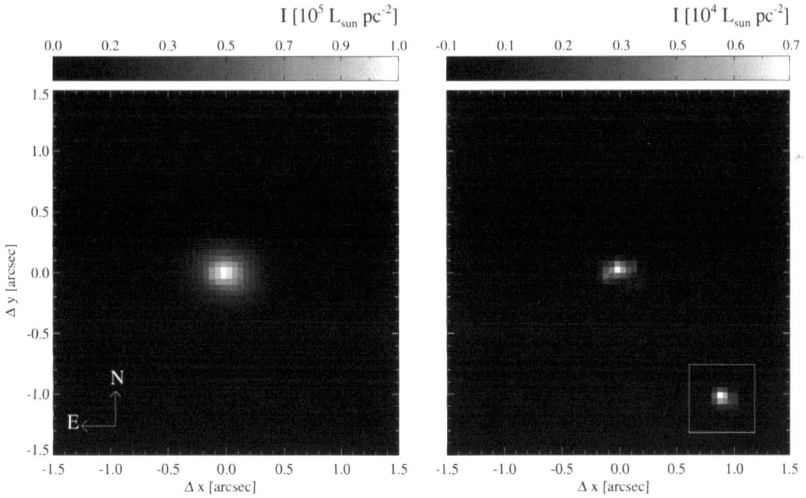

Figure 3. Same as in Fig. 2 but for the 3″ × 3″ NIFS image. The residuals in the right panel are obtained after the removal of the strong PSF contribution (see text for details).

resolved. As shown in Fig. 3, its surface brightness is on average $\sim 3.0 \times 10^3\,L_{\odot,K}\,pc^{-2}$. This is the same brightness as the one observed in the continuum image at about 1 arcsecond from the center. Although it is unlikely that its contribution would change by much more than about one order of magnitude our estimation of the bulge luminosity, an exact estimation is not possible because we are missing all emission fainter than $\sim 16\,\mathrm{mag\,arcsec}^{-2}$ – which includes the disk – in the NIFS observations.

4. Stellar kinematics

4.1. Velocity distribution

We derive the line-of-sight velocity distribution (LOSVD) of the stars in the photometric bulge, by fitting a set of high angular resolution stellar templates to the spectrum integrated over an aperture of $r \approx 0.34''$, which corresponds to the scale radius of the bulge found above. We assume that the stellar LOSVD can be described by a truncated Gauss-Hermite series parametrized by the radial velocity V, the velocity dispersion σ_*, and coefficients of the third and fourth Hermite polynomials h_3 and h_4. The deep and narrow CO absorption features at 2.294 μm and 2.323 μm indicate that the velocity dispersion of the stars in the bulge is low. This implies that, despite the high spectral resolution of the NIFS data, it is not possible to constrain significantly all the LOSVD parameters simultaneously [33]. In this cases, one would like to bias the solution towards the pure Gaussian one and, in that way, avoid the coupling of the parameters, V with h_3 and σ_* with h_4. The penalized pixel fitting routine (pPXF) of Cappellari and Emsellem [33] applies such procedure, allowing to recover the LOSVD in moderately low signal-to noise (S/N) and/or undersampled data. This routine minimizes the difference between the galaxy spectrum and the combination of the stellar templates convolved with the LOSVD. The algorithm allows different weights of the stellar templates. We use a subset of the Gemini spectroscopic library of NIR stars observed with the NIFS IFU in K-band [34]. To reduce the effects of the template mismatch, we select the stellar types that match best the absorption features in both H and K bands by comparing the galaxy spectrum with medium resolution stellar templates of Ivanov et al. [35]. The selected set of stellar spectra consists of 7 objects of the following types: K2III, K5Ib,K5II,K5III, M2III, M3III and M5III.

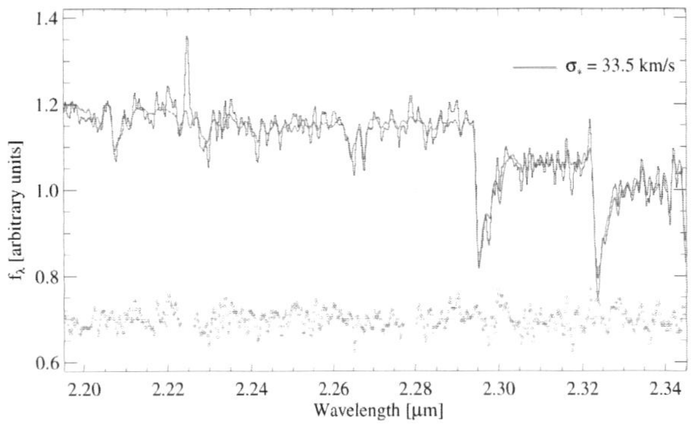

Figure 4. Fit of the stellar kinematics of the spectrum integrated over an aperture $r = r_e = 160$ pc. The observed spectrum is shown as a thin solid line, while the red solid line represents the best fit to the LOSVD. The small spectral regions delimited by the vertical green lines correspond to two H_2 emission lines, which were masked for the fit. The residuals, shown at the bottom, are shifted vertically for clarity.

Figure 4 shows the best fit of the LOSVD with the parameters: $V = (-17.1 \pm 1.6)\,\mathrm{km\,s^{-1}}$, $\sigma_* = (33.5 \pm 3.7)\,\mathrm{km\,s^{-1}}$, $h_3 = 0.06 \pm 0.05$ and $h_4 = 0.01 \pm 0.08$. The fit is obtained after adopting a bias = 0.5 and the addition of a Legendre polynomial of degree 10. In a conservative approach, the errors are calculated via Monte Carlo simulations, where 1000 realizations of an input spectrum are fitted with the pPXF routine without any penalization (bias = 0.0). The input spectrum is created by convolving a stellar template with the LOSVD and adding Poisson noise to the level of the galaxy spectrum, in this case, to reach $S/N \sim 40$. Then, the effects of undersampling are simulated allowing the velocity scale ($\mathrm{km\,s^{-1}}$ per pixel) of the input spectrum to vary in such a way that the second moment of the LOSVD, found by fitting the spectrum with the pPXF routine, varies in the range $\sim 14 - 140\,\mathrm{km\,s^{-1}}$. Given the moderate S/N of the data and that σ_* is low compared to the velocity scale (less than two pixels), the higher moments h_3 and h_4 of the LOSVD suffer from large uncertainties. Fitting only the first two moments we obtain $V = -16.1\,\mathrm{km\,s^{-1}}$ and $\sigma_* = 32.9\,\mathrm{km\,s^{-1}}$, which are within the estimated errors.

4.2. IRAS 01072+4954 bar: Kinematical evidence

Though the bar-like structure is barely detected photometrically, the kinematics of the stars very close to the center are clearly affected by its presence. To elaborate a map of the stellar LOSVD, we bin the data in apertures with three different sizes that have $S/N > 15$ per bin. Then, we fit the LOSVD of the spectra integrated in every aperture using the pPXF routine, and combine the results. This procedure gives us a first estimate of the stellar velocity field (Fig. 5). Special care has to be taken when interpreting the maps because the combination procedure smooths the velocity gradients, and the error of the values increases with the distance to the center. Given the low values of LOS velocity and velocity dispersion, the S/N of single pixel spectra, even after binning the data only information of the central square arcsecond (~ 460 pc) can be recovered.

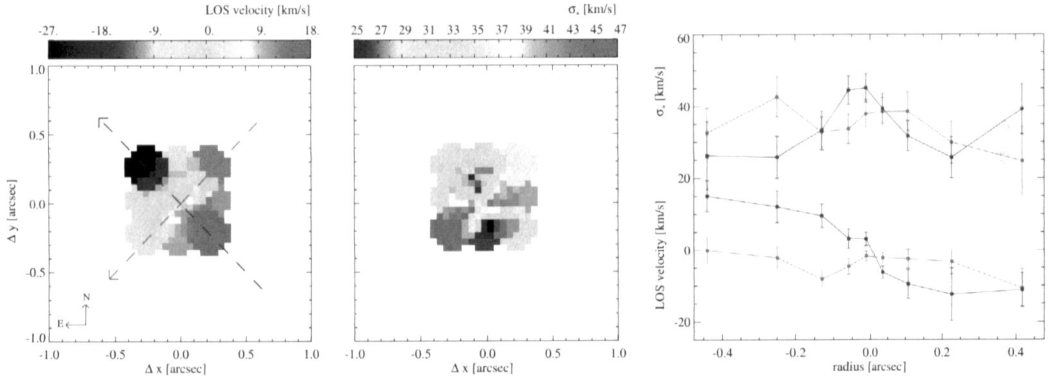

Figure 5. *Left panel:* LOS velocity. *Middle panel:* Velocity dispersion. *Right panel:* LOS velocity and velocity dispersion curves obtained from simulated long-slit spectroscopy along the two axes shown in the left panel. Blue curves+points correspond to the slit along the axis with higher LOS velocity gradient, red curves+points to a slit perpendicular to the previous one. The radius to the center increases from west to east. (These are preliminary results).

First and second panels in Fig. 5 show the stellar LOS velocity and the velocity dispersion maps constructed from the NIFS K-band data cube. The first one displays a characteristic rotation pattern consistent with the expectations for a disk galaxy nearly face-on. In the dispersion map, there is a clear evidence for an internal structure with P.A. $\approx 120°$. To get a better estimation of the rotation curve and the uncertainties, we simulate long slits of ~ 1 arcsec width that cross the center with different orientations. The data of every slit are binned to have the maximum number of points with S/N in the range 10-20. Finally, the pPXF routine is applied tuning the bias in every single bin to obtain the best fit and the uncertainties in each case. The third panel in Fig. 5 shows the derived rotation and dispersion curves for the slit oriented along the highest LOS velocity gradient (P.A. $\approx 45°$) and those for the slit perpendicular to it.

The fact that the velocity dispersion is dominated by the bar-like structure implies that the σ_* estimated above is an upper limit.

5. Discussion: Black hole mass estimations

The mass of the black hole at the center of galaxies has been found to correlate with some properties of the host galaxy bulge. However, barred galaxies show a higher scatter around the scaling relations defined by normal disk galaxies or follow a different relation [6, 7]. It seems that the vertical resonances and/or instabilities responsible for the formation of bars can affect dramatically the velocity distribution of the stars, and favor the development of pseudobulges

[15, 32, 37]. Here we use the derived upper limits of the stellar velocity dispersion and the K-band bulge luminosity to estimate the black hole mass $M_{\rm BH}$. The validity of the scaling relations used is also discussed. Table 1 summarizes the different black hole mass estimations.

5.1. $M_{\rm BH} - L_{\rm K}$

The K-band bulge luminosity, obtained by integrating the Sersic profile that best fit the surface brightness, is $L_{\rm K} = 3.24 \times 10^9\,{\rm L}_{\rm K,\odot}$. As explained previously, we expect this value to be overestimated by ~ 1 order of magnitude. In [28] we show that using the $M_{\rm BH} - L_{\rm K}$ relation from Marconi & Hunt [5], the upper limit on the black hole mass is $2.5 \times 10^6\,{\rm M}_\odot$, while with the Graham updated version of the same relation [38] it is $9.8 \times 10^6\,{\rm M}_\odot$. However, as Graham pointed out (private communication, also [39]), such a relation is not suitable for low-$M_{\rm BH}$, because it has been derived mainly based on luminous core galaxies. Core galaxies follow a luminosity-dispersion $L \propto \sigma_*^5$ relation and their $M_{\rm BH} \propto L^1$, probably as a result of a dry merging formation process [39] (see also [40]). In contrast, non-barred low-$M_{\rm BH}$ Sersic galaxies follow a $M_{\rm BH} \propto L^{2.5}$ relation. Given the lack of a study that investigates the $M_{\rm BH} - L_{\rm K}$ in such systems, we make the exercise to bend the known $M_{\rm BH} - L_{\rm K}$ relation (Eq.14 of [38]) at $\sim 7 \times 10^7\,{\rm M}_\odot$ assuming the stated proportionality. With the resulting relation, $\log(M_{\rm BH}/\,{\rm M}_\odot) \sim 2.5 \log(L_{\rm K}/{\rm L}_{\rm K,\odot}) - 18.23$, we obtain a black hole mass of $3.5 \times 10^5\,{\rm M}_\odot$.

5.2. $M_{\rm BH} - \sigma_*$

Recently Gültekin et al. [36] and Graham et al. [37] have recalibrated the $M_{\rm BH} - \sigma_*$. They use 49 objects and 18 upper limits, and 64 objects, respectively, with reliable parameter estimations. Also Xiao et al. [14] studied this relation, focusing on low-$M_{\rm BH}$ sources. We evaluate the black hole mass using their relations and report the results in Table 1.

Table 1. Black hole mass estimations from scaling relations.

$\log\left(\frac{M_{\rm BH}}{{\rm M}_\odot}\right)$	notes on the applied scaling relation	ref.
	$M_{\rm BH} - L_{\rm K}$	
6.40	27 bright, mainly early type objects. Not suitable in the case (see text).	[5]
7.00	22 bright objects. Not suitable in the case (see text).	[38]
5.54	Bending the Eq.14 of [38]. See text for details.	—
	$M_{\rm BH} - \sigma_*$	
4.83	Full sample. Intrinsic scatter 0.44 dex.	[36]
6.83	8 + 11 upper limits barred galaxies. Intrinsic scatter 0.17 dex.	[36]
3.53	Full sample. Intrinsic scatter 0.35 dex.	[37]
3.75	20 barred galaxies. Intrinsic scatter 0.31	[37]
5.10	155 low-$M_{\rm BH}$ sources. Intrinsic scatter 0.46	[14]
4.61	25 low-$M_{\rm BH}$ barred galaxies. Intrinsic scatter 0.46	[14]

From the black hole masses obtained through the $M_{\rm BH} - \sigma_*$ relation, the highest value corresponds to the one calculated from the relation found by Gültekin et al. for barred galaxies. However, when looking at the location of these sources in the $M_{\rm BH}$ vs. σ_* diagram (their Fig 1.), the 8 galaxies with dynamically detected black holes lie well within the scatter of the correlation depicted by the full sample (49 sources). Nevertheless, the $M_{\rm BH}$ calculated with the relation fitted to the latter (+ 18 upper limits) is two orders of magnitude lower. The authors also assign a probability of 0.1809 to the case of the upper limit sources not having a black hole. In contrast, the same probability stated for the full sample is 0.0004. We conclude that the difference between the two estimations is introduced by the inclusion of the upper limits to the fit which, in the case of the barred galaxies, dominate the subsample. Therefore, we do not consider that black hole mass estimation as reliable. All other estimations of the $M_{\rm BH}$ are consistent with the hypothesis of the existence of an IMBH at the center of IRAS 01072+4954. In any case, it is interesting to notice the differences in the $M_{\rm BH}$ calculated from various versions of the scaling relations. This of course is related to the sample selection, the method applied to fit the data, and the treatment of the errors.

The question of whether the emission detected in the NIFS data corresponds to a classical bulge or to a pseudobulge is still open. High angular resolution observations of the whole galaxy or a major portion of it are necessary to do a proper decomposition of the photometrical components and derive more accurate bulge parameters.

6. Summary and conclusions

IRAS 01072+4954 is a starburst/Seyfert galaxy that hosts an active nucleus detected in the NIR and in X-rays [28, 27]. Here we expose the methods employed to get photometrical and kinematical information of the bulge, a more detailed description will be published in a forthcoming paper.

We present evidence for the presence of an IMBH at the center of IRAS 01072+4954. We make use of scaling relations of the black hole mass with the K-band bulge luminosity and with the stellar velocity dispersion recently calibrated with different samples. All the reliable $M_{\rm BH}$ estimations derived from those relations are consistent with $M_{\rm BH} \lesssim 10^5\,{\rm M}_\odot$.

We also find indications for the presence of a bar-like structure with P.A. $\sim$ 100°. This implies an overestimation of the bulge luminosity, the velocity dispersion and, consequently, fn the black hole mass.

Acknowledgments

M. Valencia-S thanks J. Stern and A. Laor for constructive discussions during the AHAR Conference, and also A. Graham for his comments, in particular referring to the suitability of the scaling relations to derive black hole masses. M. Valencia-S., S. Smajic and M. Vitale are members of the International Max-Planck Research School (IMPRS) for Astronomy and Astrophysics at the Universities of Bonn and Cologne supported by the Max Planck Society. J. Zuther, S. Fischer and M. Bremer gladly acknowledge the financial support by Group of Eight Australia-Germany Joint Research Co-operation Scheme via project-ID 50753527. M. García-Marín is supported by the German federal department for education and research (BMBF) under the project No. 50OS1101. Part of this work was supported by the German Deutsche Forschungsgemeinschaft, DFG, via grant SFB 956 and fruitful discussions with members of the European Union funded COST Action MP0905: Black Holes in a violent Universe and PECS project No. 98040. Based on observations (Programm ID: GN-2008B-Q-77) obtained at the Gemini Observatory, which is operated by the Association of Universities for Research in Astronomy, Inc., under a cooperative agreement with the NSF on behalf of the Gemini partnership: the National Science Foundation (United States), the Science and Technology Facilities Council (United Kingdom), the National Research Council (Canada), CONICYT (Chile), the Australian Research Council (Australia), Ministério da Ciência, Tecnologia e Inovação (Brazil) and Ministerio de Ciencia, Tecnología e Innovación Productiva (Argentina).

References

[1] Magorrian J, Tremaine S, Richstone D, et al. 1998 *AJ* **115** 2285

[2] Ferrarese L and Merritt D 2000 *ApJ* **539** L9
[3] Gebhardt K, Bender R, Bower G, et al. 2000 *ApJ* **539** L13
[4] Graham A W, Erwin P, Caon N and Trujillo I 2001 *ApJ* **563** L11
[5] Marconi A and Hunt L K 2003 *ApJ* **589** L21
[6] Hu J 2008 *MNRAS* **386** 2242
[7] Graham A W 2008 *ApJ* **680** 143
[8] Greene J E, Peng C Y, Kim M, et al. 2010 *ApJ* **721** 26
[9] Kormendy J and Bender R 2011 *Nature* **469** 377
[10] Graham A W 2011 *Preprint*: astro-ph/1103.0525
[11] Jiang Y -F, Greene J E and Ho L C 2011 *ApJ* **737** L45
[12] Greene J E, Ho L C and Barth A J 2008 *ApJ* **688** 159
[13] Zhou H, Wang T, Yuan W, et al. 2006 *ApJS* **166** 128
[14] Xiao T, Barth A J, Greene J E, et al. 2011 *ApJ* **739** 28
[15] Kormendy J and Kennicutt Jr. R C 2004 *ARA&A* **42** 603
[16] Ryan C J, De Robertis M M, Virani S, Laor A and Dawson P C 2007 *ApJ* **654** 799
[17] Orban de Xivry G, Davies R, Schartmann M, et al. 2011 *MNRAS* **417** 2721
[18] Mathur S, Fields D, Peterson B M, and Grupe D 2011 *Preprint*: astro-ph/1102.0537
[19] Madau P and Rees M J 2001 *ApJ* **551** L27
[20] Volonteri M, Haardt F and Madau P 2003 *ApJ* **582** 559
[21] Volonteri M and Natarajan P 2009 *MNRAS* **400** 1911
[22] Greene J E and Ho L C 2005 *ApJ* **627** 721
[23] Greene J E and Ho L C 2007 *ApJ* **656** 84
[24] Dewangan G C, Mathur S, Griffiths R E and Rao A R 2008 *ApJ* **689** 762
[25] Ai Y L, Yuan W, Zhou H Y, Wang T G and Zhang S H 2011 *ApJ* **727** 31
[26] Moran E C, Halpern J P and Helfand D J 1996 *ApJS* **106** 341
[27] Panessa F, Wolter A, Pellegrini S, et al. 2005 *ApJ* **631** 707
[28] Valencia-S. M, Zuther A, Eckart A, Iserlohe C and Gillian G 2012 *Preprint*: astro-ph/1203.6063
[29] Stern J and Laor A 2012 *MNRAS* (accepted; *Preprint*: astro-ph/1203.3158)
[30] McGregor P J, Hart J, Conroy P G, et al. 2003 *Society of Photo-Optical Instrumentation Engineers (SPIE) Conf. Series* Vol. 4841, ed. M Iye and A F M Moorwood, p 15811591
[31] Skrutskie M F, Cutri R M, Stiening R, et al. 2006 *AJ* **131** 1163
[32] Athanassoula E 2005 *MNRAS* **358** 1477
[33] Cappellari M and Emsellem E 2004 *PASP* **116** 138
[34] Winge C, Riffel R A and Storchi-Bergmann T 2009 *ApJS* **185** 186
[35] Ivanov V D, Rieke M J, Engelbracht C W. et al. 2004 *ApJS* **151** 387
[36] Gültekin K, Richstone D O, Gebhardt K, et al. 2009 *ApJ* **698** 198
[37] Graham A W, Onken C A, Athanassoula E and Combes F 2011 *MNRAS* **412** 2211
[38] Graham A W 2007 *MNRAS* **379** 711
[39] Graham A W 2012 *ApJ* **746** 113
[40] Kormendy J, Fisher D B, Cornell M E and Bender R 2009 *ApJSS* **182** 216

Simulations of Nuclear Star-Forming Rings: A Case of the Milky Way

Sungsoo S. Kim, Ansun Lee

Kyung Hee University, Dept. of Astronomy and Space Science, Yongin-shi, Kyungki-do 446-701, Korea

E-mail: sungsoo.kim@khu.ac.kr

Abstract. Gas materials in the inner Galactic disk continuously migrate toward the Galactic center (GC) due to interactions with the bar potential, magnetic fields, stars, and other gaseous materials. Those in forms of molecules appear to accumulate around 200 pc from the center (the central molecular zone, CMZ) to form stars there and further inside. The bar potential in the GC is thought to be responsible for such acculmulation of molecules and subsequent star formation, which is believed to have been continous throughout the lifetime of the Galaxy. We present hydrodynamic simulations of gas clouds in the central kpc region of the Milky Way that is modeled with a three-dimensional bar potential. Our simulations consider realistic gas cooling and heating, star formation, and supernova feedback. A torus of dense gas clouds forms as a result of X_1–X_2 orbit transfer, and its size ($\sim$ 200 pc radius) coincides with the extraordinary reservoir of dense molecular clouds in the inner bulge, the Central Molecular Zone (CMZ). We also present some results from our preliminary simulations for gas transportation from the CMZ to the circumnuclear disk of molecular clouds located at a few parsecs from the GC.

1. Introduction

The CO emission survey along the Galactic plane shows that molecular gas is abundant down to a Galactocentric radius R_G of ~ 3 kpc. Inside this radius, the gas content is mostly in forms of atoms and there is no noticeable star formation activities. However, a significant amount of molecular gas, as well as various evidences of recent and current star formation, appears again inside a projected R_g of $\sim$ 200 pc. This distribution of molecular gas is called the Central Molecular Zone (CMZ). The observed longitude-velocity diagram of the molecular emission in the CMZ is generally interpreted as a result of torus-like distribution of molecules with an outer radius of $\sim$ 200 pc. The total gas mass in the CMZ is estimated to be $\sim 5 \times 10^7\,M_\odot$ [7].

The gas content in the CMZ is thought to have migrated inward to the current location from the Galactic disk. Serabyn & Morris [11] enumerate mechanisms for the inward transport of gas: shear viscosity due to the differential rotation of gas disk, compression and shocks associated with the elongated or cusped stable orbits in a non-axisymmetric potential, dynamical friction with the field stars, magnetic field viscosity, and the dilution of specific angular momentum by stellar mass loss material from outer bulge [4].

The transition from atomic to molecular status around 200 pc is believed to be responsible for the characteristics of the stable orbits in a bar potential [1]. Bar potentials have several distinct families of stable orbits, among which X_1 and X_2 orbit families are related to the discussion here: X_1, the outermost orbit family, is enlongated along the bar's major axis, wherease X_2

is enlongated along the bar's minor axis at deeper inside the potential. Hydrodynamic effects cause gas particles to generally move along stable orbits, but the innermost X_1 oribts are sharply cuspsed or even self-intersecting at the apocenters. Gas is compressed or even undergoes a shock in these regions, loses its orbital energy, and falls inward to settle onto an X_2 orbit inside. Such compression and subsequent cooling will transform the mostly atomic gas to molecular clouds.

Some of the accumulated molecular gas will keep moving inward and reach the CircumNuclear Disk (CND) of molecular clouds at R_G of few parsecs and/or eventually sink to the central supermassive black hole, while some will collapse and form stars in the CMZ. If the nonsphericity of the Galactic potential in the inner bulge has been significant enough to give rise to X_1, X_2 orbit families for the lifetime of the Galaxy, a fair amount of stars must have formed in the CMZ so far. Serabyn & Morris [11] argue that the sustained star formation in the CMZ has resulted in a cusp in the stellar number density profile with R_G of 100–200 pc.

There have been a few hydrodynamic studies on the gas inflow in the inner bulge of galaxies. Jenkins & Binney [4] were the first who numerically examed the idea of Binner et al. [1] with 2-D sticky particle simulations and confirmed that the transition of gas motion from X_1 to X_2 orbits indeed takes place in a bar potential. Englmaier & Gerhard [2] were able to observe the same pheonomenon in their larger-scale, 2-D SPH simulations of gas dynamics in the Milky Way. [8] explored a large parameter space to determine whether and at what radius a nuclear ring of gas forms with 2-D, grid-based simulations. Rodriguez-Fernandez & Combes [9] performed 2-D hydrodynamic simulations (particle motions were followed in 3-D, though) of the inner Galaxy to study the gas dynamics in that region with a secondary bar taken into account. All of these studies primarily concentrated on the gas flows and did not consider star formation in their simulations.

In the present Proceedings paper, we present 3-D hydrodynamic simualtions of gas particles in the inner bulge of the Galaxy that consider the effects of star formation, supernova feedback, and realistic gas cooling and heating. We will show that a bar potential that has a X_1–X_2 transition at $\sim$ 200 pc can indeed compress gas sufficiently enough to form stars and that the obtained star formation rates are consistent with observations. We also present some results from our preliminary simulations for gas transportation from the CMZ to the circumnuclear disk of molecular clouds located at a few parsecs from the GC.[1]

2. Simulations

We use a parallel tree SPH code ASURA (Saitoh, in preparation) that can utilize the special-purpose hardware GRAPE or an optimally-tuned gravity calculation library, Phantom-GRAPE.[2] We use an opening angle of $\theta = 0.5$ for a cell opening criterion, and the kernel size of an SPH particle is determined by imposing the number of neighbors to be 32 ± 2. We use a cooling function by [12] for gas with the solar metallicity for a temperature range of 10–10^8 K. A uniform heating from far-UV radiation is considered with a value observed in the Solar neighborhood [14].

A star particle is spawned when a gas particle satisfies all three following conditions: 1) the hydrogen number density is larger than a threshold value ($n_H > n_{th}$), 2) the temperature is lower than a threshold value ($T < T_{th}$), and 3) the flow is converging ($\nabla \cdot v < 0$). We adopt $n_{th} = 100$ cm^{-3} and $T_{th} = 100$ K. The effect of supernova feedback is implemented in a probabilistic manner. We assume that stars more massive than $8\,M_\odot$ explode as Type II supernovae and that each explosion outputs 10^{51} ergs of thermal energy into the surrounding 32 SPH particles. Detailed discussion on the choice of n_{th} and T_{th}, spawning of star particles, and supernova feedback is given in [10].

[1] The first half of this Proceedings paper (the CMZ simulations) has been published as a *Letter* [5].
[2] The library can be obtained at http://grape.mtk.nao.ac.jp/ñitadori/phantom/.

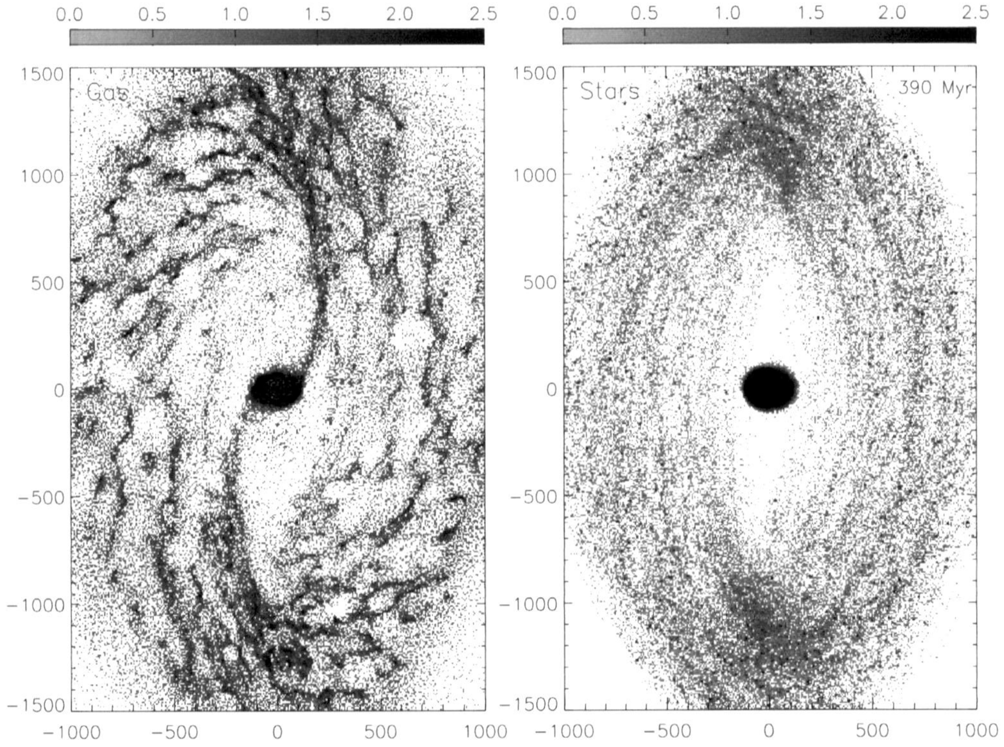

Figure 1. Distribution of gas (*left*) and star (*right*) particles in our standard run at $T = 390$ Myr. Length units ar in parsecs. Stars are predominantly formed in the central 200 pc.

For the Galactic potential, we adopt a power-law density distribution with an $m = 2$ bar for the bulge,

$$\rho = \rho_0 \left(\frac{r}{r_0}\right)^{-\alpha} [1 + b_{22} P_{22}(\cos\theta)\cos 2\phi] , \tag{1}$$

where P is the associated Legendre function. Our standard run has the following parameters: $\alpha = 1.65$, $b_{22} = 0.1$, $\rho_0 = 40\,\mathrm{M_\odot\ pc^{-3}}$, $r_0 = 100$ pc, $M_d = 4 \times 10^{10}\,\mathrm{M_\odot}$, $R_d = 3.5$ kpc, and $z_d = 40$ pc.

Initially, the simulation has gas particles only. Our standard run has the following initial distribution of particles: They are on one of the X_1 orbits whose semi-minor axes range from 300 to 1200 pc. The number of particles on each orbit is proportional to the length of the orbit, and on a given orbit, the particles are spaced over the same time-interval. The vertical distribution follows a Gaussian function with a scale height of 40 pc. The standard run has 10^5 gas particles, and the total gas mass is $5 \times 10^7\,\mathrm{M_\odot}$, thus each gas particle initially has a mass of $500\,\mathrm{M_\odot}$.

We evolve the system without cooling for the first 50 Myr to have a relaxed particle distribution.

3. Results

Figure 1 shows our standard run at $T = 390$ Myr. About a half of the gas particles that were initially on X_1 orbits have migrated inward to the CMZ, and the majority of newly formed stars remain in the central 200 pc. This implies that if gas has been supplied down to the inner bulge

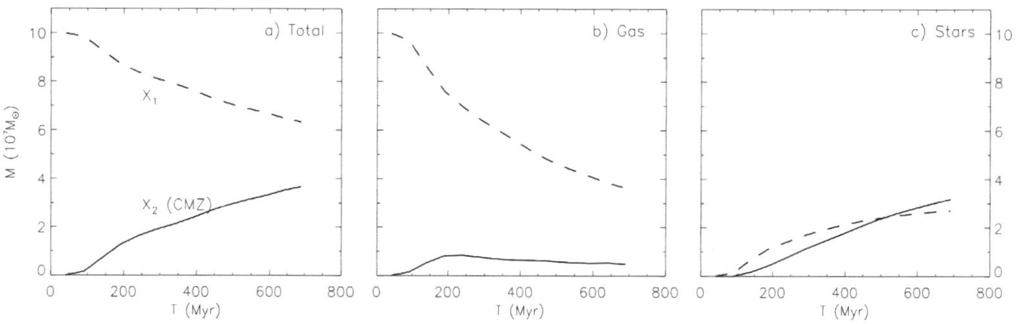

Figure 2. Evolution of total masses in gas (*left*) and stars (*right*) for our standard run. The masses are in units of initial total mass.

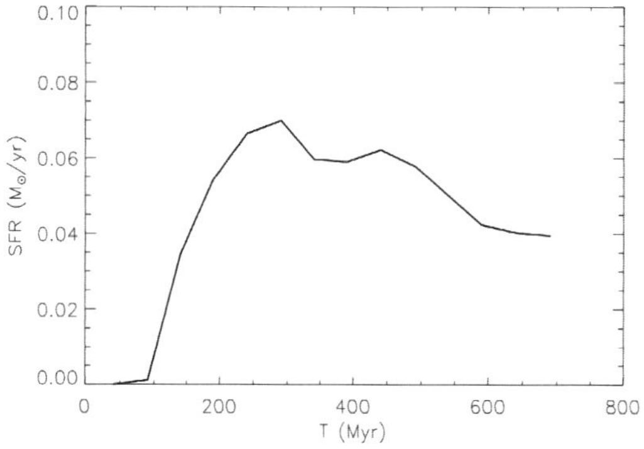

Figure 3. Star formation rate in our standard run. The obtained SFRs are very similar to the values estimated from mid-infrared observations by [13], 0.04–0.08 $M_\odot$ yr^{-1}.

region from the disk throughout the lifetime of the Galaxy, then indeed a significant amount of stars would have been born in the CMZ and a resulting stellar population would form a central cusp that is distinctive from the larger bulge.

Evolution of the masses in gas and stars for two different radial zones are plotted for our standard run in Figure 2. The evolutionary aspects of the masses do not change much after $T = 200$ Myr: the gas mass inside 250 pc (denoted by X_2) and the stellar mass outside 250 pc (denoted by X_1) are nearly constant while the gas mass outside 250 pc and the stellar mass inside 250 pc increase or decrease quite linearly. The (nearly) constant amount of gas in the CMZ implies that there is a threshold gas mass in the CMZ above which star formation becomes very efficient. On the other hand, the significantly suppressed star formation rate outside CMZ at later stages is probably a consequence of insufficient gas in that region.

Figure 3 shows the evolution of overall star formation rate (SFR) in our standard run. The SFR increases rather steeply during the first 200 Myr because there are not yet many supernova explosions that increase the temperature of nearby gas particles. During the later half of the simulation, most of the star formation takes place in the CMZ, thus the obtained SFR values of 0.04–0.06 $M_\odot$ yr^{-1} can be regarded as the SFR in the CMZ. These values are very close to the recent SFR estimated from mid-infrared observations by [13], 0.04–0.08 $M_\odot$ yr^{-1}.

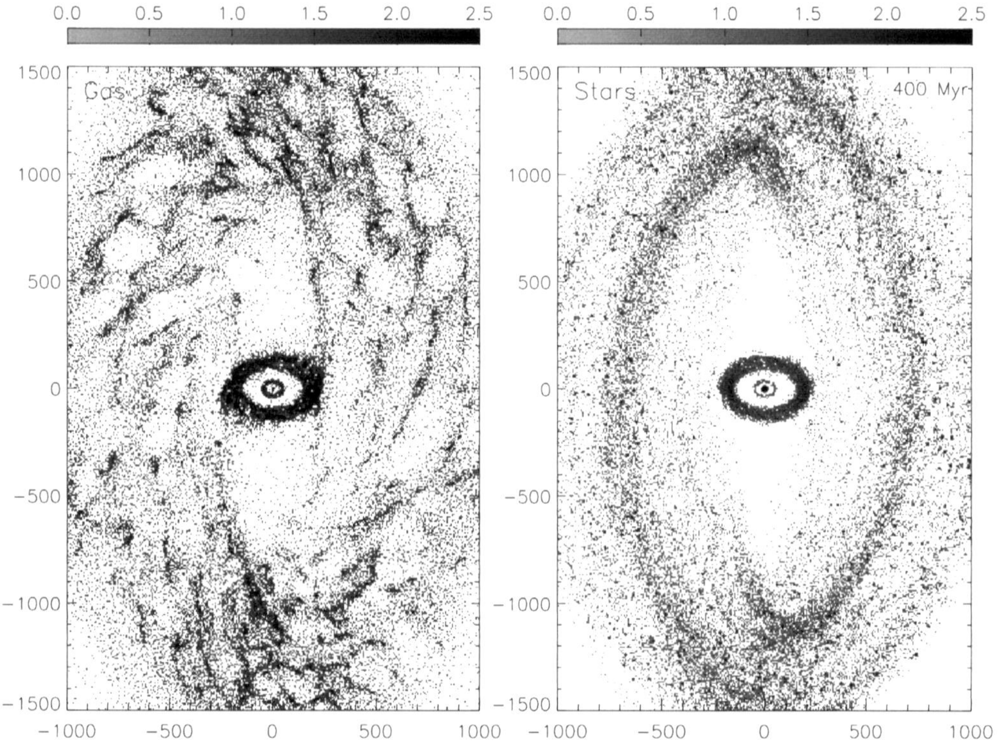

Figure 4. Distribution of gas (*left*) and star (*right*) particles in our double-bar simulation at $T = 400$ Myr. Length units ar in parsecs.

4. From CMZ to CND

Some of the accumulated gas in the CMZ will keep migrating inward and reach the CircumNuclear Disk (CND) of molecular clouds at $r = 2$–5 pc, and a part of it will sink further to form stars in the central parsec or feed the central supermassive black hole. One possible way to transfer gas from the CMZ to the CND would be a nested bar, as examined by [6]. We have performed a simple simulation of gas flow in double-bar potential as a **toy model**. Figure 4 shows a snapshot at $T = 400$ Myr from our double-bar simulation, which has an inner bar that is modeled with a $n = 1$ Ferrers density profile [3],

$$\rho = \rho_0 \left(1 - \frac{x^2}{a} - \frac{y^2}{b} - \frac{z^2}{c}\right)^n , \tag{2}$$

where $a = 180$, $b = 60$, $c = 60$ pc, and the total mass of the bar is $10^8\,M_\odot$ (the inner bar is rotating at the same pattern speed with the main bar with an angular separation of 0 degrees). Two inner rings form at $r \sim 50$ pc and ~ 10 pc, and the smaller inner ring would correspond to the CND. There is no known observed features that would correspond to our larger inner ring, but it would be not easy to observationally identify such a ring in that region if the density of the ring is much lower than the CMZ.

5. Summary

We performed 3-D hydrodynamic simulations of the CMZ in a bar potential with consideration of self-gravity, radiative cooling, and supernova feedback. Important findings are 1) that gas particles are efficiently transported from X_1 to X_2 orbits, 2) that stars are formed predominantly in the central 200 pc, 3) that the star formation rates obtained from our simulations are consistent with the recent estimates from mid-infrared observations by [13]. These results strongly support the idea that star formation has taken place in the CMZ throughout the lifetime of the Galaxy, and the resulting stellar population is the stellar cusp at the central 100-200 pc region of the Galaxy. We also find that a significant mass of gas can be transported to the CND region by the existence of an inner bar.

References

[1] Binney J, Gerhard O E, Stark A A, Bally J and Uchida K I 1991 *MNRAS* **252** 210
[2] Englmaier P and Gerhard O 1999 *MNRAS* **304** 512
[3] Ferres N 1877 *Q. J. Pure Appl. Math.* **14** 1
[4] Jenkins A and Binney J 1994 *MNRAS* **270** 703
[5] Kim S S, Saitoh T R, Jeon M, Figer D F, Merritt D and Wada K 2011 *ApJ* **735** L11
[6] Namekata D, Habe A, Matsui H and Saitoh T R 2009 *ApJ* **691** 1525
[7] Pierce-Price D, Richer J S, Greaves J S, Holland W S, Jenness T, Lasenby A N, White G J, Matthews H E, Ward-Thompson D, Dent W R F, Zylka R, Mezger P, Hasegawa T, Oka T, Omont A and Gilmore G 2000 *ApJ* **545** L121
[8] Regan M W and Teuben P 2003 *ApJ* **582** 723
[9] Rodriguez-Fernandez N J and Combes F 2008 *A&A* **489** 115
[10] Saitoh T R, Daisaka H, Kokubo E, Makino J, Okamoto T, Tomisaka K, Wada K and Yoshida N 2008 *PASJ* **60** 667
[11] Serabyn E and Morris M 1996 *Nature* **382** 602
[12] Spaans M and Norman C A 1997 *ApJ* **483** 87
[13] Yusef-Zadeh F, Hewitt J W, Arendt R G, Whitney B, Rieke G, Wardle M, Hinz J L, Stolovy S, Lang C C, Burton M G and Ramirez S 2009 *ApJ* **702** 178
[14] Wolfire M G, Hollenbach D, McKee C F, Tielens A G G M and Bakes E L O 1995 *ApJ* **443** 152

The feeding of activity in galaxies: a molecular line perspective

Santiago García-Burillo[1], **Francoise Combes**[2]

[1], OAN, Observatorio de Madrid, Alfonso XII, 3, 28014, Madrid, Spain
[2], OAN, Observatoire de Paris, LERMA, CNRS, 61, Av de lObservatoire, 75014, Paris, France

E-mail: `s.gburillo@oan.es`, `francoise.combes@obspm.fr`

Abstract.
What are the main drivers of activity in the local universe? Observations have been instrumental in identifying the mechanisms responsible for fueling activity in galaxy nuclei. In this context we summarize the main results of the NUclei of GAlaxies (NUGA) survey. The aim of NUGA is to map, at high resolution and high sensitivity, the distribution and dynamics of the molecular gas in the central kiloparsec region of 25 galaxies, and to study the different mechanisms responsible for gas fueling of low-luminosity AGNs (LLAGN). Gas flows in NUGA maps reveal a wide range of instabilities. The derived gravity torque maps show that only $\sim$1/3 of NUGA galaxies show evidence of ongoing fueling. Secular evolution and dynamical decoupling are seen to be key ingredients to understand the AGN fueling cycle. We discuss the future prospects for this research field with the advent of instruments like the Atacama Large Millimeter Array (ALMA).

1. The fueling problem

Active Galactic Nuclei (AGN) must be fueled with material which lies originally far away from the gravitational influence of the supermassive black hole. On its way to the sphere of influence of the central engine the gas must lose virtually all of its angular momentum during the process. The feeding requirements, dM/dt in $M_\odot$ yr^{-1} units, are markedly different in high-luminosity AGNs (QSOs: 10–100$M_\odot$ yr^{-1}) and low-luminosity AGNs (LLAGNS: Seyferts and LINERs: 10^{-5}–$10^{-2}M_\odot$ yr^{-1}) [1]. However, observational studies have shown that while in high-luminosity AGNs, the most demanding in terms of feeding requirements, kiloparsec-scale perturbations produced by bars and galaxy interactions are clearly related to the onset of activity [2], such correlation is marginal if any in the case of LLAGNs [3] [4] [5]. The search of a universal feeding agent in LLAGNs is seen to be challenging. Large-scale bars, bars within bars, lopsided $m = 1$ instabilities, warps, nuclear spirals, and winds from stars, among other mechanisms, have been invoked as possible feeding agents in LLAGNs. On the other hand, there is growing evidence, based on theoretical models and state-of-the-art numerical simulations that several mechanisms, and not a single mechanism, might be at work and that these operate at different spatial scales [6] [7] [8]. To further complicate this picture in the case of LLAGNs, the duty cycle of activity is expected to be very short and the associated feeding event could be of chaotic/intermittent nature [9][6]. While various models predictions are still debated in the literature, there is still ample room for improvement in the picture drawn from observations.

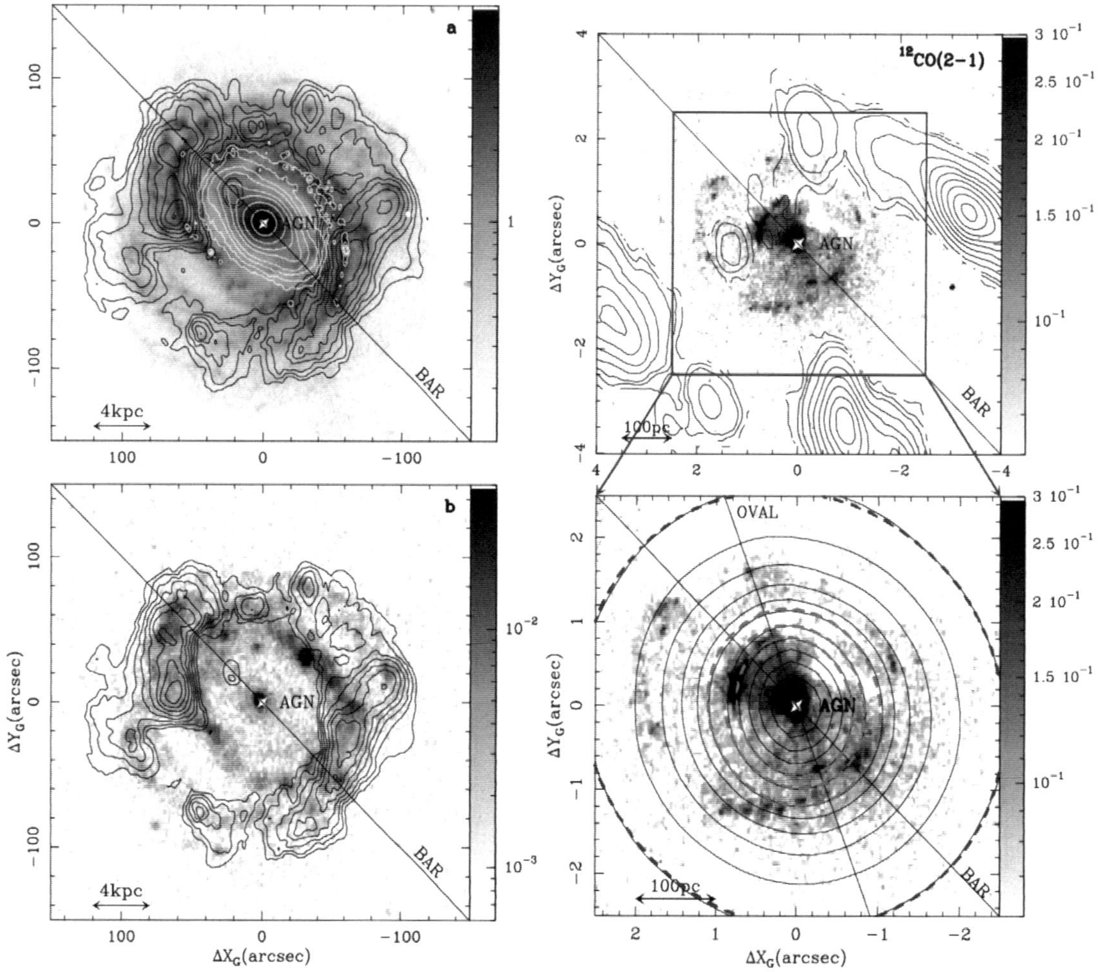

Figure 1. a) (*Left upper panel*) Overlay of the HI map (black contours), the 3.6 m Spitzer/IRAC image (white contours) and the Spitzer/IRAC image (grey scale) of NGC 4579. The large-scale bar detected in IRAC image-BAR- is identified. A pseudo-ring is detected at UHR in HI emission. b) (*Left lower panel*) Same as a) but HI intensity map (in contours) superposed on the FUV image (grey scale). c) (*Right upper panel*) The ^{12}CO(2–1) NUGA map of NGC 4579 is overlaid on the UV image of the central region (in grey scale). d) (*Right lower panel*) The K-band contours are overlaid on the UV-image of the galaxy. We show the orientation of the bar and the oval. All images have been deprojected onto the galaxy plane. Figure adapted from [14].

2. An observational perspective: The NUclei of GAlaxies (NUGA) survey

The NUclei of GAlaxies (NUGA) survey is a high-spatial resolution (~0.5"-1") and high-dynamic range interferometer CO survey of 25 nearby LLAGNs conducted with the IRAM array [10][11]. The aim of this survey is to probe the critical scales for angular momentum transfer (<10–100 pc at D~5–30 Mpc) in a significant sample of galaxies that include Seyferts, LINERs and transition objects. The high sensitivity/resolution of these observations make possible a detailed study of the distribution and kinematics of molecular gas in the circumnuclear disks of these galaxies. Molecular line maps are used to track down evidence of ongoing feeding.

Together with the CO NUGA maps, we also have access to high-resolution NIR maps obtained

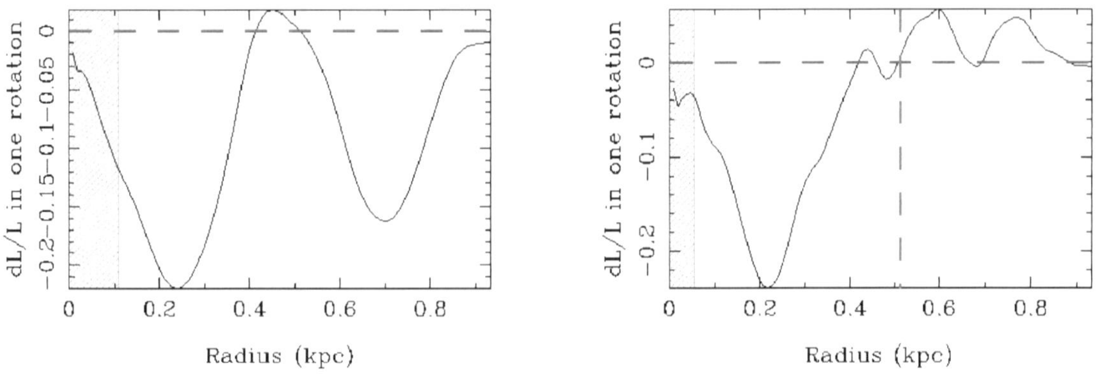

Figure 2. The radial profile of the angular momentum transfer $\Delta L/L$ in the disk of NGC 3627, computed with the HST-NICMOS F160W image is plotted for ^{12}CO(1–0) (left) and ^{12}CO(2–1) (right). The (red) dashed region identifies the radial range imposed by the resolution limit of observations. Torques are systematically negative inside $r{\sim}400$ pc. Figure adapted from [15].

by HST, Spitzer and/or ground-based telescopes for these galaxies. The NIR imaging is used to derive the stellar potentials, which are combined with the CO maps, to derive the gravity torque budget in the circumnuclear disks of the NUGA targets. We implicitly assumed that NIR maps are good estimates of mass distribution and that the overall torque budget on the gas is mostly determined by the stellar potential. Two-dimensional torque maps are used to estimate the torques averaged over the azimuth $t(r)$ using the gas column density $N(x, y)$ derived from CO as weighting function [12]. The final output of this analysis is the radial profile of the angular momentum transfer efficiency in the disks ($\Delta L/L$). With this information at hand we can quantitatively evaluate if there is ongoing AGN fueling in the disks of the galaxies analyzed, down to the spatial resolution of our observations.

3. NUGA results: statistics

About one third of the LLAGNs analyzed in NUGA show negative torques $t(r)<0$, indicative of inflow, down to typical radial distances $r<25$–100 pc. Among these galaxies, which can be considered as those showing 'smoking gun' evidence of ongoing fueling, we distinguish two types of objects:

- *Galaxies showing dynamical decoupling of embedded structures*:
  These include nuclear bars - within - bars/ovals (e.g.; NGC 2782) and nuclear ovals - within -bars (e.g.; NGC 4579). The stellar potential of NGC 2782, analyzed by [13] shows two embedded bars: an outer (weak) oval of ~6 kpc diameter and a (strong) nuclear bar of ~1.5 kpc diameter. The nuclear bar shows signs of decoupling. This configuration of the stellar potential has facilitated the inflow of molecular gas inside the Inner Lindblad Resonance (ILR) of the oval. The derived gravity torques $t(r)$ are systematically negative down to ~100–200 pc. The stellar potential of NGC 4579, studied by [14], shows two embedded bars: an outer bar of ~12 kpc diameter and an embedded weak nuclear oval of ~0.3 kpc diameter. In the outer disk, the decoupling of the spiral arms allows the gas to efficiently populate the Ultra Harmonic Resonance (UHR) region of the large-scale bar. This favors net gas inflow on intermediate scales. Furthermore, closer to the AGN, gas feels negative torques due to the combined action of the outer bar and the nuclear oval. The combination of the two $m = 2$ modes produces net gas inflow down to $r{\sim}50$ pc, providing inward gas transport on short dynamical timescales (~1–3 rotation periods) (Fig. 1).

- *Galaxies with ILR-free large-scale bars*:
  These galaxies are characterized by the apparent absence of gravity torque barriers in their nuclei (e.g.; NGC 3627). The stellar potential of NGC 3627, analyzed by [15] shows one large-scale bar of ~6 kpc diameter. The bar has no ILR barrier. Molecular gas is concentrated along the leading edges of the bar and shows no ring feature. Down to the spatial resolution of these observations (~25 pc), gravity torques $t(r)$ are systematically negative in the circumnuclear disk of NGC 3627 (Fig. 2). This scenario suggests that the bar in this galaxy is young and rapidly rotating, and thus has not yet formed an ILR.

On the contrary, about two thirds of the LLAGNs analyzed in NUGA show positive torques $t(r)>0$, indicative of outflow, down to typical radial distances $r<300$ pc. This *puzzling* gravity torque budget is found in two categories of objects in our sample:

- *Galaxies showing non-cooperative embedded structures*:
  In these galaxies, nuclear bars/ovals - within - bars, like those found in NGC 4321 and NGC 6951 are not always conducive to gas inflow at present [12][16][17]. The stellar potential of NGC 6951, analyzed by [12] and [16] shows two embedded bars: a large-scale bar of ~8 kpc diameter and an inner oval of ~0.4 kpc diameter. Molecular gas is stalled in the ILR ring of the bar. While molecular gas is also detected at the AGN locus [12][17][18], gas is gaining angular momentum inside the ILR due to oval forcing (Fig. 3).
- *Galaxies showing featureless/mostly axisymmetric potentials*:
  The absence of a clear non-axisymmetric feature in the stellar potential makes the angular momentum transfer in the nucleus a very inefficient process in some galaxies at present (e.g.; NGC 4826 [11]; NGC 7217 [16][19]; NGC 5953 [20]).

In summary, the overall results of NUGA indicate that high spatial resolution is instrumental in quantifying angular momentum transfer processes at critical scales (~10–100 pc). Gas flows in NUGA maps reveal a wide range of large-scale and embedded $m = 2$, $m = 1$ instabilities in the circumnuclear disks of LLAGNs. The derived gravity torque maps indicate that gas is frequently stalled in rings, which are the signposts of gravity torque barriers. Only ~1/3 of NUGA galaxies show negative torques down to ~50 pc. Several short-lived (<a few 10^7 yr) mechanisms are at work to drain angular momentum. These various fueling mechanisms are related to bar cycles [21][22]. Secular evolution and dynamical decoupling are key ingredients to understand the AGN fueling cycle.

4. Future surveys with ALMA

To confirm the different feeding scenarios we must improve the spatial resolution and sensitivity of millimeter line observations. The advent of the Atacama Large Millimeter Array (ALMA) will allow us to boost the spatial resolution of observations of galaxy nuclei. The full ALMA capabilities will make possible to map the emission of a set of molecular gas tracers in the inner ~1–10 pc of a significant sample of ~50 nearby (D <5–30 Mpc) AGNs in about 100 hrs of observing time with an order of magnitude higher sensitivity compared to the current NUGA survey. Improving the statistics will allow us to better explore the evolutionary sequence in AGN fueling.

The CO and NIR maps of the observed galaxies, obtained at a common spatial resolution of ~0.05"–0.1" with ALMA and HST/VLT, will provide a sharp view of the gravity torque budget in the circumnuclear disks. The efficiency of new feeding mechanisms will be observationally tested at unprecedented scales ~1–10 pc (see F. Combes' contribution to this symposium). In particular, the role of $m = 1$ modes, gas self-gravity, granularity of the stellar potential, and dynamical friction of GMCs in AGN fueling will be thus tested.

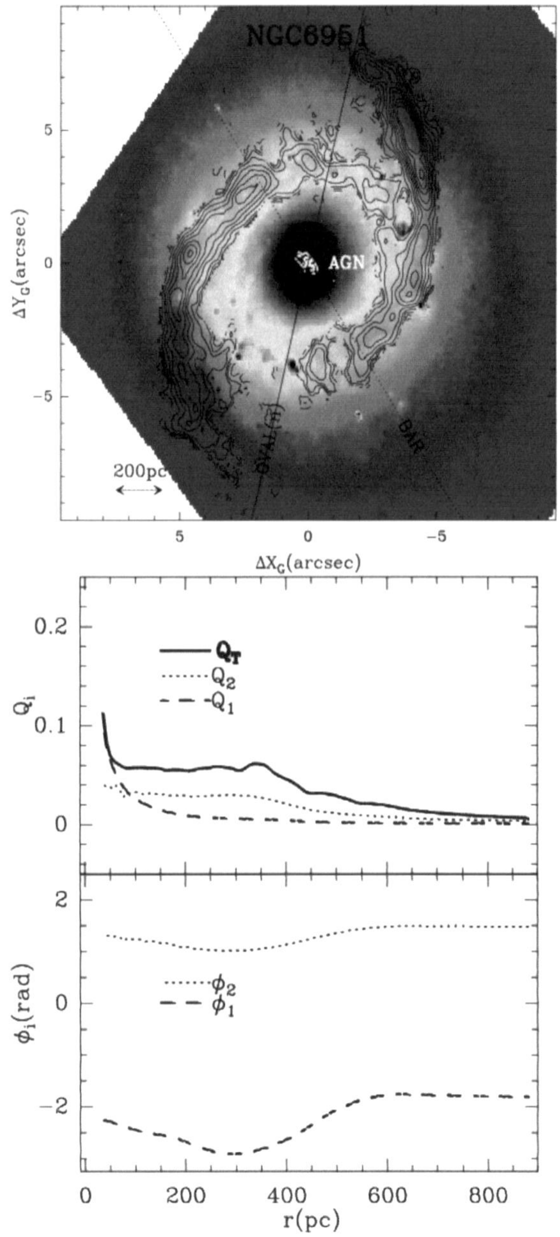

Figure 3. a) (*Upper panel*) Overlay of the $^{12}CO(2\text{–}1)$ NUGA map (contours) on the J-band HST image (grey scale) of NGC 6951; images have been deprojected onto the galaxy plane. The orientations of the large-scale bar-BAR-and of the nuclear oval-OVAL(n)-are shown. b) (*Lower panels*) We plot the strengths (Q_i, i=1,2) and phases (Φ_i, i=1,2) of the m = 1 and m = 2 perturbations inside the image field-of-view. Figure adapted from [12].

Another key ingredient in galaxy evolution is AGN feedback. ALMA enhanced capabilities will also allow us to simultaneously search for the signature of molecular gas outflows and high-velocity winds in nearby AGNs.

References

[1] Jogee, S. 2006, *Physics of Active Galactic Nuclei at all Scales* **693** 143
[2] Hutchings, J. B., & Neff, S. G. 1992 *AJ* **104** 1
[3] Mulchaey, J. S., & Regan, M. W. 1997 *ApJL* **482** L135
[4] Knapen, J. H., Shlosman, I., & Peletier, R. F. 2000 *ApJ* **529** 93
[5] Márquez, I., Durret, F., Masegosa, J., et al. 2000 *A&A* **360** 431
[6] Hopkins, P. F., & Quataert, E. 2010 *MNRAS* **407** 1529
[7] Hopkins, P. F., & Quataert, E. 2011 *MNRAS* **415** 1027
[8] Hopkins, P. F., Hayward, C. C., Narayanan, D., & Hernquist, L. 2011 *MNRAS* **2115**
[9] Hopkins, P. F., & Hernquist, L. 2006 *ApJS* **166** 1
[10] García-Burillo, S., Combes, F., Eckart, A., et al. 2003 *Active Galactic Nuclei: From Central Engine to Host Galaxy* **290** 423
[11] García-Burillo, S., Combes, F., Hunt, L. K., et al. 2003 *A&A* **407** 485
[12] García-Burillo, S., Combes, F., Schinnerer, E., Boone, F., & Hunt, L. K. 2005 *A&A* **441** 1011
[13] Hunt, L. K., Combes, F., García-Burillo, S., et al. 2008 *A&A* **482** 133
[14] García-Burillo, S., Fernández-García, S., Combes, F., et al. 2009 *A&A* **496** 85
[15] Casasola, V., Hunt, L. K., Combes, F., García-Burillo, S., & Neri, R. 2011 *A&A* **527** A92
[16] Haan, S., Schinnerer, E., Emsellem, E., et al. 2009 *ApJ* **692** 1623
[17] van der Laan, T. P. R., Schinnerer, E., Boone, F., et al. 2011 *A&A* **529** A45
[18] Krips, M., Neri, R., García-Burillo, S., et al. 2007 *A&A* **468** L63
[19] Combes, F., García-Burillo, S., Boone, F., et al. 2004 *A&A* **414** 857
[20] Casasola, V., Hunt, L. K., Combes, F., et al. 2010 *A&A* **510** A52
[21] Bournaud, F., & Combes, F. 2002 *A&A* **392** 83
[22] Bournaud, F., Combes, F., & Semelin, B. 2005 *MNRAS* **364** L18

Magnetic fields and gas flows around circumnuclear starbursts

Rainer Beck

Max-Planck-Institut für Radioastronomie, Auf dem Hügel 69, 53121 Bonn, Germany

E-mail: rbeck@mpifr-bonn.mpg.de

Abstract. Radio continuum observations of barred galaxies revealed strong magnetic fields of $\geq 50 - 100\,\mu$G in the circumnuclear starbursts. Such fields are dynamically important and give rise to magnetic stress that causes inflow of gas towards the center at a rate of several solar masses per year, possibly along the spiral field seen in radio polarization and as optical dust lanes. This may solve the long-standing question of how to feed active nuclei, and explain the relation between the bolometric luminosity of AGN nuclei and the star-formation rate of their hosts. The strong magnetic fields generated in young galaxies may serve as the link between star formation and accretion onto supermassive black holes. – Magnetic fields of $\geq 160\,\mu$G strength were measured in the central region of the almost edge-on starburst galaxy NGC 253. Four filaments emerging from the inner disk delineate the boundaries of the central outflow cone of hot gas. Strong Faraday rotation of the polarized emission from the background disk indicates a large-scale helical field in the outflow walls.

1. Introduction

Magnetic fields are a major agent in the ISM and also control the density and propagation of cosmic rays. The radio–infrared correlation indicates that turbulent fields are strongest in star-forming regions. Magnetic fields and cosmic rays can provide the pressure to drive galactic outflows. Outflows from starburst galaxies in the early Universe may have magnetized the intergalactic medium. In spite of our increasing knowledge of cosmic magnetic fields, many important questions, especially their origin, strength in intergalactic space, first occurrence in young galaxies and their dynamical importance for galaxy evolution remain unanswered.

2. Origin of galactic magnetism

The most promising mechanism to sustain magnetic fields in the interstellar medium of galaxies is the *dynamo*. In young galaxies a small-scale dynamo probably amplified seed fields from the protogalactic phase to the energy density level of turbulence within less than 10^9 yr (Schleicher et al. [24]). To explain the generation of large-scale fields in galaxies, the mean-field dynamo has been developed (Beck et al. [7]). It is based on turbulence, differential rotation and helical gas flows, driven by supernova explosions (Gressel et al. [16]) and cosmic rays (Hanasz et al. [17]). The mean-field dynamo in galaxy disks predicts that within a few 10^9 yr large-scale regular fields are excited from the seed fields (Arshakian et al. [1]), forming patterns ("modes") with different azimuthal symmetries in the disk and vertical symmetries in the halo.

3. Observation of galactic magnetism

Most of what we know about galactic magnetic fields comes through the detection of radio waves. The intensity of *synchrotron emission* is a measure of the number density of cosmic-ray electrons and of the strength of the total magnetic field component in the sky plane. The assumption of energy equipartition between the total cosmic rays and total magnetic fields allows us to calculate the total magnetic field strength from the synchrotron intensity (Beck & Krause [6]).

Polarized emission originates from ordered fields. As polarization "vectors" are ambiguous by 180°, they cannot distinguish *regular (coherent) fields*, defined to have a constant direction within the telescope beam, from *anisotropic fields*, which are generated from turbulent fields by compressing or shearing gas flows, so that their direction frequently reverses perpendicular to the flow direction. Unpolarized synchrotron emission indicates *turbulent fields* which have random directions.

The intrinsic degree of linear polarization of synchrotron emission is about 75%. The observed degree of polarization is smaller due to the contribution of unpolarized thermal emission, by *Faraday depolarization* along the line of sight and across the beam (Sokoloff et al. [26]) and by geometrical depolarization due to variations of the field orientation within the beam.

At radio wavelengths of a few centimeters and below, the orientation of the observed B–vector is parallel to the field orientation, so that the magnetic patterns of many galaxies can be mapped directly (e.g. Beck [4]). At longer wavelengths, the observed polarization vector is rotated in a magnetized thermal plasma by *Faraday rotation*. The rotation angle increases with the square of the wavelength λ^2 and with the *Rotation Measure (RM)*, which is the integral of the plasma density and the strength of the component of the field along the line of sight. As the rotation angle is sensitive to the sign of the field direction, only regular fields give rise to Faraday rotation, while anisotropic and random fields do not.

4. Magnetic fields in barred galaxies

The gravitational potential of strongly barred galaxies drives noncircular orbits and gas inflow. In many barred galaxies a circumnuclear ring is formed. Numerical models show that gas streamlines are deflected in the bar region along shock fronts, behind which the cold gas is compressed in a fast shearing flow (Athanassoula [2]). The compression regions traced by massive dust lanes develop along the edge of the bar that is leading with respect to the galaxy's rotation because the gas rotates faster than the bar pattern.

20 galaxies with large bars were observed with the Very Large Array (VLA) and with the Australia Telescope Compact Array (ATCA) (Beck et al. [8, 9]). The total radio luminosity (a measure of the total magnetic field strength) is strongest in galaxies with high far-infrared luminosity (indicating high star-formation activity), a result similar to that in non-barred galaxies. The average radio intensity, radio luminosity and star-formation activity all correlate with the relative bar length. Polarized emission was detected in 17 of the 20 barred galaxies.

NGC 1097 (Fig. 1) is one of the nearest barred galaxies and hosts a huge bar of about 16 kpc length. The total and polarized radio intensities are strongest in the downstream region of the dust lanes (southeast of the center) by compression of turbulent fields in the bar's shock. The pattern of field lines in NGC 1097 is similar to that of the gas streamlines as obtained in numerical simulations (Athanassoula [2]). This suggests that the ordered magnetic field is aligned with the flow and amplified by shear. Remarkably, the optical image of NGC 1097 shows dust filaments in the upstream region which are aligned with the ordered field (Fig. 1). Between the region upstream of the southern bar and the downstream region the field lines smoothly change their orientation by almost 90°. The ordered field is also hardly compressed, probably coupled to the diffuse gas and strong enough to affect its flow (Beck et al. [9]). The polarization pattern in barred galaxies can be used as a tracer of shearing gas flows in the sky plane and complements spectroscopic measurements of radial velocities.

Journal of Physics: Conference Series **372** (2012) 012051 doi:10.1088/1742-6596/372/1/012051

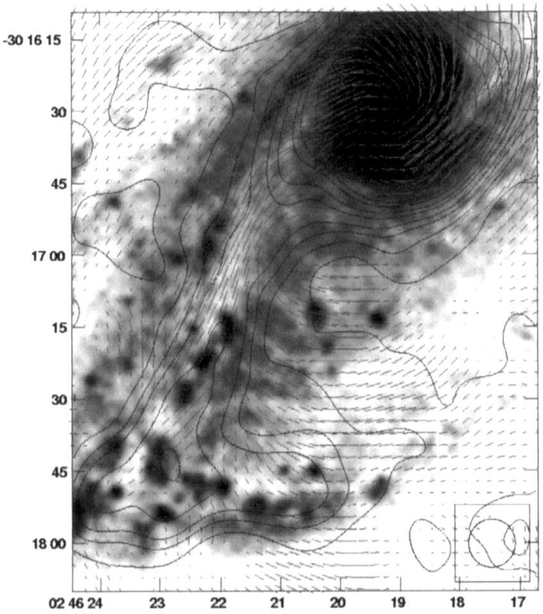

Figure 1. Total radio emission (contours) and *B*–vectors of the barred galaxy NGC 1097, observed at 6 cm wavelength with the *VLA* and smoothed to 10" resolution (Beck et al. [9]). The background optical image is from Halton Arp (*Cerro Tololo Observatory*).

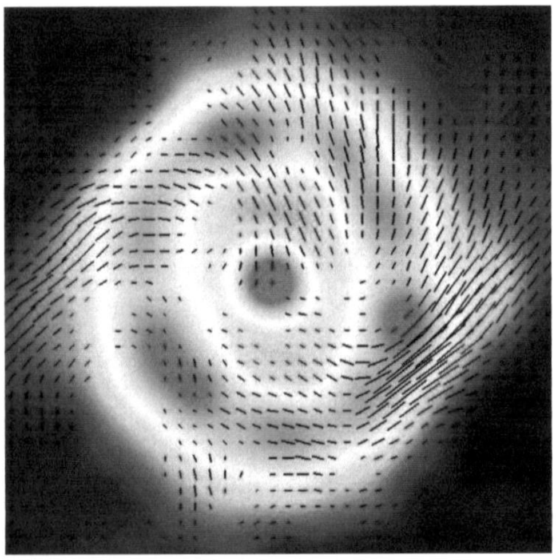

Figure 2. Total radio emission (colors) and *B*–vectors of the circumnuclear ring of the barred galaxy NGC 1097, observed at 3 cm wavelength with the *VLA* and smoothed to 3" resolution (Beck et al. [9]).

5. Magnetic fields in circumnuclear rings and gas inflows

Barred galaxies can create a gas reservoir and starburst in the central kiloparsec, fuelling an AGN needs other processes (e.g. Davies et al. [13]). The central regions of barred galaxies are often sites of ongoing intense star formation and strong magnetic fields. Bright radio emission has been observed in several galaxies with the following equipartition strengths of the total magnetic fields (Beck et al. [10]):

- NGC 1097: $\simeq 50\,\mu$G in the circumnuclear ring, $50\,\mu$G in the ring knots
- NGC 1365: $\simeq 60\,\mu$G in the dust lanes (no ring)
- NGC 1672: $\simeq 70\,\mu$G in the ring knots
- NGC 7552: $\simeq 105\,\mu$G in the ring

Note that these values are lower limits because the energy densities of cosmic-ray protons and electrons in starburst regions are reduced by various losses (Thompson et al. [27]). This is supported by an investigation of the γ-ray and radio emission from the starburst galaxies M 82 and NGC 253 which suggests that most of the radio emission is from secondaries undergoing strong bremsstrahlung and ionization losses (Lacki et al. [20]).

One of the key problems of AGN physics is how to fuel the nucleus with gas from the central region, which needs outwards transport of angular momentum (Davies et al. [12]). Magnetic fields can help here. The circumnuclear ring of NGC 1097 is bright in the optical, IR and radio spectral ranges. It has a diameter of about 1.5 kpc and an active nucleus in its center (Fig. 2). The nonthermal and weakly polarized radio emission is a signature of strong turbulent magnetic fields (Beck et al. [9]). Magnetic stress in the differentially rotating circumnuclear ring can drive mass inflow (Balbus & Hawley [3]) at a rate of:

$$dM/dt = 2\pi\,\sigma\,T_{r\phi}\,/\,\Omega \tag{1}$$

where σ is the surface mass density, Ω its angular rotation velocity and $T_{r\,\phi}$ is the magnetic stress tensor. r and ϕ denote the radial and azimuthal field components. Its dominant component can be written in terms of the Alfvén velocity v_A as $T_{r\phi} \simeq - < v_{A,r}\,v_{A,\phi} >$. This yields:

$$dM/dt = -(< b_r\,b_\phi > + B_r\,B_\phi)\,h\,/\,\Omega \tag{2}$$

where h is the scale height of the gas, b the strength of the turbulent field and B that of the ordered field. The correlation between b_r and b_ϕ is generated by shear from differential rotation. For NGC 1097, $h \simeq 100\,\mathrm{pc}$, $v \simeq 450\,\mathrm{km/s}$ at 1 kpc radius, $b_r \simeq b_\phi \simeq 50\,\mu\mathrm{G}$ and $B_r \simeq B_\phi \simeq 10\,\mu\mathrm{G}$ give an inflow rate of several $M_\odot/\mathrm{yr}$, which is sufficient to fuel the activity of the nucleus of this galaxy (Beck et al. [9]). This mechanism can be expected to operate in many galaxies.

Equipartition between the energy densities of cosmic rays and magnetic fields yields:

$$dM/dt \simeq -B_{tot}^2\;h/\,\Omega\;\propto -I_{sync}^{0.5}\,h\,/\,\Omega\;\propto -I_{IR}^{0.5}\,h\,/\,\Omega\;\propto -SFR^{0.5}\,h\,/\,\Omega \tag{3}$$

where B_{tot} is the total field strength, I_{sync} is the radio synchrotron intensity which is proportional to the infrared intensity I_{IR} (the radio-infrared correlation) and hence to the star-formation rate SFR. The relation between the bolometric luminosity of AGNs L_{bol} and the SFR of their hosts (e.g. Trakhtenbrot & Netzer [28]) can be explained by this scenario. The relation between the accretion rate and SFR (Eq. 3) should be tested with a sample of galaxies.

The ordered field in the ring of NGC 1097 has a spiral pattern and extends towards the nucleus. Its pitch angle agrees with that of the spiral filaments seen in dust and H_2 (Prieto et al. [22], Davies et al. [12]), suggesting gas inflow along the magnetic field lines.

Radio emission is a tracer of star formation also in distant galaxies. The radio–infrared correlation holds to redshifts of at least $z \simeq 3$ (Murphy [21]), indicating strong magnetic fields in young galaxies, probably generated by the turbulent dynamo mechanism (Schleicher et al. [24]). Infrared observations indicate that star formation in the hosts of the most luminous AGNs peaked at $z \simeq 3$ and decreased at later epochs (Serjeant et al. [25]), which is consistent with the evolution of the accretion rate onto supermassive black holes (Trakhtenbrot et al. [29]). The strong magnetic fields in young galaxies may serve as the link between star formation and accretion.

6. The nuclear outflow of the starburst galaxy NGC 253

NGC 253 is a prototypical starburst galaxy, hosting a starburst nucleus (Brunthaler et al. [11]). Its high inclination of 78.5° allows observing the extraplanar emission. The bright radio halo (Fig. 3) is probably a result of a galactic wind with a speed of about 300 km/s (Heesen et al. [18]).

Radio observations of the inner starburst region with high resolution revealed four filaments with widths of less than 40 pc and lengths of at least 500 pc (Fig. 4). These are interpreted as the boundaries of the outflow cone of hot gas interacting with the halo gas (Heesen et al. [19]). The equipartition strength of the total field in the filaments is at least $40\,\mu\mathrm{G}$, that in the central region $\geq 160\,\mu\mathrm{G}$.

Faraday rotation measures (RM) between 3 cm and 6 cm wavelengths are consistent with an axisymmetric, even-parity, spiral magnetic field in the disk at radii of more than 2 kpc, while

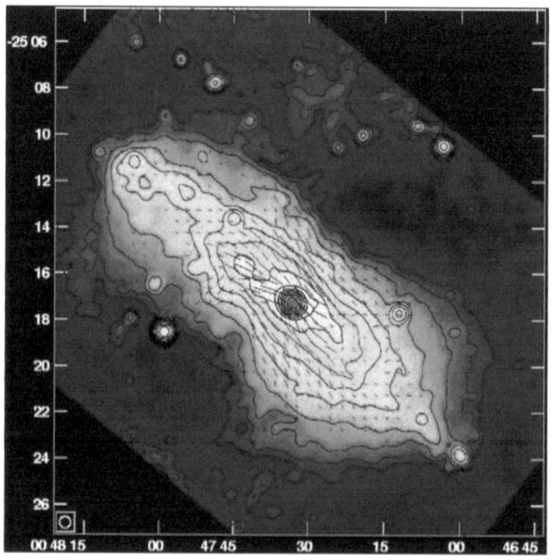

Figure 3. Total radio emission (contours and colors) and polarization B–vectors of the spiral galaxy NGC 253, combined from observations at 6 cm wavelength with the *VLA* and the *Effelsberg* telescope and smoothed to 30" resolution (Heesen et al. [18]).

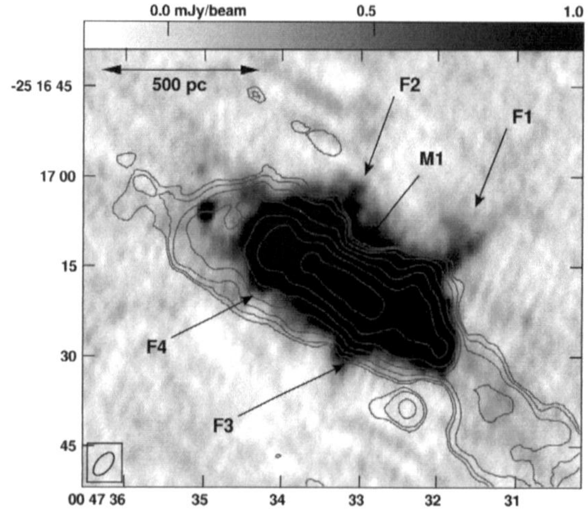

Figure 4. Total radio emission (greyscale) of the inner starburst region of NGC 253, observed at 20 cm with the *VLA* at a resolution of 1.3" x 2.2" (Heesen et al. [19]). The radio filaments (F1–4) and the minimum (M1) are indicated. The contours show the 12CO(2–1) emission, observed with the *SMA* (Sakamoto et al. [23]).

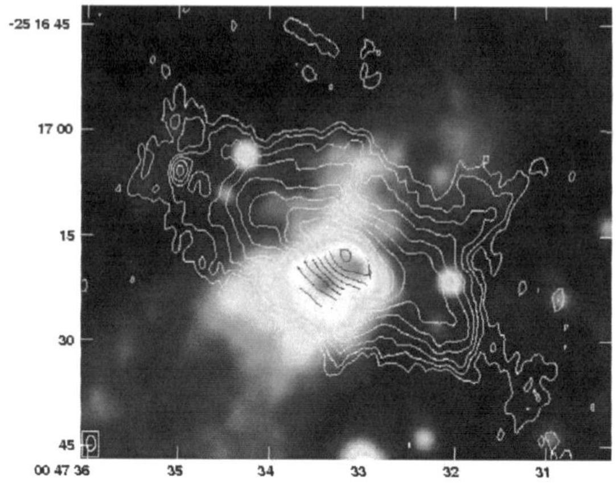

Figure 5. Total radio emission (contours) of the inner starburst region of the galaxy NGC 253, observed at 20 cm wavelength with the *VLA* at a resolution of 1.3" x 2.2" (Heesen et al. [19]). The colour image shows the X-ray emission (0.5-5 keV) measured with the *CHANDRA* satellite.

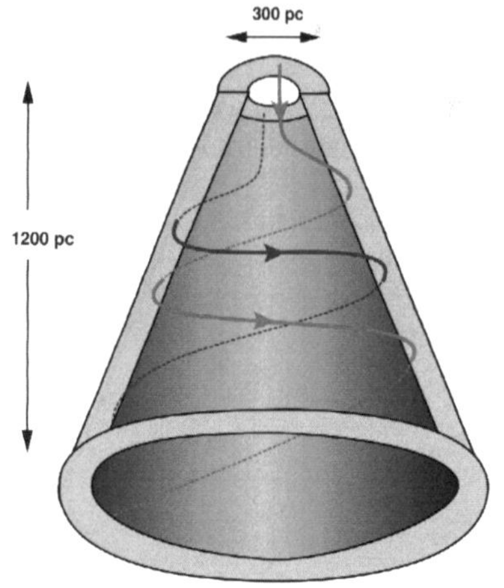

Figure 6. Model of the helical magnetic field in the walls of the outflow cone (Heesen et al. [19]).

the field in the inner disk is anisotropic (Heesen et al. [19]). Near the SE minor axis, in the region of the X-ray outflow cone (Fig. 5), RM jumps between +300 rad m^{-2} and -300 rad m^{-2}. This is interpreted as the Faraday rotation of the polarized emission from the background disk by the helical field in the walls of the outflow extending to at least 1.2 kpc from the center (Fig. 6). Only the inner part of this helical field structure is seen in emission (Fig. 4). The northern outflow cone is located behind the disk, so that its X-ray emission is mostly absorbed and cannot Faraday-rotate the radio emission.

The galaxy NGC 3079 hosts an energetic wind-driven outflow which extends 3 kpc above and below the plane (Duric & Seaquist [14]) and emits strong and highly polarized synchrotron emission, in contrast to NGC 253. Radio-bright nuclear outflows seem to be rare.

7. Summary and outlook

Magnetic fields in the central regions of galaxies are dynamically important. Magnetic stress drives gas inflow from the circumnuclear regions towards the nuclei. This process connects the star formation activity to the accretion rate and could be important for the formation of supermassive black holes. Ordered fields as observed by polarized radio emission may trace the direction of gas in- and outflows.

Results have been obtained so far only for radio-bright and nearby galaxies. A better insight into the interaction between gas and magnetic fields, especially near galactic nuclei, needs higher resolution and higher sensitivity. The *EVLA* has largely improved the sensitivity for radio continuum observations. The *EVLA* project *CHANGES* (PI: Judith Irwin) will search for outflows in a large sample of edge-on galaxies. The goal for the next decade is the *Square Kilometre Array (SKA)* with a collecting area of about 100x that of present-day telescopes. The SKA will open a new era in the observation of cosmic magnetic fields (Beck [5]). Starburst galaxies will be observable out to redshifts of $z \simeq 3$ (Murphy [21]).

8. References

[1] Arshakian T G, Beck R, Krause M and Sokoloff D 2009 *A&A* **494** 21
[2] Athanassoula E 1992 *MNRAS* **259** 345
[3] Balbus S A and Hawley J F 1998 *Rev. Mod. Phys.* **70** 1
[4] Beck R 2005 *Cosmic Magnetic Fields* ed R Wielebinski and R Beck (Berlin: Springer) p 41
[5] Beck R 2010 *PoS(ISKAF2010)*003
[6] Beck R and Krause M 2005 *Astron. Nachr.* **326** 414
[7] Beck R, Brandenburg A, Moss D, Shukurov A and Sokoloff D 1996 *Ann. Rev. A&A* **34** 155
[8] Beck R, Shoutenkov V, Ehle M, et al 2002 *A&A* **391** 83
[9] Beck R, Fletcher A, Shukurov A, et al 2005 *A&A* **444** 739
[10] Beck R, Ehle M, Fletcher A, et al 2005b *The Evolution of Starbursts* ed S Hüttemeister et al *AIP Conf. Proc.* **783** 216
[11] Brunthaler A, Castangia P, Tarchi A, et al 2009 *A&A* **497** 103
[12] Davies R I, Maciejewski W, Hicks E K S, et al 2009 *ApJ* **702** 114
[13] Davies R I, Hicks E, Schartmann M, et al 2010 *Co-Evolution of Central Black Holes and Galaxies* ed B M Peterson et al (Cambridge: Cambridge University Press) p 283
[14] Duric N and Seaquist E R 1988 *ApJ* **326** 574
[15] Fletcher A, Berkhuijsen E M, Beck R and Shukurov A 2004 *A&A* **414** 53
[16] Gressel O, Elstner D, Ziegler U and Rüdiger G 2008 *A&A* **486** L35
[17] Hanasz M, Wóltański D and Kowalik K 2009 *ApJ* **706** L155
[18] Heesen V, Beck R, Krause M and Dettmar R-J 2009 *A&A* **494** 563
[19] Heesen V, Beck R, Krause M and Dettmar R-J 2011 *A&A* **535** A79
[20] Lacki B C, Thompson T A, Quataert E, Loeb A and Waxman E 2011 *ApJ* **734**:107
[21] Murphy E 2009 *ApJ* **706** 482
[22] Prieto M A, Maciejewski W and Reunanen J 2005 *AJ* **130** 1472
[23] Sakamoto K, Ho P T P, Iono D, et al 2006 *ApJ* **636** 685
[24] Schleicher D R G, Banerjee R, Sur S, et al 2010 *A&A* **522**:A115
[25] Serjeant S, Bertoldi F, Blain A W, et al 2010 *A&A* **518** L7
[26] Sokoloff D D, Bykov A A, Shukurov A, et al 1998 *MNRAS* **299** 189 and *MNRAS* **303** 207
[27] Thompson T A, Quataert E, Waxman E, Murray N and Martin C L 2006 *ApJ* **645** 186
[28] Trakhtenbrot B and Netzer H 2010 *MNRAS* **406** L35
[29] Trakhtenbrot B, Netzer H, Lira P and Shemmer O 2011 *ApJ* **730**:7

Journal of Physics: Conference Series **372** (2012) 012052 doi:10.1088/1742-6596/372/1/012052

Molecular gas in active environments

S Mühle[1,2], C Henkel[3], T de Maio[4] and E R Seaquist[5]

[1] Argelander-Institut für Astronomie, Auf dem Hügel 71, 53121 Bonn, Germany
[2] Joint Institute for VLBI in Europe, Postbus 2, 7990 AA Dwingeloo, The Netherlands
[3] MPIfR, Auf dem Hügel 69, 53121 Bonn, Germany
[4] University of Colorado, 389 UCB, Boulder CO, 80309, USA
[5] Department of Astronomy and Astrophysics, University of Toronto, 50 St. George Street, Toronto, ON M5S3H4, Canada

E-mail: muehle@astro.uni-bonn.de

Abstract. The question whether or not the initial mass function (IMF) is universal, i.e. the same in all kinds of environments, is subject to intense debate. A number of recent observations have been interpreted as evidence for a nonstandard IMF. Hydrodynamical simulations indicate that the kinetic temperature of the collapsing molecular gas is crucial for the shape of the resulting IMF. Unfortunately, the kinetic temperature of the molecular gas in external galaxies is often not well constrained. We demonstrate the diagnostic power of a selected set of para-formaldehyde lines as tracers of the kinetic temperature as well as the gas density in external galaxies using our non-LTE radiative transfer model. With this new observational tool, we have engaged in characterizing the properties of the dense molecular gas phase in a number of nearby starburst galaxies and near AGN. Our first results suggest the existence of a dense molecular gas phase in these active environments that is significantly warmer than the dust and much warmer than dense molecular gas found in the disk of our own Galaxy.

1. Introduction

Star formation and its feedback are among the most fundamental processes that shape the evolution of galaxies throughout the universe. In the disk of our own Galaxy, groups of stars form out of dense molecular gas that is cold ($T \sim 10 - 30\,\mathrm{K}$) and the stellar initial mass function (IMF) can generally be described by a power-law with a Salpeter index ($\Gamma = 1.35$) for masses above a few solar masses. Whether this so-called "standard IMF" is universal is a matter of intense debate (see Bastian et al. 2010 for a review). Non-canonical, usually top-heavy, IMFs have been suggested for a range of objects, including the stellar disks in the Galactic Center (Bartko et al. 2010), disturbed galaxies (Habergham et al. 2010) and ultra-compact dwarf galaxies (Dabringhausen et al. 2009). In hydrodynamic simulations of the influence of star formation feedback and other environmental parameters on the IMF, the temperature of the collapsing molecular gas plays a key role in the fragmentation process, thus influencing the shape of the IMF (Klessen et al. 2007, Krumholz et al. 2010). Unfortunately, the temperature of the molecular gas outside of our own Galaxy is usually not well constrained. Observational evidence is growing, though, that a significant fraction of the molecular gas in starburst galaxies and near AGN may be warm, with kinetic temperatures of 100 K and more (Rigopoulou et al. 2002, Mauersberger et al. 2003, Krips et al. 2011), sufficient to yield a top-heavy IMF. Possible heating sources of the molecular gas, even down to the densest cores, include intense UV and X-ray radiation, cosmic rays and me-

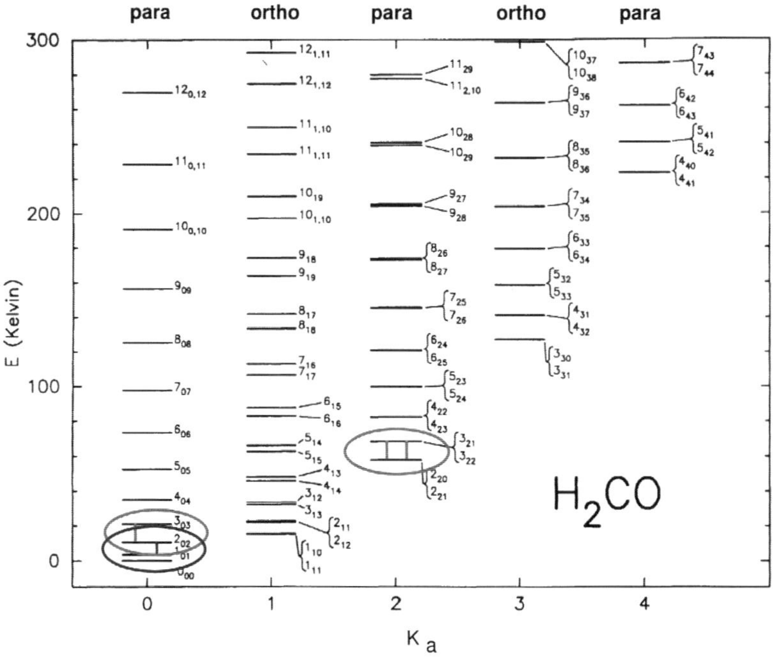

Figure 1. The lowest energy levels of ortho- and paraformaldehyde, adapted from Mangum & Wootten 1993. The selected para-H_2CO transitions at 218 GHz and 146 GHz are marked by ellipses in red and blue, respectively.

chanical heating (Hocuk & Spaans 2010, Krumholz et al. 2010, Papadopoulos 2010, Meijerrink et al. 2011). In starburst galaxies, the conversion factor between the CO(1-0) line intensity and the H_2 column density has been found to be up to an order of magnitude smaller than the Galactic conversion factor (Downes & Solomon 1998). Thus, the properties of the molecular gas in active environments with strong feedback in form of shocks, intense radiation and/or cosmic ray emission seem to differ significantly from the properties of the molecular gas in the Galactic disk.

Given the important influence of the kinetic temperature of the star-forming molecular gas on the IMF, we have developed a new method for deriving the key properties of the dominant phase of the molecular gas in external galaxies. Here we present this new method and discuss the first results of our survey of nearby active environments.

2. Tracing the properties of extragalactic molecular gas

The diagnostic power of many commonly observed gas tracers like HCN or HCO^+ suffers from a degeneracy between the kinetic temperature and the gas density in the model analysis, i.e. the intensity ratios of the observed lines are good density tracers, but only if the kinetic temperature is known. The easily thermalized CO($1 \rightarrow 0$) and CO($2 \rightarrow 1$) transitions could act as a temperature tracer, but unfortunately, the filling factor of extragalactic clouds is not well constrained. In contrast, the rich millimeter and submillimeter spectrum of formaldehyde H_2CO (Fig. 1) offers tracers of both temperature and density: line ratios involving different K_a ladders are sensitive to the kinetic temperature, whereas ratios of lines of the same K_a ladder form excellent density tracers, once the temperature is derived (Mangum & Wootten 1993). On the other hand, the spectrum of formaldehyde is not so rich that many different transition would be blended into one broad feature in extragalactic observations. In addition, the abundance of H_2CO has been found to change only within one order of magnitude in a variety of Galactic environments (Johnstone et al. 2003), whereas the abundance of ammonia, the "standard cloud thermometer", can vary by three orders of magnitude (Benson & Myers 1983, Mauersberger et al. 1987).

Because of these excellent diagnostic properties of formaldehyde, we selected three para-formaldehyde lines as molecular gas tracers in external galaxies: $H_2CO(3_{03} \rightarrow 2_{02})$ at 218.22 GHz, $H_2CO(3_{21} \rightarrow 2_{20})$ at 218.76 GHz and $H_2CO(2_{02} \rightarrow 1_{01})$ at 145.60 GHz. The paraformaldehyde lines at 218 GHz are particularly valuable from a diagnostic point of view. Within 1 GHz, i.e. usually within the same low frequency resolution spectrum, there are the three para-H_2CO transitions $3_{03} \rightarrow 2_{02}$ (218.22 GHz), $3_{22} \rightarrow 2_{21}$ (218.48 GHz) and $3_{21} \rightarrow 2_{20}$ (218.76 GHz). While the $H_2CO(3_{22} \rightarrow 2_{21})$ line may be contaminated by methanol emission, the intensity ratio of the other two lines forms an excellent temperature tracer, without any uncertainty due to calibration issues, pointing errors or different beam widths. The set of diagnostic lines is completed by $H_2CO(2_{02} \rightarrow 1_{01})$ at 145.60 GHz, an easily detectable line that together with $H_2CO(3_{03} \rightarrow 2_{02})$ forms a sensitive density tracer.

3. The analysis of the diagnostic lines

As the upper levels of all the selected para-formaldehyde transitions lie below 70 K, the excitation conditions are similar for all lines. Therefore, it is reasonable to assume that the emission of all H_2CO lines originates from the same volume. This in turn implies that the lines have the same velocity profile, thus considerably reducing the number of free parameters in the derivation of the line intensities. If the line profiles are approximated by Gaussian curves, the velocity and width of all lines should be the same, leaving the peak intensity as the only free parameter. By thus fixing the velocity and the width of the common line profile with the help of a strong line, even the intensities of weak or slightly blended lines can be derived with high reliability.

When the ratios of the observed line intensities are compared with the predicted line ratios of a model, the physical properties of the gas like the kinetic temperature and the average gas density within the molecular cloud complexes can be derived. For the analysis of the para-H_2CO lines, we have developed a non-LTE model with a spherically symmetric cloud geometry and adopted the LVG approximation. The searched parameter space covers a kinetic temperature of $T_{\rm kin} = 5$ to 300 K in steps of 5 K, a molecular gas density of $\log n_{\rm H2} = 3.0$ to 6.0 (in $\rm cm^{-3}$) in steps of 0.1 and a para-H_2CO column density per velocity interval of $\log N_{\rm pH2CO}/\Delta v = 10.5$ to 14.5 (in $\rm cm^{-2}\,km^{-1}\,s$) in steps of 0.1 (see Mühle et al. 2007 for details).

4. First results

The first targets in our survey of the molecular gas properties in active environments were the nearby galaxies M 82 and NGC 253, which are both regarded as prototypical starburst galaxies, with high infrared luminosities, super-star clusters and a prominent galactic wind. Both galaxies are viewed at a high inclination and their dense molecular gas is concentrated around the nucleus. In M 82, though, the fractional abundances of a number of molecules differ strongly from those derived for NGC 253 and other local starburst galaxies, which has been explained as M 82 being in an evolved state of a starburst (Takano et al. 2002, Wang et al. 2004). Note that the massive star cluster M 82 F is to date the best candidate to host a top-heavy IMF (Bastian et al. 2010).

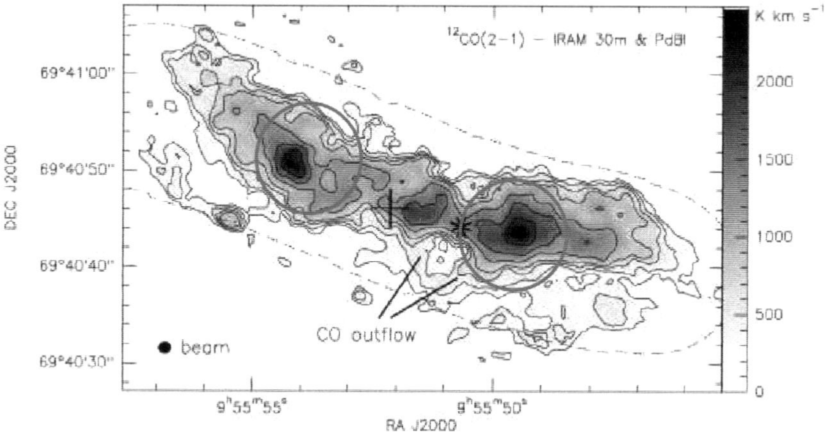

Figure 2. Positions of observed H_2CO emission in M 82 superposed on a high-resolution maps of the integrated CO($2 \rightarrow 1$) intensity from Weiß et al. (2001). The diameter of the red circles corresponds to the width of the beam of the IRAM 30-m telescope at 218 GHz.

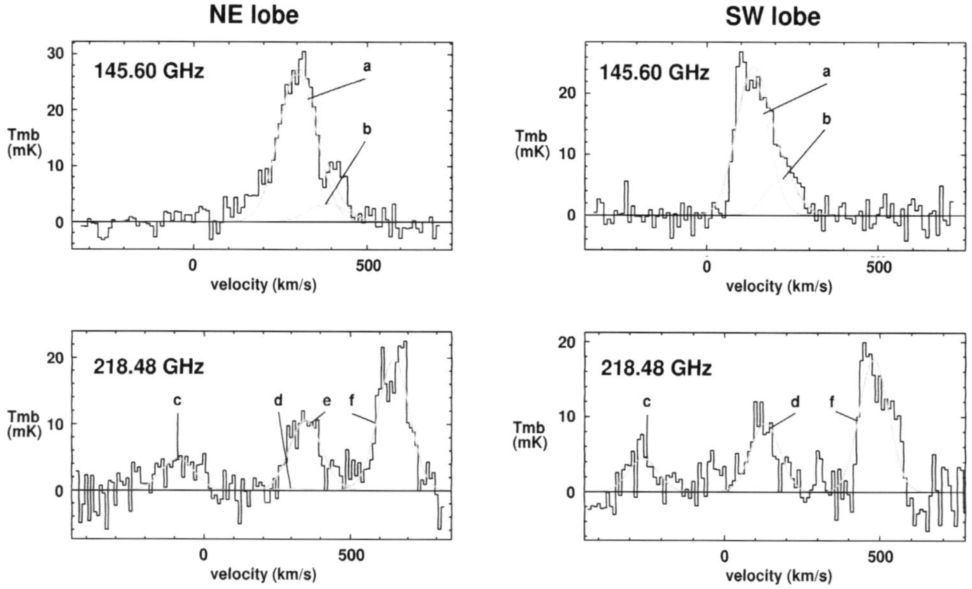

Figure 3. Observed spectra of the NE (left) and the SW lobe (right) of M 82. Each spectrum is labeled with the frequency the receiver was tuned to. Thus, the velocity scale of the 218.48 GHz spectra refers to the $H_2CO(3_{22} \rightarrow 2_{21})$ line. The $H_2CO(3_{03} \rightarrow 2_{02})$ and the $H_2CO(3_{21} \rightarrow 2_{20})$ transitions are offset by 348 km s^{-1} and -390 km s^{-1}, respectively, while the $CH_3OH(4_2 \rightarrow 3_1\ E)$ line is offset by 49 km s^{-1}. In the 146 GHz spectra, the $HC_3N(16 \rightarrow 15)$ line is offset by 86 km s^{-1} from the $H_2CO(2_{02} \rightarrow 1_{01})$ emission. The curves show the Gaussian fit to each individual line as well as the spectrum resulting from a superposition of all identified lines: a = $H_2CO(2_{02} \rightarrow 1_{01})$, b = $HC_3N(16 \rightarrow 15)$, c = $H_2CO(3_{21} \rightarrow 2_{20})$, d = $H_2CO(3_{22} \rightarrow 2_{21})$, e = $CH_3OH(4_2 \rightarrow 3_1\ E)$, f = $H_2CO(3_{03} \rightarrow 2_{02})$.

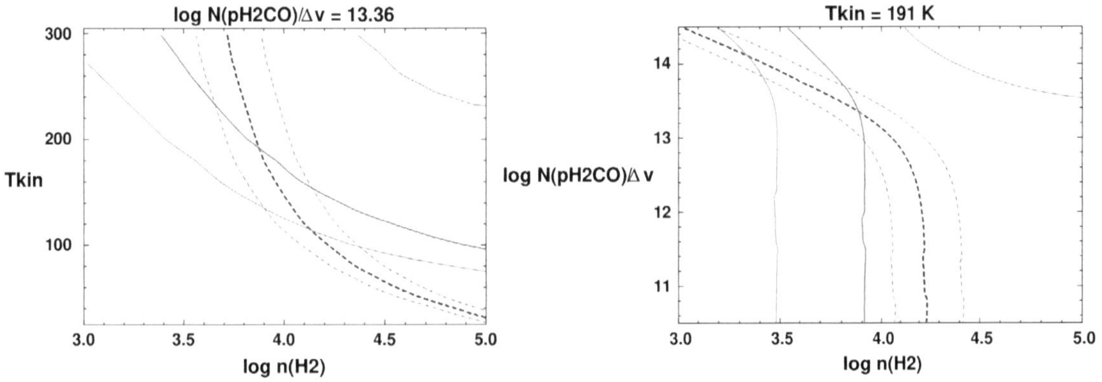

Figure 4. Results of our LVG analysis for the SW lobe of M 82 shown as cuts through the 3-D parameter space ($T_{\rm kin}$ in K, $n_{\rm H2}$ in $\rm cm^{-3}$, $N_{\rm pH2CO}/\Delta v$ in $\rm cm^{-2}\,km^{-1}\,s$) along a plane of constant para-H_2CO column density per velocity interval (left) and of constant kinetic temperature (right). The lines represent the following line ratios: $H_2CO(3_{03} \rightarrow 2_{02})/H_2CO(3_{21} \rightarrow 2_{20})$ (red), $H_2CO(2_{02} \rightarrow 1_{01})/H_2CO(3_{03} \rightarrow 2_{02})$ (blue), $H_2CO(2_{02} \rightarrow 1_{01})/H_2CO(3_{21} \rightarrow 2_{20})$ (green). The corresponding thin lines outline the uncertainties of each line ratio. The adopted source size is $\theta_s = 7\farcs5$ and the assumed para-H_2CO abundance per velocity gradient is $\Lambda = 1\times10^{-9}\,\rm km^{-1}\,s\,pc$.

In M 82, we observed the two lobes of the circumnuclear ring with the IRAM 30-m telescope both at 218 GHz and at 146 GHz to an rms noise level of 2 − 3 mK and 2 mK, respectively, at a velocity resolution of $10\,\rm km\,s^{-1}$ (Fig. 2). All H_2CO transitions have been detected at both pointing positions (Fig. 3). At 145.60 GHz, the $H_2CO(2_{02} \rightarrow 1_{01})$ line is slightly blended with the $HC_3N(16 \rightarrow 15)$ line. However, by applying the constraints discussed above when fitting Gaussian curves to the line profile, it is possible to derive the intensities of these lines with high confidence. In the 218.48 GHz spectrum of the northeastern (NE) lobe, the central line seems to be shifted compared to the frequency of that line in the southwestern (SW) lobe. This is most likely due to methanol emission from the NE lobe (see Mühle et al. 2007 for details).

The intensity ratios derived from the observed H_2CO lines are remarkably similar for the two lobes. Comparing these values to our LVG model, we derive a kinetic temperature of $T_{\rm kin} \sim 200\,\rm K$ and a gas density of $n_{\rm H2} \sim 7\cdot10^3\,\rm cm^{-3}$ for a para-H_2CO abundance per velocity gradient of $X_{\rm pH2CO}/{\rm grad}(v) = 1\cdot10^{-9}\,\rm km^{-1}\,s\,pc$, and an assumed source size of $\theta_s = 7\farcs5$ (Fig. 4). The mass of the molecular gas in the lobes is $\sim 3\cdot10^8\,M_\odot$. These values are in very good agreement with the high-excitation molecular gas component of other recent molecular line studies, which also suggest that this component may constitute more than half of the total molecular gas mass in M 82 (Mao et al. 2000, Ward et al. 2003).

In NGC 253, we observed six positions along the major axis of the circumnuclear disk at 218 GHz and at 146 GHz with the IRAM 30-m telescope (Fig. 5). In all spectra, the rms noise level at a velocity resolution of $10\,\rm km\,s^{-1}$ is a few milliKelvin. The $H_2CO(2_{02} \rightarrow 1_{01})$ line at 145.60 GHz, again slightly blended with the $HC_3N(16 \rightarrow 15)$ line, has been detected at all pointing positions. The three H_2CO lines at 218 GHz have been detected towards the inner four positions, but not at the outermost positions, most likely due to the decrease of H_2CO emission towards the edge of the circumnuclear disk. As an example, the spectra observed towards the nucleus (offset (0/0) in Fig. 5) are shown in Fig. 6.

The line intensity ratios vary significantly along the major axis of the molecular disk in NGC 253, an indication for different physical properties of the dense molecular gas across the

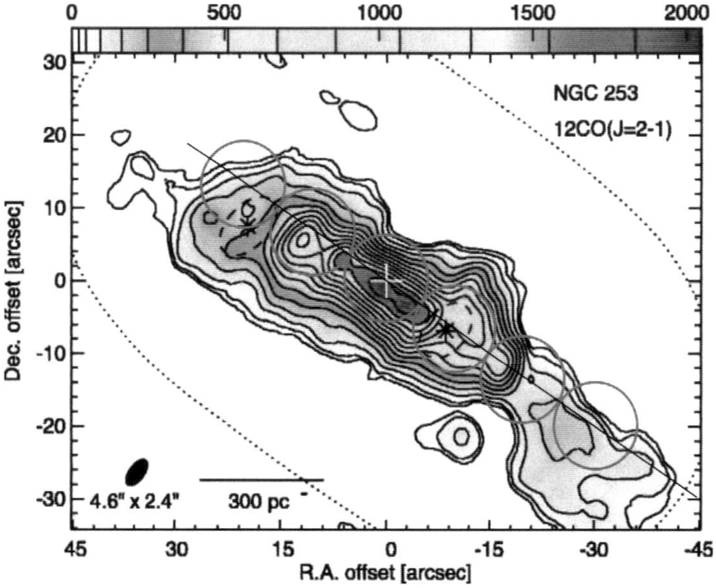

Figure 5. Positions of observed H_2CO emission in NGC 253 superposed on a high-resolution map of the integrated CO($2 \rightarrow 1$) intensity from Sakamoto et al. (2006). The diameter of the red circles corresponds to the width of the beam of the IRAM 30-m telescope at 218 GHz. The beam width of the interferometric CO($2 \rightarrow 1$) observations is shown in the lower left corner.

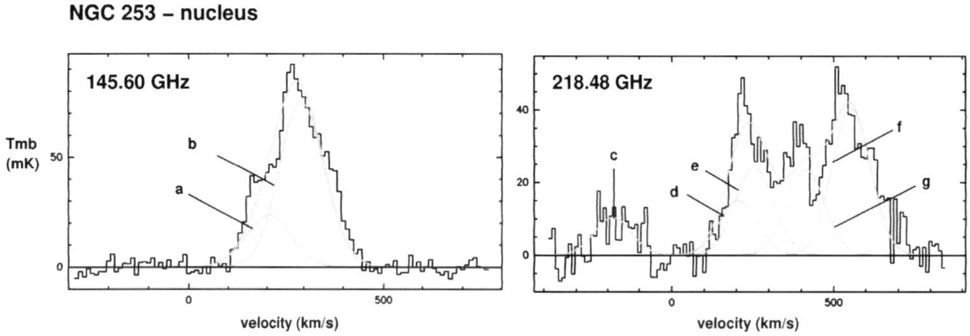

Figure 6. Observed spectra of the nucleus of NGC253 (offset (0/0) in Fig. 5). Like in Fig. 3, each spectrum is labeled with the frequency the receiver was tuned to. The velocity scale of the 218.48 GHz spectrum refers to the $H_2CO(3_{22} \rightarrow 2_{21})$ line. The $H_2CO(3_{03} \rightarrow 2_{02})$ and the $H_2CO(3_{21} \rightarrow 2_{20})$ transitions are offset by 348 km s^{-1} and -390 km s^{-1}, respectively, while the $CH_3OH(4_2 \rightarrow 3_1\,E)$ and the $HC_3N(24 \rightarrow 23)$ lines are offset by 49 km s^{-1}and 207 km s^{-1}, respectively. In the 146 GHz spectrum, the $HC_3N(16 \rightarrow 15)$ line is offset by 86 km s^{-1} from the $H_2CO(2_{02} \rightarrow 1_{01})$ emission. The curves show the Gaussian fits to each individual line as well as the spectrum resulting from a superposition of all identified lines: a = $H_2CO(2_{02} \rightarrow 1_{01})$, b = $HC_3N(16 \rightarrow 15)$, c = $H_2CO(3_{21} \rightarrow 2_{20})$, d = $H_2CO(3_{22} \rightarrow 2_{21})$, e = $CH_3OH(4_2 \rightarrow 3_1\,E)$, f = $H_2CO(3_{03} \rightarrow 2_{02})$, g = $HC_3N(24 \rightarrow 23)$. Note the strong HC_3N emission in the nucleus of NGC 253 compared to its small contribution to the spectra of M 82 (Fig. 3).

galactic core region. Unlike in M 82, there is strong HC_3N emission at both 146 GHz and at 218 GHz, most pronounced at the nucleus (Fig. 6). A preliminary LVG analysis of the line ratios derived for the different positions in NGC 253 yields kinetic temperatures of $T_{\rm kin} \sim 70 \ldots 150$ K and gas densities of $n_{\rm H2} \sim 10^4\,{\rm cm}^{-3}$ for a para-H_2CO abundance per velocity gradient of $X_{\rm pH2CO}/{\rm grad}(v) = 1 \cdot 10^{-9}\,{\rm km}^{-1}\,{\rm s\,pc}$, and an assumed source size of $\theta_s = 10''$.

Thus, in both starburst galaxies, the dense molecular gas component traced by the para-formaldehyde lines is considerably warmer than the bulk of the dense molecular gas in our Galaxy and probably warm enough to give rise to a top-heavy IMF should this gas be turned into stars.

5. Conclusions and outlook

Selected H_2CO lines can be powerful diagnostic tools for the derivation of key properties of the molecular gas in external galaxies. Para-H_2CO lines up to $K_a = 2$ can be detected in nearby starburst galaxies, providing a direct measure the the kinetic temperature of the molecular gas within one spectrum. The first results of our survey support the view that a considerable fraction of the dense molecular gas in starburst galaxies is significantly warmer than the dense molecular gas in quiescent galaxies, possibly giving rise to a nonstandard IMF in active environments.

However, the survey of the properties of dense molecular gas in active extragalactic environments is still in an early phase and many questions about the extent of the warm gas phase and the environments where it can be found are still to be investigated. The boost in sensitivity that ALMA will provide will make the para-H_2CO lines readily detectable in a variety of external galaxies. At the same time, ALMA will offer an unprecedented angular resolution, which will allow us to resolve the distribution and temperature gradients of the warm molecular gas phase in nearby galaxies and to relate them to the different possible heating sources.

Acknowledgments

We wish to thank the staff at Pico Veleta/Granada for their support during the observations. E.R.S. acknowledges a Discovery Grant from NSERC. This work has made use of the following software and resources: GILDAS, specx, Statistiklabor (FU Berlin, CeDiS), asyerr (Seaquist & Yao), the JPL Catalog of spectral lines, The Cologne Database for Molecular Spectroscopy (Müller et al. 2005) and NASA's Astrophysics Data System Bibliographic Services (ADS).

References

[1] Bartko H et al. 2010 *ApJ* **708** 834-40
[2] Bastian N, Covey K R and Meyer M R 2010 *ARAA* **48** 339-89
[3] Benson P J and Myers P C 1983 *ApJ* **270** 589-604
[4] Dabringhausen J, Kroupa P and Baumgardt H 2009 *MNRAS* **394** 1529-43
[5] Downes D and Solomon P M 1998 *ApJ* **507** 615-654
[6] Habergham S M, Anderson J P and James P A 2010 *ApJ* **717** 342-7
[7] Hocuk S and Spaans M 2010 *A&A* **522** A24
[8] Johnstone D, Boonman A M S and van Dishoeck E F 2003 *A&A* **412** 157-74
[9] Klessen R S, Spaans M and Jappsen A K 2007 *MNRAS* **374** L29-L33
[10] Krips M et al. 2011 *ApJ* **736** 37
[11] Krumholz M R, Cunningham A J, Klein, R I and McKee C F 2010 *ApJ* **713** 1120-33
[12] Mangum J G and Wootten A 1993 *ApJS* **89** 123-53
[13] Mao R Q, Henkel C, Schulz A, Zielinsky M, Mauersberger R, Störzer H, Wilson T L and Gensheimer P 2000 *A&A* **358** 433-50
[14] Mauersberger R, Henkel C and Wilson T L 1987 *A&A* **173** 352-60

[15] Mauersberger R, Henkel C, Weiß A, Peck A B and Hagiwara Y 2003 *A&A* **403** 561-71
[16] Meijerrink R, Spaans M, Loenen, A F and van der Werf P P 2011 *A&A* **525** A119
[17] Mühle S, Seaquist E R and Henkel C 2007 *ApJ* **671** 1579-90
[18] Müller H S P, Schlöder F, Stutzki J and Winnewisser G 2005 *J. Mol. Struct.* **742** 215-27
[19] Papadopoulos P P 2010, *ApJ* **720** 226-32
[20] Rigopoulou D, Kunze D, Lutz D, Genzel R and Moorwood A F M 2002 *A&A* **389** 374-86
[21] Sakamoto K, Ho P T P, Iono D, Keto E R Mao R-Q, Matsushita S, Peck A B, Wiedner M C, Wilner D J and Zhao J-H 2006 *ApJ* **636** 685-97
[22] Takano S, Nakai N and Kawaguchi K 2002 *PASJ* **54** 195-207
[23] Wang M, Henkel C, Chin Y-N, Whiteoak J B, Hunt Cunningham M, Mauersberger R and Muders D 2004 *A&A* **422** 883-905
[24] Ward J S, Zmuidzinas J, Harris A I, Isaak K G 2003 *ApJ* **587** 171-85
[25] Weiß A, Neininger N, Hüttemeister S and Klein U 2001 *A&A* **365** 571-87

A fresh look at the starburst-AGN connection

M Blank[1], W J Duschl[1,2]

[1] Institut für Theoretische Physik und Astrophysik, Christian-Albrechts-Universität zu Kiel, Leibnizstr. 15, 24118 Kiel, Germany

[2] Steward Observatory, The University of Arizona, 933 N. Cherry Ave., Tucson, AZ 85721, USA

E-mail: mblank@astrophysik.uni-kiel.de

Abstract. There is ever mounting evidence for a connection between the evolution of AGN and starburst-activity, as, for instance, suggested by a relation between the mass of the central black hole and the velocity dispersion of the bulge stars. However, the nature of this connection remains unclear, raising a number of questions, e. g.: Is the AGN triggered by the starburst or is the starburst activity caused by AGN? Which physical processes link starbursts and AGN? Here we will give a short review of crucial observations and theoretical work and describe our plans in that regard.

1. Introduction

Many galaxies show AGN and starburst activity at the same time as, e. g., revealed by the infrared astronomical satellite (IRAS) in 1984. This satellite discovered many luminous infrared galaxies (LIRGs, ULIRGs) whose luminosity is caused by dust-obscured starbursts and AGN driven by strong interacting merger systems as stated in e. g. Sanders et al. [1].

Furthermore, there exists an entire body of observations indicating a correlation, and thus, in all likelihood, a physical connection between these two phenomena. Some theoretical approaches have been undertaken as well, but as yet, however, it is not really clear how starbursts and AGN actually interact with each other, leaving a lot of open questions.

Understanding the starburst-AGN connection will yield a deeper understanding of the origin of starbursts and of AGN, in other words: the evolution of galaxies in general.

2. Observations

Numerous observations indicate a starburst-AGN connection, mostly through correlations between a starburst- and an AGN-related property. Examples are the relation between the black hole mass $M_{\rm BH}$ and the stellar velocity dispersion σ in the bulge, $M_{\rm BH} \sim \sigma^4$ (e. g., Gebhardt et al. [2]) or a relation between $M_{\rm BH}$ and the bulge mass $M_{\rm bulge}$ (e. g., Kormendy and Richstone [3]).

Wild et al. [4] find a correlation between the time since the onset of a starburst and the accretion rate of the AGN's black hole with the accretion rate steeply rising about 250 Myr after the onset of the starburst. This strongly indicates that the starburst triggers the AGN. Additionally they find that galaxies with strong star formation in their bulges are more likely to host a massive black hole than galaxies with low star-forming activity.

According to Veilleux [5] there is also strong evidence for circumnuclear starbursts in local AGN, and many ULIRGs contain an AGN coexisting with a starburst.

But whether these correlations are based on a connection between starbursts and AGN is difficult to answer, the nature of these correlations remains unknown.

3. Theory

Simulating the starburst-AGN connection has proven to be very challenging, because here very different physical scales are involved. On the one hand, mergers which occur on length scales of about 100 kpc lead to gas inflows which feed starbursts and AGN, on the other hand the accretion of material onto black holes occurs on scales of the Schwarzschild radius. Therefore it is necessary to work with parameterizations. Such a numerical framework for investigating the starburst-AGN connection was developed by Springel et al. [6]. Here, the central black holes are modeled as collisionless sink particles, which grow by accreting gas from their environment. The accretion rate $\dot{M}_{\rm BH}$ of the AGN's central black hole with mass $M_{\rm BH}$ is estimated via a Bondi-Hoyle-Lyttleton parameterization, which is limited by the Eddington rate.

$$\dot{M}_{\rm BH} = \min\left(\frac{4\pi\alpha G^2 M_{\rm BH}^2 \rho}{(c_{\rm s}^2 + v^2)^{3/2}}, \frac{4\pi G M_{\rm BH} m_{\rm p}}{\varepsilon_{\rm r}\sigma_{\rm T} c}\right)$$

Here ρ is the density and $c_{\rm s}$ the sound speed of the gas, respectively, v is the velocity of the black hole relative to the gas, $m_{\rm p}$ the proton mass, $\sigma_{\rm T}$ the Thomson cross-section, c the speed of light, $\varepsilon_{\rm r}$ the radiative efficiency, which is usually set to 0.1, and α a numerical parameter of order unity.

A feedback of the AGN is realized via a heating rate $\dot{E} = \varepsilon_{\rm f} L$ that is proportional to the luminosity $L = \varepsilon_{\rm r}\dot{M}_{\rm BH}c^2$ of the AGN and therefore to the central black hole's accretion rate. It couples the accretion process thermodynamically to the surrounding gas. The feedback efficiency $\varepsilon_{\rm f}$ is usually set to 0.05. Star formation and stellar feedback are modeled by using a subresolution multiphase model for star-forming gas, see Springel and Hernquist [7] for details.

With this model Di Matteo et al. [8] simulated galaxy mergers and were able to reproduce the observed $M_{\rm BH} - \sigma$ relation, here the feedback of the AGN quenches both, star formation and further black hole growth due to unbinding gas from the galaxy, leaving a dead elliptical galaxy and therefore explaining the short quasar lifetime.

Johansson et al. [9] used the same methods but varied mass ratios and used different progenitor morphologies finding that the timescales for star formation termination and for the mass growth of the black hole increase for unequal-mass disk mergers compared to equal-mass disk mergers. Furthermore for original galaxy mass ratios of at least 3:1 star formation is not quenched completely.

DeBuhr et al. [10] use a local viscous estimate to determine the accretion rate onto a black hole, AGN feedback is realized via depositing momentum into the surrounding gas, which is proportional to the luminosity of the AGN. This results in a self-regulated black hole accretion, but different to the results of Di Matteo et al. [8] the AGN feedback does not quench star formation because it shuts down the accretion onto the black hole before unbinding the gas of the interstellar medium.

Likewise in Kim et al. [11] AGN-feedback heats the surrounding interstellar medium and suppresses further star formation resulting in a self-regulated growth of the black hole.

4. Open questions

Some aspects of the starburst-AGN connection have been investigated yet, but many questions remain unanswered or need further investigations:

- Which mechanisms are at all capable of connecting starburst and AGN activities? (e. g. stellar winds, SN ejecta, AGN feedback: jets, heating)
- Do AGN trigger starbursts or vice versa, or are they triggered by external processes? (e. g. tidal interactions, mergers)
- Does this connection show a cosmological evolution?
- Do starbursts and AGN always occur together?
- Can and do they occur as independent processes?
- Are starbursts and AGN well synchronized (with or without a time lag)?
- Can a starburst quench further black hole growth?

5. Future work

To gain a deeper understanding of the starburst-AGN connection and to answer some of the questions raised above we will perform numerical simulations using the TreeSPH code Gadget-2 by Springel [12] with the routines implementing star formation and SN feedback developed by Xiang-Grüß [13].

For treating black hole growth and AGN feedback we will use more specific methods. For instance assuming that the mass flow onto the galaxy's center forms an accretion disk around the black hole we will solve the corresponding hydrodynamic equations with a grid code embedded into Gadget-2 to calculate the mass flow onto the black hole.

References

[1] Sanders D B, Soifer B T, Elias J H *et al.* 1988 *ApJ* **325** 74
[2] Gebhardt K, Bender R, Bower G *et al.* 2000 *ApJL* **539** L13
[3] Kormendy J and Richstone D 1995 *ARA&A.* **33** 581
[4] Wild V, Heckman T and Charlot S 2010 *MNRAS* **405** 933
[5] Veilleux S 2001 *Starburst Galaxies: Near and Far* ed L Tacconi & D Lutz p 88
[6] Springel V, Di Matteo T and Hernquist L 2005 *MNRAS* **361** 776
[7] Springel V and Hernquist L 2003 *MNRAS* **339** 289
[8] Di Matteo T, Springel V and Hernquist L 2005 *Nat* **433** 604
[9] Johansson P H, Naab T and Burkert A 2009 *ApJ* **690** 802
[10] DeBuhr J, Quataert E, Ma C P *et al.* 2010 *MNRAS* **406** L55
[11] Kim J, Wise J H, Alvarez M A *et al.* 2011 *ApJ* **738** 54
[12] Springel V 2005 *MNRAS* **364** 1105
[13] Xiang-Grüß M 2009 *Large-scale structure of disc galaxies* Ph.D. thesis Christian-Albrechts-Universität zu Kiel

Astronomy at High Angular Resolution 2011 IOP Publishing
Journal of Physics: Conference Series **372** (2012) 012054 doi:10.1088/1742-6596/372/1/012054

Optical Integral Field Spectroscopy of NGC 5850

M Bremer[1], J Scharwächter[2], A Eckart[1,3], J Zuther[1], S Fischer[1], M Valencia-S.[1,3]

[1]I.Physikalisches Institut, Universität zu Köln, Zülpicher Str.77, 50937 Köln, Germany
[2]Research School of Astronomy and Astrophysics, The Australian National University, Mount Stromlo Observatory, Cotter Road, Weston Creek 2611, Australia
[3]Max-Planck-Institut für Radioastronomie, Auf dem Hügel 69, 53121 Bonn, Germany

E-mail: mbremer@ph1.uni-koeln.de

Abstract. Here we present the preliminary results of the analysis of VIMOS observations of the central 4.5 kpc of the double-barred galaxy NGC 5850. We use optical diagnostic diagrams to study the main ionization mecahnism across the field of view confirming the LINER nature in the continuum peak location. Also a star-forming (SF) region is found close to it (0.46kpc), a second SF region is located east of the center (1.6kpc). Further the data reveals a complex nuclear gas kinematics which is likely to be dominated by the secondary bar.

1. Introduction

Scaling relations, like the one between the mass of the central black hole and the stellar velocity dispersion of the bulge, suggest the co-evolution of the central engine and its host; or at least some form of feedback between them.

NGC 5850 is a prototype double-barred early-type spiral SB(r) b galaxy. It is located at a distance of 34.2 Mpc [1] and hosts a polar gaseous disk [2]. Its two spiral arms form ring-like structure, almost resembling the morphology of the center, where the inner bar is surrounded by a (pseudo-) ring. The north-western spiral arm is distorted probably due to a recent (< 200 Myr) high velocity encounter with NGC 5846 [3], which is $10'$ to the north-west of NGC 5850. The low X-ray luminosity $L_X(0.5 - 3\text{ keV}) = 10^{40.36}$ erg s^{-1} [4] suggests NGC 5850 to be in a state of low activity. The high luminosity of [NII] lines at the center classifies it as a Low Ionisation Nuclear Emission Region (LINER) [5; 6]. Proposed by Shlosman et al. (1989) [7] and simulated by Friedli et al. (1993) [8], nested bars can be a mechanism to fuel AGN. Its morphology and proximity make NGC 5850 an ideal object to study fuelling and feedback processes. Integral Field Spectroscopy (IFS) in the visible wavelength regime allows us to study in detail the excitation mechanisms of the gas and investigate gas kinematics.

2. Observations And Data Reduction

The observations were performed with the integral field spectrometer VIMOS on the Very Large Telescope (VLT) of the European Southern Observatory (ESO) and cover the wavelength range from 4100 − 8800 Å with blue and red gratings at high spectral and spatial resolution ($R \approx 2300$ at $\lambda = 5500$ Å, $R \approx 3200$ at $\lambda = 7500$ Å, $0.67''/px \approx 0.163$ kpc, FWHM(PSF) ≈ 2 px). The source was observed in four nights in 2009, two images per night

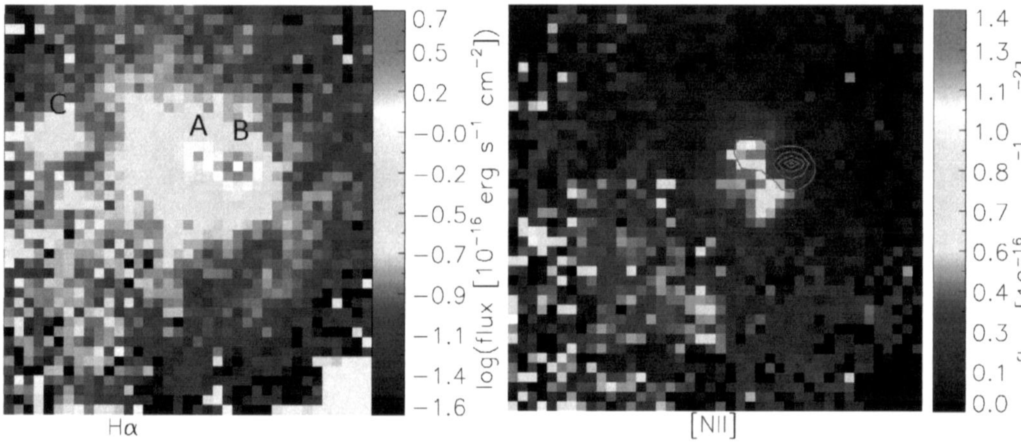

Figure 1. 27" × 27" emission maps; north is up, east is left. Left: Line map of Hα emission. The location of the continuum peak is labeled with 'A', the HII regions are labeled with 'B' and 'C'. In this and in all other images 1 px corresponds to approximately 0.163 kpc. Right: Line map of [NII] λ6585 emission with overplotted contour lines for 95, 90, 80, 60 and 40% of the Hα emission.

and per grating. The images were reduced with the VIMOS pipeline provided by ESO. We additionally applied a fringing correction. The resulting cubes were corrected for pixel-to-pixel variation and subsequently averaged for each grating separately. Flux calibrations were done by the VIMOS pipeline, but due to a number of bad pixels in the image of the standard star, the values are not reliable yet. However, the absolute flux calibration is not relevant for the preliminary results presented here, but will be improved for future analysis purposes. A sky correction was not performed, because no sky emission line lies close to those of the galaxy and the sky continuum contribution is neglible.

After correcting for intergalactic extinction we shifted the spectra to the rest frame. The stellar continuum was subtracted by application of the STARLIGHT program by [9] combined with the theoretical models by Bruzual and Charlot [10]. The emission lines were fitted in IDL with simple Gaussians using the MPFITPEAK and MPFITEXP routines. The Hα and the [NII] λ6549 lines were fitted together, though they are not blended, as well as the [SII], λλ6718 + 6732 lines. The latter were tied to each other by the constant wavelength difference between their peak values.

3. Preliminary Results

3.1. Emission Maps And Diagnostics

The Hα emission line map in the left panel of Fig. 1 shows three peaks, one located on the continuum peak (A), a second one 0.46 kpc (B) west of the center and a third one east at 1.6 kpc (C) projected distance to the core (A). The comparison of the map with the one for [NII] λ6585 line emission in the right panel of Fig. 1 implies, that location B and C are clearly distinctive in their ionisation mechanisms from the core, showing only little [NII] emission compared to the Balmer emission. This is also supported by the emission maps of Hβ and [OIII] λ5008 (left and right panel of Fig. 2, respectively), where Hβ is strong in region B, weak in the core region A and even weaker in region C. Unfortunately for region C no [OIII] fluxes could be obtained, because of low S/N. However, considering the distance of region C to the

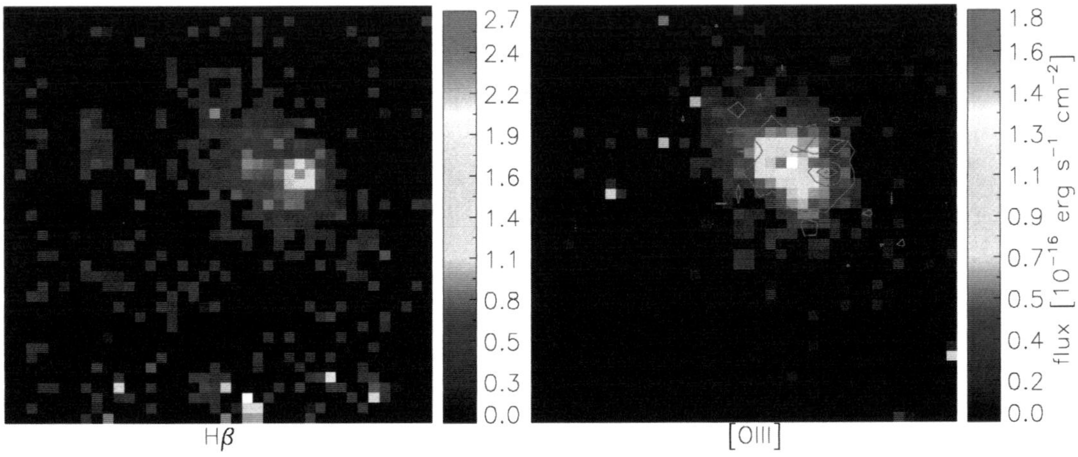

Figure 2. 27" × 27" emission maps; north is up, east is left. Left: Line map of Hβ emission. Right: Line map of [OIII] λ5008 emission with overplotted contour lines for 95, 90, 80, 60 and 40% of the Hβ emission.

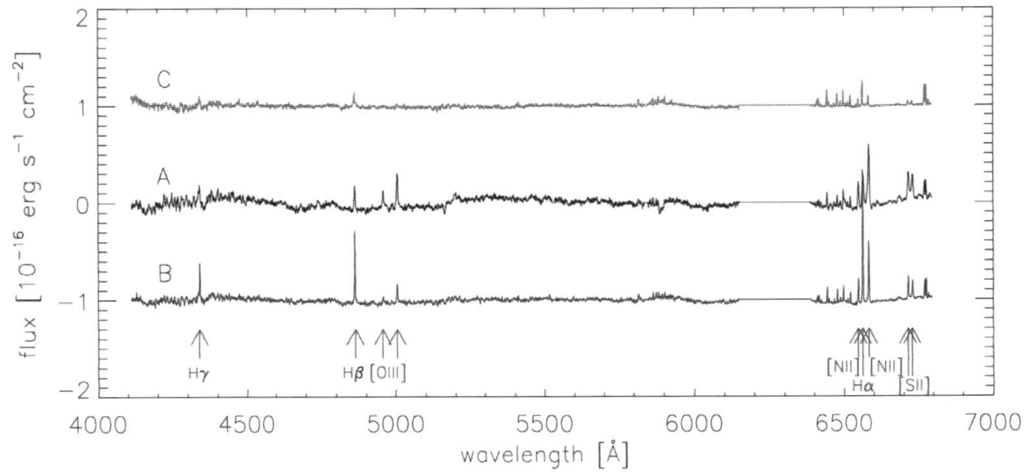

Figure 3. Spectra of the regions of interest stacked of 3 by 3 pixels centered on the respective brightest pixel. The spectra are denoted according to the regions depicted in Fig. 1. For better clarity of the plot, the values of spectrum B and C were shifted by $\pm 1 \cdot 10^{-16}$ erg sec^{-1} cm^{-2}.

center, it is likely to be a star-forming region. Also to mention is the fact, that the [NII] λ6585 emission does not appear to be confined to the core region but it extends to the south-west. In Fig. 3 the stacked spectrum (consisting of 3 by 3 pixels) of each region marked in Fig. 1 is shown.

In agreement with [6], the diagnostic diagram in Fig. 4 reveals the brightest pixel of the core (A) to be a LINER. The brightest pixel in region B is classified as an HII region. Region C cannot be classified by the means of the diagnostic diagram, because its [OIII] λ5008 emission line could not be securely detected.

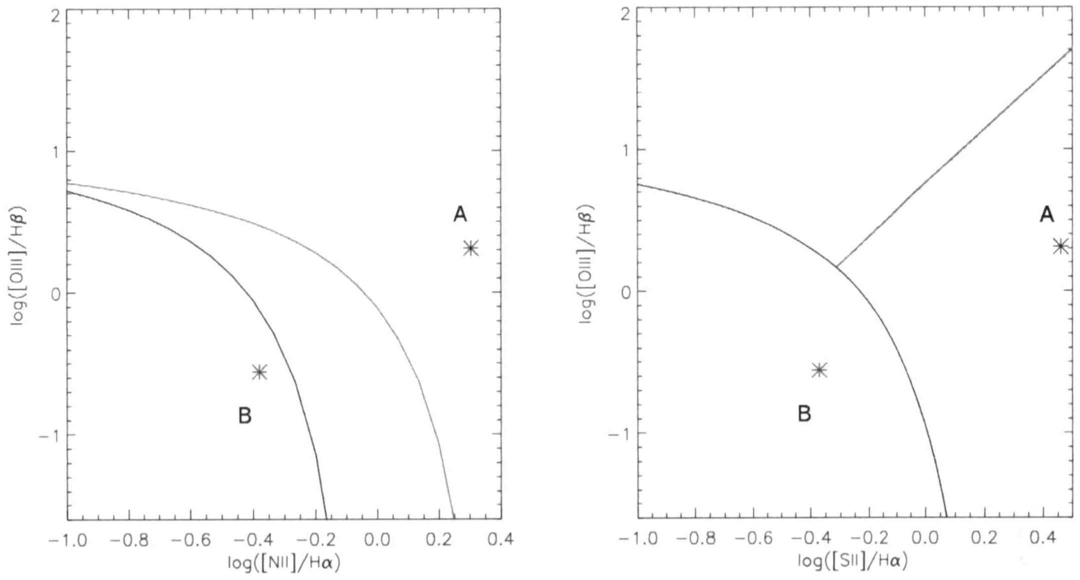

Figure 4. The locations of the brightest pixel in the regions A and B, as denoted in Fig. 1, in the diagnostic diagrams. Left: The red line marks the maximum starburst line, that represents the border between the composite (HII + AGN) and the AGN domain [11]; the black line indicates the border between the pure star-forming region and the composite region [12]. Right: The black line indicates the maximum starburst line, the red line marks the border between the Seyfert and LINER classification [13].

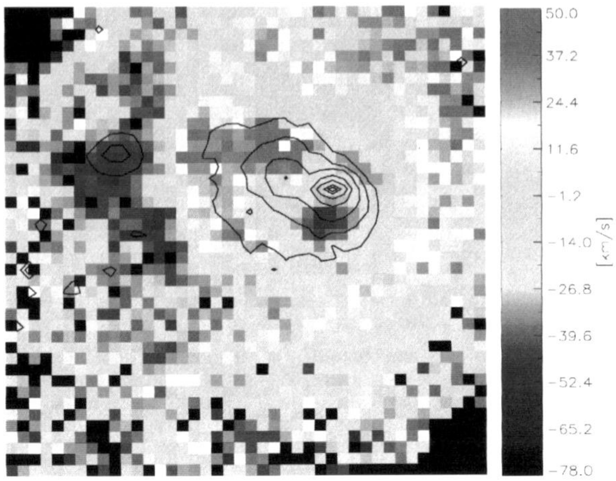

Figure 5. The line of sight velocity field derived from the Hα emission line with the same overplotted contours of Hα emission as in the right panel of Fig. 1. The field of view is 27" × 27"; north is up, east is left.

3.2. Kinematics

The Hα line of sight velocity map (Fig. 5) shows a peculiar structure, resembling the typical 'Yin-Yang' pattern and is possibly dominated by the nuclear bar. The gas kinematics is completely different from the rotational stellar velocity field published by [14]. The overplotted Hα emission contours show the HII region (B) either being located in an area of close to zero velocity or the region's intrinsic velocity being lower than that of the surrounding gas. The core (A) is also in an area of relative rest, as expected, but it is also found close to a region of high redshifted velocities to the north and east. The star-forming region in the east (C) is associated with high blueshifted velocities. Though very noisy, the eastern edge of the field of view suggests the detection of the galaxy's inner ring.

4. Conclusion And Outlook

Our observations confirm the low activity of NGC 5850 and the emission line ratios put it in the LINER area of the diagnostic diagram. The maps also show two star forming regions (denoted as B and C) at $\sim$ 1.6 kpc and $\sim$ 0.46 kpc projected distance to the core (A). The HII-region (B) close to the core is an interesting object, in particular, regarding fuelling scenarios of the central engine and its feedback into the host galaxy. The complex gas kinematics introduced by the inner bar is likely to provide further insight into fuelling scenarios for NGC5850.

However, the question whether the core of a LINER is a downscaled AGN has not been answered, yet. The extended [NII] emission around the core suggests that the LINER emission might not only be caused by the unresolved central engine. Alternatively, the LINER emission could be caused by shock ionisation. Nevertheless, central point sources are capable of creating extended line emission [15; 16], thus the presence of an AGN cannot be ruled out.

Acknowledgments

Part of this work was supported by the German Deutsche Forschungsgemeinschaft, DFG, via grant SFB 956 and fruitful discussions with members of the European Union funded COST Action MP0905: Black Holes in a violent Universe and PECS project No. 98040. M. B., J. S., J. Z. and S. F. gladly acknowledge the financial support by Group of Eight Australia-Germany Joint Research Co-operation Scheme via project-ID 50753527. M. V.-S. is member of the International Max Planck Research School (IMPRS) for Astronomy and Astrophysics at the MPIfR and the Universities of Bonn and Cologne.

References

[1] Wozniak H, Friedli D, Martinet L, Martin P and Bratschi P 1995 *Astron. And Astroph.* **111** 115
[2] Moiseev A V, Valdés J R and Chavushyan V H 2004 *Astron. And Astroph.* **421** 433–453 (*Preprint* arXiv:astro-ph/0311271)
[3] Higdon J L, Buta R J and Purcell G B 1998 *Astron. J.* **115** 80
[4] Fabbiano G, Kim D W and Trinchieri G 1992 *Astroph. J. Supplem. Series* **80** 531–644
[5] Heckman T M 1980 *Astron. And Astroph.* **87** 152–164
[6] Keel W C 1983 *Astroph. J. Supplem. Series* **52** 229–239
[7] Shlosman I, Frank J and Begelman M C 1989 *Nature* **338** 45–47
[8] Friedli D and Martinet L 1993 *Astron. And Astroph.* **277** 27
[9] Cid Fernandes R, Mateus A, Sodré L, Stasińska G and Gomes J M 2005 *Monthly Notes of the Royal Astron. Soc.* **358** 363–378 (*Preprint* arXiv:astro-ph/0412481)
[10] Bruzual G and Charlot S 2003 *Monthly Notes of the Royal Astron. Soc.* **344** 1000–1028 (*Preprint* arXiv:astro-ph/0309134)
[11] Kewley L J, Dopita M A, Sutherland R S, Heisler C A and Trevena J 2001 *Astroph. J.* **556** 121–140 (*Preprint* arXiv:astro-ph/0106324)

[12] Kauffmann G, Heckman T M, Tremonti C, Brinchmann J, Charlot S, White S D M, Ridgway S E, Brinkmann J, Fukugita M, Hall P B, Ivezić Ž, Richards G T and Schneider D P 2003 *Monthly Notes of the Royal Astron. Soc.* **346** 1055–1077 (*Preprint* arXiv:astro-ph/0304239)

[13] Kewley L J, Groves B, Kauffmann G and Heckman T 2006 *Monthly Notes of the Royal Astron. Soc.* **372** 961–976 (*Preprint* arXiv:astro-ph/0605681)

[14] de Lorenzo-Cáceres A, Falcón-Barroso J, Vazdekis A and Martínez-Valpuesta I 2008 *Astroph. J.* **684** L83–L86 (*Preprint* 0808.0517)

[15] Scharwächter J, Dopita M A, Zuther J, Fischer S, Komossa S and Eckart A 2011 *Astroph. J.* **142** 43 (*Preprint* 1106.0300)

[16] Yan R and Blanton M R 2011 *ArXiv e-prints* (*Preprint* 1109.1280)

Astronomy at High Angular Resolution 2011 IOP Publishing
Journal of Physics: Conference Series **372** (2012) 012055 doi:10.1088/1742-6596/372/1/012055

QPO in RE J1034+396: model constraints from observed trends

Bozena Czerny, Pawel Lachowicz
Copernicus Astronomical Center, Bartycka 18, 00-716 Warsaw, Poland
E-mail: bcz@camk.edu.pl

Michal Dovčiak, Vladimír Karas, Tomáš Pecháček
Astronomical Institute, Academy of Sciences, Boční II 1401, 14131 Prague, Czech Republic

Tapas K. Das
Harish Chandra Research Institute, Jhunsi, Allahabad 211 019, India

Abstract. We analyze the time variability of the X-ray emission of RE J1034+396, an active galactic nucleus with the first firm detection of a quasi-periodic oscillations (QPO). Based on the results of a wavelet analysis, we find a drift in the QPO central frequency. The change in the QPO frequency correlates with the change in the X-ray flux with a short time delay. Linear structures such as shocks, spiral waves, or very distant flares seem to be a favored explanation for this particular QPO event.

1. Introduction

The QPO appearance has been reported to occur in a bright Seyfert 1 galaxy, RE J1034+396 (Gierlinski et al. 2008) and BL Lacertae object PKS 2155-304 (Lachowicz et al. 2009). So far the QPO origin has not been well understood. The wavelet analysis can be particularly useful in the exploration of the QPO properties. Similar to Fourier spectra methods, the wavelet analysis provides fast linear operations on data vectors. Unlike the Fourier transform, the base functions of wavelets are localized, and this brings new opportunities to the data analysis. Based on the wavelet results we discuss possible constraints on the physical origin of the QPO in RE J1034+396.

2. Wavelet analysis of RE J1034+396 light curve

The source RE J1034+396 was observed by XMM-Newton satellite on 2007 May 31, and the resulting X-ray light curve analysis was reported in G08. Using the same procedure, we extract the source light curve in the 0.3-10 keV energy band, which covers 8.5×10^4 s of continuous observations sampled evenly every $t = 100$ s (see Figure 1). We use the wavelet analysis codes of Torrence & Compo (1998) to analyse the source behaviour.

The wavelet transform can be compared with the Fourier transform, which is used routinely to represent a measured light curve as a superposition of sine and cosine components. Both transforms can be classified into discrete versus continuous forms. In certain situations, wavelets

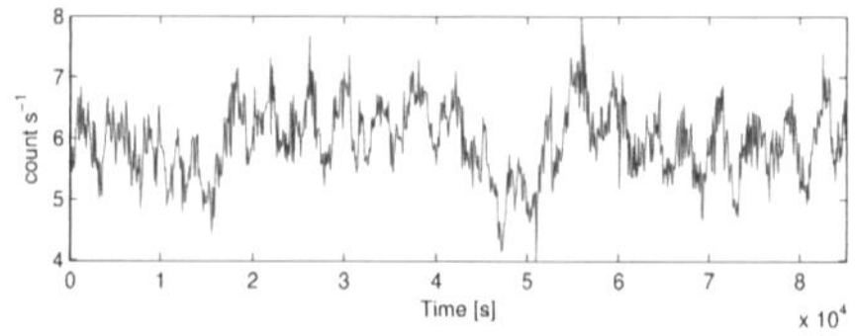

Figure 1. The X-ray light curve of RE J1034+396. The significant regions are within a period range of 3000–4500 s, where they form a rather complex pattern. The corresponding wavelet map is shown in Figure 2.

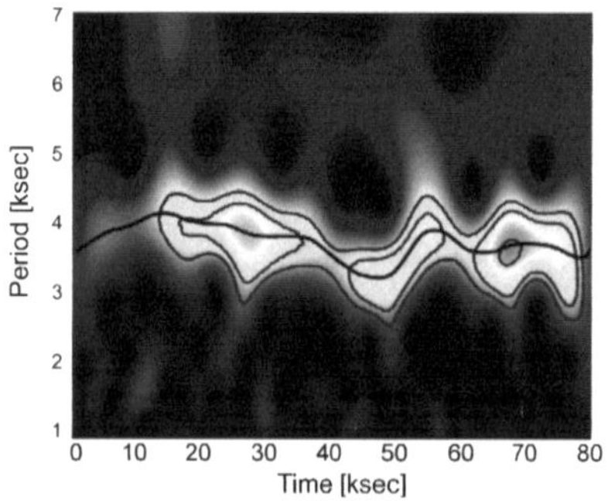

Figure 2. The increasing values of wavelet power are denoted by gradual changes of the colours, going from deep blue to red. The confidence contours of 90%, 95%, and 99% are marked. A solid black line traces the the QPO period.

can provide a more suitable representation of the signal by allowing a balanced resolution between time and frequency domains. Furthermore, the discrete wavelet transform can be less computationally expensive, as it requires $\mathcal{O}(N)$ time steps compared to $\mathcal{O}(N \log N)$ steps for the corresponding resolution in fast Fourier transform.

The wavelet map confirms the presence of the QPO in the source at the period $P_0 \sim 3733$ s through a range of peaks. In order to assign the confidence levels to those peaks we have performed Monte Carlo simulations (Czerny et al. 2010). We created 7500 simulated lightcurves and the corresponding wavelet maps. Then we built the statistics for each of the scales in the discussed range. This allowed us to assign significance levels for each wavelet scale independently, which is important in the lightcurves with underlying red noise background. The average wavelet values rise toward longer periods, as does the power density spectrum, and this drift causes the observed misalignment between the colors on the map (referring to absolute values) and confidence levels in Figure 2. In order to find whether any additional periodicity is present in the system, we analyzed a temporal variability of the QPO period. The periodogram analysis indicates a time-scale of $\sim$ 24 ks but the formal significance of the periodicity is low because of the high red-noise level.

3. Flux versus QPO period trend

The wavelet map shows a strong decrease in the marked QPO period at the time about 5×10^4 s, which is accompanied by a decrease in the X-ray flux. This motivated us to check for the overall correlation between the flux and the QPO period in the data. The correlation is present, and the time delay between the curves is 900 s, with the flux lagging behind the frequency change. The delay is by a factor 4.1 shorter than the QPO period itself. If only the second part of the data is investigated (Segment 2 in notation of G08), the measured delay is somewhat longer, 1400 s. The cross-correlation function between the X-ray luminosity and the QPO period for Segment 2 is shown in Figure 3, and the corresponding Flux-Period relation is given in Figure 4. It suggests that some underlying mechanism must be responsible for observed modulation of the QPO amplitude, the frequency and the flux.

Since the QPO of a 3733 s period smears the correlation at the longer time-scales, we extract the short-time trend from the light curve by applying the moving average approach, with a 5000 s window function. In consequence, we obtain a smoothed curve, free from temporal high-frequency flux variability. The scatter is the lower, as expected, but the overall trend is the same (see Czerny et al. 2010). Because the amplitude of the wavelet map also varies considerably, we tested for other possible correlations as well. The weak correlation between the wavelet amplitude and the QPO period was present but it was less significant than the correlation of X-ray flux vs. QPO period.

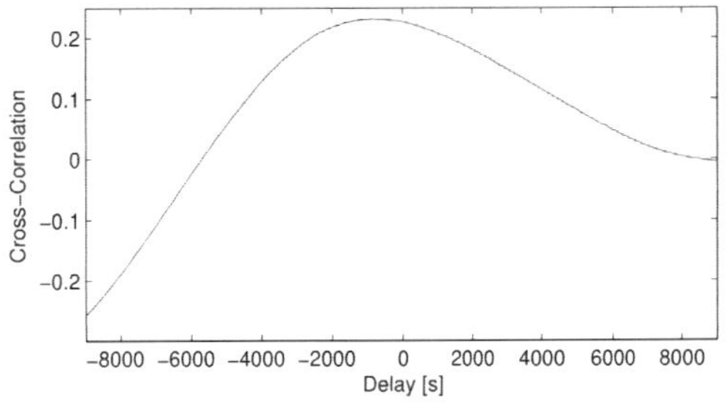

Figure 3. The normalized cross-correlation function of the QPO period and the X-ray light curve in the Segment 2.

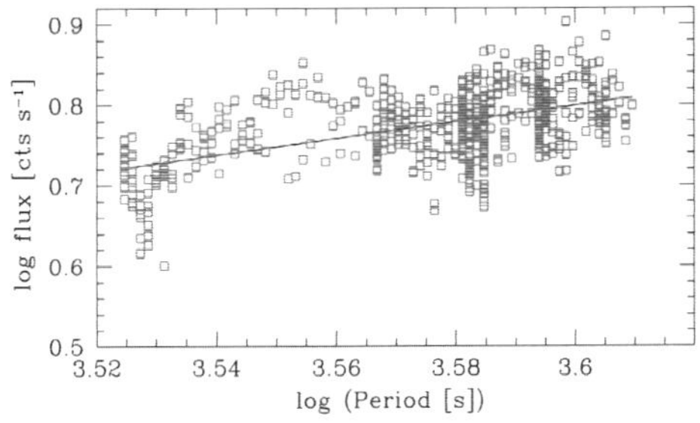

Figure 4. The flux versus QPO period relation in the same Segment 2 of the data as in Figure 3.

4. Shock oscillation scenario

We now study in detail one specific scenario of the QPO formation. In this model, QPO oscillations are due to the oscillations of the shock formed in the low angular momentum hot accretion flow, and the variation of the shock location corresponds to the observed changes in the QPO period and the X-ray flux. We calculate the shock location as a function of parameters for a stationary flow in the Kerr black hole metric. We calculate the QPO frequency from the Keplerian frequency at the shock location and the compression term, since the oscillations are not strictly dynamical, as shown by Molteni (1996). The X-ray flux attributed to the shock is estimated from the thermal energy flux across the shock. Next, we consider the variations of the flow parameters in quasi-stationary approach and study the relation between the flux and the period when either angular momentum of the matter or the Bernoulli constant vary. The examples of the Flux-Period coupling are shown in Figures 5 and 6.

We obtain that a change in the shock location caused by perturbation of the flow angular momentum is compatible with the trends observed in RE J1034+396, whereas the perturbation of the specific flow energy results in too strong flux response to the change of the oscillation period. Finally, we examined whether the results depend significantly on the spin of the black hole. However, since the shock forms not very close to the black hole, the effect is not strong. The slope of the Flux-Period relation is practically unchanged, the shock frequency itself varies with the Kerr parameter but this information cannot be used to determine the spin unless the black hole mass is known.

5. Discussion and Conclusions

Our analysis of the X-ray light curve of RE J1034+396 through the wavelet technique suggests the dependence of the QPO peak position in a single event on time, and the change of the QPO period correlates with the X-ray flux. This correlation implies a common source of perturbation. If the result indicating the proportionality between the flux and the QPO period is a more general property, then it implies strong constraints on the QPO mechanism. In particular, it appears to contradict the explanation in terms of an orbiting hot spot model and it supports the linear structures with intrinsic timescales like shocks, or spiral waves.

In Galactic sources, single QPO high-frequency events are currently unresolved, therefore this newly found correlation cannot be tested. However, future X-ray observations should bring more light curves of AGN with a quality good enough to perform a similar analysis, even if the QPO duty cycle is relatively low (Middleton et al. 2011).

There are also other approaches than the wavelet analysis which allow for a deeper insight into the QPO phenomenon. Zoghbi & Fabian (2011) examined time-lags of the signal arriving at different X-ray energy bands. In this way they reveal different reverberation properties of

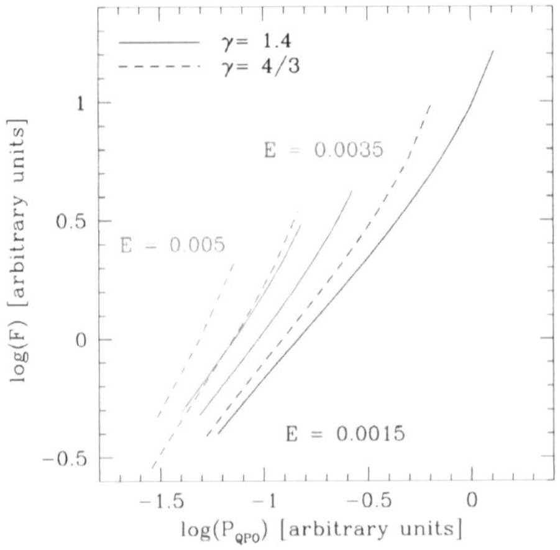

Figure 5. Flux vs. period relation for a fixed value of specific energy $\mathcal{E}$, and variable angular momentum, λ, for three pairs of solutions for three different values of $\mathcal{E}$. Each pair is characterized by two different values of the adiabatic indices, $\gamma = 4/3$ (dashed curve) and $\gamma = 1.4$ (solid curve).

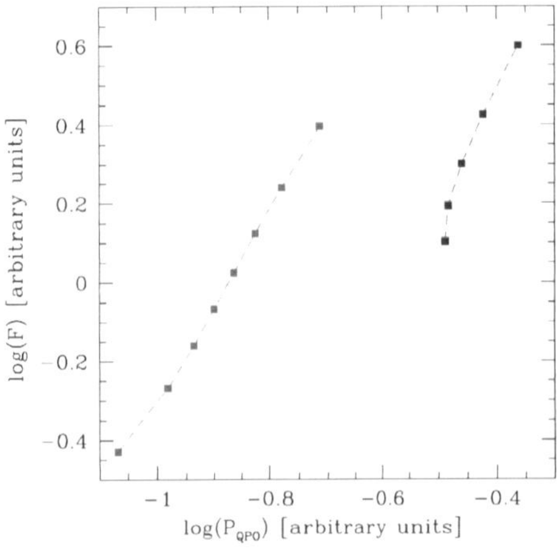

Figure 6. Flux vs. period relation for a fixed angular momentum, λ, and variable asymptotic value of the Bernoulli constant, $\mathcal{E}$. Two sequences represent two values of λ: 1.725 (left curve) and 1.775 (right curve), $\gamma = 4/3$.

the source during the period when the QPO appeared and when it was not present. Middleton et al. (2011) studied the covariance spectra and identified the soft component accompanying the QPO hotter than the soft X-ray excess seen in the average spectra. Both results imply that QPO signal is modulated at a distance of a few to few tenth of gravitational radii from the black hole.

This work has been supported in part by NN203 380136 grant in Poland, the XIth plan at HRI in India, and PECS 98040 in Czech Republic.

References

[1] Czerny B., Lachowicz P., Dovčiak M., Karas V., Pecháček T., Das Tapas K. 2010, A&A, **524**, A26
[2] Das Tapas K., Czerny B. 2011, MNRAS, **414**, 627
[3] Gierlinski M., Middleton M., Ward M., Done C. 2008, Nature, **455**, 369
[4] Lachowicz P., Gupta A. C., Gaur H., Wiita P. J.2009, A&A, **506**, L17
[5] Middleton, M., Uttley, P., Done, C., 2011, MNRAS, **417**, 250
[6] Molteni D., Sponholz, H., Chakrabarti, S.K., 1996, ApJ, **457**, 805
[7] Torrence C., Compo G. P. 1998, Bulletin of the American Meteorological Society, **79**, 61
[8] Zoghbi A., Fabian A.C. 2011, MNRAS, **418**, 2642

Astronomy at High Angular Resolution 2011 IOP Publishing
Journal of Physics: Conference Series **372** (2012) 012056 doi:10.1088/1742-6596/372/1/012056

Polarization in lamp-post model of black-hole accretion discs

M Dovčiak[1], F Muleri[2], R W Goosmann[3], V Karas[1] and G Matt[4]

[1] Astronomical Institute, Academy of Sciences, Boční II, CZ-14131 Prague, Czech Republic
[2] Istituto di Astrofisica Spaziale e Fisica Cosmica, Via del Fosso del Cavaliere 100, I-00133 Rome, Italy
[3] Observatoire Astronomique de Strasbourg, 11 rue de l'Université, F-67000 Strasbourg, France
[4] Dipartimento di Fisica, Università degli Studi "Roma Tre", Via della Vasca Navale 84, I-00146 Rome, Italy

E-mail: dovciak@astro.cas.cz

Abstract. We revisit the lamp-post geometry of the black-hole accretion disc with a primary illuminating source on the rotational axis. The primary X-ray power-law radiation is Compton reflected from the disc towards the observer. The gravitational field of a rotating black hole influences the photon properties on its way from the primary source to the disc and from the primary source and accretion disc to a distant observer. We study the polarization properties of the radiation how they would be observed in this scenario. The degree and the angle of polarization are examined as functions of the black hole spin, observers inclination angle and the position of the primary source.

1. The lamp-post geometry

The lamp-post geometry was originally proposed by [8] with the aim to provide a simple common scheme for the origin of the X-ray power-law continuum and the relativistic spectral features seen in accreting black hole sources. According to this arrangement, the primary 'lamp' moves

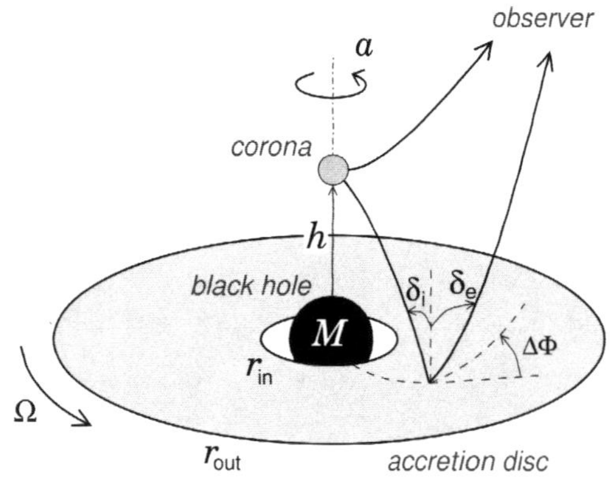

Figure 1. An elevated primary source situated on the axis of the black hole at height h illuminates the corotating accretion disc. The observer at infinity receives both the unpolarized power-law radiation from the lamp post as well as the polarized reflected light from the disc. The normal to the disc and incident and emitted light rays projected onto the disc surface are depicted by the dashed lines.

along the black hole axis, typically at heights of only up to 10 – 20 gravitational radii above the disc plane, see Fig. 1. The primary emission that is assumed to be unpolarized illuminates the accretion disc where it is reprocessed and part of it is emitted towards the observer. In this scenario the path of the light that travels from the lamp-post through some point in the disc up to the observer is set for each position on the disc unambiguously, thus defining the scattering geometry at every point in the disc plane. Here, we consider only direct photons, i.e., we assume optically thick disc and we neglect the photons that might bend near the photon orbit in such a way that they would circle around the black hole and only then either reach the observer or strike the surface of the disc.

We compute the flux and polarization as the observer at infinity would measure in this scenario, including both the reprocessed and direct emission. We use four energy bands (2–6 keV, 6–10 keV, 10–20 keV and 20–50 keV), three different inclination angles (30, 60 and 80 degrees) and various values of the height, h, of the primary source. Two different values of the angular momentum of the black hole have been explored: a static, a=0, and an extremely rotating black hole, a=1, where, as customary, a indicates the angular momentum per unit mass in units of the gravitational radius. We use these units for the remainder of this paper. The accretion disc in our computations extends from the marginally stable orbit ($r_{\rm in} = r_{\rm ms}$) up to the radius $r_{\rm out} = 1000\, GM/c^2$.

2. Local polarization of the reflected emission

We assume that the primary source of unpolarized emission radiates isotropically with the specific intensity being the usual power-law dependence on the energy, $I_{\rm lamp}(E) = N_{\rm lamp}\, E^{-\alpha}$. In our computations we use the value of the power-law index $\alpha = 1$ and the normalization factor $N_{\rm lamp}$ to be unity in the local static frame. In the equatorial plane of the system we assume a neutral cold geometrically thin and optically thick Keplerian accretion disc. Intensity of the reflected radiation in the local frame co-moving with the accretion disc was computed by the

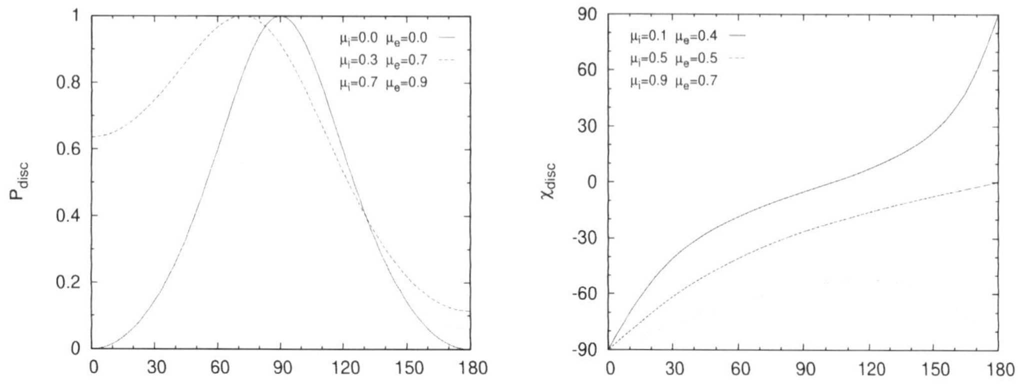

Figure 2. The properties of the local reprocessed radiation depend on the position on the disc. This is due to their dependence on the local scattering geometry. *Left:* The local polarization degree can have any value between 0 and 1. We show the dependence on the relative azimuthal angle $\Delta\Phi$ for three pairs of incident and emission cosines. *Right:* The local polarization angle obtains any value between -90° and 90°. The dependence on the relative azimuthal angle $\Delta\Phi$ for three pairs of incident and emission cosines is shown. A polarization angle of 0° represents the direction perpendicular to the disc and it is rotated counter-clockwise for positive values when looking towards the approaching photon.

Monte Carlo multi-scattering code NOAR [6]. The single scattering approximation [2] is used for the local polarization of the reflected continuum component, see [5] for the details. The polarization properties of the reflected radiation depend highly on the scattering geometry, i.e., the incident and the emission angles, δ_i and δ_e, as well as the relative azimuthal angle, $\Delta\phi$, of the incident and emitted light rays. This is illustrated in Fig. 2. We stress the dependence on the incident and relative azimuthal angles for the lamp-post geometry. This feature is additional to the pure directionality dependence on the emission angle in the case of an extended corona when the disc is usually assumed to be illuminated isotropically.

3. Overall polarization far from source

To get the resultant polarization far from the source, one has to integrate the emission over the disc surface and thus one integrates the local polarization properties. From the previous section it is obvious that the result of such integration is not easily estimated because of the complicated dependence on the local geometry of scattering. Moreover, relativistic effects enhance radiation from some parts of the disc. They also rotate the polarization angle and thus the overall observed polarization properties will be given by an interplay of the local polarization properties on the disc and relativistic change acting on photons on their way to the observer. The dependence of the polarization degree and angle at infinity on the lamp-post height and observer inclination is shown in Fig. 3 and 4.

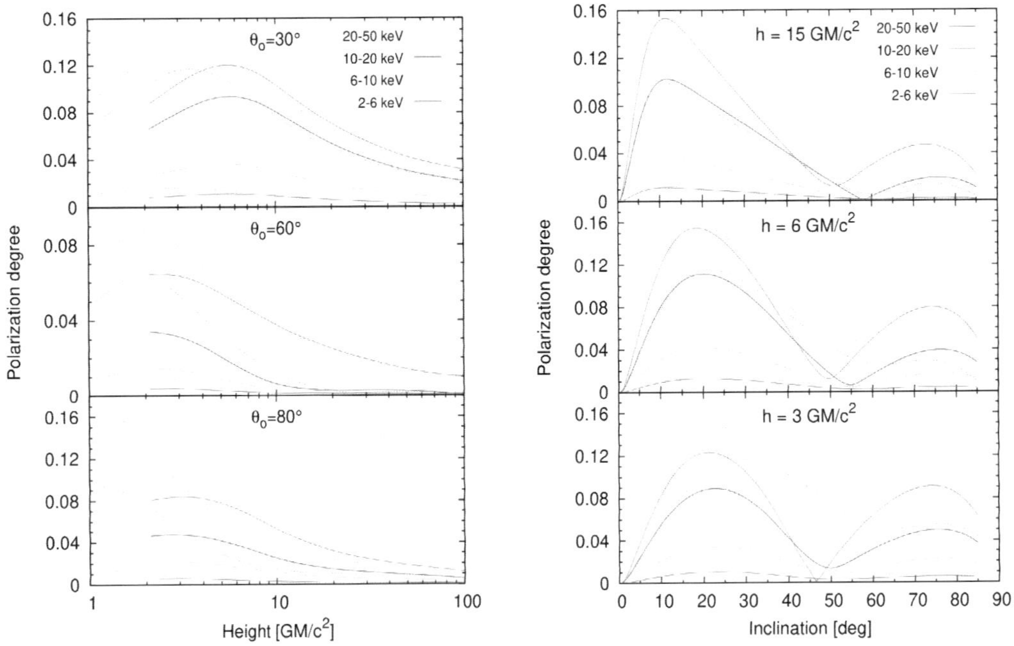

Figure 3. The dependence of the polarization degree on the height of the primary source (left) and inclination (right) as the observer at infinity would measure. The radiation from the primary source is taken into account as well. The value of the observer inclination, θ_o, the height, h, and the energy ranges are depicted inside the graphs. The solid lines show the results for the non-spinning Schwarzschild black hole, $a = 0$, while the dotted lines are for highly spinning Kerr black hole, $a = 1$. The polarization degree shown would be zero for unpolarized light and unity for totally polarized light.

Journal of Physics: Conference Series **372** (2012) 012056 doi:10.1088/1742-6596/372/1/012056

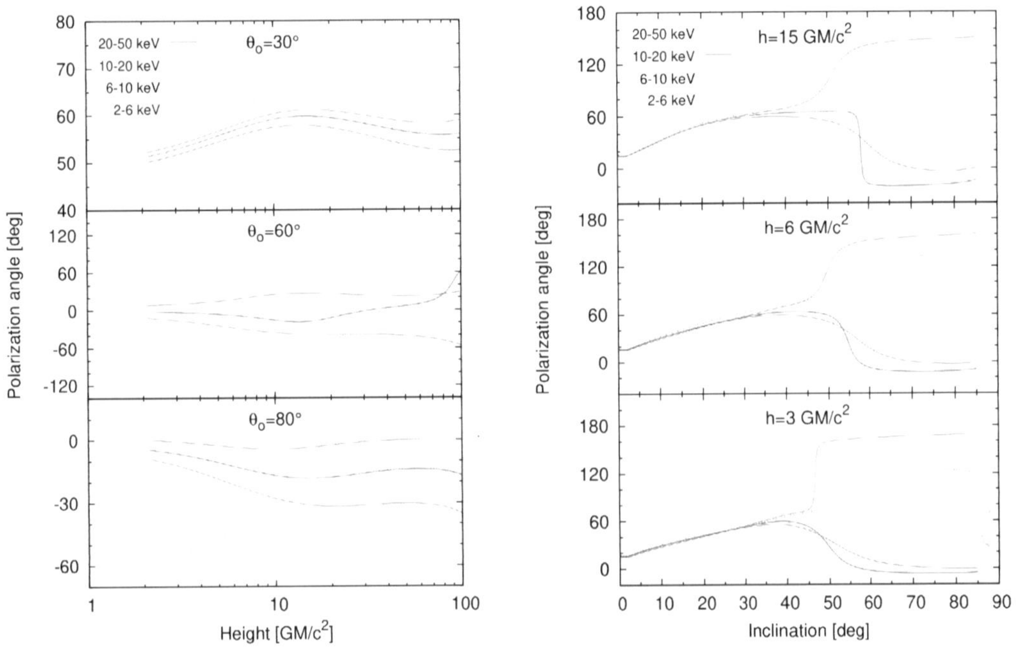

Figure 4. The same as in Fig. 3 but for the polarization angle.

The polarization degree is highest when the radiation from the inner accretion disc has a large contribution to the overall flux. This is opposite to the de-polarization one gets in the inner disc for the Comptonized thermal radiation, see [4, 9]. The cause of the thermal de-polarization is the relativistic rotation of the polarization angle because the local polarization is constant in that case. In the reflection model the local polarization angle changes due to its dependence on the local scattering geometry. When integrating the contributions from the disc, the local polarization angle adds up with the relativistic rotation in such a way that the overall polarization is much less de-polarized.

This has two consequences. Firstly, the highest polarization degrees are achieved for low lamp-post heights when the inner region of the disc is more illuminated than its more distant parts. Secondly, the radiation from the extreme Kerr black-hole accretion disc is more polarized than that in the Schwarzschild case where there is a hole in the disc below marginally stable orbit. Note, however, that there are some exceptions to the latter behaviour, see the results for some intervals of the lamp-post heights and for the observer inclination 30° and 80°.

From left panels in Fig. 3 and 4 one can see that the polarization degree and angle for different spins are almost the same for heights around $30\,GM/c^2$ and above. This is due to the fact that at these heights the innermost accretion disc already makes quite a small contribution to the overall reflected radiation.

The polarization degree has a maximum for low observer inclinations, e.g. in the highest studied energy band, $20-50\,\mathrm{keV}$, for the lamp-post height $h = 3\,GM/c^2$ it peaks in the extreme Kerr case at inclination 25° where it reaches almost 15% polarization (see the right panel in Fig. 3).

The polarization degree for zero inclination angle has to be zero because of the symmetry. As soon as the symmetry is broken by non-zero inclination the total polarization increases, determined by the region below critical radius[1] which for low inclinations is far from the center. As soon as the emission above this radius starts to contribute significantly the overall polarization starts to decrease with the inclination angle (the larger the inclination the lower the critical radius). This turn-around is at lower inclinations for larger heights because higher lamp-post illuminates better farther radii (i.e. radii above the critical one). There is another turn-around at the inclination when the critical radius moves close to the center. The contribution to the polarization from below and above critical radius cancel each other and the dependence of the polarization degree on the observer inclination reaches its minimum. For even higher inclinations the polarization again increases, it is determined mainly by the emission from far above the critical radius with lower and lower contamination from the regions around and below this radius. For very high inclinations the reflection is small and thus the overall polarization decreases again.

The polarization angle at infinity is quite sensitive to the details described in the points mentioned above and its dependence on height and inclination is rather complex. As discussed above for the polarization degree, the polarization angle for low inclinations is also determined mainly by the region below the critical radius while for high inclinations by the region above it. The change in the behaviour is very well visible by the rapid change of polarization angle with the inclination in Fig. 4. The transition happens at the inclination when the polarization above critical radius starts to dominate for higher inclinations. The transition depends on the lamp-post height and energy band and it can be either gradual or quite abrupt. Notice that there is no transition (or a very mild one) in the bottom panel in Fig. 4 for the extreme Kerr black hole and for a very low height of the primary source. This is due to the fact that the inner region below the critical point still has large impact on the polarization at infinity even for very high inclinations. In this case also the dip in the dependence of the polarization degree on inclination is not so deep, see the dotted graphs of bottom right panel in Fig. 3.

4. Predictions for the AGN case — MCG–6-30-15

In our simulations we model the polarization observations of the Seyfert 1 galaxy MCG–6-30-15. We assume that the lowest flux state of MCG–6-30-15 observed in [7] corresponds to the

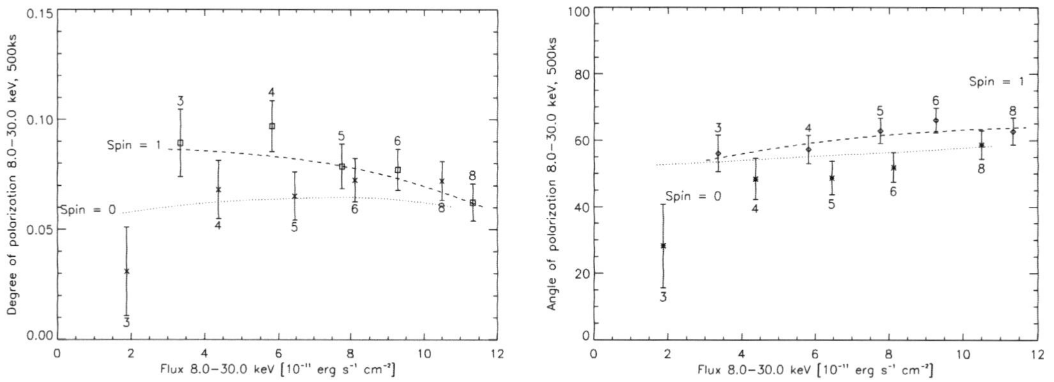

Figure 5. The correlation of the degree (left) and of the angle of polarization (right) with the flux for MCG–6-30-15. Each point corresponds to a different height of the source, specified within the figure (in units of GM/c^2). The dashed and the dotted lines are the expected dependence for spin 1 and 0, respectively.

[1] for the definition of the critical point and its role in the relativistic rotation of the polarization angle see [3]

primary source height $h = 3\,GM/c^2$. This choice, though arbitrary, is supported by conclusions of other authors, who suggest that the illuminating source is quite close to the central black hole. We assume that the observation inclination is 30° and that the black hole in MCG–6-30-15 is maximally rotating.

In the lamp-post geometry the polarization usually increases with the energy (see Fig. 3). Therefore we use the Gas Pixel Detector, see [1] for details, onboard a medium sized satellite mission sensitive in the interval 8–30 keV, as an example of the observation simulations. In Fig. 5 we show the correlations of the degree and angle of polarization with the flux. Both the flux and the polarization are integrated in the energy range 8–30 keV to avoid the iron line which is expected to be unpolarized. Each point represents an observation of 500 ks and corresponds to a different state of the source, that is a different height as indicated in the figure. Although the angle of polarization is almost constant with the flux, a certain evolution of the degree of polarization should be detectable. Moreover, we report for comparison the evolution of the polarization expected in the case of a Schwarzschild black hole to show that such evolution is somehow dependent on the spin of the central black hole.

5. Conclusions

In this paper we have discussed the observational properties of the polarization in the lamp-post geometry (the light bending model) and we used our theoretical computations to model possible future observations by next-generation X-ray satellite missions equipped with an X-ray polarimeter on board.

The polarization degree in a lamp-post geometry is higher at the highest studied energy band, 20–50 keV. This is a natural result coming from the fact that the primary source has a power-law spectrum with a negative index. Thus the best results for polarimetry of reflection spectra should be achieved in the Compton hump energy region. We also get higher polarization degree for large values of spin of the black hole, small height of the primary source and low inclination of the observer.

The behaviour of the polarization degree and polarization angle in the light bending model is quite complex. Because the result depends on the interplay of several parameters, the polarization properties may be degenerate with respect to different parameter values (e.g. for larger heights we cannot distinguish a difference between a rotating and a non-rotating black hole). Nevertheless, when combined with spectral and timing observations, polarimetry is yet another important channel that can help us to uncover the physical parameters of the black-hole accretion disc systems, such as the black hole spin, the system inclination and the height of the illuminating source.

Acknowledgments

MD and VK acknowledge the Czech Science Foundation (ref. 202/09/0772), the ESA PECS project (ref. 98040) and Academy of Sciences of the Czech Republic. GM and FM acknowledge financial support from Agenzia Spaziale Italiana (ASI I/088/06/0).

References

[1] Bellazzini R and Muleri F 2010 *NIMPA* **623** 766–770
[2] Chandrasekhar S 1960 *Radiative transfer* (Dover: New York)
[3] Dovčiak M, Goosmann R W, Karas V and Matt G 2008 J. Phys.: Conf. Series 131 012004
[4] Dovčiak M, Muleri F, Goosmann R W, Karas V and Matt G 2008 *MNRAS* **391** 32–38
[5] Dovčiak M, Muleri F, Goosmann R W, Karas V and Matt G 2011 *ApJ* **731** 75
[6] Dumont A-M, Abrassart A and Collin S 2000 *A&A* **357** 823–838
[7] Lee J C, Fabian A C, Reynolds C S, Brandt W N and Iwasawa K 2000 *MNRAS* **318** 857–874
[8] Matt G, Perola G C and Piro L 1991 *A&A* **247** 25–34
[9] Schnittman J D and Krolik J H 2009 *ApJ* **701** 1175–1187

Astronomy at High Angular Resolution 2011 IOP Publishing
Journal of Physics: Conference Series **372** (2012) 012057 doi:10.1088/1742-6596/372/1/012057

A close look at Seyfert 2 nuclei

Sebastian Fischer[1], **S. Smajic**[1,2], **M. Valencia-S.**[1,2], **A. Vitale**[1,2], **J. Zuther**[1], **and A. Eckart**[1,2]

[1] I.Physikalisches Institut, Universität zu Köln, Zülpicher Str.77, 50937 Köln, Germany
[2] Max-Planck-Institut für Radioastronomie, Auf dem Hügel 69, 53121 Bonn, Germany

E-mail: `fischer@ph1.uni-koeln.de`

Abstract. We present SINFONI adaptive optics assisted and seeing limited NIR integral field spectroscopy of the central hundreds of pc of ten z < 0.01 Seyfert 2 galaxies. The main goal of this study is to assess the significance of star formation and extinction in the circumnuclear region of Seyfert 2s. The immediate surroundings of the nuclei are resolved at linear scales of about 50-100 parsecs for most of the observed sources. The intensity and line-of-sight velocity distribution of different species is derived from the 3D SINFONI data by calculating the higher order moments of the emission lines. As part of this work in progress, the resulting maps are currently analyzed following the approach of generalized surface photometry, which allows us to identify the multiple kinematical components in the circumnuclear region of Seyfert 2s.

1. Introduction

If most galaxies harbor a central massive black hole, why are the nuclei of some galaxies more active than others? Probably, the main difference is in the available amount of fuel and the efficiency of feeding of the central engine (cf. [1], [2]). Bars or disturbed morphologies are supposed to be efficiently fuelling the nuclear region with gas from the outskirts of the galaxy. L. K. Hunt and M. A. Malkan [3] investigate the detailed circumnuclear morphologies of Seyfert 1, Seyfert 2, LINER, and starburst galaxies using HST/NICMOS. They report that H II/starburst galaxies show the strongest deviations from smooth elliptical isophotes, while the normal galaxies and LINERs have the least disturbed morphology. The Seyfert 2s have significantly more twisted isophotes than any other category, and the early-type Seyfert 2s are significantly more disturbed than the early-type Seyfert 1s, which cannot be explained by pure orientation effects. However, even the near-infrared (NIR) can be affected by dust extinction, which has to be taken into account (e.g. [4]).

On 0.1–1 kpc scales circumnuclear star formation is an important process in all types of AGN. In particular, empirical population synthesis of Seyfert 2 nuclei ([5], [6]) reveals evidence for recent (< 1 Gyr) star formation within a few hundred parsecs of the nuclei in about 40% of Seyfert 2s. Star formation models have begun to consider star formation within the torus ([7], [8]) which is proposed by the unification scheme, and it has shown that the contribution of dust is a key component in the nuclear region of such active galaxies.

It has been found that Seyfert 2 galaxies exhibit a higher contribution of star formation to the nuclear emission than Seyfert 1 galaxies, challenging the simple unified scheme ([9]). However, when compared to the well-known Kennicutt-Schmidt law, the star formation in the circumnuclear regions of Seyfert galaxies appears to be suppressed, emphasizing the importance of AGN feedback ([10]). In this context, the analysis of the velocity distribution in the circumnuclear region is of key interest (c.f. NuGa consortium work). Only a thorough decomposition of the different kinematical components in the vicinity of the AGN can help us in understanding the mechanisms behind the feeding of the central supermassive black hole.

2. The data

2.1. The sample

We obtained SINFONI $H+K$ imaging spectroscopy of the central 400 pc of 10 nearby $V^{\rm mag} < 14$ Seyfert 2 galaxies - based on the VCV catalog ([11]). Since the analysis of circumnuclear star formation and extinction is our main interest in this study, we selected only sources, where high resolution optical HST images are available. Apart from being optically classified as Seyfert 2 galaxies, no selection on the base of morphology has been imposed. Fig. 1 shows large scale visible HST images of the sample sources.

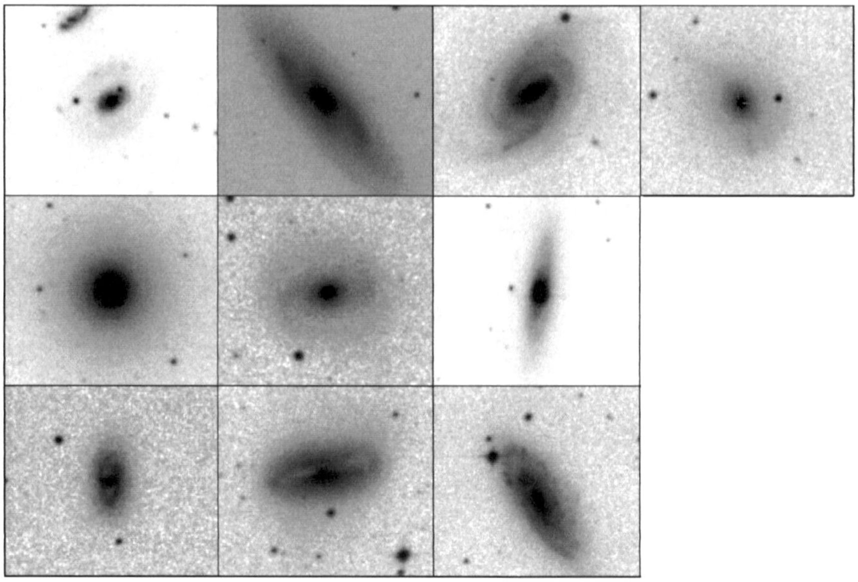

Figure 1. HST F606W images of the observed Seyfert 2 galaxies. The size of each image is approximately $3.25' \times 2.75'$. North is up, East is to the left. Note that the SINFONI field of view only covers the immediate circumnuclear region of 3" × 3" (centered on the nucleus).

2.2. Reduction and analysis

If possible, SINFONI was operated in AO mode using the 100mas scale and guiding was always performed on the nucleus itself. Here, the central 700 pc can be resolved up to linear scales of about 50-100 pc. Due to weather constraints, the performance of the AO system was limited to a PSF of approximately 0.4" (derived from the K-band continuum). For some galaxies, weather prevented the AO to close the loop and it was only possible to operate SINFONI in seeing-limited mode, where we chose the 250mas in order to compensate for the loss in spatial resolution. Telluric standards were observed prior and following the science targets and basic data reduction was applied using the ESO SINFONI pipeline.

Our analysis of one of these galaxies, NGC 7172, has recently been published in [4]. For the first time, we were able to identify broad recombination lines at the center of this extremely

Journal of Physics: Conference Series **372** (2012) 012057 doi:10.1088/1742-6596/372/1/012057

edge on galaxy. Our results suggest that in this case, significant galactic extinction might prohibit a detection of the Seyfert 1 core region at visible wavelengths, making even polarimetric measurements of reflected broad emission difficult.

As shown exemplary in Fig. 2, the core regions of our Seyfert 2 galaxies show a plentitude of emission lines such as recombination and molecular Hydrogen lines and, in certain cases, also the coronal line [SiVI]. In order to handle this analysis in a very systematic approach, we are developing an IDL analysis tool for 3D-data in general that extracts emission line maps together with the line-of-sight (LOS) velocity field and a map of the LOS velocity dispersion (Fig. 2). These information are derived from the 3D data cubes by calculation of the zeroth, first and second order moments of the respective emission line. The calculation of the higher order moments is, in particular for low S/N spectral regions, significantly more robust than an e.g. Gaussian fit to the emission line. However, since the noise in the resulting moment map increases with the order of the moment, we introduced a S/N clipping on the base of the zero order moment, i.e. the line has to be detected with a minimum of 3–5 sigma above the continuum noise. Resulting moment maps are shown in Fig. 3 for two different circumnuclear Seyfert 2 regions. Additional output of our tool yields error propagation for the three moments.

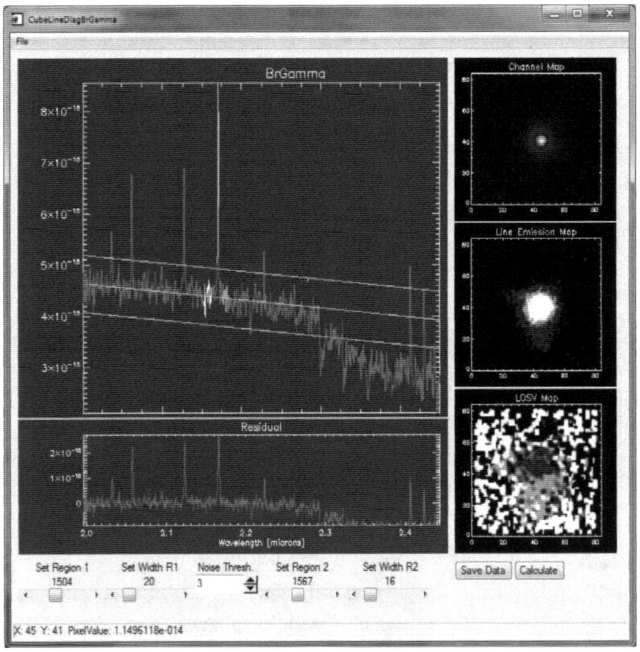

Figure 2. Dialog of the 3D-data analysis tool developed to create emission line and maps of the line-of-sight velocity field and the velocity dispersion. These maps are created by calculating higher order moments. As an example, the central spectrum is plotted in the top left window, with continuum regions defined close to the Brγ emission line and the continuum subtracted residuum on bottom left. The 3-sigma continuum noise is indicated in green. Several additional emission lines such as molecular Hydrogen can be seen in the spectrum together with prominent stellar CO absorption features. The different maps display the (currently selected) channel map, the emission line map (0th-order), and the LOS velocity map (1st order, from top to bottom right). On the bottom of the window, various adjustments to the definition of the extraction can be made. The maps can be created interactively or in a scripted way.

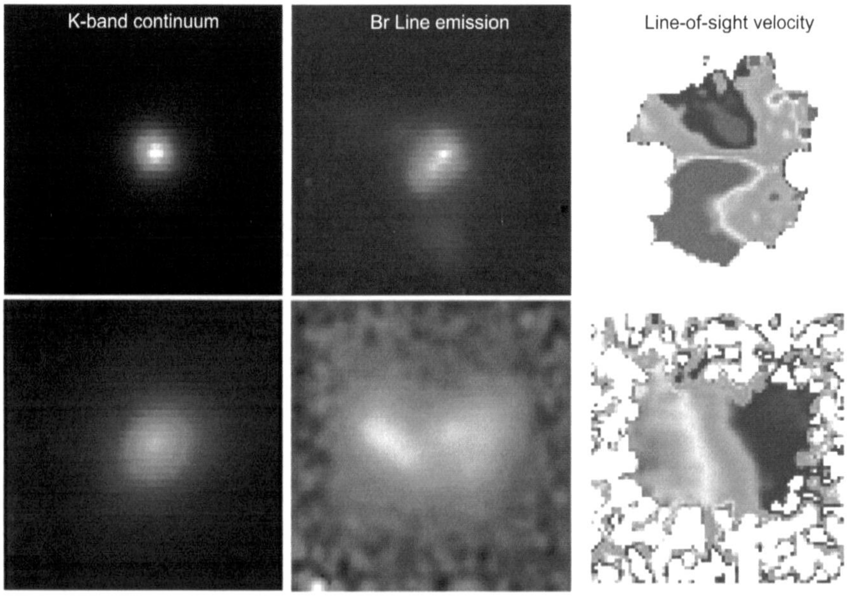

Figure 3. K-band continuum, Brγ zero-order (i.e. emission line) and first order (i.e. line-of-sight velocity) maps for two selected sample galaxies. The maps show the central 4×4 arcsecs, which represents the central 500 pc. The spatial resolution derived from the K-band continuum is, in both cases, of the order of 0.4 arcsecs (despite AO in use, weather limited its performance). Prominent star formation regions and structures suggesting a mini spiral can be identified.

3. Outlook

We are reporting preliminary results on the NIR study of the central hundreds of parsecs of a sample of Seyfert 2 galaxies. As the morphology of the recombination-line emission already implies (Fig. 3), we are able to spatially resolve different circumnuclear star formation regions. In addition to line ratios in these prominent emission line regions, which are giving us insight into the excitation mechanisms and the possible influence of the AGN (e.g. [12], [13]), the kinematic information contained in the higher order maps of the LOS velocity distribution are currently under investigation.

Similar to what is described in [14], the analysis of the different moment maps following the approach of a generalized surface photometry allows us to detect multiple co- and counter rotating components. In contrast to the recombination lines, the extended molecular hydrogen emission can in most cases be described by a single disc component. In a next step, these results have to be compared to extinction maps derived from a multi-component continuum decomposition together with the high resolution visible data, because especially when dealing with Seyfert 2 nuclear regions, extinction can play an important role. In addition, the different regions will be modeled using starburst models such that eventually, a thorough model of the circumnuclear environment of each of the Seyfert 2 nuclei will be obtained.

Acknowledgments

S. Fischer is supported by the German federal department for education and research (BMBF) under the project numbers: 50OS0502 & 50OS0801. S. Smajic, M. Valencia-S. and A. Vitale are members of the International Max-Planck Research School (IMPRS) for Astronomy and Astrophysics at the Universities of Bonn and Cologne supported by the Max Planck Society.

Part of this work was supported by the COST Action MP0905: Black Holes in a violent Universe and PECS project No. 98040.

References

[1] I. Shlosman, M. C. Begelman, J. Frank, 1990, Nature, 345, 679-686
[2] P. F. Hopkins, T. J. Cox, D. Keres, L. Hernquist, 2008, ApJS, 175,356
[3] L. K. Hunt, M. A. Malkan, 2004, ApJ, 616, 707
[4] S. Smajic, S. Fischer, J. Zuther, A. Eckart, 2012, A&A, in print.
[5] R. Cid Fernandes, L. Sodrém H. R. Schmitt, J. R. S. João, 2001, MNRAS, 325, 60
[6] R. M. González Delgado, T. Heckman, C. Leitherer, 2001, ApJ, 546, 845
[7] K. Wada, G. Meurer, C. A. Norman, 2002, ApJ, 577, 197
[8] T. A. Thompson, E. Quataert, N. Murray, 2005, ApJ, 630, 167
[9] J.-M. Wang, E.-P. Zhang, 2007, ApJ, 660, 1072
[10] J.-M. Wang, Y.-M. Chen, C.-S. Yan, C. Hu, W.-H. Bian, 2007, ApJ, 661, 143L
[11] M.-P. Véron-Cetty, P. Véron, 2006, A&A, 455, 773
[12] J. Zuther, S. Fischer, J.-U. Pott, T. Bertram, A. Eckart, C. Straubmeier, C. Iserlohe, W. Voges, G. Hasinger, 2007, ESO Msngr., 128, 44
[13] J. Scharwächter, M. A. Dopita, J. Zuther, S. Fischer, S. Komossa, A. Eckart, 2011, AJ, 142, 43
[14] D. Krajnović, M. Cappellari, P. T. de Zeeuw, Y. Copin, 2008, MNRAS, 366, 787K

Astronomy at High Angular Resolution 2011 IOP Publishing
Journal of Physics: Conference Series **372** (2012) 012058 doi:10.1088/1742-6596/372/1/012058

Continuum reverberation mapping in a $z = 1.41$ radio-loud quasar

L J Goicoechea, V N Shalyapin, R Gil-Merino and V F Braga

GLENDAMA Project Team, Facultad de Ciencias, Universidad de Cantabria, Avda. de Los Castros s/n, 39005 Santander, Spain

E-mail: goicol@unican.es, vshal@ukr.net, r.gilmerino@gmail.com, bravi4@hotmail.com

Abstract. Q0957+561 was the first discovered gravitationally lensed quasar. The mirage shows two images of a radio-loud quasar at redshift $z = 1.41$. The time lag between these two images is well established around one year. We detected a very prominent variation in the optical brightness of Q0957+561A at the beginning of 2009, which allowed us to predict the presence of significant intrinsic variations in multi-wavelength light curves of Q0957+561B over the first semester of 2010. To study the predicted brightness fluctuations of Q0957+561B, we conducted an X-ray, NUV, optical and NIR monitoring campaign using both ground-based and space-based facilities. The continuum NUV-optical light curves revealed evidence of a centrally irradiated, standard accretion disk. In this paper, we focus on the radial structure of the standard accretion disk and the nature of the central irradiating source in the distant radio-loud active galactic nucleus (AGN).

1. Introduction

Reverberation (or echo) mapping is a time-domain technique to resolve the accretion flow in AGNs. This relies on the analysis of time-delayed responses of different emitting regions to original fluctuations in an irradiating source [1]. While concurrent X-ray-UV-optical continuum monitoring campaigns of low-redshift AGNs are leading to puzzling results (e.g., [2]), there is only one AGN at $z \geq 1$ with a continuum reverberation map: Q0957+561 [3]. The $z = 1.41$ radio-loud quasar Q0957+561 suffers a strong gravitational lens effect, so one observes two images Q0957+561A and Q0957+561B from Earth. Intrinsic flux variations in Q0957+561B lag those in Q0957+561A by about 14 months [4], and we taken advantage of this time delay between images. Optical follow-up observations in late 2008 and the first semester of 2009 showed very prominent variations in Q0957+561A [5], which enabled us to know in advance the flaring behaviour of Q0957+561B during the first semester of 2010, and to use the Liverpool Robotic Telescope (Sloan *griz* bands), the UVOT on board the Swift satellite (U band) and the Chandra Space Telescope (X-rays) for such a time period.

From the Ugr light curves of Q0957+561B in 2010, we measured interband delays of several days that were consistent with a centrally irradiated, standard accretion disk [3]. In this scenario, a flaring source very close to the central supermassive black hole illuminates a standard accretion disk. The central flares are then thermally reprocessed into FUV-MUV variations in the inner disk to produce the observed NUV-optical variability. In Section 2 we present the NIR observations and the corresponding light curves, which were not discussed in [3]. These new NIR data are used to check the accretion scenario and discuss in detail the disk radial structure. In

Section 3 we infer possible reconstructions of the effective luminosity of the central irradiating source from the NUV-optical-NIR light curves. Our main conclusions appear in Section 4.

2. NIR observations and the accretion scenario

In 2010, we conducted a NIR monitoring programme of Q0957+561 with the RATCam CCD camera on the Liverpool Robotic Telescope at La Palma, Canary Islands. The exposure times were 120 and 240 s per night in the i and z bands, respectively. The pre-processing steps included in the telescope pipeline are: bias subtraction, overscan trimming and flatfielding. In addition, we interpolate over bad pixels using the bad pixel mask, clean some cosmic rays and remove fringe patterns. Our crowded-field photometry pipeline is then used to produce instrumental fluxes of both quasar images over 54 epochs. To achieve a reasonable compromise between photometric quality and sampling rate, later we apply thresholds on the signal-to-noise ratio in the i and z bands. The final processing stage consists of the conversion from instrumental magnitudes to physical fluxes (in mJy), and the final NIR fluxes are available at `http://grupos.unican.es/glendama/LQLMII_DR.htm`. We also display the NIR light curves of Q0957+561B in Fig. 1a. This includes 50 data points at λ_o = 7481 Å (i-band) and 45 data points at λ_o = 8931 Å (z-band). While the i-band flux accuracy is 1.4%, the z-band flux accuracy is only 2.0%. Fig. 1b includes all NUV-optical-NIR records of Q0957+561B in 2010.

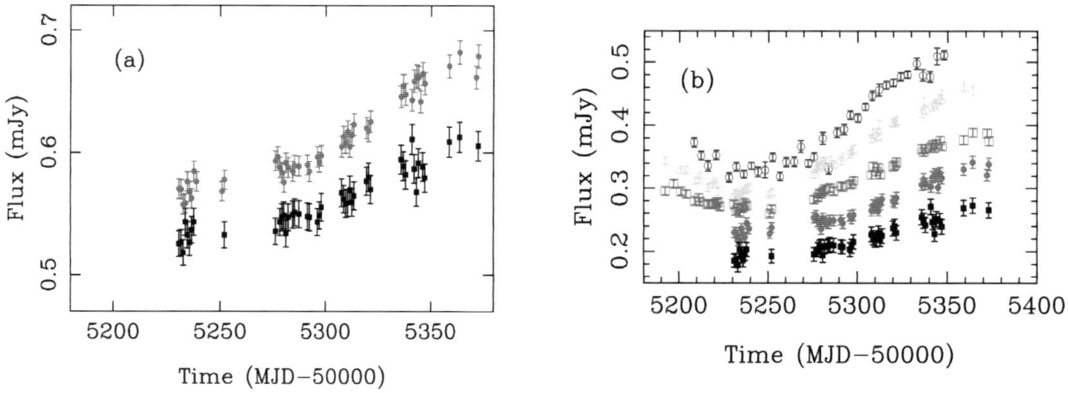

Figure 1. (a) Fluxes of Q0957+561B in the i [●] and z [■] bands of the SDSS photometric system in 2010. (b) NUV-optical-NIR light curves of Q0957+561B in 2010: U [○], g - 0.13 mJy [△], r - 0.25 mJy [□], i - 0.34 mJy [●] and z - 0.34 mJy [■] bands. This multiwavelength monitoring campaign was planned after an optical alert in 2009 (see main text).

As a result of the cosmic expansion, observed wavelengths (λ_o) are longer than emission wavelengths at the AGN (λ). Hence, our NUV-optical [3] and NIR light curves of Q0957+561B in 2010 correspond to UV continuum sources at z = 1.41 (emission at λ = 1438–3706 Å). In our recent paper [3], we have only analysed the best data from each telescope. This approach precluded the use of NIR data, since the photometric accuracy and time coverage in the iz bands are worse than those in the gr bands. Here, using a χ^2 minimization (e.g., [4]), the NIR light curves are compared to the g-band brightness record (λ = 1944 Å). We thus estimate two additional relative delays for emissions at λ = 3104 Å (i-band) and λ = 3706 Å (z-band): τ(3104 Å) - τ(1944 Å) = $9.0^{+1.3}_{-2.1}$ days and τ(3706 Å) - τ(1944 Å) = $14.0^{+3.1}_{-5.3}$ days (1σ intervals), which are used to check the inner accretion onto the supermassive black hole. These new delays confirm the scenario derived from the previous interband delays [3], i.e., all measured delays are consistent with observer-frame lags $\tau(\lambda) \propto \lambda^{4/3}$ between a central irradiating source and standard disk rings emitting at different wavelengths λ (e.g., [6]). Interestingly, the normalization based on NUV-optical data, τ(1944 Å) = 9 days, also works at λ = 3000–4000 Å (see Fig. 2).

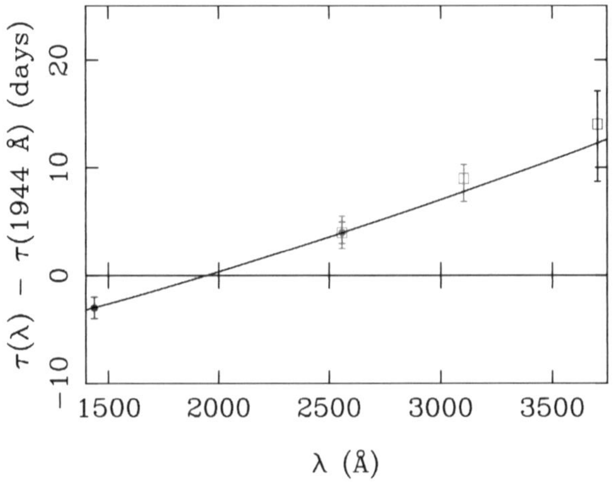

Figure 2. Relative delays using correlation functions [●] and a χ^2 minimization [□]. We show that both techniques (correlation functions and χ^2) produce similar delays at $\lambda = 2558$ Å (r-band). We also display the predictions of the centrally irradiated accretion disk with $\tau(1944$ Å$) = 9$ days [——].

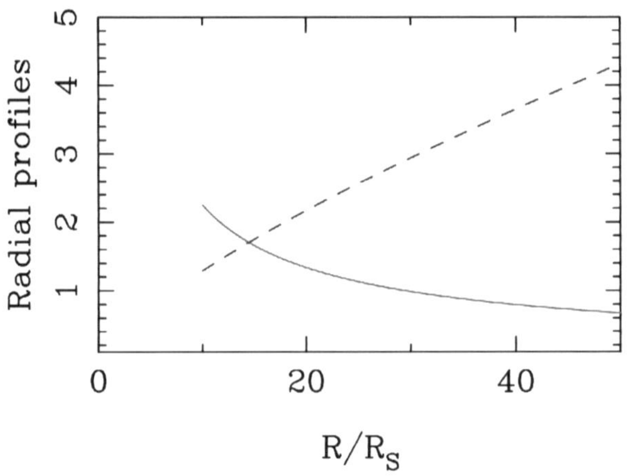

Figure 3. Radial profiles for the irradiated accretion disk in Q0957+561: temperature in 10^4 K [——] and emission peak wavelength in 10^3 Å [- - -].

The UV continuum sources (λ = 1438–3706 Å) lie within 10–30 R_S of the central black hole, where R_S is the Schwarzschild radius for a black hole mass $M = 2.5 \times 10^9 M_\odot$ [7]. If the irradiating source is just above the black hole, the disk temperature at radius $R >> R_S$ is given by (e.g., [6])

$$T(R) = [3GM\dot{M}(1+\alpha_{\rm iv})/8\pi\sigma R^3]^{1/4}, \tag{1}$$

where

$$\alpha_{\rm iv} = \frac{2(1-A)LH}{3GM\dot{M}}. \tag{2}$$

In Eqs. (1) and (2), G is the gravitation constant, σ is the Stefan constant, $\dot{M}$ is the mass accretion rate, $\alpha_{\rm iv}$ is the irradiation-to-viscosity ratio, A is the disk albedo, and L and H are the luminosity and height of the central irradiating source. Each central flare propagates out at the speed of light c and arrives at radius R after an observer-frame time $\tau = (1+z)R/c$. The temperature then rises and the peak emissivity increases, so each flare is reprocessed to roughly produce a fluctuation at $\lambda = 0.26\,[hc/kT(R)]$ (black-body emission peak), k and h being the Boltzmann constant and the Planck constant, respectively [6]. The expected $\tau - \lambda$ relationship is

$$\tau(\lambda) = 6\ (1+z)\left[\frac{3GM\dot{M}(1+\alpha_{\rm iv})}{8\pi\sigma c^3}\right]^{1/3}\left(\frac{k\lambda}{hc}\right)^{4/3}, \tag{3}$$

and we compare this Eq. (3) to the observed law $\tau(\lambda) = (9\text{ days}) \times (\lambda/1944\text{ Å})^{4/3}$. We obtain an effective mass accretion rate $\dot{M}(1 + \alpha_{\rm iv}) = 3\ M_\odot\ {\rm yr}^{-1}$ that translates into two extreme accretion-irradiation regimes:

(a) $\dot{M} = 3\ M_\odot\ {\rm yr}^{-1}$ and $L \leq 10^{45}$ erg s^{-1}, and
(b) $\dot{M} = 1\ M_\odot\ {\rm yr}^{-1}$ and $L = 10^{47}$ erg s^{-1},

assuming the reasonable constraints $4(1-A)(H/3R_{\rm S}) = 1$ and $\dot{M} \geq 1\ M_\odot\ {\rm yr}^{-1}$ [8]. Fig. 3 shows some radial profiles for the black hole mass and effective mass accretion rate of Q0957+561.

3. Central irradiating source

If central flares are thermally reprocessed in the disk at 10–30 $R_{\rm S}$ from the central supermassive black hole, what is the source of such flares? Standard simulations of X-ray reprocessing (e.g., [9]) ruled out the possibility that the disk variability is driven by a standard corona just above the black hole. The power-law X-ray emission is plausibly originated in the base of the Q0957+561 jet at a typical height of 200 $R_{\rm S}$ [3]. Moreover, the thermal source that we detect in the Chandra X-ray spectra ($kT = 0.08$ keV) can not account for the observed NUV-optical-NIR variations. A central EUV (unobservable!) source seems the best candidate to illuminate the disk and drive its variability. While standard reprocessing simulations rely on the lamppost model and use observed high-energy variations to try to reproduce observed fluctuations at lower energies (presumably originated in accretion disks; e.g., [9, 3] and the last paper in [2]), low-energy ($\lambda_{\rm o}$ = 3000–10000 Å) brightness records and inverse problem techniques (e.g., [10]) can be powerful tools for reconstructing emissivities of central irradiating sources. The so-called inverse problem in reprocessing has a very promising future, and here we concentrate on direct reconstructions of the central variable (effective) luminosity of Q0957+561 based on our NUV-optical-NIR data in 2010.

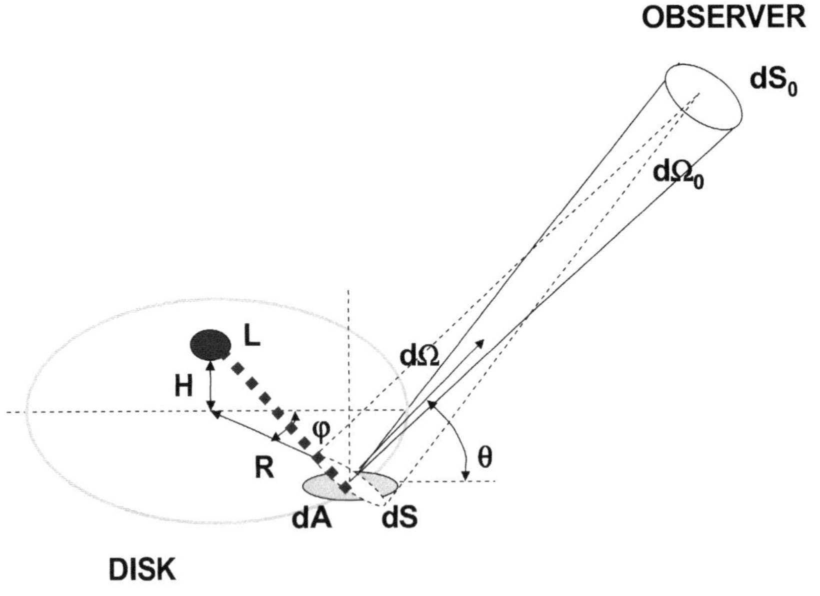

Figure 4. The standard accretion disk is illuminated by an isotropic source of luminosity L that is located at a height H above the central supermassive black hole (lamppost model). A distant observer is situated at latitude θ.

A fraction $1 - A$ of the illuminating EUV radiation would be absorbed in the disk, and then reprocessed into thermal radiation. This variable illumination would cause the variable emission of the disk. Considering the lamppost model and an observer at latitude θ (see Fig. 4), the contribution of the disk portion dA at (R, φ) to the flux received at time t_0 by the distant observer is related to the luminosity of the central source at time $t - \tau(R, \varphi; \theta)$, where $\tau(R, \varphi; \theta) = [(R^2 + H^2)^{1/2} - R\cos\theta\cos\varphi + H\sin\theta]/c$. As a result, the expected flux from the whole disk $F(\lambda_0, t_0)$ depends on different central luminosities $L(t - \tau)$ at different lags τ with respect to the emission time t. A simple approach consists of assuming an effective luminosity $L_{\rm eff}$ responsible for the disk emission by irradiation, which would correspond to an effective lag $\tau_{\rm eff}$. For each observation epoch t_0, one can minimize the difference between the modeled and observed flux: $|F_{\rm obs}(\lambda_0, t_0) - F_{\rm mod}(L_{\rm eff})|$, and thus obtain the value of $L_{\rm eff}$ for the data point at t_0. We take $M = 2.5 \times 10^9 M_\odot$ and $z = 1.41$, as well as a typical latitude $\theta = 45°$, $H = R_{\rm S}$ and $A = 1/4$. Note that the values of H and A verify the constraint $4(1 - A)(H/3R_{\rm S}) = 1$ (see above). It is not so easy to set a reasonable value of the AGN-observer transmission factor $\mu = \mu_{\rm lens} \times \mu_{\rm dust}$. This is a combination of two different contributions: the gravitational lens magnification ($\mu_{\rm lens} > 1$) and the dust extinction ($\mu_{\rm dust} < 1$) of Q0957+561B. We somewhat arbitrarily assume that both effects compensate each other, so that $\mu = 1$. First, we check the two extreme mass accretion rates that we find from the time delay analysis in Section 2, i.e., (a) $\dot{M} = 3\ M_\odot\ {\rm yr}^{-1}$ and (b) $\dot{M} = 1\ M_\odot\ {\rm yr}^{-1}$. Both extreme values lead to luminosities exceeding 10^{47} erg s^{-1}, in reasonable agreement with expected emissivities in the accretion-irradiation regime (b). Second, we select the self-consistent regime (b), and set $\dot{M} = 1\ M_\odot\ {\rm yr}^{-1}$.

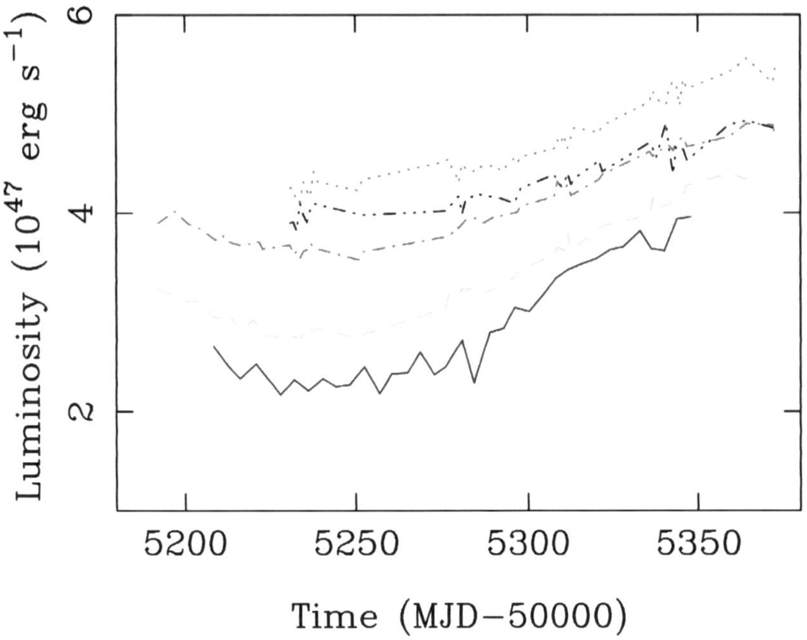

Figure 5. $L_{\rm eff} - t_0$ laws from data in the U [——], g [– – –], r [— · —], i [· · · · ·] and z [— · · —] bands.

We infer effective luminosities of a few 10^{47} erg s^{-1} from the NUV-optical-NIR records (see Fig. 5). The effective luminosity curves are smoothed versions of the actual luminosity curve. Thus, the variability from the Swift/UVOT data (U band) is more realistic than those from the optical-NIR fluxes, since the U-band source is relatively small and the smoothing is not so important. We also remark that the luminosity offsets between the $Ugri$-based reconstructions are plausibly due to wavelength-dependent dust extinction in the host galaxy, the lensing galaxy and the Milky Way. This produces a chromatic μ, which has not been taken into account in our calculations. In addition, the ordering of our reconstructions is the expected arrangement for an unrealistic achromatic extinction.

4. Conclusions

A recent multiwavelengtht monitoring of a sharp fluctuation in the trailing image of the gravitationally lensed radio-loud quasar Q0957+561 allows us to dicuss the accretion physics in the vicinity of a distant supermassive black hole at $z = 1.41$. There is evidence for an irradiated (standard) accretion disk around the black hole, where the disk heating is mostly generated from irradiation by a central source. This central irradiating source very likely emits EUV photons and produces luminosities $\geq 10^{47}$ erg s^{-1}. Our data also support the presence of an EUV luminosity variation of about 100% on a relatively short source-frame timescale of 40 days.

Acknowledgments

The authors thank Harvey Tananbaum and Neil Gehrels for granting Director's Discretionary Time for the Chandra and Swift observations, respectively. We also thank the operations group of the Liverpool Telescope (Robert Smith and Jon Marchant) for their kind interaction and support during our ground-based monitoring projects. The Liverpool Telescope is operated on the island of La Palma by Liverpool John Moores University in the Spanish Observatorio del Roque de los Muchachos of the Instituto de Astrofísica de Canarias with financial support from the UK Science and Technology Facilities Council. This research has been supported by the Spanish Department of Science and Innovation grants ESP2006-13608-C02-01 and AYA2010-21741-C03-03 (Gravitational LENses and DArk MAtter - GLENDAMA project), and University of Cantabria funds.

References

[1] Blandford R D and McKee C F 1982 *Astrophys. J.* **255** 419
Netzer H and Peterson B M 1997 *Astronomical Time Series*, ed D Maoz, A Sternberg and E M Liebowitz (Dordrecht: Kluwer) p 85
[2] Maoz D, Edelson R and Nandra K 2000 *Astron. J.* **119** 119
Gaskell C M 2006 *ASP Conf. Series* **360** 111
Arévalo P, Uttley P, Lira P, Breedt E, McHardy I M and Churazov E 2009 *Mon. Not. R. Astr. Soc.* **397** 2004
Breedt E et al 2009 *Mon. Not. R. Astr. Soc.* **394** 427
[3] Gil-Merino R, Goicoechea L J, Shalyapin V N and Braga V F 2012 *Astrophys. J.* **744** 47
[4] Kundic T et al 1997 *Astrophys. J.* **482** 75
[5] Goicoechea L J and Shalyapin V N 2009 *The Astronomer's Telegram* **2228** 1
[6] Collier S, Horne K, Wanders I and Peterson B M 1999 *Mon. Not. R. Astr. Soc.* **302** L24
Goicoechea L J, Shalyapin V N and Gil-Merino R 2010 *The Open Astronomy Journal* **3** 193
[7] Peng C Y, Impey C D, Rix H W, Kochanek C S, Keeton C R, Falco E E, Lehár J and McLeod B A 2006 *Astrophys. J.* **649** 616
[8] McLure R J and Dunlop J S 2004 *Mon. Not. R. Astr. Soc.* **352** 1390
[9] Kazanas D and Nayakshin S 2001 *Astrophys. J.* **550** 655
[10] Craig I J D and Brown J C 1986 *Inverse Problems in Astronomy* (Bristol: Adam Hilger Ltd)

Dynamical evolution of thin stellar discs

Jaroslav Haas and Ladislav Šubr

Astronomical Institute, Faculty of Mathematics and Physics, Charles University, V Holešovičkách 2, 18000 Praha, Czech Republic

E-mail: haas@sirrah.troja.mff.cuni.cz, subr@sirrah.troja.mff.cuni.cz

Abstract. We investigate, by means of direct numerical N-body integration, the orbital evolution of an initially thin stellar disc around a dominating central mass. As a perturbation to the Keplerian stellar motions, we consider the gravitational influence of an extended spherically symmetric stellar cusp and the mutual gravitational interaction of the stars within the disc. Our results suggest that the two-body relaxation of the disc, which is often neglected in similar analyses, leads to significant changes of its radial density profile on time-scales of 10^5–10^6 yr for systems with dimensions of the young stellar disc observed in the Sgr A* region.

1. Model

The individual components of the studied system are modelled in the following way:

- the dominating central mass is considered to be a source of static Keplerian potential of mass $4 \times 10^6\, M_\odot$,
- the stellar disc is treated as a group of 1000 gravitating particles on initially circular orbits with semi-major axes initially distributed according to $\mathrm{d}N/\mathrm{d}a \propto a^{-1}$, $a \in \langle 0.03, 0.3 \rangle$ pc,
- the spherical cusp is modelled either as a source of static gravitational potential, or as 5×10^4 gravitating particles on orbits with semi-major axes $a \in \langle 0.015, 0.6 \rangle$ pc.

The orbital evolution of this system is followed numerically, by means of the N-body integration code NBODY6 (Aarseth 2003).

2. Rapid two-body relaxation

Our calculations suggest that the orbital evolution of the disc leads to similar distribution of semi-major axes, regardless the considered way of modelling and characteristic mass of the spherical cusp. We thus understand that the energy transfer within the disc is determined solely by its high stellar density. This is demonstrated in Fig. 1 which shows the comparison of the initial distribution and the state after ≈1.3 Myr of orbital evolution for three models with a different treatment of the gravity of the spherical cusp — with no cusp present (solid boxes), with the cusp represented by a static potential (dotted boxes), and with the cusp treated as 5×10^4 particles (dashed boxes). In the latter two cases, both the radial density profile and characteristic mass of the cusp are equal.

The comparison with the initial state further indicates that the overall shape of the distribution evolves significantly. We thus suggest that the evolution of the radial density profile should be considered in the dynamical studies of particular observed systems, such as the young stellar disc in the Galactic Centre.

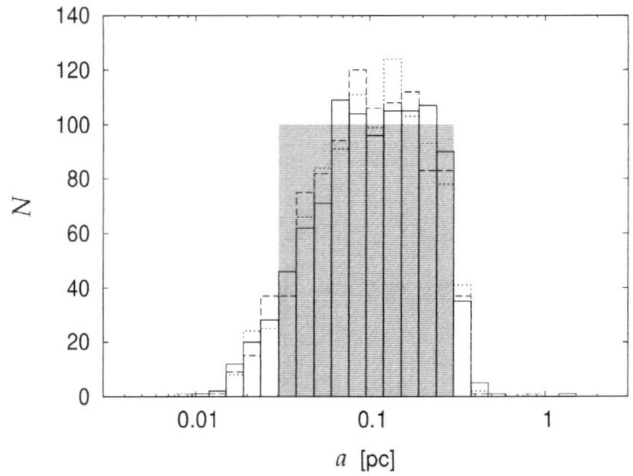

Figure 1. Distribution of the orbital semi-major axes in the disc after ≈1.3 Myr of its dynamical evolution for three models with a different treatment of the gravity of the spherical cusp: no cusp present (solid), the cusp as a source of static potential (dotted), the cusp treated as 5×10^4 particles (dashed). The initial distribution is denoted by the grey rectangle.

3. Enhanced growth of eccentricity and inclination

Fig. 2 illustrates the evolution of the mean eccentricity (left) and inclination (right) of the stellar orbits in the disc for the same three cases as in the previous section. If the spherical cusp is not included (solid lines), our results are in agreement with those obtained by Cuadra et al. (2008). The evolution is not affected, when the cusp modelled with a static potential is considered (dotted lines). On the other hand, if the cusp is treated in a full N-body way (dashed lines), the evolution of both orbital elements is significantly accelerated. This behaviour is in accord with the previous findings of Löckmann et al. (2009) who considered a somewhat simpler model of the cusp.

Acknowledgments

This work was supported from project 367611 of the Grant Agency of Charles University in Prague.

References

[1] Aarseth S J 2003 *Gravitational N-Body Simulations* (Cambridge: Cambridge University Press)
[2] Cuadra J, Armitage P J and Alexander R D 2008 *MNRAS* **388** L64
[3] Löckmann U, Baumgardt H and Kroupa P 2009 *MNRAS* **398** 429

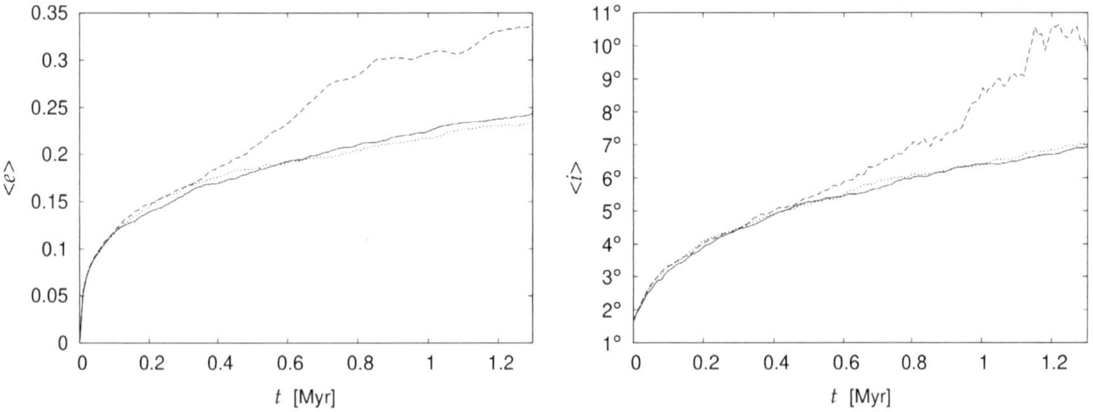

Figure 2. Evolution of the mean eccentricity (left) and inclination (right) of the stellar orbits in the disc for three different ways of modelling the gravity of the spherical cusp: no cusp present (solid), the cusp as a source of static potential (dotted), the cusp treated as 5×10^4 particles (dashed).

Probing broad-line region of the weak line quasar SDSS J094533.99+100950.1

K. Hryniewicz[1], M. Nikołajuk[2] and B. Czerny[1]

[1]Nicolaus Copernicus Astronomical Centre, Bartycka 18, 00-716 Warsaw, Poland
[2]Faculty of Physics, University of Białystok, Lipowa 41, 15-424 Białystok, Poland

E-mail: `krhr@camk.edu.pl`

Abstract.
SDSS J094533.99+100950.1 is a weak line quasar (WLQ) characterised by a low equivalent widths of the high-ionization emission lines such as CIV or HeII, typical iron emission, radio quiescence and X-ray weakness. In our work we tried to answer the question if it is possible that observed emission is intrinsically weak and come from not fully developed broad emission-line region (BELR). We also analyse the observed continuum to check if it is able to ionize a BELR gas and reproduce a weakness of emission-lines or this spectral feature can be simply explained by an intrinsic absorption. Our conclusion is that the minimal active galactic nuclei (AGN) engine consisting of an unobscured accretion disk with a partially ionized gas just expanding from the disk atmosphere is enough to explain observed features of SDSS J094533.99+100950.1.

1. Introduction

SDSS J094533.99+100950.1 (thereafter SDSS J0945+1009) is an exceptional quasar at redshift z $\simeq 1.67$ discovered by the Sloan Digital Sky Survey (SDSS) project. Its extremely weak medium-ionization (CIII], AlIII), high-ionization lines (CIV, HeII) together with typical FeII emission and near-ultraviolet power-law continuum are remarkable and puzzling. We present its spectra compared to ordinary quasar composite in Figure 1.

For PHL1811 analogues it was proposed that soft continuum is caused by shielding gas and consequences of soft continuum are low line intensities ([1], [2]). This explanation could be valid for all WLQs. In the case of SDSS J0945+1009 we checked if all observed components can be explained by a rather minimalistic AGN engine — accretion disk around black hole plus single cloud BLR without additional components.

2. Global parameters of SDSS J0945+1009

In Figure 2 we plot photometric points together with the mean spectral energy distribution (SED) of an ordinary quasar composed by [3]. The photometric points span the near-infrared (NIR), optical, and ultraviolet (UV) ranges and comes from the Two Micron All Sky Survey (2MASS) [4], SDSS [5] and the Galaxy Evolution Explorer (GALEX) [6] observational surveys. The Figure 2 shows that the optical-UV continuum of the quasar is soft, steep in UV region and dominated by quasar's accretion disk. The latest estimation of the black hole (BH) mass equals to $M = 2.9^{+5.5}_{-1.9} \times 10^9 M_\odot$ based on MgII line width and continuum luminosity measured

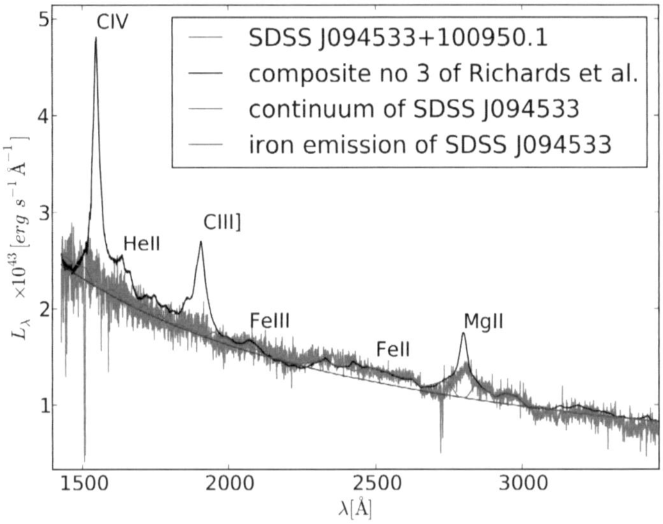

Figure 1. Spectrum of SDSS J0945+1009 (red) together with fitted continuum (blue) and iron emission (green). For comparison purposes ordinary quasar composite spectrum come from [8] is plotted in black.

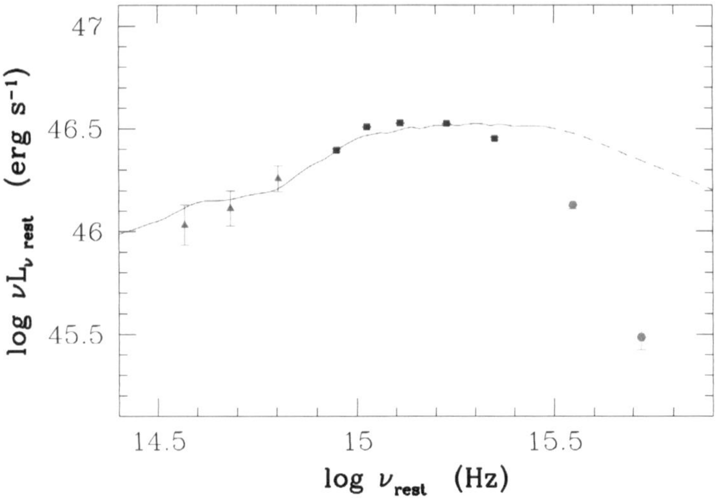

Figure 2. Photometric points of SDSS J094533.99+100950.1 together with ordinary quasar SED taken from [3]. Triangles show 2MASS IR photometric points, squares – optical points observed by SDSS, and hexagons are points which come from GALEX.

at wavelength 3000Å (see formulas in [7]). This value gives the non super-Eddington accretion rate $\dot{m} = 0.17$.

2.1. Accretion disk continuum

[9] have fitted an accretion disk continuum, where the disk's matter falls onto a spinning black hole, and they put constraints on parameters of the disk. The fitted parameters were the supermassive black hole mass, M, the Eddington accretion rate, $\dot{m}$, the dimensionless spin, a,

and the angle between the normal to the accretion disk and a direction to the observer i.e. the inclination i. The best fit parameters obtained by [9] are $M = 3 \times 10^9 M_\odot$, $\dot{m} = 0.14^{+0.11}_{-0.02}$, $a = 0.3^{+0.1}_{-0.9}$, and $\cos i > 0.7$. In work [10] authors fit SED of SDSS J0945+1009 by the cold accretion disk. They conclude that SED of this object can be produced by the cold disk and a relative low temperature of the disk can explain WLQ spectral features i.e. weakness of emission-lines.

3. Photoionization simulations

The aim of our project is to reconstruct the weakness of the emission-lines in WLQ by modification of its broad emission-line region parameters. We used Cloudy code ([11]) to perform photoionization-based fitting of the BELR assuming that the BELR is represented by a single cloud of plasma. The quasar's SED can be described by the standard Cloudy AGN two component spectrum:

$$f_\nu = A\,\nu^{\alpha_{\rm UV}} \exp\left(-\frac{h\,\nu}{k\,T_{\rm max}}\right) \exp\left(-\frac{h\,\nu}{k\,T_{\rm min}}\right) + B\,\nu^{\alpha_{\rm x}} \qquad (1)$$

where the spectral index $\alpha_{\rm UV} = 0.7$, the minimal and maximal effective cutoff temperatures $T_{\rm min} = 1.6 \times 10^2$ K and $T_{\rm max} = 4 \times 10^5$ K, respectively, fit very well the observed continuum. The second part of formula (1) illustrate the behaviour of the spectrum in the X-ray band. SDSS J0945+1009 has been unobserved by Chandra and ROSAT satellite ([12, 13]). Thus, we assume $\alpha_{\rm x} = -1$ and calculate the constant B from adopting a mean for WLQs $\alpha_{\rm ox} = -1.7$. The free parameters are (1) the local hydrogen density, n_H, (2) the radius of the BELR from the central black hole, r, and (3) the hydrogen column density, N_H. We assume the solar abundance.

3.1. Single cloud BELR model

We see in Figure 3 that all line luminosity ratios agree well in a narrow parameter range (intersection of solid lines at $\log(n_H[{\rm cm}^{-3}]) \approx 11$, $\log(r[{\rm cm}]) \approx 18$). The narrow intersection region can suggest that all emission lines can be produced in the single cloud or compact set of clouds with narrow range of densities. Absolute line luminosities agree quite well except one line (HeII, which could be underestimated). Additionally, MgII could be overestimated due to blended FeII emission under MgII line. Intersection position depends very weakly on the hydrogen column density (see panels b-e in Figure 3).

3.2. Location of the BELR

It results from our simulation of a single cloud model that the distance of the BELR from BH is approximately 10^{18} cm. This preferred location agrees very well with the value $\log(R_{\rm BLR}(L_{3000\text{Å}})[{\rm cm}]) = 17.98$ calculated with use of formula from [14] which is based on the reverberation mapping method.

3.3. Comparison to disk vertical structure

We try to locate the BELR in SDSS J0945+1009 in an environment of the supermassive BH. We independently carry out the next simulation based on the code described by [15] where the disk vertical structure (i.e. relatively cold disk and hot atmosphere) is calculated. Obtained parameters of the disk atmosphere (see Figure 4) are in agreement with those preferred by Cloudy's simulation — $\log(n_H[{\rm cm}^{-3}]) \approx 11$, $\log(r[{\rm cm}]) \approx 18$. This supports hypothesis in which the BELR is compact and not fully developed. It appears as an extension of the accretion disk atmosphere.

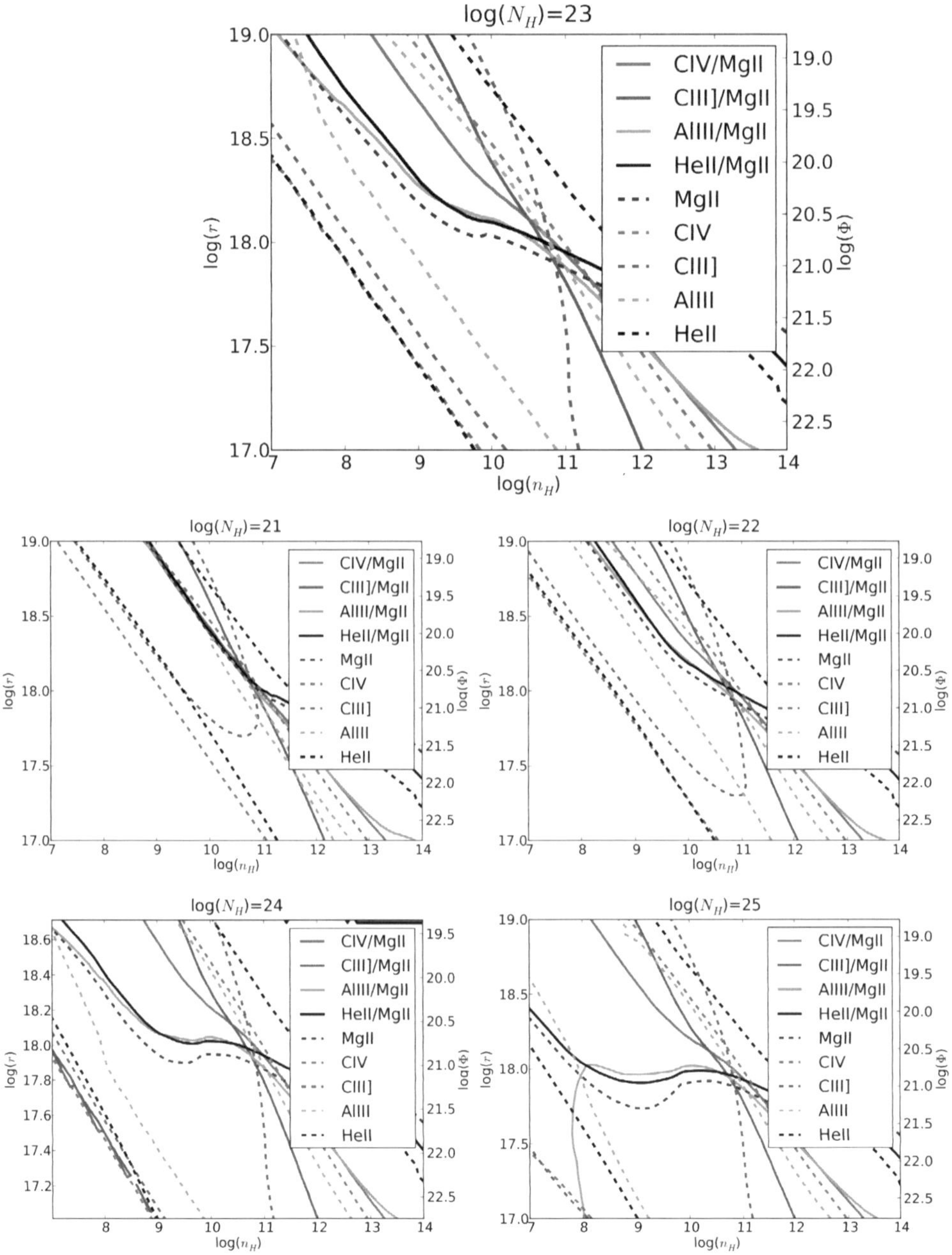

Figure 3. The observed line intensity ratios in the radius–local hydrogen density plane for the fixed column density: $N_H = 10^{23}$ cm^{-2} (panel a), $N_H = 10^{21}, 10^{22}, 10^{24}$, and 10^{25} cm^{-2} (panels b-e, respectively).

3.4. Locally Optimally Emitting Clouds

We also perform calculation of lines luminosities in the Locally Optimally Emitting Clouds (LOC) model where the BELR consists of many clouds. Therefore, each line luminosity is the integral of an emission line flux of a single cloud, $F(r, n_H)$, over a whole parametric space:

$$L_{line} \propto \int\int r^2\, F(r, n_H)\, r^{\gamma}\, n_H^{\beta}\, \mathrm{d}n_H\, \mathrm{d}r\,. \tag{2}$$

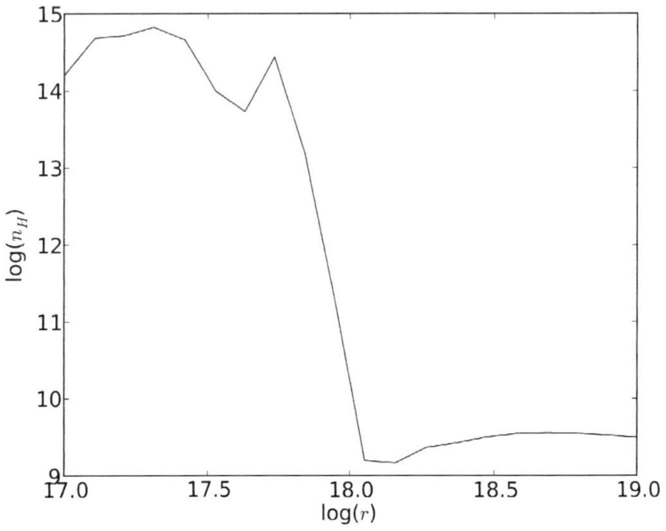

Figure 4. The local hydrogen density along radius. n_H has been calculated in the grey disk atmosphere at the effective optical depth = 2/3. Obtained results come from the disk vertical structure simulation ([15]).

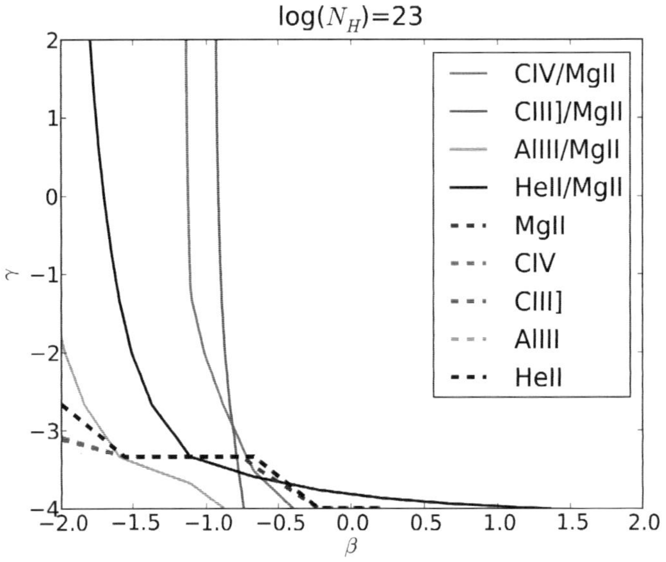

Figure 5. The dependence of the integrated lines luminosities (continuous lines) and ratios (dotted lines) on γ and β indices.

Dependence of emitted lines and their observed ratios on β and γ indices are shown in Figure 5. We were able to match observed line luminosities with calculated in LOC model. However, it demands γ and β to be more extreme than those adopted e.g. by [16]. Still line luminosity in this approach depends on the integrating range. LOC method introduces few additional free parameters thus there is many possibilities here which we do not fully explored.

4. Summary

We show preliminary results of photoionization simulations that supports compact BELR scenario in SDSS J0945+1009. The obtained preferred position of the BELR is consistent with the value which comes from the reverberation mapping formula. Additionally, this distance and the calculated local hydrogen density agree with values derived from the disk vertical structure simulations. Those remarks may encourage hypothesis that the BELR in SDSS J0945+1009 is in an early developing stage during active galaxy nuclei activation phase. This scenario is very incentive to explore this possibility. To further investigation it is necessary to answer the question how sensitive results are on simulation setup and on potential inaccuracies in line fitting.

Acknowledgments

This work was supported in part by grant NN203 380136 of the Polish State Committee for Scientific Research.

References

[1] Leighly K. M., Halpern J. P., Jenkins E. B., Casebeer D., 2007, *The Intrinsically X-Ray-weak Quasar PHL 1811. II. Optical and UV Spectra and Analysis*, ApJS, 173, 1
[2] Wu J.; Brandt W. N.; Hall P. B.; Gibson R. R.; Richards G. T.; Schneider D. P.; Shemmer O.; Just D. W.; Schmidt S. J., 2011, *A Population of X-Ray Weak Quasars: PHL 1811 Analogs at High Redshift*, ApJ, 736, 28,
[3] Elvis M., Wilkes B. J., McDowell J. C., Green R. F., Bechtold J., Willner S. P., Oey M. S., Polomski E., Cutri R., 1994, *Atlas of quasar energy distributions*, ApJS, 95, 1
[4] Skrutskie M.F. et al. 2006, *The Two Micron All Sky Survey (2MASS)*, AJ, 131, 1163
[5] Adelman-McCarthy J. K. et al. 2007, *The Fifth Data Release of the Sloan Digital Sky Survey*, ApJS, 172, 634
[6] Martin et al., 2005, *The Galaxy Evolution Explorer: A Space Ultraviolet Survey Mission*, ApJL 619, L1
[7] Wang J.-G., Dong X.-B., Wang T.-G., Ho L., Yuan W., Wang H., Zhang K., Zhang S.,Zhou H., 2009, *Estimating Black Hole Masses in Active Galactic Nuclei Using the Mg II λ2800ÅEmission Line*, ApJ, 707, 1334
[8] Richards G. T. et al., 2003, *Red and Reddened Quasars in the Sloan Digital Sky Survey*, AJ, 126, 1131
[9] Czerny B., Hryniewicz K., Nikołajuk M., Sądowski A., 2011, *Constraints on the black hole spin in the quasar SDSS J094533.99+100950.1*, MNRAS, 415, 2942
[10] Laor A. & Davis S. W., 2011, *Cold accretion discs and lineless quasars*, MNRAS, 417, 681L
[11] Ferland G. J., Korista K. T., Verner D. A., Ferguson J. W., Kingdon J. B., Verner E. M., 1998, *CLOUDY 90: Numerical Simulation of Plasmas and Their Spectra*, PASP, 110, 761
[12] Hryniewicz K., Czerny B., Nikołajuk M., Kuraszkiewicz J., 2010, *SDSS J094533.99+100950.1 - the remarkable weak emission line quasar*, MNRAS, 404, 2028
[13] Wu J., Brandt W. N., Anderson S. F., Diamond-Stanic A. M., Hall P. B., Plotkin R. M., Schneider D. P., Shemmer O., 2011, *X-ray and Multiwavelength Insights into the Nature of Weak Emission-Line Quasars at Low Redshift*, ApJ, ArXiv e-prints
[14] Kong M.-Z., Wu X.-B., Wang R., Han J.-L., 2006, *Estimating Black Hole Masses of AGNs using Ultraviolet Emission Line Properties*, Chin. J. Astron. Astrophys., 6, 396
[15] Różańska, A.; Czerny, B.; Życki, P. T.; Pojmański, G., 1999, *Vertical structure of accretion discs with hot coronae in active galactic nuclei*, MNRAS, 305, 481
[16] Baldwin, J.; Ferland, G.; Korista, K.; Verner, D., 1995, *Locally Optimally Emitting Clouds and the Origin of Quasar Emission Lines* , ApJ, 455, L119

Star Formation in the Central Kiloparsec of Nearby Active Galaxies

Chorng-Yuan Hwang and Mengchun Tsai

Graduate Institute of Astronomy, National Central University, 300 Jhongda Rd, Jungli, Taoyuan 32001, Taiwan

E-mail: hwangcy@astro.ncu.edu.tw

Abstract. We investigate star formation (SF) activity in the central kpc of a sample of nearby Active Galactic Nuclei (AGNs). AGN activities are expected to either trigger SF via accreting ISM to the central regions of the host galaxies or quench the SF via the energy feedback of the AGNs. To study the AGN-SF relation we select 113 nearby galaxies that host 8 GHz central radio sources. We use 8 GHz radio emission to represent the AGN activity and 8 micron dust emission in the central kpc regions of these galaxies to estimate the SF rate (SFR). The SFR is found to be correlated with the stellar mass for stellar mass greater than 10^{10} solar mass and looks scattered for stellar mass less than 10^{10} solar mass. There is no correlation between the specific SFR (SSFR) and the AGN activity for all sources. However, if we exclude the sources with the central stellar mass greater than 10^{10} solar mass, we find that the 8 GHz radio emission is well correlated with the SSFR. These results suggest that the AGN activity is significant in triggering SF activity only for small galaxies. Besides, we also select about 20 nearby AGN galaxies to investigate the radial variation of their surface specific star formation rate.

1. Introduction

Active galactic nuclei (AGNs) are believed to be powered by accretion onto central supermassive black holes in their host galaxies. The properties of molecular gas near an AGN can be significantly influenced by the AGN. AGN activities are expected to either trigger star formation (SF) via accreting ISM to the central regions of the host galaxies or quench the SF via the energy feedback of the AGNs. Measuring the star formation rate (SFR) of the host galaxy is thus important for studying the co-evolution of AGN and their hosts. In this paper, we select more than one hundred nearby galaxies that host 8 GHz central radio sources to study the AGN-SF relation. Besides, we also select about 20 nearby AGN galaxies to investigate the radial variation of their SFRs.

2. Methods

We select our sources from the AGN lists of [1], [2], [3], and [4]. There are total 113 nearby active galaxies in our sample. We obtained the 8 GHz fluxes of these source from the VLA archive at https://archive.nrao.edu/archive/archiveimage.html. We obtained the 3.6, 4.5, 8 and 24 μm IR images of these sources from the *Spitzer* archive. We use the 3.6 and 8 μm images to estimate the stellar mass and SFR respectively. The stellar mass and the SFR are calculated using the formulae of [5] and [6]. We use the 8 GHz fluxes to represent the strengths of the host AGNs. In studying the radial variation of the SFRs of nearby AGN galaxies, the positions of

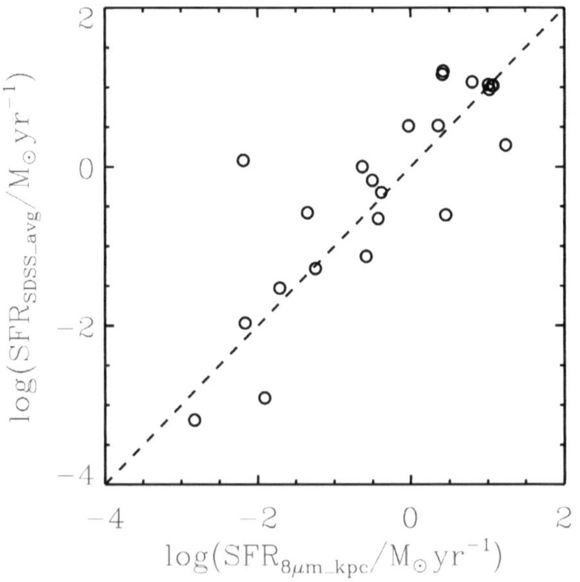

Figure 1. Comparison of the SFR within 1 kpc estimated by the *Spitzer* 8μm dust intensity with the SFR obtained from the *SDSS* DR4.

the nuclei are identified using the 8 GHz images or with fitting the bulge centers of the 3.6 μm images. The physical scales are estimated using the parameters determined in the bulge fitting of the 3.6 μm images.

3. RESULTS

We compare the *SDSS* SFR catalog with the SFR that was estimated by *Spitzer* 8 μm dust intensity. As can be seen in Figure 1, the SFRs estimated by emission line intensities and those estimated by the 8 μm dust intensities are linearly correlated, indicating that the 8 μm emission is also a good star formation tracer. However, we note that the SFR from the *SDSS* DR4 is estimated using the averaged magnitude within 3 arc second, and the SFR derived from the 8 μm dust is estimated within the central 1 kpc radius.

To study the activity of AGN, we select 8 GHz emission as our AGN activity tracer. We select the active galaxies with a compact point source from the VLA archive images to avoid contamination from the free-free emission associated with the ionized gas surrounding hot young stars.

The relation of SSFR, stellar mass surface density and 8 GHz luminosity for the central 1 kpc radius region of 113 nearby active galaxies is shown in Figure 2 and Figure 3. These AGNs have been classified into different types according to their line widths. In Figure 2, we find that the 8 GHz luminosity and the SSFR are uncorrelated. Even for Seyfert 2 galaxies, which are believed to have significant star formation activity, the correlation coefficient r between the SSFR and the 8 GHz luminosity is only 0.22 with a probability $p = 77.33\%$. In other words, the SSFR has no strong correlation with AGN activity for all nuclei types. But we note that the Seyfert 1 and other active galaxies have higher radio powers than Seyfert 2 have.

On the other hand, Figure 3 shows a strong correlation between the stellar mass surface density and the 8 GHz luminosity. The correlation coefficient r is 0.41 with a probability

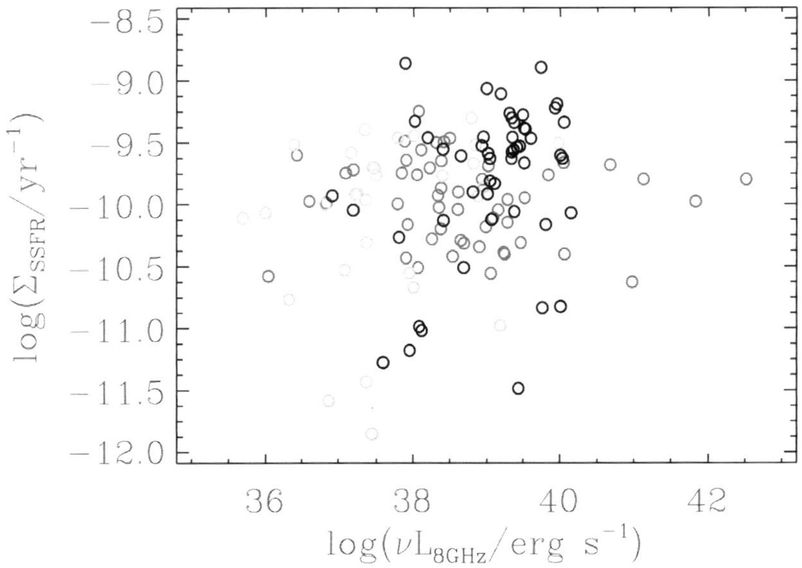

Figure 2. Relation between the 8 GHz luminosity and the SSFR surface density. Red circles are Seyfert 1.0-1.5 galaxies, green circles are Seyfert 1.8-2.0 galaxies and black circles are other active nucleus types.

p = 99.68% for Seyfert 1. This result indicates that the central radio powers of AGNs are correlated with the central stellar mass. This indicates that the AGN activity is related to the central mass or concentration of the host galaxy.

We note that the 8 GHz emission might be contaminated by free-free emission from star formation. To check whether 8 GHz do represent the activity of AGN, we compare the 8 GHz fluxes with the X-ray flux derived from *XMM* for our AGN sources. The result is shown in Figure 4. The correlation coefficient between the 8 GHz and the soft x–ray (0.5–2.0 keV) fluxes is r = 0.32 with a probability p = 99.61% for 82 AGN galaxies. However, we note that there is no correlation for Seyfert 2 galaxies; the probability of Seyfert 2 is only p = 11.84%. This suggests that the 8 GHz luminosity do represent the activity of the AGN, at least for Seyfert 1 galaxies.

To further investigate the effects caused by different environmental properties within the central regions of the active galaxies, we compare the SSFR for different types of galaxies, including both active and normal galaxies. The normal galaxies are selected from [7] and [5]. Figure 5 shows a strong correlation between the stellar mass and SFR for both active and normal galaxies within the central 1 kpc radius. Most of the active galaxies seem to locate at relatively higher SFR regions.

However, as shown in Figure 6, most of nearby galaxies, including normal and active galaxies actually have similar SSFR, $\sim 10^{-9.5} - 10^{-10.5}$ except for elliptical type galaxies. In other words, the high SFRs of the active galaxies are actually caused by their high surface stellar density instead of having enhanced SFRs triggered by AGN activity. Different SSFRs are mainly related to different morphologies of host galaxies instead of AGN activities.

The SSFR is in fact the ratio between the current SFR to the accumulated old SF and cannot represent the current star formation efficiency (SFE). In order to obtain the current SFE, we

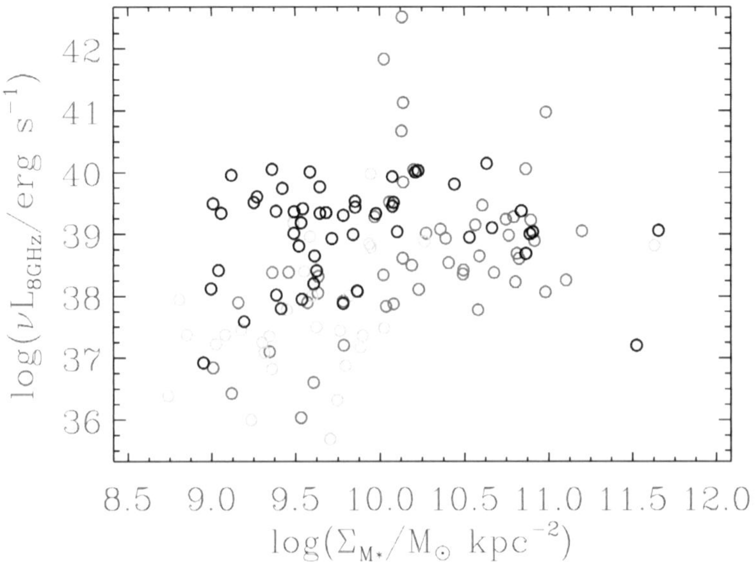

Figure 3. Correlation between the SSFR surface density and the stellar mass density within 1 kpc radius. Symbols are the same as in Figure 2. The correlation coefficient r 0.41 with a probability $p = 99.68\%$ for Seyfert 1 (red circles)

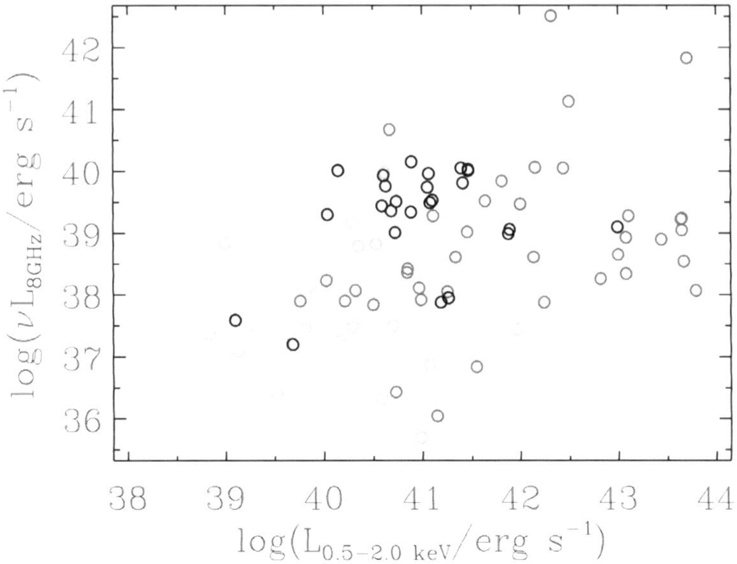

Figure 4. Comparison between the 8 GHz luminosity and x–ray luminosity (0.5–2.0 keV). Symbols are the same as in Figure 2.

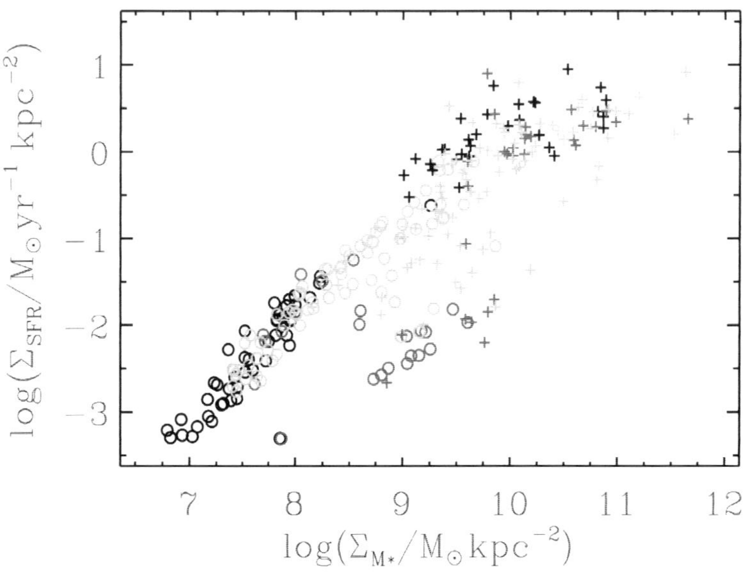

Figure 5. Relation between stellar mass surface density and star formation rate surface density within 1 kpc. Red color means elliptical-like, green means spiral-like and black means irregular or unclassified galaxies. Crosses represent 133 active galaxies and circles are 155 normal galaxies.

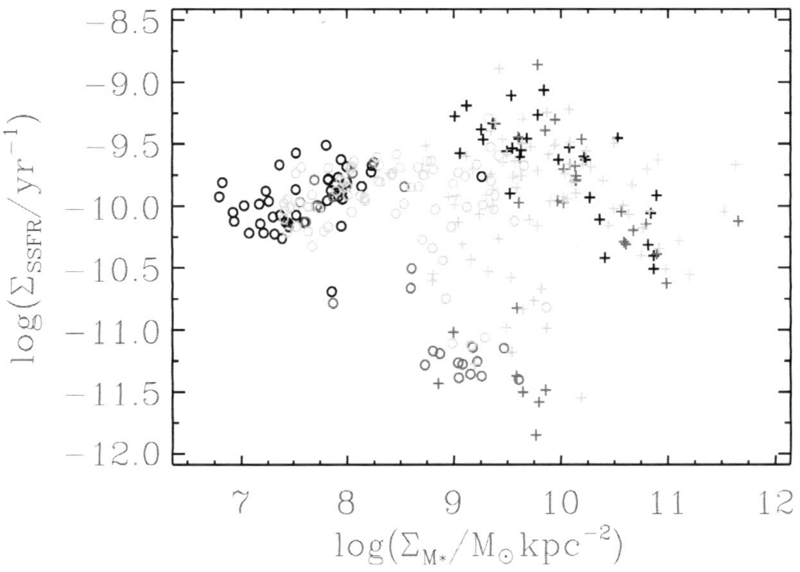

Figure 6. Relation between stellar mass surface density and specific star formation rate surface density within 1 kpc. Symbols are the same as in Figure 5.

need to have the molecular mass instead of stellar mass. We therefore try to obtain molecular gas to estimate the SFE (SFR/M_{gas}) for several normal and active galaxies. We have searched sources from the *BIMA-SONG* project, which have performed CO(1–0) observations for several active and normal galaxies. We are able to select 11 normal and 6 active galaxies with CO(1–0) observations in our sample. Although we have only a small sample, we find that the fractions of the gas to stellar mass surface density around the active galaxies are usually higher than that in the normal ones. However, the SFEs in the central regions of the active and normal galaxies are almost the same. The AGNs usually have larger central stellar masses as we have shown; therefore, they also have more gas abundance. In other words, the high SFR around AGN is actually caused by a large gas abundance instead of a high SFE.

Besides, we find that the radio powers of the AGNs seem to be related to the existence of some inner structures in the optical images of these AGNs. These structures locate within the central regions of the host galaxies. In fact, the radio power shows a strong correlation with SSFR only for AGNs with the inner structures. This suggests that the AGN activity might also be related to the central environments of the host galaxy.

Acknowledgments

We acknowledge support from the National Science Council (NSC) of Taiwan through grants NSC 100-2119-M-008-011-MY3 and NSC 99-2112-M-008-014-MY3.

References

[1] Condon, J. J., Huang, Z.-P., Yin, Q. F., Thuan, T. X., 1991, *Astrophys. J.*, 378, 65
[2] Ho, L. C., Ulvestad, J. S., 2001, *Astrophys. J. Supp.*, 133, 77
[3] Spinelli, P. F., Storchi-Bergmann, T., Brandt, C. H., Calzetti, D., 2006, *Astrophys. J. Supp.*, 166, 498
[4] Melendez, M., Kraemer, S. B., Schmitt, H. R., 2010, *MNRAS*, 406, 493
[5] Wu, H., Cao, C., Hao, C.-N., Liu, F.-S., Wang, J.-L., Xia, X.-Y., Deng, Z.-G., & Young, C. K.-S. 2005, *Astrophys. J. Lett.*, 632, L79
[6] Zhu, Y.-N., Wu, H., Li, H.-N., & Cao, C. 2010, *Research in Astronomy and Astrophysics*, 10, 329
[7] Tully, R. B., 1988, *Science*, 242, 310

Magnetically controlled accretion onto a black hole

N R Ikhsanov[1], **L A Pustil'nik**[2] **and N G Beskrovnaya**[1]

[1] Pulkovo Observatory, Pulkovskoe Shosse 65, Saint-Petersburg 196140, Russia

[2] Israel Space Weather and Cosmic Ray Center, Tel Aviv University, Israel Space Agency, & Golan Research Institute, Israel

E-mail: ikhsanov@gao.spb.ru

Abstract. An accretion scenario in which the material captured by a black hole from its environment is assumed to be magnetized ($\beta \sim 1$) is discussed. We show that the accretion picture in this case is strongly affected by the magnetic field of the flow itself. The accretion power within this Magnetically Controlled Accretion (MCA) scenario is converted predominantly into the magnetic energy of the accretion flow. The rapidly amplified field prevents the accretion flow from forming a homogeneous Keplerian disk. Instead, the flow is decelerated by its own magnetic field at a large distance (Shvartsman radius) from the black hole and switches into a non-Keplerian dense magnetized slab. The material in the slab is confined by the magnetic field and moves towards the black hole on the time scale of the magnetic field annihilation. The basic parameters of the slab are evaluated. Interchange instabilities in the slab may lead to a formation of Z-pinch type configuration of the magnetic field over the slab in which the accretion power can be converted into jets and high-energy radiation.

1. Introduction

We consider a black hole of the mass $M_{\rm bh}$, surrounded by a gas of the average density ρ_∞, which moves in the frame of the black hole with a relative velocity $v_{\rm rel} = \left(v_{\rm s}^2 + v_{\rm w}^2\right)^{1/2}$, where $v_{\rm s}$ and $v_{\rm w}$ are the sound speed and the velocity of the gas proper motion. A maximum distance at which the black hole interacts with its environment through gravitational potential is the Bondi radius $R_{\rm G} = 2GM_{\rm bh}/v_{\rm rel}^2$. It is defined by equation $v_{\rm p}(R_{\rm G}) = v_{\rm rel}$, where $v_{\rm p}(r) = (2GM_{\rm bh}/r)^{1/2}$ is the parabolic (free-fall) velocity at a distance r. The mass with which the black hole interacts in a unit time is $\dot{\mathfrak{M}}_{\rm c} = \pi R_{\rm G}^2 \rho_\infty v_{\rm rel}$. The mass capture rate by the black hole from its environment in general case is limited to $\dot{\mathfrak{M}}_{\rm a} \leq \dot{\mathfrak{M}}_{\rm c}$.

The structure of the accretion flow onto a black hole and the mode by which the accretion power is released depends on the initial conditions in the material captured by the black hole at the Bondi radius. A situation in which this material is slowly moving, i.e. $v_{\rm w} < v_{\rm s}$, and non-magnetized is discussed in the next section. It can be expected if the black hole captures material from a giant molecular cloud. If the material captured by the black hole is supplied by nearby stars its properties can significantly differ from those mentioned above. The velocity of stellar wind usually exceeds the sound speed in the outflowing material, i.e. $v_{\rm w} \gg v_{\rm s}$, and the magnetic pressure in the wind, $\mathcal{E}_{\rm m}^{(0)} = B_{\rm w}/8\pi$, is comparable to its thermal pressure, $\mathcal{E}_{\rm th}^{(0)} = \rho_\infty v_{\rm s}^2$. This is justified by studies of solar wind (see e.g. [17], and references therein) and recent results on the surface field of massive O/B-type stars (see e.g. [18], and references therein). The accretion picture which is realized under these conditions is discussed in Sect. 3. We show that accretion picture in this case is strongly affected by the magnetic field of the flow itself. It prevents the material from forming a Keplerian disk. Instead appearance of a dense magnetized non-Keplerian slab is expected. Formation of the slab can under certain conditions result in a strong electric potential in the vicinity of the black hole which might be responsible for powerful jets. Our conclusions are summarized in Sect. 4.

2. Non-magnetized accretion flow scenario

Accretion scenario in AGNs is usually built around the assumptions that the material surrounding the black hole is non-magnetized and moves with a relatively small velocity, i.e. $v_{\rm rel} \sim v_{\rm s}$. As this material is captured by a black hole it initially follows ballistic trajectories forming a quasi-spherical accretion flow. If the accreting material possesses angular momentum the flow can switch its geometry from quasi-spherical to a Keplerian disk. For the disk to form the circularization radius, $R_{\rm circ} = \dot{J}^2/GM_{\rm ns}\dot{\mathfrak{M}}_{\rm a}^2$, at which the angular velocity of the material, $\omega_{\rm en} = \xi\Omega_0\left(R_{\rm G}/r\right)^2$, reaches the Keplerian angular velocity, $\omega_{\rm k} = \left(2GM_{\rm bh}/r^3\right)^{1/2}$, should exceed a distance at which the ballistic trajectories of the material are truncated. Here $\dot{J} = \xi j_0 \dot{\mathfrak{M}}_{\rm a}$ is the angular momentum accretion rate, j_0 is the specific angular momentum of the material captured by a black hole at the Bondi radius, and ξ is the factor by which angular momentum accretion rate is reduced due to inhomogeneities (velocity and density gradients) and the magnetic viscosity in the accretion flow. The specific angular momentum carried by a material captured by the black hole from turbulent Interstellar Medium (ISM) can be evaluated as $j_0 = v_{\rm t}(R_{\rm t})R_{\rm t}^{-1/3}R_{\rm G}^{4/3}$, where $v_{\rm t}(R_{\rm t})$ is the velocity of turbulent motions at the scale of $R_{\rm t}$ and the Kolmogorov spectrum of the turbulent motions is assumed (see e.g. [12], and references therein). The circularization radius of the accretion flow in this case can be expressed as

$$R_{\rm circ} = \frac{\xi^2 v_{\rm t}^2 R_{\rm t}^{-2/3} R_{\rm G}^{8/3}}{GM_{\rm bh}}. \tag{1}$$

A formation of Keplerian accretion disk would be expected if $R_{\rm circ} > \kappa_{\rm r} R_{\rm g}$, where $R_{\rm g} = 2GM_{\rm bh}/c^2$ is the gravitational radius of the black hole, $\kappa_{\rm r} = R_{\rm st}/R_{\rm g}$, and $R_{\rm st}$ is the radius of the last stable orbit around the black hole. Solving this inequality for $v_{\rm rel}$ yields $v_{\rm rel} < v_{\rm cr}^{(0)}$, where

$$v_{\rm cr}^{(0)} \simeq 690\,{\rm km/s} \times \kappa_{\rm r}^{-3/16} m_6^{1/8} \left(\frac{v_{\rm t}}{10\,{\rm km/s}}\right)^{3/8} \left(\frac{R_{\rm t}}{10^{20}\,{\rm cm}}\right)^{-1/8}. \tag{2}$$

Here m_6 is the mass of the black hole in units of $10^6\,{\rm M}_\odot$, and parameters $v_{\rm t}$ and $R_{\rm t}$ are normalized following [14]. The derived value of $v_{\rm cr}^{(0)}$ significantly exceeds the sound speed in ISM and thus, a supermassive black hole within the non-magnetized accretion flow scenario is expected to be surrounded by a Keplerian accretion disk.

As a black hole is accreting material from the Keplerian disk its gravitation energy is converted predominantly into the kinetic and thermal energy of the accreting material, which then can be converted into the magnetic energy due to dynamo action. The pressure of the magnetic field generated in the disk? therefore, is limited to $\mathcal{E}_{\rm m}^{(\rm d)}(r) < \rho(r)v_{\rm k}^2(r)$, where $v_{\rm k}(r) = (GM_{\rm bh}/r)^{1/2}$ is the Keplerian velocity. An influence of the field on the accretion flow in this case is relatively small and does not lead to any significant changes in the flow dynamics. The disk remains Keplerian and the accreting material approaches the black hole on the viscous timescale. The scale of the magnetic field generated in the turbulent differentially rotating Keplerian disk by the dynamo action is limited to the disk thickness (which represents the largest scale of the turbulent motions in the disk). The energy of the magnetic field is released in the disk corona, which is formed as the field emerges from the disk due to the buoyancy instability. The energy release process is associated with the magnetic reconnection which leads to heating and ejection of material [5]. However, the kinetic luminosity of the ejecta is relatively small and the outflowing material is non-collimated.

Existence of collimated powerful jets cannot be explained within this scenario unless some additional assumptions are incorporated into the model. In particular, one can assume that the jets are powered by the rotational energy of the black hole, which is released as the inner radius of the disk is approaching the last stable orbit. The ejected material can be self-collimated, or (which looks more reasonable) is collimated by the large-scale magnetic field of the disk. The field in the latter case should be strong enough to collimate a powerful jet. This criterium is difficult to satisfy if the large-scale field of the disk is provided by the small-scale magnetic arches generated in the disk due to dynamo action. The pressure of the magnetic field in this situation is substantially smaller than the ram pressure of the accretion flow itself. It is more likely that the large-scale field of the disk was initially present in the material captured by the black hole and has been amplified during the accretion process. In the next Section we show that this assumption cannot be simply incorporated into the traditional accretion model. The magnetic field in the spherical accretion flow increases rapidly and prevents accreting material from forming a Keplerian disk.

3. Magnetically controlled accretion scenario

A situation in which $v_{\rm rel} \approx v_{\rm w}$ and $\beta = \mathcal{E}_{\rm th}^{(0)}/\mathcal{E}_{\rm m}^{(0)} \sim 1$ can be realized if the material captured by a black hole is supplied by the stellar wind of nearby stars. The ram pressure of the material at the Bondi radius, $\mathcal{E}_{\rm r}^{(0)} = \rho_{\infty} v_{\rm w}^2$, in this case significantly exceeds its thermal and magnetic pressure and hence, the Alfvén velocity, $v_{\rm A} = B_{\rm w}/(4\pi\rho)^{1/2}$, in the accreting material is much smaller than the free-fall velocity. The time of the magnetic field annihilation in the accretion flow,

$$t_{\rm rec} = \frac{r}{\eta_{\rm m} v_{\rm A}} = \eta_{\rm m}^{-1} t_{\rm ff} \left(\frac{v_{\rm ff}}{v_{\rm A}}\right), \tag{3}$$

under these conditions significantly exceeds the dynamical (free-fall) time, $t_{\rm ff} = \left(r^3/2GM_{\rm bh}\right)^{1/2}$, and the magnetic flux in the free-falling gas is conserved. Here $B_{\rm w}$ is the field strength in the accretion flow and the efficiency parameter of the magnetic reconnection ranges in the interval 0.01–0.15 (see e.g. [11, 10]).

3.1. Shvartsman radius

The magnetic field in the free-falling material is dominated by the radial component [19], which under the condition of the magnetic flux conservation increases as $B_{\rm r}(R) \sim B_{\rm f}(R_{\rm G})\,(r/R_{\rm G})^{-2}$ [1]. The magnetic pressure in the accreting material, therefore, increases while approaching the black hole as

$$\mathcal{E}_{\rm m}(r) = \mathcal{E}_{\rm m}(R_{\rm G}) \left(\frac{r}{R_{\rm G}}\right)^{-4}, \tag{4}$$

and the ram pressure of the free-falling spherical flow is

$$\mathcal{E}_{\rm ram}(r) = \mathcal{E}_{\rm ram}^{(0)} \left(\frac{r}{R_{\rm G}}\right)^{-5/2}. \tag{5}$$

This indicates that the magnetic energy in the free-falling gas increases more rapidly than its kinetic energy, $\mathcal{E}_{\rm m}/\mathcal{E}_{\rm ram} \propto r^{-3/2}$, and hence, the gravitational energy of the black hole in the scenario under consideration is converted predominantly into the magnetic energy of the accreting material.

A distance $R_{\rm sh}$, at which the magnetic pressure in the accretion flow reaches its ram pressure, can be evaluated by equating $\mathcal{E}_{\rm m}(R_{\rm sh}) = \mathcal{E}_{\rm ram}(R_{\rm sh})$. This yields [16]

$$R_{\rm sh} = \beta^{-2/3}\left(\frac{V_{\rm s}}{V_{\rm rel}}\right)^{4/3} R_{\rm G} = \beta^{-2/3}\,\frac{2GM_{\rm ns}V_{\rm s}^{4/3}}{V_{\rm rel}^{10/3}}. \qquad (6)$$

Parameter $R_{\rm sh}$ (hereafter Shvartsman radius) represents a minimum distance from which the accretion process is fully controlled by the magnetic field of the flow itself. The Alfvén velocity in the accretion flow at this distance reaches the free-fall velocity. An accretion of homogeneous gas in which the magnetic flux is conserved inside Shvartsman radius is impossible. Otherwise, the magnetic energy in the flow would exceed the gravitational energy, which contradicts the energy conservation law (for discussion see [16]). Further accretion, therefore, can be realized only on the timescale of the field dissipation, $t_{\rm rec}$, which is significantly larger than the free-fall time (see Eq. 3). This indicates that the initially free-falling accretion flow is decelerated by its own magnetic field at $R_{\rm sh}$. The deceleration leads to formation of a shock in which the gas is heated up to a temperature $T_{\rm s} = (3/16)T_{\rm ff}(R_{\rm sh})$, where $T_{\rm ff}(r) = GM_{\rm ns}m_{\rm p}/k_{\rm B}r$ is the proton free-fall temperature, $m_{\rm p}$ is the proton mass and $k_{\rm B}$ is the Boltzmann constant [9]. If cooling of the material in the region of flow deceleration is ineffective a hot envelope surrounding the black hole at the Shvartsman radius forms.

Rapid amplification of the magnetic field in the spherical flow as well as deceleration of the flow by its own magnetic field at the Shvartsman radius have been confirmed by the results of numerical studies of magnetized spherical accretion onto a black hole [6, 7]. These calculations have shown that the magnetized flow under the conditions of interest is shock-heated at the Shvartsman radius up to the adiabatic temperature. These authors have considered further accretion under assumptions that the flow is radiatively inefficient and heating of the flow dominates cooling. An overheating of material in this case can lead to transition of the accretion flow into convective-dominated stage in which a portion of the accreting material is leaving the system in a form of turbulent jets and the mass accretion rate inside the region of the flow deceleration significantly decreases. Here we show, however, that MCA picture can be constructed without these assumptions and a question about the nature of energy source responsible for the overheating of material (which is required for the flow to switch into the convective-dominated phase) can be avoided.

3.2. No Keplerian disk within MCA scenario
Ballistic trajectories of the free-falling spherical accretion flow within MCA scenario are truncated at the Shvartsman radius. The condition for the Keplerian disk formation in this case reads $R_{\rm circ} \geq R_{\rm sh}$. This inequality can be expressed using Eqs. (1) and (6) as $v_{\rm rel} \leq v_{\rm cr}^{\rm mag}$, where

$$v_{\rm cr}^{\rm mag} \simeq 5\times 10^{5}\,{\rm cm\,s^{-1}}\;\beta^{1/3}\xi_{0.2}m_6^{1/3}\left(\frac{v_{\rm s}}{10\,{\rm km\,s^{-1}}}\right)^{-2/3}\left(\frac{V_{\rm t}}{10\,{\rm km\,s^{-1}}}\right)\left(\frac{R_{\rm t}}{10^{20}\,{\rm cm}}\right)^{-1/3}. \qquad (7)$$

Parameter $\xi_{0.2} = \xi/0.2$ is normalized here to its maximum average value, which has been derived in numerical studies of the accretion process within the non-magnetized spherical accretion flow approximation [13]. Since the derived value of $v_{\rm cr}^{\rm mag}$ is comparable or even smaller than the sound speed in the surrounding gas the angular momentum of the accreting material at the Shvartsman radius appears to be insufficient for the Keplerian disk to form.

3.3. Cooling of the accretion flow
Cooling of the accretion flow in the considered case is dominated by the inverse Compton scattering of electrons on photons emitted in the vicinity of the black hole. The Compton

cooling time of the material at $R_{\rm sh}$ can be evaluated as [4],

$$t_{\rm c}(R_{\rm sh}) = \frac{3\pi m_{\rm e} c^2 R_{\rm sh}^2}{2\sigma_{\rm T} L_{\rm bh}}, \tag{8}$$

where $m_{\rm e}$ is the electron mass, $\sigma_{\rm T}$ is the Thomson cross-section and $L_{\rm bh}$ is the luminosity of the source associated with the black hole. Cooling is dominating heating at the Shvartsman radius if $t_{\rm c}(R_{\rm sh}) \leq t_{\rm ff}(R_{\rm sh})$. Solving this inequality yields $v_{\rm rel} \geq v_{\rm cc}$ where

$$v_{\rm cc} \simeq 2 \times 10^7 \ \beta^{-1/5} m_6^{3/5} L_{42}^{-3/5} \left(\frac{v_{\rm s}}{10\,{\rm km\,s^{-1}}}\right)^{2/3} \ {\rm cm\,s^{-1}}, \tag{9}$$

and $L_{42} = L_{\rm bh}/10^{42}\,{\rm erg\,s^{-1}}$. The value of $v_{\rm cc}$ is much smaller than typical wind velocity of massive stars and is even comparable to the velocity of stellar wind emitted by the red dwarfs. This indicates that a stationary magnetically controlled accretion at the rate $\dot{\mathfrak{M}}_{\rm a} \approx \dot{\mathfrak{M}}_{\rm c}$ is expected if the material captured by the black hole is predominantly supplied by the stellar wind of nearby stars. Compton cooling in this case prevents the flow from switching into the convective-dominated stage.

3.4. Non-Keplerian slab

The accretion picture of a cold ($T < T_{\rm s}$) magnetized ($\beta < 1$) gas has been discussed in [2, 3]. It has been shown that the material in this case tends to flow along the magnetic field lines and in the region $r \leq R_{\rm sh}$ is accumulated in a dense non-Keplerian slab (see Fig. 1 in [3]). The material in the slab is confined by the magnetic field of the flow itself and its radial motion continues as the field is annihilating. The accretion process in the slab, therefore, occurs on the timescale of $t_{\rm rec}$.

The magnetic pressure inside the Shvartsman radius increases according to the energy conservation law as $\mathcal{E}_{\rm m} = \mathcal{E}_{\rm m}(R_{\rm sh})\,(R_{\rm sh}/r)^{5/2}$, where

$$\mathcal{E}_{\rm m}(R_{\rm sh}) = \frac{\dot{\mathfrak{M}}_{\rm c}(2GM_{\rm bh})^{1/2}}{4\pi R_{\rm sh}^{5/2}} \tag{10}$$

is the magnetic pressure at the Shvartsman radius. This indicates that the field strength in the vicinity of the black hole reaches the value

$$B(R_{\rm g}) = 20 \ L_{42}^{1/2} m_6^{-1} \ {\rm kG}. \tag{11}$$

Since the material is confined by the field in the slab its thermal pressure does not exceed the magnetic pressure. The gas number density at the inner radius of the slab in this case can be estimated as

$$\rho_{\rm sl} \leq 10^{19}\,{\rm cm^{-3}} \ L_{42} m_6^{-2} T_4^{-1}, \tag{12}$$

where T_4 is the gas temperature at the inner radius of the slab in units of 10^4 K. The thickness of the slab depends on the magnetic field configuration and in the first approximation can be evaluated from continuity equation as

$$h_{\rm z}(R_{\rm g}) \simeq 10^5\,{\rm cm} \ \eta_{0.01}^{-1} m_6 T_4, \tag{13}$$

where $\eta_{0.01} = \eta_{\rm m}/0.01$. It should be noted, that the thickness of the slab strongly depends on the field annihilation time, which can be evaluated from the shortest timescale of the source variability. Furthermore, we cannot discard a possibility that the slab is disintegrated by interchange instabilities into a large number of filaments connected by the field lines as it sketched in Fig. 1 in [8]. Formation of jets in this case can be associated with interaction between the magnetic arches connecting the filaments which leads to a Z-pinch field configuration over the slab. The process of energy release and plasma ejection in this configuration will be discussed in a forthcoming paper.

4. Conclusions

Black holes within the non-magnetized accretion flow scenario are expected to be surrounded by Keplerian disks. The scale of the magnetic field generated in the disk due to dynamo action is comparable to the disk thickness and the magnetic pressure associated with this field insignificantly contributes the total power of the disk. A formation of a strong large-scale magnetic field in the accretion flow can be expected if the material captured by the black hole is magnetized. The behavior of the accreting material in this case is strongly affected by its magnetic field. The flow is decelerated by the magnetic field at the Shvartsman radius and moves towards the black hole on the timescale of the magnetic field annihilation. Formation of a homogeneous Keplerian disk within this Magnetically Controlled Accretion (MCA) scenario does not occur. Instead, the accreting material is accumulated in a non-Keplerian dense slab. The material in the slab is confined by its own magnetic field and moves towards the black hole as the magnetic field is annihilating. Interchange instabilities of the slab may lead to formation of the Z-pinch type configuration of the magnetic field in which the accretion power is released in the form of jets and high energy emission.

Acknowledgments

We would like to thank G.S. Bisnovatyi-Kogan, M.V. Medvedev and P.L. Biermann for useful discussions. Nazar Ikhsanov is grateful to Alexander von Humboldt Foundation for support in attending the conference and Michael Kramer for useful discussions and kind hospitality at Max-Planck Institute of Radio Astronomy in Bonn. The research has been partly supported by the Program of RAS Presidium N 19, NSH-3645.2010.2, and by the grant "Infrastructure" of Israel Ministry of Science.

References

[1] Bisnovatyi-Kogan, G.S., Fridman, A.M., 1970, *Soviet Astronomy*, **13**, 566–568
[2] Bisnovatyi-Kogan, G.S., Ruzmaikin, A.A. 1974, *Astrophys. and Space Sci.*, **28**, 45–59
[3] Bisnovatyi-Kogan, G.S., Ruzmaikin, A.A. 1976, *Astrophys. and Space Sci.*, **42**, 401–424
[4] Elsner R.F., Lamb F.K. 1977, *Astrophys. J.*, **215**, 897–913
[5] Galeev, A.A., Rosner, R., Vaiana, G.S. 1979, *Astrophys. J.*, **229**, 318–326
[6] Igumenshchev, I.V., Narayan, R., Abramowicz, M.A. 2003, *Astrophys. J.*, **592**, 1042–1059
[7] Igumenshchev, I.V. 2006, *Astrophys. J.*, **649**, 361–372
[8] Ikhsanov, N.R., Pustil'nik, L.A. 1994, *Astrophys. J. Suppl.*, **90**, 959–961
[9] Lamb, F.K., Fabian, A.C., Pringle, J.E., Lamb, D.Q. 1977, *Astrophys. J.*, **217**, 197–212
[10] Noglik, J.B., Walsh, R.W., Ireland, J. 2005, *Astron. Astrophys.*, **441**, 353–360
[11] Parker, E.N. 1971, *Astrophys. J.*, **163**, 279–285
[12] Prokhorov, M.E., Popov, S.B., and Khoperskov, A.V. 2002, *Astron. Astrophys.*, **381** 1000–1006
[13] Ruffert, M. 1999, *Astron. Astrophys.*, **346**, 861–877
[14] Ruzmaikin, A.A., Sokolov, D.D., and Shukurov, A.M. 1998, *Nature*, **336**, 341–347
[15] Shakura, N.I. 1973, *Soviet Astronomy*, **16**, 756–762
[16] Shvartsman, V.F. 1971, *Soviet Astronomy*, **15**, 377–384
[17] Strumik, M., Ben-Jaffel, L., Ratkiewicz, R., Grygorczuk, J. 2011, *Astrophys. J.*, **741**, 6
[18] Walder, R., Folini, D., Meynet, G. 2011, *Space Sci. Rev.*, **125**, in press
[19] Zel'dovich, Ya.B., Shakura, N.I. 1969, *Soviet Astronomy*, **13**, 175–183

The Galactic centre mini-spiral with CARMA

D Kunneriath[1], A Eckart[2,3], S N Vogel[4], P Teuben[4], K Mužić[5], R Schödel[6], M García-Marín[2], J Moultaka[7], J Staguhn[4,8,9], C Straubmeier[2], J A Zensus[3,2], M Valencia-S[2,3], V Karas[1]

[1] Astronomical Institute, Academy of Sciences, Boční II 1401, CZ-14100 Prague, Czech Republic
[2] I.Physikalisches Institut, Universität zu Köln, Zülpicher Str. 77, 50937 Köln, Germany
[3] Max-Planck-Institut für Radioastronomie, Auf dem Hügel 69, 53121 Bonn, Germany
[4] Department of Astronomy, University of Maryland, College Park, MD 20742-2421, USA
[5] Department of Astronomy and Astrophysics, University of Toronto, 50 St. George Str., Toronto ON M5S 3H4, Canada
[6] Instituto de Astrofísica de Andalucía - CSIC, Glorieta de la Astronomía S/N, 18008, Spain
[7] LATT, Université de Toulouse, CNRS, 14, Avenue Edouard Belin, 31400 Toulouse, France
[8] NASA Goddard Space Flight Ctr., Greenbelt, MD 20771, USA
[9] Department of Physics and Astronomy, Johns Hopkins University, Baltimore, MD 21218, USA

E-mail: devaky@astro.cas.cz

Abstract. The Galactic centre mini-spiral region is a mixture of gas and dust with temperatures ranging from a few hundred K to 10^4 K. We report results from 1.3 and 3 mm radio interferometric observations of this region with CARMA, and present a spectral index map of this region. We find a range of emission mechanisms in the region, including the inverted synchrotron spectrum of Sgr A*, free-free emission from the mini-spiral arms, and a possible dust emission contribution indicated by a positive spectral index.

1. Introduction

The central region of the centre of our Galaxy, the Milky Way, is dominated by the non-thermal emission from Sgr A*, the supermassive black hole of mass $\sim 4 \times 10^6\,M_\odot$ [1, 2, 3], the interstellar medium of the Circumnuclear Disk (CND) and the mini-spiral, and the nuclear stellar cluster.

The mini-spiral is believed to be a superposition of infalling streams of gas and dust from the CND, an association of molecular gas and dust of mass $\sim 10^5\,M_\odot$ [4] extending from 1.5 pc to about 4–7 pc. The infalling streams are ionized by the stars and Sgr A* in the central parsec. The main features of the mini-spiral are the Northern Arm, Western Arc, Eastern Arm and the Bar, which are believed to be in Keplerian orbits around the central SMBH, with some significant deviations possibly caused by stellar winds [5, 6]. The radio emission from the region is dominated by the thermal emission of the ionized gas from the mini-spiral arms, while the near-infrared (NIR) bands are dominated by the stellar sources. The mid-infrared (MIR) is dominated by dust emission at $\sim$200 K [7], resulting in a mixture of dust and gas with temperatures ranging from a few hundred K from the dust to up to 10^4 K radio bremsstrahlung plasma in this region.

Radio interferometric observations are a useful tool to separate the emission of Sgr A* from the surrounding mini-spiral, especially at higher frequencies with reasonably high resolution (~2″). In order to understand the emission mechanisms dominating the central parsec region of the Galaxy and to derive some of the physical characteristics of the medium, such as number density, emission measure, etc., we analysed the region at multiple wavelengths, ranging from the NIR to MIR to radio mm wavelengths. The results are presented in [8, 9]. In this paper, we present a subset of those observations, namely the radio mm observations and the spectral index map produced at these wavelengths.

2. Observations

We observed the Galactic centre region with the Combined Array for Research in mm-wave Astronomy (CARMA)[1] at 1.3 mm (230 GHz) with the C and D array configuration and at 3 mm (100 GHz) with the C configuration. The array consists of 15 telescopes, with six 10.4 m telescopes and nine 6.1 m telescopes, with five array cconfigurations, A to E, with baselines ranging from 0.25–2 km to 8–66 m, respectively. Fig.1 shows the D configuration 1.3 mm map of the central 1.5 pc (40″).

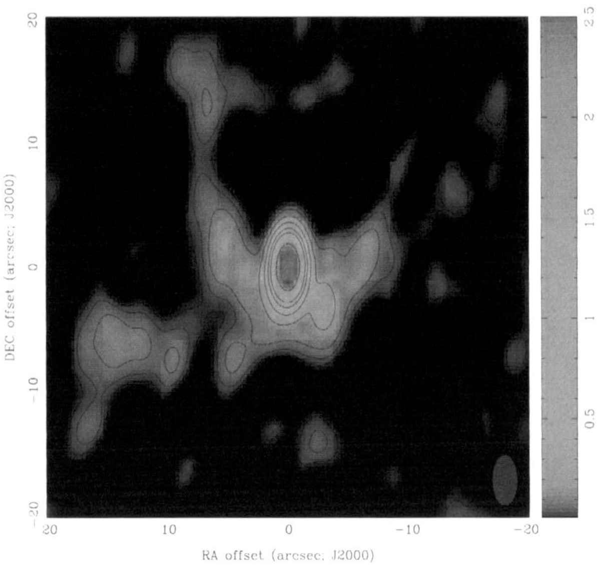

Figure 1. 1.3 mm map of the mini-spiral region at resolution 4.0″ by 2.0″ (P.A=0°). Contour levels are 0.02, 0.04, 0.08, 0.2, 0.4, 0.7, 1.0, 2.0 Jy/beam.

3. Results and Discussion

The spectral index map of the central 1.5 pc of the GC produced using the 1.3 mm and 3 mm maps convolved to the same angular resolution of 4″×2″ (P.A.=0°) is shown in Fig.2.

We obtained a spectral index of ~0.5 for Sgr A*, which indicates an inverted synchrotron spectrum, in agreement with previously published results [10, 11]. From the ionized gas in the mini-spiral arms we expect optically thin bremsstrahlung radiation with a spectral index of

[1] Support for CARMA construction was derived from the states of California, Illinois, and Maryland, the Gordon and Betty Moore Foundation, the Kenneth T. and Eileen L. Norris Foundation, the Associates of the California Institute of Technology, and the National Science Foundation. Ongoing CARMA development and operations are supported by the National Science Foundation under a cooperative agreement, and by the CARMA partner universities.

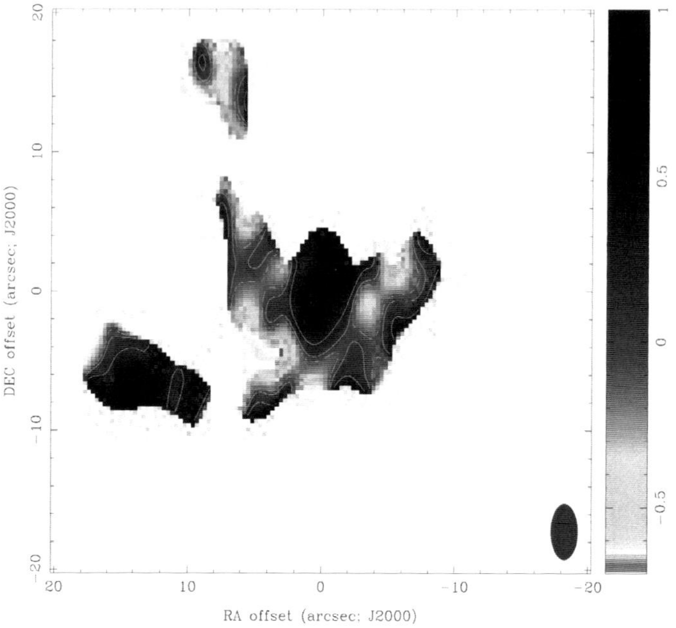

Figure 2. Spectral index map obtained from 1.3 mm and 3 mm, beam size 4″×2″ (P.A.=0°). Contour levels indicate spectral indices of -0.1 (dashed), 0.4 and 1.0.

~0.1, which is observed in most of the mini-spiral but with some exceptions. The Northern Arm shows steeper spectral index values ~-0.5, which are generally associated with optically thin synchrotron emission, but in this case they could be the result of resolved out flux at 1.3 mm.

We also obtained positive spectral index values of ~1 around the edges of the Bar and most of the Easterm Arm. At 3 mm, the mini-spiral radiation is believed to consist of mainly free-free emission. Using this assumption, we scaled our 3 mm map to 1.3 mm using the spectral index notation ($S \sim \nu^{-0.1}$), and subtracted this scaled map from our 1.3 mm map. As the bright Sgr A* dominates the spectral index in the central region, we masked the central region to exclude the emission from Sgr A*. The resultant residual map (Fig. 3) shows evidence for excess emission (compared to Bremsstrahlung) in the regions corresponding to positive spectral index regions in the spectral index map, the Eastern Arm, parts of the Bar, and a portion of the Northern Arm, which we attribute to dust emission which begins to be important at wavelengths $\leq$ 1 mm.

3.1. Uncertainty in Spectral Index

Radio interferometric maps are sensitive to the brighter emission from the compact components of the mini-spiral, and miss out on the large scale diffuse emission seen by single dish telescopes. In order to minimize effects from missing flux due to differences in uv-coverage, we have produced the spectral index map using data from array configurations which have a similar uv-coverage. The D configuration at 1.3 mm has an angular resolution of 2.18″ at 11 m to 29.74″ at 150 m, while the C configuration at 3 mm has an angular resolution of 2.16″ at 30 m to 25.16″ at 350 m. Thus we ensure that the effects of missing flux at short spacings are similar at 1.3 mm and 3 mm. Furthermore, the total flux measured by the CARMA map at 3 mm (~13 Jy) is in perfect agreement with previously published interferometric maps of the region [12, 13]. As a result it is conceivable to assume that both bands trace the same emission from the brighter small-scale features of the mini-spiral. Hence we are confident in the reliability of the spectral indices derived from these maps. There are two main aspects of the spectral index information: the emission

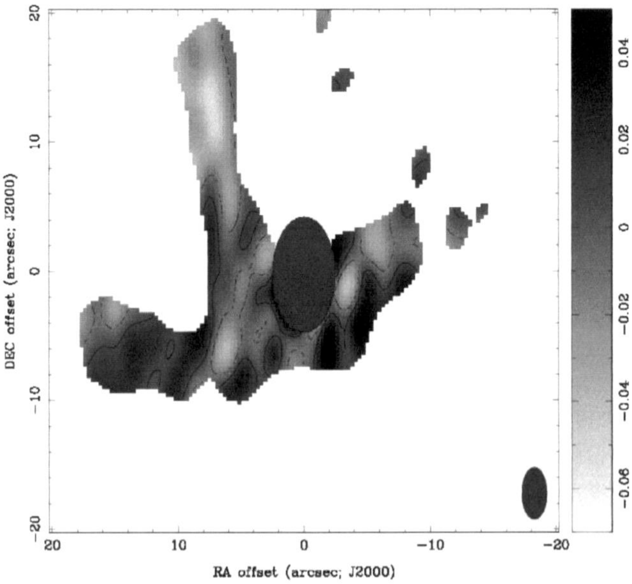

Figure 3. 1.3 mm excess emission (compared to Bremsstrahlung) at a resolution of $4'' \times 2''$ (P.A.=0°). See text for details of subtraction of free-free emission. The contour levels represent -0.01,0.01 and 0.05 Jy/beam.

mechanisms in the region, and resolution effects at different frequencies. The median spectral index obtained from the bright mini-spiral arms is ~-0.1, indicating the dominance of free-free emission. If the radio maps suffered from resolution effects, we would see steeper spectral index values, which are not observed, except in the case of the Northern Arm, and around the edges of the mini-spiral.

Acknowledgments
D. Kunneriath was supported by the International Max Planck Research School (IMPRS) for Astronomy and Astrophysics at the MPIfR and the Universities of Bonn and Cologne for part of this work. M. Valencia-S is currently a member of the IMPRS. R. Schoedel acknowledges support by the Ramón y Cajal programme by the Ministerio de Ciencia y Innovación of the government of Spain. M. Garcia-Marin is supported by the German federal department for education and research (BMBF) under the project numbers: 50OS0502 & 50OS0801. Part of this work was supported by the COST Action MP0905: Black Holes in a violent Universe and PECS project No. 98040.

4. References

[1] Eckart A and Genzel R 1996 *Nature* **383** 415–417

[2] Schödel R, Ott T, Genzel R, Hofmann R, Lehnert M, Eckart A, Mouawad N, Alexander T, Reid M J, Lenzen R, Hartung M, Lacombe F, Rouan D, Gendron E, Rousset G, Lagrange A, Brandner W, Ageorges N, Lidman C, Moorwood A F M, Spyromilio J, Hubin N and Menten K M 2002 *Nature* **419** 694–696 (*Preprint* `arXiv:astro-ph/0210426`)

[3] Ghez A, Morris M, Lu J, Weinberg N, Matthews K, Alexander T, Armitage P, Becklin E, Brown W, Campbell R, Do T, Eckart A, Genzel R, Gould A, Hansen B, Ho L, Lo F, Loeb A, Melia F, Merritt D, Milosavljevic M, Perets H, Rasio F, Reid M, Salim S, Schödel R and Yelda S 2009 *astro2010: The Astronomy and Astrophysics Decadal Survey* (*Astronomy* vol 2010) p 89

Journal of Physics: Conference Series **372** (2012) 012063 doi:10.1088/1742-6596/372/1/012063

[4] Christopher M H, Scoville N Z, Stolovy S R and Yun M S 2005 *ApJ* **622** 346–365 (*Preprint* arXiv:astro-ph/0502532)
[5] Zhao J H, Morris M R, Goss W M and An T 2009 *ApJ* **699** 186–214 (*Preprint* 0904.3133)
[6] Zhao J H, Blundell R, Moran J M, Downes D, Schuster K F and Marrone D P 2010 *ApJ* **723** 1097–1109 (*Preprint* 1009.1401)
[7] Cotera A, Morris M, Ghez A M, Becklin E E, Tanner A M, Werner M W and Stolovy S R 1999 *The Central Parsecs of the Galaxy* (*Astronomical Society of the Pacific Conference Series* vol 186) ed H Falcke, A Cotera, W J Duschl, F Melia, & M J Rieke p 240
[8] Kunneriath D, Eckart A, Zamaninasab M, Witzel G, Schödel R, García-Marín M, König S, Krichbaum T P, Lu R, Moultaka J, Mužić K, Sabha N, Sjouwerman L O, Straubmeier C, Vogel S N, Teuben P and Zensus J A 2011 *Astronomical Society of the Pacific Conference Series* (*Astronomical Society of the Pacific Conference Series* vol 439) ed M R Morris, Q D Wang, & F Yuan p 327
[9] Kunneriath D, Eckart A, Vogel S N, Teuben P, Mužić K, Schödel R, García-Marín M, Moultaka J, Staguhn J, Straubmeier C, Zensus J A, Valencia-S M and Karas V 2012 *A&A* **538** A127 (*Preprint* 1201.2362)
[10] Yusef-Zadeh F, Roberts D, Wardle M, Heinke C O and Bower G C 2006 *ApJ* **650** 189–194 (*Preprint* arXiv:astro-ph/0603685)
[11] Falcke H, Goss W M, Matsuo H, Teuben P, Zhao J and Zylka R 1998 *ApJ* **499** 731 (*Preprint* arXiv:astro-ph/9801085)
[12] Shukla H, Yun M S and Scoville N Z 2004 *ApJ* **616** 231–246 (*Preprint* arXiv:astro-ph/0408136)
[13] Wright M C H, Genzel R, Güsten R and Jaffe D T 1987 *The Galactic Center* (*American Institute of Physics Conference Series* vol 155) ed D C Backer pp 133–137

CI, [CII] and CO observations towards TNJ 1338−1942: Probing the ISM in a massive proto-cluster galaxy at z = 4.11

S König[1], T R Greve[1], N Seymour[2,3], J I Rawlings[3], P Papadopoulos[4], R J Ivison[5,6], C De Breuck[7], J Stevens[8], I Smail[9] and A Kovacs[10]

[1] Dark Cosmology Centre, Niels Bohr Institute, University of Copenhagen, Juliane Maries Vej 30, 2100 Copenhagen, Denmark
[2] CSIRO Astronomy & Space Science, PO Box 76, Epping, NSW 1710, Australia
[3] Mullard Space Science Laboratory, University College London, Holmbury St. Mary, Dorking, Surrey RH5 6NT, UK
[4] Max Planck Institute for Radio Astronomy, Auf dem Hügel 69, 53121 Bonn, Germany
[5] UK Astronomy Technology Centre, Royal Observatory, Blackford Hill, Edinburgh EH9 3HJ, UK
[6] Institute for Astronomy, University of Edinburgh, Blackford Hill, Edinburgh EH9 3HJ, UK
[7] European Southern Observatory, Karl Schwarzschild Straße 2, 85748 Garching, Germany
[8] Centre for Astrophysics Research, Science & Technology Research Institute, University of Hertfordshire, Hatfield, AL10 9AB, UK
[9] Institute for Computational Cosmology, Durham University, South Road, Durham DH1 3LE, UK
[10] University of Minnesota, 116 Church St. SE, Minneapolis, MN 55414, USA

E-mail: skoenig@dark-cosmology.dk

Abstract. A number of studies have dealt with the link between distant powerful radio sources and the most massive galaxies in the early Universe. Despite major advances in our understanding of high redshift radio galaxies (HzRGs), we still only have a rudimentary picture of the physical conditions (i.e. gas density, temperature, ambient UV-field) prevailing in the interstellar medium (ISM) of these objects. Here we report on ongoing CI, [CII] and CO observations of TNJ 1338−1942 at $z = 4.11$ with the IRAM 30m telescope, the JCMT and ATCA. With these observations we will make a first attempt at constraining the average ISM conditions in TNJ 1338−1942..

1. Introduction

Deep near-IR and mid-IR observations have shown that high-z radio galaxies ($z > 1$) are associated with large stellar populations ($M_* > 10^{11}$ $M_\odot$, e.g. [1]). This is consistent with the existence of extremely active supermassive black holes (SMBHs) at the centers of HzRGs ([2]). Observations of HI absorption ([3]), CO (e.g. [4]) and dust emission (e.g. [5]) established the presence of vast reservoirs of atomic and molecular hydrogen in some HzRGs.
TNJ 1338−1942 – the first $z > 4$ radio galaxy to be discovered in the southern sky and one of the most luminous Lyα objects in its class is most likely an extreme starburst. Both its Lyα profile and radio structure are very asymmetric ([6]), which indicates strong interaction with

dense gas. [7] showed that the high Lyα emission rate in TNJ 1338−1942 cannot be explained by AGN photo-ionization or shocks alone, but requires substantial star formation.
Not only does TNJ 1338−1942 reside at the center of an overdense distribution of both Lyα emitters ([8]) and LBGs ([9]), but it is also one of the most striking examples of an overdensity of (sub)mm selected galaxies ([10]). The latter study revealed 10 such sources in the field, corresponding to a 7σ overdensity compared to blank field surveys. Thus, TNJ 1338−1942 is unambiguously a massive galaxy at the center of a proto-cluster ([11]).
Up to now, we only have a very basic understanding of the physical conditions of the ISM in HzRGs. In an attempt to rectify this situation we have initiated a search for the most prominent ISM line tracers towards HzRGs, namely the rotational lines of carbon monoxide (CO), the FIR fine-structure lines of atomic carbon (CI($^3P_1 \rightarrow {}^3P_0$) and CI($^3P_2 \rightarrow {}^3P_1$) at 370 and 609 μm, respectively) and single ionized carbon (CII at 158 μm). The modest excitation requirement of the two lowest rotational transitions of CO make them ideal tracers of the total molecular gas mass, and the encompassed dynamical mass in galaxies. However, these low-J CO lines have been observed in only two HzRGs to date ([12], [13]). Typically, only J=3-2 CO lines or higher are detected in such systems, tracing only the dense and warm star forming gas. Compared to CO, which is routinely used to determine gas masses in galaxies, the prospect of utilizing the fine-structure lines of CI to trace molecular gas and dynamical masses at high-z was only recognized much more recently (e.g. [14], [15], [16]). The latter exploits the concomitance of CI and CO observed in molecular clouds in our Galaxy ([17]), the simple, three-level partition function, and the low optical depth and modest excitation requirements of the CI lines. Moreover, The CI(1-0) line can remain luminous even for UV-shielded, cold molecular gas (e.g. [18]) where the [CII] line luminosity is negligible. The [CII] line is the major coolant for much of the ISM and is a sensitive extinction-free probe of the gas physical conditions and ambient radiation fields. Both, CI and [CII], have been detected in high-z submillimeter galaxies and QSOs (e.g. [16],[28]).

2. Observations

2.1. ATCA

TNJ 1338−1942 was observed in the CO (2-1) emission line with the Australia Telescope Compact Array (ATCA) in the H214 hybrid array configuration in October 2010. The array included five antennas that were spread along an east-west and a north-south axis. At a redshift of 4.11 the CO (2-1) line (ν_r = 230.538 GHz) is shifted to an observed frequency of 45.115 GHz, i.e. within the 7 mm CABB receiver band (30-50 GHz). Using the CFB 1M-0.5k CABB observing mode provided us with two 2 GHz bands (total velocity range of 13610 km s^{-1} each) with a resolution of 1 MHz corresponding to a channel size of 6.65 km s^{-1}. The two 2 GHz bands were centered at 45.3 and 45.4 GHz, respectively. Only the band centered at 45.3 GHz was used in the final data analysis, since during the observations a laser diode on antenna 3 failed, therefore disabling the data transmission for the second observing frequency band. During the observations several sources were used as calibrators: the nearby source 1334-127 was used as phase calibrator to TNJ 1338−1942. The planet Uranus and 1934-638 were used as flux calibrators, 1921-293 was used for the bandpass calibration.
We reduced and analyzed the data using MIRIAD ([19]) following the standard CABB reduction procedures. The line free channels were combined to produce a continuum image that was used to subtract the continuum contribution from the spectral line data. The resulting synthesized beam was 3.99″ × 3.16″ (FWHM) with a primary beam diameter of 1.12′. The final spectrum has been binned into 100 km s^{-1} channels with in a channel-to-channel rms noise $\sigma_{\rm ch}$ of ∼1 mJy. The 3σ upper limits for the velocity-integrated line intensities were derived by $S\Delta v < 3 \times \sigma_{\rm ch} \times \sqrt{(v_{\rm FWHM} \times \Delta v_{\rm channel})}$ ([20]). Assuming a line width of 600 km s^{-1}, which is typical for CO line widths observed towards HzRGs (e.g. [20], [12]), results in a 3σ upper limit of 0.8 Jy km s^{-1} and an upper limit for the CO (2-1) luminosity $L_{CO(2-1)}$ of 5.0×10^7 $L_\odot$.

2.2. IRAM 30m

The observations of the CI (1-0) and (2-1) emission lines towards TNJ 1338−1942 were carried out with the IRAM 30m telescope[1] in early May, August and September 2011. Due to its proximity in frequency we were also able to look for the CO (7-6) emission line. The E0 and E1 EMIR (Eight Mixer Receiver) single pixel heterodyne receivers were used as frontends to observe the 3 and 2 mm bands simultaneously. E0 was centered on a frequency of 96.313 GHz and E1 on 158.384 GHz, which are the expected frequencies for CI (1-0) and CI (2-1) for our HzRG. Since VESPA provides only a small band width, the VESPA data (bandwidth of 480 MHz) will not be further taken into account. With bandwidths of 480 MHz 4 GHz (FFTS) and 16 GHz (WILMA), centered on the redshifted rest frame frequencies, we obtained data with velocity ranges of 12 450 $\mathrm{km\,s^{-1}}$ (FFTS) and 11 579 $\mathrm{km\,s^{-1}}$ (WILMA) for CI (1-0), and 7572 $\mathrm{km\,s^{-1}}$ (FFTS) and 7042 $\mathrm{km\,s^{-1}}$ (WILMA) for CI (2-1) and CO (7-6), and velocity resolutions of 0.6079 $\mathrm{km\,s^{-1}}$ (FFTS), and 6.225 $\mathrm{km\,s^{-1}}$ (WILMA) for CI (1-0) and 0.3697 $\mathrm{km\,s^{-1}}$ (FFTS) and 3.786 $\mathrm{km\,s^{-1}}$ (WILMA) for CI (2-1) and CO (7-6). The observations were done in wobbler switching mode with a symmetrical azimuthal throw of $120''$ and a frequency of 0.5 Hz. Pointing checks were performed every 90 min on the sources 1334-127 and 1253-055 from the IRAM Pointing Catalog, the focus was checked on Saturn.
We reduced and analyzed the data using the GILDAS[2] CLASS package. All spectra have been averaged, when applicable. To obtain a better signal-to-noise ratio, most of the spectra were Hanning smoothed. Baselines were fitted and subtracted from each subscan individually before averaging of the spectra. All polynomial fits to the baselines have been of the order of zero or one. The final spectra have been binned into channels with widths of 100 (CI (1-0)) and 121 $\mathrm{km\,s^{-1}}$ (CI (2-1) and CO (7-6)) with channel-to-channel rms noises of 1.8, 3.3 and 3.3 mJy, respectively. Assuming a CI line width of 600 $\mathrm{km\,s^{-1}}$ we obtain 3σ limits of 1.3 $\mathrm{Jy\,km\,s^{-1}}$ (CI (1-0)), 2.7 $\mathrm{Jy\,km\,s^{-1}}$ (CI (2-1)) and 2.7 $\mathrm{Jy\,km\,s^{-1}}$ (CO (7-6)) and upper limits for the luminosities of $\mathrm{L_{CI(1-0)}}<1.9\times10^8\ \mathrm{L_\odot}$, $\mathrm{L_{CI(2-1)}}<6.3\times10^8\ \mathrm{L_\odot}$ and $\mathrm{L_{CO(7-6)}}<6.2\times10^8\ \mathrm{L_\odot}$.

2.3. JCMT

We observed TNJ 1338−1942 in the [CII] transition at $\lambda = 158\ \mu$m with the James Clerk Maxwell Telescope (JCMT) on Mauna Kea, Hawai'i, in January 2011. We used the HARP 16 pixel single-side-band (SSB) SIS receiver as frontend and the ACSIS digital autocorrelation spectrometer as backend. With a bandwidth of 1800 MHz, centered on the redshifted rest frame frequency for TNJ 1338−1942 (372 GHz), and a channel spacing of 977 kHz we were able to obtain data with a velocity range Δv of $\sim$1450 $\mathrm{km\,s^{-1}}$ and a velocity resolution of 0.788 $\mathrm{km\,s^{-1}}$. The data were obtained in beam switching mode. The beam efficiency was $\sim$0.85 for a beam size of each receptor of $14''$. The on-source integration time was 7.4 hours. Pointing checks were performed every 50 minutes using sources from the JCMT Pointing Catalog (based on [21]). As flux calibrators we used CRL 618, IRC 10216, 16293-2422, G 34.3, CRL 2688 and Uranus, taken from the list of JCMT Heterodyne Standards.
Data reduction and analysis were performed using the STARLINK Software Package[3]. All spectra have been averaged, when applicable. To obtain a better signal-to-noise ratio, most of the spectra were Hanning smoothed. For data taken on different dates the subscans of each observation were calibrated with respect to a suitable flux calibrator. Baselines were fitted and

[1] Based on observations with the IRAM 30m Telescope.
[2] http://www.iram.fr/IRAMFR/GILDAS
[3] http://starlink.jach.hawaii.edu

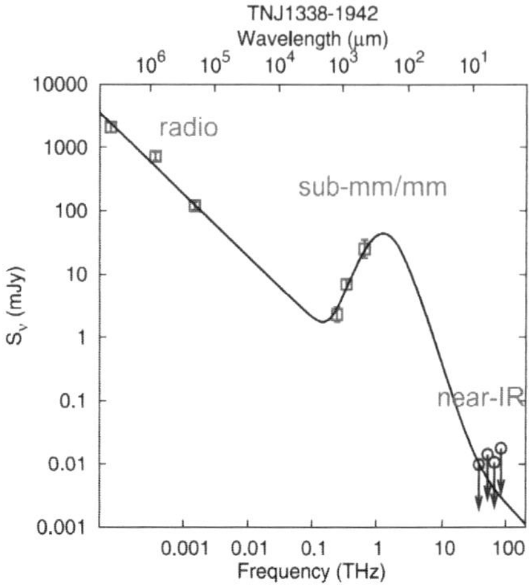

Figure 1. Full spectral energy distribution (SED) of TNJ 1338−1942. We fit a modified black body law (with an extension at radio wavelengths) to detections obtained at 74 MHz (VLA, [22]), 365 MHz (Texas Interferometer, [23]) and 1.4 GHz (VLA, [10]) in the radio, 450, 850 μm (SCUBA, JCMT, [3]) and 1200 μm (MAMBO, IRAM 30m, [10]) in the submm/mm and detections at 3.6,4.5, 5.8 and 8.0 μm (IRAC, Spitzer, [1]) in the near-IR wavelength regime. Integrating the SED over the rest-frame range 8–1000μm yields an infrared luminosity of $L_{IR}=2.4\times10^{13}$ $L_{\odot}$.

subtracted from each subscan individually before resampling and averaging of the spectra. Another baseline fit was performed after the final resampling and averaging of all available data. All polynomial fits to the baselines have been of the order of zero. The final spectrum (Fig. 3) has a spectral resolution of 40 km s^{-1} per channel with in a channel-to-channel rms noise of ~49 mJy. Assuming a typical CII line width of 600 km s^{-1} we get 3σ upper limits for integrated intensity and [CII] luminosity of $S\Delta v < 22.6$ Jy km s^{-1} and $L_{[CII]} < 1.2\times10^{10}$ $L_{\odot}$.

3. Inferred properties of TNJ 1338−1942

TNJ 1338−1942 has been detected in multiple (sub)mm bands (450, 850 and 1200 μm – [3], [10]), which, along with Spitzer IRAC detections ([1]), allows us to make a robust estimate of the IR luminosity. We fitted a power-law dust temperature distribution ($\alpha=7.2$, typical for starbursts, and emissivity index $\beta=1.5$). The resulting fit yields an IR luminosity (integrated from 8–1000 μm) of 2.4×10^{13} $L_{\odot}$ (Fig. 1), corresponding to a star formation rate (SFR) of >3800 $M_{\odot}$ yr^{-1} and a dust mass of 1.8×10^{8} $M_{\odot}$. With its IR luminosity, TNJ 1338−1942 has similar L_{IR} to other HzRGs observed at submillimeter wavelengths (e.g., L_{FIR}(TNJ 0924−2201) $= 3.4\times10^{12}$ $L_{\odot}$, [3]; L_{FIR}(4C 41.17) $= 2.0\times10^{12}$ $L_{\odot}$, [24]). TNJ 1338−1942 has a SFR that is higher than the typical SFRs of submillimeter selected galaxies (SMGs, SFR~900 $M_{\odot}$ yr^{-1} [25]). The lack of detection in the Spitzer 24 μm MIPS band suggests that an AGN is not dominating the IR energetics. In addition, a power-law extrapolation from the total 8.2 GHz flux density ([26]) predicts a negligible AGN synchrotron contribution. Therefore we assume that star formation is dominating the infrared luminosity.

Our search for CI (1-0) and (2-1), CO (2-1), CO (7-6) and [CII] line emission towards TNJ 1338−1942 has up to now yielded no detections (Figs. 2 and 3). We use the preliminary upper limit values to put more stringent constraints on the gas masses and the line transition luminosities (Table 1). Using $M_{H_2} = \alpha \times L'_{CO(2-1)}$, applying a conversion factor α of 0.8 $M_{\odot}$ (K km s^{-1}pc^{2})$^{-1}$ and assuming a CO(2-1)/CO(1-0) ratio of unity we obtain an upper limit on the molecular gas mass of 1.8×10^{11} $M_{\odot}$, which is consistent with gas masses observed for other HzRGs (e.g. TNJ 0924−2201: ~10^{11} $M_{\odot}$ [13], 4C60.07: 4×10^{11} $M_{\odot}$ [20]). The observed upper limits for $L_{[CII]}$ ($< 1.2\times10^{10}$ $L_{\odot}$) and $L_{[CII]}/L_{IR}$ ($<5\times10^{-4}$) are comparable to values found for high redshift sources detected by [27] and [28]. ALMA will provide the capabilities for much deeper observations and will therefore enable us to do much more sensitive and detailed studies of the ISM in distant radio galaxies.

Journal of Physics: Conference Series **372** (2012) 012064 doi:10.1088/1742-6596/372/1/012064

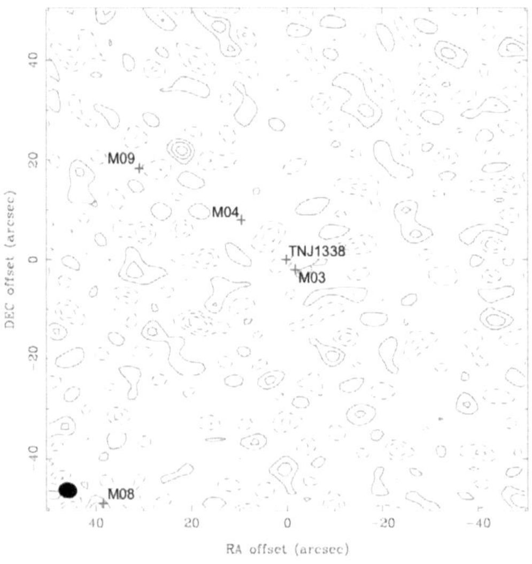

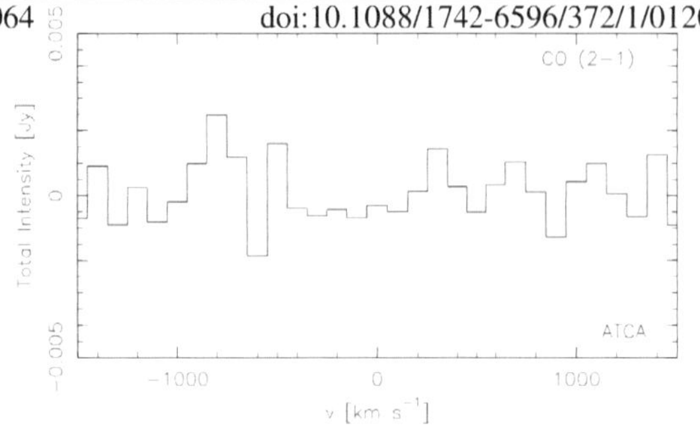

Figure 2. Moment zero map of the CO (2-1) line emission in TNJ 1338−1942 (*left*) and spectrum of the CO (2-1) emission within the central $20'' \times 20''$ (*right*). There was no line detected. The grey dashed line represents the zero intensity level. The rms noise in the map is $\sigma \sim 0.16$ Jy/beam/km s^{-1}, and the contours indicate the -1, 1, $2\times\sigma$ emission levels. The spectrum has been binned into 100 km s^{-1} channels, resulting in a channel-to-channel rms noise of ~ 1 mJy.

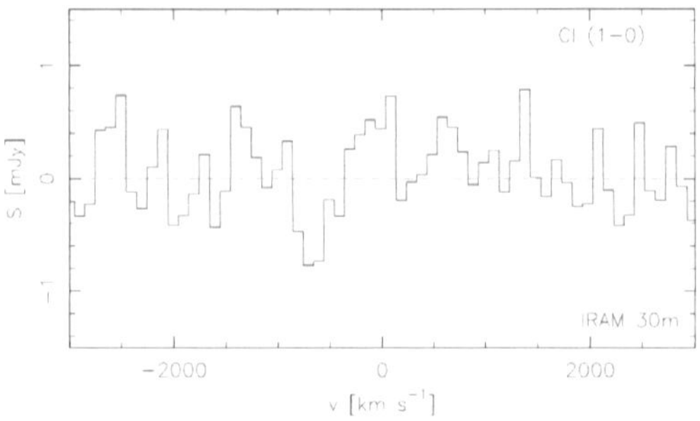

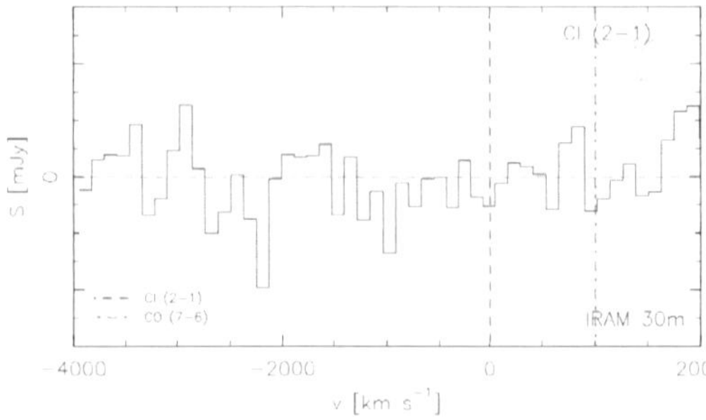

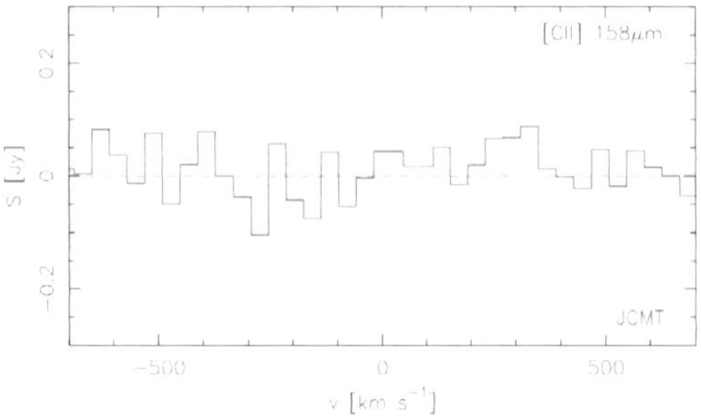

Figure 3. **Top:** IRAM spectra of the CI (1-0) (*left*), (2-1) and CO (7-6) (*right*) emission of TNJ 1338−1942. There was no line detected. The dashed (dashed-dotted) line indicate the central velocities of the CI (2-1) and the CO (7-6) emission line, the grey dashed line represents the zero intensity level. The spectral resolutions are 100 (CI (1-0)) and 121 km s^{-1} (CI (2-1)), resulting in channel-to-channel rms noise values of $\sim$ 1.8 mJy (CI (1-0)) and $\sim$ 3.3 mJy (CI (2-1)). **Bottom:** JCMT spectra of the [CII] 158 μm emission of TNJ 1338−1942. There was no line detected. The spectrum has been binned into 40 km s^{-1} channels, resulting in a channel-to-channel rms noise of $\sim$ 49 mJy.

Table 1. Properties of the line transitions observed for TNJ 1338−1942.

transition	ν_{obs} [GHz]	$t_{obs,on}$ [hrs]	$S_\nu \Delta v$ [Jy km s^{-1}]	L_{line} [10^9 $L_\odot$]
CI (1-0)	96.313	4.4	< 1.3	< 0.2
CI (2-1)	158.384	4.4	< 2.7	< 0.6
CO (2-1)	45.115	5.9	< 0.8	< 0.06
CO (7-6)	157.857	4.4	< 2.7	< 0.6
[CII] 158 μm	371.926	7.4	< 22.6	< 12.0

Acknowledgments
The Dark Cosmology Centre is funded by the Danish National Research Foundation. IRAM is supported by INSU/CNRS (France), MPG (Germany) and IGN (Spain). The James Clerk Maxwell Telescope is operated by The Joint Astronomy Centre on behalf of the Science and Technology Facilities Council of the United Kingdom, the Netherlands Organisation for Scientific Research, and the National Research Council of Canada. The JCMT data were obtained under Program ID M10BU30. The Australia Telescope Compact Array is part of the Australia Telescope National Facility which is funded by the Commonwealth of Australia for operation as a National Facility managed by CSIRO.

References

[1] Seymour N *et al.* 2007 *ApJS* **171** 353–375
[2] Blandford R D and Payne D G 1982 *MNRAS* **199** 883–903
[3] Reuland M *et al.* 2004 *MNRAS* **353** 377–390
[4] Breuck C D *et al.* 2005 *A&A* **430** L1–L4
[5] Ivison R J 1995 *MNRAS* **275** L33–L36
[6] Breuck C D *et al.* 1999 *A&A* **352** L51–L56
[7] Villar-Martín M *et al.* 2007 *MNRAS* **375** 1299–1310
[8] Venemans B P *et al.* 2002 *ApJL* **569** L11–L14
[9] Miley G K *et al.* 2004 *Nature* **427** 47–50
[10] Breuck C D *et al.* 2004 *A&A* **424** 1–12
[11] Breuck C D *et al.* 2010 *ApJ* **725** 36–62
[12] Greve T R *et al.* 2004 *A&A* **419** 99–107
[13] Klamer I J *et al.* 2005 *ApJL* **621** L1–L4
[14] Papadopoulos P P *et al.* 2004 *MNRAS* **351** 147–160
[15] Papadopoulos P P and Greve T R 2004 *ApJL* **615** L29–L32
[16] Weiß A *et al.* 2005 *A&A* **429** L25–L28
[17] Keene J *et al.* 1996 (*IAU Symposium* vol 178) ed E F van Dishoeck p 129
[18] Oka T *et al.* 2001 *ApJ* **558** 176–184
[19] Sault R J *et al.* 1995 (*Astronomical Society of the Pacific Conference Series* vol 77) ed R A Shaw, H E Payne, & J J E Hayes p 433
[20] Ivison R J *et al.* 1996 *MNRAS* **278** 669–672
[21] Loup C *et al.* 1993 *A&AS* **99** 291–377
[22] Cohen A S *et al.* 2007 *AJ* **134** 1245–1262
[23] Douglas J N *et al.* 1996 *AJ* **111** 1945
[24] Solomon P M and Vanden Bout P A 2005 *ARA&A* **43** 677–725
[25] Tacconi L J *et al.* 2006 *ApJ* **640** 228–240
[26] Pentericci L *et al.* 2002 *A&A* **396** 109–115
[27] Maiolino R *et al.* 2009 *A&A* **500** L1–L4
[28] Stacey G J *et al.* 2010 *ApJ* **724** 957–974

Journal of Physics: Conference Series **372** (2012) 012065 doi:10.1088/1742-6596/372/1/012065

Modeling the optical/UV polarization while flying around the tilted outflows of NGC 1068

Frédéric Marin[1], René Goosmann[1] and Michal Dovčiak[2]

[1]Observatoire astronomique de Strasbourg, Section Hautes Energies, 11 Rue de l'Université, F-67000 Strasbourg, France

[2] Astronomical Institute, Academy of Sciences of the Czech Republic, Boční II 1401, CZ-14131 Prague, Czech Republic

E-mail: `frederic.marin@astro.unistra.fr`

Abstract. Recent modeling of multi-waveband spectroscopic and maser observations suggests that the ionized outflows in the nuclear region of the archetypal Seyfert-2 galaxy NGC 1068 are inclined with respect to the vertical axis of the obscuring torus. Based on this suggestion, we build a complex reprocessing model of NGC 1068 for the optical/UV band. We apply the radiative transfer code STOKES to compute polarization spectra and images. The effects of electron and dust scattering and the radiative coupling occurring in the inner regions of the multi-component object are taken into account and evaluated at different polar and azimuthal viewing angles. The observed type-1/type-2 polarization dichotomy of active galactic nuclei is reproduced. At the assumed observer's inclination toward NGC 1068, the polarization is dominated by scattering in the polar outflows and therefore it indicates their tilting angle with respect to the torus axis. While a detailed analysis of our model results is still in progress, we briefly discuss how they relate to existing polarization observations of NGC 1068.

1. Introduction

According to the unified model of active galactic nuclei (AGN) [1] the accreting supermassive black hole of so-called type-2 thermal AGN lies hidden behind an optically thick, dusty torus. The obscuration prevents the broad line region to be seen and therefore type-2 AGN lack the kineticly broadened Balmer emission lines that characterize type-1 objects. Observations indicate that outflows are ejected from close to the accretion disk and it is possible that the torus funnel collimates these winds along polar directions. The launch of the wind is not yet understood, but it is a natural assumption that the ejection and collimation happen symmetrically with respect to the axis of the dusty torus and the accretion disk. In the case of the archetypical Seyfert-2 galaxy NGC 1068, the AGN is observed at a rather high inclination and the ionized winds are therefore well seen from the side.

A recent study by [2] suggests that the polar winds of NGC 1068 are inclined with respect to the torus axis. Based on this assumption, [3] and [4] started to investigate the radiative transfer effects of the misaligned outflows on the induced X-ray polarization. The modeling suggests that the position angle of the wind on the sky plane could be determined unambiguously from a soft X-ray polarization measurement. In addition to this, the tilting angle of the wind with respect to the torus axis could be constrained by the rotation of the polarization position angle between the soft and hard X-ray bands.

flared disk	**dusty torus**	**polar outflows**
$R_{\rm min} = 0.02$ pc	$R_{\rm min} = 0.1$ pc	$R_{\rm min} = 0.3$ pc
$R_{\rm max} = 0.04$ pc	$R_{\rm max} = 0.5$ pc	$R_{\rm max} = 1.8$ pc
half-opening angle = 20°	half-opening angle = 60°	half-opening angle = 40°, tilted by 18°
equat. optical depth = 1	equat. optical depth = 750	vertical optical depth = 0.03
electron scattering	Mie scattering	electron scattering

Table 1. Parameters of the three scattering regions. The polar outflows and torus half-opening angles are measured with respect to the vertical symmetry axis of the torus. The half-opening angle of the flared disk is measured from the equatorial plane.

The predictions for future X-ray polarization measurements are more solid when they can be coherently related to present-day polarization results from a different waveband. Therefore, we here present some results for the optical/UV counterpart of the same model setup. Similarly to [4], we present spectral and polarization modeling of NGC 1068 at several polar and azimuthal viewing angles. We use the new polarization mapping capacity that was recently implemented in our code STOKES. A more detailed description of the new STOKES version and more results of the current study will be published elsewhere (Marin et al., in preparation).

2. The model setup for NGC 1068

This research note focuses on the expected polarization signatures of NGC 1068 with misaligned outflows as a function of the polar and azimuthal viewing angle. We base our model on the description given in [2] assuming that the bi-conical outflows are inclined by 18° from the vertical torus axis. The model setup is very similar to the one used in [4]. Its main characteristics are summarized in Table 1. Note that here we did not include an irradiated accretion disk as it was done for the X-ray band where a lamp-post geometry was assumed. Instead, the negligibly small accretion disk produces an optical/UV continuum spectrum having a $F_*(E) \propto E^{-\alpha}$ power law shape. The spectral index is set to $\alpha = 1$. The lower and upper limits of the spectrum are set to 2000 Å and 8000 Å, respectively. The Cartesian coordinate system of the model space and its relation to the polar and azimuthal viewing angle is as described in [4]: all reprocessing regions are centered on the coordinate origin and the wind axis is tilted inside the yz-plane leaning towards the positive y-axis. The observer's inclination and the tilting angle of the outflows are measured from the z-axis. The azimuthal viewing angle, ϕ, is taken inside the xy-plane with respect to the negative y-axis. The modeling is carried out using the latest version of the Monte Carlo radiative transfer code STOKES presented in [5] and recently upgraded (Marin et al., in preparation).

We investigate the complex radiative coupling for a model composed of : (1) an optically and geometrically thick dusty torus, (2) an hourglass shaped, bi-conical polar wind filled with electrons, and (3) an equatorial, electron filled, radiation supported, flared disk. The equatorial scattering region is important in order to explain the observed polarization dichotomy between type-1 and type-2 AGN: it turns out from the observations that in type-1 AGN the polarization position angle in the optical/UV preferentially aligns with the projected symmetry axis of the polar outflows [6]. Observationally, this axis is assumed to be traced by the small-scale radio jets (note that in radio-quiet AGN these jets are much smaller and weaker than the ballistic, kpc-size jets seen in radio-loud objects). The small jets in radio-quiet AGN should indicate nuclear outflows that contribute to the larger-scale, bi-conical ionization regions seen in optical/UV imaging. When investigating the scattering-induced polarization produced by the different reprocessing regions listed above, it turns out that it is difficult to reproduce the observed aligned polarization at type-1 viewing-angles, unless an equatorial scattering region, such as a flared disk, is present [7, 8, 9].

3. Spinning around NGC 1068 : modeling results

The main goal in this research note is to revisit the X-ray polarimetry modeling of [4] while shifting the energy band to UV and optical wavelengths and without doing any morphological changes to the model (with the exception of leaving aside the lamp-post irradiation geometry used only in the X-ray modeling).

3.1. Spectropolarimetric signature

The spectropolarimetric modeling results are shown in Fig.1. We consider four inclinations at which we investigate the spectral and polarization response of the system. The angle $i \sim 26°$ features a type-1 object (face-on view), $i \sim 60°$ refers to a viewing angle passing through the upper layers of the torus, $i \sim 73°$ is around the expected inclination of NGC 1068 [10] and, finally, $i \sim 84°$ features an extreme type-2 object (edge-on view). We cover half a round in azimuthal angle at $\phi = 35°$, $\phi = 80°$, $\phi = 125°$ and $\phi = 170°$. We extend the coverage to a full round for the spectra of the polarization position angle ψ. We define $\psi = 0°$ as "perpendicular" polarization with respect to the z-axis. It rotates clock-wise until $\psi = 90°$ ("parallel" polarization).

Similarly to the X-ray results presented in [4], the spectral flux is rather ϕ-independent at face-on view because the model does not deviate much from axis-symmetry. Since at low inclination the primary source is directly visible, the total flux is dominated by the power law continuum spectrum. Nevertheless, an additional scattered component contributes by more than 20% to the normalized flux. The flux level decreases rapidly as the observer's line of sight changes from type-1 to type-2 viewing directions. Heavy obscuration arises when the line of sight passes through the dusty torus, and the central source is no longer visible. Instead, scattered radiation dominates the spectrum.

The spectral shape of the polarization percentage, P, depends on the inclination: at $i \sim 26°$ and $i \sim 60°$, P has low values due to strong dilution by the unpolarized primary radiation coming from the directly visible source. The low polarization is nearly wavelength-independent in the UV, which reveals electron scattering inside the flared disk as the main polarization mechanism. The polar outflows are seen either in transmission (forward-scattering in the near cone) or in straight-line reflection (backscattering in the far cone) and they thus do not contribute much to the polarization. A dim contribution of dust scattering to the polarization appears toward increasing wavelengths when the Mie scattering phase function becomes less anisotropic so that multiply scattered photons have a higher probability to escape from the torus funnel. When the line of sight crosses the torus horizon and moves towards the highest inclination, the effects of dust scattering are less prominent as the torus funnel becomes more and more obscured. Electron scattering in the polar outflows is dominant and therefore P rises with i and remains constant with wavelength. This is in agreement with the observations [11, 12, 13]. The spectral shape of P is nearly independent of azimuth, but not its normalization, which varies with ϕ.

The polarization position angle ψ shows a systematic behavior as the virtual observer travels around the object in the ϕ-direction. Despite the tilt of the outflows, the modeled ψ confirms the observed polarization dichotomy at all ϕ-angles: for type-1 objects, we obtain $\psi \sim 90°$ and therefore the polarization roughly aligns with the axis of the outflow. For type-2 objects, the polarization is oriented perpendicularly to this axis. This picture would change, though, for more extreme tilting angles of the wind. The parallel polarization at type-1 viewing angles can only be preserved when the outflows are seen along a line of sight that mainly favors forward and backscattering. The polarization angle is symmetric in ϕ, which is expected because the axis of the inclined outflows lies in the yz-plane. The variation of ψ with ϕ is broader at type-2 than at type-1 view. By comparing the sign of ψ at azimuthal viewing angles that are symmetric with respect to the yz-plane, we can recover the wind tilting direction: If the polarization angle is positive (negative), the polar outflows should be inclined in a clock-wise (counter clock-wise) direction from the z-axis.

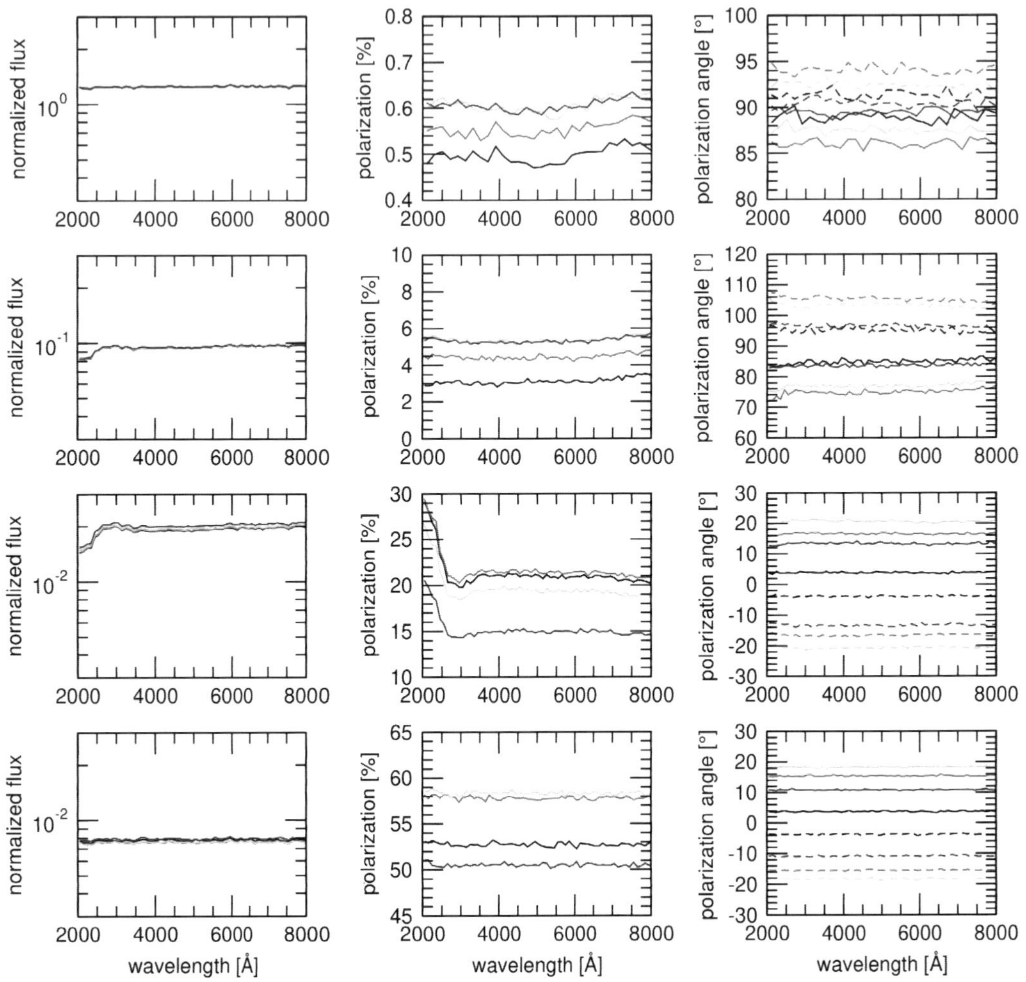

Figure 1. Spectral and polarization results in the optical/UV for the NGC 1068 model with inclined polar outflows. From top to bottom the panels represent the four inclinations $i \sim 26°$, $i \sim 60°$, $i \sim 73°$ and $i \sim 84°$. *Left:* spectral flux F/F_{cent} normalized to the source flux F_{cent}. *Middle:* percentage of polarization P. *Right:* polarization position angle ψ. The different curves denote four azimuthal viewing angles (half a round) measured from the negative y-axis : $\phi = 35°$ (blue), $\phi = 80°$ (green), $\phi = 125°$ (red) and $\phi = 170°$ (black). For the polarization position angle, ψ, the spectra at the symmetric azimuth (ϕ = -ϕ) are also shown using dashed lines. For $\psi = 0°$ the polarization is perpendicular to the z-axis; it rotates clock-wise with rising ψ.

The position of the outflow axis in NGC 1068 was investigated by [14] conducting kinematic modeling of the narrow line region. This region is assumed to be the natural continuation of the ionized winds. The kinematic modeling suggests that the axis does not lie in the plane of the sky but is indeed azimuthally rotated. From spectroscopy alone, however, this conclusion is difficult to verify and the polarization modeling carried out here can give an additional tracer of the scattering geometry. The combination of the polarization percentage and position angle adds two more observables that are sensitive to the position of the wind axis.

3.2. Polarization imaging

Results of our simulated polarization cartography are given in Fig. 2. We show the maps at an inclination of $i \sim 73°$, which is in agreement with the expected viewing angle towards NGC 1068 as derived by [10]. The maps simultaneously show the polarized flux, PF_ν, and the polarization position angle, ψ. The angle ψ is represented by black bars drawn in the center of each spatial bin. A vertical bar indicates a polarization of $\psi = 90°$, a bar leaning to the right denotes $90° > \psi > 0°$ and a horizontal bar stands for $\psi = 0°$. For each pixel, the Stokes parameters are integrated over the full wavelength range.

The images illustrate how the net polarization of the AGN is determined by the polarized flux integrated over the projected extension of the object. Independently of the azimuthal angle, strong polarized flux with a polarization position angle geared towards a vertical direction is spotted at the center. The polarization here is dominated by the combined scattering in the flared electron disk and the funnel of the relatively flat torus. It was shown in [5] that both such scattering regions produce parallel polarization. Even though the central polarized flux predominates over the average polarized flux from the outflows by a factor of 5, the large spatial extension of the latter determines the net polarization.

It is instructive to compare our model images with the observed UV polarization maps shown in Fig. 2 of [15]. Our model reproduces the arch-like distribution of polarization position angles across the extended outflows rather well. In the center, where according to [15] the main scattering source is located, the observed behavior still differs from our modeling. Although the observed polarization position angle near the central source shows some deviation from its behavior in the surroundings, this local turn is less prominent than our model suggests. It is possible that foreground matter on the observer's line of sight leads to additional polarization and can reduce the difference. Therefore, our analysis and modeling of the observed polarization images of NGC 1068 still needs some refinement, in particular in the central region of the object. It may also be that the inner surfaces of the torus are indeed steeper than assumed here so that the resulting polarization from the torus funnel is perpendicular and not parallel. Also, we have not investigated the impact of the electron optical depth in the polar outflows or the equatorial scattering disk yet. Such investigations are currently carried out and will be presented in a forthcoming paper (Marin et al., in preparation).

References

[1] Antonucci R 1993 *ARAA* **31** 473–521
[2] Raban D, Jaffe W, Röttgering H, Meisenheimer K and Tristram K R W 2009 *MNRAS* **394** 1325–1337 (*Preprint* 0901.1306)
[3] Goosmann R W and Matt G 2011 *MNRAS* **415** 3119–3128 (*Preprint* 1012.4652)
[4] Goosmann R W and Matt G 2011 *SF2A-2011: Proceedings of the Annual meeting of the French Society of Astronomy and Astrophysics* ed G Alecian, K Belkacem, R Samadi, & D Valls-Gabaud pp 583–586
[5] Goosmann R W and Gaskell C M 2007 *AAP* **465** 129–145 (*Preprint* arXiv:astro-ph/0507072)
[6] Antonucci R R J 1982 *Nature* **299** 605
[7] Chandrasekhar S 1960 *Radiative transfer*
[8] Antonucci R R J 1984 *ApJ* **281** 112–125
[9] Smith J E, Young S, Robinson A, Corbett E A, Giannuzzo M E, Axon D J and Hough J H 2002 *MNRAS* **335** 773–798 (*Preprint* arXiv:astro-ph/0205204)
[10] Hönig S F and Kishimoto M 2010 *AAP* **523** A27 (*Preprint* 0909.4539)
[11] Miller J S and Antonucci R R J 1983 *ApJL* **271** L7–L11
[12] Antonucci R R J and Miller J S 1985 *ApJ* **297** 621–632
[13] Antonucci R, Hurt T and Miller J 1994 *ApJ* **430** 210–217
[14] Das V, Crenshaw D M, Kraemer S B and Deo R P 2006 *AJ* **132** 620–632 (*Preprint* arXiv:astro-ph/0603803)
[15] Capetti A, Macchetto F, Axon D J, Sparks W B and Boksenberg A 1995 *ApJL* **452** L87

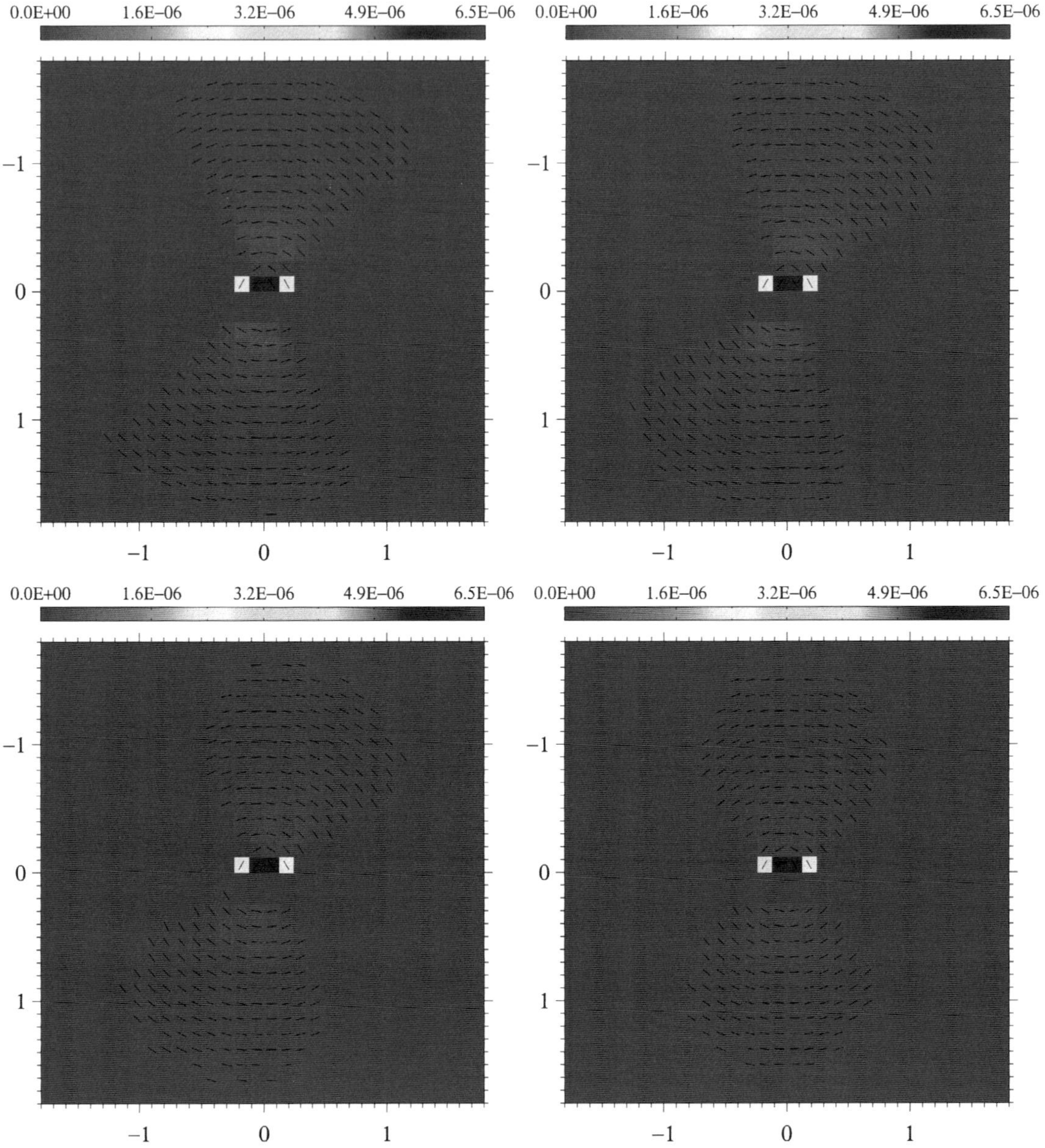

Figure 2. Modeling the spatial distribution of the polarized flux, PF_ν, of our model for NGC 1068 with inclined polar outflows. The PF_ν is denoted by the color-coded grid and normalized to the total flux. The polarization angle ψ is represented by black bars (see text). The model region is observed at the inclination $i \sim 73°$ for the same ϕ-angles as used in Fig.1. *Top, left*: $\phi = 35°$, *Top, right*: $\phi = 80°$, *Bottom, left*: $\phi = 125°$, *Bottom, right*: $\phi = 170°$.

Star Formation Timescale in the Circumnuclear Starburst Ring of Barred Galaxy NGC 7552

Hsi-An Pan[1,2,3,4], Jeremy Lim[2,5], Satoki Matsushita[2] and Tony Wong[6]

[1]Department of Earth Sciences, National Taiwan Normal University, Taipei 11677, Taiwan
[2]Institute of Astronomy & Astrophysics (ASIAA), Academia Sinica, Taipei 10617, Taiwan
[3]Department of Astronomical Science, The Graduate University for Advanced Studies, Shonan Village, Hayama, Kanagawa 240-0193, Japan
[4]Nobeyama Radio Observatory, NAOJ, Minamimaki, Minamisaku, Nagano 384-1305, Japan
[5]Department of Physics, University of Hong Kong, Pokfulam, Hong Kong
[6]Department of Astronomy, University of Illinois, 1002 W. Green Street, Urbana, IL 61801, USA

E-mail: pan@nro.nao.ac.jp

Abstract. We present the multiple line observations carried out with interferometry. We observed the circumnuclear region of the barred galaxy NGC 7552 in molecular lines tracing diffuse molecular gas (^{12}CO J = 2 - 1; observed by SMA) and relatively dense molecular gas (HCN J = 1 - 0; observed by ATCA). We also reprocessed a published HI image which covers the entire galaxy to gain an analytical image with higher resolution.

The displacement between HCN (J = 1 – 0) and radio knots (3 cm continuum; ATCA archive data) is clearly seen in the circumnuclear starburst ring of NGC 7552. The propagation time derived from ^{12}CO J = 2 - 1 and HI based rotation curve between the HCN (J = 1 – 0) and radio knots implies the timescale between the formation and death of massive stars. The timescale of NGC 7552 is about an order of magnitude shorter than 5 – 10 Myr timescale for OB stars to become supernovae. It is possible that star formation in the ring is top heavy, resulting in a shorter timescale.

1. Introduction

Over half of all spiral galaxies in the Local Universe exhibit large-scale bars [1]. Theory predicts that a galactic bar is able to transport disk molecular gas toward the center via dust lanes. In addition, observation in ^{12}CO (1 – 0) also shows a correlation between the degree of central concentration of molecular gas and the bar strength [2]. The accumulation of molecular gas in central region of galaxies gives the chance to form stars. Therefore, it is not surprising that circumnuclear starburst rings are commonly seen in the centers of barred galaxies.

In this proceeding, we present the results of multi-line interferometric observations of NCC 7552. Since our lines trace different stages of star formation, the results allow us to calculate the star formation timescale in the circumnuclear ring.

2. Target

NGC 7552 is a face-on ($i = 25°$) barred galaxy with a distance of 22.3 Mpc at 23° 16′ 10.8″ -42^{d} 35^{m} 05^{s}, corresponding to 1″ = 21 Mpc. The galaxy is a member of the galaxy group known as the Southern Grus Quartet. Figure 1 shows the optical image of NGC 7552 with different

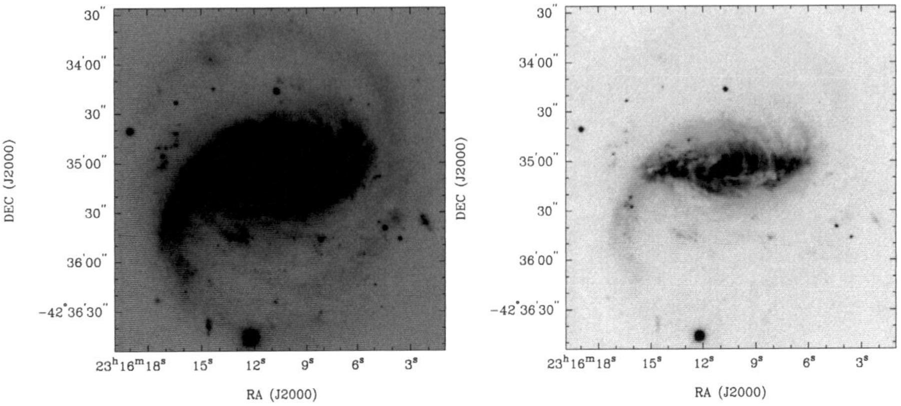

Figure 1. Optical B-band images of NGC 7552 taken at the CTIO, and retrieved from the NASA Extragalactic Database(NED). The left panels shows the image with more saturation, showiing the spiral arms. The right panel with less saturation indicates the bar and dust lanes.

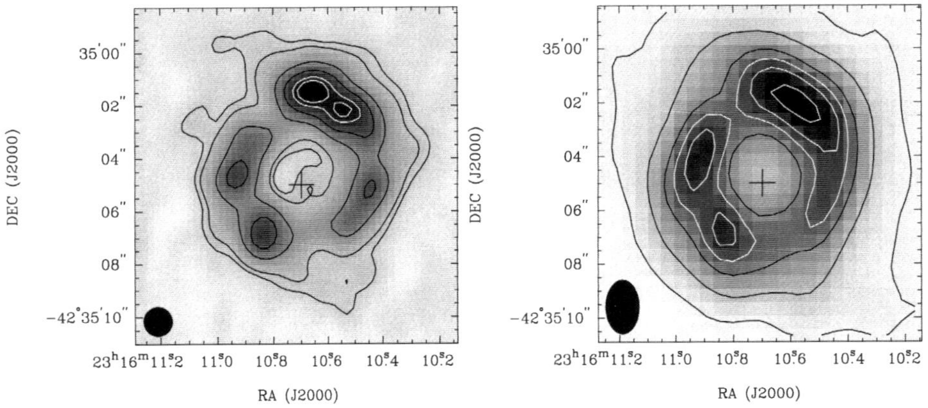

Figure 2. Radio continuum images of the circumnuclear starburst ring in NGC 7552 at 6 cm (left panel) and 3 cm (right panel), re-created from the data taken by and published in [3]. The contour levels of 6 cm image are 3, 5, 15, 25, 35, and 40 σ, where 1 σ = 72 μJy beam^{-1}. On the other hand, contour levels of 3 cm continuum are plotted in 3, 15, 35, 50, 65, and 80 σ, with 1 σ = 66 μJy beam^{-1}. The cross indicates the center of the ring as defined by [3]. The synthesized beam is $2.0'' \times 1.3''$ at 6 cm $1.1'' \times 1.1''$ at 3 cm, and shown at the lower left corner of the respective panels.

saturation levels. As can be seen, NGC 7552 has a galactic bar with radius of ~5 kpc along with spiral arms emerging from the bar ends.

Forbes et al (1994)[3] have presented the observation of the center of NGC 7552 in 3 cm and 6 cm continuum (Figure 2). The ring-like structure is completely resolved in both wavelengths. There are several knots in the ring, indicating the regions where the star formation activity is intense.

3. Observation

We observed two molecular lines ^{12}CO (J = 2 − 1) and HCN (J = 2 − 1) toward the central region of NGC 7552. ^{12}CO (J = 2 − 1) observation is carried with compact configuration of the Submillimeter Array [4] on 4 August 2006. Seven out of the eight antennas of the SMA were available during our observation, with baseline lengths ranging from 16 m to 69 m. The data reduction was done by using SMA-specific MIR tasks, which was originally developed for the Owens Valley Radio Observatory [5]. The final integrated intensity and velocity maps have a resolution of $7.0'' \times 2.8''$ at a position angle (PA) of $-11.9°$. We do not show the maps of ^{12}CO ($J = 2-1$) in this proceeding but the inferred rotation curve. HCN ($1-0$) was observed by the Australian Telescope Compact Array (ATCA) between May and September 2005 in the H75, H168, and H214 configurations, providing baseline lengths ranging from 30 m to 200 m, The data was reduced using MIRIAD. Resolution of finale maps are $2.6'' \times 2''$ with PA = -80°.

We re-created the 21-cm neutral atomic hydrogen (HI) maps of [6]. By using robust weighting, we obtained higher angular resolution ($20'' \times 20''$) than the images in [6]. The HI maps cover not only the central region but also the outer disk of NGC 7552, which allows us to derive a large scale rotation curve.

4. Results and Discussion

Figure 3 shows the integrated intensity of HCN ($1-0$) emission (grey scale and white contour). The resolution of the HCN ($1-0$) observation is $2.6'' \times 2.0''$. A ring-like structure is seen, with two knots at the northern and southern ring. The 3 cm continuum shown in §2 is overlaid in contours but now convolved to the same angular resolution as the HCN ($1-0$) map. At this resolution, 3 cm continuum shows two knots in HCN ($1-0$) emission. We therefore confirm that HCN ($1-0$) and 3 cm continuum originate from the same source, *i.e.* the circumnuclear ring. Note that there are small displacements between the centroids of the dense molecular gas (HCN) and supernova emission (3 cm continuum). We will use these displacements to compute star formation timescale in circumnuclear ring later.

If we assume that the spiral arms in NGC 7552 are trailing, then the radio continuum knots are located downstream of the HCN knots in both NGC 7552. Accordingly, the aforementioned displacements reflects the timescale between the formation of massive stars from dense molecular gas at or close to the contact points, and the eventual demise of these stars as supernova explosions, a schematic explanation of this idea is shown in Firgure 6. We then check this argument by calculating the time it would take.

We derived a rotation curve of NGC 7552 based on our measurements of the kinematics of the galaxy in ^{12}CO (from 0 to 1.2 kpc) as well as the HI line (2 to 20 kpc). We fit the sum of two functions to the derived rotation curve. A Brandt curve reflects the gravitational potential of the galactic bulge. An exponential curve reflects the gravitational potential of the galactic disk and dark matter halo. With standard methods (Figure 5) and the widely accepted assumption that the Corotation Radius (CR) is located atthe end of the, the pattern speed is therefore derived as 20 km s^{-1} kpc^{-1}.

The northern radio continuum and HCN knots have deprojected angular offset $0.47'' \pm 0.05''$, while the offset between southern radio continuum and HCN knots is $0.67'' \pm 0.08''$. The rotational velocity of the circumnuclear starburst ring is about 170 km s^{-1}, and the pattern speed of the bar about 20 km s^{-1} kpc^{-1} (translate to $\sim$10 km s^{-1} at ring, which has radius of $\sim$ 0.5 kpc); their net velocity is therefore about 160 km s^{-1}. The propagation time between the northern radio and HCN knots is therefore $(3.0 \pm 0.4) \times 10^5$ years, and that between the southern radio and HCN knots $(5.0 \pm 0.5) \times 10^5$ years.

The timescale is about one order of magnitude shorter than the timescale for OB stars to explode as supernovae. There are two possible reason for the shorter timescale. First of all, the insufficient resolution in ^{12}CO and HI might leads to underestimation in pattern speed and/or overestimation in the rotational velocity of the ring. Besides, star formation in a starburst environment may be top heavy compared to recently accepted Salpeter-like IMF, resulting in a shorter than usual timescale.

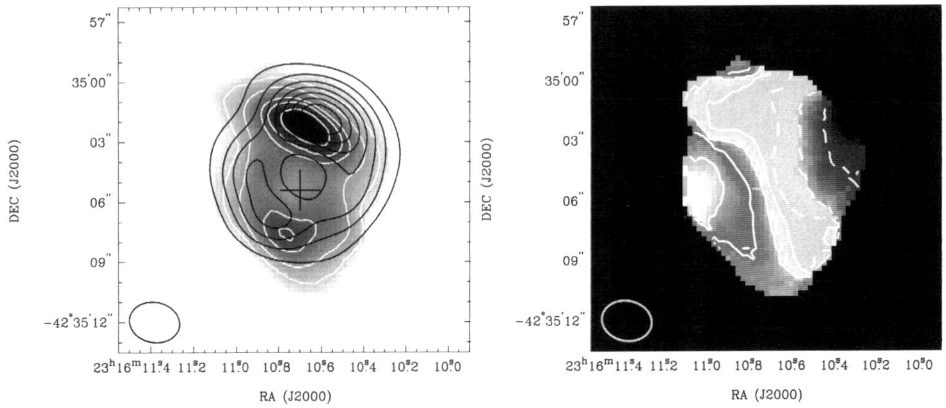

Figure 3. Integrated HCN (1 – 0) intensity (left panel) and intensity-weighted mean HCN (1 – 0) velocity (right panel). In the left panel, white contours indicate the intensity of HCN (1 – 0) plotted in 3, 6, 10, 12, and 15 σ, while 1 σ = 0.3 Jy beam^{-1} km s^{-1}. Black contours correspond to the 3 cm continuum image shown in Figure 1 but convolved to the same angular resolution as in HCN (1 – 0). The contour interval is 13% of peak intensity 9.4 mJy beam^{-1}. In the right panel, contours correspond to the intensity-weighted mean HCN (1 – 0) velocity just like the color map and are plotted in steps of 20 km s^{-1} starting from -60 km s^{-1} and ending at 60 km s^{-1} with respect to the systemic velocity. A cross is plotted at the center of the galaxy. The synthesized beam is shown at the lower left corner of each panel. The synthesized beam is $2.6'' \times 2.0''$ at PA = 80°.

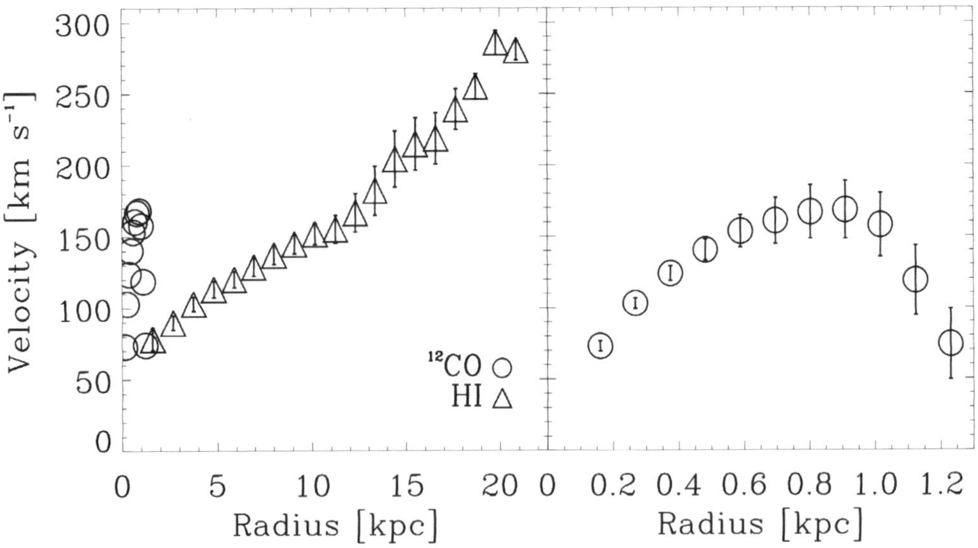

Figure 4. Rotation curve derived for NGC 7552 from our ^{12}CO (2 - 1) maps, plotted as circles, and HI maps, plotted as triangles. We assumed a constant inclination of 25° and position angle for the kinematic major axis of 110°. The error bars of rotation curve in central region are ignored in left panel to make the figure neat. Th right panel shows the rotation curve in center region of NGC 7552, along with the uncertainties of data points.

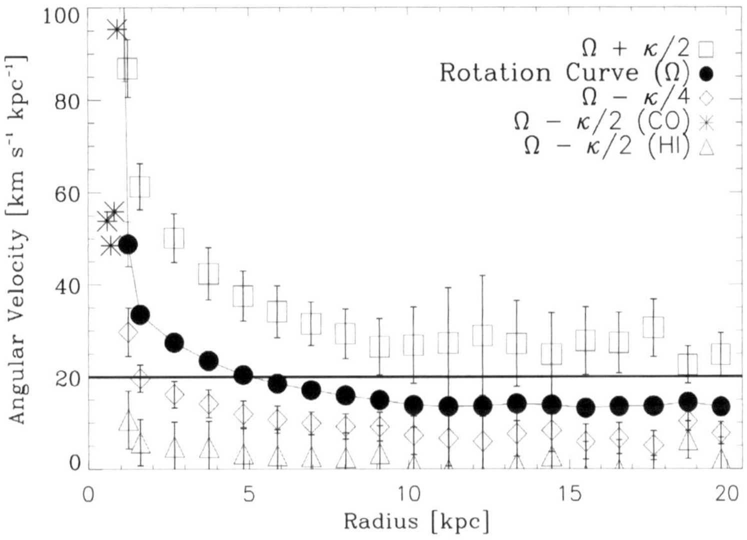

Figure 5. Diagram used to infer the locations of dynamical resonances and pattern speed of a barred galaxy. The angular velocity of the galaxy, Ω, based on an interpolation of the rotation curve shown in Figure 4 is plotted as the black solid line and solid circles. Assuming that CR is at the end of the bar (5 kpc), the pattern speed is therefore the angular velocity (20 km s^{-1} kpc^{-1}) at the bar end. The individual measurements that define the $\Omega - \kappa/2$ (ILR) curve are plotted as asterisks (from ^{12}CO (2 – 1)) and triangles (from HI), where κ is the epicyclic frequency of the orbit. The $\Omega - \kappa/4$ (UHR) curve is plotted as diamonds, and the $\Omega + \kappa/2$ (OLR) curve with squares. Resonances occur where the values of these curves are equal to the pattern speed of the bar, assumed to be equal to the rotational velocity at the ends of the bar as indicated by the red horizontal line.

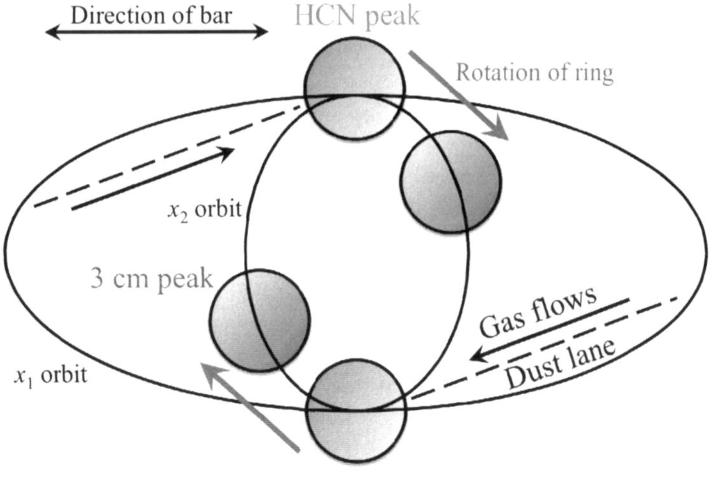

Figure 6. Schematic explanation of the displacement seen in HCN (1 – 0) and 3 cm continuum images.

5. Summary

We have observed the center of NGC 7552 in multiple lines. Both HCN (1 – 0) and 3 cm continuum show ring-like two peaks structures. The displacement between knots of HCN (1 – 0) and 3 cm continuum reflect the star formation timescale between dense molecular gas and supernova explosion. For calculating the timescale, we first derive a rotation curve for the diffuse gas tracer ^{12}CO (2 – 1) and 21 cm Atomic Hydrogen (HI), then derive the pattern speed of bar with standard manner. The pattern speed is 20 km s^{-1} kpc^{-1}. After subtracting the pattern speed from the rotational velocity derived from the rotation curve, we estimate the timescale between dense molecular gas and supernova explosion. The timescale is at the shorter end of 5–10 Myr for OB stars to explode as supernovae. This may reflect a top-heavy IMF, or uncertainties in the inferred angular velocities of the ring and bar.

References

[1] Eskridge, P. B. et al. 2000 *AJ* **119** 536
[2] Kuno, N. et al. 2007, 2007 *PASJ* **59** 117
[3] Forbes, D. A., Kotilainen, J. K. & Moorwood, A. F. M. 1994s *ApJ* **433** 13
[4] Ho, P. T. P., Moran, J. M. & Lo, K. Y. 2004 *ApJ* **616** 1
[5] Scoville, N. Z. et al. 1993 *PASP 105* 1482
[6] Dahlem, M. 2005 *A&A* **429** 5

Astronomy at High Angular Resolution 2011 IOP Publishing
Journal of Physics: Conference Series **372** (2012) 012067 doi:10.1088/1742-6596/372/1/012067

Searching for Molecular Gas in Southern Radio Galaxies

I Prandoni[1], R A Laing[2], H R de Ruiter[1], and P Parma[1]
[1] INAF - Istituto di Radioastronomia, Bologna, Italy
[2] European Southern Observatory, Garching b. Munchen, Germany
E-mail: prandoni@ira.inaf.it

Abstract. It has recently been proposed that the jets of low-luminosity radio galaxies are powered by direct accretion of the hot phase of the IGM onto the central black hole. Cold gas remains a plausible alternative fuel supply, however. The most compelling evidence that cold gas plays a role in fueling radio galaxies is that dust is detected more commonly and/or in larger quantities in (elliptical) radio galaxies compared with radio-quiet elliptical galaxies. On the other hand, only small numbers of radio galaxies have yet been detected in CO (and even fewer imaged), and whether or not all radio galaxies have enough cold gas to fuel their jets remains an open question. If so, then the dynamics of the cold gas in the nuclei of radio galaxies may provide important clues to the fuelling mechanism. The only instrument capable of imaging the molecular component on scales relevant to the accretion process is ALMA, but very little is yet known about CO in southern radio galaxies. Our aim is to measure the CO content in a complete volume-limited sample of southern radio galaxies, in order to create a well-defined list of nearby targets to be imaged in the near future with ALMA. APEX has been equipped with a receiver (APEX-1) able to observe the 230 GHz waveband. This allows us to search for CO(2-1) line emission in our target galaxies. Here we present the results of CO(2-1) APEX-1 spectroscopy taken in 2008 and 2010 for our southern sample. The experiment was successful with nearly all targets detected, and several indications for double-horned CO line profiles, consistent with ordered rotation.

1. Introduction

It is generally accepted (e.g. [1, 2]) that *powerful* radio galaxies (RG; i.e. radio-loud early-type galaxies) are triggered by cold gas transported to the center through merging or collisions with gas-rich galaxies. During the merger event, part of the gas can be transported to the central region of the merging system, where it may trigger a starburst, and, if it loses sufficient angular momentum, feed the AGN.

The most compelling evidence that cold gas can play a role in fueling *low power* RGs as well, is that in such systems cold gas (often in form of nuclear dusty disks) is detected more often and in larger quantities than in radio-quiet ETGs. ([3, 4]). In addition, the

first systematic CO observing campaigns of nearby (mostly low power) RGs (e.g. [5, 6, 7]) have revealed higher CO detection rates than in radio-quiet ETGs ($45 - 65\%$ against $\sim 20\%$, [5, 8]). This provides clear support for the idea that the jets are powered by the accretion of cold gas, rather than by Bondi accretion of hot, X-ray emitting plasma ([9]). Large quantities of warm, but still molecular gas have recently been found in radio-loud *Molecular Hydrogen Emission Galaxies* (MOHEGs), including some with high cold gas masses ([10]), and this is also

a plausible candidate for the fuel supply. Incidentally, the H_2 lines do not appear related to star-formation, but are excited by shocks, related to AGN feedback. Whatever its temperature, if molecular gas is the dominant fueling mechanism in low power radio galaxies, as well as in their brighter counterparts, we expect to detect CO in virtually all radio galaxies, provided that CO spectroscopy goes deep enough.
Only high-resolution imaging of CO or other molecular lines would allow us to probe the dynamics of the cool gas directly, to determine if and how much of it is infalling, and what kind of motions (rotation, non-circular, radial in- or out-flows) are present. Interferometry with mm-instrumentation is challenging and so far has been possible for only a few objects (see e.g. [11]).
ALMA, thanks to its combination of angular resolution and sensitivity, will give us a unique opportunity to assess the cold accretion scenario in low luminosity radio galaxies, by directly imaging the gas on scales $\sim 10 - 100$ pc and determining its dynamical state. Very little is known about cold gas in Southern galaxies, however, with very few objects observed in CO.
In this paper we report the result of an APEX CO(2-1) detection campaign of a volume-limited sample of Southern low power radio galaxies, to be followed up with ALMA on due time.

Table 1. Southern Sample

Source	z	HST dust features
PKS 0007-325 [IC 1531]	0.0256	-
PKS 0131-31 [NGC 612]	0.0298	disk[*]
PKS 0320-37 [NGC 1316, Fornax A]	0.0059	clouds
PKS 0336-35 [NGC 1399]	0.0048	No dust
PKS0718-34	0.0284	-
PKS 0958-314 [NGC 3100]	0.0088	-
PKS 1107-372 [NGC 3557]	0.0102	disk
PKS1258-321	0.0170	-
PKS 1333-33 [IC 4296]	0.0125	disk
PKS 2128-388 [NGC7075]	0.0185	-
PKS 2254-367 [IC1459]	0.0060	clouds

[*] Ground-based optical observations

2. Sample Description

The parent sample is selected from the Parkes 2.7-GHz survey, in the declination range $-17^o < d < -40^o$, as described by Ekers et al. ([12]). It has a radio flux-density limit of 0.25 Jy at 2.7 GHz and an optical limit of $m_V \leq 17.0$, and consists of 191 radio galaxies. We extracted those sources with redshifts z¡0.03 which are associated with elliptical galaxies. The resulting 11 objects (listed in Table 1) thus form a small, but volume-limited and complete subset of the Ekers et al. sample. All eleven sources of the volume-limited sample are of the FRI type ([13]). Five galaxies (NGC1316, NGC1399, NGC3557, IC4296 and IC1459) have HST imaging available ([14]); as an example, three of them are reported in Figure 1, where we show the radio source (left) and the HST image showing, when present, dust features in the cores (right, see also last column of Table 1). A sixth galaxy (NGC612) show a very prominent dust

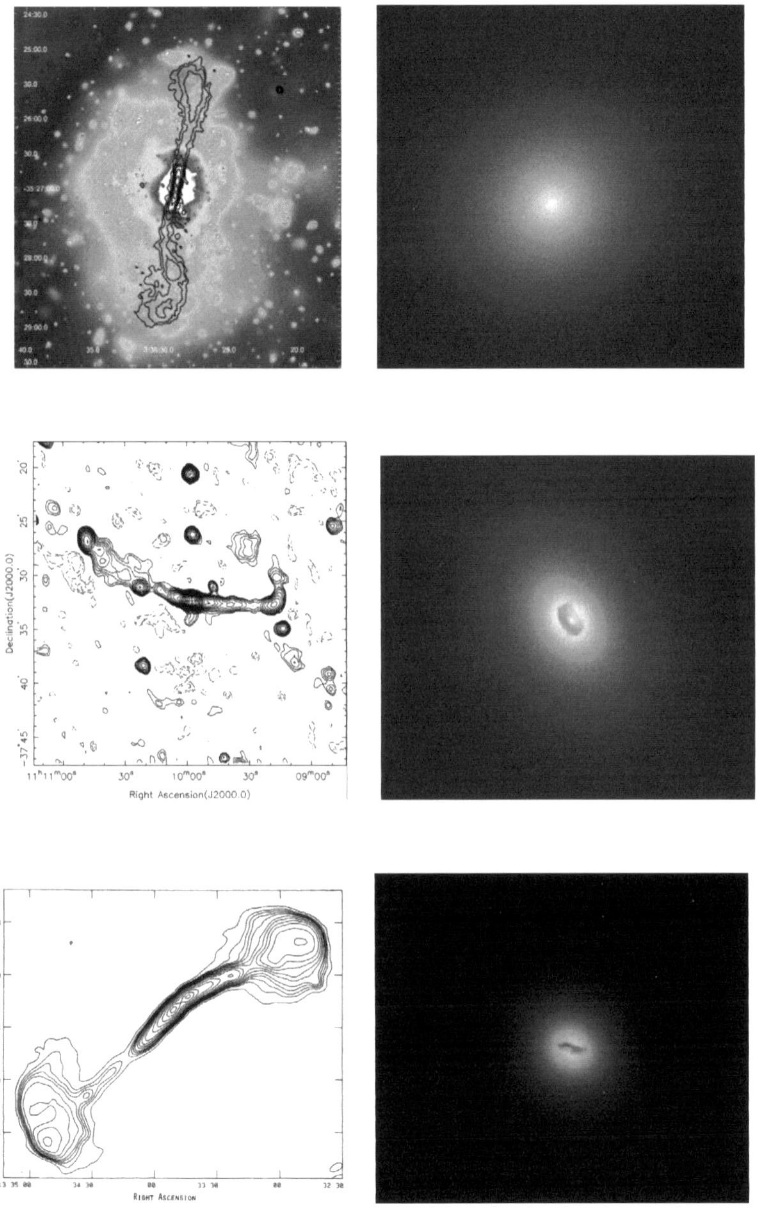

Figure 1. 1.4 GHz VLA radio continuum contours (*left*) and HST images (*right*) for radio galaxies NGC 1399 (*top*), NGC 3557 (*middle*), IC 4296 (*bottom*). 1.4 GHz radio continuum contour images are taken respectively from Shurkin et al. ([15]) for NGC 1399 (here overlaid to a Chandra X-ray color image); from the NVSS database ([16]) for NGC 3557; and from Killeen et al. ([17]) for IC 4296. The HST images are taken from Lauer et al. ([14]).

lane perpendicular to the radio jet in ground optical images (marked [*] in last column of Table 1). NGC1316 (the well known Fornax A), which shows a complex dust structure on HST scale

([18]), was imaged in 12CO(1-0) and 12CO(2-1) by Horellou et al. ([19]). No CO imaging was available for the remaining sample members. Two of the five sources with available Spitzer spectra are classified as MOHEGs (work in progress).
Our southern sample is quite similar to the sample of 3C and B2 radio galaxies in the Northern Hemisphere, which consists of 23 low luminosity radio sources associated with elliptical galaxies at z¡0.03; 16 of these have been observed in CO with the IRAM 30m telescope ([5]).

3. CO(2-1) Line Observations with APEX

We have been using APEX to search for emission in the (2-1) transition of CO in the ten radio galaxies with no available CO observations. The observing campaign started in Summer 2008, and preliminary results for three sources (those shown in Figure 1) are presented in Prandoni et al. ([6]). In December 2010 the remaining sources were observed, together with IC 4296, that was not detected during the first observing session.
For most sources the APEX 230-GHz 30 arcsec FWHM beam probes the presence of molecular gas in the galaxy cores (the inner $3-7$ kpc, depending on the redshift, see last column of Table 2), allowing a direct comparison with dust structures on similar scales imaged with HST. The data were reduced with the CLASS package. Scans severely affected by ripples and/or noisy baselines were removed and first-order baselines were subtracted. Line fluxes were measured by numerically integrating over the channels in the line profile. A source was considered detected when the $^{12}CO(2-1)$ emission line had $T_a^{peak} > 3T_a^{rms}$, and tentatively detected when $T_a^{peak} > 2T_a^{rms}$. H_2 molecular masses were derived for a λ-CDM cosmology and $H_0 = 70$ km/s/Mpc. All targets were detected or tentatively detected at a channel resolution of v_{res} 40 km/s. However in five cases a degraded channel resolution was needed (80 or 120 km/s) to get secure detections ($T_a^{peak} > 3T_a^{rms}$). These latter sources have been re-observed in December 2011 to get better CO line parameter estimates and data reduction is in progress. Preliminary results of our CO line measurements are given in Table 2.

Table 2. APEX CO(2-1) line measurements

Source	z	v_{res} (km/s)	T_a^{rms} (mK)	S/N	Δv (km/s)	$T_a dv$ (K km/s)	logM(H_2) (M_{sun})	size (kpc)
IC 1531	0.0256	40	1.0	7.3	895	4.9	9.6	14
NGC 612	0.0298	40	0.8	18.8	774	7.2	9.9	17
NGC 1316 [**]	0.0059	-	-	-	-	-	8.7	-
NGC 1399	0.0048	120	0.4	3.3	365	0.4	6.4	3
PKS0718-34	0.0284	40	0.7	3.7	292	0.5	8.4	16
NGC 3100	0.0088	80	0.8	5.8	489	1.7	8.2	5
NGC 3557	0.0102	40	2.0	3.5	248	1.5	7.6	5
PKS1258-321	0.0170	80	0.2	12.5	683	1.1	8.6	7
IC 4296	0.0125	80	0.4	6.5	904	1.3	8.4	6
NGC7075	0.0185	40	0.8	6.3	650	2.2	9.0	7
IC1459	0.0060	80	0.6	3.2	492	0.7	7.4	4

[**] from Horellou et al. ([19]

4. Preliminary Results

The H_2 masses of the low-luminosity southern radio galaxies are plotted against redshift in Figure 2, together with the 23 B2 $z < 0.03$ radio galaxies ([5]), the brighter $z < 0.03$ 3C radio galaxies studied by Lim et al. ([20]) and the $z < 0.0233$ UGC galaxies with radio jets studied by Leon et al. ([21]). We find that molecular gas masses for our southern sample are in line with those found in the other samples, even though two high redshift galaxies show significantly larger masses. We are now exploring whether this is due to different physical properties (one of the two galaxies is a MOHEG), or whether this is due to technical reasons. Our APEX observations can probe larger line widths (see Δv in Table 2) and we cannot exclude that H_2 masses in the northern samples may be (at least in some cases) underestimated, due to a different observing strategy.

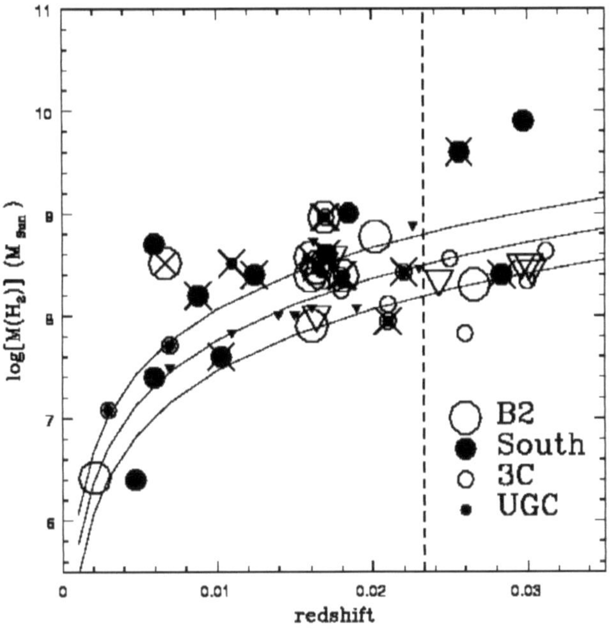

Figure 2. H_2 mass vs redshift for the $z < 0.03$ B2 radio galaxies (large empty symbols); for $z < 0.03$ 3C radio galaxies (small empty symbols); for $z < 0.0233$ UGC galaxies (small filled symbols); and for the eleven (including Fornax A) $z < 0.03$ Southern radio galaxies (large filled symbols). Circles refer to detections and triangles to upper limits. Crosses indicate sources which display a double-horned CO line profile. Solid lines indicate the M_{H_2} 3σ detection limit for σ ($= T_{\rm a}^{\rm rms}$) = 0.1, 0.2 and 0.4 mK respectively (assuming a 500 km s^{-1} wide CO line). The vertical dashed line at $z = 0.0233$ indicates the redshift limit to which all three Northern samples overlap.

Seven examples of CO detections without visible dust, including NGC 1399, are plotted in Figure 2 with inferred H_2 masses range from 8×10^7 to 4×10^8 M_{sun}. However CO line detections are frequent in radio galaxies with dust, even more when the dust is in form of nuclear disks. We also find that several southern objects with dusty disks detected in CO have double-horned line profiles (indicated by the crosses in Figure 2). Since many of the other detections are still marginal, it is possible that the molecular gas in the majority of low-luminosity radio galaxies is in the form of kpc-scale nuclear disks or rings, co-spatial with the dust and in ordered rotation. The fact that we are finding CO in radio galaxies with no detectable dust may suggest that in (at least some) objects the bulk of the gas is in a cold (molecular) phase, supporting the idea that virtually all radio galaxies may be fuelled by cold gas.

References

[1] Heckman T M, Smith E P, Baum S A, et al. 1986 *ApJ* **311** 526
[2] Baum S A, Heckman T M and van Breugel W 1992 *ApJ* **389** 208
[3] de Ruiter H R, Parma P, Capetti A, Fanti R and Morganti R 2002 *A&A* **396** 857
[4] de Koff S, Best P, Baum S A, et al. 2000 *ApJS* **129** 33
[5] Prandoni I, Laing R A, Parma P, et al. 2007 *NewAR* **51** 43
[6] Prandoni I, Laing R A, Parma P, et al. 2010 *A&A* **523** A38
[7] Ocaña Flaquer B, Leon T S, Combes F and Lim J 2010 *A&A* **518** 90
[8] Young L M, Bureau M, Davis T A, et al. 2011 *MNRAS* **414** 940
[9] Allen S W, Dunn R J H, Fabian A C, Taylor G B, Reynolds C S 2006 *MNRAS* **372** 21
[10] Ogle P, Boulanger F, Guillard P 2010 *ApJ* **724** 1193
[11] Okuda T, Kohno K, Iguchi S and Nakanisji K 2005 *ApJ* **620** 673
[12] Ekers R D, Wall J V, Shaver P A, et al. 1989 *MNRAS* **236** 737
[13] Fanaroff B L and Riley J M 1974 *MNRAS* **167** 31P
[14] Lauer T R, Faber S M, Gebhardst K, et al. 2005 *AJ* **129** 2138
[15] Shurkin K, Dunn R J H, Gentile G, Taylor G B and Allen S W 2008 *MNRAS* **383** 923
[16] Condon J J, Cotton W D, Greisen E W, et al. 1998 *AJ* **115** 1693
[17] Killeen N E B, Bicknell G V and Ekers R D 1986 *ApJ* **302** 306
[18] Grillmair C J, Forbes D A, Brodie J P and Elson R A W 1999 *AJ* **117** 167
[19] Horellou C, Black J H, van Gorkom J H, et al. 2001 *A&A* **376** 837
[20] Lim J, Leon S, Combes F and van-Trung D 2003 *ASP Conf. Ser.* **290** 525
[21] Leon S, Lim J, Combes F and van-Trung D 2003 *ASP Conf. Series* **290** 525

Astronomy at High Angular Resolution 2011 IOP Publishing
Journal of Physics: Conference Series **372** (2012) 012068 doi:10.1088/1742-6596/372/1/012068

[FeII] as a tracer supernova rate

M J F Rosenberg, P P van der Werf and F P Israel

Sterrewacht Leiden, Universiteit Leiden, P.O. Box 9513, NL-2300 RA Leiden, The Netherlands

E-mail: `rosenberg@strw.leidenuniv.nl`

Abstract. Supernovae play an integral role in the feedback of processed material into the ISM of galaxies and are responsible for most of the chemical enrichment of the universe. The rate of supernovae can also reveal the star formation histories. Supernova rate is usually measured through the non-thermal radio continuum luminosity, but in this paper we establish a quantitative relationship between the $[FeII]_{1.26}$ luminosity and supernova rate in a sample of 11 near-by starburst galaxies. SINFONI data cubes are used to perform a pixel pixel analysis of this correlation. Using Brγ equivalent width and luminosity as the only observational inputs into Starburst 99, the supernova rate is derived at each pixel and a map of supernova rate is created. This is then compared morphologically and quantitatively to $[FeII]_{1.26}$ luminosity map. We find a strong linear and morphological correlation between supernova rate and $[FeII]_{1.26}$ on a pixel-pixel basis:

$$log\frac{\nu_{SNrate}}{yr^{-1}pc^{-2}} = (1.01 \pm 0.2) * log\frac{[FeII]_{1.26}}{ergs^{-1}pc^{-2}} - 41.17 \pm 0.9$$

The Starburst 99 derived supernova rates are also in good agreement with the radio derived supernova rates, which further demonstrates the strength of [FeII] as a tracer of supernova rate. With the strong correlation found in this sample of galaxies, we now qualitatively use $[FeII]_{1.26}$ to derive supernova rate on either a pixel-pixel or integrated galactic basis.

1. Introduction

Supernova rate (SNrate) is typically estimated by the integrated non-thermal radio continuum emission. The connection between radio continuum emission and SNrate comes from the infrared-radio relation that shows a tight correlation between star formation rate and both radio and infrared emission (1), and references therein). In addition, near-infrared (NIR) observations of supernova remnants (SNR) often show strong [FeII] emission line flux coincident with the radio peak. Iron atoms are typically locked in dust grains and shock fronts of the SNR cause efficient grain destruction through thermal sputtering. This releases the iron into the gas-phase where it is singly ionized by the interstellar radiation field, making [FeII] a strong tracer of shocks, which are often caused by expanding SNR. Imaging and spectroscopy point to [FeII] emission being coincident with known radio SNR, indicating that [FeII] can provide at least a relative measure of SNrate.

2. Sample

We have selected 11 bright near-by starburst galaxies at distances ranging from 10-100 Mpc. A summary of the galaxies included in this study can be found in Table 1. All galaxies are spiral galaxies ranging from S0-a to Sd with the exception of NGC 520, which is a merger galaxy. There are two edge-on galaxies in our sample, NGC 520 and NGC 3628. The far-infrared luminosities range from $1.4 \times 10^{10} L_{\odot}$ to $3.2 \times 10^{12} L_{\odot}$, where Arp 220 is considered an Ultra Luminous Infrared Galaxy (ULIRG), with $L > 10^{12} L_{\odot}$, and NGC 6240, NGC 1614 and NGC 7552 are classified as Luminous Infrared Galaxies

Journal of Physics: Conference Series **372** (2012) 012068 doi:10.1088/1742-6596/372/1/012068

Table 1. Summary of galaxy parameters.

Galaxy	Morph. J2000	Activity	Dist. [Mpc]	i [°]	log L_{FIR} [$L_\odot$]	
NGC 3628	Sb pec edge-on	HII LINER	12.8	847±2	79.29	10.14
NGC 4536	SAB(rs)bc	HII Sbrst	15.4	1802±3	58.9	10.17
NGC 1792	SA(rs)bc	HII	13.2	1210±5	62.78	10.22
NGC 1084	SA(s)c	HII	16.6	1409±4	46	10.42
NGC 1808	(R)SAB(s)a	HII	12.3	1001±4	83.87	10.55
NGC 520	pec	Merger Sbrst	30.5	2162±4	77.49	10.81
NGC 7552	(R')SB(s)ab	HII LINER	22.5	1611±6	23.65	11.03
NGC 7632	(R')SB(s)0^0	HII	19.3	1535±15	82.44	11.43
NGC 1614	SB(s)c pec	Sbrst	64.2	4778±6	41.79	11.43
NGC 6240	S0-a	LINER	108.8	7242±45	73.0	11.73
Arp 220	Sd	LINER Sy2	82.9	5420±6	57	12.50

(LIRGs), with $L > 10^{11} L_\odot$,. Our sample excludes galaxies with very active nuclei, however we have four Low-Ionization Nuclear Emission-line Region (LINER) galaxies and one weak Seyfert 2 (Sy2) nucleus.

3. Methods

3.1. Extinction Correction

In order to study the true emission line strengths, it is crucial to determine the amount of dust extinction obscuring each nucleus. The extinction is calculated at each pixel and an extinction map is created showing the regions most effected by dust. The A_V is determined using the extinction law at NIR wavelengths of $A_\lambda \propto \lambda^{-1.8}$ (2).

To grasp the true morphology and magnitude of the line emission, extinction-corrected $[FeII]_{1.26}$ and Brγ linemaps are created using the Paβ/Brγ line ratio. The [FeII] map is used directly to find the [FeII] luminosity at each pixel. The Brγ linemap, on the other hand, is used along with the K band continuum to calculate both the Brγ equivalent width and the Brγ luminosity, which are used as the *only* observational inputs to calculate SNrate from the Starburst 99 model, described in detail in Section 3.2 below.

3.2. Calculating SNrate

Starburst 99 (from here on referred to as SB99) is a tool that models spectrophotometric properties of star forming galaxies (3) such as spectral energy distributions (SEDs), luminosities, equivalent widths, supernova rates and colors. It includes predictions of the variations in these properties as a function of starburst age. These models are calculated for two extreme star formation modes: continuous, where star formation proceeds continuously at a constant rate, and instantaneous (delta function) starburst.

A near solar metallicity ($Z = 0.02$), Salpeter IMF ($\alpha = 2.35, M_{low} = 1 M_\odot, M_{up} = 100 M_\odot$) and an instantaneous star formation mode are used. For the case of an instantaneous starburst, the burst is normalized to an initial starburst mass of 10^6 $M_\odot$. Using the observed Brγ equivalent width (EW(Brγ)), which is independent of this normalization, the average age of the population dominating the emission in each pixel is calculated. From age, SB99 then provides the expected SNrate. Because of the normalization, the SNrate must be appropriately scaled by comparing the SB99 age-dependent prediction of ionizing photon flux (N(H°)) to the observed ionizing photon flux, which scales with Brγ luminosity. This comparison allows a scaling factor that is directly proportional to the initial mass and

Table 2. Average equivalent width of Brγ and the SB99 derived average ages and integrated SNrates using the instantaneous starburst model.

Galaxy	EW(Brγ) Å	Age_{inst} (Myr)	$SNrate_{inst}$ yr^{-1}
NGC 7632	5.9	7.0	0.03
NGC 3628	7.5	6.9	0.01
NGC 4536	10.6	6.7	0.3
NGC 1792	4.1	7.9	0.007
NGC 1084	4.8	7.3	0.009
NGC 1808	7.2	6.9	0.06
NGC 520	11.7	6.6	0.3
NGC 7552	13.7	6.7	0.3
NGC 7632	5.9	7.0	0.03
NGC 1614	22.9	6.4	0.9
NGC 6240	3.6	7.7	3.6
Arp 220	8.0	6.8	0.7

initial star formation rate, allowing the SNrate to be scaled based on the true initial conditions of the region represented in each pixel.

4. Analysis

4.1. Age and SNrate

Table 2 presents the EW(Brγ) followed by the SB99 results, including the average age and integrated SNrate using the instantaneous burst model for each galaxy. Although the analysis was done on a pixel-pixel basis, the numbers presented in Table 2 are averaged over the galaxy, in the case of age and equivalent width, and integrated over the galaxy in the case of SNrate.

4.2. Qualitative Correlation

To demonstrate the morphological correlation between the [FeII] and SNrate, a side-by-side comparison of the K band continuum, extinction corrected Brγ flux, extinction corrected $[FeII]_{1.26}$ flux, and SNrate is shown in Figure 1 for NGC 4536, NGC 1808, and NGC 7552. These galaxies were selected for presentation because of the high quality of the observations. Visually comparing the morphologies of the four different maps, we find that the SNrate most closely resembles the [FeII] emission.

Focusing on NGC 4536, the Brγ emission has a bright knot directly north of the nucleus with secondary emission peaks to the north-west and south-east of the nucleus. The [FeII] emission is concentrated around the galactic center with a long plume of emission coming from the nucleus and extending towards the north-west corner of the galaxy. There is a small knot of faint [FeII] emission at the location of the Brγ peak. Similar to the [FeII] emission, the SNrate peaks around the center with an arm extending towards the north-east corner.

NGC 1808 displays a clear starburst ring in Brγ surrounding the bright nucleus. The [FeII] reveals a much less prominent ring configuration with more extended emission around the nucleus spanning the south-east to north-west plane. The SNrate map shows more ring structure than the [FeII] but it is quite diffuse and lacks the strength of the Brγ peak in the north-east region of the ring.

The face-on spiral, NGC 7552, also reveals a starburst ring, with little Brγ emission in the nucleus. Focusing on the northern Brγ knots, we see three 3 main knots, two slightly north-west of the nucleus, and one elongated knot directly north of the nucleus. Comparing the Brγ emission to the [FeII] emission, we discuss only one of the two north-western Brγ knots. In addition, the morphology in the elongated knot differs from that of [FeII] and appears as two distinct knots. As shown by the SNrate morphology,

Journal of Physics: Conference Series **372** (2012) 012068 doi:10.1088/1742-6596/372/1/012068

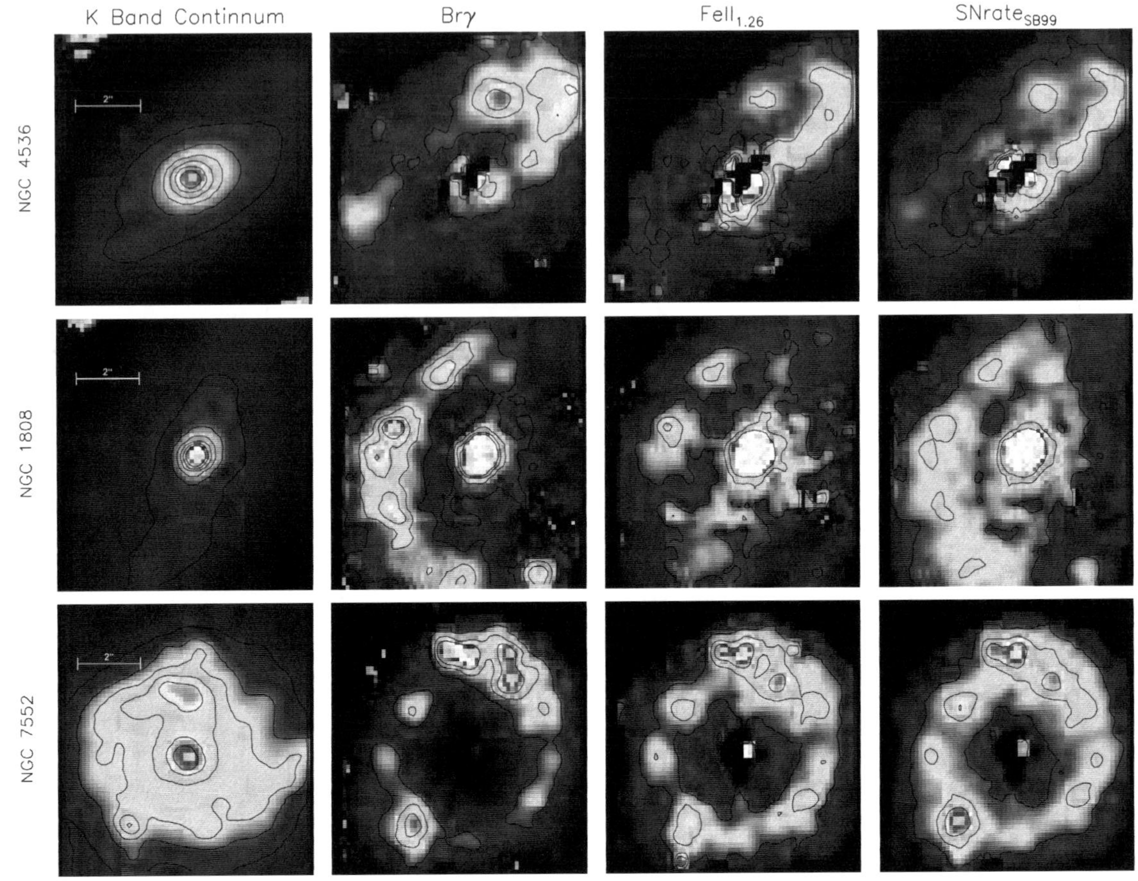

Figure 1. A comparison between the K-band continuum, Brγ linemap, $[FeII]_{1.26}$ linemap, and the SNrate morphologies for NGC 4536, NGC 1808, and NGC 7552.

the northern emission knots resemble the realtive flux and morphologies of the [FeII] emission much more closely than the Brγ emission. Specifically, the top north-western knot is also missing in the SNrate map and the elongated northern Brγ peak is resolved into two peaks, resembling the relative flux ratios seen in [FeII].

The comparison of the Brγ and [FeII] emission to the morphology of the SNrate per pixel reveals a general similarity between the [FeII] emission and SNrate, a much closer correlation than the [FeII] and Brγ or the SNrate and Brγ. We emphasize again the *only* observational input into the SNrate calculation is the Brγ luminosity and EW(Brγ), the [FeII] flux is never used. However, the SNrate correlates more with the [FeII] emission than the Brγ. This strongly supports [FeII] as a robust tracer of SNrate.

4.3. Quantitative Correlation

To test if the [FeII] to SNrate correlation holds quantitatively, we compare the $[FeII]_{1.26}$ luminosity to the SNrate, pixel by pixel. The linemaps were not additionally filtered with the exception of NGC 6240, which was filtered in regions where the Brγ flux is insufficient to determine the true extinction. The pixel-pixel plot of SNrate against [FeII] luminosity for each galaxy is shown in Figure 2. Both the SNrate and [FeII] luminosity are scaled per square parsec such that each pixel represents the same physical area. A least squares regression was performed in log space combining the points from all the galaxies excluding

NGC 1792, due to the poor signal-to-noise linemaps. This combined linear regression is plotted by a solid black line in Figure 2 for direct comparison and the relation is shown below.

$$log\frac{\nu_{SNrate}}{yr^{-1}pc^{-2}} = (1.01 \pm 0.2) * log\frac{[FeII]_{1.26}}{ergs^{-1}pc^{-2}} - 41.17 \pm 0.9 \quad (1)$$

The above equation is a power-law with a slope of nearly 1.0, which represents a linear relationship. The error is calculated from finding the slope and intercept for each individual galaxy, and finding the standard deviation of that spread. Therefore the error represents the variations between individual galactic fits.

The main uncertainty in the SNrate calculation is caused by the observational errors in Brγ, [FeII] and K band continuum. The calibration uncertainty dominates the error, this is 10% of the flux for all cases except NGC 1792, where the uncertainty is dominated by noise and is estimated to be 20% of the flux. If we alter the Brγ flux by $\pm$10%, this translates to a change in SNrate of 10%. In addition, a major systematic source of error is the choice of burst model. Since our calculation of SNrate depends heavily on SB99, we are limited to only two star formation models, instantaneous and continuous. In the next section, the dependency on the burst model is discussed in detail.

5. Comparison to Continuous Star Fromation and Radio SNrate

An initial assumption used to perform this analysis was that our galaxies are best represented by an instantaneous burst of star formation. The other extreme case, also modeled by SB99, is the continuous star formation scenario. It is unlikely that our galaxies are represented by either of these extreme cases, but rather somewhere in between a starburst that is extended in time. Since radio continuum is the classical tracer of SNrate, the radio flux densities provide an independent measurement of the accuracy of the SB99 derived SNrate. To estimate the radio SNrate, the equation given by (4), relating non-thermal radio luminosity and SNrate, is used along with VLA observations. The integration area is matched to that of SINFONI. Figure 3 compares the SB99 instantaneous SNrate to both the radio SNrate and the SB99 continuous SNrate.

For all galaxies, except NGC 1792, NGC 4536, NGC 6240 and Arp 220, the radio SNrate closely matches those given by the SB99 instantaneous-burst model. As noted before, the NIR-emission line-maps of NGC 1792 are too noisy to be reliable. However, NGC 4536 is a nearly face-on galaxy with relatively low extinction and high signal-to-noise spectra, and the NIR data should be reliable. Thus, NGC 4536 is undergoing star formation that is closer to instantaneous but extended in time. Arp 220 and NGC 6240 have radio SNrates even higher than the SB99 instantaneous SNrates. Also, in Figure 2, the points for these galaxies lie above the best fit line, demonstrating a [FeII] excess. This excess is caused by merger-related shocks that are exciting the [FeII] in addition to the SNRs. These powerful merger (U)LIRGs may demonstrate a limit to the validity of the direct relationship between SNrate and [FeII] luminosity. In addition, determining an accurate extinction in (U)LIRGs presents a challenge. The under or over estimation of extinction leads to inaccurate [FeII] luminosities as well as Brγ luminosities used to scale the SNrate. However, this can only be established by studying a larger sample of (U)LIRGs. In any case, it appears that the majority of (modest) starburst galaxies is well-represented by the assumption of a (nearly) instantaneous burst of star formation. In addition, the very good agreement between SNrates derived from the radio continuum, and from NIR data provides added confidence in the diagnostic strength of [FeII] as a tracer of SNrate.

6. Conclusion

Using SINFONI observations of 11 near-by galaxies, we performed a pixel-by-pixel analysis of the correlation between $[FeII]_{1.26}$ emission and SNrate. First, Brγ, Paβ, $[FeII]_{1.26}$, $[FeII]_{1.64}$, and $H_{2,2.12}$ line fluxes were remeasured. The Brγ line flux along with the K band continuum was used to find the equivalent width of Brγ, which was then used as input into the SB99 model to estimate the age of each

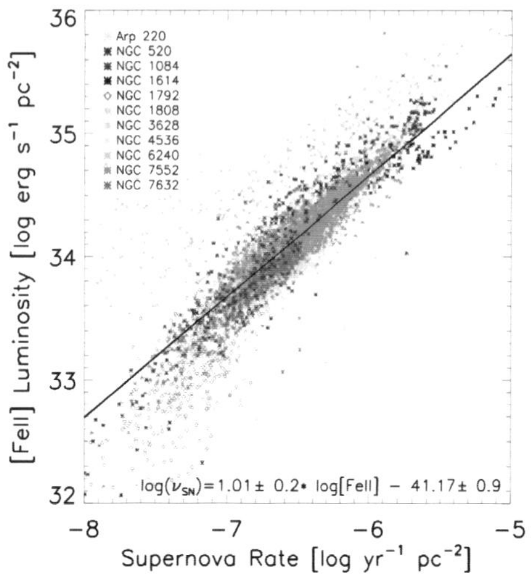

Figure 2. A pixel-pixel plot of SNrate, as derived from SB99, compared to [FeII] luminosity. Each galaxy is represented with a different color and the values are normalized to 1 square parsec. The black line represents the best fit powerlaw excluding NGC 1792.

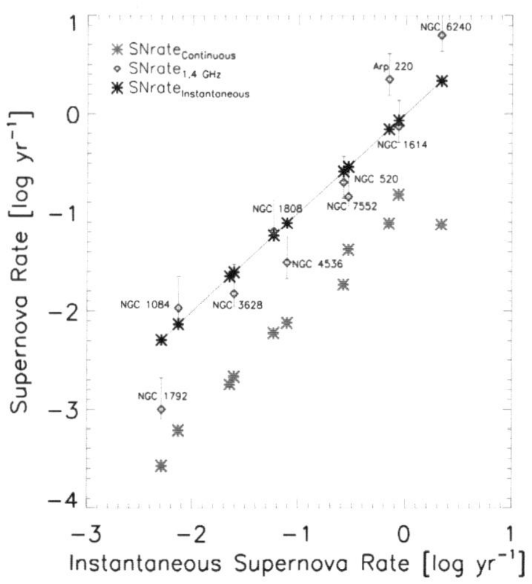

Figure 3. Comparison of the SB99 instantaneous SNrate (x-axis) to the radio SNrate (blue diamond), the SB99 continuous SNrate (red asterisk) and for comparison the SB99 instantaneous SNrate (black asterisk).

pixel. From age, SB99 provides a SNrate, normalized to an initial mass of $10^6 M_\odot$. The normalized SNrate is scaled by Brγ luminosity to derive the true SNrate in each pixel.

Comparison of the $[FeII]_{1.26}$ luminosity to the SNrate (derived only from Brγ equivalent width and luminosity) reveals a nearly linear correlation. This relationship is valid both on a pixel-pixel basis and for the integrated galaxy. For the integrated [FeII] luminosity and SNrate, the fit is remarkably tight with very little spread. However, to correctly determine the absolute SNrates, it is still critical to determine whether star formation has occurred in a (nearly) instantaneous burst, or has proceeded in a continuous fashion. SNrates derived from radio continuum observations may be used to distinguish these scenarios. Most of the modest starburst galaxies in our sample are best fitted assuming instantaneous star formation. However, we find that the relationship breaks down for (U)LIRGs in our sample.

The use of [FeII] as a robust tracer of SNrate is a useful diagnostic tool. Specifically, it would allow measurements of SNrates from NIR observations of distant galaxies, where individual SNR can not be resolved. With the strong correlation found in this sample of galaxies, we can confidently say that $[FeII]_{1.26}$ is quantitatively correlated to SNrate and can be used to derive SNrate on either a pixel-pixel or integrated galactic basis.

References

[1] Condon J J 1992 *ARA&A* **30** 575–611

[2] Martin P G and Whittet D C B 1990 *ApJ* **357** 113–124

[3] Leitherer C, Schaerer D, Goldader J D, González Delgado R M, Robert C, Kune D F, de Mello D F, Devost D and Heckman T M 1999 *ApJS* **123** 3–40 (*Preprint* arXiv:astro-ph/9902334)

[4] Huang Z P, Thuan T X, Chevalier R A, Condon J J and Yin Q F 1994 *ApJ* **424** 114–125

Journal of Physics: Conference Series **372** (2012) 012069 doi:10.1088/1742-6596/372/1/012069

Near infrared hydrogen emission line ratios as diagnostics of the broad emission line region

Andrea J. Ruff[1], David J. E. Floyd[2], Kirk T. Korista[3], Rachel L. Webster[1], Ryan L. Porter[4] and Gary J. Ferland[4]

[1] School of Physics, University of Melbourne, Parkville, VIC 3010, Australia
[2] School of Physics, Monash University, Clayton, VIC 3800, Australia
[3] Department of Physics, Western Michigan University, Kalamazoo, MI 49008, USA
[4] Department of Physics, University of Kentucky, Lexington, KY 40506, USA

E-mail: aruff@unimelb.edu.au

Abstract.
Broad emission line flux ratios are a powerful diagnostic of the physical conditions of the broad-line region gas in Active Galactic Nuclei. With recent advances in infrared spectroscopy, previously unstudied emission lines provide a new means to investigate the physical nature of the BELR gas. The hydrogen emission lines are particularly sensitive to the upper limits of both the radius from the central ionising source and the number density of the gas. Using an existing subset of near-infrared quasar spectra from the Glikman et al. (2006) sample [1] together with Cloudy photoionization simulations, we confirm the Locally Optimally emitting Cloud (LOC) model's ability to reproduce observed emission line flux ratios. The model is then used to constrain physical conditions for individual sources. The photoionization models show that high number density, low incident flux gas is required to reproduce observed near-infrared hydrogen emission line ratios. We also find that comparison to individual sources, rather than composites, is vital.

1. Introduction

Broad emission lines are a prominent feature in quasar spectra, yet the physical state of this gas is not well understood. The locally optimally emitting cloud (LOC) model has been successful in reproducing UV and optical emission line ratios [2, 3], but has not been extensively tested on the hydrogen emission lines. The hydrogen emission lines are interesting because the physics is relatively simple, and there is a negligible dependence on several free parameters of the model: metallicity, column density and slope of the ionising continuum.

A subset of quasars from the Glikman et al. (2006) [1] sample was used to compare to photoionization simulations. The subset has a mean redshift of $z = 0.2$. Deconvolution of the broad and narrow components of the emission lines is not necessary, as the narrow contribution is small in quasars. Using existing near-infrared spectra of quasars together with Cloudy simulations of the broad line region, limits on the physical conditions of the hydrogen emission line region have been inferred.

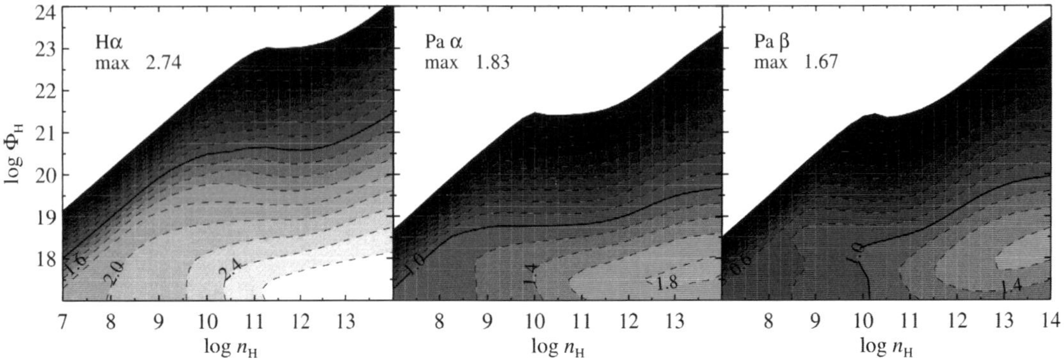

Figure 1. Logarithmic contours of equivalent width in angstroms (referenced to the incident continuum at 1216Å) in the plane formed by $n_{\rm H}$ ($\rm cm^{-3}$) and $\Phi_{\rm H}$ ($\rm cm^{-2}\,s^{-1}$). Only $W_{1216} \geq$ 0.1Å contours are shown. The solid lines show 1 decade intervals and the dashed lines show 0.2 decade intervals. $W_{1216} = 0.1$Å is plotted with a very thick solid contour for each transition. For each each transition, the maximum $\log(W_{1216})$ is quoted in the legend.

2. Photoionization simulations

Following the standard LOC approach [2, 4, 3], large grids of photoionization models were computed using the photoionization code, Cloudy v08.00, last described by [5]. Models that predict hydrogen emission line strengths were generated for a range of hydrogen number density ($n_{\rm H}$) and hydrogen ionizing flux ($\Phi_{\rm H}$) values. The plane formed by these parameters represents the range of cloud densities and distances from the ionizing continuum that are expected to exist within the BELR. For each of these simulations, the output emission is computed as an equivalent width (W_{1216}) in angstroms relative to the incident continuum at 1216Å. The equivalent width, W_{1216}, describes how efficiently the line is produced from the incident continuum. Full geometric coverage was assumed and a constant column density of $10^{23}\,\rm cm^{-2}$ was used. Full detail of these simulations will be given by [6].

Figure 1 shows logarithmic contours of W_{1216} in the density-flux plane for three NIR emission lines: Hα, Pa$\,\alpha$ and Pa$\,\beta$. Light colours indicate a high reprocessing efficiency, i.e. large W_{1216} values. The hydrogen lines are emitted efficiently over a large range in the density-flux plane, and emit particularly efficiently in the high density, low incident ionising flux regime. However, the entire range of parameter space shown in this plot might not be physical. Although the lower limit on the number density of the broad emission line gas can be inferred from the presence of semi-forbidden and absence forbidden transitions, constraining the upper limit is more difficult. Dust grains will condense at low values of incident ionising flux ($\Phi_{\rm H} \sim 10^{18}\,\rm cm^{-2}\,s^{-1}$) [7], which will clobber any line emission. The hydrogen emission lines emit strongly near these limits. As the predicted hydrogen emission line ratios are senitive to the upper $n_{\rm H}$ and lower $\Phi_{\rm H}$ limits, they can be used to constrain these limits.

3. Predicting broad emission line ratios

Ratios of hydrogen emission lines are a powerful diagnostic of the physical conditions of the BELR gas, as the strength of emission is very dependent on the incident ionising flux and number density, but not on the covering factor. As show by [2], the line luminosity is given by:

$$L_{\rm line} \propto \int_{r_{\rm min}}^{r_{\rm max}} \int_{n_{\rm min}}^{n_{\rm max}} W_{1216}(r, n_{\rm H})\, f(r)\, g(n_{\rm H})\, dn_{\rm H}\, dr, \qquad (1)$$

where $W_{1216}(r, n_{\rm H})$ is the equivalent width of a particular line (relative to the continuum at 1216Å) from a single cloud at radius r with number density $n_{\rm H}$. The cloud covering fractions as functions of radius from the ionising source and hydrogen number density are given by $f(r)$ and $g(n_{\rm H})$, respectively. For a more complete description of the LOC parameters, see [8]. In the simplest LOC integration model, the covering fractions are given by $f(r) \propto r^{-1}$ and $g(n_{\rm H}) \propto n_{\rm H}^{-1}$. The line luminosity can then be simplified to:

$$L_{\rm line} \propto \int_{\log \Phi_{\rm min}}^{\log \Phi_{\rm max}} \int_{\log n_{\rm min}}^{\log n_{\rm max}} W_{1216}(\Phi, n_{\rm H})\, d\log n_{\rm H}\; d\log \Phi_{\rm H}, \tag{2}$$

where the integration limits are free parameters of the model and $\Phi_{\rm H} \propto r^{-2}$. This is simply a sum over each point in parameter space between the integration limits, as the grid simulations are weighted evenly per decade. Since only three independent ratios were measured, the model can only have two free parameters. The free parameters of the model are $n_{\rm max}$ and $\Phi_{\rm min}$, which the hydrogen ratios are particularly sensitive to. As shown in Figure 1, the hydrogen line emission is only weakly dependent on $n_{\rm min}$ and $\Phi_{\rm max}$(varying these limits casues the predicted hydrogen ratios to change by less than 1%). These parameters are fixed in this model.

4. Data

Very few quasars have been observed in the infrared with high signal-to-noise. Glikman et al. (2006) [1] published a set of 27 quasars, presented as a composite quasar spectrum. Rather than using the measured emission line flux ratios from the composite, individual spectra with the highest signal-to-noise ratios were used. Emission line fluxes were measured for the strongest hydrogen lines. Glikman et al. [1] found the Pa α/Pa β composite ratio to be 0.64±0.01. However, values measured from individual objects are significantly higher and consistently above 1.0. Further detail of the measurements will be given by [9].

5. Fit to data

A set of 'best fit' $\Phi_{\rm min}$ and $n_{\rm max}$ integration limits was found for each object by comparing simulated and measured flux ratios. The most likely set of $\Phi_{\rm min}$ and $n_{\rm max}$ limits was found by minimising χ^2:

$$\chi^2 = \sum_{i=1}^{N} \frac{(R_{{\rm obs},i} - R_{{\rm model},i})^2}{\sigma^2_{{\rm stat},i} + \sigma^2_{{\rm sys},i}}, \tag{3}$$

where $R_{\rm obs}$ and $R_{\rm model}$ are the observed and predicted ratios, and $\sigma_{\rm stat}$ and $\sigma_{\rm sys}$ are the statistical and systematic uncertainties on each measured ratio. The sum is over each emission line flux ratio used to find the best model: Pa α/Pa β, Pa γ/Pa β and Hα/Pa β. $R_{\rm model}$ is the predicted ratio for each combination of $\Phi_{\rm min}$ and $n_{\rm max}$ values considered.

Figure 2 shows χ^2 contours for two objects in the $\Phi_{\rm min}$–$n_{\rm max}$ plane. The point where χ^2 is minimised (shown with a red diamond) is at high density and low ionising flux. Results for all measured objects will be presented by [9].

6. Summary

A simple LOC model can reproduce measured near-infrared hydrogen broad emission line ratios. The measured Pa α/Pa β ratio for individual sources was inconsistent with the value presented by [1], which was measured from the composite spectrum.

Limits on the physical conditions of the BELR gas were inferred by comparing measured line ratios to simulations. The predicted outer radius of the BELR is consistent with the radius of dust sublimation calculated by [7]. The predicted upper limit on the gas number density is high compared to previous calculations, however, all predictions stated here should be considered

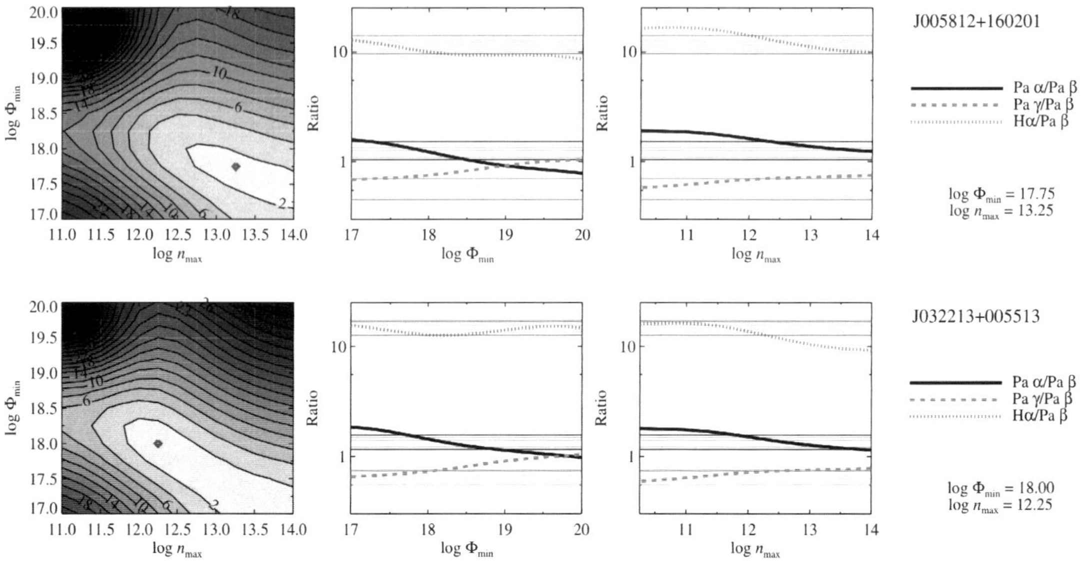

Figure 2. Measured ratios are compared to the best fit models for two quasars. **Left:** Contours of reduced χ^2. The location of the minimum χ^2 in the $n_{\rm max}$-$\Phi_{\rm min}$ plane is shown with a red diamond. **Right:** The emission line flux ratio plotted as a function of $\Phi_{\rm min}$ and $n_{\rm max}$. The solid, dashed, dot dashed and dotted lines show the simulated Pa α/Pa β, Pa γ/Pa β and Hα/Pa β ratios, respectively. The measured values are shown as a band with 1σ uncertainties in a corresponding colour. The best fit $\Phi_{\rm min}$ and $n_{\rm max}$ parameters are given in the legend for each object.

estimates, because of the simplicity of the model and limitations in the quality and number of spectra. This model can be extended to predict other emission line ratios, such as Hα/Hβ. This ratio can also be compared to measured values and can be used to estimate the amount of intrinsic dust.

References

[1] Glikman E, Helfand D and White R 2006 *ApJ* **640** 579
[2] Baldwin J, Ferland G, Korista K and Verner D 1995 *ApJL* **455**
[3] Korista K and Goad M 2000 *ApJ* **536** 284
[4] Korista K, Baldwin J, Ferland G and Verner D 1997 *ApJS* **108** 401
[5] Ferland G, Korista K, Verner D, Ferguson J, Kingdon J and Verner E 1998 *PASP* **110** 761
[6] Ruff A, Korista K, Porter R and Ferland G 2012, *in prep*
[7] Nenkova M, Sirocky M, Nikutta R, Ivezić Z and Elitzur M 2008 *ApJ* **685** 160
[8] Bottorff M, Baldwin J, Ferland G, Ferguson J and Korista K 2002 *ApJ* **581** 932
[9] Ruff A, Floyd D, Webster R and Korista K 2012, *submitted*

Journal of Physics: Conference Series **372** (2012) 012070 doi:10.1088/1742-6596/372/1/012070

The long-term evolution of the parsec scale jet of the quasar 3C 345

F. K. Schinzel, A. P. Lobanov, and J. A. Zensus

Max-Planck-Institut für Radioastronomie, Auf dem Hügel 69, 53121 Bonn, Germany

E-mail: frank@schinzels.de

Abstract. The quasar 3C 345 is one of the best examples of an active galactic nucleus (AGN) showing structural and flux variability on parsec scales around a compact unresolved radio core. Over the past 30 years, it has been followed up closely from radio to gamma-ray wavebands with a special focus on very long baseline interferometry (VLBI) observations in the range of 1–100 GHz. The complex parsec-scale jet of 3C 345 exemplifies an archetypical 'superluminal' jet with an apparent helical morphology. Here we present first results from a study of the long-term jet evolution, especially focusing on the evolution of trajectories, kinematics, and emission in more than 20 enhanced emission regions embedded in the jet. A "closeup" on physical properties of individual features implies that the outer jet is most likely dominated by Kelvin-Helmholtz instability. Studies of general trends in properties of those features provide certain evidence for their apparent trajectories to result from an underlying (slowly evolving) pattern lit up by passages of plasma condensations ejected during the nuclear flares. The long-term evolution of this pattern indicates its possible relation either to the elliptical mode of Kelvin-Helmholtz instability or precession of the jet direction.

1. Introduction

The archetypical quasar 3C 345 is one of the best studied "superluminal" radio source, with its parsec-scale radio emission monitored for over 30 years, in particular with very long baseline interferometry (VLBI). On parsec scales, the jet of 3C 345 exhibits a remarkably complex and a highly variable structure around a compact unresolved radio core. At high angular resolution, the radio emission of 3C 345 has been closely monitored since 1979 and it was among the first four quasars claimed to exhibit superluminal motion (cf. [1, 2, 3, 4, 5, 6, 7, 8]). Figure 1 shows a stacked image of all VLBI observations at 15.4 GHz conducted by NRAO's Very Long Baseline Array between 1994 and 2010 (98 epochs). It shows that the jet of 3C 345 is highly collimated, with an initial direction of the flow in Eastern direction and turning northward at a distance of $\sim$5 mas from the core. At a redshift of $z = 0.5928$ [9], 1 mas corresponds to 6.64 pc ($\sim$76 pc in the source frame) and a proper motion scale of 1 mas year^{-1} corresponding to 34.5 c.

An unprecedented database of high angular resolution observations of 3C 345 was established covering a frequency range from 1 – 43 GHz and a time range of 31 years (1979 – 2010). This database comprises a total of 319 observations and combines the data of VLBI observations from the days of the US VLBI network up to the construction and operation of the VLBA. To interpret the data, the method of fitting the observed visibility data with a model (model fitting) was used, implemented in the software package "Difmap" [10]. This method provides a parametric description of the observed source brightness distribution. Brightness peaks in the

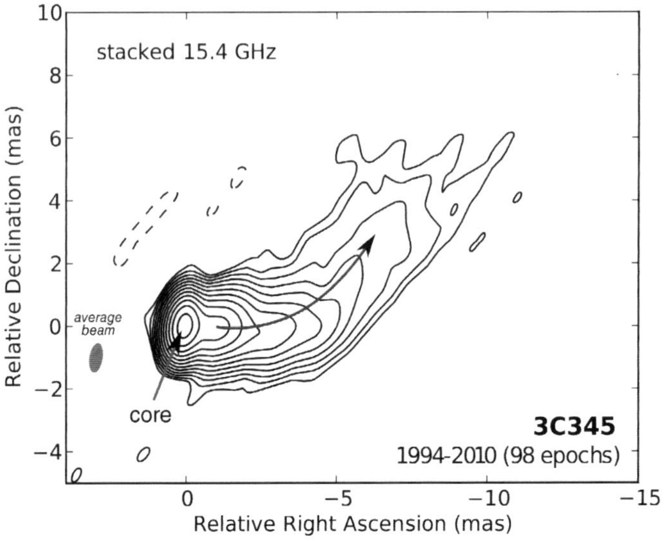

Figure 1. Cleaned Stokes I map from stacking of 98 observations at 15.4 GHz observed between 1995 and 2010, representing the observed brightness distribution and outlining the average channel of the flow. The "core" marks the origin and the optically thick end of the jet of 3C 345. The brown arrow indicates the general direction of the flow. Contour levels are (-0.03, 0.03, 0.06, 0.12, 0.24, 0.48, 0.96, 1.9, 3.8, 7.7, 15, 31, 61) % of the peak flux density with a value of 4.0 Jy beam^{-1} and an image RMS noise of 0.40 mJy beam^{-1}, corresponding to a dynamic range of 10000:1.

obtained images are represented by 2D-Gaussian components, which are Fourier transformed and fitted to the observed visibility data. Their parameters, flux density, size (full width at half maximum of the Gaussian component), position, and ellipticity are varied, minimizing the χ^2 value using a least squares implementation of the maximum likelihood method [11]. The errors for modelfit parameters are estimated from the image plane, following analytical approximations with modifications to account for the strong side-lobe case inherent to VLBI observations (see [12] and references therein). Jet components are labeled following the naming scheme developed for 3C 345 in the early analysis of the Caltech group [1, 4] and continued by Lobanov [5] and Schinzel [13].

The full data-set of high resolution multi-wavelength observations of the parsec scale jet of 3C 345 is combined to study the variability of the jet structure, covering a time range of 31 years. In Section 2 we present our first results from the investigation of the jet structural variability of 3C 345 over three decades. Section 3 discusses the long-term kinematics of 18 distinct emission regions embedded in the jet. In Section 4 we summarize and draw conclusions from the results presented.

2. Structural Variability

Parametrized trajectories and observed component positions were combined to study structural variability of the parsec scale jet in 3C 345. Component angles measured at 0.5 mas radial distance from the core offer a way to represent time dependent variations of the observed trajectories of jet features. A significant change in this jet position angle (at 0.5 mas) is observed between 1995 and 2005, where the position angles changed from -120° to -70° and back to about -100° by 2010. Previous works have interpreted this trend as oscillation, fit by a sinusoidal function [8]. If this observed variation were due to a precession in the jet, a period of ~14.5

years could be inferred.

However, an alternate explanation for the oscillation in the component position angle of the jet could be the existence of an underlying pattern in the jet that is lit up by passages of plasma condensations introduced at the base of the jet. The left-side plot in Figure 2 shows the 2D-trajectory of component C9. This component was first detected in October 1996 and since then has been followed by VLBI observations, with a reliable identification in a total of 157 observations so far. The overall trajectory of this feature traces the outline of the jet and is well represented by a polynomial fit, reporting an average apparent speed of (0.381 ± 0.034) mas year^{-1} and an average direction of the motion of $(-88.6 \pm 1.8)^{\circ}$. In addition, the position of C9 oscillates around this general motion in a more or less regular pattern. Such an oscillation is seen in many of the observed component trajectories.

The size of the emission region observed at 15.4 and 22.2 GHz for components C8, C9, and C10 is shown in the bottom plot in Figure 2. The sizes resemble a step-like evolution or in the cases of C8 and C9 a clear oscillation. The times at which the first two steps occur (0.4 and 1.5 mas) match the oscillation period observed in the trajectories of C8, C9 (Figure 2 *top*), and C10. Such pattern is expected if the jet is changing its position angle to our line of sight, possibly caused by Kelvin-Helmholtz instabilities [14]. The observed size evolution resembles two threads resulting from the elliptical surface mode of a Kelvin-Helmholtz instability. During maxima, the threads are apart and a double humped (edge brightened) transverse profile of the jet at this position is expected, during minima the threads overlap (in projection) leading to a single peaked transverse jet profile, which is indicated through inlets in the right-hand plot in Figure 2. With this interpretation the observed oscillation in the jet position angle measured at constant distance in the jet, discussed above, is a consequence of the rotation of this Kelvin-Helmholtz instability induced pattern and a possible time evolution of the instability modes.

3. Long-term Kinematics

Polynomial fits to the trajectories of individual jet components provide not only information about the temporal evolution of the general jet flow, but also give insights into the actual dynamics of each of the superluminal features.

The values for the apparent speeds are directly inferred from the polynomial fits to the component trajectories. Physical parameters like the Doppler factor, Lorentz factor, and viewing angle were determined under the assumption of dominating radiative losses (cf. [15]) and the presence of optically thin shocked gas. This requires the determination of the flux variability time scale. The flux variability time scale is the ratio of the measured maximum component flux density to the minimum component flux density. The time of the minimum component flux density is selected to maximize the absolute value of the time derivative of the flux density variation. This flux variability time scale was determined from 15 GHz observations only. Thus, the obtained average kinematic parameters, derived from the trajectories using the flux variability timescale, provide median values for the apparent speed of 12.5c, Lorentz factor 14.1, Doppler factor 15.6, and viewing angle of 3.8°. Figure 3 provides an overview of the distribution for each of these parameters.

The evident spread in the observed apparent speeds shown in Figure 3 indicates a wide range of observed component speeds that can be explained by non-constant speeds. The fit to the observed trajectories of most components requires second order polynomials in at least one direction. With the parametrized description of the general component motion, it is possible to reconstruct the intrinsic motion pattern in the source frame. In order to do so, at first we assumed a constant viewing angle. This leads for most of the components to a dramatic increase in the required Lorentz factor and in some cases leave the parameter space beyond which no valid solutions are found. For accelerating components the assumption of constant viewing

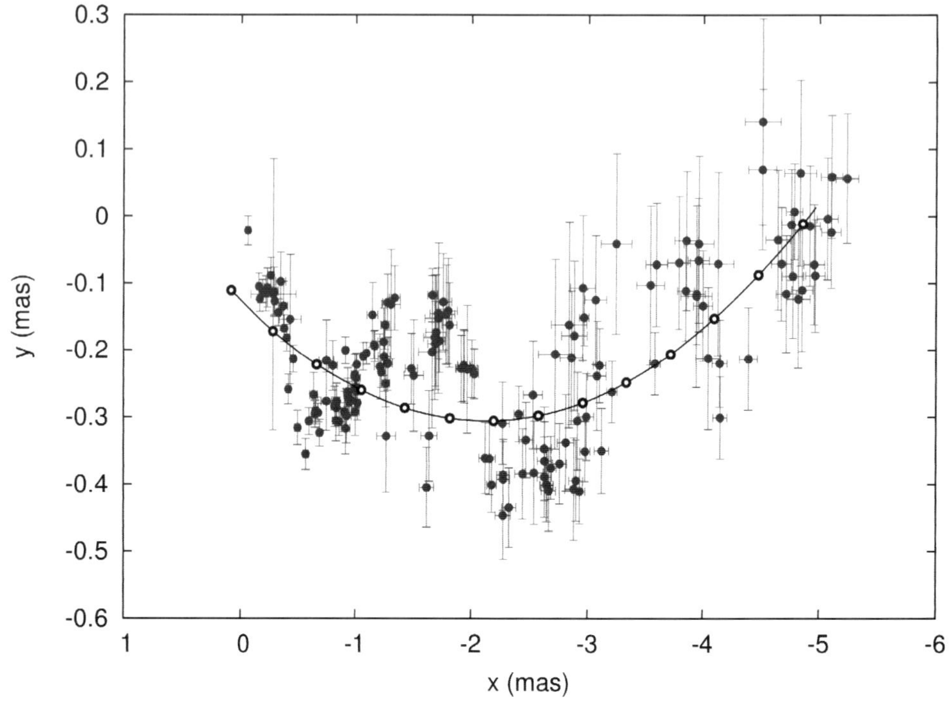

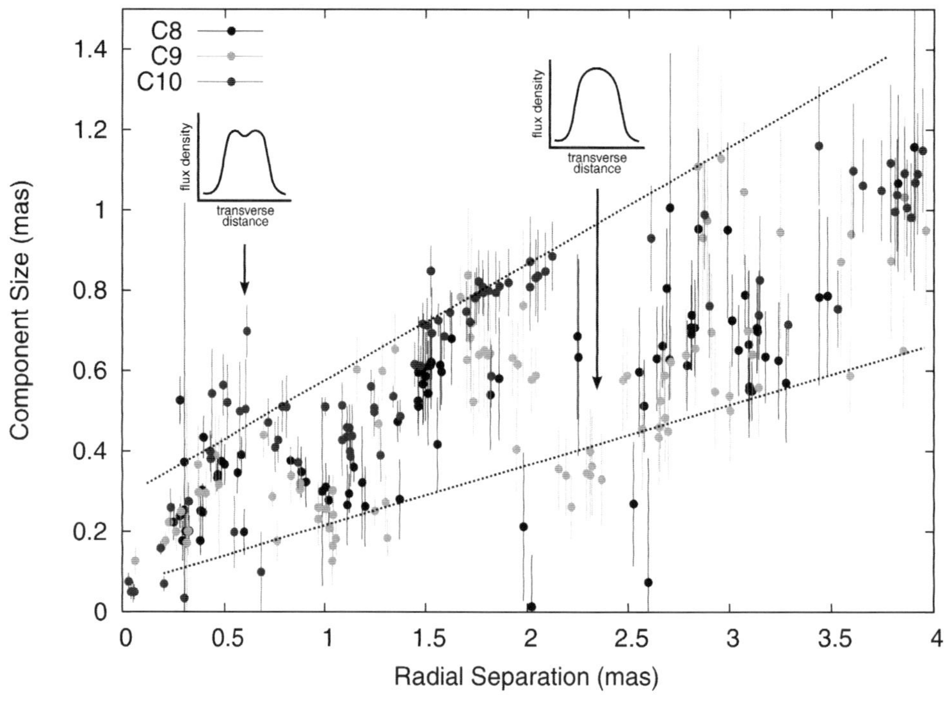

Figure 2. *top:* Two-dimensional trajectory of component C9. The thick black-line is reconstructed from a polynomial fit in x and y direction. White circles are spaced at intervals of one year. Note the oscillating pattern of the component evolution. *bottom:* Component size evolution (15.4 and 22.2 GHz). Predicted transverse brightness profiles induced by Kelvin-Helmholtz instability are indicated by inlets, denoting the expected shape.

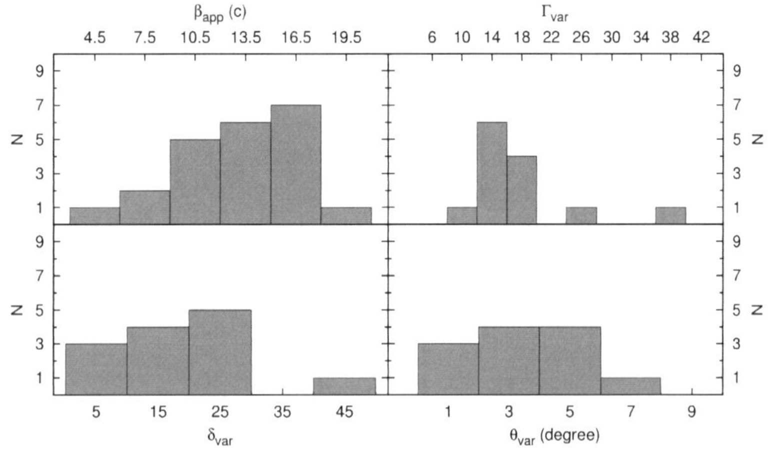

Figure 3. Distribution of average kinematic parameters, derived from the trajectories using the flux density decay timescale. The top left panel shows the distribution of the apparent component speeds, the top right panel shows the distribution of the Lorentz factors, the bottom left panel the distribution of the Doppler factors, and the bottom right panel the distribution of the viewing angles. The median values for the respective panels are 12.5 c, 14.1, 15.6, 3.8°.

angle seems too simple. In a second attempt we investigated the more realistic possibility of a curved jet, fixing the Lorentz factor to a constant value. For slow jet components this yields unrealistically small viewing angles, also the distances jet components travel in the source frame are on the order of hundreds of pc. Finally, a combination of a curved jet in combination with a varying Lorentz factor yields satisfying results. The required change in the Lorentz factor is relatively large, but remains in a plausible range of 4 – 24, whereas the viewing angles range between 2 and 10° and the distances traveled in the source frame are on the order of 2 – 5 pc. Altogether, it appears that the combination of acceleration and viewing angle change produces the best results and provides a satisfactory interpretation of the observed component motion.

4. Summary

A compilation of over thirty years of VLBI observations of the quasar 3C 345 provides us with new insights into the long-term properties of its parsec-scale jet. A first systematic analysis of an unprecedented database of jet kinematics for a simple object was presented. This data-set revealed a highly collimated and stable outflow over a period of over three decades, supporting a slow sheath/fast spine jet flow model, providing the conditions for Kelvin-Helmholtz instabilities. Standing oscillating patterns are observed in component motions and size evolution. The pattern resembles small changes in the viewing angle and a rotation of the internal structure of the jet. This supports the interpretation of elliptical surface modes from K-H instabilities being imprinted on the observed jet. We have found the observed apparent speeds are not constant. From the investigation of different scenarios we conclude that two effects acting together are able to reproduce the observed variability in speeds. To the one hand a change in the jet viewing angle leads to a change in observed speeds, on the other hand an intrinsic acceleration of the jet flow over parsec-scales is required.

Acknowledgments
Frank Schinzel was supported for this research through a stipend from the International Max Planck Research School (IMPRS) for Astronomy and Astrophysics at the Universities of Bonn and Cologne. This research has made use of NASAs Astrophysics Data System and used data from the MOJAVE database that is maintained by the MOJAVE team [16]. The authors acknowledge the efforts of the many people who made the study of such a vast dataset possible. This research has made extensive use of data collected by the Very Long Baseline Array, operated by the National Radio Astronomy Observatory (NRAO). The NRAO is a facility of the National Science Foundation operated under cooperative agreement by Associated Universities, Inc.

References
[1] Biretta J A, Moore R L and Cohen M H 1986 *Astrophys. J.* **308** 93–109
[2] Unwin S C and Wehrle A E 1992 *Astrophys. J.* **398** 74–86
[3] Matveenko L I, Graham D A, Pauliny-Toth I I K, Sherwood W A and Baath L B 1992 *Soviet Astronomy Letters* **18** 379
[4] Zensus J A, Cohen M H and Unwin S C 1995 *Astrophys. J.* **443** 35–53
[5] Lobanov A P 1996 *Physics of the Parsec-Scale Structures in the Quasar 3C 345* Ph.D. thesis New Mexico Institute of Mining & Technology, Socorro, NM, USA
[6] Ros E, Zensus J A and Lobanov A P 2000 *A&A* **354** 55–66 (*Preprint* arXiv:astro-ph/9911454)
[7] Klare J 2003 *Quasi-Periodicity in the Parsec-Scale Jet of the Quasar 3C 345* Ph.D. thesis Rheinische-Friedrich-Wilhelms-Universität Bonn
[8] Lobanov A P and Roland J 2005 *A&A* **431** 831–846 (*Preprint* arXiv:astro-ph/0411417)
[9] Marziani P, Sulentic J W, Dultzin-Hacyan D, Calvani M and Moles M 1996 *ApJS* **104** 37
[10] Shepherd M C, Pearson T J and Taylor G B 1995 *Bulletin of the American Astronomical Society* (*Bulletin of the American Astronomical Society* vol 27) ed B J Butler & D O Muhleman p 903
[11] Pearson T J 1999 *Synthesis Imaging in Radio Astronomy II* (*Astronomical Society of the Pacific Conference Series* vol 180) ed G B Taylor, C L Carilli, & R A Perley p 335
[12] Schinzel F K, Lobanov A P, Taylor G B, Jorstad S G, Marscher A P and Zensus J A 2012 *A&A* **537** A70 (*Preprint* 1111.2045)
[13] Schinzel F K 2011 *Physics and Kinematics of the Parsec Scale Jet of the Quasar 3C345* Ph.D. thesis Universität zu Köln, Germany
[14] Hardee P E and Eilek J A 2011 *Astrophys. J.* **735** 61 (*Preprint* 1104.4480)
[15] Jorstad S G, Marscher A P, Lister M L, Stirling A M, Cawthorne T V, Gear W K, Gómez J L, Stevens J A, Smith P S, Forster J R and Robson E I 2005 *Astron. J.* **130** 1418–1465 (*Preprint* arXiv:astro-ph/0502501)
[16] Lister M L, Aller H D, Aller M F, Cohen M H, Homan D C, Kadler M, Kellermann K I, Kovalev Y Y, Ros E, Savolainen T, Zensus J A and Vermeulen R C 2009 *Astron. J.* **137** 3718–3729 (*Preprint* 0812.3947)

Journal of Physics: Conference Series **372** (2012) 012071 doi:10.1088/1742-6596/372/1/012071

Charged Balanced Black Rings in Five Dimensions

Kirsten Schnülle
MPIA Heidelberg, Königstuhl 17, D-69117 Heidelberg
E-mail: schnuelle@mpia.de

Abstract. Balanced black ring solutions of pure Einstein-Maxwell theory in five dimensions are presented here. Those solutions are asymptotically flat, and their tension and gravitational self-attraction are balanced by the repulsion due to rotation and electrical charge. Hence they are free of conical singularities and possess a regular horizon which exhibits the topology $S^1 \times S^2$ of a torus. The global charges and the horizon properties of the solutions are discussed, and it is shown that they satisfy a Smarr relation. These black ring solutions are constructed numerically, and are restricted to the case of black rings with a rotation in the direction of the S^1.

1. Introduction

When looking for a theory that unifies all fundamental interactions including gravity, string theory is one of the most promising candidates. For mathematical consistency however, it requires spacetime to consist of more than four dimensions. Thus, higher-dimensional black holes have received much interest in recent years, opening up the possibility of direct observation in future high energy collisions.
The presence of extra dimensions can affect the properties of black holes dramatically: In four-dimensional spacetimes, black holes are restricted to have a spherical horizon topology, and are uniquely characterized by their global charges. In five dimensions however, topologies like black strings and black rings can be found. In a certain parameter range, spherical black holes and black rings coexist, so black hole uniqueness is violated in five dimensions.

2. The Vacuum Black Ring

A fascinating development was the discovery of the black ring as a topologically different solution of the five-dimensional Einstein-Maxwell equations by Emparan and Reall [2,3]. This is a black hole with ring horizon topology $S^1 \times S^2$ and the metric:

$$ds^2 = -\frac{F(y)}{F(x)}\left(dt + CR\frac{1+y}{F(y)}d\psi\right)^2 + \frac{R^2}{(x-y)^2}F(x)\left[-\frac{G(y)}{F(y)}d\psi^2 + \frac{dy^2}{G(y)} + \frac{dx^2}{G(x)} + \frac{G(x)}{F(x)}d\phi^2\right],$$

$$F(\xi) = 1 + \lambda\xi, \qquad G(\xi) = (1-\xi^2)(1+\nu\xi), \qquad C = \sqrt{\lambda(\lambda-\nu)\frac{1+\lambda}{1-\lambda}}.$$

Static solutions always suffer from the presence of a conical singularity, preventing the ring from collapsing under its tension and gravitational self-attraction. Their horizon is not smooth however (Fig. 1(a)), so no regular solutions exist. When including rotation, the tension and gravitational self-attraction can be balanced by the centrifugal repulsion, leading to balanced solutions without conical singularity (Fig. 1(b)).

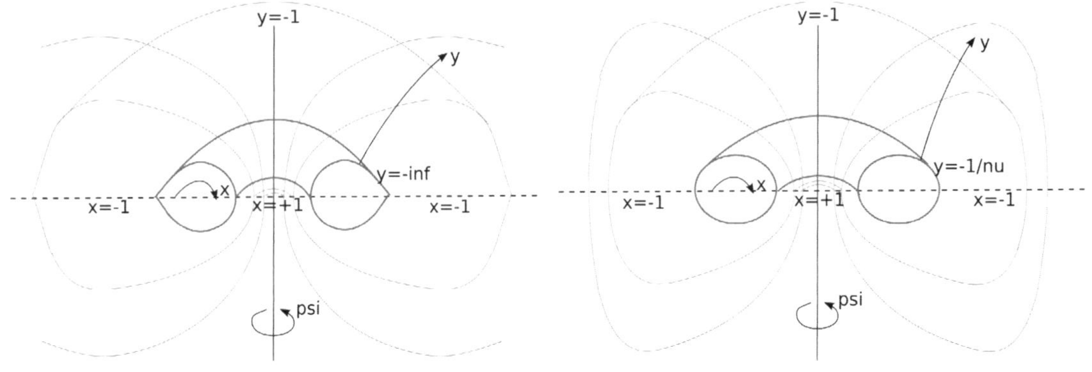

Figure 1. Spatial sections of the static (a) and rotating (b) black ring described by surfaces of constant y. The ψ part of the metric is suppressed.

3. Charged Balanced Black Rings – Action and Ansatz

A charged stationary black ring solution of pure Einstein-Maxwell theory is not known in closed from yet. However, we have constructed such a solution numerically.

From the five-dimensional Einstein-Maxwell action

$$S = \frac{1}{16\pi G_5}\int d^5x\sqrt{-g}(R - F_{ab}F^{ab}),$$

we obtain the Einstein-Maxwell equations

$$G_{ab} = 2T_{ab}, \qquad \nabla_b F^{ab} = 0,$$

with the curvature scalar R, the five-dimensional Newton constant G_5 and the field strength tensor $F_{ab} = \partial_a A_b - \partial_b A_a$, where A_a denotes the gauge potential. G_{ab} and T_{ab} represent the Einstein tensor and the energy-stress tensor, respectively. For the numerical work, we choose an ansatz in isotropic coordinates

$$ds^2 = -f_0(r,\theta)dt^2 + \frac{1}{f_1(r,\theta)}(dr^2 + r^2 d\theta^2) + f_2(r,\theta)d\psi^2 + f_3(r,\theta)(d\varphi - \frac{\omega(r,\theta)}{r}dt)^2,$$

$$A_a(r,\theta)dx^a = A_0(r,\theta)dt + A_\varphi(r,\theta)d\varphi.$$

Here, the rotation occurs in the direction of the S^1 of the ring. The resulting system of seven coupled non-linear elliptic partial differential equations is solved numerically.

4. Charged Balanced Black Rings – Results

The solutions depend on four parameters, i.e. the horizon radius r_0, the horizon angular velocity Ω_H, the charge parameter α and the parameter b representing a rough measure for the radius of the S^1.

In order to analyze the physical properties of the solution, it is convenient to work with dimensionless quantities, i.e. the scaled horizon area a_H and the scaled temperature t_H as well as the scaled squared angular momentum j^2 and the scaled charge q

$$a_H = \frac{3}{16}\sqrt{\frac{3}{\pi}}\frac{\mathcal{A}_H}{(G_5 M)^{3/2}}, \qquad t_H = T_H\sqrt{G_5 M}, \qquad j^2 = \frac{27\pi}{32 G_5}\frac{J^2}{M^3}, \qquad q = \frac{Q}{M}.$$

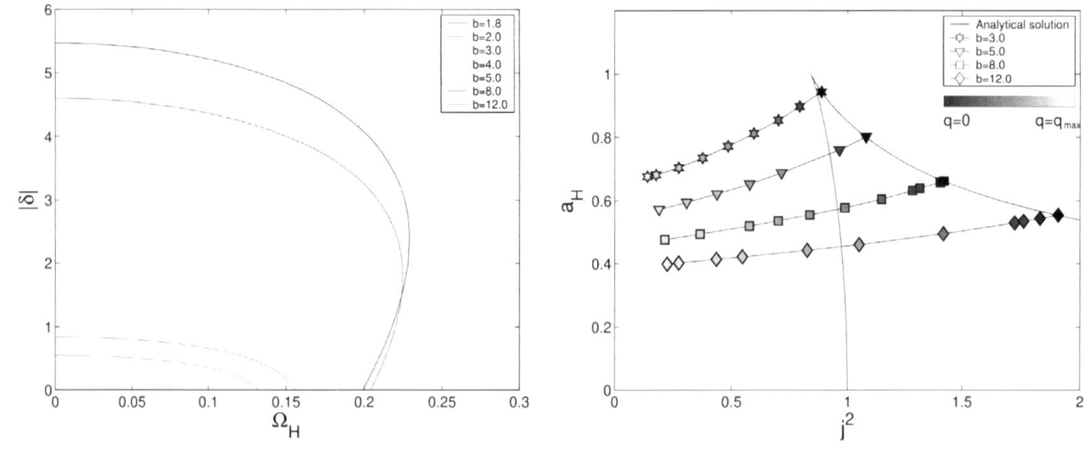

Figure 2. (a) Dependence of the conical singularity δ on the parameters b and Ω_H for neutral black rings ($r_0 = 1$, $\alpha = 0$). (b) Scaled horizon area a_H versus scaled squared angular momentum j^2 for the neutral rotating black ring (see [5] for the neutral analytical curve) and for the electrically charged balanced black ring (with $r_0 = 1$).

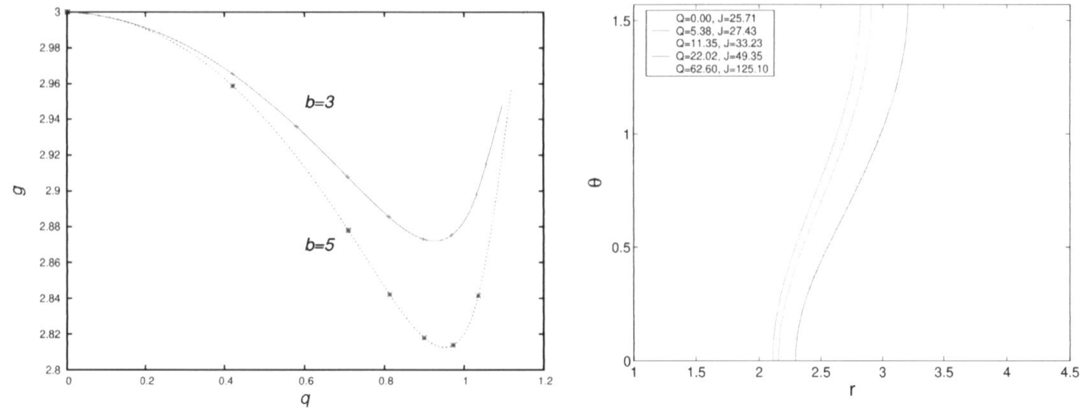

Figure 3. (a) Dependence of the gyromagnetic ratio g on the scaled charge q for charged balanced black ring solutions with $b = 3$ and $b = 5$ ($r_0 = 1$). The red and blue dots represent the calculated values, the lines are interpolated resp. extrapolated into the vicinity of the corresponding extremal values; the set of extremal values appears to approach $g = 3$ for q_{max} (green dot). (b) Location of the ergosurface $g_{tt} = 0$ (for $r_0 = 1$, $b = 5$).

We find that all solutions satisfy a Smarr relation

$$M = \frac{3}{16\pi G_5}\kappa\mathcal{A}_H + \frac{3}{2}\Omega_H J + \Phi_H Q,$$

where $\kappa = 2\pi T_H$ is the surface gravity.

In the static case as well as for generic values of the rotation parameter, the solutions suffer from a conical deficit. By adjusting the parameter values, however, we have succeeded in obtaining regular solutions. Our solutions agree excellently with the static analytical solution by

Yazadjiev [4] and the neutral analytical solution by Emparan and Reall [3]. In comparison with the neutral rotating solutions, we observe that for the electrically charged balanced solutions an increasing value of the scaled charge q makes the conical deficit vanish for lower values of the rotation parameter Ω_H and hence for lower values of the scaled squared angular momentum j^2. The phase diagram is shown in Fig. 2(b).

While the gyromagnetic ratio $g = \frac{2M\mathcal{M}_\varphi}{QJ}$ for weakly charged rings has been found to be $g = 3$, in accordance with Ortaggio and Pravda [6], a negative deviation from this value has been observed for larger magnitudes of the electric charge (Fig. 3(a)), similar to the case of black holes with a spherical horizon topology ([7]).

The ergosurface of the balanced solutions moves farther away from the horizon with increasing scaled charge (Fig. 3(b)).

For further reading confer the peer-reviewed article of this work ([1]).

References

[1] Kleihaus B., Kunz J. and Schnülle K., 2011, *Phys. Lett.* B **699**, 192
[2] Emparan R. and Reall H. S., 2002, *Phys. Rev.* D **65**, 084025
[3] Emparan R. and Reall H. S., 2002, *Phys. Rev. Lett.* **88**, 101101 (2002)
[4] Yazadjiev S. S., 2005, arXiv: hep-th/0507097
[5] Emparan R. and Reall H. S., 2008, *Living Rev. Rel.* **11**, 6
[6] Ortaggio M. and Pravda V., 2006, *JHEP* **0612**, 054
[7] Kunz J., Navarro-Lerida F. and Petersen A. K., 2005, *Phys. Lett.* B **614**, 104

Astronomy at High Angular Resolution 2011 IOP Publishing
Journal of Physics: Conference Series **372** (2012) 012072 doi:10.1088/1742-6596/372/1/012072

Spectral Evolution of Field Galaxies

Banafsheh Shahzamanian[1,2,3], **Sadollah Nasiri**[3], **Sepehr Arbabi**[4]

[1]I.Physikalisches Institut, Universitaet zu Koeln, Zuelpicher Str.77, 50937 Koeln, Germany
[2]Max-Planck-Institut fuer Radioastronomie, Auf dem Huegel 69, 53121 Bonn, Germany
[3]Physics Group, Science Faculty, Zanjan University, Zanjan, Iran
[4]Iran National Observatory, Tehran, Iran

E-mail: shahzaman@ph1.uni-koeln.de

Abstract. For the purpose of analyzing the spectral evolution of stellar population, we derive spectral energy distributions of simple stellar population in different ages from GALAXEV. Then, from stellar populations spectra, physical parameters are obtained for observational spectra of galaxies in the wavelength range from 3200 Å to 9500 Å. We use spectral data of Coleman and Wu for plotting the spectra of four types of galaxies. These observational data are combined and averaged from different sources; also they are extrapolated to high redshifts. In addition, we plot 16 synthesized spectra by considering Salpeter Initial Mass Function (IMF), exponential and constant Star Formation Rate (SFR) and solar metallicity for all spectra. Comparing the observational spectra with stellar population spectra, we analyze the age distribution for elliptical, irregular and two types of spiral field galaxies and obtain age and time scale parameters for each.

1. Model description

Bruzual, Charlot (2003) [hereafter BC03] evolutionary model is used for computing the spectral evolution of stellar populations at ages between $1\times10^{5}yr$ and $2\times10^{10}yr$ at a resolving power($\frac{\Delta\lambda}{\lambda}$) of 2000 for a wide range of metallicities. Also, the spectral evolution across the larger wavelength range from 91Å to 160μm at lower resolving power can be computed. Isochrone synthesis method has been used for evaluating the spectral evolution of stellar populations. BC03 stellar population synthesis model consists of two components: stellar evolution prescription (Padova 1994 Library, Padova 2000 Library, Geneva Library) and stellar spectral library (BaSeL, STELIB, Pickles). The mass distribution functions in BC03 model are Salpeter and Chabrier. [1, 2, 3]

1.1. GALAXEV

GALAXEV is a library of evolutionary stellar population synthesis models computed using the isochrone synthesis code of BC03. We computed the Spectral Energy Distributions (SEDs) of Stellar Populations in different ages by using this library.

2. Data Analysis

We used spectral data of Coleman,Wu and Weedman (1980) for plotting the four types of field galaxies spectra. Observational mean data are extrapolated to high redshifts. These data are empirical data from various sources that have been combined and averaged. In Figure 1, we

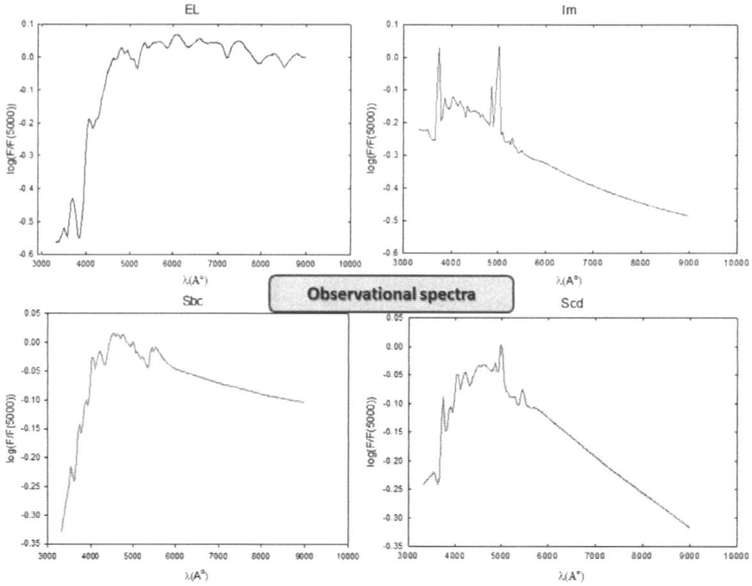

Figure 1. Observational spectra for four types of field galaxies.

plotted four empirical spectra obtained from empirical data. In these plots, we made the flux of galaxy dimensionless according to its flux in 5000 Å.

Moreover, we plotted 16 synthesized spectra in Figure 2 by considering Saltpeter IMF, exponential and constant SFR, Padova1994 evolutionary prescription, Pickles spectral library and solar metallicity for all spectra. We have 12 spectra by considering exponential function of star formation and different time scales: $\tau = 1Gyr,4Gyr,15Gyr$. Also, we obtained four spectra by considering constant function of star formation and different ages: $t = 0, 1Gyr, 1Gyr,5Gyr,10Gyr$. In each plot of Figure 2, time scale is constant and the different colours show the different ages of stellar populations. In these plots, we made the flux of galaxy dimensionless according to its flux in 5000 Å.

3. Result

Comparing the observational spectra with the stellar population model spectra, we computed the age distribution of stellar population (t) and obtained the time scale parameter τ of Elliptical, Spiral and Irregular galaxies in a field region. In Table 1, the amounts of t and τ for each type of galaxy are shown. For fitting the model and observed spectrum the relation bellow (Mean Square Error) must be minimized:

$$R_K^2 = \sum_{i=1}^{N}[\Delta \log F_{\lambda_i}]^2 = \sum_{i=1}^{N}[\log F_\lambda^{obs}(\lambda_i) - \log F_\lambda(\lambda_i, t_k)]^2. \quad (1)$$

In this relation the first flux is the flux of observed spectrum in wavelength i and the second one is the flux of stellar population model in wavelength λ_i and age t_k.

4. Conclusion

The obtained time scale is similar to result of Kauffmann and Charlot (1998) for same types of galaxies. Having a shorter time scale in Elliptical galaxies in comparison with Spiral ones shows that they have experienced a short time of star burst in their early evolution. Also, we obtained

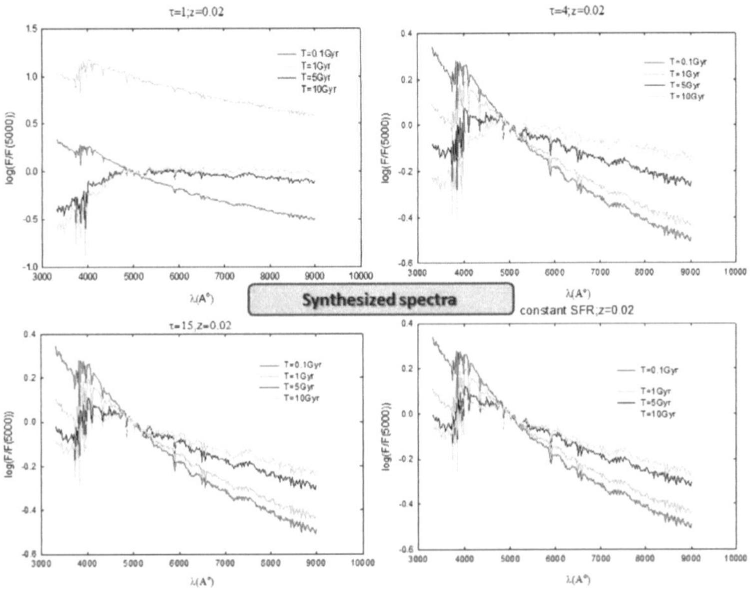

Figure 2. Synthesized spectra calculated by GALAXEV.

Table 1. Age distribution and time scale for different types of galaxies.

Type	$\tau(Gyr)$	$t(Gyr)$
Sbc	4	10
Scd	4	5

Type	$\tau(Gyr)$	$t(Gyr)$
El	1	10
Im	∞	5

the constant SFR for highly Irregular galaxies. Furthermore, we estimated the age of elliptical galaxies to be 10 Gyr, which shows this type of galaxies has old stellar populations. The age of irregular galaxies was found 5 Gyr, so they have intermediate stellar populations. For spiral galaxies we were not able to get a good estimation due to the lack of observational data.

References

[1] Bruzual A G 1983 *ApJ* **273** 105
[2] Bruzual A G and Charlot S 2003 *MNRAS* **344**
[3] Charlot S and Bruzual A G 1991 *ApJ* **367** 126
[4] Kauffman G and Charlot S 1998 *MNRAS* **297** L23

Astronomy at High Angular Resolution 2011 IOP Publishing
Journal of Physics: Conference Series **372** (2012) 012073 doi:10.1088/1742-6596/372/1/012073

Modeling polarization signatures of NIR radiation from the Sgr A* black hole environs

M Valencia-S.[1,2], M Bursa[3], G Karssen[1], M Dovčiak[3], A Eckart[1,2], J Horák[3] and V Karas[3]

[1] I.Physikalisches Institut, Universität zu Köln, Zülpicher Str.77, 50937 Köln, Germany
[2] Max-Planck-Institut für Radioastronomie, Auf dem Hügel 69, 53121 Bonn, Germany
[3] Astronomical Institute, Czech Academy of Sciences, Boční II 1401, 141 00 Praha 4, Czechia

E-mail: `bursa@astro.cas.cz`

Abstract. The near-infrared (NIR) emission of Sagittarius A* (Sgr A*), the source associated with the supermassive black hole ($4.4 \times 10^6\,\mathrm{M}_\odot$) at the center of our galaxy, is polarized and highly variable. Correlations between intensity and polarimetric parameters of the observed light curves compared with the predicted ones for different configurations, allow us to extract information about the geometry of the radiating region. Here we present the theoretical polarimetric light curves expected in the case of optically thin NIR emission from overdense regions close to the marginal stable orbit. Using a numerical code we track the time evolution of detectable polarization properties produced by synchrotron emission of compact sources in the vicinity of the black hole. We show that the different setups lead to very distinctive patterns in the time profiles of polarized flux and the orientation of the polarization vector and as such may be used for determining the geometry of the accretion flow around Sgr A*.

1. Introduction

At the position of Sgr A*, at the Milky Way's center, stellar orbits have convincingly proven the existence of a supermassive black hole of mass $\sim 4.4 \times 10^6\,\mathrm{M}_\odot$. Sgr A* is a source of variable radio, near-infrared, and X-ray emission. In the NIR, flux-increases of about one order of magnitude and variations in the polarized emission have been observed (e.g., [1, 2, 3]). Some bright NIR and X-ray flares seem to be simultaneous ([4] and references therein), which points to a common physical origin of the observed phenomena. The short time scale variations seen during the NIR and X-ray flares also argue for an emitting region not bigger than about ten Schwarzschild radii ($r_\mathrm{S} = 2GM/c^2 = 2\,r_\mathrm{g}$). The nature of this emission is of a non-thermal origin, most likely Synchrotron and/or Synchrotron self-Compton radiation from overdense regions very close to the black hole (see e.g., [4], and also Eckart et al. in this proceedings). The observed flux variations have been interpreted to be due to spots on relativistic orbits close to the inner most marginal stable orbit (ISCO) on an accretion disk ([5, 6, 7, 3]) or as variations in a short jet component([8, 9]).

A number of decades ago it was realized that polarization studies could provide firm clues to the physics of accreting black holes [10]. In strong gravitational field, the observed polarization is affected by light-bending, aberration and Doppler boosting [11]. In the context of the Galactic Center, the polarized NIR emission of SgrA* can be used to constrain the essential parameters of the system like spin and inclination by fitting the predicted light-curves to the observed ones [12, 13].

In this short paper we summarize results from our modeling of optically thin emission that arise from the close proximity of a black hole and that is a result of two setups: a spot model and an ejected plasmoid. We use a numerical code for our simulations that is able to ray-trace in Kerr metric and track polarization properties of produced synchrotron emission, it handles all relativistic effects present in strong gravity regime not excluding the parallel transport of the polarization vector along the photon trajectory that plays a role in this environment and in this kind of study. We show that the different setups that we are covering here all lead to very distinctive patterns in the time profiles of polarized flux and the orientation of the polarization vector. As such, those patterns can be used for matching the observed variations and can help in determining the geometry of the accretion flow around Sgr A*.

2. Modeling of polarized emission

In the following, we shall compare polarization signatures and their time variations produced by magnetically threaded hot spots/blobs as overdensities in the accretion disk or as ejected plasmoids.

2.1. Orbiting spot

This model has been studied and used in the context of Sgr A* emission by several authors (see e.g. [5, 14, 6, 15, 13]), and we show it here mainly as a reference. We model a single spot/blob as an optically thin 3D-Gaussian with $\sigma = 1\,r_{\rm g}$, on an optically thick accretion disk, that is orbiting at radius of $1.1 \times$ ISCO. In this case, we kept the structure of the emitting region not allowing it to deform due to differential rotation in the disk. The orientation of the system is such that the symmetry axis of the disk points towards the north (up) and the observer sees the equatorial plane under inclination of 70 degrees from this axis. Figure 1 presents snapshots from the photon ray-tracing, for the case of a black hole with dimensionless spin parameter $a = 0.5$. The color gradient corresponds to relative flux density (in arbitrary units) and the short lines indicate the orientation of the polarization angle. Figure 2 shows the characteristic light-curve pattern of the orbiting spot for $a = 0.0, 0.5$ and 0.9 – periodically rising and lowering flux which is the

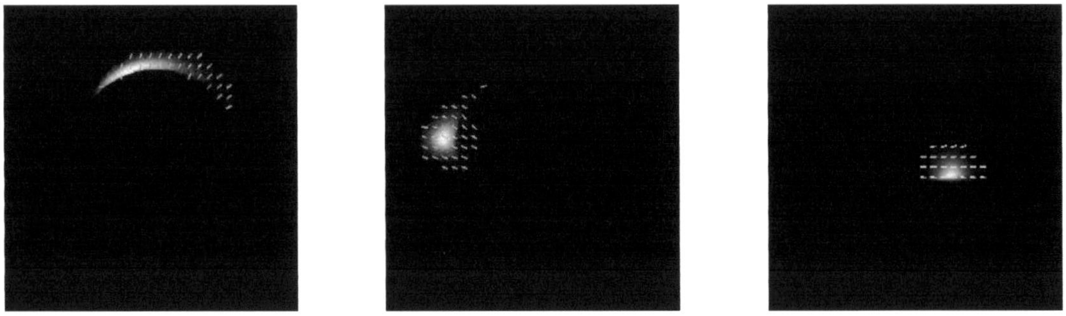

Figure 1. Snapshot images of $10\,r_{\rm g}$ side length taken from the numerical simulation of an orbiting blob for an inclination 70 deg and spin $a = 0.5$. The flux in every snapshot has been normalized to its maximum in order to enhance the low-flux images. The actual flux modulation is presented in the next figure. Color coding corresponds to a linear gradient of the flux emission levels and the green dashes show the polarization (orientation and length of the green lines represent the polarization angle and degree, respectively).

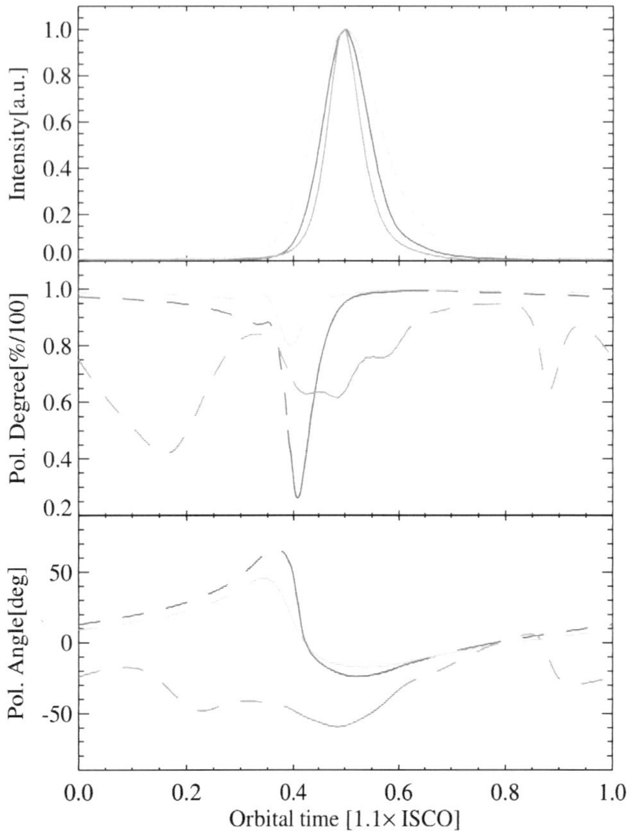

Figure 2. Time dependence of the total intensity (*top*), polarization degree (*middle*) and polarization angle (*bottom*) over one orbital period of a Gaussian blob orbiting in the disk at a radius $1.1 \times$ ISCO. The colors indicate the parameter behavior for different black hole spin: $a = 0.0$ (*cyan*), $a = 0.5$ (*magenta*) and $a = 0.9$ (*orange*). In all cases we have assumed a vertical magnetic field and the emission from the disk has been subtracted. The observer is inclined 70 degrees with respect to the symmetry axis of the disk. Dashed lines correspond to sections where the total flux is less than 5% of the maximum, and therefore, even during strong flares, they are not measurable with the current detectors.

result of relativistic Doppler beaming as well as the gravitational bending as the spot describes one orbit around the black hole. Here we have subtracted the disk emission, and therefore we present only the variable and polarized component. We assumed that a vertical magnetic field is present in the local frame of the spot. Figure 2 also shows the changes in polarization degree (as percentage of the polarization degree at the source) and angle. Such polarized light-curve profile displays the main expected relativistic effects: decrease of polarization degree and rapid change of polarization angle just before the maximum of intensity, though this pattern depends strongly on the local polarization at the source reference frame.

2.2. Blob ejected from the disk

The next example assumes a plasma blob that might be produced by a magnetic loop on the disk corona and is ejected out of the disk surface at $r = 1.3 \times$ ISCO. The plasmoid keeps its orbital motion and shape while moving away. As a first approximation we considered the radial component of the velocity to be zero and assumed the presence of a helical magnetic field with a higher azimuthal component (10 times the vertical one). Figures 3 and 4 illustrate the situation for an observer at 70 degrees with respect to the disk symmetry axis. Figure 3 shows snapshot images in the case of a black hole with spin $a = 0.5$, while Fig. 4 displays the behavior of the intensity, polarization degree and angle for three different spin parameters. The light curve in Fig. 4 covers the evolution of the blob over almost two full orbits and it has two peaks that arise from the Doppler boosting. The intensity modulations are similar to those in the previous

Journal of Physics: Conference Series **372** (2012) 012073 doi:10.1088/1742-6596/372/1/012073

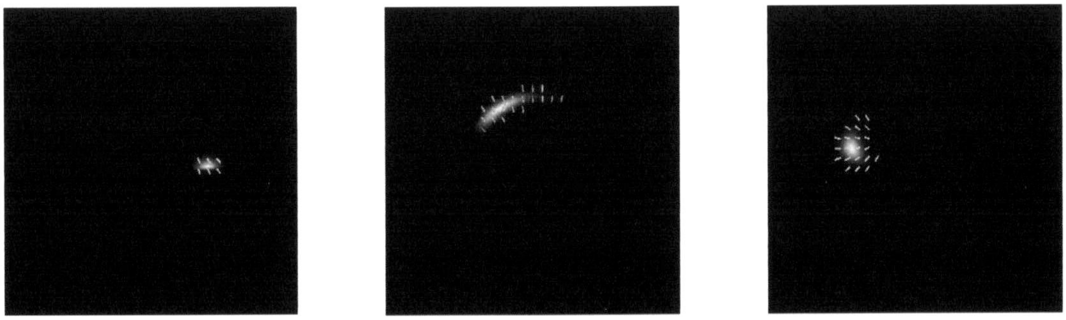

Figure 3. Snapshot images of $20\,r_g$ side length taken from the numerical simulation of a blob that originates in the accretion disk and is ejected away from it. The inclination of the system is 70 deg and the black hole spin $a = 0.5$. After the ejection event, the blob material keeps its angular momentum and follows a helical trajectory. Color coding as in Fig. 1.

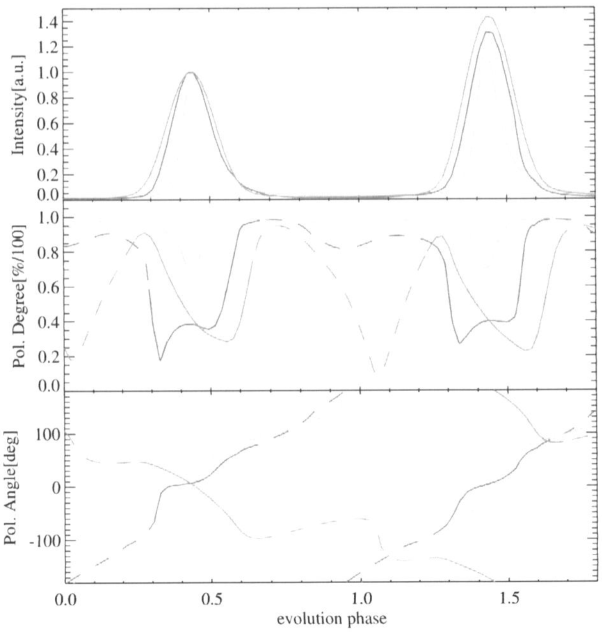

Figure 4. Time dependence of the total intensity (*top*), polarization degree (*middle*) and polarization angle (*bottom*) over the evolution of the plasmoid (a blob) that is ejected out of the disk surface at a radius 1.3 × ISCO. The colors correspond to different black hole spin parameters: $a = 0.0$ (*cyan*), $a = 0.5$ (*magenta*) and $a = 0.9$ (*orange*). The magnetic field is assumed to be helical with a higher toroidal component. The disk contribution has been subtracted in all cases (disk emission is unpolarized).As in Fig. 2, dashed lines correspond to sections where the total flux is less than 5% of the maximum.

case. Given the assumed magnetic field, the major depolarizations in the light curves coincide with the maxima in flux (compare figures 2 and 4). We can see that this scenario produces a different pattern and could be easily distinguished from the orbiting spot scenario.

3. Conclusions

We have shown preliminary results of two possible models for polarized emission from Sgr A* that can be matched against observational data to better guess what types of processes and geometries could be the sources of the observed NIR flares. Each setup is characterized by a distinct profile of polarized flux and changes in the angle of the polarization vector. This is a second step on modeling the Sgr A* activity (the first approach has been done using the 2-dimensional KY-code). Further efforts to account for radiative transfer effects in optically thick environments will allow us to describe multiwavelength observations. Ultimately this would lead to constraining the spin of the Sgr A* black hole and its orientation with respect to the galactic plane.

These results can be compared with today's data as well as with data from future near-infrared polarimetric detectors that will extend our observational capabilities and provide a higher sensibility and resolution than it is available today.

Acknowledgments
M. Valencia-S. is member of the International Max-Planck Research School (IMPRS) for Astronomy and Astrophysics at the Universities of Bonn and Cologne supported by the Max Planck Society. Part of this work was supported by the COST Action MP0905: Black Holes in a violent Universe and PECS project No. 98040.

References
[1] Baganoff F K, Bautz M W, Brandt W N, et al. 2001 *Natur.* **413** 45
[2] Eckart A, Baganoff F K, Morris M, et al. 2004 *A&A* **427** 1
[3] Zamaninasab M, Eckart A, Witzel G, Dovčiak M, et al. 2010 *A&A* **510** 3
[4] Eckart A, Garcia-Marin M, Vogel S N, Teuben P, et al. 2012 *A&A* **537** 52
[5] Broderick A E and Loeb A 2005 *MNRAS* **363** 353
[6] Meyer L, Eckart A, Schödel R, et al. 2006 *A&A* **460** 15
[7] Pecháček T and Karas V 2008 *A&A* **487** 815
[8] Falcke H and Markoff S 2008 *A&A* **362** 113
[9] Markoff S, Bower G C and Falcke H 2007 *MNRAS* **379** 1519
[10] Rees M J 1975 *MNRAS* **171** 457
[11] Connors P A, Stark R F and Piran T 1980 *ApJ* **235** 224
[12] Meyer L, Schödel R, Eckart A, Duschl W J, Karas V and Dovčiak M 2007 *A&A* **473** 707
[13] Zamaninasab M, Eckart A, Dovčiak M, Karas V, et al. 2011 *MNRAS* **413** 322
[14] Dovčiak M, Karas V and Matt G 2006 *AN* **327** 993
[15] Noble S C, Leung P K, Gammie C F and Book L G 2007 *CQG* **24** 259

Astronomy at High Angular Resolution 2011 IOP Publishing
Journal of Physics: Conference Series **372** (2012) 012074 doi:10.1088/1742-6596/372/1/012074

Classifying Active Galactic Nuclei from the zCOSMOS 20k spectroscopic survey. Looking for a mass trend.

M Vitale[1,2,3], M Mignoli[3] and A Cimatti[3]

[1]I. Physikalisches Institut - Universitaet zu Koeln, Zuelpicher Strasse 77, 50937 Koeln, Germany
[2]Max Planck Institute for Radio Astronomy, Auf dem Huegel 69, 53121 Bonn, Germany
[3]Dipartimento di Astronomia - Bologna University, via Ranzani 1, 40127 Bologna, Italy

E-mail: `vitale@ph1.uni-koeln.de`

Abstract. We analyze the evolution of galaxy spectral properties up to z=1 by using galaxies from the zCOSMOS-Bright 20k spectroscopic survey. We divide the sample in several mass-redshift bins to obtain stacked galaxy spectra and measure emission lines with a higher S/N. The aim of this work is to investigate whether it is possible to highlight some evidence of AGN activity in suppressing star formation during the transitional phase of galaxies from late to early type, with a particular look on a possible trend of the galaxy total stellar mass.
Our main finding is a clear role of the total stellar mass in leading galaxy evolution. As the diagnostic diagrams indicate, the objects show a progressive shift towards the AGNs region with increasing total stellar mass. Additionally, the composites follow, for a fixed redshift bin, evolutionary tracks that move from the upper part of the SF galaxies sequence down to the same metallicity sequence. For stellar masses greater than 1.6×10^{10} solar masses, the objects start being classified as composites or AGNs. This indicates that AGNs mostly act at the highest mass bins, where our sample is firstly populated by passive galaxies that have experienced the quenching.
The evidence that older stellar populations inhabit more massive galaxies, together with the increasing AGN detection rate with increasing masss, suggest a scenario in which AGNs might act to quench the star formation and, then, contribute to the transformation from young blue late-type galaxies to old red early-type galaxies.

1. Introduction

In recent years, a better overview of galaxy formation and evolution has been developing, with a great contribution coming from large cosmic surveys. The colour function of galaxies has been discovered to be a bimodal function [1]. The two dominant sequences can be associated with the ellipticals+S0s population (*red cloud*) and the spirals+irregulars population (*blue sequence*). Galaxies showing intermediate properties between the red and the blue sequences are considered as transitional objects.
The transition from late to early-type galaxies is supposed to be very fast in suppressing the star formation activity, a phenomenon that is known as *quenching*. It is currently thought to be largely responsible for the growing abundance of galaxies on the red sequence by preventing further episodes of star formation and activating a passive galaxy evolutionary phase. There

are several hypotheses to explain the quenching, but all of them still require observational counterparts.
A role of Active Galactic Nuclei (AGNs) in quenching the star formation inside hosts has been suggested [2], [3]. The discovery of how the black hole mass strongly correlates with the properties of the hosting galaxies [4], [5], has generated new hypotheses in linking the black hole growth with galaxy formation and evolution. Since every massive galaxy seems to host a central super massive black hole, which releases a large amount of energy during the active phase, we can imagine a scenario where a fraction of this energy is absorbed by the host galaxy to quench the star formation (by heating or blowing away part of the gas content of the galaxy [6]).
One of the most promising ways to test the scenario in which AGNs firstly initiate the quenching event and then prevent hot gas in passive galaxies from cooling to form stars is to investigate the properties of AGN host galaxies, searching for the signature of this feedback.
Spectra are known to contain a large amount of information on age and metallicity of the stellar populations, but also on the ionizing source of galaxies. We analyze the evolution of galaxy spectral properties up to $z \sim 1$ by using galaxies from the zCOSMOS-Bright 20k sample [7]. With the large number ($\sim 2\times10^4$) of observed objects, we are able to investigate galaxy evolution by measuring emission lines from stacked spectra of galaxies placed in relatively small bins of redshift and mass (Fig. 1).
The line-emitting gas in AGNs and star-forming galaxies is powered by two different ionizing mechanisms that produce different emission line ratios: accretion around the black hole (non-thermal) in the former and photoionization by hot massive OB stars (thermal) in the latter. It is important to notice that emission lines in narrow-line AGNs have considerably greater ionization range than in star-forming (SF) galaxies. In particular, high-ionization lines, such as [OIII]λ5007 Å, and low-ionization lines, like [OI]λ6300 Å, are stronger in narrow-line AGNs than in normal star-forming galaxies. Photoionization by a central non-thermal source is the ideal candidate to explain the excitation mechanism in galaxies that display strong emission lines as a sign of unusual activity in their nuclei [8].

Emission-line diagnostic diagrams [9], which make use of these emission lines, represent a powerful tool for probing the nature of the dominant ionizing source in galaxies.

2. Results and discussion

In this work we have analyzed the spectral properties of galaxies from the 20k zCOSMOS-Bright spectroscopic survey, looking for a trend of the total stellar mass in the classification of objects preferentially as star-forming galaxies or AGNs.
Kauffman et al. [10] showed that SDSS galaxies hosting an AGN are more likely to have a larger stellar mass compared to SF galaxies that do not show optical AGN signatures. The transition between the blue and the red sequence seems to occur around $3 \times 10^{10} M_{\odot}$, suggesting a mass dependence in the observed change of galaxies properties.
In the stacked zCOSMOS spectra the optical continuum is quite flat and featureless at low redshifts, especially for the objects having low masses, whereas it displays a drop in the bluest part of the spectrum when we move to higher redshifts. This means that the galaxies are spectroscopically more evolved at high redshifts, according to the downsizing scenario [11]. There are also some notable differences among the mass bins: aside from being flatter, the spectra at lower masses exhibit strong and well detectable lines, which become easily visible in most of the cases.

The stellar continuum subtraction - which has been performed by using our set of Bruzual & Charlot templates (BC03, [12]) plus reddened stellar spectra - helps to stress the spectroscopic features that allow us to identify AGNs (Fig. 2). It also gives more detailed information on the age and the kind of stellar population that inhabits a specific mass-z bin.

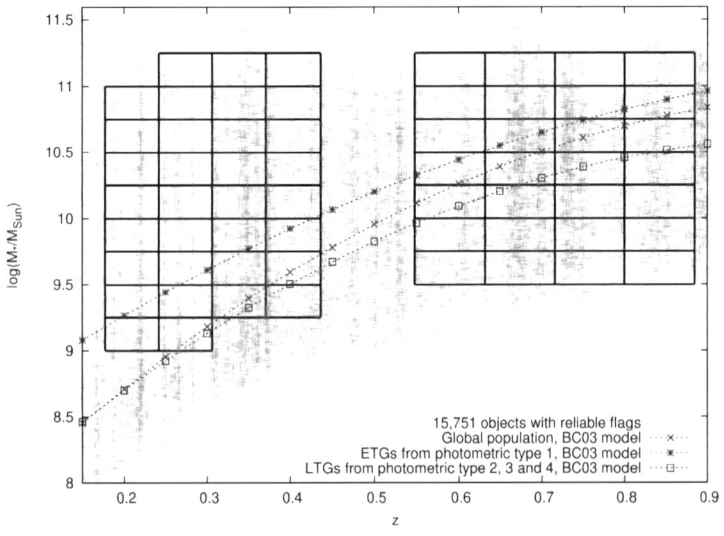

Figure 1. Original sample of zCOSMOS 20k galaxies in the range $0.15 \leq z \leq 0.9$. The blue dashed lines represent, respectively, the mass completeness limits of the global galactic population (crosses), the ETGs population (asterisks) and the LTGs population (squares). The black boxes are the 61 mass-redshift bins that define the stacked galaxies we study with the diagnostic diagrams.

Another point is the general increase of the age of the stellar populations with decreasing redshift. In addition we found older stellar populations in more massive galaxies. The effect of having younger populations at higher redshifts was expected from galaxy evolution scenarios. The evidence of older stellar population with increasing mass and fixed redshift indicates that, inside these objects, a mechanism acted to quench the star formation. Considering a hierarchical scenario in which young low mass galaxies are undergoing star formation and high mass galaxies are the product of a later mass assembly, the latter are passively-evolving galaxies with no star formation episodes. In this way, AGNs might have just suppressed star formation with their feedback, by means of transforming blue galaxies into red ones that continue to accrete mass.
We concentrated on the study of different types of diagnostic diagrams: BPT diagrams [9] and MEx diagram [13]. These kinds of diagrams are used to display the differences between the ionizing mechanisms responsable for the production of low-ionization emission lines (stars or accretion around the central black hole). Therefore, diagrams show if our sample is dominated by SFGs or if there is trace of some AGN contamination.
We found a clear role for the total stellar mass in leading galaxy evolution. As the classical diagnostic diagram with [NII] indicates (Fig. 3), the objects show a progressive shift towards the AGNs region - on the right-hand side of the diagrams - with increasing stellar mass. The objects with the highest mass are almost always placed on the right-hand side of the demarcation curves that separate the region where the ionization mechanism that originates the emission lines is nuclear activity (black hole accretion). Additionally, the composites follow, for a fixed redshift bin, evolutionary tracks that move from the upper part of the SF galaxies sequence, defined by the SDSS galaxies [14], down to the so called metallicity sequence. For $\log(M_*/M_\odot) = 10.2$

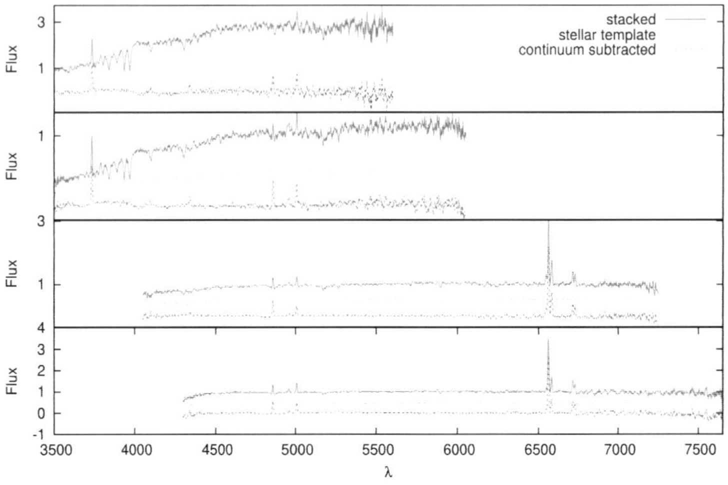

Figure 2. Some examples of stellar continuum subtraction. For each panel we show the original composite spectrum (red), the most suitable stellar continuum to be subtracted (green) and the resulting zCOSMOS spectrum after the stellar continuum subtraction (blue).

the objects start being classified as composites or AGNs, in agreement with Kauffmann et al. 2003 [10]. Due to the fringing and the lines blending, the [SII] diagnostic diagram is heavily affected by errors and leads to results that are not as reliable as the ones we obtain with the [NII] diagram.

At higher redshift, in the [OII] diagram, all the objects are placed in the SF zone and they do not show a particular trend with the mass. This diagram is known to be the most sensitive to shocks and this might be the reason why some objects are likely to be misclassified as star-forming galaxies rather than AGNs.

The trend we observe in the [NII] diagram is clearer in the MEx diagram (Fig.4), which makes use of the galaxy stellar mass instead of the emission line ratio [NII]λ6583/Hα (lines only visible up to $z \sim 0.4$).

The dependency of the stellar population evolution on stellar mass, which has been observed during the continuum subtraction procedure, together with the increasing AGNs detection rate with increasing mass represented in the diagnostic diagrams, suggest a scenario in which AGNs could act to quench the star formation and, then, contribute to the transformation from young blue late-type galaxies to old red early-type galaxies.

3. Summary and conclusions

In this work we have used the zCOSMOS 20k catalogue to create a sample of galaxies with reliable spectroscopic redshift in order to investigate the evolution of galaxy spectral properties up to $z \sim 1$. This sample has been divided in 61 bins, taking into account the completeness in mass, the galaxies density (also related to selection effects) and the gap in redshift, this latter due to the impossibility of measuring all the emission lines involved in classical diagnostic diagrams. Every bin includes galaxies with various mass and redshift intervals. In order to improve the

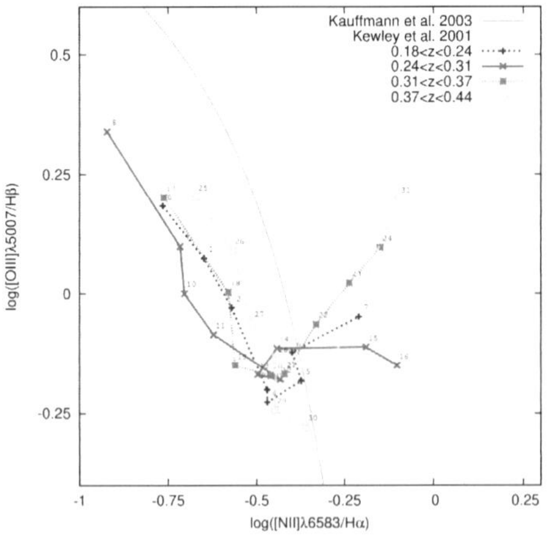

Figure 3. Evolutionary tracks for the [NII] diagnostic diagram. Each color represents the mass trend (increasing numbers for increasing mass) for a fixed redshift bin. The progressive shift of the objects towards the AGN region is shown.

signal to noise (S/N) of a single spectrum, we have combined all the spectra in each bin to obtain stacked spectra, allowing accurate flux measurements.
After obtaining the spectra, we have applied the continuum subtraction procedure to the stacked galaxy spectra by using templates of stellar population synthesis. This provides new spectra that are free from contaminations by stars. It also provides information on the age of the stellar populations and their composition (metallicity). Our main findings are:

- Stellar populations hosted in galaxies are older at low redshift and for more massive objects. This is in agreement with an evolutionary scenario that accounts for a faster mass assembling and stellar population ageing in case of massive objects.
- Galaxies are more likely to be classified as AGNs rather that star-forming galaxies if their total stellar mass is high. In particular, we found that galaxies with $\log(M_*/M_\odot) > 10.2$ start to be classified as transitional objects (where both photoionization and nuclear activity contribute to ionization of the emission lines) and move towards the AGN region of the diagnostic diagrams. This is well visible in the [NII] and in the MeX diagram, whereas it has not been observed in the [OII] diagram.
- The galaxy composites seem to follow some evolutionary tracks that can be traced if we fix the redshift interval and move along the mass sequence. The trend follows the left hand branch of the Sloan Digital Sky Survey (SDSS) classification, defined by the galaxy metallicity evolution, down to the bulk of the population in the SF region of the diagram. Then the tracks continue and end in the composite or AGN region.
- The MeX diagram represents an useful tool to investigate the ionizing mechanisms inside galaxies, especially at high redshift. The classical [OII] diagram is indeed not effective in distinguishing between the contribution from an AGN or star-forming regions to the lines emission, due to its particular sensitivity to shocks and the need to be in a specific redshift interval to measure all the emission lines therein.

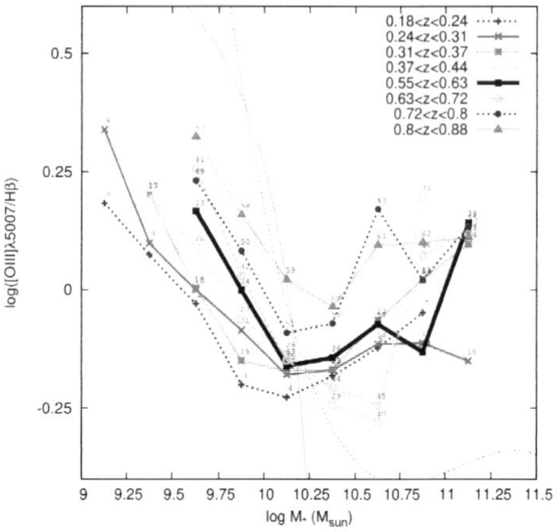

Figure 4. The MEx diagram makes use of the stellar mass instead of the emission line ratio [NII]λ6583/Hα. In this plot, every color represents the evolution in mass for a fixed redshift bin. The numbers increase with mass along the same redshift bin. At low masses the composites are classified as SF galaxies, while for $\log(M_*/M_\odot) > 10.2$ they start being classified as transitional objects and move towards the AGNs region, on the right-hand side.

Acknowledgments

This work has been carried out at the University of Bologna, Italy, thanks to the support of the zCOSMOS team. M. Vitale is currently a member of the International Max Planck Research School for Astronomy and Astrophysics in Bonn, Germany.

[1] Baldry I K, Glazebrook K, Brinkmann J, Ivezić Ž, Lupton R H, Nichol R C and Szalay A S 2004 *ApJ* **600** 681–694 (*Preprint* `arXiv:astro-ph/0309710`)
[2] Croton D J, Springel V, White S D M, De Lucia G, Frenk C S, Gao L, Jenkins A, Kauffmann G, Navarro J F and Yoshida N 2006 *MNRAS* **365** 11–28 (*Preprint* `arXiv:astro-ph/0508046`)
[3] Narayanan D, Cox T J, Kelly B, Davé R, Hernquist L, Di Matteo T, Hopkins P F, Kulesa C, RobeThe Demography of Massive Dark Objects in Galaxy Centersrtson B and Walker C K 2008 *ApJ* **176** 331–354 (*Preprint* `0710.0384`)
[4] Magorrian J, Tremaine S, Richstone D, Bender R, Bower G, Dressler A, Faber S M, Gebhardt K, Green R, Grillmair C, Kormendy J and Lauer T 1998 *AJ* **115** 2285–2305 (*Preprint* `arXiv:astro-ph/9708072`)
[5] Ferrarese L and Merritt D 2000 *ApJ* **539** L9–L12 (*Preprint* `arXiv:astro-ph/0006053`)
[6] Cattaneo A, Faber S M, Binney J, Dekel A, Kormendy J, Mushotzky R, Babul A, Best P N, Brüggen M, Fabian A C, Frenk C S, Khalatyan A, Netzer H, Mahdavi A, Silk J, Steinmetz M and Wisotzki L 2009 *Nature* **460** 213–219 (*Preprint* `0907.1608`)
[7] Lilly S J, Le Fèvre O and Renzini A 2007 *ApJ* **172** 70–85 (*Preprint* `arXiv:astro-ph/0612291`)
[8] Ho L C 2008 *ARAA* **46** 475–539 (*Preprint* `0803.2268`)
[9] Baldwin J A, Phillips M M and Terlevich R 1981 *PASP* **93** 5–19
[10] Kauffmann G, Heckman T M, Tremonti C, Brinchmann J, Charlot S, White S D M, Ridgway S E, Brinkmann J, Fukugita M, Hall P B, Ivezić Ž, Richards G T and Schneider D P 2003 *MNRAS* **346** 1055–1077 (*Preprint* `arXiv:astro-ph/0304239`)
[11] Cimatti A, Daddi E and Renzini A 2006 *A&A* **453** L29–L33 (*Preprint* `arXiv:astro-ph/0605353`)
[12] Bruzual G and Charlot S 2003 *MNRAS* **344** 1000–1028 (*Preprint* `arXiv:astro-ph/0309134`)
[13] Juneau S, Dickinson M, Alexander D M and Salim S 2011 *ApJ* **736** 104 (*Preprint* `1105.3194`)
[14] Kewley L J, Groves B, Kauffmann G and Heckman T 2006 *MNRAS* **372** 961–976 (*Preprint* `arXiv:astro-ph/0605681`)

AUTHOR INDEX

AUTHOR INDEX

AUTHOR INDEX

Institute of Physics
Dirac House, Temple Back
Bristol BS1 6BE UK

proceedings.com

International Workshop on Dirac Electrons in Solids 2015

Journal of Physics: Conference Series Volume 603

Tokyo, Japan
14 – 15 January 2015

ISBN: 978-1-5108-0263-6
ISSN: 1742-6588

IOP Institute of Physics